Design and Application of Modern Synchronous Generator Excitation Systems

Design and Application of Modern Synchronous Generator Excitation Systems

Jicheng Li
Tsinghua University
China

This edition first published 2019 by John Wiley & Sons Singapore Pte. Ltd under exclusive licence granted by China Electric Power Press for all media and languages (excluding simplified and traditional Chinese) throughout the world (excluding Mainland China), and with non-exclusive license for electronic versions in Mainland China.
© 2019 China Electric Power Press

All rights reserved. No part of this publication may be reproduced, stored in a retrieval system, or transmitted, in any form or by any means, electronic, mechanical, photocopying, recording or otherwise, except as permitted by law. Advice on how to obtain permission to reuse material from this title is available at http://www.wiley.com/go/permissions.

The right of Jicheng Li to be identified as the author of this work has been asserted in accordance with law.

Registered Offices
John Wiley & Sons, Inc., 111 River Street, Hoboken, NJ 07030, USA
John Wiley & Sons Singapore Pte. Ltd, 1 Fusionopolis Walk, #07-01 Solaris South Tower, Singapore 138628

Editorial Office
1 Fusionopolis Walk, #07-01 Solaris South Tower, Singapore 138628

For details of our global editorial offices, customer services, and more information about Wiley products visit us at www.wiley.com.

Wiley also publishes its books in a variety of electronic formats and by print-on-demand. Some content that appears in standard print versions of this book may not be available in other formats.

Limit of Liability/Disclaimer of Warranty
While the publisher and authors have used their best efforts in preparing this work, they make no representations or warranties with respect to the accuracy or completeness of the contents of this work and specifically disclaim all warranties, including without limitation any implied warranties of merchantability or fitness for a particular purpose. No warranty may be created or extended by sales representatives, written sales materials or promotional statements for this work. The fact that an organization, website, or product is referred to in this work as a citation and/or potential source of further information does not mean that the publisher and authors endorse the information or services the organization, website, or product may provide or recommendations it may make. This work is sold with the understanding that the publisher is not engaged in rendering professional services. The advice and strategies contained herein may not be suitable for your situation. You should consult with a specialist where appropriate. Further, readers should be aware that websites listed in this work may have changed or disappeared between when this work was written and when it is read. Neither the publisher nor authors shall be liable for any loss of profit or any other commercial damages, including but not limited to special, incidental, consequential, or other damages.

Library of Congress Cataloging-in-Publication Data

Names: Li, Jicheng, 1930- author.
Title: Design and application of modern synchronous generator excitation
  systems / Jicheng Li, Tsinghua University, China.
Description: Hoboken, NJ, USA : Wiley-IEEE Press, 2019. | Includes
  bibliographical references and index. |
Identifiers: LCCN 2018049587 (print) | LCCN 2018056673 (ebook) | ISBN
  9781118841020 (Adobe PDF) | ISBN 9781118841051 (ePub) | ISBN 9781118840870
  | ISBN 9781118840870 (hardcover)
Subjects: LCSH: Electric machinery, Synchronous. | Electronic excitation.
Classification: LCC TK2731 (ebook) | LCC TK2731 .L5 2019 (print) | DDC
  621.31/34–dc23
LC record available at https://lccn.loc.gov/2018049587

Cover design: Wiley
Cover image: © jauhari1/iStock.com

Set in 10/12pt WarnockPro by SPi Global, Chennai, India
Printed in Singapore by C.O.S. Printers Pte Ltd

10 9 8 7 6 5 4 3 2 1

Contents

About the Author *xxi*
Foreword *xxiii*
Preface *xxvii*
Introduction *xxix*
Acknowledgement *xxxi*

1	**Evolution and Development of Excitation Control** *1*	
1.1	Overview *1*	
1.2	Evolution of Excitation Control *1*	
1.2.1	Single Variable Control Based on Classical Control Theory *1*	
1.2.1.1	Proportional Control *2*	
1.2.1.2	PID Control *4*	
1.2.2	Linear Multivariate Control Based on Modern Control Theory *7*	
1.2.2.1	Linear Optimal Control Principle *7*	
1.2.2.2	Quadratic Performance Index *8*	
1.2.2.3	Linear Optimal Controller *10*	
1.3	Linear Multivariable Total Controller *11*	
1.3.1	Overview of TAGEC *11*	
1.3.2	TAGEC Control *12*	
1.3.2.1	TAGEC-I *12*	
1.3.2.2	TAGEC-II *12*	
1.3.2.3	TAGEC-III *13*	
1.3.3	Mathematical Model of TAGEC Control System *14*	
1.3.4	Composition of TAGEC System *15*	
1.3.4.1	Sampling Operation *15*	
1.3.4.2	Matrix Operation *17*	
1.3.4.3	Stability Margin Monitoring Control *17*	
1.3.4.4	TAGEC Input and Output *18*	
1.3.5	Digital Simulation Test of Power Transmission System *18*	
1.3.5.1	Composition of Power System *18*	
1.3.5.2	Generators *18*	
1.3.5.3	Excitation and Speed Governing System *19*	
1.3.5.4	Power System Stability Test *19*	
1.3.5.5	Power System Stability Test Result *19*	
1.4	Nonlinear Multivariable Excitation Controller *20*	
1.5	Power System Voltage Regulator (PSVR) *25*	
1.5.1	Overview *25*	
1.5.2	Effect of SVR on Improvement of System Voltage Characteristics *25*	

1.5.3	Composition of PSVR	29
1.5.3.1	Basic Control of PSVR	29
1.5.3.2	Basic Equation of PSVR Control	29
1.5.4	Comparison between PSVR and AVR in Control Characteristics	31
1.5.5	Basic Functions of PSVR	31
1.5.5.1	Basic Control	31
1.5.5.2	Program Voltage Setting	31
1.5.5.3	Output Limit	31
1.5.5.4	Anomaly Self-Detection	31
1.5.5.5	Control Stability	31
1.5.6	PSVR Simulation Test	33
2	**Characteristics of Synchronous Generator**	**35**
2.1	Electromotive Force Phasor Diagram of Synchronous Generator	35
2.1.1	Non-salient Pole Generator	35
2.1.1.1	Electromotive Force Phasor Diagram of Non-salient Pole Generator	35
2.1.1.2	Emf Equation of Non-salient Pole Generator	35
2.1.2	Salient Pole Generator	36
2.1.2.1	Emf Phasor Diagram of Salient Pole Generator	36
2.1.2.2	Emf Equation of Salient Pole Generator	37
2.2	Electromagnetic Power and Power Angle Characteristic of Synchronous Generator	38
2.2.1	Power and Torque Balance Equations	38
2.2.2	Electromagnetic Power and Power Angle Characteristic Expressions	38
2.2.2.1	Power Angle Characteristic Curve of Non-salient Pole Generator	40
2.2.2.2	Power Angle Characteristic Curve of Salient Pole Generator	40
2.2.2.3	Power Angle Expression of Reactive Power	40
2.3	Operating Capacity Characteristic Curve of Synchronous Generator	41
2.3.1	Operating Capacity Diagram of Non-salient Pole Generator	41
2.3.1.1	Power Diagram of Non-salient Pole Generator	41
2.3.1.2	Operating Capacity Curve of Non-salient Pole Generator	42
2.3.2	Operating Capacity Curve of Salient Pole Generator	44
2.4	Influence of External Reactance on Operating Capacity Characteristic Curve	45
2.5	Operating Characteristic Curves of Generator	50
2.5.1	Operating Characteristic Curve of Hydro Generator	51
2.5.1.1	No-Load Saturation Characteristic Curve	51
2.5.1.2	Short Circuit Characteristic Curve	51
2.5.1.3	Air Gap Line	52
2.5.1.4	Load Characteristic Curve ($\cos\varphi = 0.9$)	52
2.5.1.5	Load Characteristic Curve ($\cos\varphi = 0$)	52
2.5.2	Capacity Characteristics Curve of Hydro Generator	52
2.5.3	V-Shaped Curve of Hydro Generator	52
2.5.3.1	Normal Operating Area	53
2.5.3.2	Unstable Operating Area	53
2.5.3.3	Phase-Modulated Operation Capacity	53
2.6	Transient Characteristics of Synchronous Generator	54
2.6.1	Transient Reactance X'_d	54
2.6.2	Transient Emf E'_q	57
2.6.3	Transient Equation of Rotor Excitation Circuit	59
2.6.4	Change in Transient Emf in the Case of Three-Phase Short Circuit	60
2.6.5	Influence of Change of Excitation Voltage on Transient Emf	62

3	**Effect of Excitation Regulation on Power System Stability** 67
3.1	Definition and Classification of Power System Stability 67
3.1.1	Steady-State Stability 67
3.1.2	Dynamic Stability 67
3.1.3	Transient Stability 67
3.2	Criterion of Stability Level 68
3.3	Effects of Excitation Regulation on Power System Stability 68
3.3.1	Effect of Excitation Regulation on Steady-State Stability 68
3.3.2	Influence of Excitation Regulation on Transient Stability 70
3.3.2.1	Enhancement of the Steady-State Stability of Power System 72
3.3.2.2	Improving Transient Stability 74
3.3.2.3	Improving Dynamic Stability 74

4	**Static and Transient State Characteristics of Excitation Systems** 77
4.1	Static Characteristics of Excitation System 77
4.1.1	Block Diagram of Static Characteristics 77
4.1.2	Natural Ratio of Generator Voltage to Reactive Current of Generator 77
4.1.2.1	Formula Relating Generator Voltage and Reactive Current 77
4.1.2.2	Determination of Natural Ratio of Generator Voltage to Reactive Current of Generator 78
4.1.3	Static Voltage Droop Ratio of Generator 80
4.1.3.1	A Setting Voltage Static Error 80
4.1.3.2	Static Error Caused by the Load Disturbance I_Q 81
4.2	Ratio and Coefficient of Generator Voltage to Reactive Current of Generator 81
4.2.1	Ratio of Generator Voltage to Reactive Current of Generator 81
4.2.2	Determination of Ratio and Coefficient of Generator Voltage to Reactive Current of Generator at Different Operation Connection Modes 81
4.2.3	Natural Ratio and Additional Coefficient of Generator Voltage to Reactive Current of Generator 83
4.2.3.1	Natural Regulation Ratio of Generator Voltage to Reactive Current of Generator 83
4.2.3.2	Additional Regulation Coefficient of Ratio of Generator Voltage to Reactive Current of Generator 83
4.2.4	Formation of Additional Coefficient of Reactive Current Compensation 85
4.2.4.1	Analog AVR 85
4.2.4.2	Digital AVR 86
4.2.5	Conclusions 86
4.3	Transient State Characteristics of Excitation System 87
4.3.1	Transient Response of Large Disturbance Signal 87
4.3.1.1	Ceiling Voltage Factor of Excitation System 87
4.3.1.2	Voltage Response Ratio of Excitation System 88
4.3.2	Transient Response of Small Deviation Signal 90
4.3.3	Unit Step Response 91
4.3.4	Frequency Response 92
4.4	Stability Analysis of Excitation System 94

5	**Control Law and Mathematical Model of Excitation System** 97
5.1	Basic Control Law of Excitation System 97
5.1.1	Basic Terms of Excitation Control System 97
5.1.2	Transfer Function of Control System 98
5.1.3	Basic Control Law of Excitation Controller 98
5.1.3.1	Proportional Control Law (P) 99

5.1.3.2	Integral Control Law (I)	99
5.1.3.3	Proportional-Integral Control Law (PI)	100
5.1.3.4	Differential Control Law (D)	101
5.1.3.5	Proportional-Differential Control Law (PD)	101
5.1.3.6	Proportional-Integral-Differential Control Law (PID)	103
5.1.4	Numerical Description of the PID Control Law	105
5.1.4.1	Position Algorithm	105
5.1.4.2	Increment Algorithm	106
5.1.4.3	Practical Algorithm	106
5.1.5	Parallel Feedback Correction of Excitation Control System	107
5.1.5.1	Parallel Correction – Proportional Feedback	107
5.1.5.2	Parallel Correction – Dynamic Feedback	108
5.2	Mathematical Model of the Excitation System	108
5.2.1	Static Self-Excitation System	109
5.2.1.1	Series Compensation Self-Excitation System	109
5.2.1.2	Feedback Compensation Self-Excitation System	109
5.2.1.3	Proportional-Integral (PI) Self-Excitation System	109
5.2.2	AC Exciter System	110
5.2.2.1	AC Shunt-Exciter-Type Self-Excitation System	110
5.2.2.2	AC Separate-Exciter-Type Excitation System	113
5.2.2.3	Feedback Compensation Brushless Excitation System	113
5.2.2.4	PI Type Brushless Excitation System	114
5.2.3	Rectifier	115
5.3	Mathematical Model of Excitation Control Unit	118
5.3.1	Load Current Compensator (LCC)	118
5.3.2	Automatic Reactive Power Regulator (AQR)	118
5.3.3	Automatic Power Factor Regulator (APFR)	119
5.3.4	Under-Excitation Limiter (UEL)	120
5.3.5	Over-Excitation Limiter (OEL)	121
5.3.6	Power System Stabilizer (PSS)	123
5.3.6.1	ΔP Type PSS	123
5.3.6.2	$\Delta \omega$ Type PSS	123
5.3.6.3	Δf Type PSS	123
5.4	Parameter Setting of Excitation System	124
5.4.1	Setting of Proportional AVR Parameters	124
5.4.1.1	AVR Steady-State Open-Loop Gain K_A	124
5.4.1.2	Setting of AVR Phase Compensation and Damping Loop Parameters	124
5.4.2	Frequency Response and Step Response Calculation	125
5.4.3	PI AVR Parameter Setting	126
5.4.3.1	AVR Gain K_A	128
5.4.3.2	Integral Time Constant	129
5.4.4	PSS Parameter Setting	131
5.4.4.1	Setting of Time Constant T_g and Gain K_{PSS} for the Washout Filter	131
5.4.4.2	Phase Compensation	131
5.4.4.3	Gain and Output Limit	133
5.4.4.4	Setting of PSS Gain and Phase Compensation	133
6	**Basic Characteristics of Three-Phase Bridge Rectifier Circuit**	**137**
6.1	Overview	137
6.2	Operating Principle of Three-Phase Bridge Rectifier	137

6.3	Type I Commutation State	*139*
6.4	Commutation Angle	*144*
6.5	Average Rectified Voltage	*144*
6.5.1	Zero Rectifier Control Angle α and Commutation Angle γ	*144*
6.5.2	$\alpha > 0, \gamma = 0$, for Zero Rectifier Load	*145*
6.5.3	$\alpha > 0, \gamma = 0$, in Controllable and Loaded State	*145*
6.6	Instantaneous Rectified Voltage Value	*147*
6.7	Effective Element Current Value	*147*
6.8	Fundamental Wave and Harmonic Value for Alternating Current	*152*
6.8.1	Fundamental Wave for Alternating Current	*152*
6.8.2	AC Harmonic Value	*153*
6.9	Power Factor of Rectifying Device	*156*
6.10	Type III Commutation State	*161*
6.11	Type II Commutation State	*167*
6.12	External Characteristic Curve for Rectifier	*168*
6.13	Operating Principle of Three-Phase Bridge Inverter Circuit	*170*

7	**Excitation System for Separately Excited Static Diode Rectifier**	*175*
7.1	Harmonic Analysis for Alternating Current	*175*
7.2	Non-distortion Sinusoidal Potential and Equivalent Commutating Reactance	*177*
7.2.1	Non-distortion Sinusoidal Potential and Equivalent Commutating Reactance	*177*
7.2.2	Non-distortion Sinusoidal Potential Determined by Simplification	*178*
7.2.3	Calculation of Equivalent Commutating Reactance	*179*
7.3	Expression for Commutation Angle γ, Load Resistance r_f, and Commutating Reactance X_γ	*182*
7.4	Rectified Voltage Ratio β_u and Rectified Current Ratio β_i	*184*
7.4.1	Rectified Voltage Ratio β_u	*184*
7.4.2	Rectified Current Ratio β_i	*185*
7.5	Steady-State Calculations for AC Exciter with Rectifier Load	*186*
7.6	General External Characteristics of Exciter	*189*
7.7	Transient State Process of AC Exciter with Rectifier Load	*191*
7.8	Simplified Transient Mathematical Model of AC Exciter with Rectifier Load	*193*
7.8.1	Simplified Computation Method of External Characteristics of Rectifier	*193*
7.8.2	Simplified Computation Method of External Characteristics of Exciter	*194*
7.9	Transient State Process of Excitation System in Case of Small Deviation Change in Generator Excitation Current	*196*
7.10	Influence of Diode Rectifier on Time Constant of Generator Excitation Loop	*200*
7.11	Excitation Voltage Response for AC Exciter with Rectifier Load	*201*
7.11.1	Excitation Voltage Responses of AC Exciter without Load	*201*
7.11.2	Excitation Voltage Response of AC Exciter with Load	*201*
7.11.3	Field Voltage Response When Three-Phase Short-Circuit Occurs in the Generator	*204*
7.12	Short-Circuit Current Calculations for AC Exciter	*205*
7.12.1	Abrupt Short Circuit on DC Side of Rectifier	*205*
7.12.2	Determination of Transient Rotor Current-Free Component in Case of Abrupt Generator Terminal Three-Phase Short Circuit	*209*
7.13	Calculations for AC Rated Parameters and Forced Excitation Parameters	*211*
7.13.1	Determination of Distortion Sinusoidal Potential E_x and Rated Power P_N	*211*
7.13.2	Determination of Rated Parameters for AC Pilot Exciter	*212*

8 Brushless Excitation System 215

- 8.1 Evolution of Brushless Excitation System 215
- 8.1.1 Diode Brushless Excitation System 215
- 8.1.2 SCR Brushless Excitation System 218
- 8.2 Technical Specifications for Brushless Excitation System 219
- 8.2.1 Ratings 219
- 8.2.2 AC Exciter 220
- 8.2.3 AC Pilot Exciter 221
- 8.2.4 Rotating Rectifier 221
- 8.3 Composition of Brushless Excitation System 221
- 8.3.1 Wiring 221
- 8.3.2 Composition of AC Excitation Unit 222
- 8.3.2.1 AC Main Exciter 222
- 8.3.2.2 Rotating Rectifier 224
- 8.4 Voltage Response Characteristics of AC Exciter 224
- 8.4.1 Excitation Voltage Response when Generator Is Subject to Three-Phase Short Circuit 226
- 8.5 Control Characteristics of Brushless Excitation System 227
- 8.5.1 Time Constant Compensation in Case of Small Deviation Signals 227
- 8.5.2 Time Constant Compensation in Case of Large Disturbance Signal 230
- 8.5.2.1 Connect Additional Resistors in Series 230
- 8.5.2.2 Increasing the Excitation Voltage Multiple of the Exciter 230
- 8.5.2.3 Comparison of the Above Two Methods 231
- 8.6 Mathematical Models for Brushless Excitation System 232
- 8.6.1 Determination of Saturation Coefficient 233
- 8.6.2 I-Type Model 234
- 8.6.2.1 Block Diagram of I-Type Model 234
- 8.6.2.2 Derivation of I-Type Model 235
- 8.6.2.3 Linearization of Small Deviation Signal AC Exciter Model 237
- 8.6.3 AC-I Model 238
- 8.6.3.1 The Block Diagram 238
- 8.6.3.2 Saturation Coefficient S_E 238
- 8.6.3.3 Armature Reaction Coefficient K_D 238
- 8.7 AC2 Model 243
- 8.8 Generator Excitation Parameter Detection and Fault Alarm 246
- 8.8.1 Determination of Excitation Voltage and Current of Rotor Excitation Winding 246
- 8.8.2 Protection, Fault Monitoring, and Alarm of Brushless Exciter 248
- 8.8.2.1 Fault Types 248
- 8.8.2.2 Fault Monitoring 249
- 8.8.2.3 Telemetry Detection Technology 254

9 Separately Excited SCR Excitation System 255

- 9.1 Overview 255
- 9.2 Characteristics of Separately Excited SCR Excitation System 255
- 9.2.1 Characteristics of Hydro-generator SCR Excitation System 255
- 9.2.2 Characteristics of SCR Excitation System for Steam Turbine Generator 258
- 9.3 Influence of Harmonic Current Load on Electromagnetic Characteristics of Auxiliary Generator 260
- 9.3.1 Harmonic Current mmf 260
- 9.3.1.1 Harmonic Current Value 260
- 9.3.1.2 mmf Generated by Harmonic Current 261

9.3.2	Influence of Harmonic mmf on Voltage Waveform of Generator without Damping Winding	263
9.3.3	Influence of Harmonic mmf on Voltage Waveform of Generator with Damping Winding	264
9.3.4	Damping Winding Loss and Allowable Value	265
9.3.5	Stator Winding and Core Loss	266
9.3.5.1	Stator Winding Loss	266
9.3.5.2	Stator Core Loss	267
9.3.6	Armature Reaction and Power Factor	267
9.4	Parameterization of Separately Excited SCR Excitation System	268
9.4.1	Auxiliary Generator	268
9.4.1.1	Preselected Commutation Reactance Value X_γ	269
9.4.1.2	Calculated emf and Actual AC Stator Current of Auxiliary Generator	270
9.4.1.3	Rated Phase Current of Auxiliary Generator	270
9.4.1.4	Rated Phase Voltage of Auxiliary Generator	270
9.4.1.5	Capacity of Auxiliary Generator	270
9.4.2	Calculation of SCR Control Angle in Different Operating States	270
9.4.2.1	No-Load Operating Status	270
9.4.2.2	Rated Operating Status	271
9.4.2.3	Forced Excitation Status	271
9.4.3	Current-Sharing Reactor	271
9.5	Separately Excited SCR Excitation System with High-/Low-Voltage Bridge Rectifier	272
9.6	Parameterization of High-/Low-Voltage Bridge Rectifier	276
9.7	Transient Process of Separately Excited SCR Excitation System	281
10	**Static Self-Excitation System**	**285**
10.1	Overview	285
10.2	Characteristics of Static Self-Excitation System	288
10.2.1	Main Circuit Connection	288
10.2.2	Field Flashing Circuit	289
10.2.3	Stable Operating Point of Static Self-Excitation System	290
10.2.4	Short-Circuit Current of Self-Excited Generator	291
10.2.4.1	Relevant Expressions in Case of Three-Phase Short Circuit	292
10.2.4.2	Transient Characteristics of Three-Phase Short Circuit	294
10.2.5	Critical External Reactance	296
10.2.6	Influence of Short-Circuit Mode on Short-Circuit Current	297
10.2.7	Terminal Voltage Recovery of Self-Excited Generator after Short-Circuit Fault is Cleared	299
10.2.8	Inversion De-excitation of Self Thyristor Excitation System	300
10.2.9	Excitation Transformer Protection Scheme	302
10.3	Shaft Voltage of Static Self-Excitation System	307
10.3.1	Overview	307
10.3.2	Source and Protection of Shaft Voltage	307
10.3.3	Shaft Grounding System for Large Steam Turbine Generator	309
10.3.4	New Shaft Grounding Device	309
10.4	Coordination between Low Excitation Restriction and Loss-of-Excitation Protection	311
10.4.1	Generator Operating Limit Diagram	312
10.4.2	Admittance Measurement Principle and Derivation	316
10.4.3	Comparison of Impedance and Admittance Measurement	319
10.5	Electric Braking of Steam Turbine	321
10.5.1	Selection of Braking System	321
10.5.1.1	Mechanical Braking	321
10.5.1.2	Electric Braking	321

10.5.2	Basic Expression for Electric Braking	*323*
10.5.3	Selection of Braking Transformer	*324*
10.5.4	Key Points of Design of Electric Braking Circuit	*325*
10.6	Electric Braking Application Example at Pumped-Storage Power Station	*326*
10.6.1	Composition of Electric Braking System	*326*
10.6.2	Control of Electric Braking System	*326*
10.6.2.1	Setting of Control Logic of Monitoring System	*326*
10.6.2.2	Excitation System Electric Braking Execution Procedure	*327*
10.6.2.3	Short-Circuit Switch Hard Wiring Latching	*328*
10.6.2.4	Protection Latching at Time of Electric Braking Put-In	*328*

11 Automatic Excitation Regulator *329*

11.1	Overview	*329*
11.2	Theoretical Basis of Digital Control	*330*
11.2.1	Digital Discrete Technology	*330*
11.2.2	Z-Transform	*331*
11.2.3	Discretization of Continuous System	*332*
11.2.4	Transfer Function of Holder	*333*
11.2.5	Discrete Similarity Model and Bilinear Transformation	*336*
11.3	Digital Sampling and Signal Conversion	*337*
11.3.1	AC Sampling	*337*
11.3.2	Fourier Algorithm for AC Sampling	*338*
11.3.3	Three-Phase One-Point Algorithm	*339*
11.3.4	Speed Measurement Algorithm	*340*
11.4	Control Operation	*340*
11.5	Per-Unit Value Setting	*345*
11.6	Digital Phase Shift Trigger	*346*
11.6.1	Digital Phase Shift Trigger	*346*
11.6.2	Characteristics of Digital Phase Shift	*347*
11.7	External Characteristics of Three-Phase Fully Controlled Bridge Rectifier Circuit	*348*
11.7.1	Three-Phase Fully Controlled Bridge Rectifier Circuit	*348*
11.7.2	Linear Phase Shift Element	*348*
11.7.3	Cosine Phase Shift Element	*350*
11.7.4	Mathematical Model of Three-Phase Fully Controlled Rectifier	*350*
11.8	Characteristics of Digital Excitation Systems	*351*
11.8.1	Overview	*351*
11.8.2	Characteristics of Typical Digital Excitation System	*353*
11.8.3	Operation Modes of Digital Excitation System	*354*
11.8.4	Functions of Digital Excitation System	*355*

12 Excitation Transformer *365*

12.1	Overview	*365*
12.2	Structural Characteristics of Resin Cast Dry-Type Excitation Transformer	*367*
12.2.1	Core	*368*
12.2.2	Winding Insulation Structure	*368*
12.2.3	Winding Material	*368*
12.2.4	Heat Dissipation and Cooling	*368*
12.3	Application Characteristics of Resin Cast Dry-Type Excitation Transformer	*369*
12.4	Specification for Resin Cast Dry-Type Excitation Transformer	*369*
12.4.1	Rated Capacity of Excitation Transformer	*369*

12.4.2	Connection Set	370
12.4.3	Insulation Class and Temperature Rise	370
12.4.4	Impedance Voltage	371
12.4.5	Short-Time Current Overload Capacity	372
12.4.6	Resistance to Sudden Short-Circuit Current	375
12.4.6.1	Calculation of Sudden Short-Circuit Current	375
12.4.6.2	Thermal Stability	375
12.4.6.3	Dynamic Stability under Action of Sudden Short-Circuit Current	376
12.4.7	Overcurrent Protection	376
12.4.8	Overvoltage Suppression	376
12.4.8.1	Closing Surge Overvoltage	376
12.4.8.2	Opening Overvoltage	377
12.4.8.3	Atmospheric Overvoltage	377
12.4.9	AC R-C Protection	377
12.4.9.1	Role of R-C Protection	377
12.4.9.2	Energy Conversion Caused by Commutation Voltage Effect	377
12.4.9.3	Energy Conversion Caused by Reverse Current Recovery Effect	378
12.4.9.4	Calculation of Magnetic Energy Stored	379
12.4.10	Test Voltage	381
12.4.11	Noise	382
12.4.11.1	Overview	382
12.4.11.2	Sound Wave, Sound Pressure, Sound Intensity, and Sound Intensity Level	382
12.4.11.3	Noise of Epoxy Dry-Type Excitation Transformer	382
12.4.11.4	Noise Reduction Measures	382
12.4.11.5	Structure Carrier Noise Reduction Design	385
12.4.12	Electrostatic Shield	387
12.4.13	Operational Environmental Impact	388
12.4.13.1	Environmental Class	388
12.4.13.2	Climate Class	388
12.4.13.3	Fire Class	388
12.4.13.4	Operating Environment	388
12.5	Harmonic Current Analysis	389
12.5.1	Overview	389
12.5.1.1	Harmonic Current Analysis	389
12.5.1.2	Influence of Harmonic Current	391
12.5.1.3	Harmonic Loss Calculation	392
12.5.2	Excitation Transformer Capacity Calculation Checking Example: 740 MVA ABB Hydro-Turbine of Three Gorges Hydro Power Station	393
12.5.2.1	Specification for Technical Parameters of Excitation Transformer	394
13	**Power Rectifier**	**395**
13.1	Specification and Essential Parameters for Thyristor Rectifier Elements	395
13.1.1	Overview	395
13.1.2	Specification for Thyristor Rectifier Elements	395
13.1.3	Operating Characteristics of Thyristor	395
13.1.3.1	Nominal Voltage	398
13.1.3.2	Rated Current	398
13.2	Parameterization of Power Rectifier	400
13.2.1	Thyristor Element Parameters	400
13.2.2	Calculation of Reverse Repetitive Peak Voltage (U_{RRM})	400

13.2.3 Calculation of Thyristor Element Loss *401*
13.2.4 Selection of Radiator *401*
13.2.5 Selection of Fan Model *403*
13.2.5.1 Calculation of Ventilation Quantity *403*
13.2.5.2 Selection of Fans *404*
13.2.6 Selection of Quick Fuse Parameters *404*
13.2.7 Calculation of Rectifier Loss *406*
13.3 Cooling of Large-Capacity Power Rectifier *407*
13.3.1 AN Radiator *407*
13.3.2 AF Radiator *407*
13.3.3 Heat Pipe Radiator *408*
13.3.3.1 Working Principle of Heat Pipe *408*
13.3.3.2 Loop Heat Pipe (LHP) Radiator *409*
13.3.3.3 Selection of Power Radiator *410*
13.3.3.4 Design of High-Power Heat Pipe Rectifier *411*
13.4 Current Sharing of Power Rectifier *413*
13.4.1 Background *413*
13.4.2 Factors Influencing Effect of Current Sharing *414*
13.4.2.1 AC-Side Inlet Line *414*
13.4.2.2 DC-Side Outlet Line *414*
13.4.2.3 Commutation Process *414*
13.4.2.4 Thyristor Parameters *415*
13.4.3 Digital Current Sharing *415*
13.4.4 Comparison Between Digital and Conventional Current Sharing *415*
13.4.5 Conclusions *416*
13.5 Protection of Power Rectifier *416*
13.5.1 Power Rectifier Anode Overvoltage Suppressor *417*
13.5.2 R-C Protection of Power Rectifier Thyristor Elements *421*
13.5.2.1 Working Principle of R-C Protection *421*
13.5.2.2 Principle for Parameter Selection for Commutation Snubber *422*
13.5.2.3 Energy Conversion of Commutation Process *426*
13.5.2.4 Commutation Snubber Parameter Calculation Program *426*
13.5.2.5 R-C Snubber Parameter Verification *427*
13.6 Thyristor Damage and Failure *429*
13.6.1 Factors Affecting Thyristor Operation Safety in Lectotype Design *429*
13.6.2 Thyristor Damage and Failure Analysis and Identification *430*
13.6.2.1 Thyristor Overvoltage Breakdown *430*
13.6.2.2 Thyristor Overcurrent *431*
13.6.2.3 Too Large On-State Current Critical Rise Rate (di/dt) *431*
13.7 Capacity of Power Rectifiers Operating in Parallel *433*
13.7.1 Verification and Calculation of Thyristor Power Cabinet Capacity *433*
13.7.1.1 Calculation of Power Rectifier Load *433*
13.7.1.2 Calculation of Power Rectifier Thyristor Element Loss *434*
13.7.1.3 Calculation of Thyristor Junction Temperature and Radiator Temperature Rise *434*
13.7.2 Influence of Altitude on Output Capacity of Power Rectifier *437*
13.8 Uncertainty of Parallel Operation of Double-Bridge Power Rectifiers *437*
13.8.1 Background *437*
13.8.2 Analysis of Uncertainty of Protection Action of Fast Fuses in Double-Bridge Parallel Scheme *437*
13.9 Five-Pole Disconnector of Power Rectifier *439*

14	**De-excitation and Rotor Overvoltage Protection of Synchronous Generator** *441*	
14.1	Overview *441*	
14.2	Evaluation of Performance of De-excitation System *443*	
14.2.1	Equivalent Generator Time Constant Method *443*	
14.2.2	Valid De-excitation Time Method *446*	
14.2.3	Generator Voltage-Based De-excitation Time Determination Method *446*	
14.3	De-excitation System Classification *447*	
14.3.1	Linear Resistor De-excitation System *447*	
14.3.1.1	Expression for Linear De-excitation Time *447*	
14.3.1.2	Linear Resistor De-excitation Commutation Conditions *448*	
14.3.1.3	Stepped Linear Resistor De-excitation System *450*	
14.3.2	Nonlinear Resistor De-excitation System *452*	
14.3.2.1	Expression for Nonlinear De-excitation Time *453*	
14.3.2.2	Nonlinear Resistor De-excitation Commutation Conditions *454*	
14.3.3	Crowbar De-excitation System *455*	
14.3.3.1	Overview *455*	
14.3.3.2	Functions of Crowbar *456*	
14.3.4	AC Voltage De-excitation System *459*	
14.4	Influence of Saturation on De-excitation *463*	
14.5	Influence of Damping Winding Circuit on De-excitation *465*	
14.6	Field Circuit Breaker *467*	
14.6.1	DC Field Circuit Breaker *468*	
14.6.1.1	CEX Series of Modular DC Contactors from France's Lenoir Elec *468*	
14.6.1.2	HPB Series of Fast DC Circuit Breakers from Switzerland's Secheron *469*	
14.6.1.3	UR Series of DC Circuit Breakers from Switzerland's Secheron *469*	
14.6.1.4	Gerapid Series of Fast DC Circuit Breakers *471*	
14.6.2	AC Field Circuit Breaker *475*	
14.6.2.1	AC Air Circuit Breaker *475*	
14.6.2.2	AC Disconnector *475*	
14.6.2.3	DC Disconnector *475*	
14.7	Performance Characteristics of Nonlinear De-excitation Resistor *477*	
14.7.1	U-I Characteristic Expression for Single Valve *478*	
14.7.2	U-I Characteristic Expression for Elements *479*	
14.7.3	Temperature Coefficient of SiC Nonlinear Resistor *480*	
14.7.4	Calculation of Temperature Rise *480*	
14.7.5	De-excitation Time *481*	
14.7.6	Parameter Selection for SiC Nonlinear Resistor *481*	
14.7.7	Timeliness of SiC Nonlinear Resistor *482*	
14.7.8	Damage and Failure Forms for SiC Nonlinear Resistor *482*	
14.7.9	Specifications *484*	
15	**Excitation System Performance Characteristics of Hydropower Generator Set** *485*	
15.1	Overview *485*	
15.2	Static Self-Excitation System of Xiangjiaba Hydro Power Station *485*	
15.2.1	Overview *485*	
15.2.2	Generator Parameters and Excitation System Configuration *487*	
15.2.3	Function and Component of Excitation System *489*	
15.2.4	Composition of Excitation System *489*	
15.2.4.1	Automatic Voltage Regulator *489*	
15.2.4.2	Thyristor Rectifier *497*	

15.2.4.3 De-excitation and Overvoltage Protection Device *501*
15.2.4.4 Excitation Transformer *507*
15.2.4.5 Electric Braking *508*
15.2.5 System Control Logic *510*
15.2.5.1 Control Logic Mode *510*
15.2.5.2 Excitation System Start and Stop Condition *512*
15.2.5.3 Special Running Modes *512*
15.2.5.4 Limiters *514*
15.2.5.5 PSS *519*

16 Functional Characteristics of Excitation Control and Starting System of Reversible Pumped Storage Unit *521*
16.1 Overview *521*
16.2 Operation Mode and Excitation Control of Pumped Storage Unit *521*
16.2.1 Operation Modes of Pumped Storage Unit *521*
16.2.2 Excitation Control Characteristics of Pumped Storage Unit *522*
16.2.3 Selection of Main Circuit of Excitation System of Pumped Storage Unit *523*
16.3 Application Example of Excitation System of Pumped Storage Unit *525*
16.3.1 Excitation System of Units of Tianhuangping Pumped Storage Power Station *525*
16.3.2 Composition of Excitation System *527*
16.3.2.1 Excitation Transformers *527*
16.3.2.2 Power Rectifiers *527*
16.3.2.3 Initial Field Flashing Circuit and De-excitation System *527*
16.3.2.4 Overvoltage Protection Devices *527*
16.3.2.5 Excitation Regulator *528*
16.3.3 Excitation Regulator Software *529*
16.3.4 Excitation Regulator Software Flow Diagram *530*
16.3.4.1 Overview *530*
16.3.4.2 Switching Between Automatic Mode and Manual Mode *531*
16.3.4.3 Reactive Power/Power Factor Regulator *531*
16.3.4.4 Reactive Power Capacity Characteristic Curves and Limiters of Generator *532*
16.3.4.5 Limiters *532*
16.3.4.6 Stator Current Limiter *535*
16.3.4.7 Rotor Power Angle Limiter *536*
16.3.4.8 Voltage/Frequency Limiter Generating Additional Frequency Signals *537*
16.3.4.9 PSS *538*
16.3.5 Additional Functions of Excitation Regulator *540*
16.3.5.1 Limitation Protection *540*
16.3.5.2 Operation and Switching of Excitation Regulator *540*
16.3.5.3 Debugging and Self-Test of Excitation Regulator *540*
16.3.6 Excitation Regulation in Different Operation Modes *540*
16.3.6.1 Working Conditions Including GO, GCO, PMO, and PCO Provided for Pumped Storage Units of Tianhuangping Pumped Storage Power Station *540*
16.3.6.2 Simultaneous Start under PMO Conditions *541*
16.3.6.3 Pump Operation Mode *541*
16.3.6.4 Electric Braking *541*
16.3.6.5 SFC or Back-To-Back Operation *541*
16.3.6.6 Test Operation Mode *541*
16.4 Working Principle of SFC *542*
16.4.1 Connection of SFC of Pumped Storage Unit *542*

16.4.2	Composition of SFC *542*	
16.4.2.1	Power Unit *543*	
16.4.2.2	Control and Protection Units *544*	
16.4.2.3	SFC Control Mode *545*	
16.4.3	SFC Start Procedure *548*	
16.4.3.1	SFC Auxiliaries Start Process *548*	
16.4.3.2	SFC Preparation Process *548*	
16.4.3.3	SFC Start Process *548*	
16.4.3.4	SFC Standby Process *549*	
16.4.3.5	SFC Auxiliaries Stop Process *549*	
16.4.4	Establishment of SFC Electrical Axis *549*	
16.4.5	Starting System of SFC of Units of Tiantang Pumped Storage Power Station *551*	
16.4.5.1	Composition of SFC System *551*	
16.4.5.2	Back-To-Back Start *553*	
16.4.6	Electromagnetic Induction Method for Identification of Initial Position of Rotor *554*	
16.4.7	Capacity Calculation of SFC *557*	
16.4.8	SFC Power Supply Connection *559*	
16.5	SFC Current and Speed Dual Closed-Loop Control System *560*	
16.5.1	Speed Control of Pump Motor Unit during SFC Start *560*	
16.5.2	Unit Voltage Control during SFC Start of Pump Motor Unit *561*	
16.5.3	Synchronization Control of Pump Motor Unit *561*	
16.6	Influence of SFC Start Current Harmonic Components on Power Station and Power System *562*	
16.6.1	Harmfulness and Characteristics of Harmonics *562*	
16.6.2	Simplified Calculation of Harmonic Components *563*	
16.6.3	Improvement of Harmonic Operation State of SFC *565*	
16.7	Local Control Unit (LCU) Control Procedure for Pumped Storage Unit *566*	
16.7.1	LCU Configuration *566*	
16.7.2	LCU Control Program *566*	
16.7.3	GO Control Process *567*	
16.7.4	PMO Control Flow *567*	
16.8	Pumped Storage Unit Operating as Synchronous Condenser *568*	
16.8.1	GCO Mode *568*	
16.8.2	PCO Mode *568*	
16.8.3	PCO Start Mode and Coordination *569*	
16.9	De-excitation System of Pumped Storage Unit *569*	
16.9.1	Composition of AC De-excitation System *569*	
16.9.2	Functional Characteristics of AC De-excitation System *570*	
16.9.2.1	AC Field Circuit Breaker Application Characteristics *570*	
16.9.2.2	Coordination of Thyristor Rectifier Bridge Trigger Pulse AC Field Circuit Breaker *570*	
16.9.2.3	De-excitation System Composed of AC Switch and Linear Resistor Crowbar *570*	
16.9.2.4	Characteristics of AC De-excitation System *571*	
16.9.3	Operation Time Sequence of AC De-excitation System *571*	
16.9.3.1	De-excitation for Stop in Normal Case *571*	
16.9.3.2	De-excitation for Stop in Case of Fault *572*	
16.10	Electric Braking of Pumped Storage Unit *572*	
16.11	Shaft Current Protection of Pumped Storage Unit *574*	
16.11.1	Configuration of Shaft Current Protection *574*	
16.11.2	Problems and Handling *576*	
16.11.2.1	Shaft Current Protection Action Caused by Metal Lap Joint Between Upper Bushing Back and Oil Cooler Upper Cover Plate *576*	

16.11.2.2	Decrease in Upper Insulation of Thrust Bearing Caused by Impurities in Lubricating Oil	*576*
16.11.2.3	Equipment Grounding in Insulation Measurement Area	*577*
16.12	Application Characteristics of PSS of Pumped Storage Unit	*577*

17 Performance Characteristics of Excitation System of 1000 MW Turbine Generator Unit *579*

17.1	Introduction of Excitation System of Turbine Generator of Malaysian Manjung 4 Thermal Power Station	*579*
17.2	Key Parameters of Turbine Generator Unit and Excitation System	*581*
17.2.1	Key Parameters of Generator Unit	*581*
17.2.2	Key Parameters of Excitation System	*581*
17.2.2.1	Data of Generator Unit	*581*
17.2.2.2	Data of Excitation System	*584*
17.3	Parameter Calculation of Main Components of Excitation System	*585*
17.3.1	Excitation Transformer	*585*
17.3.1.1	Excitation Transformer Output Current	*585*
17.3.1.2	Transformer Output Voltage	*585*
17.3.1.3	Excitation Transformer Capacity	*586*
17.3.1.4	Design Values of Excitation Transformer	*586*
17.3.2	DC Field Circuit Breaker	*586*
17.3.2.1	Rated Operating Voltage of Field Circuit Breaker	*586*
17.3.2.2	Rated Current of Main Contact of Field Circuit Breaker	*587*
17.3.2.3	Design Values of Field Circuit Breaker	*587*
17.3.3	Field Discharge Resistor	*587*
17.3.4	Crowbar	*588*
17.3.5	Field Flashing	*588*
17.3.5.1	Excitation Data	*588*
17.3.5.2	AC Flashing – Design of the Transformer	*588*
17.3.6	Current Transformer	*589*
17.3.6.1	Current Transformer on Primary Side of Excitation Transformer	*589*
17.3.6.2	Current Transformer on Secondary Side of Excitation Transformer	*589*
17.3.7	Parameters and Configuration of Excitation System	*589*
17.4	Block Diagram of Automatically Regulated Excitation System	*592*
17.4.1	AVR Reference Voltage	*592*
17.4.2	AVR Reference Voltage Limitation	*592*
17.4.3	Main Voltage Control Loop for Static Excitation System - VREG	*593*
17.4.4	Under-Excitation Limit – UEL	*593*
17.4.5	Overexcitation Limitation for Static Excitation System – OEL	*593*
17.4.6	Dual-Input PSS	*595*
17.4.7	Stator Current Limitation – STCL	*595*
17.4.8	Reactive Power or Power Factor Regulators – RPPF	*596*
17.4.9	Droop	*596*
17.4.10	Potential – Source Excitation System ST7B	*596*

18 Performance Characteristics of 1000 MW Nuclear Power Steam Turbine Excitation System *601*

18.1	Performance Characteristics of Steam Turbine Generator Brushless Excitation System of Fuqing Nuclear Power Station	*601*
18.1.1	Basic Parameters	*601*
18.1.1.1	Generator Parameters	*601*

18.1.1.2	Exciter Parameters	*601*
18.1.1.3	Excitation Transformer Parameters	*601*
18.1.2	Description of Excitation System	*601*
18.1.2.1	Automatic Regulation Excitation System	*602*
18.1.2.2	External Communication Network	*603*
18.1.2.3	Excitation System Monitoring and Protection	*603*
18.2	Structural Characteristics of Brushless Excitation System	*608*
18.2.1	Basic Parameters	*608*
18.2.2	Structural Characteristics	*608*
18.2.3	Rotary Diode Rectifier	*610*
18.3	Analysis of Working State of Multi-Phase Brushless Exciter	*612*
18.3.1	Overview	*612*
18.3.2	Working State Analysis of Odd-Numbered Diodes' Rectifier Circuit	*612*
18.3.3	Double-Layer Fractional Slot Variable Pitch Wave Winding Connection	*614*
18.3.4	Phase Angle of 39-Phase Outgoing Lines Emf	*615*
18.3.5	Phase Angle of Emf in Single Winding	*615*
18.3.6	Phase Analysis of Single Winding Emf	*616*
18.3.7	Phase of Phase Emf	*616*
18.3.8	Armature Winding Phase Emf Phase Diagram	*617*
18.3.9	Phasor Diagram of Total Emf Composed of 39-Phase Windings	*617*
18.4	Calculation of Excitation System Parameters of Fuqing Nuclear Power Station	*618*
18.4.1	Ceiling Parameters	*618*
18.4.2	Excitation Transformer	*618*
18.4.3	Thyristor Bridge	*619*
18.4.4	Other Thyristor Protections	*622*
18.4.5	Flashing Circuit	*622*
18.4.6	De-excitation Circuit	*623*
18.4.7	Field Circuit Breaker	*623*
18.5	Static Excitation System of Sanmen Nuclear Power Station	*624*
18.5.1	Introduction	*624*
18.5.2	Basic Parameters	*624*
18.5.2.1	Available Synchronous Machine Data	*624*
18.5.2.2	Excitation System Rated Output Value	*626*
18.5.2.3	Excitation Transformer Parameters	*626*
18.5.2.4	Thyristor Rectifier Parameters	*626*
18.5.2.5	DC Short Circuit	*627*
18.5.2.6	Nonlinear Field Suppression	*627*
18.5.3	Functional Description of Static Excitation System	*627*
18.5.3.1	Overview	*627*
18.5.3.2	Control Electronics	*628*

References *639*

Index *643*

About the Author

Jicheng Li
Principal Engineer
 Part-time Researcher in the State Key Laboratory of Power System of Tsinghua University.
 Senior Technical Advisor for the excitation system of Three Gorges Hydropower Station.
 Technical Advisor of Electric Power Systems Automation Committee of China Society for Hydropower Engineering.
 Chief Technical Advisor, working group on excitation of the nuclear power plant, China Nuclear Energy Association.
 Entitled to the special allowance for experts issued by the State Council.

- Graduated from the Department of Electrical Engineering of Northeastern University in September 1953, and engaged in the research and development of synchronous generator excitation systems, first at Harbin Electrical Machinery Plant and then at Harbin Institute of Electrical Engineering and Ministry of Machine-Building Industry. Designed China's first excitation regulator and the first set separated excited diode rectifier excitation system.
- Transferred to Harbin Power Plant Equipment Design Institute of Ministry of Machine-Building Industry in 1979.
- Studied the design of brushless excitation systems for 600 MW turbo-generator in the Orlando and Pittsburgh branches of Westinghouse Electric Corporation in September 1982.
- As a representative of Huaneng International Power Development Co., Ltd., supervised and inspected four sets of 350 MW turbo-generators of Huaneng Shangan Power Plant and Huaneng Nantong Power Plant in General Electric Company in the United States in June 1987.
- Participated in the academic exchange visits to the electricity testing institutes in different areas of the former Soviet Union as a member of the Chinese electrical expert delegation formed by China Electricity Council of Ministry of Energy in September 1990.
- Responsible for the design, installation, and commissioning of the complete excitation equipment for hydroelectric projects in the Malaniangate Hydropower Station in the Philippines in July 1997.
- Appointed as a part-time researcher in the National Key Laboratory of The Power System of Tsinghua University in 1999.
- Conducted technical support and consultancy work of the training, on-site commissioning, and implementation of the 1000 MW large turbo-generator excitation system in Daya Bay Nuclear Power Plant, 600 MW large turbo-generator excitation system in Yangzhou Second Power Plant, and 740 MW large turbo-generator excitation system in Zhuhai Power Plant in 2000.
- Hired as senior technical adviser for the Three Gorges hydropower station in 2002, and participated in the implementation and commissioning work of the 700 MW hydro-generator excitation system in the left bank.

- Hired by Tianwan Nuclear Power Plant of Jiangsu Nuclear Power Co. Ltd. to conduct technical support and consultancy work of staff training and commissioning of the 1000 MW brushless turbo-generator excitation system in 2004.
- Participated in the bidding and evaluation of many hydroelectric excitation systems in the Three Gorges Hydropower Station, Longtan Hydropower Station, Jinghong Hydropower Station, and Laxiwa Hydropower Station since 2005.
- Visited the Excitation and Product & Technology Department of Asea Brown Boveri Ltd. in Switzerland and the Technical Department of Basler Electrical Branch in France for technical exchanges in September 2013.
- Visited United Kingdom–based M&I Materials Ltd. in Manchester for technical exchanges about SIC Metrosil Varistor and MIDEL synthetic ester in May 2015.

Foreword

Modern Synchronous Generator Excitation System Design and Application has finally come out, and this is the fifth excitation monograph I have authored in my career of more than 60 years as an excitation professional.

The first excitation monograph, *Automatic Adjustment Excitation Devices*, which was written in 1958, was published by Water Resources and Electric Power Press. I was studying Russian at the preparation department of study abroad in the USSR of Beijing Foreign Studies University. My initial plan was to use my spare time to summarize the results of my engagement in the field of excitation in Harbin Electric Machine Factory over the past five years. But later, I got the idea of writing a book with the support, encouragement, and recommendation of Professor Shi Cheng of Department of Electrical Engineering in Tsinghua University, who studied in Germany. From my technical career perspective, the reason why this book has been published is because Professor Shi Cheng was my first mentor.

Automatic Adjustment Excitation Devices was the first excitation monograph published in China that discusses excitation systems with the direct-current exciter and magnetic amplifier excitation regulator. Its success was unexpected effects were gained after publishing, and it almost became the most influential monograph in the water and electricity industry for more than ten years. No matter where I would go, even in a small hydropower station on a remote mountain, I would get a warm reception. It was such an honor for a young man in his twenties to become a recognized author. Cases like this, there are still more.

The publication of the second excitation monograph, *Modern Synchronous Generator Rectifier Excitation Systems*, was much bumpier. Due to well-known historical circumstances, the manuscript was written in 1968 and published only in 1988, a gap of twenty years. Four times during this period, the manuscript was rewritten, and the hard work it entailed is beyond the reach of the pen. Some materials in the manuscript were written on road trips, some were conceived in crowded train stations, in the shuttling lounges, while some chapters were written during holidays.

In particular, I remember some of the chapters were written when I worked in the rural production team in Liuhaogou, Lixin commune, Acheng district, in Harbin suburb in the early 1970s. In the severe winter in the northern rural areas, the night is quiet and the wind is cold and chilly. In the rural cottage where I lived, I often used to burn clay or fill a small indoor furnace with a pile of corn sticks to keep warm. The raging flames filled me with hope; however, the warmth was only momentary, just like the matches in the hands of the "little match girl" of Hans Christian Andersen's, it flashed! A wisp of moonlight gleaming through the window, flashing against the freezing frost of the corner of the wall. The scene, the situation, was really "a window of moon, a curtain of frost." Loneliness, depression, and tears welled up in my heart, and this was often the case with writing at night.

As time passes, my nib has been creeping on paper, like a silkworm crawling on mulberry leaves. Over the decades of my youth, I have been steadfastly trying to realize my dream of writing to give play to my personal value.

There are many happy moments associated with writing in the journey of my life, and the social identity of being a writer has given me great encouragement and inspiration. I remember that when I visited the Xinjiang Dushanzi power plant in 1995, I wanted to return to Urumqi on that very day, but when the factory leader came to know that I was the author of *Modern Synchronous Generator Rectifier Excitation System*, he said, "Teacher

Li is here! Teacher Li is here ...." I was invited to stay overnight immediately, with five leaders in the factory accompanying me, and a special banquet was arranged for Xinjiang flavor in the luxurious Bachman hotel. What shocked me the most was that even in the long northwest desert, a book can have such wide-ranging influence. It should be mentioned that Professor Jingde Gao, master of Tsinghua University, wrote the preface for my second monograph. I will always remember his care and encouragement for the rest of my life.

Design and Application of Modern Synchronous Generator Excitation Systems was my third excitation monograph, conceived at the end of 1995 and published in July 2002; its writing took more than seven years. The manuscript was written in all parts of the country; some were written in the library of Urumqi Petrochemical Plant in Xinjiang Urumqi, and some parts were written in Fulaerji Thermal Power Plant in Jinan Shandong, Chaozhou Guangdong and the Northeast, Daya Bay Nuclear Power Station, Tianwan Nuclear Power Station, Yangzhou Second Power Plant, Zhuhai Power Plant, and other places.

The third monograph was published at a time of rapid economic and technological development in China. Due to the needs of the times, the publication of the third excitation monograph received a favorable response.

It is particularly worth noting that the publication of the third monograph received the support and help of many colleagues in the industry, especially my best friend, Qing LU, academician of Chinese Academy of Science and professor of Tsinghua University, who wrote the preface for the monograph. I would like to express my heartfelt thanks to him.

My fourth excitation monograph, *Design and Application of The Modern Synchronous Generator Excitation System (second edition)*, was published in 2009 and was conceived in 2003; it took me six years to write it. The writing background of the fourth monograph is that I served as the chief technical adviser of the 700 MW hydropower unit excitation system of the Three Gorges Hydropower Station in 2003, and participated in the commissioning and commissioning of the left bank unit excitation system. At the construction site, I personally feel hydropower builders' love for the motherland and their selfless dedication of the great mind. This vigorous and enterprising spirit of the ages has given me great momentum to keep pace with the times! Through two years on-site work in the Three Gorges hydropower station excitation system, I gained a better understanding and mastery of the performance characteristics of the excitation system in large units.

After that, I also worked at the Tianwan Nuclear Power Station as a technical support consultant for the Russian 1000 MW nuclear power plant excitation system, participated in on-site commissioning, special research and personnel training, translated the relevant Russian technical documents as well as organizing personnel training.

These business experiences, as well as the training work in Zhuhai Power Plant 740 MW excitation system, Yangzhou Second Power Plant 600 MW excitation system, Westinghouse Thermal Power Plant excitation system, and Daya Bay Nuclear Power Station 1000 MW Russian nuclear power plant excitation system; and the research and special project investigation on the excitation system of the pumped storage unit, all these gave me the idea of creating a monograph to discuss excitation systems across disciplinary boundaries. For example, in the field of excitation of hydropower pumped storage units, while discussing the excitation system, the switching of operation mode and frequency conversion starting feature are supplemented, giving the reader a complete and systematic understanding of the topic. At the same time, in the discussion of the basic chapters, for example, more emphasis was put on the application of the excitation system during the discussion of the basic characteristics of the synchronous motor and the description of the characteristics of the rectifying line. That is the reason why *Design and Application of The Modern Synchronous Generator Excitation System (Second Edition)* was developed.

Design and Application of Modern Synchronous Generator Excitation System is my fifth excitation monograph, which was written in 2013 and took more than four years to finalize. The biggest difference between the third edition and the second edition is that the second edition is about the excitation systems of Three Gorges Hydropower Station and Tian Wan Nuclear Power Plant, all of which belong to the design level of around 2000, while the third edition discusses mainly the 850 MW hydroelectric power unit excitation system in Xiangjiaba Hydropower Station with the world's largest single machine capacity, the 1000 MW steam turbine thermal power unit excitation system of Malaysia Manjung 4 Thermal Power Station, and the excitation systems of the 1278 MVA half-speed turbo-generator unit of Fuqing Nuclear Power Plant and the 1407

MVA half-speed turbo-generator unit of Sanmen Nuclear Power Plant, which are all excitation systems of world-class design level in 2015.

In addition to focusing on the development of contemporary cutting-edge control technology of excitation, the content of the third edition pays more attention to its novel and comprehensive requirements, laying emphasis on the close integration of engineering applications and basic theory. For example, the premise of a discussion of the basic theory of the synchronous motor should be to help analyze the characteristics of excitation systems, and the discussion of the three-phase bridge rectifier line should be closely related to the analysis of the operational characteristics of the power rectifier.

The third edition, based on a discussion of excitation systems, focuses on the integration with other related disciplines. For example, in the chapter on hydroelectric excitation systems, the excitation method of the pumped storage unit is also treated, and variable frequency starting is also extended. In the discussion of the underexcitation limit of the generator, the interdisciplinary fields of loss of field failure and relay protection of the generator are also organically integrated. All of these enable readers to broaden their horizons and understand the essence of the thematic content on a deeper level.

In the writing process of the third edition, many of the domestic and international excitation industry colleagues have given great support and help for the writing of this book. At the time of publication of this book, I would like to express my sincere thanks to Xiluodu Hydropower Station, Xiangjiaba Hydropower Station, Three Gorges Hydropower Station, Guangzhou China Electric Apparatus Research Institute, ABB Engineering (Shanghai) Co. Ltd., SIEMENS Power Pant Automation Co., Ltd. (Nanjing), and Alstom China.

On the publication of the third edition, it should also be noted that according to the copyright output agreement signed by China Electric Power Press and United States John Wiley & Sons Inc. (which has a publishing history of 210 years), Wiley Press will issue the English version of the book around the world based on the Chinese version model of *Modern Synchronous Generator Excitation System Design and Application (Third Edition)* at the beginning of 2018. This English translation work has also been supported and helped by colleagues from home and abroad, and I express my heartfelt gratitude to all of them.

December 23, 2018 Jicheng Li
Snowy winter in Harbin

Preface

Power systems are the most complex structures and technical systems ever made by humans. China has experienced amazing growth in electric energy production and consumption over the last few decades. Today, China is the world's largest electricity producer. The famous Three Gorges hydropower station at Yangtze River is only one example of the incredible development of China's power system.

Stable and reliable operation of the electric power system is a precondition for all modern societies and a challenge to power engineers. A deep understanding of the system with all its components is crucial to analyze and maintain power system stability.

Synchronous generators are utilized to produce the majority of electrical power. The required field current of the machine is provided by the excitation system. Moreover, the equipment includes all the control elements needed for the generator. It is essential to control the voltage and to ensure that the generator remains in synchronism. Hence, the design of excitation systems is a very important subject.

Professor Jicheng Li is one of the most experienced experts in power engineering, especially in excitation systems. Beyond his comprehensive expertise of power engineering, his practical experience in numerous projects is remarkable. In this textbook, all aspects are covered: synchronous generator, brushless and static excitation system, stability, control systems, and so on. In addition, various applications help illustrate this sophisticated subject.

It is my great honor and pleasure to write a foreword for this book. It will be of great value to power engineers and all students willing to meet the challenge of this fascinating field.

University of Applied Sciences and Arts Hannover *Prof. Dr. Ruediger Kutzner*
Faculty of Electrical Engineering and Information Technology

Technical Biography of Prof. Dr. Ruediger Kutzner
Since 2004 Professorship for Control Systems at the Faculty of Electrical Engineering and Information Technology at the University of Applied Sciences and Arts, Hannover, Germany. He gives lectures to bachelor's and master's students of electrical engineering and power engineering programs.
- 1996–2004 Siemens AG, Germany
- He was in charge of the development and design of control systems for power plants. Together with his team he developed real-time simulators to check and optimize the performance of excitation control systems.
- 1997 He received the Dr.-Ing. degree (equivalent to PhD) in Electrical Engineering from the Technical University of Braunschweig, Germany. Earlier, he studied Electrical Engineering at the same University and received the Dipl.-Ing. degree (equivalent to MSc).

- Main focus Excitation systems, power system stabilizers, turbine governors, electrical machines, simulation, and closed-loop control of generators.
- Memberships IEEE PES: Excitation System and Controls Subcommittee
- VDE (Association for Electrical, Electronic & Information Technologies): Chair of the working group "Control of Synchronous Machines and Transformers."

Introduction

Keeping pace with the times, this book presents an original and comprehensive description of the excitation systems featuring world-class designs on the hydroelectric unit of Xiangjiaba Hydro Power Station, the 1,000 MW thermal power unit of Malaysia Manjung 4 Coal-Fired Power Station, and the nuclear power units of Fuqing and Sanmen Nuclear Power Stations.

The contents of this book are as follows: evolution and development of excitation control, characteristics of synchronous generators, effect of excitation regulation on power system stability, static and transient state characteristics of excitation systems, control law and mathematical models of excitation systems, basic characteristics of the three-phase bridge rectifier circuit, excitation system for the separately excited static diode rectifier, brushless excitation system, separately excited SCR excitation system, static self-excitation system, automatic excitation regulator, excitation transformer, power rectifier, de-excitation and rotor overvoltage protection of synchronous generators, performance characteristics of the excitation system of hydro-turbines, functional characteristics of excitation control and starting system of the reversible pumped storage unit, performance characteristics of the excitation system of the 1,000 MW steam turbine, and performance characteristics of the excitation system of the 1,000 MW nuclear power steam turbine.

This book can be used as a reference for excitation-related professionals from electric power test institutes and motor manufacturing plants, power station designers, debuggers, operators, and maintainers or as a learning/reference material for college students studying electric power systems.

Acknowledgement

The book is rich in content and novel in materials. It is a best-seller with a circulation of 15,000 in China. The English version of the book will surely play a powerful role in promoting the development of world excitation technology.

Mr. Giles Salt
CEO Chief Executive Director, M&I Materials.
Manchester, United Kingdom.

1

Evolution and Development of Excitation Control

1.1 Overview

In the modern power system, improving and maintaining the stability of synchronous generator operation is fundamental to the safe and economic operation of the power system. Of the numerous measures to improve the stable operation of synchronous generators, the use of modern control theory for improved excitation control is recognized as an economic and effective way.

Since the 1950s, with the passage of time, significant progress has been made both in control theory and in the development and application of electronic devices, further accelerating the development of the excitation control technology.

This chapter briefly describes the evolution of excitation control in different historical periods over the past half century, and develops useful conclusions from the main principles recognized from industry practice rather than from derivations based on mathematical logic.

1.2 Evolution of Excitation Control

In the early 1950s, the main function of an automatic voltage regulator (AVR) was to maintain the generator voltage at a given value, and most voltage regulators were mechanical. Following that, electronic and electromagnetic types were introduced.

In the late 1950s, with the increase in the size of the power system and the growth in the unit capacity of generators, the function of AVRs was no longer limited to maintaining the generator voltage constant; they were designed to improve the static and dynamic stability of generators for improved stability of the power system. This marked a fundamental shift in the functional requirements of exciter regulators.

In the 1950s, there were different opinions about the role of forced excitation. There was a view that in the event of a system fault the role of excitation should be limited to preventing generator stator current overload. However, Soviet scholars concluded through experiments and practice that the use of forced excitation could accelerate voltage recovery after the removal of the system fault, and could shorten the time of stator current overload, which would be extremely favorable for shortening the recovery time of system voltage and for maintaining system stability after the fault.

Since the 1950s, great progress has been made in excitation control. In a nutshell, the evolution of excitation control has undergone several stages, including proportional control for single variable input and output, multivariable feedback control for linear multivariable input and output, and nonlinear multivariable control supported by control theory; these are separately described below.

1.2.1 Single Variable Control Based on Classical Control Theory

In the early 1950s, with the development of power and electronic technology, the power system continued to impose new requirements on the control function of the generator excitation system, mainly reflected in the shift of the functional requirement on automatic excitation regulators from maintaining constant voltage at the generator side to increasing the steady-state stability limit of generator operation [1]. In this historical period, generators usually used DC exciter excitation, the regulation of excitation was usually effected on the field winding side of the DC exciter, and the regulation of generator excitation could only be enabled through the exciter power element with considerable inertia, namely, a slow excitation control system. During this period, in terms of excitation control, the following types of excitation regulation were usually used:

(1) Proportional excitation regulation based on generator terminal voltage deviation.
(2) Duplex excitation compensation regulation based on the generator stator current as a disturbance variable.
(3) Phase compensation excitation regulation based on signals such as the generator terminal voltage and stator current and power factor angle. As the DC exciter was the main mode at the time, the excitation regulator was composed of magnetic elements and was virtually able to meet the requirements of the operation mode.

In this period, Soviet researchers made many important discoveries in research on power system stability. For example, in the 1950s, C. A. Lebedjew and M. M. Botvinick first proposed the concept of a synchronous generator operating in the artificial stable zone in their research on power system stability. They pointed out that, as long as the automatic excitation regulator exhibits inert-zone-free functional performance, the generator stable operation zone can be extended to the rotor power angle $\delta > 90°$ zone even under the law of simple control based on generator voltage deviation negative feedback. Given the generator operation power angle limit $\delta = 90°$ in the absence of excitation regulation, the power angle $\delta > 90°$ extended operation zone was called the "artificial stable zone."

In terms of the excitation control law, most excitation regulators of the period adopted proportional regulation based on generator voltage deviation negative feedback or proportional–integral–differential (PID) regulation based on generator voltage deviation.

1.2.1.1 Proportional Control

The expression for the transfer function of proportional control is

$$\frac{U}{\Delta U_i} = K_P$$
$$\Delta U_i = U_{ref} - U_t(t) \tag{1.1}$$

where

U – output;
ΔU_i – input;
K_P – proportional regulation factor;
U_{ref} – reference voltage;
$U_t(t)$ – average of real-time three-phase effective values of generator terminal voltage.

The expression for the transfer function of PID regulation based on generator voltage deviation is

$$\frac{U}{\Delta U_i} = (K_P + K_D s) \frac{1}{1 + K_I s} \tag{1.2}$$

where

K_P, K_I, K_D – proportional, integral, and differential regulation factors, respectively.

The block diagrams of the transfer functions of the closed-loop system corresponding to Eqs. (1.1) and (1.2) are shown in Figures 1.1 and 1.2, respectively.

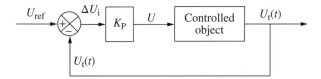

Figure 1.1 Block diagram of transfer function of univariate proportional control.

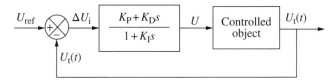

Figure 1.2 Block diagram of transfer function of PID control.

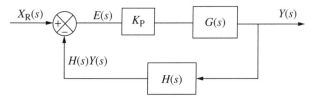

Figure 1.3 SISO closed-loop regulation system.

The physical concept of PID control shown in Figure 1.2 is hereby further elaborated. As can be seen from Eq. (1.2), the transfer function of PID control is composed of the sum of the proportional element K_P and the differential element $K_D s$ in series with the inertia element $\frac{1}{1+K_I s}$. If the time constant of the inertia element is large enough, or $K_I s \gg 1$, the value 1 can be ignored. Here, the inertia element is similar to an integral element $\frac{1}{K_I s}$. Thus, the control mode can be called the PID control system based on generator voltage deviation.

The performance characteristics of the single-variable input and output (SISO) PID control system shown in Figure 1.3 are discussed below.

In Figure 1.3, $X_R(s)$, $Y(s)$, and $E(s)$ respectively represent the Laplace transform function of the input $X_R(t)$, the output $y(t)$, and the regulation error $e(t)$. K_P and $G(s)$ respectively represent the transfer function of the forward paths. $H(s)$ represents the transfer function of the feedback channel.

As can be seen from the classical regulation principle, for the closed-loop control system shown in Figure 1.3, as the gain K_P increases, the dominant root of the closed-loop system characteristic equation will move to the right of the complex plane. When the gain K_P exceeds its critical value K_C, a pair of closed-loop system characteristic roots will appear on the right half of the complex plane. At this point, the closed-loop system will be unstable, and the dynamic response of the system will show an increased oscillation. Therefore, the gain K_P of the proportional regulation system will be limited to the range of $K_P < K_C$ to ensure system stability. At this point, if only generator voltage deviation control is adopted, for a long-distance transmission system, the weaker the electrical connection between the generator and the system, the smaller the allowable critical gain K_C will be, usually 5–20.

However, requirements for excitation regulation performance involve not only stability but also accuracy. For the closed-loop system shown in Figure 1.3, its static error is

$$\varepsilon(\infty) = \lim_{t \to \infty} e(t)$$

According to the standard applicable in China, the static error $\varepsilon(\infty)$ of generator terminal voltage regulation shall not be greater than 0.5%.

1 Evolution and Development of Excitation Control

For the system shown in Figure 1.3, its closed-loop transfer function is

$$\frac{Y(s)}{X_R(s)} = \frac{K_P G(s)}{1 + K_P H(s) G(s)} \tag{1.3}$$

The transfer function between the static error $\varepsilon(t)$ and the input $X_R(t)$ is

$$\frac{E(s)}{X_R(s)} = \frac{X_R(s) - H(s)Y(s)}{X_R(s)} = \frac{1}{1 + K_P H(s) G(s)} \tag{1.4}$$

With Eq. (1.4), the following can be obtained:

$$E(s) = \frac{1}{1 + K_P H(s) G(s)} X_R(s) \tag{1.5}$$

It is assumed that the input $X_R(t)$ is a unit step function and its Lagrangian transform function is $X_R(s) = \frac{1}{s}$. At this point, for the closed-loop regulation system shown in Figure 1.3, the static error Laplace transform expression under the action of the unit step function is

$$E(s) = \frac{1}{1 + K_P H(s) G(s)} \times \frac{1}{s} \tag{1.6}$$

As can be seen from the final value theorem in the regulation principle, the steady-state value of the static error is

$$e(\infty) = \lim_{t \to \infty} e(t) = \lim_{s \to 0} s E(s)$$

If Eq. (1.6) is substituted into the above equation, $H(s)G(s)$ can be written in the polynomial form of s:

$$e(\infty) \lim_{s \to 0} \frac{1}{1 + K_P \frac{b_m s^m + \cdots + b_1 s + 1}{a_n s^n + \cdots + a_1 s + 1}} = \frac{1}{1 + K_P} \tag{1.7}$$

As can be seen from Eq. (1.7), for a SISO closed-loop regulation system, its static error $e(\infty)$ approximates the reciprocal of the closed-loop gain K_P under the action of the unit step function, because $K_P \gg 1$ in the general case. The following can be obtained in turn:

$$e(\infty) \approx \frac{1}{K_P}$$

It can be concluded from the above equation that the open-loop gain K_P of the excitation system should be no smaller than 200, so that the static error of the generator voltage can be kept less than 0.5% under the action of the unit step function. However, in proportional excitation control, an excessive open-loop gain may cause unstable operation of the excitation system. Therefore, both requirements should be taken into account in selection of the gain K_P.

1.2.1.2 PID Control

When consideration is given to both the static error and the transient stability of the system, the structure of the transfer function of the excitation regulator can be changed to divide the gain of the excitation regulator into two parts. One part is the delay-free transient gain K_D, and the other part is the delay steady-state gain K_S. The block diagram of the corresponding excitation regulator transfer function is shown in Figure 1.4a. Figure 1.4b is an equivalent simplified block diagram.

If the unit step function $E(s) = \frac{1}{s}$ is added to the system input side at $t = 0^+$, as can be seen from the initial value theorem, the transient output of the control side is

$$u(0^+) = \lim_{t \to 0^+} u(t) = \lim_{s \to \infty} s U(s) = \lim_{s \to \infty} \left[s \frac{(K_D + K_P) + K_D T s}{1 + T s} \times \frac{1}{s} \right]$$

$$= \lim_{s \to \infty} \frac{(K_D + K_P) + K_D T s}{1 + T s} = K_D \tag{1.8}$$

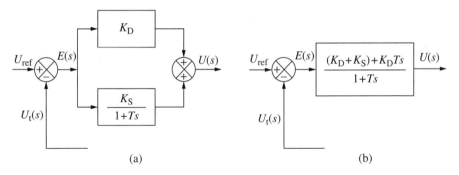

Figure 1.4 Block diagrams of static and transient gain transfer functions of excitation regulator: (a) transfer function, (b) equivalent transfer function.

Meanwhile, as can be seen from the final value theorem, the control output of the unit step output $E(s) = \frac{1}{s}$ in the steady state is

$$u(\infty) = \lim_{t \to \infty} u(t) = \lim_{s \to 0} sU(s) = \lim_{s \to 0} \left[s \frac{(K_D + K_P) + K_D Ts}{1 + Ts} \times \frac{1}{s} \right] = K_D + K_P \quad (1.9)$$

From Eq. (1.9), if the gain of the excitation regulator is divided into two parts (steady-state and transient), its transient gain is equivalent to the proportional regulation whose gain is K_D at the beginning of the transition process, and its steady-state gain is equivalent to the proportional regulation whose gain is $K_D + K_S$.

The transient gain reduction can be used for the coordination between precision and stability regulation. That is the basic role of PID excitation regulators.

It is emphasized that since the 1950s the control law of most univariate excitation regulators has been based on analysis of the performance of the excitation system in the s frequency domain under the classical adjustment principle. Under the condition of linearized small deviation, the excitation control law can be regulated on the basis of generator voltage deviation or the PID rule. On the basis of the frequency characteristics of the excitation system transfer function, the Bode plot is given. In turn, the excitation system's amplitude-frequency and phase frequency margin are obtained, and the appropriate corrective actions are selected.

Determining the no-load stability of the generator excitation system with the excitation system's open-loop characteristics and determining the parameters of the power system stabilizer (PSS) with the excitation system's closed-loop characteristics is the basic analysis method of the classical regulation principle applied so far.

It should be noted that, in the 1950s, with the application of high initial rapid ion excitation systems, the so-called dynamic stability problem appeared. When the system restores the original mode of operation after the disturbance of a major fault, a fast excitation system is helpful for the rotor swing braking in the rotor's swing period. However, in the subsequent dynamic stabilization process, the fast excitation system may extend the rotor swing time, increase power angle oscillation, and even cause oscillation and loss of synchronization in special cases.

In this regard, Soviet researchers realized that to suppress such power oscillation and loss of synchronization, a generator-power-related additional quantity should be added to the control law of the excitation regulator to improve the transient stability of the generator in operation.

In the 1950s, the so-called "strong excitation regulator" developed by the Soviet Union was multi-parameter regulator based on the above basic idea. However, since the measurement of the generator rotor power angle was difficult, the power angle δ signal was replaced by a value approximately equal to the action angle δ. The signals in the strong excitation regulator included such parameters as generator voltage deviation, voltage derivative, frequency deviation, frequency derivative, and generator rotor current. Under the condition of small deviation disturbance, a strong excitation regulator could make the system stability power limit 10%–12% higher than that in the case of a proportional excitation regulator. As can be seen, although the strong excitation regulator applied classical regulation theory to parameter setting, it adopted the so-called

D domain division method. The relation between two parameters could be studied on an s frequency plane, so that the common stability domain of each parameter such as ΔU, Δf, and f' could be determined. Thus, a multi-parameter setting could be realized. But that method was too complicated. When any system structure parameters changed, its modification was complex. Therefore, the method was not applied internationally. The strong excitation regulator primarily took the voltage deviation ΔU and the differential U' as regulation signals and introduced the rotor voltage soft negative feedback $\Delta U'_f$ to reduce the transient gain and maintain a high steady-state gain. In addition, it took the frequency deviation Δf and the differential f' as power damping signals. So, in essence, the Soviet strong excitation regulator was the equivalent of a PID excitation regulator with the stable power signal Δf.

Subsequently, in the mid-1960s, with the wide application of fast excitation systems, low-frequency power oscillation and oscillation and loss of synchronization in the process of dynamic stability restoration after major disturbance faults were frequently seen on some large power transmission systems. In this regard, American researcher F. D. Demello and C. Concordia probed the causes for deterioration of dynamic stability under the condition of a fast excitation system and specific power system parameters, starting with an analysis of the mechanism of low-frequency oscillation. Using the mathematical model based on the linearized small deviation theory proposed by R. A. Phillips and W. G. Ephron, they concluded that deterioration of the positive damping torque of a single generator on an infinite system was primarily caused by the hysteresis characteristics of the excitation system and the generator excitation winding.

Under normal operating conditions, the generator closed-loop excitation regulation system with the generator terminal voltage ΔU_t as negative feedback is stable. As can be seen from Figure 1.5, there is a certain phase lag between the two phases ΔU_t and $\Delta E'_q$ (transverse axis transient electromotive force deviation), and the lagged phase is related to the frequency. Therefore, when the rotor power angle oscillates, $\Delta E'_q$ lags behind ΔU_t. That is, the phase of the excitation current that the excitation system provides lags behind the rotor power angle. At a certain frequency, when the lag angle reaches 180°, the original negative feedback will become positive feedback. The change in the excitation current will in turn lead to oscillation of the rotor power angle. That is, the so-called "negative damping" will be generated.

If an excitation system adopts PID control, excitation regulation with the generator voltage deviation signal is helpful in improving the dynamic and static stability of the generator voltage. Meanwhile, the leading phase output provided for the excitation system compensates to some extent for the lag phase of the excitation current and the negative damping torque. But PID regulation is primarily designed for the voltage deviation signal; the leading phase frequency it generates is not necessarily the same as the low-frequency oscillation frequency. That is, it may not necessarily meet the need for the phase required for negative damping compensation. Besides, in order to control the voltage in the PID regulation system, the voltage deviation should be continuously regulated. Thus, it is impossible to distinguish between positive and negative damping torque

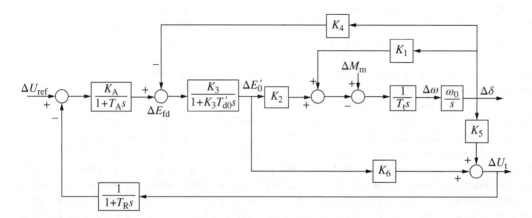

Figure 1.5 Block diagram of synchronous generator excitation control system under the condition of small deviation.

and difficult to simultaneously regulate the generator voltage and ensure a positive damping torque. So, the role of PID regulation in suppressing the low-frequency oscillation of the system is limited.

The PSS based on the theory by F. D. Demello and C. Concordia is designed to just suppress the low-frequency oscillation and improve the power system dynamic stability. At present, it has been widely used in excitation systems.

In the PSS excitation control mode described above, the excitation control law retains PID control based on generator voltage deviation and adds a power-related signal, for example, generator power, frequency, speed, or the rotor power angle.

Since PSS parameter selection is usually determined by specific operating conditions, when the system parameters change, the effective suppression frequency band set by the PSS will deviate from the actual system oscillation frequency band, and the control effect will significantly decrease. Therefore, in recent years, multi-parameter and adaptive PSSs have emerged to address a wider range of application when the power system parameters change.

A promoted new technology should be perfected and improved in practice. PSS applications once had some negative effects. For example, the PSSs developed by the US companies Westinghouse and GE take the generator speed ω as the input signal. In the case of torsional vibration of the unit, the inherent torsional vibration frequency of a certain order in the generator shaft system will be amplified by the PSS gain component that takes the speed ω as the signal. This further stimulates resonance of the torsional vibration, eventually damaging the generator shaft system. In December 1970 and October 1971, respectively, Mohave Power Station located in Nevada encountered two torsional vibration resonance faults, which caused breakage of the exciter coupled to the generator shaft. On another large steam turbine manufactured by Westinghouse, torsional vibration resonance as a result of the PSS's amplification of the shaft system torsional vibration signal caused breakage of the low-pressure cylinder blades. In 1987, a GE-made 800 MW steam turbine installed at Taiwan Ma'anshan Power Station also encountered breakage of the low-pressure cylinder blades caused by torsional vibration resonance. In view of that, Westinghouse and GE added the trap component to their PSSs that take the speed ω as the signal. Based on the difference in the order of the inherent torsional vibration frequency of the generator shaft system, when the set torsional vibration frequency order appears, the trap will prevent the speed signal from passing it through, thus avoiding intensification of the torsional vibration.

Any other PSS line that takes a physical variable other than the speed $\Delta\omega$, for example, the generator terminal power P or the rotor power angle δ, as input signal does not need to add a trap.

1.2.2 Linear Multivariate Control Based on Modern Control Theory

1.2.2.1 Linear Optimal Control Principle

Since P. E. Kalman laid the foundation for the modern control theory in 1960, some well-known researchers have put forward the application of modern control theory in power system research.

Since a power system is a complex nonlinear multivariate system, there are many restrictions on analysis of the system with classical linear univariate control theory. Linear multivariate modern control theory based on the state space description method can address these restrictions more easily.

In 1970, Canadian Dr. Yu Yaonan conducted the first research on the linear multivariate optimal control law for power systems with modern control theory.

In China, a group of researchers on power system stability represented by Professor Lu Qiang from Tsinghua University achieved much success in applying linear multivariate optimal control theory to synchronous generator excitation system control.

Optimal control theory is a developed and more widely applied key branch of modern control theory. Its goal is to select the optimal control law that will make performance of the control system under certain conditions optimal.

The state equation of a time-invariant system can generally be expressed as

$$\dot{X} = AX + BU \tag{1.10}$$

Where,

X – n-dimensional phasor;
U – r-dimensional state control phasor;
A – state coefficient matrix;
B – control coefficient matrix.

The eigenvalues of the system state equation are determined by the matrix A. To change its characteristics, the feedback of the state phasor can be introduced to form a closed-loop system, as shown in Figure 1.6.

The state phasor of the feedback system is

$$U = V - KX \tag{1.11}$$

where

K – state feedback gain matrix.

If Eq. (1.11) is substituted into Eq. (1.10), the following is obtained:

$$\dot{X} = AX + B(V - KX)$$
$$= (A - BK)X + BV \tag{1.12}$$

At this point, the eigenvalues of the closed-loop system will be determined by the matrix $A\text{-}BK$. Therefore, the optimal control law is in essence to select a K value and make the performance of the controlled system optimal under the given control rule under certain conditions.

1.2.2.2 Quadratic Performance Index

System performance indicators are usually determined on the basis of the project's actual requirements. Different performance indicators lead to different control laws.

The SISO control system shown in Figure 1.7 is taken as an example.

When the unit step function is entered in the system shown in Figure 1.7, the output is that shown in Figure 1.8b.

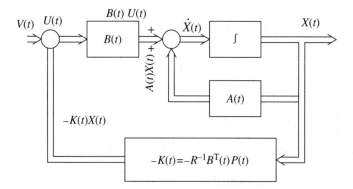

Figure 1.6 Block diagram of the feedback closed-loop system of a normal system linear.

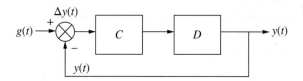

Figure 1.7 SISO control system.

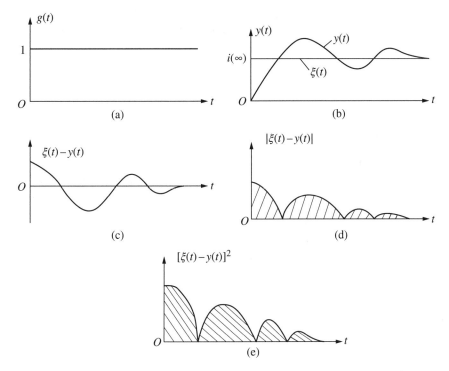

Figure 1.8 Error response under the action of unit step function: (a) unit step input, (b) ideal dynamic response and actual dynamic response, (c) error response, (d) absolute value of error response, (e) square of error response.

It is assumed that $y(t)$ is the actual response of the system, and $\xi(t)$ is the expected dynamic response. The optimal control performance indicators minimize the deviation $y(t) - \xi(t)$. There are three common forms of performance indicators represented by mathematical expressions:

$$J = \xi(t) - y(t) = J_{\min} \tag{1.13}$$

$$J = \int_0^\infty [\xi(t) - y(t)] dt = J_{\min} \tag{1.14}$$

$$J = \int_0^\infty [\xi(t) - y(t)]^2 dt = J_{\min} \tag{1.15}$$

Figures 1.8c–e show the graphs of the deviation value $\xi(t)-y(t)$, the deviation integral value, and the deviation quadratic integral value.

Equation (1.14) shows that the definite integral of the absolute value of the deviation of the actual response $[y(t)]$ from the expected response $[\xi(t)]$ is minimal. J is a generic function that changes with the function $[y(t)]$. Equation (1.15) shows that the expected value for the definite integral value of $[\xi(t)-y(t)]^2$ is minimal. In addition, Eq. (1.15) also shows equal treatment of positive and negative deviations and greater emphasis on large deviations.

Since Eq. (1.15) obtains the definite integral of the square of the deviation in the $0 \sim \infty$ time interval, it is called the quadratic performance index.

For a multivariate system, the above concept can also be used to determine its performance indicators.

If the actual state phasor is expressed as $X(t)$ and the expected state phasor is expressed as $\widehat{X}(t)$, the quadratic performance index that requires the state phasor deviation to be minimal is

$$J = \int_0^\infty [\widehat{X}(t) - X(t)]^T [\widehat{X}(t) - X(t)] dt = J_{\min} \tag{1.16}$$

However, when the above-mentioned optimal control performance criterion is met, it may be hard to achieve because of the need for excessively controlled variable. Therefore, the control phasor $U(t)$ is also limited. It is expressed as follows:

$$J = \int_0^\infty \{[\widehat{X}(t) - X(t)]^T Q[\widehat{X}(t) - X(t)] + U^T(t)RU(t)\} \, dt = J_{min} \tag{1.17}$$

where

Q, R – weight matrix corresponding to the state phasor and the control phasor, respectively.

Equation (1.17) expresses the degree of emphasis on the phase phasor deviation and the control phasor amplitude, indicating that the sum of the phasors accumulated over the whole control process and the generalized control energy weights consumed will be minimal.

In engineering, in order to facilitate analysis, the balance of the control system is often placed at the origin of the state space. When the system is disturbed, the state phasor will deviate from the origin. If the system is asymptotically stable, the state phasor will eventually tend to the origin. Under such conditions, the expected state phasor is just the origin, that is, $\widehat{X}(t) = 0$. At this point, Eq. (1.17) can be rewritten as

$$J = \frac{1}{2} \int_0^\infty [X^T(t)QX(t) + U^T(t)RU(t)]dt = J_{min} \tag{1.18}$$

Equation (1.18) is just the quadratic performance index of the optimal control of the linear time-invariant system.

1.2.2.3 Linear Optimal Controller

In order to improve the static and dynamic stability of generators under small interference conditions, China's scientists once applied linear multivariate optimal control theory to the control of excitation systems. The linear optimal excitation control (LOEC) regulator developed by Tsinghua University was once applied at some large hydropower stations.

The purpose of designing a linear optimal control system is to identify the optimal control phasor in all possible control phasors.

It is assumed that a controllable linear time-invariant system is expressed as

$$\dot{X} = AX + BU \tag{1.19}$$

where

X – n-dimensional state phasor;

U – r-dimensional state control phasor;

A, B – $n \times n$ and $n \times r$ constant matrix, respectively.

If the optimal control system is designed on the basis of the quadratic performance index shown in Eq. (1.18), it can be proved that the optimal control law exists and is unique. It is expressed as follows:

$$U = -R^{-1}B^T PX = -KX \tag{1.20}$$

where

P – $n \times n$-dimensional symmetric constant matrix, which is the positive definite solution of the Riccati algebraic matrix.

In order to minimize the generalized function J of Eq. (1.18), the necessary condition should satisfy the Riccati algebraic matrix equation, as follows:

$$-PA - A^T P + PBR^{-1}B^T P - Q = 0 \tag{1.21}$$

Figure 1.9 Block diagram of linear optimal control system.

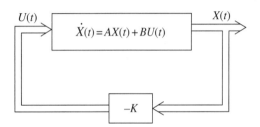

Figure 1.9 shows that the optimal control phasor U is the linear negative feedback of the state phasor X.

As can be seen from the above points, optimal control means making the deviation of the control process smallest, the time required to reach the expected value of the final value shortest, the final value optimal, and the control energy minimal under certain specific conditions.

Compared with classical control theory, optimal control theory has the following characteristics:

(1) Classical control theory synthesizes primarily in the complex frequency domain, whereas modern optimal control theory directly synthesizes primarily in the time domain and meanwhile balances dynamic quality and stability better.
(2) Classical control theory employs the concept of the transfer function in the complex frequency domain and applies it to SISO systems, whereas optimal control theory employs the state space analysis method in the time domain and applies it to multivariable systems as well as to simulation calculation with the digital computer.
(3) Classical control theory only applies to time-invariant systems, whereas optimal control theory can be extended to time-varying systems.

1.3 Linear Multivariable Total Controller

In recent years, some countries have made rapid progress in the development of linear multivariable optimal controllers based on modern control theory. For example, in the late 1980s, Japan's Fuji Electric developed a multivariable generator total controller, called Total Automatic Generation Controller (TAGEC).

1.3.1 Overview of TAGEC [2]

The total controller developed by Fuji Electric was designed to integrate a generator's excitation and speed control systems in one unit. In the traditional control mode, the reactive power of the generator is controlled by the excitation regulator and the active power by the governor. Because of the large inertia time constant of the mechanical governor, when the active power oscillates, generator excitation control is usually adopted. For example, the oscillation is suppressed with the PSS's additional power signal, while the speed governing system fails to govern the speed in a timely manner.

In recent years, due to the digitization of control systems, their inertia time constant has been greatly reduced. This provides the possibility of directly suppressing the oscillation of active power with the speed governing system.

TAGEC takes the minimum linear quadratic integral LQI deviation in modern control theory as the performance index. Conventional univariate excitation regulators usually take the generator voltage deviation ΔU as the feedback control variable, while governors take the speed deviation $\Delta \omega$ as the control variable. The two feedback deviations ΔU and $\Delta \omega$ act independently.

In TAGEC, the feedback variables include the generator's state variable U, power P_e, current I, magnetic field flux Φ_{fi}, speed ω, and power angle δ as well as the prime mover's state variable servomotor opening P_m and mechanical torque T_m. In addition, the power-system-related state variables are taken as feedback variables.

The schematic diagram of the TAGEC control system is shown in Figure 1.10.

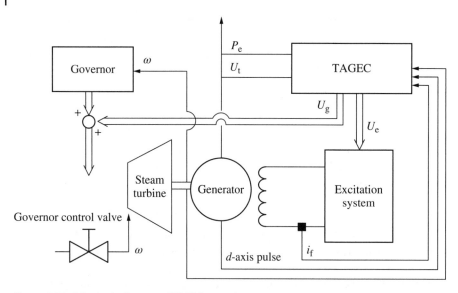

Figure 1.10 Schematic diagram of TAGEC control system.

The main characteristics of the system are as follows:

(1) It is a multivariable control system based on modern control theory.
(2) It is capable of total control of excitation and the governor.
(3) It is able to calculate the internal state of the generator and the gain value adapted to the operation in advance in case of any great change in the operating state.
(4) It has a regulating function for hydropower stations that is capable of considering the dynamic characteristics of hydraulic systems.
(5) It is able to control judgment and processing of start, stop, and load shedding programs.
(6) It has the fail-safe function.
(7) It is able to record generator state changes in transient processes such as system faults.

With the above functions, it can automatically compensate for negative damping, improve the system dynamic stability, and inhibit long-time power swing.

In addition, if a major disturbance change occurs in the state of the system connected to the generator, for example, failure of one of the circuits of the two-circuit transmission line, the control system can obtain the optimal gain, thus delivering good robustness.

It is able to control multi-computer systems well. Besides, it has the self-test function, which is capable of an automatic switch if a hardware failure occurs.

1.3.2 TAGEC Control

1.3.2.1 TAGEC-I
As shown in Figure 1.11, all of the regulation functions before the phase shift trigger circuit for the excitation and before the servomotor for the governor are replaced by the adaptive TAGEC.

This mode is applicable to new hydro turbines that pose no special requirements for governor control.

1.3.2.2 TAGEC-II
As shown in Figure 1.12, generator excitation control is the same as that in the TAGEC-I mode. For governor control, while the conventional governor control system functions are retained, a compensation signal is added by TAGEC, with its amplitude limited. This is similar to the function of PSS's additional signal, which acts on the excitation control's feeding point and achieves synthesis.

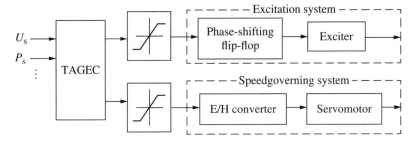

Figure 1.11 Block diagram of TAGEC-I control system.

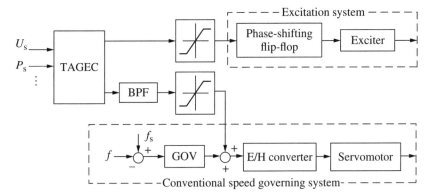

Figure 1.12 Block diagram of TAGEC-II control system.

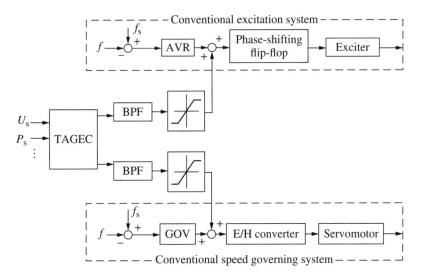

Figure 1.13 Block diagram of TAGEC-III control system.

1.3.2.3 TAGEC-III

As shown in Figure 1.13, in this control mode, TAGEC provides an additional signal and acts on the given points of the excitation regulator and the governor accordingly to obtain the total compensation function without changing the original excitation regulator and governor. This mode is applicable to modernization of units that have been in operation.

1.3.3 Mathematical Model of TAGEC Control System

Under the condition that the generator and the system are operating in parallel, the related Park equation is listed as the basis for establishing the mathematical model of the TAGEC control system.

Generally, the TAGEC's sampling operation time is about 20 ms. Under the condition that the stator resistance and the damping circuit are neglected, the following basic equation of the generator can be obtained on the basis of the basic mathematical models corresponding to the TAGEC-I and TAGEC-II control systems:

$$\dot{\Phi}_{fd} = [(r_f/X_{ad})U_e - r_f(\Phi_{fd} - \Phi_{ad})/X_{lfd}]\omega_0 \tag{1.22}$$

The equation operating in parallel with the system is

$$\dot{\omega} = [(T_m - T_e) - D(\omega - \omega_0)]/M \tag{1.23}$$

$$\dot{\delta} = \omega - \omega_0 \tag{1.24}$$

The generator output equation is

$$U_t = K_6 \Delta\Phi_{fd} + K_5 \Delta\delta + U_{tb} \tag{1.25}$$

$$P_e = K_2 \Delta\Phi_{fd} + K_1 \Delta\delta + P_{eb} \tag{1.26}$$

The water turbine and governor system equations are

$$\dot{P}_m = (1/T_g)(U_g - P_m) \tag{1.27}$$

$$\dot{q} = (2/T_\omega)(P_m - q) \tag{1.28}$$

$$T_m = 3q - 2P_m \tag{1.29}$$

Equations (1.27)–(1.29) express the relation between the governor opening P_m and the flow q and the mechanical torque T_m when the hydraulic system model is considered.

Under ideal conditions, the transfer function of the hydraulic system model can be written as

$$\omega(s) = \frac{1 - T_\omega s}{1 + 0.5 T_\omega s}$$

For the steam turbine governor, there are the following relations:

$$\dot{X}_g = \frac{-K_g[K_f(\omega - \omega_0) + P_e - P_s]}{T_g} \tag{1.30}$$

$$\dot{T}_{mh} = \frac{X_g - K_g[K_f(\omega - \omega_0) + P_e - P_s + U_g - T_{mh}]}{T_h} \tag{1.31}$$

$$\dot{T}_{ml} = (T_{mh} - T_{ml})/T_{rh} \tag{1.32}$$

$$\dot{T}_m = T_h T_{mh} + K_1 T_{ml} \tag{1.33}$$

where

Φ_{fd}	–	excitation flux;
ω	–	rotation angular velocity;
r_f	–	excitation winding resistance;
U_e	–	excitation voltage control;
Φ_{ad}	–	d-axis interlocking flux;
X_{lfd}	–	excitation winding leakage reactance;
T_m	–	mechanical torque;
T_e	–	electrical torque;

D	–	damping coefficient;
M	–	inertia coefficient;
δ	–	power angle;
U_t	–	terminal voltage;
U_{tb}	–	voltage linearization base value;
P_e	–	output power;
P_{eb}	–	output power linearization base value;
K_1, K_2, K_5, K_6	–	W. G. Ephron mathematical model coefficients;
U_g	–	governor opening instruction;
P_m	–	governor opening;
T_g	–	governor primary inertia time constant;
T_ω	–	hydraulic system time constant;
q	–	water flow;
X_g	–	governor PI regulator integrator output;
K_g, T_g	–	integrator gain and time constants;
K_f	–	frequency deviation measurement circuit gain;
P_s	–	output setpoint;
T_{mh}	–	high-pressure cylinder output torque;
T_{ml}	–	medium- or low-pressure cylinder output torque;
T_h	–	time constant of high-pressure cylinder with governor time lag;
T_{rh}	–	time constant of reheater with medium- or low-pressure cylinder time lag.

On the basis of the above mathematical model equation, the state equation is linearized and discretized. Through operation with the following quadratic performance index function, the optimal gain value of each variable can be calculated:

$$J = \sum_{j=0}^{\infty}[Q_V(\Delta U_{tj})^2 + Q_P(\Delta P_{ej})^2 + R_e(\Delta U_{ej})^2 + R_P(\Delta U_{gj})^2] \quad (1.34)$$

where

Q_v, Q_p, R_e, R_p	–	weight coefficients;
ΔU_{tj} and ΔP_{ej}	–	deviations of real-time value from setpoint at the point j;
ΔU_{ej} and ΔU_{gj}	–	deviations of real-time value from excitation and governor control variable at the point *j*.

1.3.4 Composition of TAGEC System

The block diagram of the TAGEC system based on a high-speed microcomputer is shown in Figure 1.14.

1.3.4.1 Sampling Operation
As shown in Figure 1.15, the control input of the excitation and governor system is operated, and a judgment is made and the data logging function is performed on the basis of the program conditions.

1 Evolution and Development of Excitation Control

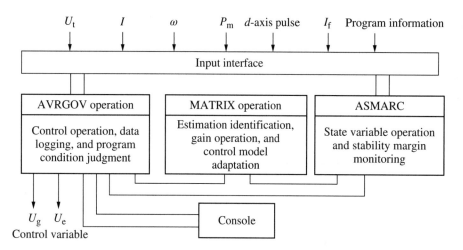

Figure 1.14 Block diagram of TAGEC system.

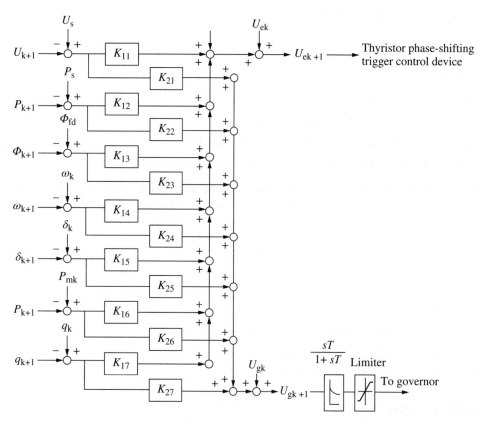

Figure 1.15 Block diagram of total control system.

1.3.4.2 Matrix Operation

On the basis of the equivalent system reactance, the optimal control model is calculated, and the optimal gain at the time of changes in the generator's and the system's operating state is determined.

1.3.4.3 Stability Margin Monitoring Control

The vertical and horizontal axis synchronous reactance of the generator and the vertical axis component of the generator terminal voltage are calculated, so that the stability margin's control of excitation increase or output power reduction for the generator can be ensured.

Conventionally, monitoring of generator stability margin is achieved with a leading phase reactive power monitoring relay. The action principle is that when a single generator is operating in parallel with an infinite system, if the excitation of the generator remains constant and the output of the generator is gradually increased, as the vertical axis component U_d of the generator terminal voltage U_t is increased to the maximum value U_{dmax}, beyond the point the generator will be in the acceleration state till it loses synchronization. U_{dmax} is the value when the excitation remains unchanged, as shown in Figure 1.16a.

The expression for U_{dmax} can be written as

$$U_{dmax} = \frac{X_q}{X_q + X_{ex}} U_b \qquad (1.35)$$

where

X_q – horizontal axis reactance of generator in operating state;

X_{ex} – external reactance;

U_b – infinite system line voltage.

It can be seen from Figure 1.16b that when the excitation of the generator is increased, the terminal voltage changes from U_t (U_d, U_q) to U'_t (U'_d, U'_q) and the rotor power angle changes from δ_0 to δ'_0.

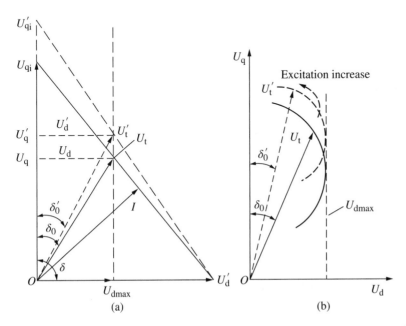

Figure 1.16 Principle of U_{dmax}: (a) basic phasor, (b) U_d and U_q phasor characteristics of generator. U_{qi} – X_q post voltage, U_t – generator terminal voltage, I – generator current, U_d – U_t vertical axis component, U_q – U_t horizontal axis component, and δ – power angle.

Table 1.1 TAGEC input and output physical quantities.

Measured value	TAGEC internal operation processing capacity	State variable	Remark
d-axis pulse	δ	U_t	TAGEC-I and II General
U_t three-phase	Φ_{ad}	P_e	
I_t three-phase	$U_{t(rms)}$	Φ_{fd}	
I_f	$P_{L(rms)}$	ω	
ω	U_e (excitation output)	δ	
P_m	$U_s, P_s,$ and U_g (governor output)	P_m and q	TAGEC-I For hydropower stations
P_s	U_g (governor output) band-pass filter, amplitude limit $\pm a\%$	$X_g, P_m, T_h,$ and T_1	TAGEC-II For thermal power stations

For stability margin monitoring, it is assumed that the maximum change in the external reactance is ΔX_{ex}. It can be seen from Eq. (1.35) that, at this point $U'\text{dmax} = \frac{X_q}{X_q + X_{ex} + \Delta X_{ex}} U\text{b}$. The stability margin after the excitation of the generator is increased is determined by the real-time value $U_d < U'_{d\text{max}}$. As can be seen from the above discussion, in the TAGEC control system, operation of the stability margin depends on X_q and X_{ex} corresponding to the operating output state, especially the value X_q. As can be seen from actual measurement, the value X_q will significantly decrease with an increase in the load. On a 700 MW steam turbine, for example, the design value X_q is 1.65 p.u. and will decrease to 1.40 p.u in the rated load state.

1.3.4.4 TAGEC Input and Output

The input and output physical variables of TAGEC are shown in Table 1.1.

The following physical variables for the monitoring values are entered directly from the outside into the TAGEC control system:

(1) d-axis pulse (the excitation magnetic pole position signal measured from the steam turbine shaft end or an equivalent signal).
(2) Generator terminal voltage signal (TV secondary three-phase voltage) U_t.
(3) Generator output current (TA secondary three-phase current) I_t.
(4) Generator excitation current I_f.
(5) Speed ω.
(6) Speed output measured by the turbine or the generator shaft end sensor.
(7) Governor opening P_m.
(8) Output setpoint (for TAGEC-II only) P_s.

1.3.5 Digital Simulation Test of Power Transmission System

Fuji Electric conducted a stability test for two generators to an infinite system with thermal power 200 kVA and hydropower 30 kVA simulated transmission line equipment.

1.3.5.1 Composition of Power System

The power system consisted of a 500 kV/300 km (or 600 km) two-circuit line of a simulated transmission system.

1.3.5.2 Generators

Two simulated four-pole non-salient pole 100 kVA/90 kW generators were used: voltage: 220 V; speed: 1500 r min^{-1} (or 1800 r min^{-1}); $X_d = 167\%$, $X'_d = 43\%$, $X''_d = 37\%$, $T_a = 0.3$ s, and $T'_{d0} = 3.0$ s.

1.3.5.3 Excitation and Speed Governing System

In order to compare the performance of the TAGEC control system, only three modes (AVR, AVR-PSS, and TAGEC-II) are adopted in excitation control.

The turbine speed governing system is electric. However, for the TAGEC-II control mode, the opening of the speed governing system is controlled by the compensation correction signal of TAGEC.

1.3.5.4 Power System Stability Test

The two generators, G1 and G2, adopt different excitation control modes, such as AVR-AVR, AVR-PSS, PSS-PSS, and PSS-TAGEC (or TAGEC-AVR/TAGEC-PSS).

The stability limit test is carried out under the condition of constant G2 output power and gradually increasing G1 output power. For the base value of the stability limit, the corresponding value of AVR-PSS is taken as the basis for the comparison. The added stability limit power of G1 is expressed as the per-unit value.

1.3.5.5 Power System Stability Test Result

Under the condition of two generators in an infinite system, the static, dynamic, transient, and long-cycle dynamic stability tests are conducted. The static stability test result in the AVR-PSS mode is shown in Figure 1.17 as a basic value for the comparison. The load of G2 remains unchanged during the test.

Figure 1.18 shows the result of the dynamic stability test of the two-generator system. The test condition is that one of the circuits of the transmission line is cut off.

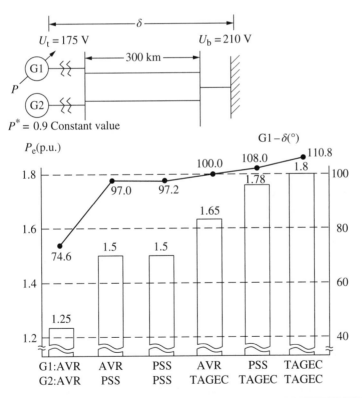

AVR/AVR	AVR/PSS	PSS/PSS	AVR/TAGEC	PSS/TAGEC	TAGEC-TACEC
—	Base value	0.0	10.0%	18.7%	20.0%

Figure 1.17 Static stability limit of two generators on an infinite system.

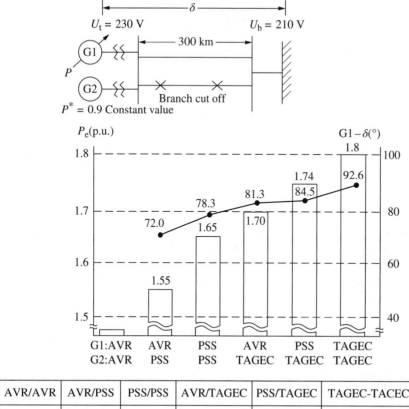

Figure 1.18 Dynamic stability test of two-generator system when one of the circuits of transmission line is cut off.

Figure 1.19 shows the transient stability test when one of the circuits of the transmission line is three-phase shorted out for four cycles and then cut off.

Figure 1.20 shows the long-cycle stability evaluation when a 300 km transmission line of one of the branches of the two-circuit transmission line is cut off in two generators in an infinite system. At this point, the squared value of the generator voltage is used as the stability evaluation index.

It should be emphasized that, as can be seen from Figure 1.20, the ability to maintain the generator terminal voltage with TAGEC is lower than with the corresponding value of the conventional excitation mode AVR-AVR or AVR-PSS.

In addition to the above simulation test, a prototype industrial operation test was conducted by Fuji Electric on a 32 MW hydro turbine in 1990, which delivered a good result.

1.4 Nonlinear Multivariable Excitation Controller [3]

Over the past two decades, with the continuous path-breaking progress in research on modern differential geometry, numerous achievements have also been made in studies on the application of differential geometry to nonlinear control systems. Moreover, a new discipline – the differential geometry theoretic system of nonlinear control systems – has been formed on this basis. In this regard, Professor A. Isdori from Sapienza University of Rome pointed out: "Just as the introduction of the Laplace transform and the transfer function before the 1950s and the introduction of the linear algebraic method in the 1960s brought major achievements

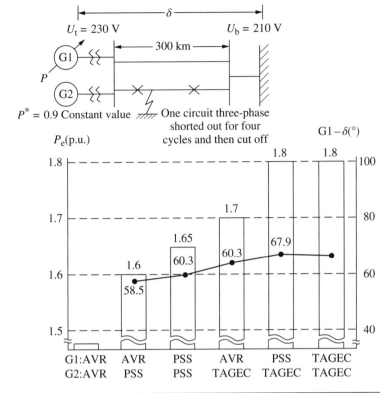

Figure 1.19 Three-phase short circuit test of two-generator system.

in control theory in terms of SISO and multivariable linear systems, respectively, introduction of the differential geometry method into nonlinear control systems will also bring breakthroughs in control theory." A branch of the new modern differential geometry theoretic system – the nonlinear system state feedback accurate linearization theory – has been developed very rapidly and applied to projects.

Professor Lu Qiang from Tsinghua University applied nonlinear system control theory based on the differential geometric method to complex power systems, gave the first rigorous mathematical proof of the optimal properties of the nonlinear control solution internationally, and further developed a decentralized nonlinear optimal excitation controller (NOEC) for power systems. A large number of simulation study results and field tests show that the nonlinear excitation control law can significantly improve the transient stability of power systems and the power transmission limit of transmission corridors, thereby maximizing the capacity of generator units.

However, a real-world engineering control system is always more or less affected by various uncertainties or external disturbances. This must be considered in research on controllers for power systems which are typical nonlinear systems. Without exceptions, some controllers applied to power systems including PID, PSS, LOEC, and NOEC adopt models with fixed structure and parameters in modeling. That is, the impact of uncertainties (e.g., external interferences and unmodeled dynamics) is not taken into account.

The nonlinear differential geometry control method is also built upon an accurate mathematical model of the controlled object. Although it provides an analytical method for design of nonlinear control, the controller design suffers from inherent defects due to ignorance of the uncertainties of the model in modeling. In the case of disturbances due to the uncertainties, it is difficult to achieve the desired performance indicators. Given this,

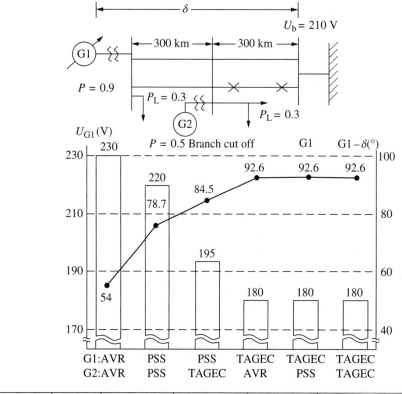

Figure 1.20 Long-cycle stability of two-generator system when one branch is cut off.

a branch of modern control theory called robust control was formulated. In system modeling and controller design, the impact of uncertainties on system performance is considered, and the actual control system is regarded as a system family. On this basis, the analytical method is used to design controllers, enabling the controlled object to meet the desired performance indicators as far as possible, even in the case of changes in the model.

For a nonlinear system subject to external uncertain disturbances, its nonlinear H_∞ control law can be obtained by solving an HJI inequation. However, the HJI inequality is a first-order partial differential inequation, and its general analytical solution cannot be obtained mathematically. However, in linear cases, the inequation can be reduced to a Riccati inequation. Solving the algebraic inequation is not mathematically difficult.

On the basis of the above idea, from a summary of NOEC, Professor Lu conducted the pioneering methodological integration of the differential geometry control theory, the dissipation system theory, the differential strategy, and the H_∞ method based on nonlinear robust control theory and exported the exact analytic expression of the nonlinear robust power system stabilizer (NR-PSS) control strategy from that, thus completing the creation of the nonlinear robust control theoretical system of power systems.

Besides, since the control strategy expression contains only the local parameters and the local and unit state variables and does not contain the power network parameters, robust excitation controllers have higher adaptability to network structure and parameter changes and better ability to suppress various external interferences to make the control law highly robust.

Figure 1.21 Coordination of NR-PSS and AVR.

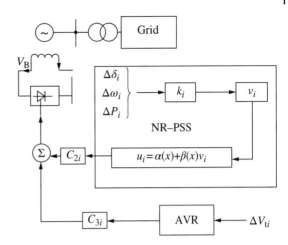

The establishment of this scientific theoretical system is of great significance in academia.

Since American researchers put forward the classical PSS excitation control theory in the 1960s, the NR-PSS excitation control strategy and the complete theoretical system created by Chinese researchers have undoubtedly set a new milestone in the field of excitation system control.

Under the basic design philosophy of NR-PSS, the nonlinear model of a multi-generator excitation system considering interferences is built, and the original nonlinear system is accurately linearized with the linearization technique. Then the linear H_∞ control law is designed for the linearized model, and the control law is brought back to the nonlinear feedback law set in the precise linearization process. Thus, the NR-PSS control law of the original system is obtained. This makes the designed controller strongly robust theoretically, thus ensuring the ability to suppress such uncertainties as external interferences and internal unmodeled dynamics and higher practicability in engineering applications.

Similar to conventional PSS applications, a NR-PSS applied to a project superimposes its output on the output point of the AVR in the form of additional signals. The outputs of the two systems are independent of each other. The connection is shown in Figure 1.21.

The NR-PSS nonlinear robust excitation control law obtained on the basis of the nonlinear robust control design principle is

$$V_{fiPSS} = E_{qi} - \frac{T'_{d0i}}{i_{qi}}[E'_{qi} + (X_{qi} - X'_{di})(i_{qi}\dot{i}_{di} + i_{di}\dot{i}_{qi})]$$
$$+ C_{1i}\frac{T_{ji}T'_{d0i}}{\omega_0 i_{qi}}\left(k_{1i}\Delta\delta + k_{2i}\Delta\omega - k_{3i}\frac{\omega_0}{T_j}\Delta P_e\right) \quad (1.36)$$

where

i	– parameters and state variables of the generator i (subscript i);
E'_{qi}, E_{qi}	– transient potential and no-load potential of synchronous generator respectively (p.u.);
T'_{d0i}	– time constant of excitation winding at the time of stator open circuit s;
i_{di} and i_{qi}	– d-axis and q-axis components of armature current;
$\dot{i}_{di}, \dot{i}_{qi}$	– differential variable of i_d and i_q, respectively;
X_{qi}, X'_{di}	– q-axis synchronous reactance and d-axis transient reactance (p.u.);
T_{ji}	– rotational inertia s;
ω_0	– synchronous angular velocity rad s^{-1};
$\Delta\delta$	– rotor operating angle deviation rad;

$\Delta\omega$ — angular velocity deviation rad s^{-1};

ΔP_e — electromagnetic power deviation (p.u.);

$k_{1i}, k_{2i}, k_{3i}, C_{1i}$ — accommodation coefficients.

As can be seen from Eq. (1.36), the NR-PSS control law has the following characteristics:

(1) As suppression of interferences is fully considered in the system design, the designed control law has a significant inhibitory effect on external interferences. Since the control law contains only locally measurable parameters (parameters of the generator), which are thus independent of the network parameters. So, it is adaptable to network structure changes, thereby ensuring robustness.
(2) The parameters in the control law are all locally measured, and they are not directly related to the state or output variables of other units, which ensures applicability to decentralized coordination control of multi-generator systems.
(3) The control law takes into account the transient salient pole effect of the generator on the basis of the generator's biaxial model and eliminates the original assumption that the NOEC $X'_d = X_q$, thus making the designed control law more accurate. This extends the scope of application of the controller theoretically.

When a NR-PSS and an AVR are used in combination, the total control law is

$$V_{fi} = C_{3i} V_{iAVR} + C_{2i} V_{iNR\text{-}PSS}(C_{1i}) \tag{1.37}$$

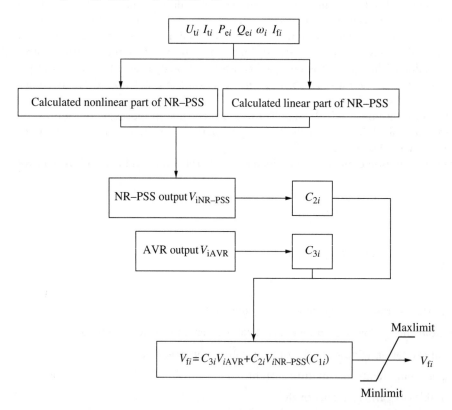

Figure 1.22 NR-PSS nonlinear robust excitation control engineering algorithm routine.

where

$$V_{fiPSS} = E_{qi} - \frac{T'_{d0i}}{i_{qi}}[E'_{qi} + (X_{qi} - X'_{di})(i_{qi}\dot{i}_{di} + i_{di}\dot{i}_{qi})]$$
$$+ C_{1i}\frac{T_{ji}T'_{d0i}}{\omega_0 i_{qi}}\left(k_{1i}\Delta\delta + k_{2i}\Delta\omega - k_{3i}\frac{\omega_0}{T_j}\Delta P_e\right)$$
$$V_{iAVR} = (k_P + k_D s)\frac{1}{1 + k_1 s}\Delta V_{ti}$$

The nonlinear robust excitation control engineering algorithm routine is shown in Figure 1.22.

1.5 Power System Voltage Regulator (PSVR) [4]

1.5.1 Overview

A conventional AVR is designed to keep the generator voltage constant. A system voltage drop caused by a transmission line failure will result in an increase in the reactive power loss of the entire system, thus reducing the system voltage stability.

Moreover, in recent years, in some countries such as some metropolises in Japan, with the popularity of refrigerating units, loads have had a voltage characteristic that is close to constant power. Meanwhile, due to the application of cables to power supply systems, there is an obvious downward trend in the trunk line system voltage in power supply to downtown areas with long-distance 275 kV high-voltage transmission systems. Therefore, the maintenance of trunk line system voltage stability has become important. In the 1990s, Japan set up a new PSVR, which can improve high-voltage transmission system voltage and maintain it at a certain value on generator units connected with 500 kV and 275 kV systems.

1.5.2 Effect of SVR on Improvement of System Voltage Characteristics

Figure 1.23a shows the conventional AVR mode. In the system, the voltage deviation of the generator is regulated as the feedback variable, and the voltage of the generator is maintained at a given value.

Under such conditions, any transmission line failure will cause a drop in the system voltage U_s. When U_s drops to the voltage at the intersection with the leading edge of the P–U curve, the generator's ability to provide reactive power will be determined by the intersection of the AVR's control characteristics and the leading line of P–U, as shown in Figure 1.24b.

If the PSVR mode shown in Figure 1.23b is adopted to maintain the high-voltage side voltage of the power station at a high level, it can improve not only the generator output reactive power limit but also the system voltage stability.

It should be noted that the PSVR just applies the generator's potential limit reactive power capacity, but it is not an overload condition in which the generator exceeds the allowable reactive capacity. A simulated system of the PSVR is shown in Figure 1.24a. The reactive power range that can be regulated by the PSVR is shown in the shaded area in Figure 1.24b.

To explain the effect of PSVR on the improvement of the system voltage characteristics, the compensation effect is described when the power system is equipped with a power capacitor, a synchronous rotary condenser, and a PSVR, respectively.

It is assumed that the model of a single generator on a single load system is shown in Figure 1.25.

1 Evolution and Development of Excitation Control

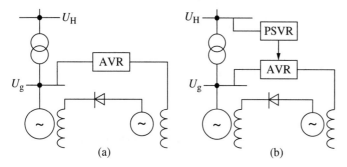

Figure 1.23 Generator excitation control modes: (a) AVR, (b) PSVR.

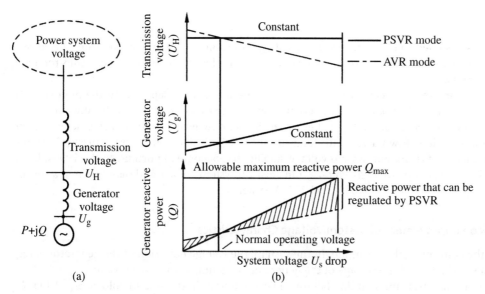

Figure 1.24 PSVR and AVR control characteristics when system voltage drops: (a) simulated system, (b) PSVR and AVR control characteristics.

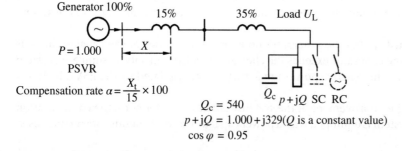

$Q_c = 540$
$p + jQ = 1.000 + j329$ (Q is a constant value)
$\cos \varphi = 0.95$

Figure 1.25 Model of a single generator on a single load system.

The parameters of the power system include $X_t = 15\%$, $X_e = 35\%$, $P + jQ = 1000 + j329$ (Q is a constant value), and $\cos\varphi = 0.95$.

(1) Compensation characteristic when a power capacitor is installed

It is assumed that the power capacitor's admittance is Y_{SC}. At this point, the generator power expression is

$$P^2 + \left[\left(\frac{1}{X_e + X_t} - Y_e - Y_{SC}\right)U_L^2 + Q\right]^2 = \frac{U_g^2 U_L^2}{(X_e + X_t)^2} \tag{1.38}$$

where

P and Q — active and reactive power of load, respectively;
U_g — generator terminal voltage;
U_L — load terminal voltage;
X_e — transmission line reactance;
X_t — boosting transformer reactance;
Y_c — admittance of power capacitor on load side before fault.

(2) Compensation characteristic when a synchronous rotary condenser is installed (compensation capacity Q_{RC})

The expression for the compensation characteristic when a synchronous rotary condenser is installed is

$$P^2 + \left[\left(\frac{1}{X_e + X_t} - Y_c\right)U_L^2 + (Q - Q_{RC})\right]^2 = \frac{U_g^2 U_L^2}{(X_e + X_t)^2} \tag{1.39}$$

(3) Compensation characteristic when a PSVR is installed

At this point, the reactance of the boosting transformer is compensated by the PSVR to $X_t' = \alpha X_t$, where α is the compensation factor. Then

$$P^2 + \left[\left(\frac{1}{X_e + \alpha X_t} - Y_c\right)U_L^2 + Q\right]^2 = \frac{U_g^2 U_L^2}{(X_e + \alpha X_t)^2} \tag{1.40}$$

If the power system parameter values shown in Figure 1.25 are substituted into Equations (1.38)–(1.40), the corresponding P–U curves can be obtained, as shown in Figure 1.26.

As can be seen from Figure 1.26, when a power capacitor or synchronous rotary condenser is installed, with the increase in installed capacity, the leading edge voltage of the P–U curve shows an obvious upward trend. That is particularly obvious when a power capacitor is installed. The voltage stabilization limit is defined by the leading voltage becoming lower than the load terminal voltage rating.

When the PSVR is set, the reactance of the boosting transformer can be partially compensated to decrease the leading edge voltage. This can improve the system voltage stability. Its effect is equivalent to that of an addition of transmission system and transformer capacity.

The same effect can be achieved for multi-generator systems. Figure 1.27 shows the simulation test result for a 500 kV system when the generator adopts a PSVR.

Table 1.2 shows the reactive load balance when the one of the circuits of the 500 kV transmission system is cut off.

As can be seen from Figure 1.27, when one of the circuits of the 500 kV transmission system is cut off, the PSVR mode can make the system voltage of the central part drop by 12–14 kV. This can improve the system's voltage maintenance level to a certain extent.

In addition, this can reduce the system's reactive power loss by about 1.7 Mvar.

28 | *1 Evolution and Development of Excitation Control*

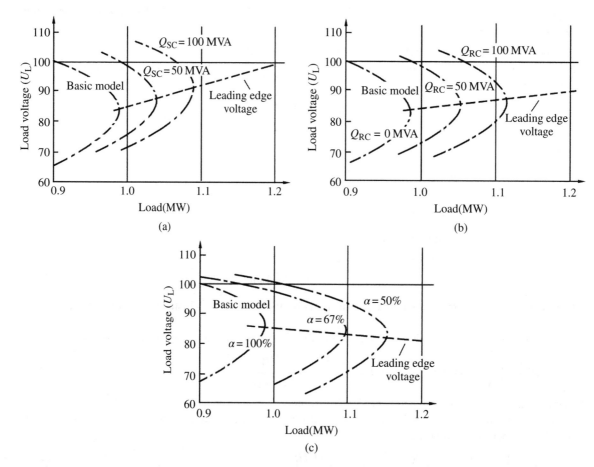

Figure 1.26 P–U curve for model of a single generator on a single load system: (a) P–U curve when power capacitor is installed, (b) P–U curve when rotary condenser is installed, (c) P–U curve when PSVR is installed.

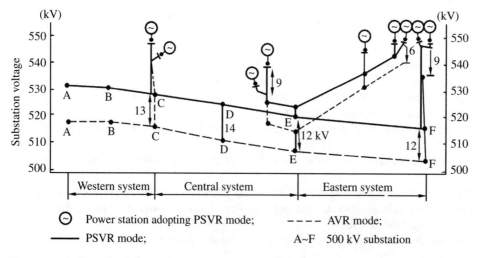

Figure 1.27 Voltage distribution when one of the circuits of 500 kV transmission system is cut off.

1.5 Power System Voltage Regulator (PSVR)

Table 1.2 System-wide reactive power balance when one of the circuits of 500 kV transmission system is cut off.

Control mode	Balance	PSVR mode (Mvar)	Conventional AVR mode (Mvar)	Difference (Mvar)
Mains side	Generator	14.080	16.540	−2.460
	Rotary condenser	13.730	13.010	+0.720
Load side	Load	12.140	12.140	0
	Transmission loss	15.670	17.410	−1.740
	Total	27.810	29.550	−1.740

1.5.3 Composition of PSVR

1.5.3.1 Basic Control of PSVR

There are multiple schemes of PSVR control. Figure 1.28a shows a simple control scheme. In the scheme, additional signals are supplied by the transmission high voltage side to compensate for the generator voltage benchmark value U_g. The degree of compensation, that is, the slope of the generator voltage on the external characteristic curve of the reactive power is $(0.5-5)\% U_H/Q_{gmax}$, where Q_{gmax} is the maximum allowable reactive power, as shown in Figure 1.28b.

1.5.3.2 Basic Equation of PSVR Control

The PSVR control system block diagram is shown in Figure 1.29.

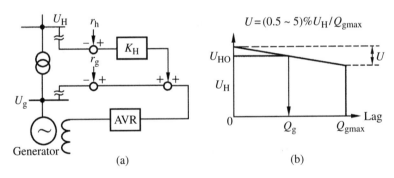

Figure 1.28 PSVR control mode: (a) diagram of control system, (b) generator voltage control characteristics.

Figure 1.29 PSVR control system block diagram.

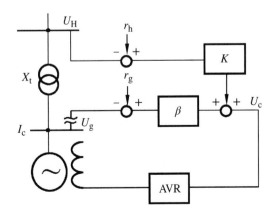

For the PSVR control system shown in Figure 1.29, it is assumed that the normal gain of the AVR is infinite and decreases to $\beta = 1$ after the compensation. The following equations can be obtained:

$$n(U_g - X_t I_q) = U_H \tag{1.41}$$

$$(r_h - U_H)K_H = U_c \tag{1.42}$$

$$U_g = r_g + U_c \tag{1.43}$$

If U_c in Eq. (1.42) is substituted into Eq. (1.43) and then U_g is substituted into Eq. (1.41), the basic equation into which U_H is substituted can be obtained:

$$n[r_g + (r_h - U_H)K_H - X_t I_q] = U_H \tag{1.44}$$

Through calculation, the following is obtained:

$$U_H = \frac{n(r_g + r_h K_H)}{1 + nK_H} - \frac{nX_t}{1 + nK_H}I_q = \frac{n(r_g + r_h K_H)}{1 + nK_H} - \alpha_v I_q$$

$$\alpha_v = \frac{nX_t}{1 + nK_H} \tag{1.45}$$

where

α_v — voltage slope coefficient;

K_H — voltage gain coefficient.

If U_H in Eq. (1.45) is substituted into Eq. (1.41), the expression for U_g can be obtained:

$$U_g = \frac{r_g + r_h K_H}{1 + nK_H} + \frac{K_H n X_t}{1 + nK_H}I_q = \frac{r_g + r_h K_H}{1 + nK_H} + K_H \alpha_v I_q \tag{1.46}$$

As can be seen from Eq. (1.44), a drop in the system voltage will cause an increase in the generator reactive current I_q. The degree of the drop depends on α_v.

As can be seen from Eq. (1.45), as I_q increases, the generator terminal voltage will increase. The magnitude of the increase depends on K_H and α_v. If the gain reduction set for the generator AVR measurement circuit is taken into account, that is, the gain reduction factor β is not equal to 1, Eq. (1.43) can be rewritten as

$$U_g = r_g + U_c/\beta \tag{1.47}$$

With simultaneous Eqs. (1.41), (1.42), and (1.47), U_H and U_g can be obtained:

$$U_H = \frac{n(\beta r_g + r_h K_H)}{\beta + nK_H} - \frac{\beta n X_t}{\beta + nK_H}I_q = \frac{n(\beta r_g + r_h K_H)}{\beta + nK_H} - \alpha_{v\beta} I_q$$

$$\alpha_{v\beta} = \frac{\beta n X_t}{\beta + nK_H} \tag{1.48}$$

$$U_g = \frac{\beta r_g + r_h K_H}{\beta + nK_H} + K_H \alpha_{v\beta} I_q \tag{1.49}$$

where

$\alpha_{v\beta}$ — voltage gain slope coefficient with gain reduction coefficient β factored in.

1.5.4 Comparison between PSVR and AVR in Control Characteristics

For the AVR mode, the high-voltage-side feedback coefficient (K_H) can be made equal to 0. In this case, Eqs. (1.45) and (1.46) can be rewritten as

$$U_{HA} = nr_g - nX_t I_q = nr_g - \alpha_v(1 + nK_H)I_q \tag{1.50}$$

$$U_{gA} = r_g \tag{1.51}$$

For the PSVR mode, when $K_H \neq 0$, Eqs. (1.45) and (1.46) can be rewritten as

$$U_{HP} = nr_g \frac{1 + r_h K_H / r_g}{1 + nK_H} - \alpha_v I_q \tag{1.52}$$

$$U_{gP} = r_g \frac{1 + r_h K_H / r_g}{1 + nK_H} + K_H \alpha_v I_q \tag{1.53}$$

If it is assumed that $n = r_h / r_g$, Eqs. (1.52) and (1.53) can be written as

$$U_{HP} = nr_g - \alpha_v I_q \tag{1.54}$$

$$U_{gP} = r_g + K_H \alpha_v I_q \tag{1.55}$$

As can be seen from these two equations, compared with the AVR mode, the PSVR control can reduce the voltage drop on the high-voltage side to $1/(1 + nK_H)$ while the generator-side voltage increases only by $K_H \alpha_v$.

1.5.5 Basic Functions of PSVR

Figure 1.30 shows the basic composition of the PSVR control circuit. In order to achieve AVR control by the high-voltage-side line voltage, PSVR can automatically determine the target value of the output voltage by the time program and detect anomalies. The relevant functions are as follows.

1.5.5.1 Basic Control

PSVR can quickly and accurately take samples at a sampling frequency of 600 Hz or more an accuracy of ±0.2% or less and measure the generator three-phase voltage RMS and average. In order to ensure the generator excitation system regulation stability, it provides phase compensation and gain reduction circuits. It can achieve generator voltage stability and reactive power balance settings with the regulation gain K_H.

1.5.5.2 Program Voltage Setting

It can achieve control of up to 16 ladder benchmark values with different models of working and rest days.

1.5.5.3 Output Limit

It provides a voltage limiting circuit at the PSVR output terminal.

1.5.5.4 Anomaly Self-Detection

It can self-detect operation anomalies.

1.5.5.5 Control Stability

In the conventional AVR control mode, its deviation signal is added with the system voltage-related deviation and amplified for operation control. The result is that the gain selected by the AVR is too large, so that the

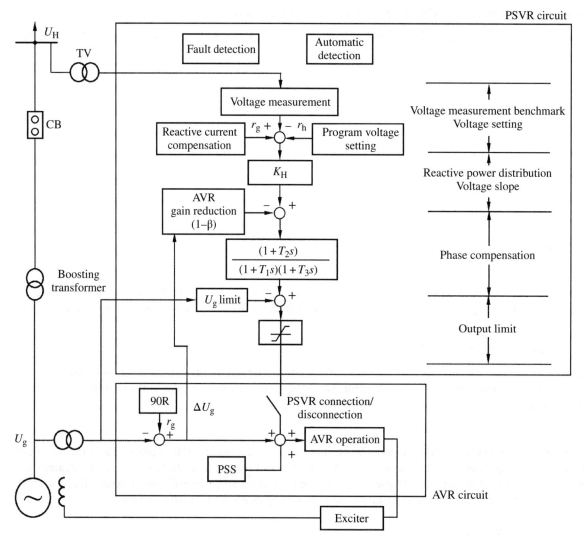

Figure 1.30 Basic composition of PSVR control circuit.

damping system power oscillation capacity declines if any system transmission line failure occurs. To improve the damping capacity, the following measures can be taken:

(1) Provide a gain reduction (1-β) circuit in the AVR control circuit;
(2) Provide phase compensation (advanced first order and lagged second order);
(3) Increase the limit on the PSS.

Figure 1.31 shows signal operation and phase compensation coordination in the PSVR control system.

The generator voltage is determined by the sum of the reference voltage value r_g of the AVR and the output voltage of the PSVR. When the output voltage of the generator exceeds the allowable value, the output voltage of the PSVR will be limited. If only the gain of the AVR is reduced and connected to the limited output terminal of the PSVR, the transient gain of the AVR will be reduced when a major disturbance event occurs on the transmission line. For this reason, if the gain reduction $\beta(\beta \leq 1)$ of the AVR is connected to the input terminal of the phase compensation circuit with a greater lag time as shown in Figure 1.31, under the action of a large transient disturbance, the gain reduction of the AVR can give rise to the same response ability as before.

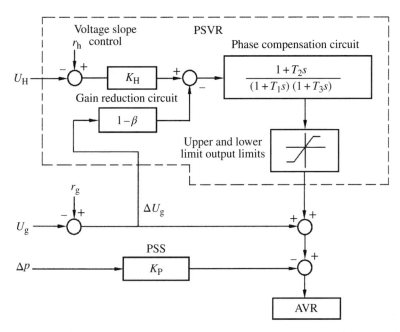

Figure 1.31 Signal operation and phase compensation coordination in PSVR control system.

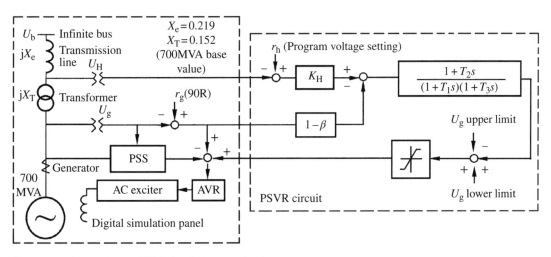

Figure 1.32 Composition of PSVR simulation test circuit.

1.5.6 PSVR Simulation Test

The PSVR simulation test is shown in Figure 1.32. The digital simulation test is conducted by a single generator in an infinite system.

The test includes a steady-state test and a transient test, with the results shown in Figure 1.33. The test in Figure 1.33a is carried out without phase compensation. During the test, the generator voltage and the active power both produced slight oscillations. The test in Figure 1.33b is carried out with phase compensation, without a generator voltage and active power oscillation. In Figure 1.33c, the gain is increased, so that $\beta = 1$. The voltage closed circuit system unstably oscillates. Figure 1.33d shows the limit action under the generator inverse time limit voltage limit. It maintains the stability of the voltage closed-loop system.

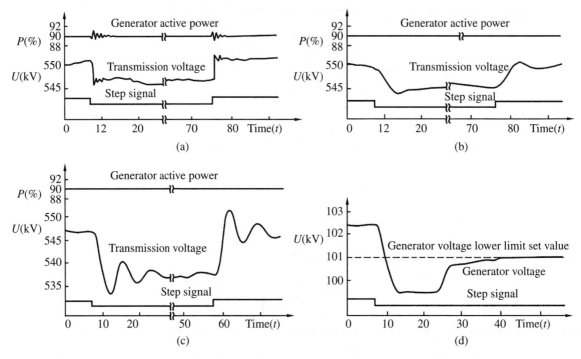

Figure 1.33 PSVR and generator simulation test results: (a) without phase compensation, (b) with phase compensation, (c) gain increase, $\beta = 1$, (d) inverse time limit voltage limit.

PSVRs can comprehensively improve the voltage stability of trunk lines of transmission systems. In Japan, PSVRs have been applied to 500 kV transmission systems of hydro, thermal, and nuclear power generating units since the early 1990s. Nowadays their effectiveness has been achieved in actual operation, and their effect in improvement of system stability has been confirmed.

2

Characteristics of Synchronous Generator [5]

2.1 Electromotive Force Phasor Diagram of Synchronous Generator

2.1.1 Non-salient Pole Generator

2.1.1.1 Electromotive Force Phasor Diagram of Non-salient Pole Generator

When a non-salient pole synchronous generator operates with a symmetrical inductive load, its electromotive force (emf) phasor diagram without taking the influence of core saturation into account is shown in Figure 2.1.

In Figure 2.1a, the sum of the excitation flux $\dot{\Phi}_0$ phasor and the armature reaction flux $\dot{\Phi}_a$ phasor is just the air gap flux ($\dot{\Phi}_\delta = \dot{\Phi}_0 + \dot{\Phi}_a$) which actually exists in on-load operation. Obviously, compared with the main flux $\dot{\Phi}_0$ at no load, both the magnitude and the distribution of $\dot{\Phi}_\delta$ have changed. $\dot{E}_\delta$ is just the emf of the air gap flux induction. If the stator winding resistance and leakage reactance are taken into account, the sum of the air gap emf $\dot{E}_\delta$ phasor and the leakage emf $\dot{E}_s$ phasor can be regarded as the phase emf that actually exists in the stator windings. When the resistance voltage drop of the stator current on the one-phase winding is subtracted, the generator stator phase voltage $\dot{U}$ can be obtained. Obviously, it not only decreases from the no-load emf $\dot{E}_0$ numerically but also changes in phase.

Figures 2.1b and c respectively show the reduced phasor diagram taking the stator winding voltage drop into account and that without taking the stator winding voltage drop into account.

2.1.1.2 Emf Equation of Non-salient Pole Generator

A non-salient pole generator rotor is cylindrical. Its air gap is generally uniform. Since the change in the magnetic circuit magnetoresistance with the change in the rotor position is slight, it is not necessary to decompose the armature reaction into two components. So, its emf equation is

$$\dot{E}_0 + \dot{E}_a + \dot{E}_s = \dot{U} + \dot{I}R_a \tag{2.1}$$

On the vertical and horizontal axes, $X_{ad} \approx X_{aq} = X_a$ is called the armature reaction reactance. The armature reaction emf $\dot{E}_a = -j\dot{I}\,X_a$. So, Eq. (2.1) can be written as

$$\dot{E}_0 = \dot{U} + \dot{I}R_a + j\dot{I}X_a + j\dot{I}X_s \tag{2.2}$$

or

$$\dot{E}_0 = \dot{U} + \dot{I}R_a + j\dot{I}X_d \tag{2.3}$$

where $X_d = X_a + X_s \approx X_q$ is called the synchronous reactance of the non-salient pole generator, usually expressed as X_d only. Its value reflects the influence of the armature magnetic field generated by the three-phase symmetrical stator current on the stator phase emf.

Equation (2.3) is just the stator winding one-phase emf balance equation in symmetrical stable operation of the non-salient pole generator.

For a large generator, the stator winding resistance is so much smaller than the synchronous reactance that it can be ignored. The stator winding resistance voltage drop is smaller than 1% of the stator terminal voltage. So, the above equation can be simplified in practical application.

Design and Application of Modern Synchronous Generator Excitation Systems, First Edition. Jicheng Li.
© 2019 China Electric Power Press. Published 2019 by John Wiley & Sons Singapore Pte. Ltd.

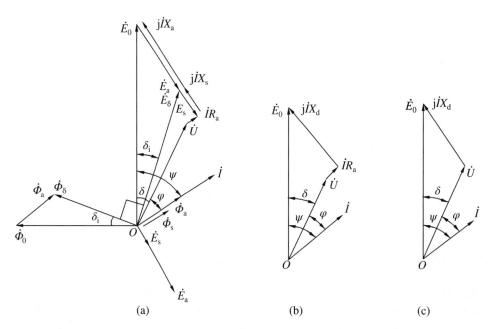

Figure 2.1 Emf phasor diagram of non-salient synchronous generator ($\varphi > 0$): (a) emf phasor diagram, (b) reduced phasor diagram, (c) reduced phasor diagram without taking the stator winding voltage drop into account.

For a non-salient generator, Eq. (2.3) can be reduced to

$$\dot{E}_0 = \dot{U} + j\dot{I} X_d \tag{2.4}$$

2.1.2 Salient Pole Generator

2.1.2.1 Emf Phasor Diagram of Salient Pole Generator

When a salient pole synchronous generator stably operates with a symmetrical inductive load, its emf phasor diagram without taking the influence of core saturation into account is shown in Figure 2.2.

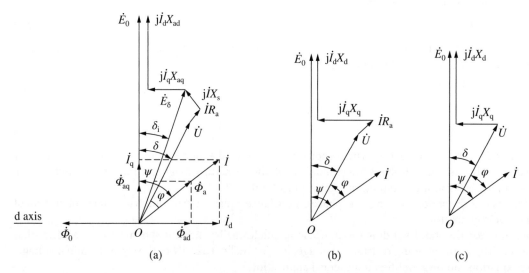

Figure 2.2 Emf phasor diagram of salient pole synchronous generator ($\varphi > 0$): (a) emf phasor diagram, (b) reduced phasor diagram, (c) reduced phasor diagram without taking the stator winding voltage drop into account.

In Figure 2.2, φ is the power factor angle, which depends on the load impedance, reflecting the nature of the load. ψ is the internal power factor angle, which is determined by the magnitude and nature of the load of the generator but is also related to the generator stator winding impedance. In the figure, $0° < \psi < 90°$. δ is the rotor power angle, δ_i is the angle between $\dot{E}_0$ and $\dot{E}_\delta$, $\dot{E}_\delta$ is the air gap emf, and $\dot{\Phi}_a$ is the armature reaction flux.

At any instant when the salient pole generator operates under the condition of lagging phase, the following relations between the various parts' fluxes and induced emfs hold:

(1) The excitation magnetomotive force (mmf) F_0 produced by the rotor excitation current I_f generates the main flux $\dot{\Phi}_0$, inducing the no-load emf and $\dot{E}_0$.
(2) The vertical axis armature mmf F_{ad} produced by the vertical axis component of the stator three-phase current ($\dot{I}_d$, i.e., the resultant of $\dot{I}_{dA}$, $\dot{I}_{dB}$, and $\dot{I}_{dC}$) generates the vertical axis armature reaction flux $\dot{\Phi}_{ad}$ and its induced vertical axis armature reaction emf $\dot{E}_{ad}$.
(3) The horizontal axis armature mmf F_{aq} produced by the horizontal axis component of the stator three-phase current ($\dot{I}_q$, i.e. the resultant of $\dot{I}_{qA}$, $\dot{I}_{qB}$ and $\dot{I}_{qC}$) generates the horizontal axis armature reaction flux $\dot{\Phi}_{ad}$ and its induced horizontal axis armature reaction emf $\dot{E}_{aq}$.
(4) The stator leakage flux $\dot{\Phi}_s$ induces the leakage emf $\dot{E}_s$ in the stator one-phase windings.

2.1.2.2 Emf Equation of Salient Pole Generator

As stipulated, the positive direction of the emf and the positive direction of the phase current are the same. Thus, the sum of the emfs less the stator windings' resistance voltage drop $\dot{I}R_a$ equals $\dot{U}_0$. The expression is

$$(\dot{E}_0 + \dot{E}_{ad} + \dot{E}_{aq} + \dot{E}_s) - \dot{I}R_a = \dot{U}$$

and can be rewritten as

$$\dot{E}_0 + \dot{E}_{ad} + \dot{E}_{aq} + \dot{E}_s = \dot{U} + \dot{I}R_a \tag{2.5}$$

Equation (2.5) is the stator winding one-phase emf equation when the salient pole generator is in lagging-phase symmetrical stable operation.

Due to the differences in the magnetic circuit and magnetoresistance, the reactance values corresponding to the magnetic fluxes of the various parts of the generator are different. What correspond to the vertical and horizontal axis armature reaction fluxes are the vertical axis armature reaction reactance X_{ad} and the horizontal axis armature reaction reactance X_{aq}. What corresponds to the leakage flux is the leakage reactance X_s. The relation between the voltage drop of the current on the reactance and the induced emf of the magnetic flux of the same part in the winding is

$$\dot{E}_{ad} = -j\ \dot{I}_d X_{ad} \tag{2.6}$$
$$\dot{E}_{aq} = -j\ \dot{I}_q X_{aq} \tag{2.7}$$
$$\dot{E}_s = -j\dot{I} X_s \tag{2.8}$$

Thus, Eq. (2.5) can also be written as

$$\dot{E}_0 = \dot{U} + \dot{I}R_a + j\ \dot{I}_d X_{ad} + j\ \dot{I}_q X_{aq} + j\dot{I} X_s \tag{2.9}$$

If the leakage reactance voltage drop is also divided into two components (a vertical axis and a horizontal axis), the above equation can be written as

$$\dot{E}_0 = \dot{U} + \dot{I}R_a + j\ \dot{I}_d(X_{ad} + X_s) + j\ \dot{I}_q(X_{aq} + X_s) \tag{2.10}$$

or

$$\dot{E}_0 = \dot{U} + \dot{I}R_a + j\ \dot{I}_d X_d + j\ \dot{I}_q X_q \tag{2.11}$$

where

X_d — vertical axis synchronous reactance of salient pole generator,
$X_d = X_{ad} + X_s$;

X_q — horizontal axis synchronous reactance of salient pole generator,
$X_q = X_{aq} + X_s$

2 Characteristics of Synchronous Generator

Because $X_{ad} > X_{aq}$, $X_d > X_q$. The ratio of X_d to X_q reflects the characteristic unevenness of the air gaps of the salient pole synchronous generator. The magnitude reflects the influence of the armature magnetic field on the stator phase emf along the vertical and horizontal axes, respectively.

2.2 Electromagnetic Power and Power Angle Characteristic of Synchronous Generator

2.2.1 Power and Torque Balance Equations

For a synchronous generator, the mechanical power P_1 supplied by the prime mover is transmitted from the rotor side to the stator side via the air gap resultant magnetic field to convert the mechanical energy into electrical energy. The power of this part is called the electromagnetic power P_e. After the stator winding copper loss P_c and the no-load mechanical loss P_0 are subtracted from the electromagnetic power, the residual power is the generator output active power P. In the case of stable operation, the balance equation between the powers is

$$P = P_e - (P_c + P_o) \tag{2.12}$$

For a large generator unit, at the rated load, the proportion of the stator winding copper loss P_c in the rated power is very small and is thus approximately negligible. At this point, it can be considered that the generator's electromagnetic power is approximately equal to its active power. That is,

$$P_e \approx P = 3UI \cos\varphi \tag{2.13}$$

The relation between the various powers and torques of the generator is expressed as

$$P = \omega T \tag{2.14}$$

where

ω – mechanical angular velocity of the rotor, $\omega = n_c/60$;

n_c – synchronous speed

The torque balance equation that relates the drag torque T_1 of the generator, the mechanical loss braking torque T_0, and the electromagnetic braking torque T_e is as follows:

$$T_1 = T_0 + T_e \tag{2.15}$$

2.2.2 Electromagnetic Power and Power Angle Characteristic Expressions

As mentioned above, when the stator loss is ignored, the active power output by the unit will be equal to its electromagnetic power.

With Eq. (2.13) and Figures 2.2a and c, the power expression of a salient pole generator can be obtained:

$$P_e = 3UI \cos\varphi = 3UI \cos(\psi - \delta) = 3UI (\cos\psi \cos\delta + \sin\psi \sin\delta) \tag{2.16}$$

Since

$$I_d = I \sin\psi, \quad U_d = U \cos\delta = E_0 - I_d X_d$$
$$I_q = I \cos\psi, \quad U_q = U \sin\delta = I_q X_q$$

the following can be obtained:

$$P_e = 3(U\cos\delta I_q + U\sin\delta I_d)$$

$$= 3\left(U\cos\delta\frac{U\sin\delta}{X_q} + U\sin\delta\frac{E_0 - U\cos\delta}{X_d}\right)$$

$$= 3\left[\frac{E_0 U}{X_d}\sin\delta + \frac{U^2}{2}\left(\frac{1}{X_q} - \frac{1}{X_d}\right)\sin 2\delta\right]$$

Expressed as per-unit value, the above equation can be written as

$$P_e = \frac{E_0 U}{X_d}\sin\delta + \frac{U^2}{2}\left(\frac{1}{X_q} - \frac{1}{X_d}\right)\sin 2\delta = P_{e1} + P_{e2} \tag{2.17}$$

where

P_{e1} – electromagnetic power related to excitation;

P_{e2} – additional electromagnetic power.

The additional electromagnetic power is unrelated to the excitation but is related to the grid voltage and the vertical and horizontal axis synchronous reactances. That is, when the rotor excitation winding has no excitation current, as long as $U \neq 0$ and $\delta \neq 0$, P_{e2} will be generated. Since it is caused wholly by the inequality between the d- and q-axis magnetoresistances, it is also known as the magnetoresistive power or salient pole power. Its amplitude increases as the difference between X_d and X_q increases. The torque corresponding to P_{e2} is known as the reluctance torque or the salient pole torque.

For a non-salient generator, because $X_d \approx X_q$, at this point $P_{e2} \approx 0$. So,

$$P_e = \frac{E_0 U}{X_d}\sin\delta \tag{2.18}$$

As can be seen from Eq. (2.18), for a non-salient pole generator, its electromagnetic power varies with the angle δ as a sine function. The relation between the electromagnetic power and the angle δ is called the power angle characteristic or the power characteristic. This characteristic indicates that when the grid voltage is constant and the excitation current of the generator operating on the grid is kept constant (i.e., E_0 and U are constant), the magnitude of the electromagnetic power depends only on the power angle δ value. The power angle characteristic curves of synchronous motors are shown in Figure 2.3.

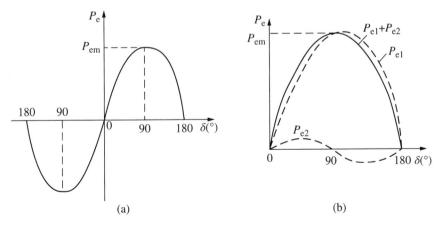

Figure 2.3 Power angle characteristic curves of synchronous motor: (a) non-salient pole generator, (b) salient pole generator.

2.2.2.1 Power Angle Characteristic Curve of Non-salient Pole Generator

The power angle characteristic curve of a non-salient pole generator is shown in Figure 2.3a.

(1) When $0° < \delta < 90°$, the electromagnetic power P_e increases as the angle δ increases. When $90° < \delta < 180°$, P_e decreases as the angle δ increases. When $\delta = 90°$, the electromagnetic power reaches the limiting value P_{em}, which is

$$P_{em} = \frac{E_0 U}{X_d} \tag{2.19}$$

(2) When the angle δ is positive, P_e is positive, indicating that the synchronous motor supplies active power to the grid and works as a generator. On the other hand, when the angle δ is negative, P_e is negative, indicating that the motor absorbs active power from the grid and works as an electric motor.

2.2.2.2 Power Angle Characteristic Curve of Salient Pole Generator

The power angle characteristic curve of a salient pole generator is shown in Figure 2.3b. In addition to a sinusoidal wave, the power angle characteristic curve has a second harmonic that is two times the sine wave frequency. The power component makes the salient pole generator reach the power limit before $\delta < 90°$.

In the time phase, δ represents the angle of the leading terminal voltage $\dot{U}$ of the induced emf $\dot{E}_0$, and varies with the unit's load (see Figures 2.1 and 2.2).

When the stator leakage resistance and the resistance of the generator are small, it can be approximately considered that $\delta = \delta_i$ (see Figures 2.1a and 2.2a). Besides, in the spatial position, δ represents both the spatial angle of the rotor main flux Φ_0 in the rotor rotation direction ahead of the air gap resultant magnetic flux Φ_δ and the spatial angle of the rotor pole axis as a prime mover in the rotor rotation direction ahead of the air gap resultant magnetic field pole axis, as shown in Figure 2.4.

2.2.2.3 Power Angle Expression of Reactive Power

As mentioned earlier, the power angle is an important parameter reflecting energy conversion within a generator. A change in the power angle will produce changes of the active power and the reactive power. The power angle characteristic of the reactive power of the generator can be obtained with the derivation method similar to that used for the active power.

(1) For a salient pole generator, since reactive power $Q = 3UI \sin\varphi$, the following can be obtained from Figure 2.2c:.

$$Q = \frac{3E_0 U}{X_d} \cos\delta - \frac{3U^2}{2} \cdot \frac{X_d + X_q}{X_d X_q} + \frac{3U^2}{2} \cdot \frac{X_d - X_q}{X_d X_q} \cos 2\delta \tag{2.20}$$

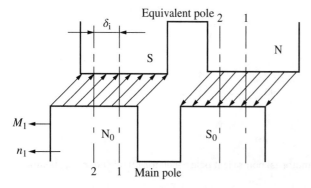

Figure 2.4 Schematic diagram when power angle is spatial angle (generator state): 1 – equivalent pole axis of air gap resultant magnetic field, 2 – rotor pole axis, S and N – polarity of stator equivalent poles, S_0 and N_0 – polarity of rotor poles, M_1 and n_1 – rotor torque and speed.

Figure 2.5 Curve of relation $Q = f(\delta)$ of non-salient pole generator.

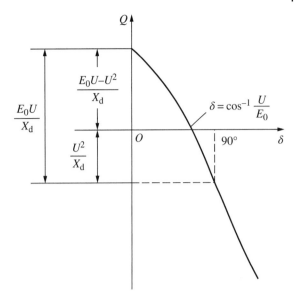

(2) For a non-salient pole generator, since $X_d \approx X_q$, reactive power can be written as

$$Q = \frac{3E_0 U}{X_d} \cos\delta - \frac{3U^2}{X_d}$$

With per-unit value, the relation between the output reactive power and the power angle for a non-salient pole generator can be obtained:

$$Q = \frac{E_0 U}{X_d} \cos\delta - \frac{U^2}{X_d} \tag{2.21}$$

Equation (2.21) shows that, for a non-salient pole generator in operation, since X_d has been determined, $Q = f(\delta)$ is a cosine function when the grid voltage is constant and the excitation current of the non-salient pole generator operating on the grid is kept constant. The shape of the curve is shown in Figure 2.5.

2.3 Operating Capacity Characteristic Curve of Synchronous Generator

The generator operating capacity diagram expresses the relation between the generator's active power and reactive power under the condition of rated terminal voltage and cooling medium temperature. The diagram shows the long-term safe operation range of the generator under different power factor operating conditions. The principle used to determine the operating capacity curve is introduced here.

2.3.1 Operating Capacity Diagram of Non-salient Pole Generator

2.3.1.1 Power Diagram of Non-salient Pole Generator

The emf phasor diagram of a non-salient pole generator under the rated operating conditions is shown in Figure 2.1c. If each side of the emf triangle in Figure 2.1c is multiplied by $3U/X_d$, the power diagram of the non-salient pole generator can be obtained, as shown in Figure 2.6. As can be seen the figure:

(1) The rated capacity of the generator is $S_N = \overline{OA} = 3UI$.
(2) The projection of $\overline{OA}$ on the vertical axis and that on the horizontal axis respectively represent the rated active power and reactive power; that is,

$$P_N = 3UI \cos\varphi_N$$
$$Q_N = 3UI \sin\varphi_N$$

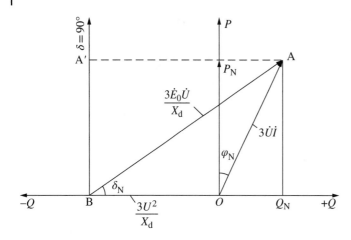

Figure 2.6 Power diagram of non-salient pole generator.

(3) Point A corresponds to the rated operating conditions for the generator (i.e., the stator rated voltage U_N, the stator rated current I_N, the rated power factor $\cos\varphi_N$, and the rated cooling medium parameters). At the point, the rated power angle is δ_N.
(4) The vertical axis (+P) corresponds to $\cos\varphi = 1$. The horizontal axis (+Q) corresponds to $\cos\varphi = 0$ (lagging) and −Q to $\cos\varphi = 0$ (leading).
(5) As the power angle increases, when $\delta = 90°$, the capacity line moves to OA′, and the limiting power is BA′. OB represents the maximum reactive power that can be absorbed when P = 0, as the generator is operating in parallel with an infinite power grid.

2.3.1.2 Operating Capacity Curve of Non-salient Pole Generator

If the power triangle shown in Figure 2.6 is expressed as the per-unit value based on rated capacity, $\overline{OA} = 1$ can be obtained, and $\overline{OA}$ can represent the stator rated current. When the generator is not saturated, the excitation current is in direct proportion to the potential. So, $\overline{AB}$ also represents the excitation rated current. The change in φ can reflect the change in the $\cos\varphi$ value and the generator capacity. Thus, the operating capacity diagram of the generator can be obtained.

(1) *Lagging-phase capacity curve.* When the non-salient pole generator operates in lagging phase and keeps the temperature of the cooling medium constant, the stator and rotor currents shall not exceed the rated values in order to ensure that the stator and rotor winding temperature rises do not exceed the allowable value and cause overheating. As shown in Figure 2.7, the rotor rated current arc $\widehat{AC}$ is plotted with point B as its center and $\overline{AB}$ as its radius, and the stator rated current arc ($\widehat{ADGP}$) is plotted with point O as its center and $\overline{OA}$ as its radius. The stator and rotor currents corresponding to the point of intersection of the two arcs A reach the rated values simultaneously. When $\cos\varphi$ decreases (angle φ increases), due to the rotor current constraint, the generator's operating point should not exceed Arc $\widehat{AC}$. Point C is the maximum reactive power output by the generator when $\cos\varphi = 0$. When $\cos\varphi$ increases (angle φ decreases), the generator's armature reaction decreases, and the required excitation current decreases. So, the excitation current is not a limiting factor. But at this point the stator current is a constraint. The generator's operating point should not exceed Arc $\widehat{ADGP}$, and its active power should not exceed the rated power of the turbine. Therefore, after point D, when $\cos\varphi$ continues to increase, there will be the limitation of the prime mover output limit line $\widehat{DF}$. That is, the capacity of the generator is limited.

Thus, the lagging-phase operation range of the generator is the closed area demarcated by the points (O, E, D, A, C, and O).

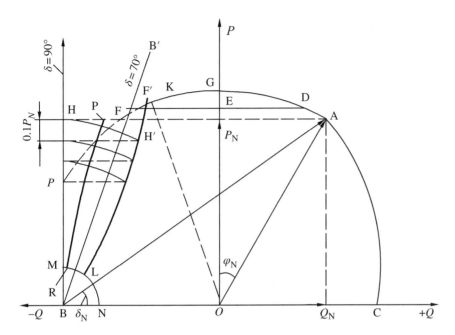

Figure 2.7 Operating capacity diagram of synchronous non-salient pole generator.

(2) *Leading phase operating capacity curve.* When the active power is constant and the generator switches to leading phase operation, the generator is in the low excitation state, with reduced internal emf and electromagnetic torque and increased power angle and static stability margin, and is prone to loss of static stability. Meanwhile, the stator end leakage flux tends to deteriorate and the loss increases. The leading phase operating capacity (i.e., P and Q) is determined by the stator core end overheat or the static stability limit or the dynamic stability limit, whichever is the lowest.

When the capacity is determined by the generator stator end temperature rise, due to the differences in its end structure and material properties and cooling conditions, accurate calculation of the temperature rise of the various parts of the end is difficult. So, the allowable actual value of the leading phase capacity when the local high temperature does not exceed the limit cannot be determined. Generally, the capacity of the end heating limit is estimated.

The determination of the leading phase operation static stability limit capacity employs the power limit value determined when the generator has no automatic excitation regulator, outputs various active power values and operates in leading phase and the operating power angle $\delta \leqslant 90°$ ($X_e = 0$) as the basis and takes the appropriate static stability reserve factor into account. Usually 10% of P_N is taken as a static stability reserve. An arc is plotted, with point B in Figure 2.7 as its center and $\overline{BH}$ as its radius. Then, a straight line is plotted at $0.9P_N$, parallel to the horizontal axis and intersecting the arc at point H′. In this way, one intersection can be obtained every $0.1P_N$. The line connecting these points, $\overline{F'L}$, is just the static stability capacity limit line.

The straight line with power angle $\delta = 70°$ can also be taken as the static stability capacity limit line, like BB′ shown in Figure 2.7. At this point, the static overload capacity of the generator $K = 1/\sin 70° = 1.06$, with a 6% static stability reserve.

If the generator is connected to the grid through the contact reactance X_e, its static stability limit capacity is reduced. Since $X_e \neq 0$, the static stability limit line in the P,Q plan is no longer a straight line but an arc, like Arc $\widehat{RP}$ shown in Figure 2.7. The equation of this circle is

$$P^2 + \left[Q - \frac{U^2}{2}\left(\frac{1}{X_e} - \frac{1}{X_d}\right)\right]^2 = \left[\frac{U^2}{2}\left(\frac{1}{X_e} + \frac{1}{X_d}\right)\right]^2 \tag{2.22}$$

The circle's center coordinates are on the reactive power axis Q. Its coordinate point is $\left[0, \frac{U^2}{2}\left(\frac{1}{X_e} - \frac{1}{X_d}\right)\right]$, and its radius is $\frac{U^2}{2}\left(\frac{1}{X_e} + \frac{1}{X_d}\right)$. Thus, under the condition of constant active power P and voltage U, the reactive power that can be absorbed by the generator at this point can be calculated. When the voltage changes, the static stability limit line will be arcs with different centers and radiuses. In the figure, Curve $\overarc{RP}$ is the static stability limit capacity curve when $X_e \neq 0$ and $U = 1$. This indicates that, due to the influence of X_e, the generator's ability to absorb reactive power will decrease in the case of the same active power. It can be determined that the leading phase operation range of the non-salient pole generator is OEF′LN.

Curve $\overarc{MN}$ is the minimum allowable excitation current limit line, usually considered to be 10% of the rated current.

2.3.2 Operating Capacity Curve of Salient Pole Generator

The expression for the electromagnetic power of a salient pole generator is shown in Eq. (2.17). When the stator winding resistance copper loss is ignored, the generator's electromagnetic power can be considered equal to its output active power P. That is,

$$P = \frac{3UE_0}{X_d}\sin\delta + \frac{3U^2}{2}\left(\frac{1}{X_q} - \frac{1}{X_d}\right)\sin 2\delta = P_1 + P_2$$

Figure 2.8 is the power diagram of a salient pole generator. On the $-Q$ axis of the P,Q plan, a circle is plotted with $3U^2\left(\frac{1}{X_q} - \frac{1}{X_d}\right)$ as its diameter. Thus, its radius is $\frac{3U^2}{2}\left(\frac{1}{X_q} - \frac{1}{X_d}\right)$. Then, a straight line $\overline{BD}$ is plotted from Point B, intersecting the horizontal axis at an angle equivalent to the rated power angle δ_n, and extending to the rated operating point A. Since $\angle DO_1C = 2\angle DBC = 2\delta_N$, the following can be obtained:

$$P_2 = \frac{3U^2}{2}\left(\frac{1}{X_q} - \frac{1}{X_d}\right)\sin 2\delta_N$$

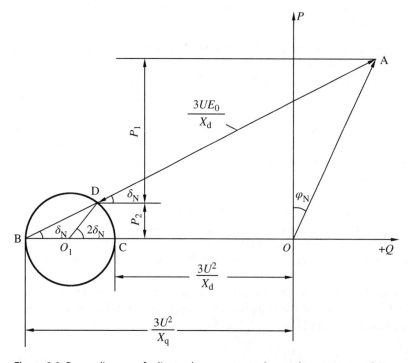

Figure 2.8 Power diagram of salient pole generator under rated operating conditions.

Since

$$\overline{AD} = \frac{3UE_0}{X_d}$$

the following can be obtained:

$$P_1 = \frac{3UE_0}{X_d} \sin \delta_N$$

The projection of the straight line $\overline{AB}$ on the active power axis is

$$P = P_1 + P_2 = \frac{3UE_0}{X_d} \sin \delta_N + \frac{3U^2}{2}\left(\frac{1}{X_q} - \frac{1}{X_d}\right) \sin 2\delta_N = 3UI \cos \varphi_N$$

Similarly, the straight line $\overline{AD}$ represents the excitation rated current and the straight line $\overline{OA}$ the stator rated current. Thus, the operating capacity curve of the stator or rotor current limit can be plotted.

2.4 Influence of External Reactance on Operating Capacity Characteristic Curve [6]

The connection and phasor diagrams of the connection between a generator and a system are shown in Figure 2.9. First, the generator's active power P and reactive power Q are determined on the basis of the phasor diagram in Figure 2.9.

The d- and q-axis components of the generator voltage U_G are respectively

$$U_{Gd} = I_q X_q \tag{2.23}$$
$$U_{Gd} = E_q - I_d X_d \tag{2.24}$$

The system voltage U and the generator terminal voltage U_G are related as follows:

$$U_{Gd} - U_d = -X_e I_q \tag{2.25}$$
$$U_{Gq} - U_q = X_e I_d \tag{2.26}$$

The angle between the generator's internal emf E_q and the infinite system's voltage U is the rotor angle δ. Thus, we obtain the following:

$$U_d = U \sin \delta \tag{2.27}$$
$$U_q = U \cos \delta \tag{2.28}$$

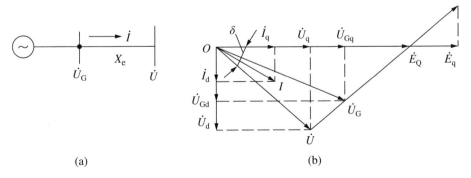

Figure 2.9 Connection and phasor diagrams of connection between hydroelectric generating and system: (a) connection diagram, (b) phasor diagram.

With Eqs. (2.23)–(2.28), U_{Gd} and U_{Gq} are eliminated, and we obtain:

$$U \sin \delta = (X_q + X_e) I_q \tag{2.29}$$

$$U \cos \delta = E_q - (X_d + X_e) I_d \tag{2.30}$$

It is assumed that the active and reactive powers at the receiving terminal voltage U are respectively P' and Q'. That is,

$$\begin{aligned} P' + jQ' &= \dot{U}\hat{I} = U(\cos \delta - j \sin \delta)(I_q + j\, I_d) \\ &= (UI_q \cos \delta + UI_d \sin \delta) + j(UI_d \cos \delta - UI_q \sin \delta) \end{aligned} \tag{2.31}$$

The generator output power $P + jQ$ is given by

$$P + jQ = P' + j(Q' + X_e I^2) \tag{2.32}$$

Thus,

$$\begin{aligned} P &= P' = UI_q \cos \delta + UI_d \sin \delta \\ &= U \frac{U \sin \delta}{X_q + X_e} \cos \delta + U \frac{E_q - U \cos \delta}{X_d + X_e} \sin \delta \\ &= \frac{U^2 \sin \delta \cos \delta}{X_q + X_e} + \frac{E_q U \sin \delta - U^2 \sin \delta \cos \delta}{X_d + X_e} \\ &= \frac{E_q U \sin \delta}{X_d + X_e} + \left(\frac{1}{X_q + X_e} - \frac{1}{X_d + X_e} \right) U^2 \sin \delta \cos \delta \\ &= \frac{E_q U}{X_d + X_e} \sin \delta + \frac{U^2(X_d - X_q)}{2(X_d + X_e)(X_q + X_e)} \sin 2\delta \end{aligned} \tag{2.33}$$

The reactive power Q is

$$\begin{aligned} Q &= Q' + X_e I^2 \\ &= UI_d \cos \delta - UI_q \sin \delta + X_e(I_d^2 + I_q^2) \\ &= U \frac{E_q - U \cos \delta}{X_d + X_e} \cos \delta - U \frac{U \sin \delta}{X_q + X_e} \sin \delta + X_e \left(\frac{E_q - U \cos \delta}{X_d + X_e} \right)^2 + X_e \left(\frac{U \sin \delta}{X_q + X_e} \right)^2 \\ &= \frac{E_q U \cos \delta}{X_d + X_e} - \frac{U^2 \cos^2 \delta}{X_d + X_e} - \frac{U^2 \sin^2 \delta}{X_q + X_e} + \frac{E_q^2 X_e}{(X_d + X_e)^2} - \frac{2 E_q U X_e \cos \delta}{(X_d + X_e)^2} \\ &\quad + \frac{U^2 X_e \cos^2 \delta}{(X_d + X_e)^2} + \frac{U^2 X_e \sin^2 \delta}{(X_q + X_e)^2} \\ &= \frac{E_q U X_d \cos \delta + E_q U X_e \cos \delta - U^2 X_d \cos^2 \delta - U^2 X_e \cos^2 \delta}{(X_d + X_e)^2} \\ &\quad + \frac{E_q^2 X_e - 2 E_q U X_e \cos \delta + U^2 X_e \cos^2 \delta}{(X_d + X_e)^2} + \frac{U^2 X_e \sin^2 \delta - U^2 X_q \sin^2 \delta - U^2 X_e \sin 2\delta}{(X_q + X_e)^2} \\ &= \frac{(X_d - X_e) E_q U \cos \delta - U^2 X_d \cos^2 \delta + E_q^2 X_e}{(X_d + X_e)^2} - \frac{U^2 X_q \sin^2 \delta}{(X_q + X_e)^2} \end{aligned} \tag{2.34}$$

For a steam turbine, in Eqs. (2.33) and (2.34), $X_d = X_q$, which yields the following form:

$$P = \frac{E_q U}{X_d + X_e} \sin \delta \tag{2.35}$$

$$Q = \frac{(X_d - X_e)E_q U \cos\delta - U^2 X_d + E_q^2 X_e}{(X_d + X_e)^2} \qquad (2.36)$$

Next, the relationship between P and Q under the condition of static stability limit is determined. When the power reaches the static stability limit, $\frac{dP}{d\delta} = 0$. With

$$P = \frac{E_q U}{X_d + X_e} \sin\delta$$

the following can be obtained:

$$\frac{dP}{d\delta} = \frac{E_q U}{X_d + X_e} \cos\delta$$

With Eq. (2.36), $\cos\delta$ is obtained as follows:

$$\cos\delta \frac{\frac{E_q^2 X_e}{(X_d+X_e)^2} - \frac{U^2 X_d}{(X_d+X_e)^2} - Q}{\frac{(X_e-X_d)}{(X_d+X_e)^2} E_q U} = \frac{E_q^2 X_e - U^2 X_d - Q(X_d + X_e)^2}{(X_e - X_d) E_q U} \qquad (2.37)$$

$$\frac{dP}{d\delta} = \frac{E_q U}{X_d + X_e} \left[\frac{E_q^2 X_e - U^2 X_d - Q(X_d + X_e)^2}{(X_e - X_d) E_q U} \right]$$

$$= \frac{X_d}{X_d^2 - X_e^2} U^2 - \frac{X_e}{X_d^2 - X_e^2} E_q^2 + \frac{X_d + X_e}{X_d - X_e} Q \qquad (2.38)$$

When $\frac{dP}{d\delta} = 0$, the following is obtained:

$$Q = \frac{X_d - X_e}{X_d + X_e} \left(\frac{X_e}{X_d^2 - X_e^2} E_q^2 - \frac{X_d}{X_d^2 + X_e^2} U^2 \right)$$

$$= \frac{X_e}{(X_d + X_e)^2} E_q^2 - \frac{X_d}{(X_d + X_e)^2} U^2 \qquad (2.39)$$

For the active power P, when $\delta = \frac{\pi}{2}$ and $\frac{dP}{d\delta} = 0$, we obtain the following:

$$P = \frac{E_q U}{X_d + X_e}$$

The corresponding phasor diagram is shown in Figure 2.10.
Figure 2.10 yields the following:

$$U_G^2 = (X_e I_d)^2 + (X_d I_q)^2$$

$$= \frac{X_e^2}{(X_d + X_e)^2} E_q^2 + \frac{X_d^2}{(X_d + X_e)^2} U^2 \qquad (2.40)$$

Figure 2.10 Static stability limit phasor diagram of non-salient pole generator.

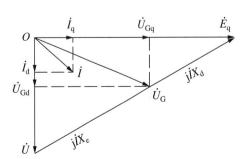

Substituting $P = \frac{E_q U}{X_d + X_e}$ into Eq. (2.40) and eliminating E_q, we obtain

$$U_G^2 = \frac{X_e^2}{(X_d + X_e)^2} \times \frac{P^2}{U^2}(X_d + X_e)^2 + \frac{X_d^2}{(X_d + X_e)^2}U^2$$

$$= \frac{X_e^2}{U^2}P^2 + \left(\frac{X_d}{X_d + X_e}\right)^2 U^2 \tag{2.41}$$

The reactive power expression, Eq. (2.39), enables E_q to be eliminated:

$$Q = \frac{X_e}{(X_d + X_e)^2} \times \frac{P^2}{U^2}(X_d + X_e)^2 - \frac{X_d}{(X_d + X_e)^2}U^2$$

$$= \frac{X_e}{U^2}P^2 - \frac{X_d}{(X_d + X_e)^2}U^2 \tag{2.42}$$

Calculation yields the following:

$$\frac{P^2}{U^2}X_e^2 = X_e Q + \frac{X_e X_d}{(X_d + X_e)^2}U^2 \tag{2.43}$$

Substituting Eq. (2.43) into Eq. (2.41) and eliminating $\frac{X_e^2}{U^2}P^2$ yields the following:

$$U_G^2 = X_e Q + \frac{(X_e X_d + X_d^2)U^2}{X_e + X_d} \tag{2.44}$$

Through calculation, the following is obtained:

$$U^2 = \frac{(X_d + X_e)^2}{X_e X_d + X_d^2}(U_G^2 - X_e Q) \tag{2.45}$$

Substituting Eq. (2.45) into Eq. (2.42) and eliminating the term U^2, the following is obtained:

$$Q = \frac{X_e X_d P^2}{(U_G^2 - X_e Q)(X_d + X_e)} - \frac{U_G^2 - X_e Q}{X_e + X_d} \tag{2.46}$$

Calculation yields the following:

$$(U_G^2 - X_e Q)(X_d + X_e)Q = X_e X_d P^2 - (U_G^2 - X_e Q)^2 + X_d X_e P^2$$
$$+ [X_e(X_d + X_e) - X_e^2]Q^2 + [2U_G^2 X_e - U_G^2(X_d + X_e)]Q = U_G^4 \tag{2.47}$$

or

$$P^2 + Q^2 + \frac{X_e - X_d}{X_d X_e}QU_G^2 = \frac{U_G^4}{X_e X_d}$$

$$P^2 + \left(Q + \frac{X_e - X_d}{2X_d X_e}U_G^2\right)^2 = \frac{U_G^4}{X_e X_d} + \frac{(X_e - X_d)^2}{4X_d^2 X_e^2}U_G^4 \tag{2.48}$$

This can be rewritten as

$$P^2 + \left[Q + \frac{1}{2}\left(\frac{1}{X_d} - \frac{1}{X_e}\right)U_G^2\right]^2 = \frac{1}{4}\left(\frac{1}{X_e} + \frac{1}{X_d}\right)^2 U_G^4$$

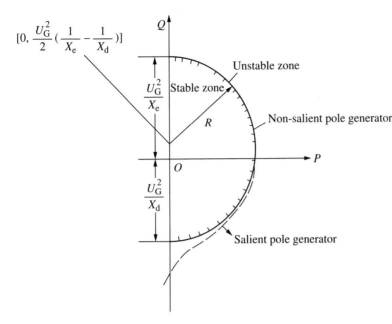

Figure 2.11 Determination of static stable zone of non-salient pole generator.

If both sides of the above equation are multiplied by $\left(\frac{X_d}{U_G^2}\right)^2$, the following is obtained:

$$\left(\frac{PX_d}{U_G^2}\right)^2 + \left[\frac{QX_d}{U_G^2} + \frac{1}{2}\left(1 - \frac{X_d}{X_e}\right)\right]^2 = \frac{1}{4}\left(1 + \frac{X_d}{X_e}\right)^2 \tag{2.49}$$

When $U_G = 1.0$,

$$P^2\left[Q + \frac{1}{2}\left(\frac{1}{X_d} - \frac{1}{X_e}\right)\right]^2 = \left[\frac{1}{2}\left(\frac{1}{X_d} + \frac{1}{X_e}\right)\right]^2 \tag{2.50}$$

The corresponding static stability limit curve is shown in Figure 2.11.

The power expression of a salient pole generator under the condition of external reactance is

$$P = \frac{E_q U}{(X_d + X_e)} \sin\delta + \frac{U^2(X_d - X_q)}{2(X_d + X_e)(X_q + X_e)} \sin 2\delta \tag{2.51}$$

The static stability limit is

$$\frac{dP}{d\delta} = \frac{E_q U}{(X_d + X_e)} \cos\delta$$
$$+ \frac{U^2(X_d - X_q)}{(X_d + X_e)(X_q + X_e)} \cos 2\delta = 0 \tag{2.52}$$

or

$$E_q \cos\delta + \frac{U(X_d - X_q)}{(X_q + X_e)}(2\cos^2\delta - 1) = 0 \tag{2.53}$$

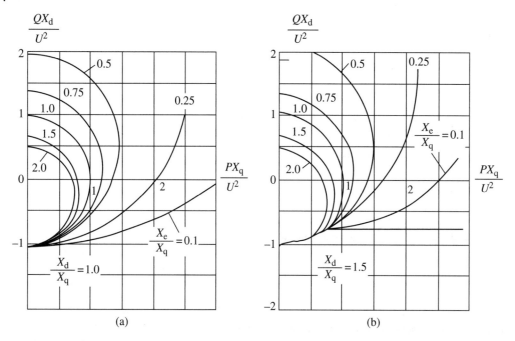

Figure 2.12 Influence of transformer and line reactance on static stability of synchronous generator: (a) non-salient pole generator, (b) salient pole generator.

E_q, $\cos\delta$ and U in Eq. (2.53) are eliminated and expressed as parameters such as P, Q, X_d, X_q, and X_e. Through calculation, the equation of the static stability zone of the salient pole generator is determined:

$$\left(\frac{PX_q}{U^2}\right)^2 \left[\frac{QX_q}{U^2} - \frac{1-\frac{X_e}{X_q}}{2\frac{X_e}{X_q}}\right]^2 + \frac{\left(\frac{X_d}{X_q}-1\right)\left(1+\frac{X_e}{X_q}\right)^2}{\frac{X_e}{X_q}\left(\frac{X_e}{X_q}+\frac{X_d}{X_q}\right)}$$

$$\times \frac{\left(\frac{PX_q}{U^2}\right)^2}{\left[\left(1+\frac{QX_q}{U^2}\right)^2 + \left(\frac{PX_q}{U^2}\right)^2\right]} = \left[\frac{1+\frac{X_e}{X_q}}{2\frac{X_e}{X_q}}\right]^2 \quad (2.54)$$

Given different X_e/X_q values, the static stability limits of non-salient and salient pole generators can be obtained, as shown in Figure 2.12.

As can be seen from Eq. (2.49), on the coordinate plane that takes $\frac{PX_d}{U^2}$ as its active per-unit value and $\frac{QX_d}{U^2}$ as its reactive per-unit value, for a non-salient pole generator, the static stability limit curve is a circle whose center is on the $\frac{QX_d}{U^2}$ axis and coordinate point at ordinate $\frac{1}{2}\left(1-\frac{X_d}{X_e}\right)$, with $\frac{1}{2}\left(1+\frac{X_d}{X_e}\right)$ as its radius. For a salient pole generator, the center and radius coordinates can be obtained from Eq. (2.54).

2.5 Operating Characteristic Curves of Generator

To fully understand the characteristics of a generator in different operating modes, operators should be familiar with the characteristic curves provided by the manufacturer in various operating modes. The following is a brief description of the relevant operating characteristic curves of the 777.8 MVA hydro turbine of Three Gorges Hydropower station as an example.

2.5.1 Operating Characteristic Curve of Hydro Generator

2.5.1.1 No-Load Saturation Characteristic Curve

The no-load saturation characteristic curve expresses the relation $U_G = f(I_f)$ between the generator voltage U_G and the excitation current I_f, as shown in Curve 1 in Figure 2.13.

Usually, the no-load characteristic curve of a generator is expressed in terms of the per-unit value as follows:

$$U_G = U_{GN} U_{G(p.u.)} \tag{2.55}$$

where

U_G – actual value of voltage of generator;

U_{GN} – actual value of rated voltage of generator;

$U_{G(p.u.)}$ – per-unit value of voltage of generator.

In Figure 2.13, correspondingly,

(1) No-load rated excitation current I_{f0}: 2351 A;
(2) Short circuit rated excitation current I_{fc}: 1964 A;
(3) Excitation current of air gap line corresponding to rated voltage I_{fA}: 2092 A;
(4) Stator rated voltage: 20 000 V;
(5) Stator rated current: 22 453 A;
(6) Excitation current: 2092 A.

2.5.1.2 Short Circuit Characteristic Curve

The short circuit characteristic curve (see Curve 2 in Figure 2.13) expresses the relation $I_{GC} = f(I_f)$ between the stator steady-state short circuit current I_{GC} and the excitation current I_f when the stator windings are three-phase shorted out.

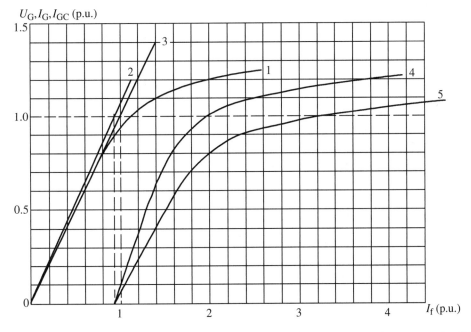

Figure 2.13 Characteristic curve of 777.8 MVA hydro turbine of Three Gorges Hydropower station (20 000 V, 50 Hz, 75 r min^{-1}, cosφ = 0.9): 1 – no-load characteristic curve, 2 – short circuit characteristic curve, 3 – air gap line, 4 – load characteristic curve (cosφ = 0.9), 5 – load characteristic curve (cosφ = 0).

2.5.1.3 Air Gap Line

The excitation current I_{fA} checked on the air gap line when the stator current is rated is defined as air gap line rated excitation current (see Curve 3 in Figure 2.13). The air gap line expresses the relation $I_G = f(I_{fA})$.

2.5.1.4 Load Characteristic Curve ($\cos\varphi = 0.9$)

The load characteristic curve (see Curve 4 in Figure 2.13) expresses the relation $U_G = f(I_f)$ between the stator voltage and the excitation current when the stator current and the power factor are constant, where $\cos\varphi = 0$ and I_G are constant.

2.5.1.5 Load Characteristic Curve ($\cos\varphi = 0$)

The load characteristic curve (see Curve 5 in Figure 2.13) expresses the relation $U_G = f(I_f)$ between the stator voltage and the excitation current when the stator current is constant and the power factor is zero, where $\cos\varphi = 0$ and I_G is a constant.

2.5.2 Capacity Characteristics Curve of Hydro Generator

The capacity characteristic curve of a hydro generator (see Figure 2.14) expresses the range that ensures that the generator operates safely in lagging and leading phase areas under the given operating conditions and the relation between the active and reactive powers. The restricted operating parameters include stator voltage, rotor excitation current, and so on.

2.5.3 V-Shaped Curve of Hydro Generator

When the active power of the hydro generator remains constant, the shape of the curve expressing the relation between the stator current and the excitation current is similar to the letter V. So, the curve is called a V-shaped curve.

Figure 2.15 shows the V-shaped curve. When the active power is constant, the angle δ increases as the excitation current and the emf E_0 decrease. The changes in I_G, $\cos\varphi$, and Q are bounded by $\cos\varphi = 1$. In

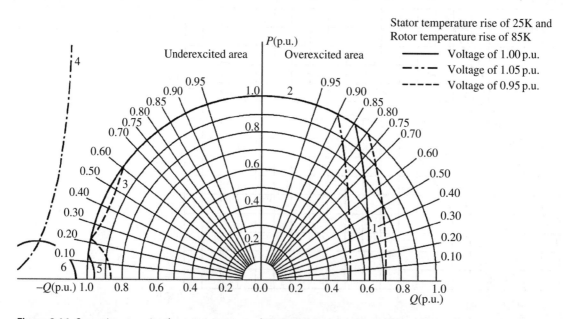

Figure 2.14 Operating capacity characteristic curve of 777.8 MVA hydro turbine of Three Gorges Hydropower station (20 000 V, 50 Hz, 75 r min^{-1}, $\cos\varphi = 0.9$): 1 – maximum excitation current limit, 2 – stator current limit, 3 – actual static stability limit, 4 – theoretical static stability limit, 5 – minimum excitation current limit, 6 – reluctance power or salient pole power.

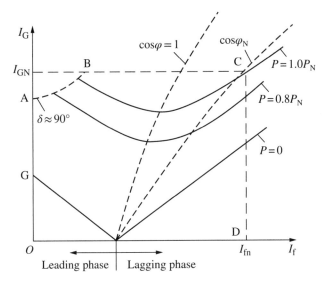

Figure 2.15 V-curve of synchronous generator.

the lagging-phase operation, the reactive power fed by the generator into the grid increases as E_0 and the stator current I_G increase. In the leading phase operation, as E_0 decreases, the stator current I_G increases, and the reactive power taken from the grid also increases. The figure shows the V-shaped curve expressing the relation between the stator current I_G and the excitation current (emf, i.g. E_0). Corresponding to a constant active power value, a V-shaped curve can be obtained. When the active load increases, the curve will move up. Thus, a cluster of V-shaped curves as shown in Figures 2.15 can be obtained.

Figure 2.15 shows the various operating areas of a generator when it has different active loads.

2.5.3.1 Normal Operating Area

In Figure 2.15, the area enclosed by the curves connecting the points (O, A, B, C, and D) is the normal operating area of the generator. In the curve cluster, the lowest point of each curve is the state where the generator operates with $\cos\varphi = 1$. When the lowest points are connected, a virtual curve tilting to the right is obtained. When the active load increases, the excitation current I_f increases, so that $\cos\varphi = 1$ can be maintained. The area on the right side of the line is the lagging-phase operating area, where the excitation current is high and the excitation potential is higher than the generator terminal voltage, which is the overexcited state. The area on the left side of the line is the leading phase operating area, where the excitation current and excitation potential are low, which is the underexcited state. When $\cos\varphi = 1$, we have the normally excited state. Point C is the rated operating point of the generator.

2.5.3.2 Unstable Operating Area

The dotted line AB in Figure 2.15 is a line connecting the operating points when the generator with different active loads reaches the critical steady state. At this point, the rotor power angle $\delta = 90°$. Thus, the curve boundary determines the unstable operating area of the generator.

2.5.3.3 Phase-Modulated Operation Capacity

In Figure 2.15, when $P = 0$, the V-shaped curve of the generator in phase-modulated operation is obtained. At this point, the reactive power is determined by the rotor rated current I_{fN}. The lowest point of the curve is the excitation current when the rotary condenser operates at no load. Point G is the stator current value when the phase-modulated operation excitation current is zero.

Figure 2.16 shows the V-shaped curve of a 777.8 MVA hydro turbine of the Three Gorges Hydropower station, where point A is the rated operating point of the generator.

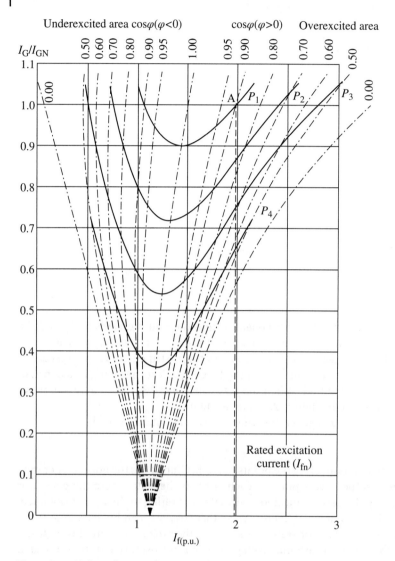

Figure 2.16 V-shaped curve of 777.8 MVA hydro turbine of Three Gorges Hydropower station: 20 000 V, 50 Hz, 75 r min^{-1}, cosφ = 0.9, rated current = 22,453 A (1 p.u.), rated stator voltage gap line excitation current = 2,092 A (1p.u.), P_1 = 100% P_N, P_2 = 80% P_N, P_3 = 60% P_N, P_4 = 40% P_N.

2.6 Transient Characteristics of Synchronous Generator

2.6.1 Transient Reactance X'_d

This section discusses the transient process when a synchronous generator's operation is disturbed and the resulting parameter changes [7]. For a normally loaded synchronous generator, there are four main flux components in the magnetic circuit: the effective main flux Φ_d, the armature reaction flux Φ_{ad}, the stator winding leakage flux Φ_s, and the excitation winding leakage flux Φ_{fs}, as shown in Figure 2.17.

In the event of a disturbance of the operation of the synchronous generator, such as a short circuit accident, the above-mentioned magnetic flux distribution will change transiently. The flux linkage change in the motor winding and the mechanical swing of the rotor determine characteristics of the transient process.

Figure 2.17 Flux components of synchronous generator.

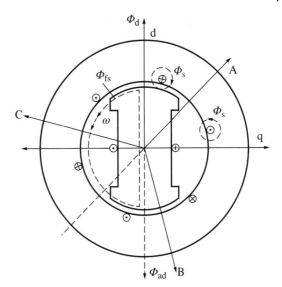

The flux linkage change can be determined with the law of conservation of flux linkage. That is, at the instant $t = 0$ of the disturbance, the change process is analyzed on the basis of the principle that the flux linkage linked to the excitation winding will remain unchanged. Then, it will embark on a gradual transition to a new stable value.

It should be noted that the law of conservation of flux linkage does not apply to a single flux component but to the total resultant flux when several flux components are linked to one winding.

Besides, in the research on the transient process of a generator short circuit, the stator current's aperiodic components attenuate faster than its periodic components. So, in the discussion on the transient process, as an approximation, the stator current's aperiodic components can be ignored once. Similarly, the current change in the damping winding circuit that attenuates faster is negligible.

First, consider the transient process when the generator is in no-load operation. If there is no leakage flux in the rotor windings, the main flux in the magnetic circuit Φ_d is equal to the total flux of the linked excitation winding Φ_f.

In the event of an abrupt disturbance of the generator (e.g., an abrupt short circuit or load addition/reduction), the vertical axis armature reaction flux Φ_{ad} will appear. From the law of conservation of flux linkage, at the instant of the abrupt disturbance, the total flux of the excitation winding Φ_f will remain constant. So, when the armature reaction flux Φ_{ad} appears, the excitation winding flux Φ_d will increase by $\Delta \Phi_d = \Phi_{ad}$ to offset Φ_{ad} and keep the resultant flux linkage unchanged at the instant $t = 0$, as shown in Figure 2.18.

As the air gap flux does not decrease at the instant of the short circuit, the armature reaction effect is not shown. In the steady state, the synchronous generator's vertical axis reactance is equal to the sum of the stator winding leakage reactance and the armature reaction reactance. That is,

$$X_d = X_s + X_{ad} \tag{2.56}$$

At the instant of the short circuit, since the armature reaction is not shown, that is, $X_{ad} = 0$, the vertical axis of the synchronous reactance corresponding to the transient state X'_d will be equal to X_s. That is,

$$X'_d = X_s \tag{2.57}$$

where

X'_d – vertical axis transient synchronous reactance.

Then, as the armature reaction potential gradually increases, X'_d gradually increases to $X'_d = X_d$. In the above discussion, it is assumed that the rotor winding leakage flux is zero. If there is leakage flux, the situation

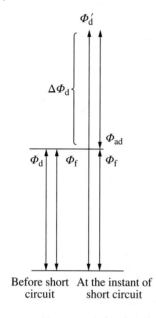

Figure 2.18 Rotor flux changes of synchronous generator in the case of short circuit.

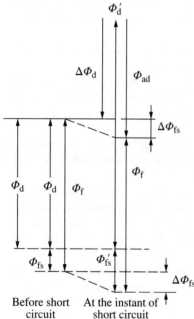

Figure 2.19 Rotor flux changes when rotor leakage flux is considered.

becomes complex. As shown in Figure 2.19, when the rotor excitation winding has a leakage flux Φ_{fs}, although the flux is still Φ_d in the air gap, the total flux of the excitation winding Φ_f will not be equal to Φ_d but will be equal to the sum of the effective flux Φ_d and the rotor winding leakage flux Φ_{fs}. That is,

$$\Phi_f = \Phi_d + \Phi_{fs} \tag{2.58}$$

At the instant of the short circuit, the armature reaction flux Φ_{ad} is shown. In order to keep the total flux of the excitation winding Φ_f constant, the fluxes Φ_d and Φ_{fs} are added to the new Φ'_d and Φ'_{fs} to compensate for the armature reaction flux Φ_{ad}. That is,

$$\Phi_f = \Phi'_d + \Phi'_{fs} - \Phi_{ad} = \Phi_d + \Phi_{fs} \tag{2.59}$$

At the instant after the short circuit, the increase in the flux component is

$$\Delta\Phi_d = \Phi'_d - \Phi_d$$
$$\Delta\Phi_{fs} = \Phi'_{fs} - \Phi_{fs} \qquad (2.60)$$

Obviously,

$$\Delta\Phi_d + \Delta\Phi_{fs} = (\Phi'_d + \Phi'_{fs}) - (\Phi_d + \Phi_{fs}) = \Phi_{ad}$$

However, in the above equation, only $\Delta\Phi_d$ passes through the air gap. $\Delta\Phi_{fs}$ is a rotor leakage flux which does not enter the air gap. That is, part of the armature reaction flux $\Delta\Phi_{ad}$ is not compensated, and its value is equal to the rotor winding leakage flux increase $\Delta\Phi_{fs}$. At this point, the transient reactance includes part of the armature reaction reactance in addition to the stator winding leakage reactance X_s. That is,

$$X'_a = X_s + K_a X_{ad} \qquad (2.61)$$

The coefficient K_a can be obtained from the following equation:

$$K_a = \frac{X_{fs}}{X_{fs} + X_{ad}} \qquad (2.62)$$

where

X_{fs} – excitation winding leakage resistance.

At the instant of the short circuit, the currents in the stator and rotor windings will increase simultaneously in order to maintain the flux linkage balance in the air gap ΔI_f. Then, with the attenuation of the increase in the free component current in the excitation winding circuit, the armature reaction effect gradually strengthens. When the free component current of the rotor current disappears completely, the armature reaction flux Φ_{ad} will be fully shown. At this point, the vertical axis reactance of the synchronous generator will equal X_d.

In the short circuit process, if the excitation system supplies the forced excitation current I_{f1}, although the free component current of the excitation windings I_{f2} is attenuated, the attenuation is compensated by the increase in the forced excitation current. If the increase in the forced excitation current can completely compensate for the decrease in the free component excitation current of the rotor windings I_{f2}, the total excitation current I_f and the stator current I can still be determined by the transient reactance X'_a at any time, just as they can at the instant of the short circuit. The changes in the stator and rotor currents in the case of non-complete compensation are shown in Figure 2.20.

2.6.2 Transient Emf E'_q

As mentioned above, the three flux components linked to the rotor excitation winding are Φ_d, Φ_{fs}, and Φ_{ad}; see Figure 2.21. The excitation winding leakage flux Φ_{fs} can be written as

$$\Phi_{fs} = \sigma_f \Phi_d \qquad (2.63)$$

Figure 2.20 Increase in stator and rotor currents of synchronous generator in case of forced excitation.

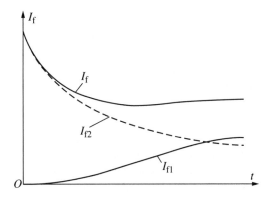

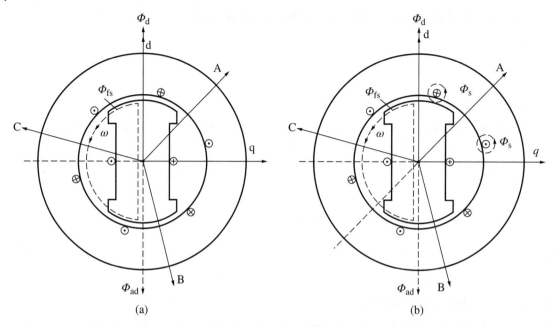

Figure 2.21 Flux components linked to excitation and stator windings.

where

σ_f – leakage flux coefficient of rotor excitation winding.

The total flux linked to the rotor windings is

$$(1 + \sigma_f)\Phi_d - \Phi_{ad} = I_f X_f \tag{2.64}$$

The flux linkage of any flux ψ can be expressed as the product of the number of turns of the winding W and the flux Φ. That is, $\psi = W\Phi$. Besides, the flux linkage can be expressed as the product of the current generating the flux I and the winding's self-inductance L or mutual inductance X. That is, $\psi = LI$. For the latter expression, as the per-unit value is adopted, and $\omega = 1$, $L = X$. The corresponding flux can be written as $\psi = IX$. In this case, the flux linkage term $(1 + \sigma_f)\Phi_d$ in Eq. (2.64) that is generated by the rotor current and linked to the rotor excitation winding can be expressed as $I_f X_f$.

Similarly, the armature reaction flux Φ_{ad} generated by the vertical axis component of the stator current I_d is linked to the excitation winding, and the flux can also be written as $I_d X_{ad}$. Thus, the total flux linkage of the excitation winding is

$$\psi_{fd} = I_f X_f - I_d X_{ad} \tag{2.65}$$

For the stator windings, the flux linkage also consists of three components, as shown in Figure 2.21b; that is, the leakage flux Φ_s generated by the vertical axis component of the stator current I_d, the armature reaction flux Φ_{ad}, and the excitation winding effective flux Φ_d. The expression for the total flux linkage of the vertical axis of the stator is

$$\psi_d = I_f X_{ad} - I_d X_d \tag{2.66}$$

where X_{ad} and X_d are the mutual inductive reactances between the stator and rotor windings and the self-inductive reactance of the stator windings, respectively.

If the stator windings' flux linkage ψ_d is divided by the number of its turns, a stator resultant flux which does not include the stator flux leakage will be obtained. When the motor rotates, the flux will induce the corresponding stator winding emf. Since the flux Φ_d does not include the stator leakage flux, what is induced is not the internal emf but U_q directly corresponding to the generator terminal voltage. That is,

$$\psi_d = U_q \tag{2.67}$$

Besides, when the motor operation abruptly changes, the stator vertical axis current and the rotor excitation current will simultaneously increase in order to keep the flux linkage ψ_{fd} unchanged. With Eq. (2.65), the following can be obtained:

$$I_f = \frac{\psi_{fd} + I_d X_{ad}}{X_f} \tag{2.68}$$

If Eq. (2.68) is substituted into Eq. (2.66), the relation between I_f and I_d can be obtained:

$$\psi_d = U_q = \frac{X_{ad}}{X_f}(\psi_{fd} + I_d X_{ad}) - I_d X_d$$

or

$$U_q = \psi_{fd}\frac{X_{ad}}{X_f} - I_d\left(X_d - \frac{X_{ad}^2}{X_f}\right) \tag{2.69}$$

Equation (2.69) can also be written as

$$U_q = E_q' - I_d X_d' \tag{2.70}$$

where

E_q' — emf after transient reactance, $E_q' = \psi_{fd}\frac{X_{ad}}{X_f}$;

X_d' — transient reactance of generator, $X_d' = X_d - \frac{X_{ad}^2}{X_f}$

The emf E_q' is not the real emf of the synchronous generator. It just represents a fictitious variable proportional to the excitation winding flux linkage ψ_{fd}. Since the rotor flux is unchanged at the instant when the operation of the generator is disturbed, the emf E_q' proportional to ψ_{fd} will be unchanged as well. This is a very important characteristic. The changes in the transient emf of the synchronous generator before and after the short circuit are shown in Figure 2.22.

Then, the excitation winding flux linkage ψ_{fd} is attenuated as the free component current of the rotor attenuates. The law of attenuation for the corresponding transient emf E_q' is the same as that for ψ_{fd}.

2.6.3 Transient Equation of Rotor Excitation Circuit

As mentioned earlier, when the operation of the generator is disturbed, the rotor excitation current will abruptly change and eventually tend to the initial stable value $I_{f0} = \frac{U_{f0}}{R_f}$.

With the excitation circuit of the generator, the following equation is obtained:

$$U_f = I_f R_f + \frac{d\psi_{fd}}{dt} \tag{2.71}$$

Dividing both sides of the equation by R_f yields the following:

$$I_{fe} = I_f + \frac{1}{R_f} \times \frac{d\psi_{fd}}{dt} \tag{2.72}$$

where

I_{fe} — steady-state excitation current corresponding to the excitation voltage U_f, $I_{fs} = \frac{U_f}{R_f}$.

Multiplying both sides of Eq. (2.72) by X_{ad}, the following is obtained:

$$I_{fe}X_{ad} = I_f X_{ad} + \frac{X_{ad}}{R_f} \times \frac{d\psi_{fd}}{dt} \tag{2.73}$$

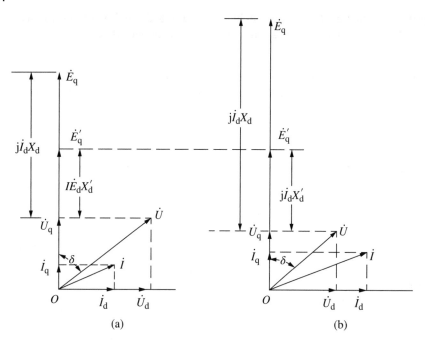

Figure 2.22 Changes in transient emf of synchronous generator in the case of short circuit: (a) before short circuit; (b) after short circuit.

The left-hand side of Eq. (2.73), $I_{fe}X_{ad}$, represents the no-load emf stability value, expressed as E_{qe}. The first term on the right-hand side, $I_f X_{ad}$, represents the no-load emf E_q induced by the stator winding flux linkage generated by the excitation current I_f. Equation (2.73) can be expressed in terms of the per-unit value as follows:

$$E_{qe} = E_q + \frac{X_{ad}}{R_f} \times \frac{d\psi_{fd}}{dt} \tag{2.74}$$

Since

$$E'_q = \psi_{fd}\frac{X_{ad}}{X_f}$$
$$= \frac{X_f}{R_f} = \frac{L_f}{R_f} = T'_{d0}$$

by substitution into the above equation, the following is obtained:

$$E_{qe} = E_q + T'_{d0}\frac{dE'_q}{dt} \tag{2.75}$$

This equation expresses the transient process of the synchronous generator without damping winding.

2.6.4 Change in Transient Emf in the Case of Three-Phase Short Circuit

In the case of a three-phase short circuit of the generator, the change in the transient emf E'_q can be obtained from the following procedure. As can be seen from Eq. (2.75), there are two unknown variables, E_q and E'_q, in it. Another equation is added in order to obtain the solution. With Figure 2.22, the following is obtained:

$$U_q = E'_q - I_d X'_d = E_q - I_d X_d \tag{2.76}$$

Eliminating I_d in Eq. (2.76) yields the following:

$$E'_q = \frac{X'_d}{X_d}E_q + \frac{X_d - X'_d}{X_d}U_q \tag{2.77}$$

2.6 Transient Characteristics of Synchronous Generator

In the case of a three-phase short circuit of the generator terminal, the voltage U_q equals 0, yielding the following:

$$E'_q = \frac{X'_d}{X_d} E_q \tag{2.78}$$

Substituting Eq. (2.78) into Eq. (2.75) and eliminating E'_q, the equation containing only the variable E_q is obtained:

$$E_{qe} = E_q + T'_d \frac{dE_q}{dt} \tag{2.79}$$

where

T'_d – rotor winding time constant in case of short circuit of stator windings. $T'_d = \frac{X'_d}{X_d} T'_{d0}$

With Eq. (2.79), the following is obtained:

$$E_q = E_{q0} + (E_{q(0)} - E_{q0}) e^{-\frac{t}{T'_d}} \tag{2.80}$$

where

E_{q0} – horizontal axis emf component before short circuit (equal to inner emf before short circuit when excitation voltage is constant);

$E_{q(0)}$ – horizontal axis emf component at the instant of short circuit.

With Eq. (2.78), $E_{q(0)}$ is obtained:

$$E_{q(0)} = \frac{X_d}{X'_d} E'_{q0}$$

where

E'_{q0} – horizontal axis invariant transient emf at the instant of short circuit.

During the three-phase short circuit, the curves of changes in the emfs E_q and E'_{q0} over time are shown in Figure 2.23.

As can be seen from Figure 2.23, the no-load emf E_q changes abruptly from E_{q0} to $E_{q(0)}$ after the short circuit and then is attenuated to the initial value based on the time constant T'_d.

Since $U_q = E_q - I_q X_d$, the stator short circuit current is obtained in the case of the short circuit where $U_q = 0$:

$$I_d = \frac{E_q}{X_d}$$

Figure 2.23 Changes in horizontal axis emfs E_q and E'_q over time.

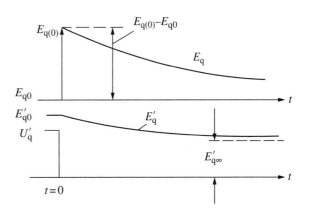

The following is a discussion on the change in the transient emf E'_q during the three-phase short circuit. Substituting Eq. (2.80) into Eq. (2.78) yields the following:

$$E'_q = \frac{X'_d}{X_d}E_{q0} + \left(E_{q(0)}\frac{X'_d}{X_d} - E_{q0}\frac{X'_d}{X_d}\right)e^{-\frac{t}{T'_d}} \tag{2.81}$$

In Eq. (2.81), the factor of the first term $\frac{E_{q0}}{X_d}$ is the steady value of the short circuit current. The product of the item and X'_d equals the transient emf in the case of a steady-state short circuit:

$$\frac{E_{q0}}{X_d}X'_d = E'_{d\infty}$$

$\frac{E_{q(0)}}{X_d}X'_d$ is the value at the instant of the short circuit. It equals E'_{q0}. Thus, Eq. (2.81) can be written as

$$E'_q = E'_{q\infty} + (E'_{q0} - E'_{q\infty})e^{-\frac{t}{T'_d}} \tag{2.82}$$

As can be seen from Eqs. (2.80) and (2.82), both E_q and E'_q get attenuated based on the time constant T'_d, but the magnitudes of their decreases are different. The transient emf at the instant of the short circuit E'_{q0} remains unchanged, while the no-load emf E'_{q0} increases.

2.6.5 Influence of Change of Excitation Voltage on Transient Emf

The last section presents the change in the emf when the excitation voltage is constant. When the excitation voltage changes, the transient emf change law will be different. Suppose the excitation voltage changes in accordance with the following law:

$$U_f = U_{f\infty} - \Delta U_f e^{-\frac{t}{T_e}}$$

where

$U_{f\infty}$ — excitation ceiling voltage stability value;
ΔU_f — excitation voltage change value;
T_e — time constant of exciter.

The steady-state value of the no-load E_{qe} of the synchronous generator will change in accordance with the same law as that for the excitation voltage U_f:

$$E_{qe} = E_{q\infty} - \Delta E_q e^{-\frac{t}{T_e}} \tag{2.83}$$

where

$E_{q\infty}$ — no-load emf corresponding to excitation voltage steady-state ceiling.

The internal emf changes with the change of the excitation voltage, as shown in Figure 2.24. E_{qe} in Eq. (2.83) is substituted into Eq. (2.79). That is,

$$E_{q\infty} - \Delta E_q e^{-\frac{t}{T_e}} = E_q + T'_d + \frac{dE_q}{dt} \tag{2.84}$$

The following solution is obtained:

$$E_q = E_{q\infty} + (E_{q(0)} - E_{q\infty})e^{-\frac{t}{T'_d}} + \Delta E_q \frac{T_e}{T'_d - T_e}\left(e^{-\frac{t}{T_e}} - e^{-\frac{t}{T'_d}}\right) \tag{2.85}$$

Figure 2.24 Change in internal emf when excitation voltage changes.

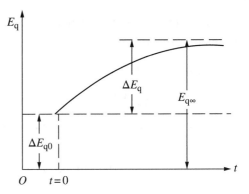

Figure 2.25 Changes in emfs (E_q and E'_q) in the case of three-phase short circuit.

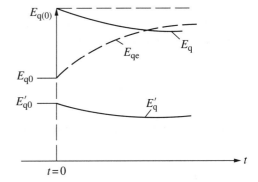

As mentioned in the previous section, the expression for the transient emf E'_q can be obtained:

$$E'_q = E'_{q\infty} + (E_{q(0)} - E'_{q\infty})e^{-\frac{t}{T'_d}} + \Delta E_q \frac{X'_d}{X_d} \times \frac{T_e}{T'_d - T_d}\left(e^{-\frac{t}{T_e}} - e^{-\frac{t}{T_d}}\right) \tag{2.86}$$

Figure 2.25 shows the curves of changes in E_q and E'_q plotted on the basis of Eqs. (2.85) and (2.86).

As can be seen from Figure 2.25, when there is excitation regulation, the attenuation of E_q in the process of approaching the steady value is much lower than that in the case of constant excitation. If the ceiling voltage is a sufficiently high multiple of the excitation voltage, E_q or E'_q may not even decrease but instead increase. E_q and E'_q follow the same change law.

Example 2.1 Try to determine the change in the emf E_q and the transient emf E'_q when the hydro-generator experiences a three-phase short circuit under the no-load rated conditions. When the excitation voltage is constant and the excitation voltage U_f equals $1.0 + 2.5\,t$, the parameters of the generator are as follows:

$$X_d = 1.05, X_q = X'_q = 0.69$$

$$X'_d = 0.29, T'_{d0} = 5\text{ s}$$

With Eq. (2.75), the following can be obtained:

$$E_{qe} = E_q + T'_{d0}\frac{dE'_q}{dt}$$

Then, the following can be obtained:

$$\frac{dE'_q}{dt} = \frac{E_{qe} - E_q}{T'_{d0}}$$

If the numerator and the denominator of the right-hand side of the above equation are multiplied by $\frac{E'_q}{E_q}$ simultaneously, we obtain:

$$\frac{dE'_q}{dt} = \frac{\left(\frac{E'_q}{E_q}\right) E_{qe} - E'_q}{\left(\frac{E'_q}{E_q}\right) T'_{d0}} \tag{2.87}$$

Equation (2.87) shows that the change of the transient emf E'_q is inversely proportional to the product of the time constant $\left(\frac{E'_q}{E_q}\right) T'_{d0}$.

Besides, as can be seen from Eq. (2.78),

$$\frac{E'_q}{E_q} = \frac{X'_d}{X_d} = \text{Constant}$$

In order to improve the calculation accuracy, in the calculation of the transient emf E'_q with the time quantum method, an appropriate Δt interval and E_{qe} corresponding to $t + \frac{\Delta t}{2}$ is selected. That is,

$$\frac{\Delta E'_q}{\Delta t} = \frac{\left(\frac{E'_q}{E_q}\right) E_{qe\left(t+\frac{\Delta t}{2}\right)} - E'_q}{\left(\frac{E'_q}{E_q}\right) T'_{d0} + \frac{\Delta t}{2}} \tag{2.88}$$

If Eq. (2.75) is directly adopted, the corresponding transient increment expression is

$$\frac{\Delta E'_q}{\Delta t} = \frac{E_{qe\left(t+\frac{\Delta t}{2}\right)} - E_q}{T'_{d0} + \left(\frac{E_q}{E'_q}\right)\frac{\Delta t}{2}} \tag{2.89}$$

Before the no-load short circuit,

$$E'_q = E_q = U = E_{qe} = 1.0$$
$$I_d = I_q = 0$$

At the instant after the short circuit, the following is obtained from the law of conservation of flux linkage:

$$E'_q = E_{qe} = 1.0, U = 0$$
$$I_d = \frac{E'_q}{X'_d} = \frac{1.0}{0.29} = 3.45, \quad I_q = 0$$
$$E_q = E_{q(0)} = E'_q + (X_d - X'_d)I_d$$
$$= I_d X'_d + (X_d - X'_d)I_d$$
$$= I_d X_d = 1.05 \times 3.45 = 3.62$$
$$\frac{E_q}{E'_q} = \frac{3.62}{1.0} = 3.62$$

During the short circuit, the ratio $\frac{E_q}{E'_q}$ is constant. $\Delta t = 0.2$ s is taken from the computational domain. With Eq. (2.89), the following is obtained:

$$\frac{\Delta E'_q}{\Delta t} = \frac{E_{qe\left(t+\frac{\Delta t}{2}\right)} - E_q}{T'_{d0} + \left(\frac{E_q}{E'_q}\right)\frac{\Delta t}{2}} = \frac{1.0 - 3.62}{5.0 + 3.62 \times 0.1} = -0.489$$

When the excitation voltage is constant, $E_{qe} = 1.0$ is constant. With the above equation, the first time interval $\Delta E'_q$ is

$$\Delta E'_q = \frac{\Delta E'_q}{\Delta t} \Delta t_1 = -0.489 \times 0.2 = -0.098$$

E'_q at the end of the first time interval Δt_1 is

$$E'_q = 1.0 - 0.098 = 0.902$$

At the second time interval Δt_2:

$$\frac{\Delta E'_q}{\Delta t_2} = \frac{1.0 - E_q}{5.36}$$

At this point:

$$I_d = \frac{E'_q}{X'_d} = \frac{0.902}{0.29} = 3.11$$

$$E_q = 3.11 \times 1.05 = 3.26$$

$$\frac{\Delta E'_q}{\Delta t_2} = \frac{1.0 - 3.26}{5.36} = -0.422$$

Thus,

$$\Delta E'_q = -0.422 \times 0.2 = -0.0844$$

E'_q at the end of the second time interval Δt_2 is

$$E'_q = 0.902 - 0.0844 = 0.818$$

In this way, the curves $E_q = f(t)$ and $E'_q = f(t)$ can be calculated point by point. See the curves labeled "a" shown in Figure 2.26.

For curves labeled "a," $U_f =$ constant; for curves labeled "b," $U_f = 1.0 + 2.5\,t$

Figure 2.26 Curves of changes of emfs $E_q = f(t)$ and $E'_q = f(t)$ with different excitation voltage change laws.

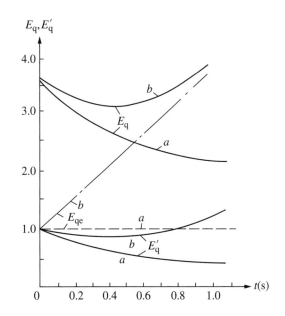

2 Characteristics of Synchronous Generator

When the excitation voltage changes in accordance with the law $U_f = 1.0 + 2.5\,t$, that is, $E_{qe} = 1 + 2.5\,t$, the average of E_{qe} of the first time interval is obtained, as mentioned earlier:

$$E_{qe}\left(t + \frac{\Delta t}{2}\right) = 1.0 + 2.5 \times 0.1 = 1.25$$

$$\frac{\Delta E'_q}{\Delta t_2} = \frac{1.25 - 3.62}{5.36} = \frac{-2.37}{5.36} = -0.441$$

$$\Delta E'_q = -0.441 \times 0.2 = -0.0882$$

E'_q at the end of the first time interval is

$$E'_q = 1.0 - 0.0882 = 0.9118$$

The calculation procedure for any other point is the same as that described above. See the curves labeled "b" in Figure 2.26 for the calculation results.

3

Effect of Excitation Regulation on Power System Stability

3.1 Definition and Classification of Power System Stability

In the 1960s, the definition of power system stability was classified into steady-state stability, dynamic stability, and transient stability by the United Kingdom, the United States, Western Europe, Japan, and so on. These terms now described.

3.1.1 Steady-State Stability

This term applies to the situation in which the system itself maintains stable transmission when the load or voltage of the power system is affected by a small disturbance,. This form of stability mainly concerns the condition of synchronism reduction in terms of excessive generator rotor angle displacement.

3.1.2 Dynamic Stability

This term applies to the ability of a synchronous generator to maintain and restore steady-state operation when a large disturbance occurs in the system. The forms of dynamic stability loss are mainly expressed as power angle oscillation among generators and other variables increasing with time, or maintaining equi-amplitude oscillation due to the effect of a nonlinear system. This process can be spontaneous, and the duration is long.

It should be emphasized that, after large disturbances, the spontaneous oscillation stability whose oscillation frequencies lie between 0.2 and 3 Hz is caused by the fast and high-gain excitation system. This kind of stability belongs to the category of dynamic stability.

3.1.3 Transient Stability

This term applies to the ability of the power system to maintain synchronism when the power system is subjected to a severe transient disturbance, such as short circuits, grounding, line breakage faults, faulty line disconnections, and so on. The duration of such disturbances is usually short, and mainly occurs within the first swing cycle of the rotor after the disturbance. After long-term research and debate, currently electrical engineers all over the world define stability in terms of large disturbances and small disturbances, as follows.

The first category is small-disturbance stability, which refers to the ability of the generator in the power system to maintain synchronism under infinitely small disturbances; linearized differential equations can be used to analyze this. When a small disturbance occurs, the process of step-out synchronism can monotonically increase (with the generator's excitation constant), as creeping step out and oscillatory step out (under the excitation regulation condition). This definition corresponds to conventional steady-state stability.

The second category is large-disturbance stability, which involves problems under conditions like short circuits, ground faults, phase failures, and so on. The stability of synchronous-generator-related synchronization capability is traditionally called transient stability.

Design and Application of Modern Synchronous Generator Excitation Systems, First Edition. Jicheng Li.
© 2019 China Electric Power Press. Published 2019 by John Wiley & Sons Singapore Pte. Ltd.

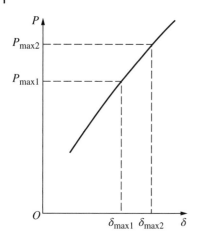

Figure 3.1 Static stability of generator.

The transient process of large-disturbance stability is short, and mainly occurs within the first swing cycle of the rotor. The descriptions of its behavior have involved nonlinear problems of the system.

Because the research scope of large-disturbance stability involves the transient state after a large disturbance and the subsequent behavior, this definition includes the traditional transient and dynamic stability problems. Besides, the method involves the nonlinear characteristics of the system.

3.2 Criterion of Stability Level

The critical indicators which determine the generator stability level under small disturbance effects are the electromagnetic power limit P_{max} and the rotor operative angle limit δ_{max}. In Figure 3.1, the micro-dynamic stability level is shown within the same generator sets under different control forms. While the generator electromagnetic power exceeds P_{max1} or P_{max2} shown in the figure, the micro-stability will be disrupted.

There are two criteria for the transient stability level under large disturbances. The first criterion is represented by the transient stability power limit P_{max}, and P_{max} is defined as follows. If the generating sets operate in parallel with the system under normal operation mode, we assume that the active power $P_e = P_{e1}$, in which case the system can still be stable even if the supposed accident point K is under some kind of short circuit. But under the condition of normal operation, $P_e = P_{e1} + \Delta P_e$ (ΔP_e is a tiny increment which is much smaller than P_{e1}), the system will be unstable when the K point is under the same fault. P_{e1} is called the transient stability power limit, which is represented by $P_{e1} = P_{max1}$. Under the condition of normal operation, the rotor power angle that corresponds to P_{e1} is called the transient stability limit angle δ_{max}.

Another method of representing the transient stability level is to compare the maximum allowable fault clearance time during the short circuit under the same default point and the same default form. When the transmission line is under a certain transmission power and the K point is under some kind of short circuit, the system is stable if the clearance time satisfies $t = t_1$. However, the system cannot be stable when $t = t_1 + \Delta t_1$ (Δt_1 is a tiny increment which is much smaller than t_1), and $t_1 = t_{max}$ is called the maximum allowable fault clearing time. Obviously, a greater value of t_1 marks a higher level of transient stability of the system.

3.3 Effects of Excitation Regulation on Power System Stability

3.3.1 Effect of Excitation Regulation on Steady-State Stability

Under normal operating conditions, the mechanical input power and the electromagnetic output power of the synchronous generator are in balance [8]. The synchronous generator operates at synchronous speed.

Figure 3.2 Turbine generator power characteristic curve.

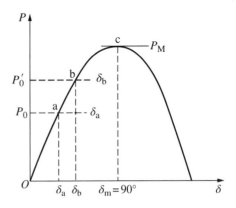

The characteristic is usually represented by the characteristic of the power angle. For turbine generators, the characteristic of the power angle is represented by

$$P = \frac{E_q U_s}{X_\Sigma} \sin \delta \tag{3.1}$$

where
- E_q — generator internal potential;
- U_s — infinite bus voltage;
- X_Σ — total reactance between generator and grid.

Figure 3.2 shows the corresponding power-angle characteristic, and this curve is also called the internal power characteristic curve. Without the regulation of excitation, E_q is constant, and the maximum output power P_M $\left(\frac{dP}{d\delta} \big|_{\delta=\delta_m} = 0 \right)$ is called the steady-state stability critical power, the value of which is $P_M = \frac{E_q U_s}{X_\Sigma}$. Under normal operation, the equilibrium point is at point a. Assuming that the mechanical input power rises from P_0 to P_0', the residual power will accelerate the rotor rotation speed, the power angle of the internal potential E_q (which is related to the receiving-end voltage system U_s) will shift from δ_a to δ_b, and thus the new working point will move from point a to point b. Under the condition of constant excitation, which means E_q is constant, the steady-state stability critical power limit can be expressed as $P_M = \frac{E_q U_s}{X_\Sigma}$ and the criterion of steady-state stability is represented by $\frac{dP}{d\delta} \geq 0$ or $\delta \leq 90°$.

If the generator can automatically adjust the excitation in the run, then at this point E_q will be a variable value, and the corresponding transmission power can be improved significantly. We assume that the automatic excitation regulation is inertia-less and the transient potential of the generator E_q' is approximately constant when the load changes. The inner potential E_q will change with the change in the load. The power characteristic is not a sine curve at this time, but a line consisting of working points corresponding to the sine curve clusters whose E_q values equal different constant values constituting a group. Curves 1–4 in Figure 3.3a show this. In order to differentiate it from the characteristic curve of the inner power when E_q is a constant, the 1–4 power characteristic curve a, b, c, d which changes with the load is called the external power characteristic curve. At the same time, because the external power characteristic curve works in the section of the curve that is aided by excitation regulation, the corresponding working segment is also called the artificial region. The difference between curve a–e of the external power characteristic in the I group and curves 1–5 in the II group is that the excitation regulator has a higher voltage amplification factor K_{OU}. Therefore, it can maintain a higher voltage level.

Besides, as shown in Figure 3.3b, E_q increases with the load increment (assuming an approximately constant E_q'). Meanwhile, in terms of the external power characteristics, the maximum power value appears not at $\delta = 90°$, but at $\delta > 90°$. The specific value depends on the condition of steady-state stability.

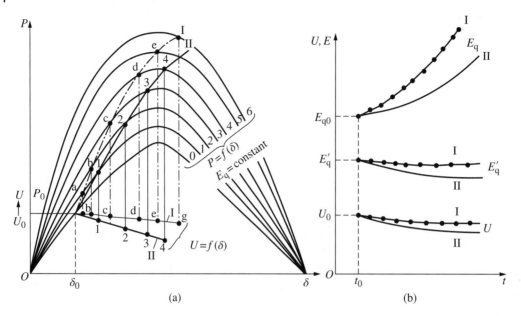

Figure 3.3 Generator power characteristic curves when the excitation regulation maintains the transient's inner potential E'_q constant: (a) power characteristic curves, (b) generator voltage U and the changes in inner potential E_q and E'_q.

Figures 3.4e and 3.4d show that at a given parameter, the corresponding rotor angle reaches δ_{max} when the external power characteristic reaches its maximum stability operation power. Afterward, as the power increases, the generator terminal voltage and output power oscillate, as shown in Figures 3.4c–f; it is caused by inappropriate parameter selection of the excitation system.

3.3.2 Influence of Excitation Regulation on Transient Stability

The above discussions concern only the micro-dynamic stability issue under a small disturbance; we will now discuss the effect of excitation regulation on the transient stability under a large disturbance. In the circuit shown in Figure 3.5a, the changes in the power characteristic under the short-circuit fault condition are discussed.

In Figure 3.5b, curve 1 represents the power characteristic with a two-circuit power supply, the amplitude of which is expressed as

$$P_M = \frac{E_q U_s}{X_\Sigma}$$

$$X_\Sigma = X_d + X_T + \frac{X_e}{2}$$

where Curve 2 represents the power characteristic curve after the clearance of a faulty line affected by a short-circuit fault. Because the line impedance increases from $\frac{X_e}{2}$ to X_e, the power characteristic curve amplitude reduces to $\frac{E_q U_s}{X_\Sigma}$, where $X_\Sigma = X_d + X_T + X_e$. Curve 3 shows the power characteristic during the fault.

If the operating point of the generator is initially located at point "a" on power characteristic curve 1, it will be determined by power characteristic curve 3 after the short-circuit. At the instant of the fault, due to inertia, with the operating point moving from point "a" to point "b," the speed is unchanged and the power angle δ remains as δ_0. Afterward, the reduced output electromagnetic power characteristic will cause the rotor to speed up and the power angle to increase. When it reaches δ_1 the fault is cleared, the power characteristic is curve 2, and the operating point moves from point c to point e. Due to inertia, the rotor keeps accelerating

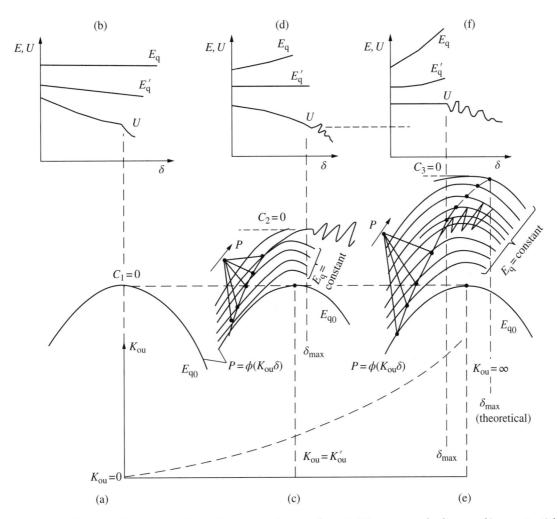

Figure 3.4 Changes in the generator internal and external power characteristic curves, end voltage, and inner potential: (a) E_q constant, (b) when E_q is constant, the changes in E_q' and U, (c) E_q' constant, (d) when E_q' is constant, the changes in E_q and U, (e) U constant, (f) when U is constant, the changes in E_q and E_q'.

along power curve 2 until it reaches point f, and the corresponding rotor power angle changes to δ_2. After repeated oscillation, it finally becomes steady at point g. As previously described, the transient stability depends on whether the acceleration area abcd is less than or equal to the deceleration area dfed. Obviously, δ_1 and the accelerated area abcd will increase if the fault clearance is slow. If the deceleration area is smaller than or equal to the acceleration area, the acceleration process will continue, therefore transient stability will be lost.

There are two methods to enhance transient stability: reduction of the acceleration area and increase of the deceleration area. One of the effective ways to reduce the acceleration area is to shorten the fault clearance time. An effective way to increase the deceleration area is to, in the excitation system, increase the excitation voltage response ratio and the ceiling voltage multiple simultaneously, so that after the clearance of the fault, the generator internal potential E_q will be rapidly boosted and the output power will increase, thus increasing the deceleration area. The corresponding changes are shown in Figure 3.6.

As shown in Figure 3.6, in the normal operation mode, the generator operating point is located at point a in power characteristic curve 1; when subjected to a short-circuit fault, the corresponding curve is shown as curve 3. If forcing excitation is provided to rapidly increase the generator internal potential E_q, the curve

3 Effect of Excitation Regulation on Power System Stability

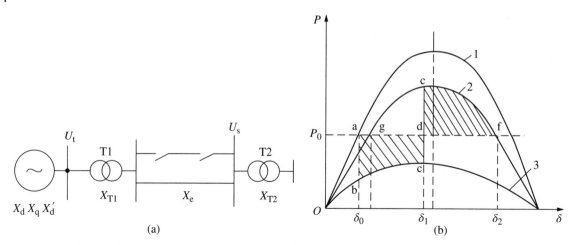

Figure 3.5 Changes in the power characteristic curve under short-circuit fault: (a) single machine infinite bus system, (b) changes in the power characteristic curve under short-circuit fault.

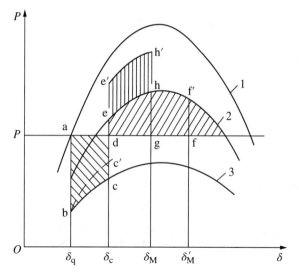

Figure 3.6 Influence of excitation regulation on transient stability.

will increase from the bc segment to the bc′ segment, thus reducing the acceleration area (from area abcd to abc′d) before the clearance of the fault. After the clearance of the fault, when $\delta = \delta_c$, the deceleration area can also be increased from dehg to de′h′g. If area de′h′g equals def′f, the maximum value of the rotor angle can be reduced from δ'_M to δ_M, which will enhance transient stability. Evidently, with a higher excitation ceiling voltage value and a faster excitation voltage response, the more obvious effect of the excitation regulation will be improvement of transient stability. Usually, it is appropriate to choose a forcing excitation voltage multiple of 2.

An example to explain the effect of excitation regulation on power system stability follows [9].

3.3.2.1 Enhancement of the Steady-State Stability of Power System

As previously described, regardless of whether a small or large disturbance occurs, the excitation control will play an important role in improving the steady-state (micro-dynamic) and transient stability of the system.

Now, the effect of the excitation control on the generator's steady-state (micro-dynamic) stability is described.

In Figure 3.5a, assume a single generator with single circuit and infinite bus, and the per-unit generator and line parameters are expressed as follows;

$$X_d = X_q = 1.5, X_{T1} = X_{T2} = 0.1, \quad X'_d = 0.3, X_e = 0.8$$

The expression for the generator power characteristic curve can be written in the following forms:

(1) Without implementing excitation regulation, E_q is constant, and the generator power transmission formula can be represented by

$$P_e = \frac{E_q U_s}{X_{d\Sigma}} \sin \delta_{Eq}$$
$$X_{d\Sigma} = X_d + X_{T1} + X_{T1} + X_e \tag{3.2}$$

where
$X_{d\Sigma}$ — total impedance;
δ_{Eq} — power angle between generator potential E_q and system voltage U_s at the receiving end.

(2) When $\delta_{Eq} = 90°$, the transmission power reaches a maximum, and the power limit of steady-state stability is

$$P_{emax(Eq)} = \frac{E_q U_s}{X_{d\Sigma}} \tag{3.3}$$

(3) With excitation regulation, the voltage gain of the regulator K_{OU} can only maintain the internal potential E'_q constant, and at this time, the power transmission formula is represented by

$$P_e = \frac{E'_q U_s}{X'_{d\Sigma}} \sin \delta_{E'q}$$
$$X'_{d\Sigma} = X'_d + X_{T1} + X_{T2} + X_e \tag{3.4}$$

where
$X'_{d\Sigma}$ — total impedance;
$\delta_{E'q}$ — power angle between generator transient inner potential E'_q and system voltage U_s at the receiving end.

(4) When $\delta_{E'q} = 90°$, the limit of the steady-state stability power is

$$P_{emax(E'q)} = \frac{E'_q U_s}{X'_{d\Sigma}} \tag{3.5}$$

(5) Due to the strong influence of automatic voltage regulation (AVR), when the transient load changes, the generator terminal voltage can still be maintained at a constant value. At this time, the expression for the transmission power is

$$P_e = \frac{U_1 U_s}{X_\Sigma} \sin \delta_{Ut} \tag{3.6}$$
$$X_\Sigma = X_{T1} + X_e + X_{T2}$$

where
X_Σ — total impedance;
δ_{Ut} — power angle between the generator terminal voltage and the receiving end system voltage U_s.

(6) When $\delta_{Ut} = 90°$, the limit of the steady-state stability critical power is

$$P_{emax(Ut)} = \frac{U_1 U_s}{X_\Sigma} \tag{3.7}$$

Substituting the given parameters into Eqs. (3.3), (3.5), and (3.7), the limits of the steady-state stability power under three states can be calculated as

$$P_{emax(Eq)} = \frac{1}{1.5 + 0.1 + 0.8 + 0.1} = 0.4 \text{ (p.u.)}$$

$$P_{emax(E'q)} = \frac{1}{0.3 + 0.1 + 0.8 + 0.1} = 0.77 \text{ (p.u.)}$$

$$P_{emax(Ut)} = \frac{1}{0.1 + 0.8 + 0.1} = 1.0 \text{ (p.u.)}$$

From the above calculated results, due to the influence of AVR, it can be seen that without excitation regulation, E_q is constant and the steady-state stability critical power is 0.4 (p.u.). When E'_q = constant with excitation regulation, the steady-state stability critical power is 0.77 (p.u.). When using fast excitation regulation, U1 = constant, and the steady-state stability critical power is 1.0 (p.u.).

This shows that the automatic excitation control system plays an important role in maintaining the generator voltage level and improving the steady-state stability of power systems.

3.3.2.2 Improving Transient Stability

Transient stability is defined as the stability of when power system is subjected to a severe transient disturbance, which refers to the first oscillation cycle of the rotor after the accident. In term of excitation control system, its effects are mainly determined by three factors:

(1) *Excitation system ceiling voltage multiple*. Increasing the field-forcing ceiling voltage multiple can enhance power system transient stability. However, the higher field-forcing multiple may in turn increase the manufacturing cost of the excitation system and requirement for the generator insulation level. Therefore, it is not necessary to consider increasing the field-forcing multiple while the fault clearance time is sufficiently short.
(2) *Ceiling voltage response ratio of excitation system*. The excitation system ceiling voltage response ratio is also known as the excitation voltage rise speed. The larger the ratio, the shorter the duration of the excitation voltage increase to the ceiling value will be, which is beneficial for the enhancement of transient stability. The ceiling voltage response ratio of excitation system is determined by the performance of the excitation system.
(3) *Utilization extent of excitation system field-forcing multiple*. Making full use of the field-forcing multiple is an important factor to improve the transient stability of an excitation system. If a power system fault occurs in the vicinity of a power station, the output voltage of excitation systems will not be able to reach the ceiling value or they may reach the ceiling value for a short time. The field-forcing has stopped before the generator voltage is restored to the pre-fault level, with the result that the field-forcing effect of the excitation system cannot be exerted completely. One of the measures to take full advantage of the excitation system ceiling voltage is to increase the open-loop gain of the excitation control system. The larger the open-loop gain, the higher the voltage regulation accuracy, and the field-forcing multiple will be utilized more fully, which will help improve the transient stability of the power system.

3.3.2.3 Improving Dynamic Stability

Dynamic stability refers to the maintenance of equilibrium when the power system is subjected to a disturbance, by restoration to the original equilibrium point or transition to a new equilibrium point (after a large disturbance). The assumption is that the original or new equilibrium point concerns steady-state stability, and the process of large disturbance concerns transient stability.

The power system dynamic stability issue can be understood as the problem of electromechanical oscillation damping in the system. When the damping is positive, the dynamic system is stable, and when it is negative, the dynamic system is unstable; when the damping is zero, the system is in the critical state. Zero or small positive damping is the uncertain factor during power system operation, and measures should be taken to enhance it.

Analysis shows that automatic voltage regulation of the excitation control system is one of the important reasons why the damping of a power system's electromechanical oscillations becomes weak or even negative. With a certain operating mode and excitation system parameters, the voltage regulator works to maintain the generator voltage at a constant value and at the same time causes negative damping.

Many studies show that within the normal application range, the negative damping effect of the excitation voltage regulator will be reinforced as the open-loop gain increases. Therefore, the requirement for higher voltage regulation accuracy conflicts with the enhancement of dynamic stability. This problem can be solved as follows:

(1) Reducing the voltage regulation accuracy requirement, and decreasing the open-loop gain of the excitation control system. From the above analyses, we know that it is inappropriate to apply this method because of the negative effect on steady-state stability and transient stability.
(2) Adding a dynamic gain reduction element in the voltage regulation channel; using this method can not only maintain the voltage regulation accuracy, but also reduce the negative damping effect of the voltage regulation channel. However, this dynamic gain reduction link has a large inertia link which reduces the response ratio of the excitation voltage, thus affecting the utilization of the field-forcing multiple, and it is also inappropriate for enhancement of transient stability. Therefore, the advantages and disadvantages of dynamic gain reduction should be considered comprehensively in practical applications.
(3) One of the effective methods of improving the excitation control system is to increase the number of excitation control channels and adopt a power system stabilizer. This additional signal could impart a positive damping action to the whole excitation system in the low-frequency oscillation range through phase adjusting.
(4) Linear and nonlinear excitation control theory can be applied to improve the dynamic performance of excitation system.

4

Static and Transient State Characteristics of Excitation Systems

4.1 Static Characteristics of Excitation System

4.1.1 Block Diagram of Static Characteristics

The static characteristic of an excitation system is obtained in the model of an excitation control system when the parameter K is a linearized value and the Laplace factors in each transfer function are zero. Now we use the excitation system static block diagram, shown in Figure 4.1, as the base of our discussions.

In Figure 4.1, the forward transfer function $G(s)$ is

$$G(s) = \frac{\Delta U_G}{\Delta U_2} = K_A K_B K_C K_G$$

The backward transfer function $H(s)$ is

$$H(s) = \frac{\Delta U_1}{\Delta U_G} = K_1$$

The open-loop transfer function $H_{0(s)}$ is

$$\begin{aligned} H_{0(s)} &= \frac{\Delta U_1}{\Delta U_2} = K_1 K_A K_B K_C K_G \\ &= K_{AVR} K_G \\ &= K_\Sigma \end{aligned} \tag{4.1}$$

where
K_Σ – open-loop gain of the excitation control system, $K_\Sigma = K_{AVR} K_G$;
K_{AVR} – gain of the excitation regulator, $K_{AVR} = K_1 K_A K_B K_C$

4.1.2 Natural Ratio of Generator Voltage to Reactive Current of Generator

The natural ratio of the generator voltage to the reactive current of a generator refers to the curve $U_G = f_{(IQ)}$ that relates the generator terminal voltage U_G to the reactive load current I_Q, when the voltage adjusting unit of AVR is not operational and under constant voltage conditions.

4.1.2.1 Formula Relating Generator Voltage and Reactive Current
From Figure 4.1, we obtain the following:

$$\Delta U_1 = K_1 \Delta U_G \tag{4.2}$$

$$\Delta U_A = K_A \Delta U_2 \tag{4.3}$$

$$\Delta U_B = K_B \Delta U_A \tag{4.4}$$

Design and Application of Modern Synchronous Generator Excitation Systems, First Edition. Jicheng Li.
© 2019 China Electric Power Press. Published 2019 by John Wiley & Sons Singapore Pte. Ltd.

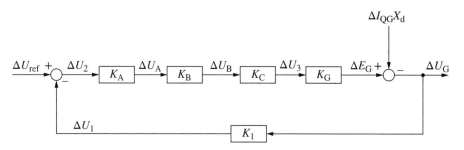

Figure 4.1 Static block diagram of the excitation control system.

$$\Delta U_3 = K_C \Delta U_B \tag{4.5}$$

$$\Delta E_G = \Delta U_3 K_G \tag{4.6}$$

Under the discussed condition of constant supplied voltage, the following formula can be obtained:

$$\Delta U_2 = -\Delta U_1 \tag{4.7}$$

And:

$$\Delta U_A = -K_A \Delta U_1 \tag{4.8}$$

The EMF increment of the generator is obtained by Eqs. (4.2)–(4.7) above:

$$\begin{aligned}\Delta E_G &= -K_1 K_A K_B K_C K_G \Delta U_G \\ &= -K_\Sigma \Delta U_G\end{aligned} \tag{4.9}$$

The disturbance values are given by $\Delta I_Q X_d$:

$$\Delta U_G = \Delta E_G - \Delta I_Q X_d$$

Substituting Eq. (4.9) into the above equation:

$$\Delta U_G = -K_\Sigma \Delta U_G - \Delta I_Q X_d$$

Simplification yields

$$\Delta U_G = -\frac{\Delta I_Q X_d}{1 + K_\Sigma} \tag{4.10}$$

Formula (4.10) indicates that the generator voltage drop ΔU_G decreases by a factor of $\frac{1}{1+K_\Sigma}$ under the automatic excitation regulator. However, in the absence of the automatic excitation regulator, when the generator reactive current increases, the generator terminal voltage will drop because of the lack of excitation regulation:

$$\Delta E_G = 0$$

From Eq. (4.9), we obtain

$$\Delta U_G = -\Delta I_Q X_d \tag{4.11}$$

4.1.2.2 Determination of Natural Ratio of Generator Voltage to Reactive Current of Generator

The voltage regulation ratio of a generator is defined as the ratio of the absolute value of the generator terminal voltage variation versus the reactive load current variation, when the voltage adjusting unit exits and the voltage setting is constant. It is obvious that this ratio expresses the slope of the natural voltage adjusting characteristic curve $\Delta U_G = f\Delta(I_Q)$ of the generator, as shown in Figure 4.2.

Figure 4.2 Generator natural voltage adjusting characteristic curve.

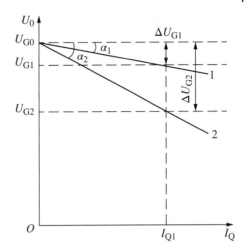

The natural ratio of the reactive current compensation is expressed in mathematical form. The AVR status of Curve 1 in Figure 4.2 can be expressed as follows by Eq. (4.10):

$$\delta_{N1} = \left| \frac{\Delta U_{G1}}{\Delta I_{Q1}} \right| = \tan \alpha_1 = \frac{X_d}{1 + K_\Sigma} \quad (4.12)$$

As shown in Figure 4.2, Curve 2 without AVR can be expressed as follows by Eq. (4.11):

$$\delta_{N2} = \left| \frac{\Delta U_{G2}}{\Delta I_{Q1}} \right| = \tan \alpha_2 = X_d \quad (4.13)$$

Eq. (4.12) is expressed in per-unit value:

$$\delta_{N1} = \frac{X_d}{1 + K_\Sigma} \times \frac{U_{GN}}{I_{GN}} \quad (4.14)$$

$$\delta_{N2} = X_d \times \frac{U_{GN}}{I_{GN}} \quad (4.15)$$

Note: In the above discussion of the natural ratio of the reactive current compensation of a generator, we used the ratio of the generator voltage variation versus the reactive load current variation. Besides, the ratio of the reactive current compensation can also expressed as the ratio of the change in the generator terminal voltage within the change range of reactive current; for instance, when the generator reactive current I_Q increases from zero to the rated value I_{QN}, the generator terminal voltage changes from the no-load value to U_{Gt}, and at this time the natural ratio of the reactive current compensation can be expressed as

$$\delta N = \frac{U_{G0} - U_{Gt}}{U_{GN}} \quad (4.16)$$

where
U_{GN} – generator rated voltage value.

Under the same condition, if the ratio of the generator terminal voltage variation to the reactive load current variation is used to express the ratio:

$$\delta_{N1} = \frac{\Delta U_G}{\Delta I_Q} = \frac{U_{G0} - U_{Gt}}{I_{QN}} \quad (4.17)$$

If I_{QN} is used as the base per-unit values of the current and generator voltage, respectively, substituting them into Eq. (4.17) yields

$$\delta_{N1} = \frac{\Delta U_G^*}{\Delta I_Q^*} = \frac{\dfrac{U_{G0} - U_{Gt}}{U_{GN}}}{\dfrac{I_{QN}}{I_{QN}}} = \frac{U_{G0} - U_{Gt}}{U_{GN}} = \delta_N \quad (4.18)$$

It can be seen that both definitions – Eq. (4.16) or Eq. (4.17) – are equivalent: the calculation result of the natural ratio of the reactive current compensation is the same in both cases.

4.1.3 Static Voltage Droop Ratio of Generator

In automatic control theory, "offset" refers to the concept of static error or steady-state error. In terms of a poor excitation control system that has negative feedback control according to the generator voltage deviation, the role of a close-loop control system is "measuring error and correcting error," regardless of the open-loop amplification factor value of the excitation control system. In theory, there is always static error or bias in working systems with the above control modes, unless the steady-state gain is infinity and no static error occurs.

The static voltage droop ratio is usually divided into two forms: the static error caused by a given input signal of the voltage and the static error caused by the disturbance signal of the excitation system. The former corresponds to the excitation system regulation accuracy, which is called the static voltage regulation accuracy, and the latter, such as the reactive current load of a generator, can be regarded as the disturbance signal. Therefore, the generator terminal voltage changes caused by this are called the disturbance signal error.

4.1.3.1 A Setting Voltage Static Error

Now, we take Figure 4.3, a simplified block diagram of an excitation control system, as an example of calculating the voltage static error. The relevant excitation system parameters are as follows:

$$X_d' = 0.346, \quad X_d = 1.693, \quad T_{d0}' = 6.85s, \quad T_e = 0.015s, \quad T_A = 0.02S, \quad K_A = 200$$

Consider the static error caused by the setting voltage signal, for steady state, $s \to 0$, $t \to \infty$. If there is no integral link between the transfer function from U_{ref} to U_G, it is a zero-order system. The static error is

$$e_{ss} = \frac{1}{1 + K_A} = \frac{1}{201} = 0.005 = 0.5\%$$

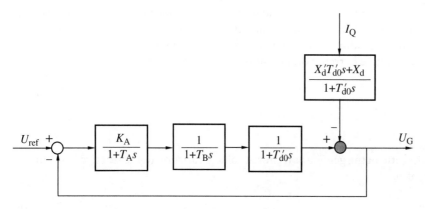

Figure 4.3 Simplified block diagram of excitation control system.

4.1.3.2 Static Error Caused by the Load Disturbance I_Q

According to Figure 4.3, first, we obtain the transfer function $G_N(S)$ from the disturbance signal I_Q to the output terminal U_G, and then calculate the static error:

$$e_{ssN} = \lim_{s \to 0} sG_N(s)\frac{1}{s}$$

$$e_{ssN} = \lim_{s \to 0} \frac{-\dfrac{X'_d T'_{do} s + X_d}{T'_{do} s + 1}}{1 + \dfrac{K_A}{(1+T_A s)(1+T_E s)(1+T'_{do} s)}}$$

$$= -\frac{X_d}{1 + K_A} = -\frac{1.693}{201} \times 100\% = -0.742\% \tag{4.19}$$

Form the above discussion, the static error of the excitation system caused by the disturbance of the load current is −0.742%.

The disturbance static error can also be obtained by the following procedure.

Corresponding to a steady state $s \to 0$, $t \to \infty$, under this condition, corresponding to Figure 4.3, the factor containing s is zero, and when the given voltage value is unchanged $U_{ref} = 0$, the disturbance static error caused by the load current is

$$\Delta U_G = -K_A \Delta U_G - X_d \Delta I_Q$$

$$\frac{\Delta U_G}{\Delta I_Q} = \frac{-X_d}{1 + K_A} = -0.742\%$$

Obviously the calculation results of these two methods are consistent, and because the latter calculation method is the same as with Eq. (4.12), we can consider natural adjustment as a static disturbance error which is caused by the load current.

4.2 Ratio and Coefficient of Generator Voltage to Reactive Current of Generator

4.2.1 Ratio of Generator Voltage to Reactive Current of Generator

The ratio of the voltage regulation of a generator can be defined by the ratio of the generator voltage variation to the reactive current variation, by the derived results of Eqs. (4.16)–(4.18). It can also be expressed by the ratio of the generator voltage variation to the reactive current variation, and the calculation results of both definitions are the same.

Now, the change in the generator's active current has very little impact on the generator terminal voltage change, but the reactive current change has a greater impact. Hence, for the definition of the voltage ratio of the reactive current compensation for a generator, it is appropriate to use the reactive current as the base when the power factor is zero.

4.2.2 Determination of Ratio and Coefficient of Generator Voltage to Reactive Current of Generator at Different Operation Connection Modes

Figure 4.4 shows the different operation connection modes of a generator.

In Figure 4.4a, the operation characteristics of the generator unit whose buses on the low-voltage side are parallel are as follows:

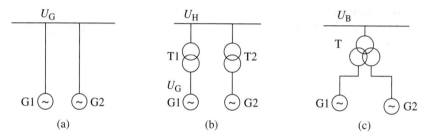

Figure 4.4 Connection of generator in different operating modes: (a) Parallel at the generator terminal, (b) parallel at the high-voltage-side connection at the generator expanding unit, (c) connection at the generator expanding unit.

For two generators with the same capacity, if an excitation device is set values $U_{ref1} = U_{ref2}$, then the two generators have the same terminal voltage value $U_1 = U_2$. The reactive power between the two generators $\Delta Q = 0$. But, in fact, due to the different characteristics of the units, different loads and temperatures, and so on. $U_1 \neq U_2$, so that the voltage difference $U_1 - U_2 = \Delta U$ exists. Under the effect of this voltage difference, it will generate a reactive power ΔQ_r between the two units of magnitude:

$$\Delta Q_r = \frac{U_1(U_1 - U_2)}{X_1 + X_2} \tag{4.20}$$

Because the reactance value $X_1 + X_2$ between the two units is very small, in terms of Eq. (4.20), under the effect of the limited voltage difference, the value of $X_1 + X_2$ is close to zero, and the quotient ΔQ_r is nearly infinite. If the general circulation is too large, the safe operation of the unit will be affected. To avoid the above-mentioned situation, the reactive current compensation droop device of the excitation regulator can be used to equivalently increase the reactance between the units, give the units a more drooping external characteristic, and improve the inter-unit when they operate in parallel.

Figure 4.5 shows the low-voltage-side parallel-connection generator unit; setting the bus voltage as U, we can see from Figure 4.6 that the reactive current shared between parallel units is

$$\sum I_Q = I_{Q1} + I_{Q2}$$

When the bus voltage drops from U to U', these two units will reassign the reactive power according to their external characteristics. In this case, the total increased reactive power is

$$\Delta I_Q = \Delta I_{Q1} + \Delta I_{Q2}$$

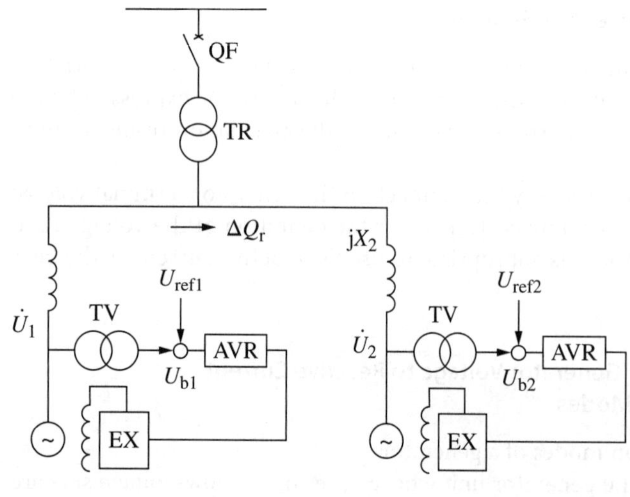

Figure 4.5 Wiring diagram of low-voltage-side parallel-connection units.

Figure 4.6 Parallel operation between two units with voltage difference.

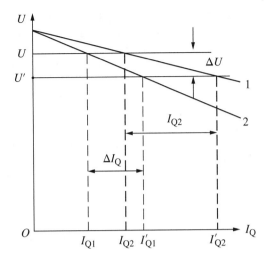

The corresponding bus voltage change can be obtained according to the ratio of the reactive current compensation definition of Eq. (4.17):

$$-\Delta U = \delta_1 \Delta I_{Q1} = \delta_2 \Delta I_{Q2}$$

From the above formula, the reactive current load borne by the units is proportional to the voltage difference ΔU, and is inversely proportional to the coefficient of reactive current compensation δ. Similarly, the above conclusion can be extended to multiple operational units. In order to stabilize the reactive power distribution among the various units, the coefficient of reactive current compensation δ should be 3%–4%.

For the generating units which are parallel on the low-voltage side, in order stabilize the reactive power distribution of the parallel operation among the generating units, positive adjustment wiring should be employed.

4.2.3 Natural Ratio and Additional Coefficient of Generator Voltage to Reactive Current of Generator

4.2.3.1 Natural Regulation Ratio of Generator Voltage to Reactive Current of Generator

As mentioned above, the natural regulation ratio related to the voltage and reactive current of a generator refers to the ratio between the generator voltage variation and the reactive power load current variation when the condition of a constant supplied voltage value no longer applies to the adjustment unit, and this ratio is equal to the slope of the generator external characteristic curve $U_G = f(I_Q)$.

For the excitation system with differential regulation, for the excitation regulator to provide the needed excitation increment, it should maintain a certain terminal voltage deviation when the generator changes from no-load to rated load, and the percentage value of the generator terminal voltage deviation is the natural ratio of the reactive current compensation, which is approximately equal to the reciprocal of the gain of the excitation system.

For a non-differential regulation system, the natural ratio of the reactive current compensation is zero.

4.2.3.2 Additional Regulation Coefficient of Ratio of Generator Voltage to Reactive Current of Generator

When the natural ratio of the reactive current compensation of the excitation regulator does not meet the requirements of reactive power distribution stability between the parallel units, additional adjustment measures have to be adopted, and the function of the additional coefficient of reactive current compensation is to introduce a voltage that is proportional to the reactive current in the AVR measuring circuit. Therefore, the terminal voltage will change as the reactive load changes.

Operational experience shows that, for analog AVR, to ensure stability of reactive power distribution between the parallel operation generator units, the minimum ratio of the reactive current compensation value is 3%–4%; and for digital AVR, the adjustment precision and resolution have been improved greatly, allowing the use of a smaller ratio of the reactive current compensation during generator operation, for example, 2%–3%. The ratio of the reactive current compensation needs to be further reduced in order to maintain the stability of operation: when employing the additional voltage adjustment, it should be noted that the voltage ratio of the reactive current compensation of the generator is defined by taking the generator rated reactive power Q_N as the per-unit value, and the additional adjustment is mainly used for compensating the voltage drop of the step-up transformer. The setting of the additional coefficient of reactive current compensation is calculated by taking the rated apparent power S_N as the per-unit value. If the apparent powers S_N of the transformer and generator are different, in order to unify the unit, per-unit value conversion will need to be done.

The following explains the setting of the additional coefficient of reactive current compensation by giving examples.

(1) The generator operates in parallel on the low-voltage side (as shown in Figure 4.4a). If the voltage regulation ratio of the AVR is zero, that is, the slope ratio of the generator voltage to the reactive current is zero, to ensure the stability of reactive power distribution between the parallel units, it should adopt the additional positive coefficient of reactive current compensation mentioned above. The ratio of the reactive current compensation of the generator is calculated according to the rated per-unit value of Q_N, and the additional coefficient of reactive current compensation is calculated according to the rated per-unit value of S_N, so that the calculation results for the ratio and coefficient must be converted.

For example, when the synthesized voltage ratio of the reactive current compensation of the generator is to be $\delta_C = 3\%$, the rated power factor $\cos\phi = 0.85$, and in order to meet this requirement, the corresponding additional coefficient of reactive current compensation is

$$\delta = \delta_C \frac{S_N}{Q_N} = \delta_C \frac{U_N I_N}{U_N I_N \sin\phi}$$

$$= \frac{\delta_C}{Q_N} = \frac{3\%}{0.526} = 5.7\% \tag{4.21}$$

(2) The generator transformer unit operates in parallel on the high-voltage side (as shown in Figure 4.4b). For the generator step-up transformer units which are connected in parallel on the high-voltage side, because the short-circuit impendence of the step-up transformer is bigger, negative adjustment is needed to partially compensate for the reactance drop of the step-up transformer when the reactive loads of the generator are increased. However, at the parallel operating point on the high-voltage side, the synthesis adjustment should still be positive adjustment. In other words, when the reactive load current of the generator is increased, the parallel running voltage on the high-voltage side is still decreased to guarantee the stability of reactive power distribution among the parallel units.

For the step-up transformer, negative adjustment is needed to partially compensate for the reactance of the step-up transformer. If the short-circuit impendence of the transformer is X_T, the synthesis ratio of the reactive current compensation is δ_C, and the additional coefficient of reactive current compensation will be

$$\delta = \frac{\delta_C}{\sqrt{1-\cos^2\phi}} - X_T \tag{4.22}$$

Because X_T takes S_N as the per-unit value, δ should also take S_N as the per-unit value.

Now, the setting of the additional coefficient of reactive current compensation is demonstrated by the following examples: Assuming $X_T = 11\%$, $\delta_C = 3\%$, $\cos\phi = 0.85$, from Eq. (4.22):

$$\delta = 3\%/\sqrt{1-0.85^2} - 11\%$$
$$= -5.3\%$$

Basic value conversion is needed if the S_N of the generator and the S_N of the step-up transformer are different. In this case, the additional negative coefficient of reactive current compensation is −5.3%, so the compensation rate of the transformer impendence should be (5.3/11) × 100% = 48%.

When the reactive power of the generator changes from zero to the rated value, the ratio of the reactive current compensation on the transformer high-voltage side should be $\delta_C = 3\%$, and this value takes the rated reactive power as the per-unit value. If it is converted to take S_N as its per-unit value, then the corresponding ratio of the reactive current compensation should be $\delta'_C = \frac{S_N}{Q_N}\delta_C$

$$\delta'_C = \frac{\delta_C}{\sin\phi} = \frac{3\%}{0.526} = 5.7\%$$

According to this, when the load of the generator is S_N, the high-voltage side of the transformer decreases to 3%, and when the generator is in the negative adjustment setting, the corresponding voltage rise of the generator is

$$= \Delta U_G = \delta\frac{Q_N}{S_N} = 5.3\% \times \sin\phi$$
$$= 5.3\% \times 0.526$$
$$= 2.6\%$$

The synthesis of each ratio of the reactive current compensation is shown in Figure 4.7.

(3) A generator transformer unit operating in parallel via a section of transmission lines, apart from compensating the transformer reactance voltage drop, must also compensate for the voltage drop of the transmission line, and the resistance and reactance of the transmission line are on the same order of magnitude, which cannot be ignored. Therefore, the reactance and resistance drops of the transmission line should be compensated for at the same time. The voltage drop of the active current on the resistor is in the same direction as the terminal voltage, and the active current or active power can be employed as the resistor voltage drop compensation.

(4) For generators connected as an expanding unit (Figure 4.4c), the reactance between generators G1 and G2 is the reactance between two low-voltage windings of transformers which has a smaller value compared with the reactance between the transformer high and low windings. If the excitation adjustment of G1 and G2 generating units is independent, then there should be a certain positive adjustment among G1 and G2 generating units, and the ratio of the reactive current compensation on the high-voltage side of the transformer should be fairly big; if G1 and G2 generating units are taken as one unit for excitation adjustment, or G1 and G2 generating units have the group excitation adjustment mode, negative adjustment could be used, which could help maintain the ratio of the reactive current compensation on the high-voltage side at the requested level.

4.2.4 Formation of Additional Coefficient of Reactive Current Compensation

4.2.4.1 Analog AVR

For analog AVR, the first side winding of the voltage transformer (VT) connected to the AVR measuring unit is connected to a class of fixed resistance, and the current component with a phase difference of 90° of the phase voltage is introduced, producing a voltage drop; for instance, the B phase resistor is connected to the AC phase current, and the voltage drop of the reactive current on the resistor is in the same direction as the voltage. For the positive voltage adjustment (slope $\Delta U_G \Delta I_Q > 0$), the reactive current increases and the resistor voltage and generator voltage are in the same direction, reflecting the artificial increase in the terminal voltage. Thus, the excitation current is regulated, reducing the generator terminal voltage. At negative adjustment wiring, the generator terminal voltage rises due to the increase in the reactive current.

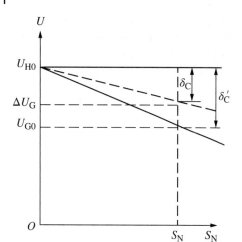

Figure 4.7 Synthesis of the ratio of reactive current compensation.

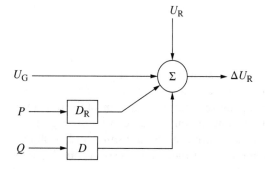

Figure 4.8 Formation of D-AVR adjustment.

4.2.4.2 Digital AVR

A digital voltage regulator (D-AVR) usually uses software to calculate the generator active power P and reactive power value Q. Figure 4.8 shows the adjustment functions performed by the software.

In Figure 4.8, D is the reactive adjustment, except the step-up transformer reactance; the circuit-compensated transmission line reactance is also needed, and after compensation the ratio of the reactive current compensation is determined according to the power system requirement. D_R is the circuit resistance compensation, which can be slightly less than the circuit resistance value, and D_R is zero when the circuit compensation is not considered.

4.2.5 Conclusions

(1) Most large-scale generator units adopt the generator transformer unit wiring. When AVR is put into negative adjustment, the reactance voltage drop of the step-up transformer can be compensated for properly to improve the operational stability of the power system, and the power system should correspondingly regulate the voltage adjustment of each generator in the system.

(2) In order to make the concept clearer, the ratio of the reactive current compensation for generator takes the generator Q_N as the per-unit value. The main role of the additional adjustment is to compensate for the corresponding reactance voltage drop of the transformer; and the voltage adjusting coefficient S_N in per-unit value is convenient to use. If the S_N values of the transformer and generator are different, the per-unit values need to be converted.

(3) The reactive ratio of the reactive current compensation δ_C at the generator parallel operation point is generally around 3%. The additional adjusting coefficient in AVR of the generator-transformer sets is expressed as $\delta = \delta_C / \sin \phi - X_T$.

(4) The ratio of the reactive current compensation of the parallel generators which connect the same bus should be identical, and their steady-state gain and transient gain should also be consistent. Before the new generators are put into operation, the excitation parameters of the former generators should be known and have uniformly adjusted.
(5) The generator unit should have a smaller ratio of the reactive current compensation, in order to increase the generator high-voltage-side bus voltage regulation precision and improve the power system's steady-state stability and voltage stability level. For example, there are a few generators in various power plants within an area of the US Western power system. They increased the capability of voltage support in the high-voltage bus and avoided a voltage collapse in one accident, whose effect is equal to the installation of a 500 kV, 400 MVA capacitor.
(6) Adopting a negative voltage adjustment can partially compensate for the reactance voltage drop of the step-up transformer, and it equivalently reduces the contact reactance between the generator unit and power system, relatively increases the system damping and helps suppress system low-frequency power oscillations. Therefore, this could increase the dynamic stability of the power system.

4.3 Transient State Characteristics of Excitation System

The dynamic characteristic of the excitation control system refers to the transient process when, in response to an external disturbance (such as a sudden change in the reactive load) or a change in the setting value U_{set}, the system transits from a stable state to another state at this moment.

The excitation control system consists of a generator, exciter, and other inertial elements. These inertial elements have their time constants, and that is why when the input value (external disturbance or setting value) changes, the output value cannot immediately reflect such behavior, but needs a certain time. The variation of the output value U_G versus time t within this time is called the transient process.

To evaluate the transient response characteristic of the excitation system, it is important to first examine the mode of the disturbance signals that acts on the excitation system. Basically, the disturbance signal is divided into two modes: the large disturbance signal and the small deviation signal, and the former signal concerns power system transient stability, whereas the latter concerns steady-state and dynamic stability of the system.

4.3.1 Transient Response of Large Disturbance Signal

In the transient response characteristic of a large disturbance signal, each element of the excitation system presents a nonlinear saturation state.

The large disturbance signal criterion is used to evaluate the transient characteristics of the excitation system that involve and affect the power system transient stability.

The evaluation standards for the above property vary from country to country. At present, the standards that are widely applied are the 421A-1978 and 421.1-1986 standards proposed by the American IEEE Society. The relative standards established by China are based on these two standards.

When evaluating the transient performance of a large disturbance signal in the excitation system, several excitation parameters are involved: ceiling voltage multiple, excitation voltage response ratio, and voltage response time of the excitation system, and so on, which are described as follows [6].

4.3.1.1 Ceiling Voltage Factor of Excitation System
Because the system can be disturbed by some severe faults, such as short-circuit, grounding, and so on, from the standpoint of enhancing a power system's transient stability, the excitation system is required to provide a sufficient ceiling voltage multiple. This is very beneficial in inhibiting the mechanical runaway of the generator rotor.

4.3.1.2 Voltage Response Ratio of Excitation System

For the same reasons discussed above, in order to increase the accuracy of relay protection action, reinforce electrical braking on the generator rotor within a short-circuit fault period, and accelerate the restoration of system voltage after the clearance of the fault, a fast excitation system needs to be adopted that has a high parameter excitation voltage response ratio, which is necessary for improving the system transient stability.

It should be stated that when the time needed by the excitation system to reach 95% of the difference between the ceiling excitation voltage and the rated excitation voltage is less than or equal to 0.1 s, this kind of fast speed excitation system is defined as a high initial response (HIR) excitation system. This definition is adopted to take full account of the growth rate of the excitation system in the 0.1 s initial period, which is of the utmost importance for the first swing vibration of the damping rotor. The current voltage response ratio is determined on the basis of the excitation voltage within 0.5 s average time, which may make the average response ratio very high, but the rising speed in the initial period is not high. Therefore, its effect obviously cannot be equivalent to the HIR excitation system.

For separate excitation and self-excitation thyristor rectifier excitation systems, because the change in the excitation voltage is instantaneous, it has the characteristic of HIR performance. Therefore, such an excitation system is called an inherent HIR excitation system. When there is a difference in the excitation voltage change pattern, its excitation voltage response ratio is different as well, for example:

(1) The excitation voltage curve changes following an exponential law. The response curve of the excitation voltage is shown in Figure 4.9, and the area encircled by the excitation voltage within a 0.5 s range is expressed as

$$S_{eac} = \int_0^{0.5} \left[(U_{fc} - U_{f1}) \left(1 - e^{-\frac{t}{T_e}} \right) \right] dt$$

$$= \left[0.5 + T_e \left(e^{-\frac{0.5}{T_e}} - 1 \right) \right] (U_{fc} - U_{f1}) \tag{4.23}$$

The area of the equivalent triangle eac is

$$\Delta S_{eac} = \frac{1}{2} \times 0.5 (U_{f2} - U_{f1}) \tag{4.24}$$

With the consistent conditions of Eqs. (4.23) and (4.24) above, it can be calculated that the excitation voltage response ratio within 0.5 s is

$$R = \frac{(U_{f2} - U_{f1})}{0.5 U_{f1}}$$

$$= \left(\frac{U_{fc}}{U_{f1}} - 1 \right) \left[8 T_e \left(e^{-\frac{0.5}{T_e}} - 1 \right) + 4 \right] \tag{4.25}$$

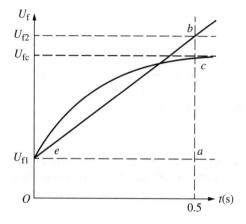

Figure 4.9 Excitation voltage curve changes exponentially.

Figure 4.10 Excitation voltage curve changes linearly.

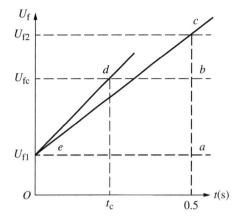

Figure 4.11 Excitation voltage curve changes with step change.

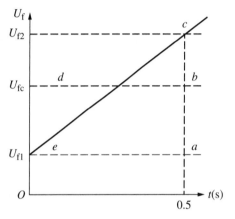

(2) The excitation voltage curve changes following a linear law. As discussed above, we can obtain the result from Figure 4.10:

$$S_{eabd} = \frac{1}{2}t_c(U_{fc} - U_{f1}) + (0.5 - t_c)(U_{fc} - U_{f1})$$
$$= \frac{1}{2}(U_{fc} - U_{f1})(1 - t_c) \tag{4.26}$$

$$S_{eac} = \frac{1}{2} \times 0.5 \ (U_{f2} - U_{f1}) \tag{4.27}$$

From $S_{eabd} = \Delta S_{eac}$, we obtain

$$R = \frac{(U_{f2} - U_{f1})}{0.5 U_{f1}} = 4\left(\frac{U_{fc}}{U_{f1}} - 1\right)(1 - t_c) \tag{4.28}$$

(3) The excitation voltage curve changes following a step change pattern, as shown in Figure 4.11, according to the equivalent areas condition, there is:

$$S_{eabd} = \Delta S_{eac}$$
$$\frac{1}{2} \times 0.5 \ (U_{f2} - U_{f1}) = 0.5 \ (U_{fc} - U_{f1})$$

We get

$$R = \frac{(U_{f2} - U_{f1})}{0.5 U_{f1}} = 4\left(\frac{U_{fc}}{U_{f1}} - 1\right) \tag{4.29}$$

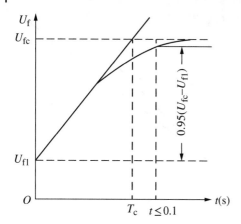

Figure 4.12 Change curve of the high initial response (HIR) excitation system.

(4) HIR excitation system
  (1) *Exponential change.* If the ultimate ceiling voltage is the saturated excitation voltage instead of the limited voltage, as shown in Figure 4.12, from the definition, the time constant T_e of HIR excitation system is calculated as

  $$0.95(U_{fc} - U_{f1}) = (U_{fc} - U_{f1})\left(1 - e^{-\frac{0.1}{T_e}}\right)$$

  The solution yields

  $$T_e = 0.0333\text{s}$$

  The response ratio of the excitation system can be obtained by Eq. (4.25):

  $$R = \left(\frac{U_{fc}}{U_{f1}} - 1\right)\ [8 \times 0.0333 \times (-1) + 4]$$
  $$= 3.73\left(\frac{U_{fc}}{U_{f1}} - 1\right) \tag{4.30}$$

  (2) *Linear change.* If the ceiling excitation voltage is limited when $U_{fc} = U_f$, as shown in Figure 4.13, according to the definition $0.95 t_c = 0.1\text{s}$, we obtain $t_c = \frac{0.1}{0.95} = 0.1053\text{s}$, and Eq. (4.28) yields

  $$R = 3.58\left(\frac{U_{fc}}{U_{f1}} - 1\right) \tag{4.31}$$

4.3.2 Transient Response of Small Deviation Signal

The small deviation signal is defined as each element of excitation system is still within the linear operating range when the excitation system is under the action of the signal. The characteristics criterion of the small deviation signal provides a basis for evaluating the transient character of the excitation system under the conditions of micro-load and voltage change of steady-state stability and the initial stage of dynamic stability. The small deviation signal also provides a method of developing or validating the performance of the excitation system model. The two most common methods for the transient response analysis of the excitation system under the action of a small deviation signal are the unit step response method (in the time domain) and the frequency response method (in the frequency domain).

Figure 4.13 Excitation voltage changes linearly in the high initial response (HIR) excitation system.

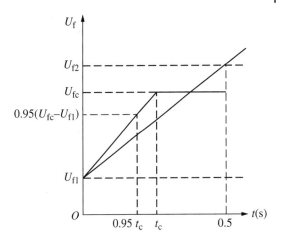

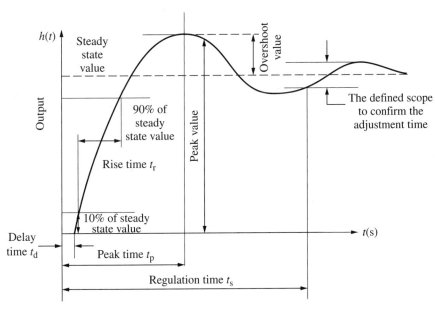

Figure 4.14 Transient response characteristics of the excitation system under the action of unit step function.

4.3.3 Unit Step Response

The transient response characteristics of the excitation system under the action of the unit step function are shown in Figure 4.14. Among them, the described performance indicators for the excitation system's transient response mainly include the following:

(1) *Delay time t_d*: The period from the time the step signal is input into the excitation system until the time the system starts to respond.
(2) *Rise time t_r*: The time period in which the response value changes from 10% of the steady-state value up to 90% of it.
(3) *Peak time t_p*: The time for the response value to surpass the steady-state value up to the first peak value.
(4) *Regulation time t_s*: The minimum time for the response value to reach the ±5% error range of the steady-state value.
(5) *Overshoot value $\sigma_\%$*: During the period of response, the system overshoot value is defined as the difference between the response output value and the steady-state value of the system peak value corresponding to

the peak time t_p. If it is expressed in per-unit terms, the value should be divided by the steady-state output base value.

The transient characteristics of the unit step response are expressed in the time domain.

4.3.4 Frequency Response

The evaluation of the transient performance of the excitation system by the frequency response method consists of applying varying sinusoidal signals with different frequencies at the input end of the excitation system, measuring the corresponding output amplitudes and the corresponding phase angle shift, and drawing the frequency characteristics of the amplitude's phase or the logarithmic amplitude's phase (known as the Bode diagram), after which the transient response characteristics of the excitation system can be evaluated. This method is widely used in engineering design.

There are two methods for the evaluation of the frequency response characteristics of the small deviation signal: the open loop and closed loop of synchronous generator excitation systems.

(1) When the synchronous generator is in the no-load state and the excitation system is open loop, the typical open loop frequency response characteristics are as shown in Figure 4.15. The performance characteristics shown in the figure include the low-frequency gain G, cut-off frequency ω_c, phase margin ϕ_m, and gain margin G_m. The phase margin ϕ_m and gain margin G_m determined by the open-loop frequency characteristics are the necessary parameters to distinguish the operation stability of the excitation system when the generator is at no-load. Generally, for most excitation control systems, in a good design the phase margin is no less than 40° and the gain margin no less than 6 dB.

(2) When a synchronous generator is in the no-load state, the closed-loop characteristics of the excitation system are as shown in Figure 4.16. Among the characteristic values of the frequency response characteristics for a closed-loop excitation system, the peak value M_p of the amplitude response is a threshold measure pertaining to stability. A high value (>1.6) represents an oscillating system with a larger overshoot in the transient response. For the majority of excitation control systems, $1.1 \leqslant M_p \leqslant 1.6$ can be considered to be an excellent design.

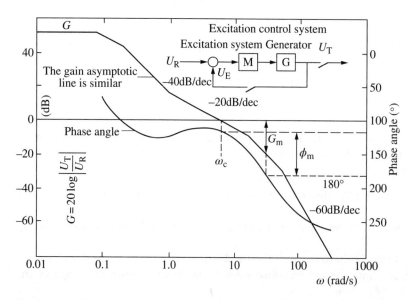

Figure 4.15 Characteristics of frequency response when the synchronous generator is in no-load state and the excitation system is open loop.

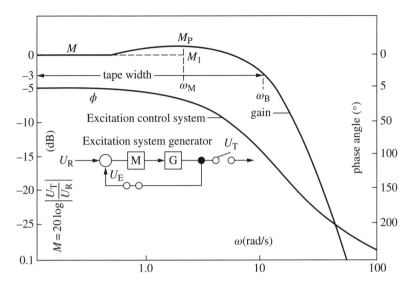

Figure 4.16 Frequency response characteristics of synchronous generator in no-load state with closed-loop excitation system.

The frequency bandwidth, ω_B, is an important indicator of the frequency response characteristics of the closed-loop system. This parameter represents the frequency bandwidth of the rise time t_r or the transient response speed when the response gain of the closed-loop frequency characteristic drops to 3 dB below the zero frequency value; the corresponding frequency range $0 \leqslant \omega \leqslant \omega_B$ is called the system bandwidth.

In feedback control systems with step input and overshoot less than 10%, the arithmetic phase angle product of the rise time t_r and the bandwidth cut-off frequency ω_c has the following relationship:

$$t_r \omega_c = 0.30 - 0.45.$$

The transient response overshoot will increase with the increase in the arithmetic product $t_r \bullet \omega_C$. The overshoot value can be ignored when $t_r \omega_c = 0.30 - 0.35$. If a PSS (power system stabilizer) is implemented in the excitation system, as the bandwidth increases, the excitation system can provide positive damping to suppress the oscillation of the power system over a wider frequency range. To ensure safety, the frequency response characteristic of the excitation system must be validated within a higher frequency range (for example, $f > 5$ Hz). If the check is found that, at a certain frequency which includes the resonant frequency of the generator mechanical torsional vibration system, the excitation system may provide negative damping and may even cause the main shaft stress to reach damaging stress.

Tables 4.1 and 4.2 show the relevant parameter indexes, as a reference for design and adjustment.

It should be stated that, under the small signal deviation condition, each index of the system is unlikely to be at its optimum value. For example, the selection of parameters for a low response peak value M_P, high value

Table 4.1 Excitation system open-loop frequency response characteristic indexes.

Performance indexes	Reference values
Gain margin G_m	$\geqslant 6$ dB
Phase margin ϕ_m	$\geqslant 40°$
Overshoot amount	(5–15%)
Amplitude response peak value M_p	1.1–1.6 (0.8–4) dB

Table 4.2 Excitation system closed-loop response characteristic indexes.

Performance indicators	Reference values
Gain of excitation system G	50–800 (per unit value)
Gain margin G_m	2–20 dB
Phase margin ϕ_m	20°–80°
Amplitude response peak value M_p	1.0–4.0 (0–12) dB
Bandwidth	0.3–12 HZ
Overshoot amount	0%–80%
Rise time	0.1–2.5 s
Adjustment time	0.2–10 s

gain margin G_m, phase margin ϕ_m, and maximum bandwidth should be determined by the comprehensive requirements of the feedback control system.

Generally, the excitation system no-load stability is validated by the open-loop frequency characteristics of the excitation system, while the PSS parameters are set by the closed-loop frequency response characteristics of the excitation system.

4.4 Stability Analysis of Excitation System

The differential equation could be used to express the performance characteristics of the generator excitation control system. Therefore, to analyze the stability of the excitation system in essence is to analyze the stability of its differential equation solutions. The methods of analyzing stability can be classified into two categories: obtain the solution of the differential equation, whether analytically or numerically, or (2) find the substitution amount of the differential equation solution – the root of the characteristic equation, and this method is called the indirect method. However, in the indirect method, it is difficult to solve higher-order systems with manual computation, especially when the computation of the stability in a complex power system can only be solved by the computer. The other method does not need to obtain the solution of a differential equation, but it needs to distinguish the stability directly, so it is called the direct method, such as the Routh criterion, Nyquist criterion, Lyapunov function, and the graphing method of classical control theory (root locus and frequency characteristics method), and so on.

Generally, the excitation control system is the system above the third order, but from the perspective of engineering approximation, it can be simplified as a second-order system, as shown in Figure 4.17. The stability of the system can be described by the damping ratio ξ and the natural frequency ω_n under the no-damping condition of this system. In other words, it analyzes the root locus of the closed-loop transfer function and the characteristic equation of this system in order to determine the stability of the system. Figure 4.17 shows

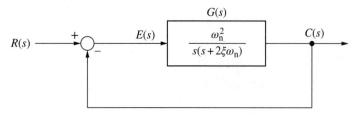

Figure 4.17 Block diagram of the second-order control system.

Table 4.3 Response characteristics of the second-order system with different damping ratios ξ.

Number	Damping ratio		Closed-loop poles of the system	Response characteristics of the system	Explanation
1	Damping ratio	$\xi = 0$	$s_{1,2} = \pm j\omega_n$		Persistent oscillation (unstable)
2	Under-damped	$0 < \xi < 1$	$s_{1,2} = -\xi\omega_n \pm j\omega_n\sqrt{1-\xi^2}$		Damping oscillation (stable)
3	Critical damping	$\xi = 1$	$s_{1,2} = -\omega_n$		Monotonous decay exponentially (stable)
4	Over-damped	$\xi > 1$	$s_{1,2} = -\xi\omega_n \pm \omega_n\sqrt{\xi^2-1}$		Monotonous decay exponentially (stable)
5	$\xi < 0$	$\xi > -1$	$s_{1,2} = \xi\omega_n \pm j\omega_n\sqrt{1-\xi^2}$		Increasing oscillation (unstable)
6		$\xi < -1$	$s_{1,2} = \xi\omega_n \pm \omega_n\sqrt{\xi^2-1}$		Monotonous growth exponentially (unstable)

that the closed-loop transfer function of the system is $\frac{C(s)}{R(s)}$, because

$$G(s) = \frac{\omega_n^2}{s(s + 2\xi\omega_n)}$$

So,

$$H(s) = 1$$
$$\frac{C(s)}{R(s)} = \frac{G(s)}{1 + G(s)H(s)}$$
$$\frac{\omega_n^2}{s^2 + 2\xi\omega_n s + \omega_n^2} \tag{4.32}$$

4 Static and Transient State Characteristics of Excitation Systems

The denominator $R(s)$ of Eq. (4.34) is a polynomial, and this formula is the characteristic equation of the closed-loop transfer function. The obtained root of the characteristic equation determines the stability of the dynamic process of the second-order-system, and it could be obtained from the following secular characteristic equation:

$$s^2 + 2\xi\omega_n s + \omega_n^2 = 0 \qquad (4.33)$$

The roots s_1 and s_2 in the above formula are

$$s_{1,2} = \frac{-2\xi\omega_n \pm \sqrt{(2\xi\omega_n)^2 - 4\omega_n^2}}{2} \qquad (4.34)$$

When the damping ratio ξ has different values, the location of these two roots (s_1 and s_2) in the s-plane will be different, as shown in Table 4.3. One can see the following:

When $\xi < 0$, the system is unstable; inappropriate to apply;

When $\xi = 0$, no damping; inappropriate to apply;

When $\xi \geqslant 1$, although it is exponentially monotonic decay, the response is slow; inappropriate to apply.

The excitation control system should be designed to respond stably and swiftly (that is, the transient process should be short). Normally, an ideal damping ratio within the range $0 < \xi < 1$ should be taken. Generally, when ξ is equal to 0.5–0.8, the response curve can quickly achieve the stable value. When $\xi = 0.7$, the overshoot is about 5%. The response characteristics of the second-order system with different damping ratios ξ are shown in Table 4.3.

5

Control Law and Mathematical Model of Excitation System

5.1 Basic Control Law of Excitation System

5.1.1 Basic Terms of Excitation Control System

To further understand and comprehend the structure, effect, and function of an excitation control system, a brief description of basic terms follows:

(1) *Manual control*: A controlled quantity is always affected by many factors during operation, and thus it deviates from the required value. Therefore, an operator should control the quantity at any moment according to his/her observation, which is called manual control or artificial control.
(2) *Automatic control*: This adopts a mechanical or electrical device to replace the artificial control, which is called automatic control. No manual intervention is involved in automatic control.
(3) *Controlled object*: A device that is controlled is called a controlled object. In an excitation system, a generator is the controlled object.
(4) *Controlled quantity*: A controlled quantity is also known as the regulated quantity, which is the output quantity of the controlled object.
(5) *Set point value*: A value is expected for a controlled variable, which is called the set point value or target value. A set point value may be a constant or vary freely over time.
(6) *Controlling quantity*: A physical quantity changed by a control action to make the controlled quantity follow the physical quantity of a set point value is called the controlling quantity.
(7) *Disturbance*: A disturbance is a signal generating a counteractive effect on the output quantity of the system. If the disturbance is generated inside the system, it is called an internal disturbance; if the disturbance is generated outside the system, it is called an external disturbance. An external disturbance can be regarded as the input quantity of the system.
(8) *Controller (regulator)*: A device that can generate a control signal to change the controlling quantity as expected is called the controller.
(9) *System*: The term "system" has been widely used in all fields, so it is quite hard to clarify its definition. On the one hand, the definition can be sufficiently generalized for all applications; on the other hand, it can also be narrowly defined for a specific application. A system is an organic combination of interconnected and interacting objects. An automatic excitation control system consists of a generator and an excitation system; that is, the control system includes a controller and a controlled object.
(10) *Feedback*: In case the output quantity of the system is fully or partly fed back to the input terminal, it shall affect the system output jointly with the input quantity.
(11) *Closed-loop control*: Closed-loop control is a control process which can minimize the deviation between the output quantity and the reference set quantity of the system. Besides, its control principle is also realized on the basis of such a deviation. From this point of view, it is important to note that closed-loop control is designed only for an unexpected disturbance. An expected or known disturbance can always be corrected in the system, and thus it is unnecessary to measure it.

Design and Application of Modern Synchronous Generator Excitation Systems, First Edition. Jicheng Li.
© 2019 China Electric Power Press. Published 2019 by John Wiley & Sons Singapore Pte. Ltd.

(12) *Remote operation*: Remote operation is also called long-distance operation, which refers to the process of operating a device that is far away.
(13) *In situ operation*: In situ operation is called local operation, which is a manually operated process.

5.1.2 Transfer Function of Control System

When determining the dynamic performance of an automatic control system, first, the mathematical model representing the dynamic performance of the control system has to be determined. A control system that is described in the time domain usually involves the solution of a differential equation of the system based on the external action and the initial conditions, to work out its output response. However, if the system parameters have changed, the differential equation must be solved again.

When solving the differential equation of a linear system with the Laplace transform, a mathematical model in the complex frequency domain can be obtained for the control system: a transfer function. The transfer function can not only represent the dynamic characteristics of the system, but also can be used to study the influence on system performance when the system structure or parameter changes.

The transfer function is a mathematical model in the complex frequency domain which is derived when solving a linear ordinary differential equation by using the Laplace transform.

For a linear ordinary differential system, its transfer function is defined as the ratio between the Laplace transform of the system's output quantity and the Laplace transform of the system's input quantity under the zero initial condition.

The transfer function and the differential equation share common characteristics. The numerator polynomial coefficient and denominator polynomial coefficient of a transfer function correspond to the differential operator polynomial coefficients on the right and left of the differential equation separately. Therefore, the transfer function can be obtained by replacing the operator d/dt of the differential equation with the complex number s; otherwise, the differential equation can be obtained by replacing the variable s in the transfer function with d/dt, for instance, based on the transfer function:

$$G(s) = \frac{C(s)}{R(s)} = \frac{b_1 s + b_2}{a_0 s^2 + a_1 s + a_2}$$

The algebraic equation of s can be solved:

$$(a_0 s^2 + a_1 s + a_2) C(s) = (b_1 s + b_2) R(s)$$

A differential equation of the corresponding system can be solved by replacing s with the differential d/dt.

$$a_0 \frac{d^2}{dt^2} C(t) + a_1 \frac{d}{dt} C(t) + a_2 C(t) = b_1 \frac{d}{dt} R(t) + b_2 R(t) \tag{5.1}$$

It is important to note that the transfer function is defined according to the ratio between the Laplace transform of the system output quantity and the Laplace transform of the system input quantity under the zero initial condition. The zero initial condition of a control system has two meanings: one is that the input quantity can act on the system only when $t \geq 0$; therefore, when $t=0$, the input quantity and its coefficients at all orders shall be zero. It also means that the system is in a stable state before adding the input quantity.

5.1.3 Basic Control Law of Excitation Controller

The basic control law of the excitation controller has three basic control actions such as proportional (P), integral (I), and differential (D) action as well as their combinations.

5.1.3.1 Proportional Control Law (P)

For a controller based on the proportional control law, the relationship between its output signal $u(t)$ and input signal $e(t)$ can be expressed as follows:

$$u(t) = K_p\, e(t) \tag{5.2}$$

where

K_p – proportional gain or proportional amplification factor of the regulator.

The transfer function of a proportional controller is

$$G_c(s) = \frac{U(s)}{E(s)} = K_p \tag{5.3}$$

where

$U(s)$ – Laplace transform of the output;

$E(s)$ – Laplace transform of the input;

K_p – Amplification factor of the controller.

A proportional regulator is actually an amplifier with an adjustable gain (amplification factor). K_p is the ratio between the output variable $u(t)$ and the input variable $e(t)$ of the regulator. $e(t)$ is the deviation between the set point value and the measured value of the control system, which is also known as the input deviation signal or control error. Since the input and output of the controller can be various physical quantities, K_p may be dimensional. The output variable quantity of the proportional controller is directly proportional to the deviation input, with no time delay.

5.1.3.2 Integral Control Law (I)

The control law of an integral controller can be expressed as follows:

$$u(t) = K_I \int_0^t e(t)\, dt \tag{5.4}$$

where

K_I – integral rate constant of the integral controller.

The transfer function of an integral controller is

$$G_c(s) = \frac{U(s)}{E(s)} = \frac{K_I}{s} \tag{5.5}$$

Obviously, Eq. (5.5) describes a straight line with a fixed value slope. The slope of the output straight line is directly proportional to the integral rate constant of the controller K_I.

The integral controller can eliminate the steady-state error of the system. The output signal magnitude is not only related to that of the input deviation signal, but is also based on the duration of the deviation. As long as a deviation exists, the controller output will change constantly, and the longer it exists, the greater the variable component of the output signal can become, until the controller output reaches the limit. Only when the deviation signal e is zero can the output signal of the integral controller be constant, which is an outstanding feature of integral control.

The defects of a pure integral controller are that, unlike proportional control, it cannot keep the output u and the input e synchronized and quick to respond, with the output changes always lagging behind the deviation changes. In this way, it is hard to overcome the influence of a disturbance in a timely and effective manner to stabilize the system. Therefore, in the excitation control process, proportional control and integral control are usually combined into proportional-integral (PI) control during the application.

5.1.3.3 Proportional-Integral Control Law (PI)

The control law of the PI controller can be expressed as follows:

$$u(t) = K_p \left[e(t) + \frac{1}{T_I} \int_0^t e(t)\,\mathrm{d}t \right] \tag{5.6}$$

where

$K_p e(t)$ — proportional term;
$\frac{K_p}{T_I} \int_0^t e(t)\,\mathrm{d}t$ — integral term;
T_I — integration time constant.

The integration time constant T_I and the integral rate constant K_I are the inverse of each other, that is, $T_I = 1/K_I$. Obviously, the PI control law is a combination of the proportional and integral control actions. It combines the advantages of the quick response of proportional control with the elimination of the steady-state error by integral control, and thus can be widely applied in practice. The transfer function of a PI controller is

$$G_c(s) = \frac{U(s)}{E(s)} = K_p \left[1 + \frac{1}{T_I s} \right] \tag{5.7}$$

With a step deviation signal input, the output characteristic of the PI controller is shown in Figure 5.1. When the amplitude of the step deviation input is A, the proportional action output immediately jumps to $K_p A$, and then the integral action output will increase linearly over time; therefore, the output characteristic of the controller is a straight line with intercept $K_p A$ and slope $K_p A/T_I$. When K_p and A are determined, the straight-line slope depends on the magnitude of integral time T_I. The smaller T_I is, the greater the integral rate K_I and the steeper the straight line, which implies the greater the integral action; the greater T_I is, the slower the integral rate K_I and the gentler the slope of the straight line, which means the smaller the integral action. When T_I gradually becomes infinite, the controller has actually become a pure proportional controller. Therefore, T_I is an important parameter for describing the intensity of the integral action. When the input is $e(t) = A$, Eq. (5.6) yields:

$$u(t) = K_p A + \frac{K_p}{T_I} A t \tag{5.8}$$

Obviously, when $t = T_I$, the output is $u = 2K_p A$. Therefore, one can define T_I as follows: with the step deviation input, the time during which the controller output achieves twice the proportional output is the integration time T_I. According to this definition, T_I can be measured with a test.

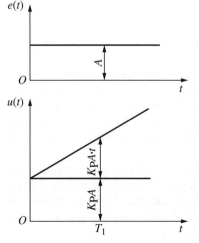

Figure 5.1 Output characteristics of the proportional-integral controller under the step deviation input.

The PI controller can be widely applied in most situations. Only when the controlled process greatly lags behind is the PI regulation time perhaps longer; or when the load fluctuation is especially intense, the PI regulation action is too slow in response, and in such a case, the differential control action must be added.

5.1.3.4 Differential Control Law (D)

For a controlled object with greater inertia, advanced regulation measures are usually adopted according to the variation trend of the controlled variable to avoid generating much greater deviation during the regulation, thus leading to the differential control law.

An ideal differential control law is expressed as follows:

$$u(t) = T_D \frac{de(t)}{dt} \tag{5.9}$$

It can be seen that the controller output $u(t)$ and the rate of change in the input deviation $de(t)/dt$ are in proportion, and the proportionality factor T_D is called the differential time constant.

From Eq. (5.9), when $t = t_0$, if inputting a step-changing deviation signal $e(t) = A$, the controller output at this moment is infinite, and the output is zero for the rest of the time. That is to say, an ideal differentiator is represented as a sharp pulse with an infinite magnitude and a width tending to zero, under a step deviation input signal. The differential controller output is only related to the rate of the change in deviation, independent of whether a deviation exists or not. When a deviation is fixed, the differential action has no output regardless of its value. This feature indicates that the differential control action is incapable of regulating a constant deviation. Therefore, the differentiator cannot simply be used as a separate controller. As a matter of fact, the differential control action is always combined and used with the proportional action or the PI control action. The transfer function of a differential controller is

$$G_c(s) = T_D s \tag{5.10}$$

5.1.3.5 Proportional-Differential Control Law (PD)

The ideal proportional-differential control law can be expressed as follows:

$$u(t) = K_p \left[e(t) + T_D \frac{de(t)}{dt} \right] \tag{5.11}$$

where

K_p — proportional gain, which can be a positive or negative value depending on the situation;

T_D — differential time constant.

The transfer function of a PD controller is

$$G_c(s) = K_p(1 + sT_D) \tag{5.12}$$

The PD controller output is the sum of proportional output and ideal differential output. The proportional term is $K_p e(t)$, and the ideal differential item is $K_p T_D \frac{de(t)}{dt}$. Under the step deviation input, its output response is shown in Figure 5.2. In the figure, a sharp pulse with an infinite magnitude and a width tending to zero that occurs at t_0 means that infinite power is difficult to realize physically. Obviously, it is difficult to realize ideal proportional-differential control in application.

A real PD controller is often adopted for an excitation control system, with its transfer function as

$$G_c(s) = K_p \frac{T_D s + 1}{\frac{T_D}{K_D} s + 1} \tag{5.13}$$

The differential gain K_D in the above real PD controller generally lies in the range 5–10. The proportional-differential control action actually is a combination of the proportional action and the

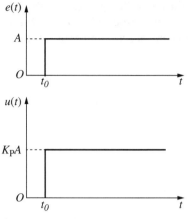

Figure 5.2 Output characteristics of ideal PD control.

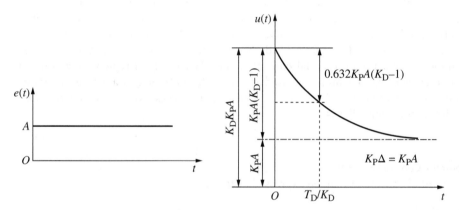

Figure 5.3 Output characteristics of the real PD control.

approximate differential action. When the amplitude of the input deviation $e(t)$ is the step signal for A, the output signal can be expressed as

$$u(t) = K_p A + K_p A (K_D - 1) e^{-\frac{K_D}{T_D}t} \tag{5.14}$$

See Figure 5.3 for the output characteristic curve of a proportional-differential controller.

When $t = T_D/K_D$, Eq. (5.14) can be expressed as

$$u(t) = K_p A + 0.368 K_p A (K_D - 1) \tag{5.15}$$

It can be seen that, within the time interval $t = T_D/K_D$, the PD controller output drops by 63.2% of the maximum output for differential action from the peak of its step pulse (or "drops to 36.8% of the maximum output for differential action"). Let $T = T_D/K_D$ be termed a time constant. From this, one can measure the time constant T via a test and solve for the differential time constant T_D accordingly.

Since the differential action always tries to prevent the controlled variable from changing (either increasing or decreasing), effective differential control action will have an oscillation suppression effect. Appropriate selection of the differential time constant will help improve system stability. If the differential time constant T_D is too large, then the differential control action will become too strong, which will undermine system stability. The differential time constant for the PD controller in an excitation system can be set within a certain range.

5.1.3.6 Proportional-Integral-Differential Control Law (PID)

The proportional-integral-differential (PID) control law can be expressed as follows:

$$u(t) = K_p \left[e(t) + \frac{1}{T_I} \int_0^t e(t)\,dt + T_D \frac{de(t)}{dt} \right] \qquad (5.16)$$

In the equation, the definitions of the parameters K_p, T_I, and T_D are the same as those of PI and PD controllers.

The transfer function of a PID controller is

$$G_c(s) = K_p \left[1 + \frac{1}{T_I s} + T_D s \right] \qquad (5.17)$$

It is clear that the ideal control law of the controller as shown in Eq. (5.16) is difficult to realize physically.

Taking into consideration the possibility of physical realization and the control requirements, a real PID controller design will be adopted in most applications. The differential equation for its input and output relationship is quite complicated and will not be discussed in detail here. However, its basic principle is the same as that for the ideal PID control law. Under the step deviation action with magnitude A, the output signal of the real PID controller can be regarded as the superposition of proportional output, integral output, and differential output; that is,

$$u(t) = K_p A + \frac{K_p A}{T_I} t + K_p A (K_D - 1) e^{-\frac{t}{T_D}} \qquad (5.18)$$

From Figure 5.4, for the PID controller based on step output, the output variation in the differential action is maximum at the beginning, which makes the overall controller output vary greatly. Then, the differential action dies out gradually, with the integral output prevailing gradually. As long as a deviation exists, the proportional action will increase gradually until the deviation disappears completely. In PID control, the proportional action always corresponds to the deviation. It is always the most essential control action.

In the PID controller, if the gain K_p, integral time constant T_I, and differential time constant T_D are properly set, higher control performance can be obtained If the gain (K_p), integral time constant (T_I), and differential time constant (T_D) are properly set, higher control quality can be achieved. Therefore, the PID controller is highly adaptable and can be widely used.

In the PID controller, mutual interference exists among the proportional, integral, and differential actions to some extent. A mutual interference coefficient F is usually adopted for correction. The value of the interference coefficient F will vary with different controller structures.

Figure 5.4 Output characteristics of PID regulation under the step deviation actions.

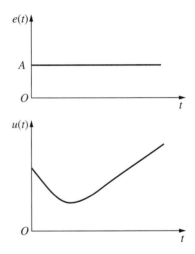

Now that descriptions of the control performance and functions of the PID have been given, to help understand the overall characteristic influence of the relevant parameters of the PID controller, we give some numerical instances based on Eq. (5.19) to describe the overall performance of the PID controller. The equation is

$$G(s) = K_p \frac{1+T_1 s}{1+T_2 s} \times \frac{1+T_3 s}{1+T_4 s} = 200 \times \frac{1+2s}{1+10s} \times \frac{1+0.3s}{1+0.01s} \tag{5.19}$$

where

K_p — proportional amplification factor;

$\frac{1+T_1 s}{1+T_2 s}$ — (Term 2) because $T_1 < T_2$ are the lagging phases, that is, the integral part;

$\frac{1+T_3 s}{1+T_4 s}$ — (Term 3) because $T_3 > T_4$ and are the leading phases, that is, the differential part.

$$T_1 = 2, T_2 = 10, T_3 = 0.3, T_4 = 0.01$$

The proportional part P in the PID controller:

When $t_1 \to \infty$, $s \to 0$, the corresponding static amplification factor is $K_1 = G(0) = K_P = 200$;

When $t \to 0$, $s \to \infty$, the corresponding transient amplification factor is $K_2 = G(\infty) = K_P \frac{T_1}{T_2} \cdot \frac{T_3}{T_4} = 1200$;

When $0 < t < \infty$, the corresponding dynamic amplification factor is $K_3 = G(s) = K_P \frac{T_1}{T_2} = 40$.

See Figure 5.5 for the corresponding amplitude/frequency characteristics.

In general, the low-frequency stage for the open-loop frequency characteristic of the control system represents stable performance of the closed-loop system, the medium-frequency stage represents the dynamic performance of the closed-loop system, and the high-frequency stage represents the noise suppression performance. Therefore, the essence of designing a control system by using the frequency method is to add a correction device with appropriate frequency characteristics in the control system to give the system the required frequency characteristic shape. In the low-frequency stage, the gain should be great enough to meet the requirement of steady-state error; the slope of the gain response in the medium-frequency stage is generally −20 dB dec^{-1} and covers a sufficient frequency bandwidth to ensure that the system can have an adequate phase margin; the gain in the high-frequency stage should be minimized as soon as possible to strengthen noise suppression.

From the above instance analysis, in the PID excitation control law, we have the following:

(1) The proportional part, P, is the basis for negative feedback control. An increased proportional amplification factor P can reduce the steady-state error in the excitation control system and accelerate the response speed; however, if the proportional part is too large, system stability will be undermined.
(2) The integral part, I, is used to improve the static amplification factor of the excitation system, reduce the steady-state error, and decrease the dynamic amplification factor, which can help control system stability.

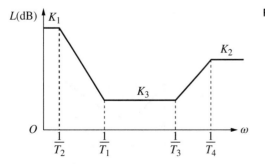

Figure 5.5 Amplitude/frequency characteristics of the PID controller.

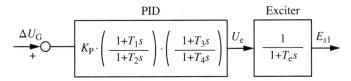

Figure 5.6 Transfer function expression of the PID control system.

(3) The differential part, D, is used to improve the response rate of the excitation control system, reduce the overshoot, and compensate for the greater inertia time constant of the system; for instance, the compensation for the time constant T_e of the exciter can help control dynamic system stability. This action can be clarified via the transfer function expression for the PID control system as shown in Figure 5.6.

In Figure 5.6, K_P is the proportional part, $T_1 < T_2$ of $\left(\frac{1+T_1 s}{1+T_2 s}\right)$ is the lagging phase integral control part, and $T_3 > T_4$ of $\left(\frac{1+T_3 s}{1+T_4 s}\right)$ is the leading phase differential control part. When setting the parameters, if $T_3 = T_e$, the transfer function of the excitation control system can be simplified as

$$G(s) = K_P \times \frac{1+T_1 s}{1+T_2 s} \times \frac{1}{1+T_4 s}$$

The time constant T_e of the exciter is reduced equivalently.

5.1.4 Numerical Description of the PID Control Law

One can obtain the discrete control algorithm used for digital computing control by changing the analog PID control algorithm. An ideal analog PID algorithm can be expressed as

$$u = K_P \left[e + \frac{1}{T_I} \int e \, dt + T_D \frac{dt}{dt} \right] \quad (5.20)$$

where

u – output signal of the regulator;

e – the difference between the set point value r and the measured value y, $e = r - y$;

K_P, T_I, T_D – proportional constant, integral time constant, and differential time constant of the controller.

There are several algorithms based on the different controller output forms:

5.1.4.1 Position Algorithm
By substituting the integral term and the differential term in Eq. (5.20) separately with a summation and an incremental ratio, the output value u_n at the sampling instant n can be obtained as

$$u_n = K_P \left[e_n + \frac{T_S}{T_I} \sum_{i=0}^{n} e_i + \frac{T_D}{T_S}(e_n - e_{n-1}) \right] \quad (5.21)$$

where

T_S – sampling period;

e_n – the sampled deviation value at time n, $e_n = r_n - y_n$.

Obviously, the computer output value u_n mutually corresponds to the sampling value; therefore, Eq. (5.21) is generally known as the position algorithm of PID.

According to Eq. (5.21), the output corresponding to the nth sampling is related to all the previous e_i values; therefore, the computer should be capable of storing massive amounts of data. Once a failure occurs, the sampling value will change dramatically.

5.1.4.2 Increment Algorithm

From Eq. (5.21), the control algorithm for the $(n-1)$th sampling can be obtained:

$$u_{n-1} = K_P \left[e_{n-1} + \frac{T_S}{T_I} \sum_{i=0}^{n-1} e_i + \frac{T_D}{T_S}(e_{n-1} - e_{n-2}) \right] \tag{5.22}$$

The computer output increment within the time interval of the two samplings is

$$\begin{aligned} \Delta u_n &= u_n - u_{n-1} \\ &= K_P(e_n - e_{n-1}) + \frac{T_S}{T_I} e_n + \frac{T_D}{T_S}(e_n - 2e_{n-1} + e_{n-2}) \\ &= K_P(e_n - e_{n-1}) + K_I e_n + K_D(e_n - 2e_{n-1} + e_{n-2}) \end{aligned} \tag{5.23}$$

where

K_I — integral coefficient, $K_I = K_P T_S / T_I$;
K_D — differential coefficient, $K_D = K_P T_D / T_S$.

Equation (5.23) represents the changed relationship between the deviation and the sampling value; that is, according to the arithmetical result, the sampling value will increase or decrease depending on the original position, and therefore Eq. (5.23) is known as the increment algorithm of PID.

5.1.4.3 Practical Algorithm

When it is specifically required to implement the PID algorithm on a computer, as an aid to programming, many practical algorithms can be derived. The following is one of them.

Equation (5.23) is written in the following form:

$$\Delta u_n = K_P \left(1 + \frac{T_S}{T_I} + \frac{T_D}{T_S}\right) e_n - K_P \left(1 + \frac{2T_D}{T_S}\right) e_{n-1} + \frac{K_P T_D}{T_S} e_{n-2} \tag{5.24}$$

When: $A = K_P \left(1 + \frac{T_S}{T_I} + \frac{T_D}{T_S}\right)$, $B = K_P \left(1 + \frac{2T_D}{T_S}\right)$, $C = \frac{K_P T_D}{T_S}$, then Eq. (5.24) can be written as

$$\Delta u_n = A e_n - B e_{n-1} + C e_{n-2} \tag{5.25}$$

The above equation is an algorithm without the traditional PID concept; therefore, the practical algorithm adopted is

$$\begin{aligned} u_n &= A e_n + q_{n-1} \\ q_n &= u_n - B e_n + C e_{n-1} \end{aligned} \tag{5.26}$$

The initial value is generally taken as $q_{n-1} = 0$, $e_{n-1} = 0$.

According to Eq. (5.26), if e_n is known, one can work out u_n, q_n under the above initial conditions; then u_{n+1} can be calculated based on q_n and e_{n-1}; by repeated iteration, u_n, u_{n+1}, and u_{n+2} …can be obtained separately. In this way, the increment output Δu_n of the computer can be obtained.

5.1.5 Parallel Feedback Correction of Excitation Control System

To improve the control system performance, apart from series correction, parallel feedback correction is also a correction method that has been widely applied.

The basic principle of parallel feedback correction is that the feedback correction device encircles the elements and exerts a major influence by improving the dynamic performance of the uncorrected system by forming a local feedback loop to make the frequency response of the corrected system meet a given requirement.

We take the transfer function block diagram of the parallel feedback correction shown in Figure 5.7 as an example to discuss the characteristics of the parallel feedback correction action.

$H(s)$ shown in Figure 5.7 is the forward path transfer function, which includes the elements that need to be corrected; $F(s)$ is the feedback path transfer function. After feedback correction, the closed-loop transfer function can be obtained by the following equation:

$$Y(s) = H(s)[X(s) - Y(s)F(s)]$$

$$[1 + H(s)F(s)]Y(s) = H(s)X(s)$$

$$G(s) = \frac{Y(s)}{X(s)} = \frac{H(s)}{1 + H(s)F(s)} \tag{5.27}$$

5.1.5.1 Parallel Correction – Proportional Feedback

We shall adopt the proportional feedback parallel correction link. See Figure 5.8 for the transfer function block diagram of the control system.

According to Figure 5.8 and Eq. (5.27), the transfer function equations of the control system are

$$G(s) = \frac{Y(s)}{X(s)} = \frac{H(s)}{1 + H(s)F(s)}$$

$$G(s) = \frac{\frac{1}{1+T_e s}}{1 + \frac{K_f}{1+T_e s}} = \frac{1}{1 + K_f + T_e s}$$

$$G(s) = \frac{\frac{1}{1+K_f}}{1 + s\frac{T_e}{1+K_f}} \tag{5.28}$$

Figure 5.7 Transfer function block diagram of the parallel feedback correction.

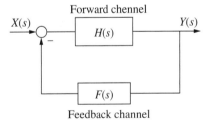

Figure 5.8 Transfer function block diagram when adopting the proportional feedback parallel correction.

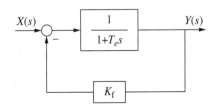

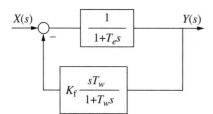

Figure 5.9 Transfer function block diagram when adopting the dynamic feedback parallel correction.

From Eq. (5.28), it is seen that after the control system employs the proportional feedback parallel correction link, the time constant T_e of the original system can be reduced by $\frac{1}{1+K_f}$, and the steady-state gain of the original system can also be reduced by $\frac{1}{1+K_f}$.

5.1.5.2 Parallel Correction – Dynamic Feedback

We shall adopt the dynamic feedback parallel correction link. See Figure 5.9 for the transfer function block diagram of the control system.

From Figure 5.9, the transfer function equations of the control system are

$$G(s) = \frac{\frac{1}{1+T_e s}}{1 + \frac{1}{1+T_e s} \times K_f \times \frac{T_w s}{1+T_w s}}$$

$$G(s) = \frac{1 + T_w s}{(1 + T_e s)(1 + T_w s) + K_f T_w s}$$

$$G(s) = \frac{1 + T_w s}{1 + [T_e + (1 + K_f) T_w] s + T_e T_w s^2}$$

$$G(s) \approx \frac{1 + T_w s}{1 + \left[\frac{T_e}{1+K_f} + (1 + K_f) T_w\right] s + T_e T_w s^2}$$

$$G(s) = \frac{1}{1 + \frac{T_e s}{1+K_f}} \times \frac{1 + T_w s}{1 + (1 + K_f) T_w s} \tag{5.29}$$

From Eq. (5.29), it is seen that after the control system employs the dynamic feedback parallel correction links, the time constant T_e of the original system can be reduced by $\frac{1}{1+K_f}$, and the dynamic-state gain can also be reduced by $\frac{1}{1+K_f}$, with the static-state gain remaining unchanged.

5.2 Mathematical Model of the Excitation System

In our study of power system stability, the operating characteristics of synchronous generators will be precisely described. We will first establish the mathematical models which can precisely explain an excitation system in various operating modes [10].

Since 1968, some famous companies and manufacturers worldwide have kept launching various excitation system mathematical models, developing them separately and perfecting them in practice. These models have proved to be a great convenience for the power sector in terms of studying power system stability. The most widely applied excitation system mathematical models are those proposed by Institute of Electrical and Electronic Engineers (IEEE) and International Electro-technical Commission (IEC).

In the meantime, some famous electric companies have created for their users of dedicated excitation system mathematical models by combining the characteristics of the excitation systems they have developed. For instance, Westinghouse Electric Corp from the United States has established mathematical models, including model I, AC-1, and AC-2 for a brushless excitation system. Some other national companies have also established corresponding excitation system mathematical models. For example, some Japanese motor manufacturers have also proposed a related mathematical model. These mathematical models are characterized by many modes excitation system. This section will focus on the excitation system multimode mathematical model that was proposed in 1995 by the Japanese Institute of Electrical Engineers (IEEJ). Since these multimode mathematical models have included some representative parameter values in the form of numerical examples, which are of great reference values in practice.

In general, regarding the application scope of the excitation system mathematical models, the highest system oscillation frequency should not exceed 3 Hz, and the system frequency deviation should not exceed ±5% of the rated values. The sub-synchronous oscillation, and so on, with a greater frequency variation range are not applicable.

The last point to be stated is that normalized excitation system mathematical models used for studying stability in the power sector have been elaborately specified in international, national, and industrial standards, for example, IEEE std. 421.5–2005. Therefore, they will not be covered in this book. The excitation system mathematical models covered in this section will be only further specified from the standpoint of the basic functions of the excitation system on the basis of engineering applications, which can help the reader better understand the excitation system models in the application context.

It is important to stress that all the excitation devices that can achieve the generator excitation control action can be considered. However, their mathematical models may not meet the standard model requirements. Therefore, there are three cases when using a specific excitation system model: (1) the original excitation system model that conforms to the fixed excitation system model in the standards or in the power system stability calculation application program; (2) the original excitation system model is converted into the fixed excitation system model in the standards or in the power system stability calculation application program; and (3) a model that conforms to the real excitation system is developed by using the custom function in the power system calculation application program.

5.2.1 Static Self-Excitation System

The transfer function model block diagrams are shown in Figures 5.10–5.12 for a static self-excitation thyristor having its excitation source from the generator terminal.

5.2.1.1 Series Compensation Self-Excitation System
The series compensation mode means that the correction element of the excitation system is in the forward path of the excitation regulating loop in series, as shown in Figure 5.10.

Table 5.1 shows sample values for the transfer function mathematical model parameters of the series compensation self-excitation system, which can be referred to in the specific applications.

5.2.1.2 Feedback Compensation Self-Excitation System
In this system, the correction action is compensated based on the parallel negative feedback mode, as shown in Figure 5.11. When carrying out the operation at each level after the amplifier K_A, the gains and limiting values in all components should be taken, including PSVR and the power system stabilizer (PSS), and so on. See Table 5.2 for the corresponding parameters.

5.2.1.3 Proportional-Integral (PI) Self-Excitation System
The PI self-excitation system with no deviation regulation can achieve no steady-state deviation of the first order when the open-loop gain of the excitation regulation system is low.

See Figure 5.12 for the transfer function model of the PI self-excitation system. See Table 5.3 for the relevant sample parameter values.

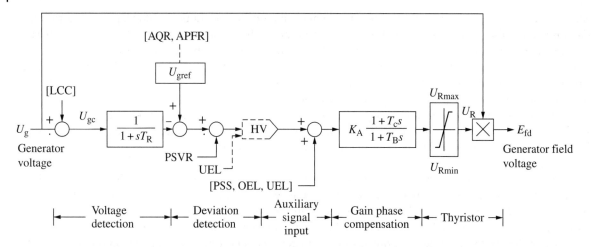

Figure 5.10 Transfer function model instance for the series compensation self-excitation system. U_{gref} – reference voltage; HV – high-pass value selection: the high-pass value will get priority to pass at the summing point for each input signal of the summator; U_{Rmax} – the max. output voltage of the thyristor rectifier under the rated generator voltage; U_{Rmin} – the min. output voltage of the thyristor rectifier under the rated generator voltage, which can be negative for a full-control-bridge rectifying circuit and zero for a half-control circuit; E_{fd} – the excitation voltage value in the actual operating status, which is proportional to the generator voltage U_g and the field regulator output voltage U_R, thus resulting in $U_g \times U_R$.

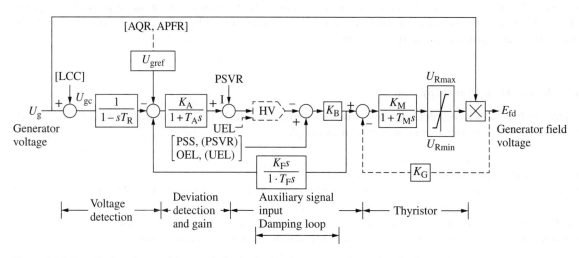

Figure 5.11 Transfer function model example for the feedback compensation self-excitation system.

5.2.2 AC Exciter System

For a system with the AC exciter as the excitation source, the corresponding transfer function model is as follows.

5.2.2.1 AC Shunt-Exciter-Type Self-Excitation System

In this system, the AC shunt exciter is excited by the excitation transformer at the generator terminal, as shown in Figure 5.13, and hence the name self-excitation system.

See Table 5.4 for the sample parameter values for the transfer function models of the AC shunt-exciter-type self-excitation system.

Table 5.1 Sample of transfer function model parameters for series compensation self-excitation system (full-control rectifying circuit).

Symbol (unit)	Parameter value	Meaning	Symbol (unit)	Parameter value	Meaning
$T_R(s)$	0.01	Time constant for the voltage transducer	$T_C(s)$	0.22	Time constant for the leading phase compensation
K_A	150	AVR gain	U^*_{Rmax} (p.u.)	4	Max. output voltage of the thyristor
$T_B(s)$	0.5	Time constant for the lagging phase compensation	U^*_{Rmin} (p.u.)	−3.2	Min. output voltage of the thyristor

Table 5.2 Sample of transfer function model parameters for the feedback compensation self-excitation system (full-control rectifying circuit).

Symbol (unit)	Parameter value	Meaning	Symbol (unit)	Parameter value	Meaning
$T_R(s)$	0.01	Time constant for the voltage transducer (secondary amplification)	T_F	0.5	Time constant for the damping loop
			K_M	25	Thyristor gain
K_A	24.0	AVR gain	$T_M(s)$	0.02	Time constant for the thyristor controller
$T_A(s)$	0.02	Time constant for the AVR amplifier	U^*_{Rmax} (p.u.)	5.0	Max. output voltage of the thyristor
K_B	5.0	Auxiliary input signal gain	U^*_{Rmin} (p.u.)	−2.5	Min. output voltage of the thyristor
K_F	0.0075	Damping loop gain	K_G	0.8	Field voltage feedback gain

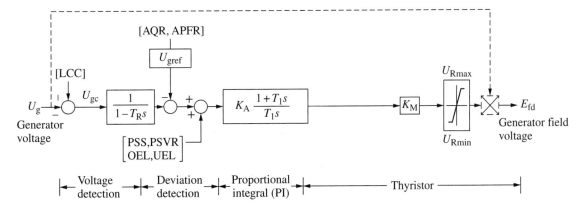

Figure 5.12 Transfer function model example for the proportional-integral self-excitation system.

Table 5.3 Samples of transfer function model parameters for the proportional-integral self-excitation system (full-control rectifying circuit).

Symbol (unit)	Parameter value	Meaning	Symbol (unit)	Parameter value	Meaning
$T_R(s)$	0.02	Time constant for the voltage transducer (secondary amplification)	K_M	4.2	Thyristor gain
K_A	7.5	AVR gain	U^*_{Rmax} (p.u.)	4.2	Max. output voltage of the thyristor controller
$T_i(s)$	0.5	Integral time constant	U^*_{Rmin} (p.u.)	−4.2	Min. output voltage of the thyristor

Table 5.4 Samples of transfer function model parameters for AC shunt-exciter-type self-excitation system.

Symbol (unit)	Parameter value	Meaning	Symbol (unit)	Parameter value	Meaning
$T_R(s)$	0.02	Time constant for the voltage transducer (secondary amplification)	$T_A(s)$	0.044	Time constant for the AVR amplifier
K_A	122	AVR gain	U^*_{Rmax} (p.u.)	5.1	Max. AVR output voltage
K_F	0.09	Damping loop gain	U^*_{Rmin} (p.u.)	−5.1	Min. AVR output voltage
$T_F(s)$	1.5	Time constant for the damping loop			

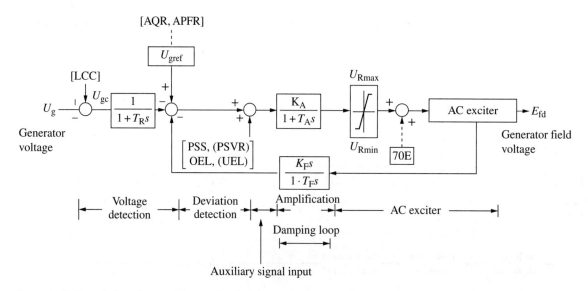

Figure 5.13 Transfer function model example of AC shunt-exciter-type self-excitation system. 70E – manually set value.

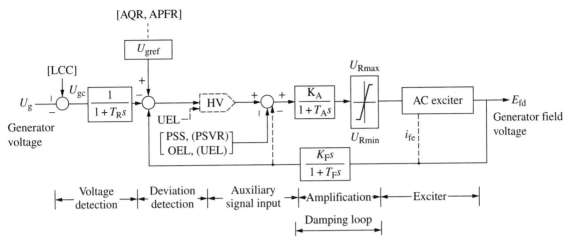

Figure 5.14 Instances of transfer function model parameters for AC separate-exciter-type excitation system (including brushless excitation system).

Table 5.5 Samples of transfer function model parameters for AC separate-exciter-type excitation system.[a]

Symbol (unit)	Parameter value	Meaning	Symbol (unit)	Parameter value	Meaning
$T_R(s)$	0.013	Time constant for the voltage transducer (secondary amplification)	$T_F(s)$	1.0	Time constant for the damping loop
K_A	200	AVR gain	U^*_{Rmax} (p.u.)	5.92	Max. AVR output voltage
$T_A(s)$	0.01	Time constant for the AVR amplifier	U^*_{Rmin} (p.u.)	−2.63	Min. AVR output voltage
K_F	0.061	Damping loop gain			

a) The input signal of the damping loop can either be the excitation voltage of the generator or the excitation current of the exciter. For the brushless excitation system, the input signal of the damping loop can only be the excitation current i_{fe} of the exciter.

5.2.2.2 AC Separate-Exciter-Type Excitation System

In the AC separate-exciter-type excitation system, the excitation power is taken from the generator axle head. See Figure 5.14 for the typical transfer function model block diagram. See Table 5.5 for sample parameter values.

5.2.2.3 Feedback Compensation Brushless Excitation System

To reduce the time constant of the AC exciter in the brushless excitation system, usually feedback compensation measures are employed. By adopting various feedback compensation modes, the transfer function model of the corresponding excitation system will also be different. For instance, Figure 5.15 shows the brushless excitation system model as recommended by IEEJ. Some sample parameter values are listed in Table 5.6.

When calculating the gain at each level after K_A, the gain for each limiting value should be taken into consideration and K_A corrected accordingly.

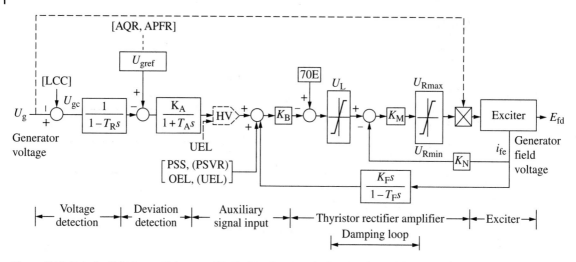

Figure 5.15 Transfer function model instance for feedback-compensation-type brushless excitation system.

Table 5.6 Transfer function model parameter sample for feedback-compensation-type brushless excitation system.

Symbol (unit)	Parameter value	Meaning	Symbol (unit)	Parameter value	Meaning
T_R (s)	0.01	Time constant for the voltage transducer (secondary amplification)	U^*_{Rmax} (p.u.)	86.0	Max. AVR output voltage
K_A	25.0	AVR gain	U^*_{Rmin} (p.u.)	−40.0	Min. AVR output voltage
T_A (s)	0.02	Time constant for the AVR amplifier	U^*_L (p.u.)	8.6	Output limit
K_B	5.0	Auxiliary input signal gain	K_F	0.32	Damping loop gain
K_M	37.0	Thyristor gain	T_F (s)	1.0	Time constant for the damping loop
K_N	0.8	Field current feedback gain of the exciter			

5.2.2.4 PI Type Brushless Excitation System

The PI type brushless excitation system can be integrated according to the deviation value, so that the first-order steady-state deviation of the system is zero and independent of the open-loop gain of the system.

Figure 5.16 shows the typical transfer function block diagram of the above system. Typical sample parameter values are listed in Table 5.7.

In addition, for the AC exciter system, there are also more commonly used transfer function models such as standard Types I, II, and III excitation system modes, as shown separately in Figures 5.17–5.19. See Tables 5.8–5.10 for the relevant symbol meanings and sample parameter values.

In Figure 5.18b, U_{e0} and U'_{e0} separately represent the exciter no-load and load voltage values corresponding to the exciter rated no-load excitation current I_{fe0}. T_{dze} represents the exciter load time constant, which is

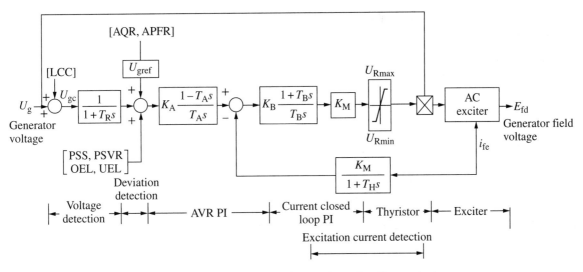

Figure 5.16 Transfer function model instance for proportional-integral-type brushless excitation system.

Table 5.7 Transfer function model parameter samples for proportional-integral type brushless excitation system.[a]

Symbol (unit)	Parameter value	Meaning	Symbol (unit)	Parameter value	Meaning
$T_R(s)$	0.02	Time constant for the voltage transducer (secondary amplification)	K_H	0.2	Current transducer gain
K_A	1.2	AVR gain	$T_H(s)$	0.006	Time constant for the current transducer
$T_i(s)$	0.8	AVR integral time constant	K_M	4.6	Thyristor gain
$T_B(s)$	1.8	ACR Integral time constant	U^*_{Rmax}(p.u.)	4.6	Upper AVR output limit
K_B	1.5	ACR gain	U^*_{Rmin}(p.u.)	−4.6	Lower AVR output limit

a) It is applicable to the self-excitation system with power supplied from the generator terminal in the hydraulic generator without a pilot exciter. Sometimes the ACR integral loop will not be adopted.

related to the exciter parameters and the load power factor; for example, when the load power factor is 0.9, the value T_{dze} is $T_{dze} = T_{doe} \times \dfrac{0.81+(X'_{de}+0.436)}{0.81+(X_{de}+0.436)} \dfrac{(X'_{qe}+0.436)}{(X_{qe}+0.436)}$

5.2.3 Rectifier

In the thyristor rectifier excitation system, the following characteristics are considered for rectifier models:

The models for the half-control rectifier and the full-control rectifier are the same, but the output voltage for only the full-control rectifier can be negative; for the half-control circuit, the output voltage can only be zero.

In addition, the following assumptions apply for the rectifier in the models:

(1) The commutating reactance and the forward voltage drop of the elements can be ignored.
(2) The phase commutation state does not change.
(3) The load is highly inductive.

5 Control Law and Mathematical Model of Excitation System

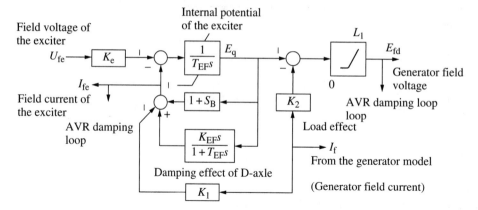

Figure 5.17 Standard Type I transfer function model S_E – saturation coefficient.

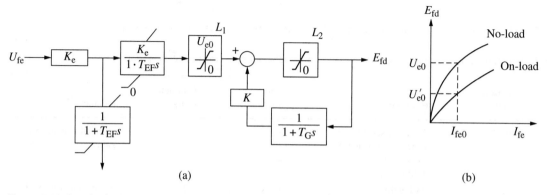

Figure 5.18 Standard Type II transfer function model: (a) function model, (b) function curve.

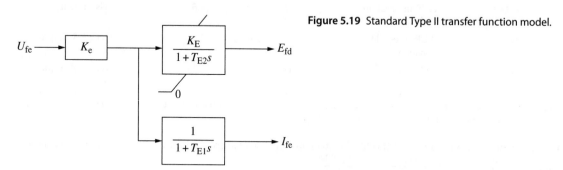

Figure 5.19 Standard Type II transfer function model.

U_{AC} represents the supply voltage of the thyristor rectifier and U_{DC} the DC output voltage. The full-control circuit is then expressed as

$$U_{DC} = 1.35 U_{AC} \cos \alpha \tag{5.30}$$

For the half-control circuit:

$$U_{DC} = \frac{1.35 U_{AC}(1 + \cos \alpha)}{2} \tag{5.31}$$

Equations (5.30) and (5.31) can be approximately expressed by Eqs. (5.24) and (5.25):

$$U_{DC} \approx 1.35 U_{AC} U_{AVR} \tag{5.32}$$

Table 5.8 Symbol meaning and parameters for standard Type I model.

Symbol (unit)	Meaning	Symbol (unit)	Meaning
K_e	$1/R_{fe}$, R_{fe} are the excitation winding resistance for the exciter	S_E	Saturation coefficient of the exciter
K_2	The load effect coefficient, values in Type I phase commutation state for the external characteristics of the rectifier	K_{EF}	Damping effect of d-axis $K_{EF} \approx T_{EF} \dfrac{(X_{de} - X'_{de})(X'_{de} - X''_{de})}{(X_{de} - X_{fe})^2}$
$T_{E1}(s)$	Exciter time constant, equivalent to T_{doe}	$T_{EF}(s), K_1$	$T_{EF} \approx T_{doc}, K_1 = (X_{de} - X'_{de})$

Table 5.9 Symbol meaning and parameters for standard Type II model.

Symbol (unit)	Parameter value	Meaning	Symbol (unit)	Parameter value	Meaning
K_e	1.0	$1/R_{fe}$, R_{fe} are the excitation winding resistance for the exciter	U^*_{CE} (p.u.)	5.0	Ceiling voltage of the excitation system
K_E	1.5	U_{e0}/I_{fe0}	K	0.5	$\dfrac{U_{e0} - U'_{e0}}{U'_{e0}}$
$T_{E2}(s)$	0.8	$\dfrac{(T_{doe} + T_{dze})}{2}$	$T_G(s)$	4.0	The generator voltage changes within a scope of ±5%, the correction value of the generator open circuit time constant T'_{d0}
$T_{E1}(s)$	0.5	$\dfrac{T_{doe}}{2}$			

Table 5.10 Symbol meaning and parameters for standard Type III model.

Symbol (unit)	Meaning
K_e	$1/R_{fe}$, R_{fe} are the excitation winding resistances for the exciter
K_E	U_{e0}/I_{fe0}
$T_{E2}(s)$	$\dfrac{T_{d0e} + T_{dze}}{2}$
$T_{E1}(s)$	$\dfrac{T'_{doe}}{2}$

$$U_{DC} \approx \frac{1.35 U_{AC}(1 + U_{AVR})}{2} \qquad (5.33)$$

Simplification of the above equations shows that U_{DC} and U_{AC} have a linear relationship. To avoid a commutation failure, the control angle should have a certain margin, generally 10° – 20°, and the variation range of the control angle α is 20° – 160°.

5.3 Mathematical Model of Excitation Control Unit

The control device of the excitation system includes various limiting functions, which will be represented in the mathematical models of the excitation system. This gives a description of the mode characteristics for each unit of the control device.

5.3.1 Load Current Compensator (LCC)

The LCC loop of the generator is used to set different coefficients of the excitation regulator:

(1) If LCC only compensates for the reactance capacity of the generator, it is also known as the cross-current compensator (CCC).
(2) Active current resistance droop $R_C I_C$.
(3) Reactive current reactance droop $jX_C I_C$.

To achieve different compensations, the above compensation quantity and the resultant phasor of the generator terminal voltage U_g are taken as the feedback quantity U_{ga} for AVR voltage detection:

$$U_{ga} = | U_g + (R_C + jX_C)\dot{I}_C | \tag{5.34}$$

where

$\dot{I}_C$ — compensating current;

R_C — compensation rate for the resistive component, which is negative in the case of LCC and positive in the case of CCC;

X_C — compensation rate for the reaction component, which is negative in the case of LCC and positive in the case of CCC.

To compensate for the voltage drop due to the machine transformer impedance, LCC negative droop is usually applied.

For more than one generator set operating in parallel at the generators' terminals, CCC positive difference regulation connection is employed in order to maintain the stability in the reactive distribution.

The current used for the compensation has the following two modes:

(1) Reactive current (cross-current) compensation between the independent generator sets, $\dot{I}_C$, is taken as the stator current I_g of each generator.
(2) Two-generator one-transformer connection, $\dot{I}_C$ is better taken as the difference between the local stator current and the average value of the total current of the two generators. If the stator current of the No. 1 Generator Set is $\dot{I}_1$ and that of the No. 2 Generator Set is $\dot{I}_2$, the compensating current of each generator separately is

$$\dot{I}_{C1} = \dot{I}_1 - \dot{I}$$
$$\dot{I}_{C2} = \dot{I}_2 - \dot{I}$$

where: $\dot{I} = \frac{\dot{I}_1 + \dot{I}_2}{2}$

5.3.2 Automatic Reactive Power Regulator (AQR)

In regard to the different operation modes, automatic reactive power regulator (AQR) has various connection modes, and only the transfer function model of its general expression will be discussed here.

The expression is

$$Q = a + bP + cP^2 \tag{5.35}$$

where

a, b, c – coefficients, determined on the basis of the operational requirements;

P, Q – active and reactive power.

See Figure 5.20 for the corresponding transfer function model. See Table 5.11 for sample parameter values.

5.3.3 Automatic Power Factor Regulator (APFR)

The APFR is used to control the power factor of the generator as the set point value or limit it within a set range. See Figure 5.21 for its transfer function model. See Table 5.12 for sample parameter values.

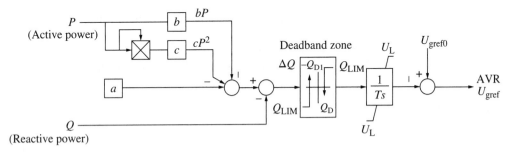

Figure 5.20 AQR transfer function model: U_{gref0} – initial value of the set voltage.

Table 5.11 AQR transfer function model parameter samples.

Symbol (unit)	Parameter value	Meaning	Symbol (unit)	Parameter value	Meaning
T (s)	30–200	Rate of changes for the voltage setting device	Q^*_{LIM} (p.u.)	1	AQR output
Q_D	0.01	Deadband zone	$\pm U^*_L$ (p.u.)	±0.1	Integral limit

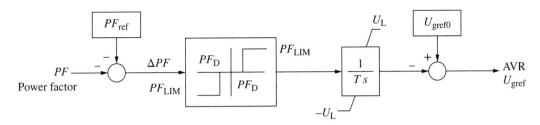

Figure 5.21 APFR transfer function model: PF_{ref} – set value for the power factor.

Table 5.12 APFR transfer function model parameter samples.

Symbol (unit)	Parameter value	Meaning	Symbol (unit)	Parameter value	Meaning
T (s)	30–200	Rate of changes for the voltage setting device	PF^*_{LIM} (p.u.)	1	APFR output
PF_D	0.005	Deadband zone	$\pm U^*_L$ (p.u.)	±0.1	Integral limit

When the APFR limits the power factor of the generator within the set point range, it must set a deadband zone to prevent the regulator from being always in the alternation mode.

Besides, when in the AQR control mode as represented in Eq. (5.35), when $a = c = 0$, one can obtain $Q = bP$, which is the APFR control mode.

In the meantime, as in the AQR mode, the APFR output also acts on the AVR voltage setting device.

5.3.4 Under-Excitation Limiter (UEL)

The UEL is used to prevent the generator from losing synchronism because the excitation current is too low. The UEL uses the stator voltage and stator current of the generator. The limiting characteristic can be a linear function or an arc limiting curve that is similar to the static stability characteristic curve.

(1) Similar to AQR, the UEL linear expression is $Q = a + bP$ where a and b are parameters determined by the limit characteristic.
(2) The static stability limit curve is solved using the power circle equation.
(3) If the static stability is limited by the linear expression, the linear equation is rewritten as

$$\frac{Q}{U_g^2} = \frac{a}{U_g^2} + b \times \frac{P}{U_g^2}$$

For the linear limit characteristics, see Figure 5.22 for the transfer function model. See Table 5.13 for sample parameter values.

For the arc limit characteristics, see Figure 5.23 for the transfer function model. See Table 5.14 for sample parameter values.

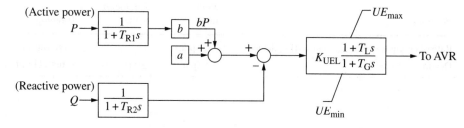

Figure 5.22 UEL transfer function model: ($Q = a + bP$ linear characteristics).

Table 5.13 UEL transfer function model parameter samples ($Q = a + bP$ linear characteristics).

Symbol (unit)	Parameter value	Meaning	Symbol (unit)	Parameter value	Meaning
$T_{R1}(s)$	0.06	Time constant for the P transducer	K_{UEL}	2	UEL gain
$T_{R2}(s)$	0.06	Time constant for the Q transducer	T_L (s)	0.87	Time constant for the leading phase compensation
UE^*_{max} (p.u.)	0.3	Upper output limit	T_G (s)	8.5	Time constant for the lagging phase compensation
UE^*_{min} (p.u.)	0	Lower output limit			

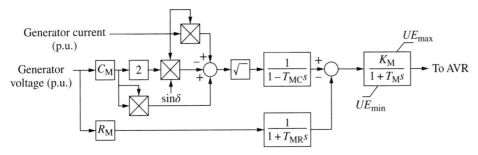

Figure 5.23 UEL transfer function model (arc limit characteristics): δ – power angle, C_M – central point of the boundary circle on Q-axle, R_M – radius of the boundary circle.

Table 5.14 UEL transfer function model parameter samples (arc limit characteristics).

Symbol (unit)	Parameter value	Meaning	Symbol (unit)	Parameter value	Meaning
T_{MC} (s)	0.02	Time constant for the filter	T_{MR} (s)	0.02	Time constant for the filter
UE^*_{max}	0.3	Upper output limit	T_M (s)	0.01	Time constant for the amplifier
UE^*_{min}	−0.3	Lower output limit (HV loop)	K_M	0.2	UEL gain

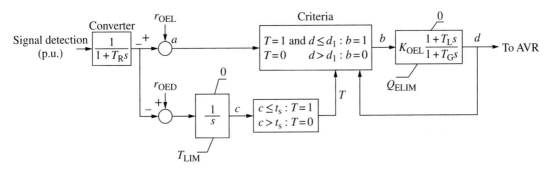

Figure 5.24 OEL transfer function model (Model 1).

5.3.5 Over-Excitation Limiter (OEL)

The OEL is used to prevent the generator rotor excitation winding from exceeding the heat capacity, with its detection signal varying with different excitation modes. These excitation modes include the following:

(1) Generator excitation current (static excitation mode, AC exciter mode).
(2) Generator excitation voltage (AC exciter mode).
(3) Exciter excitation current (AC exciter brushless excitation mode).

Figures 5.24 and 5.25 show two separate OEL models. See Tables 5.15 and 5.16 for the corresponding sample parameter values.

Due to the variety of available OEL circuits, when calculating a specific model, the real model should be obtained from the manufacturer if necessary.

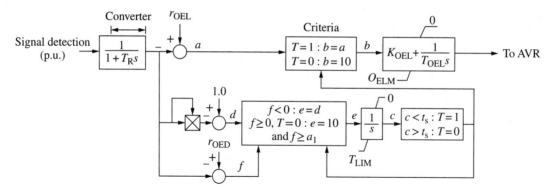

Figure 5.25 OEL transfer function model (Model 2).

Table 5.15 OEL transfer function model parameter samples (Model 1).

Symbol (unit)	Parameter value	Meaning	Symbol (unit)	Parameter value	Meaning
T_R (s)	0.04	Time constant for the transducer	a_1^* (p.u.)	–	Return difference value for operation reset
r_{OEL}^* (p.u.)	1.02	Limited target value	T_{LIM}^* (p.u.)	–3	Integral limit value
r_{OED}^* (p.u.)	1.05	Initial value for OEL operation	Q_{ELIM}^* (p.u.)	–0.6	Max. limit value
t_s	–0.75	Timing value for OEL operation	T_G (s)	5	Time constant for the lagging phase compensation
K_{OEL}	16	OEL gain	T_{OEL}	–	Integral time constant
T_L (s)	0.4	Time constant for the leading phase compensation	d_1	–	Min. detection value for guaranteed operation

Table 5.16 OEL transfer function model parameter samples (Model 2).

Symbol (unit)	Parameter value	Meaning	Symbol (unit)	Parameter value	Meaning
T_R (s)	0.02	Time constant for the converter	a_1^*	0.08	Return difference value for operation reset
r_{OEL}^*	1.0	Limited target value	T_{LIM}^*	–10	Integral limit value
r_{OED}^*	1.05	Initial value for OEL operation	Q_{ELIM}^*	–0.6	Max. limit value
t_s	–5	Timing value for OEL operation	T_G (s)	–	Time constant for the lagging phase compensation
K_{OEL}	4	OEL gain	T_{OEL}	0.25	Integral time constant
T_L (s)	–	Time constant for the leading phase compensation	d_1	–	Min. detection value for guaranteed operation

5.3.6 Power System Stabilizer (PSS)

At present, the input signals for the PSS as developed by each country have the following modes.

5.3.6.1 ΔP Type PSS

It utilizes the active power as the input signal for the PSS. When the active power decreases, the additional signal will regulate the generator excitation toward an increment. See Figure 5.26 for the circuit transfer function model of the ΔP type PSS. See Table 5.17 for the transfer function model sample parameter values.

For the PSS taking the shaft speed $\Delta\omega$ or the frequency Δf as input signal, in order to avoid resonance caused by the torsional vibration of the shaft system, a torsional notch filter should be employed, with its order given via mechanical computation by the host.

5.3.6.2 $\Delta\omega$ Type PSS

The PSS taking the generator rotor speed $\Delta\omega$ as the input signal has been applied in China's 300 MW and 600 MW steam turbine generator units produced by acquiring the technology from Westinghouse Electric Corp. The PSS developed by General Electric has also adopted $\Delta\omega$ as the input signal. When the generator speed decreases, the PSS will regulate excitation toward a decrement.

5.3.6.3 Δf Type PSS

The generator's terminal voltage or the internal electromotive force frequency Δf is taken as the input signal for the PSS. In this case, excitation decreases when the frequency drops.

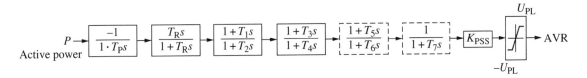

Figure 5.26 ΔP type transfer function model.

Table 5.17 PSS transfer function model parameter samples.[a]

Symbol (unit)	Parameter value	Meaning	Symbol (unit)	Parameter value	Meaning
T_P (s)	0.01	Time constant for the active power P detector	T_3 (s)	0.02–1	Time constant for the leading phase compensation
T_R (s)	1–10	Time constant for the isolation filter	T_4 (s)	0.02–1	Time constant for the lagging phase compensation
T_1 (s)	0.1–2	Time constant for the leading phase compensation	K_{PSS}	0.1–1	PSS gain
T_2 (s)	0.1–2	Time constant for the lagging phase compensation	U_{PL}^* (p.u.)	0.05–0.1	PSS output limit

a) It is applicable to the case in which Level-3 phase compensation is adopted and the case in which a noise blocking filter is adopted in the high-frequency stage.

5.4 Parameter Setting of Excitation System

5.4.1 Setting of Proportional AVR Parameters

We now describe how to set the parameters for the self-excitation system adopting series compensation that is shown in Figure 5.10 as an example.

5.4.1.1 AVR Steady-State Open-Loop Gain K_A

When the total voltage change ratio value for the generator is given, the corresponding steady-state gain K_A is

$$K_A \geq \frac{U_{fN} - U_{f0}}{U_{f0}} \times \frac{1}{\xi} \tag{5.36}$$

where

U_{f0} — the no-load rated excitation voltage of the generator;
U_{fN} — the rated excitation voltage of the generator;
ξ — the total voltage change ratio, which is generally 1%.

For the steam turbine generator unit, it is assumed that $U_{fN}/U_{f0} = 2.5$, $\xi = 1\%$, and K_A can be solved from Eq. (5.36): $K_A \geq 150$.

5.4.1.2 Setting of AVR Phase Compensation and Damping Loop Parameters

For the excitation control system shown in Figure 5.27, first, its parameters are determined on the basis of the frequency response characteristics.

(1) *Frequency response characteristics*: When assessing the excitation system stability by using the frequency response characteristics represented by the open-loop transfer function of the excitation control system including the generator, the corresponding index range is as follows (at this time the generator is in no-load operation):
Gain margin; 10–20dB;
Phase margin: 20°–80°
In addition, with respect to the dynamic performance of the excitation system, its assessment mode is as follows: when the open-loop frequency response gain is 0 dB, the cutoff frequency is ω_c. When the ω_c value increases, the dynamic performance of the control system will also increase, which is unfavorable for ensuring system stability.
(2) *Transient gain K_T*: The transient gain represented by the excitation voltage variation caused by an abrupt change in the generator voltage differs from the normal steady-state gain.

The thyristor excitation mode is fast, and the voltage response time of the excitation system is closely related to the transient gain. For the AC exciter mode, the voltage response ratio of the excitation system is controlled by the AVR ceiling voltage U_{Rmax} and the characteristics of the AC exciter.

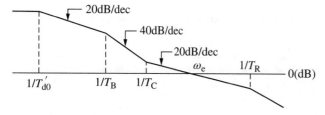

Figure 5.27 Bode plot of series compensation self-excitation system.

For the series compensation self-excitation system shown in Figure 5.10, if the time constant for the voltage detection loop is ignored, at this time the transient gain K_T can be expressed by the following equation:

$$K_T = K_A \times \frac{T_C}{T_B} \tag{5.37}$$

On the other hand, the value of K_T can be determined by the ratio of the ceiling voltage U_{Rmax} of the AVR output when the generator voltage drops by ΔU (generally 5%–10%), with its set point value as

$$K_T \geqslant U_{Rmax}/\Delta U \tag{5.38}$$

By determining the K_T value based on Eq. (5.38) and substituting it into Eq. (5.37), the T_C/T_B value can be obtained.

(3) Phase compensation time constant: Figure 5.27 shows the Bode plot of the excitation system which includes the generator; the relationship between the cutoff frequency ω_C in the figure and the transient gain K_T is

$$\frac{K_T}{T'_{d0}} \approx \omega_C \tag{5.39}$$

In Eq. (5.39), ω_C and T'_{d0} are known, from which one can obtain the K_T value. The value is compared to the K_T value obtained on the basis of Eqs. (5.37) and (5.38) to arrive at a proper value.

For a system without a PSS and with a bigger ω_c, in case an unstable case exists, K_T can also be determined on the basis of a reverse process; that is, K_T is first solved for using Eq. (5.39).

For a stable system, the amplitude response around ω_C must go through the 0 dB point with a slope of 20 dB dec^{-1}. In this case, the necessary phase and gain margin will be ensured, and the T_C value can be solved for from the following equation:

$$\frac{1}{T_C} \leqslant \frac{\omega_c}{n} \ (n \geqslant 2) \tag{5.40}$$

Specific values for T_C and T_B can be obtained from Eqs. (5.40) and (5.39). These three important factors are involved in the foregoing parameter selection and calculation such as the ceiling voltage, ω_c, and the transient gain, which affect the static and transient stability of the system as a whole.

As a matter of fact, after determining the major loop, first the ceiling voltage of the excitation system is determined (based on the power system stability requirements), then the excitation system stability is taken into consideration, based on which the coordination between the transient gain and phase is determined.

For the feedback compensation self-excitation system shown in Figure 5.11, the time constant T_A of the AVR amplification part is quite small. Therefore, it can equivalently be converted into the form shown in Figure 5.10. The damping loop constant can also be determined by the method specified in this section.

The phase compensation for the AC exciter mode is determined on the basis of excitation system stability, without necessarily considering the influence of the other factors.

However, for the AC exciter system with a high initial response, the configuration of ω_C is considered. After determining the applicable transfer function form, the target value for the open-loop transfer function ω_c is set. The stability requirements can be met when the slope for phase compensation is 20 dB dec^{-1} at ω_c.

5.4.2 Frequency Response and Step Response Calculation

See Table 5.18 for calculation examples of the thyristor self-excitation mode and the AC exciter mode. See Figures 5.28–5.35 for the relevant figures.

Table 5.18 Frequency response and step response calculation example.

SN	Excitation mode	Frequency response			Step response	Control constant setting instance
		Gain margin	Phase margin	Bode plot		
1	Thyristor self-excitation mode	23dB	60°	Figure 5.30	Figure 5.31 (±2% step response) Figure 5.32 (±10% step response)	Figure 5.28
2	AC exciter mode	16dB	44°	Figure 5.33	Figure 5.34 (±2% step response) Figure 5.35 (±10% step response)	Figure 5.29

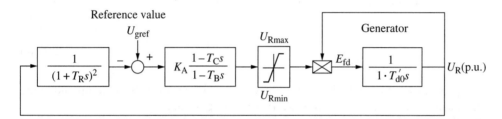

Figure 5.28 Control parameter setting example of thyristor self-excitation system.
$T_R = 0.013s$, $K_A = 150$, $T_B = 0.5s$, $T_C = 0.22s$,
$U^*_{Rmax} = +5.5$ (p.u), $U^*_{Rmin} = -4.8$ (p.u), $T'_{d0} = 6.0s$.

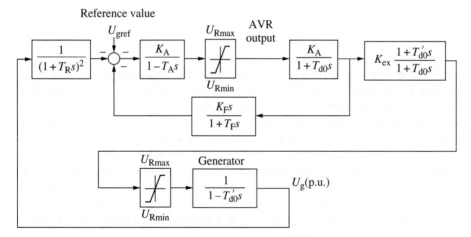

Figure 5.29 Control parameter setting example of AC exciter excitation system.
$T_R = 0.13s$, $K_A = 37.5$, $T_A = 0.01s$, $U^*_{Rmin} = 1.0$ (p.u), $U^*_{Rmin} = -0.8$ (p.u), $T_{ex} = 1.0s$,
$K_{ex} = 4$, $T'_{d0} = 6.0s$, $T_{ga} = 3.0s$, $U^*_{fRmin} = 5$ (p.u), $U^*_{fRmin} = 0$, $T_F = 0.5s$, $K_F = 0.12$.

5.4.3 PI AVR Parameter Setting

See Figure 5.36 for the typical transfer function block diagram of the PI self-excitation system.

The system has a larger time constant T_1 (the generator's time constant T'_{d0}) and a smaller time constant T_2 (the time constant of voltage measurement), and takes the primary inertial element in series as the controlled object to determine the parameters for the PI control system.

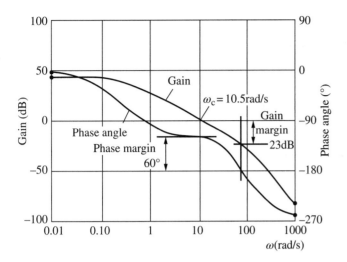

Figure 5.30 Frequency response characteristics of generator no-load thyristor self-excitation system (Bode plot).

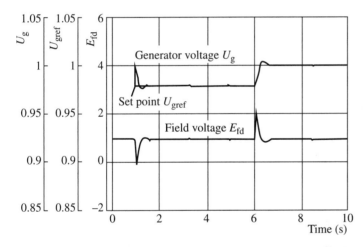

Figure 5.31 ±2% Step response characteristics of generator no-load thyristor self-excitation system.

The integral time is set as $T_i = 4T_2$ and the steady-state gain as $K_A = T_1/(2K_2T_2)$.

See Figure 5.37 for the Bode plot solved from the result of this setting. The proper gain margin and phase margin can be solved by correcting the T_i value.

When adopting a PSS, the ω_c value is generally set less than 10 rad s^{-1}.

For the thyristor self-excitation system shown in Figure 5.36, T_1, T_2, and K_2 in the figure separately correspond to T'_{do}, the time constant $T_R(0.02\,\text{s})$ for the voltage detection loop, and the thyristor rectifier gain $K_M(4.2)$.

Assume in Figure 5.36 that when $T_1 = T'_{do} = 5.2$s, the integral time constant for AVR T_i and the gain K_A for AVR will separately be

$$T_i = 4 \times T_2 = 0.08(\text{s})$$

$$K_A = \frac{T_1}{2K_2T_2} = \frac{5.2}{2 \times 0.02 \times 4.2} = 30.95 \tag{5.41}$$

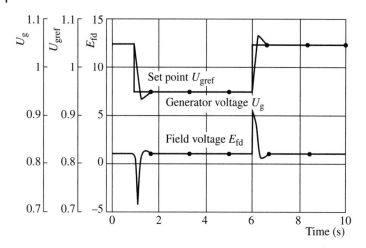

Figure 5.32 ±10% Step response characteristics of generator no-load thyristor self-excitation system.

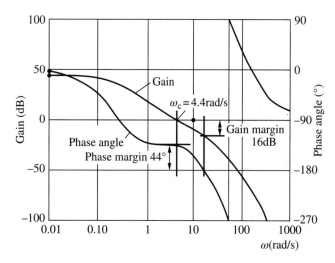

Figure 5.33 Frequency response characteristics of generator no-load AC exciter excitation system (Bode plot).

In this case, ω_c will be 22 rad s^{-1}. If a PSS is not used, a negative damping torque may be generated. Therefore, the ω_c value must be adjusted. See Figure 5.38 for the block diagram of the PI AVR system. The following are procedures relating to parameter setting.

5.4.3.1 AVR Gain K_A

The gain characteristic near ω_c is generally be determined by the generator and other gains on the basis of the following equation:

$$K_A \approx \omega_c \times \frac{T'_{do}}{K_2} \tag{5.42}$$

Setting $\omega_c = 6$ rad s^{-1}, the following is obtained by substitution into Eq. (5.42):

$$K_A = 6 \times \frac{5.2}{4.2} = 7.43$$

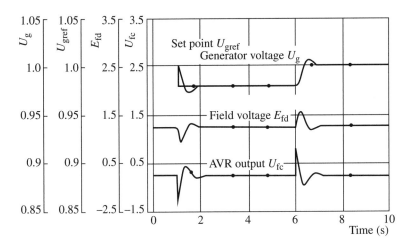

Figure 5.34 ±2% Step response characteristics of generator no-load AC exciter excitation system.

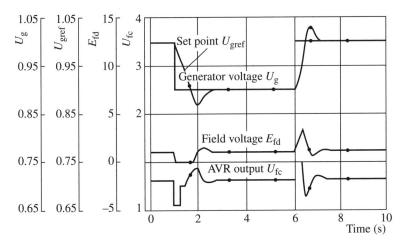

Figure 5.35 ±10% Step response characteristics of generator no-load AC exciter excitation system.

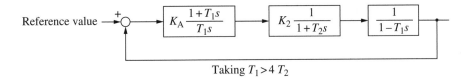

Figure 5.36 Typical transfer function block diagram of proportional-integral self-excitation system.

5.4.3.2 Integral Time Constant

In a frequency band less than ω_c, in order to ensure that 10 dB is taken as the slope for 20 dB dec^{-1}, according to

$$20\log(T_I \omega_c) = 10$$

From this, one can obtain $T_I = 0.527$ s.

For the brushless excitation system shown in Figure 5.39, the AVR's PI time constant and the relevant constants for the AC exciter can be set with the following procedures.

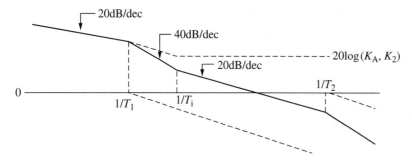

Figure 5.37 Bode plot of proportional-integral AVR system.

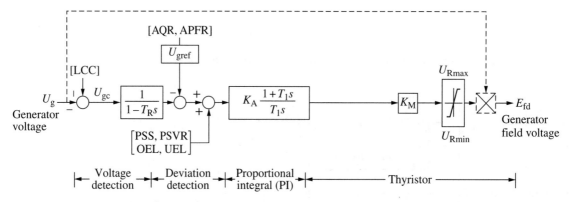

Figure 5.38 Proportional-integral AVR system block diagram.

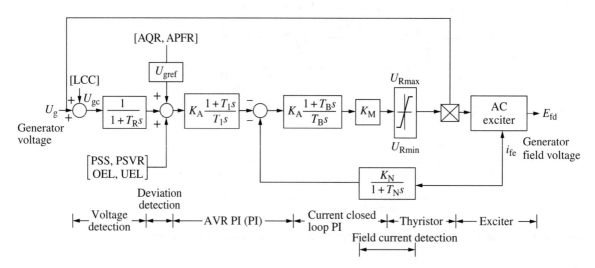

Figure 5.39 Proportional-integral type brushless excitation system block diagram.

(1) The setting of the current circuit constant K_B and T_B for the AC exciter: The constant setting in the AC exciter current circuit can still be used with the method specified in this section. However, in the closed-loop transfer function solved from the current loop, the sum of its equivalent first-order lag element time constant and first-order lag element time constant T_R (equivalent to T_2 in Figure 5.36), should be less than 1/4 of the generator time constant T'_{do} (with the target value of 1/10).

(2) AVR time constant T_I and gain K_A: The transfer function of the AC exciter current circuit can be approximately regarded as the first-order lag element. Its gain and time constant are equal to T_2 and K_2 in Figure 5.36. Therefore, the gain K_A and the integral time constant T_I of the AVR can be determined by using the same method that was applied to the thyristor self-excitation system.

5.4.4 PSS Parameter Setting

When the generator power swings, the PSS has the task of damping the power oscillation. The PSS input signals are diversified. In Japan, the active power ΔP is widely used as the input quantity for generators, and the rotational speed $\Delta \omega$ is widely adopted in the United States.

We will discuss the parameter setting mode here by taking ΔP as the input quantity of the PSS for our example.

To facilitate the discussion, we will now consider the single machine versus the infinite system shown in Figure 5.40a as an example.

Figure 5.40a represents a system model with a single synchronous generator on an infinite bus bar with a capacity much greater than that of the generator. Figure 5.40b shows the synchronous generator linearization block diagram when the model in Figure 5.40a changes slightly. Figure 5.40c refers to the change in the synchronizing torque and damping torque that the excitation system generates in case of frequency variation. Figure 5.40d refers to the synchronous generator block diagram after the synchronous generator and the excitation system are transformed equivalently.

In essence, it would be ideal to take the accelerating torque ΔT_a in Figure 5.40b as the PSS input signal. However, it is quite difficult to measure ΔT_a. Therefore, when the mechanical input ΔT_m is not changed, the electrical output torque $\Delta T_e \approx \Delta P$ is usually taken as the PSS input signal. By regulating the PSS gain and phase and taking its output as the AVR's additional signal, the damping coefficient $K_d(\omega)$ can be increased to damp power oscillation.

The PSS transfer function is set as follows:

The PSS part, indicated by the dashed line in Figure 5.40, is represented by the real transfer function, shown in Figure 5.41.

In Figure 5.41, T_p is the time constant for the detector, the signal washout filters the DC component, and the alternating component can pass through the isolation filter.

The alternating component of the active power ΔP compensates for the phase via the lead-lag phase regulation loop, which is then added to the AVR combiner via the gain and PSS output limit loop.

5.4.4.1 Setting of Time Constant T_g and Gain K_{PSS} for the Washout Filter

The PSS output is proportional to the rate of change of the generator output power over time, the washout time constant T_g, and the product of each parameter for the gain K_{PSS}. The generator output is subject to changes in the prime mover output. However, in the case of regular operation, the prime mover output will remain unchanged, from which T_g and K_{PSS} can be determined.

We shall set a small value for the time constant T_g of the washout filter. For the oscillation frequency of the power system, the rather small corresponding gain and large phase change are very beneficial for forming an arresting damping torque, and T_g generally is set between 1–5 s.

An unstable state will occur if K_{PSS} is too large, and thus generally a value less than 1.0 is selected.

5.4.4.2 Phase Compensation

From Figure 5.40: From the standpoint of phase compensation, it will be most beneficial if the electrical torque ΔT_{ex} in the output quantity of the PSS that is connected to the excitation system and $\Delta \omega$ are in the same phase.

When the oscillation frequency of the power system is ω_n and the PSS input signal is $-\Delta P$, the electrical torque connected to the excitation system will be ΔT_{ex}. See Figure 5.42 for the relationship of each phasor diagram.

In Figure 5.42, the PSS signal phase is regulated as the leading phase, with the excitation system signal θ_{AVR} being the lagging phase. The generator excitation system loop signal is also known as the lagging phase angle.

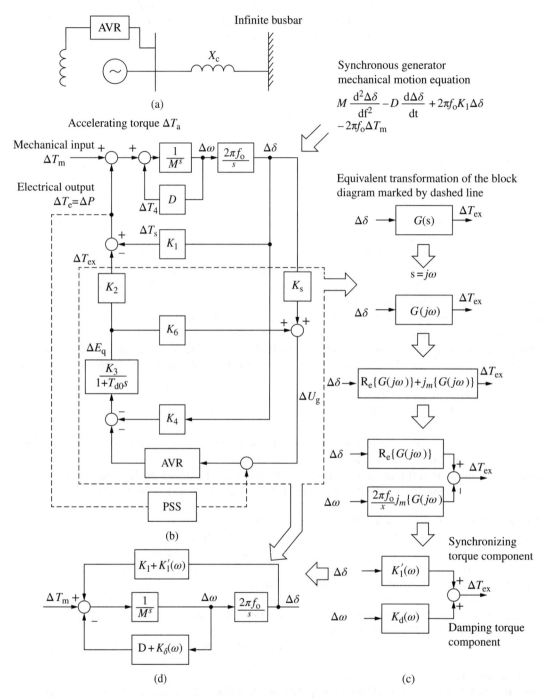

Figure 5.40 Synchronous generator excitation control system block diagram: (a) the single machine infinite bus bar configuration, (b) the synchronous generator linear block diagram, (c) the synchronizing and damping torques provided by the excitation system, (d) the synchronous generator block diagram upon equivalent transformation.

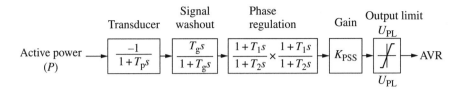

Figure 5.41 PSS transfer function diagram.

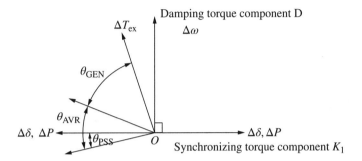

Figure 5.42 Excitation system phasor diagram when $\omega = \omega_n$.

Figure 5.43 System parameters. $X_d = 1.6$ p.u., $X'_d = 0.24$ p.u., $X''_d = 0.2$ p.u., $X_q = 1.6$ p.u., $X''_q = 0.2$ p.u., $X_t = 0.16$ p.u.

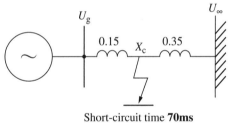

Short-circuit time **70ms**

$P_g = 0.9$ p.u.
$Q_g = 0.0$ p.u.
$U_g = 1.0$ p.u.

5.4.4.3 Gain and Output Limit

The value of the upper PSS output limit is ±5% to ±10% when reduced to the generator voltage end. Obviously, the PSS output limit can reduce its influence on power oscillation damping, which will be considered as a whole.

5.4.4.4 Setting of PSS Gain and Phase Compensation

In general, by using the computer to calculate the single machine infinite bus bar system, the damping torque can be solved for on the basis of the load frequency response. Sometimes it can calculate each analog quantity in the time domain, and it will also carry out operations for a multi-machine system if necessary.

Figure 5.43 shows the configuration of the single machine versus the infinite bus bar system. Figure 5.44 shows the calculation examples of the PSS loop constant setting. See Table 5.19 for the calculation results.

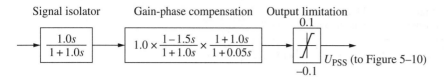

Figure 5.44 PSS transfer function parameter setting.

Table 5.19 PSS calculation results.

SN	Calculation content	Result	Thyristor self-excitation system constant	PSS constant	Model and generator constant
1	Frequency response in case of load	Figure 5.45	Figure 5.28	Figure 5.44	Figure 5.43
2	Time response	Figure 5.46	Figure 5.28		

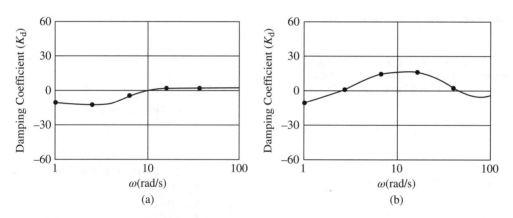

Figure 5.45 Analog system frequency response (determination of damping coefficient): (a) thyristor self-excitation mode (without PSS), (b) thyristor self-excitation mode (with PSS).

Figure 5.45a shows the calculation results for the load frequency response characteristics when the PSS is disabled. Figure 5.45b shows the calculation results for the load frequency response characteristics when the PSS is used. According to the calculation results, when the power system oscillation frequency is around 7 rad s^{-1}, the damping coefficient K_d can be greatly improved.

Figure 5.46a shows the corresponding results in the time domain when the PSS is disabled. Figure 5.46b shows the corresponding results in the time domain when the PSS is used. According to the figures, the oscillation of the power system is divergent when the PSS is not used, and is rapidly convergent when the PSS is used.

It is important to note that in a large hydroelectric generating set, if the PSS utilizes a single signal (such as the active power), when the active power of the generator changes, anti-regulation will occur with the introduction of reactive power during this process. That is, when the active power increases, the reactive power will drop

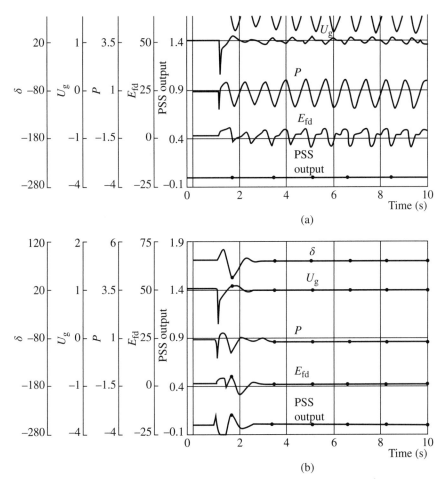

Figure 5.46 Time domain response of analog system: (a) thyristor self-excitation mode (without PSS), (b) thyristor self-excitation mode (with PSS).

accordingly. On the other hand, when the active power drops, the reactive power will increase. To avoid this reactive anti-regulation, the power system stabilizers that are used for a large hydroelectric generating set usually utilize a dual-input control signal; for instance, apart from the active power, they will also be supplied a rotational speed or frequency signal, which in the typical PSS-2A and PSS-3B models are the dual-input signal control system.

6

Basic Characteristics of Three-Phase Bridge Rectifier Circuit

6.1 Overview

In a synchronous generator excitation system, generally a power converter must be employed to convert AC into DC or to convert DC to AC.

When the power converter consists of diode rectifying elements, it can convert only AC into DC; the corresponding power converter is known as a rectifier.

If the power converter utilizes thyristor rectifying elements, the power conversion of the device will be two-way; that is, it can either convert AC to DC or convert DC to AC. The former is known as the rectifying state, and the latter is known as the inverting state. In this case, the power converter is also called a current converter since it is compatible with both rectification and inversion.

Among the various current converter types, the three-phase bridge is the most universally applied. In modern synchronous generator excitation systems, the three-phase bridge has currently become the only selected converter type. As compared to other converter types, the three-phase bridge has the following advantages:

(1) Under the same DC voltage conditions, the peak reverse voltage (PRV) withstood by the rectifying element in the off state is 1.05 times the DC voltage, which is only half those of the other types.
(2) If the commutation power is fixed, the primary winding capacity of the converter transformer is no more than the corresponding capacities of the other types; however, the secondary winding capacity is less than the corresponding capacities of the other types.
(3) The converter transformer has a simple connection, which can facilitate the insulating treatment of the transformer.
(4) If the same DC power conditions are guaranteed, the elements needed by the three-phase bridge will require the minimum volt-ampere capacity.
(5) It is characterized by a smaller DC voltage ripple value.

6.2 Operating Principle of Three-Phase Bridge Rectifier

Figure 6.1 shows the equivalent circuit of a three-phase bridge rectifier [11]. In the diagram, e_a, e_b, and e_c are the equivalent potentials of the three-phase AC power supply system, and L_γ refers to the equivalent commutation inductance for each phase of the AC system, which is from the power supply to the AC input of the rectifier bridge. To facilitate the analysis, the equivalent resistance of the AC system can be ignored.

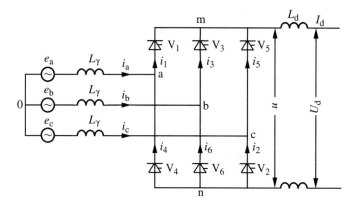

Figure 6.1 Equivalent circuit diagram for the three-phase bridge rectifier.

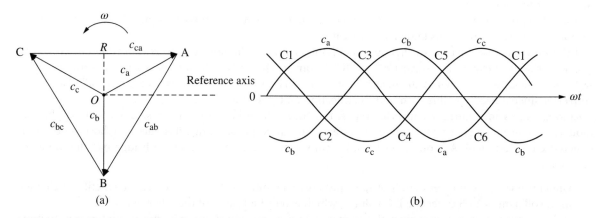

Figure 6.2 Three-phase equivalent potential change: (a) phasor diagram, (b) oscillogram.

If the AC potential phasor e_{ca} of the system is taken as the reference, according to Figure 6.2a, the instantaneous value of power phase potential is

$$\left. \begin{array}{l} e_a = e_{0a} = \sqrt{\dfrac{2}{3}} E \sin(\omega t + 30°) \\[4pt] e_b = e_{0b} = \sqrt{\dfrac{2}{3}} E \sin(\omega t - 90°) \\[4pt] e_c = e_{0c} \sqrt{\dfrac{2}{3}} E \sin(\omega t + 150°) \end{array} \right\} \quad (6.1)$$

where

E – RMS value of power line potential.

Figure 6.2a shows the phasor diagram for each phase potential in the above equation. Figure 6.2b shows the waveform changes in the corresponding phase potentials, with the ordinate representing the potential at the neutral point of the AC system, when all the six rectifying elements are in the off state. The waveform shown in Figure 6.2b refers to the potential changes at the phase potential "a," "b," and "c" at the a–c terminal of the rectifier bridge versus the neutral point.

The line potential corresponding to the phase potential in Eq. (6.1) is expressed as

$$\left.\begin{array}{l}e_{ca} = e_{c0} + e_{0a} = e_a - e_c = \sqrt{2}E\sin\omega t \\ e_{ab} = e_{a0} + e_{0b} = e_b - e_a = \sqrt{2}E\sin(\omega t - 120°) \\ e_{bc} = e_{b0} = e_{0c} = e_c - e_b = \sqrt{2}E\sin(\omega t + 120°)\end{array}\right\} \quad (6.2)$$

Figure 6.2a also shows the phasor diagram of the line potential. The zero crossing point (ZCP) of each line potential is determined by the crossing point of the two adjacent phase potentials. For example, a ZCP with the line potential e_{ca} changing from a negative value to a positive value will be at the crossing point C1 of the phase potentials e_a and e_c in the positive half-wave. In the meantime, the ZCPs for the other line potentials e_{ab} and e_{bc} will be at C2 and C3, respectively. In the case of symmetrical system potential, the phase interval between each line potential ZCP will be 60°.

6.3 Type I Commutation State

Under normal operating conditions, the rectifying element of the rectifier cathode connected to the maximum potential and the rectifying element with the anode connected to the minimum potential can form a closed circuit. When the potential of the conducting phase is the same as the potential of the adjacent phase to be operated, the commutation process will start, and the rectified current will be changed from a conducting phase to the adjacent phase, until the end of the commutation. Depending on the different time intervals for commutating current transfer, the external characteristics expressing the relationship between the rectified voltage and the rectified current can be divided into three operating states, namely, Types I, II, and III commutation states. In the case of the Type I phase commutation state, the operating process of the rectifier in each cycle can be divided into two zones: a non-commutating interval and a commutating interval.

In the non-commutating interval, both the top half and bottom half of the bridge will have one element conducting the rectified current. In the commutation state, the two adjacent elements of the top or bottom half will commutate, changing the rectified current from the original conducting phase to the adjacent phase, with the other element that is in the non-commutating state still conducting the rectified current. This cycle will repeat; two or three elements always work alternately. Such a state is known as the Type I commutation state for rectifiers, or the 2–3 commutation state. See Figure 6.3 for the working process of the Type I commutation state.

Under the initial conditions, the rectifying element V5, which is located in the c-phase of the top half, and the rectifying element V6, which is located in the b-phase bottom half, are in the on state, as shown in Figure 6.3a. See Figure 6.3a for its equivalent circuit.

Under non-commutating conditions, the rectified circuit outputs stable DC current I_d, which cannot generate a voltage drop on the equivalent commutating reactance I_γ. Therefore, the rectified voltage u_d is the same as the AC line potential of the rectifier bridge in terms of waveform, as shown in Figure 6.5.

Figure 6.5 shows the voltage waveform from the positive and negative terminals of the rectified voltage when the control angle of the thyristor rectifier is α and the commutation angle is γ to m, n, the neutral point, indicated by the solid-line-shaded part in Figure 6.5a.

The longitudinal length between u_m and u_n represents the instantaneous value of the rectified voltage u_d, as shown by the shaded part in Figure 6.5b, with the area ΔA referring to the rectification area loss due to the rectifying element commutation.

Figures 6.5c,d individually refer to the oscillograms for the rectified current and the AC-side current of the element.

By taking the three-phase bridge rectifier circuit as an example, we will discuss the Type I commutation process when $0 < \gamma < 60°$.

For the three-phase bridge circuit as shown in Figure 6.3a, when the rectifier is working in the non-commutating interval, both the top and bottom half have one element that is in the on state; for example, two elements such as V5 and V6 are in the on state, with its equivalent circuit shown in Figure 6.4a.

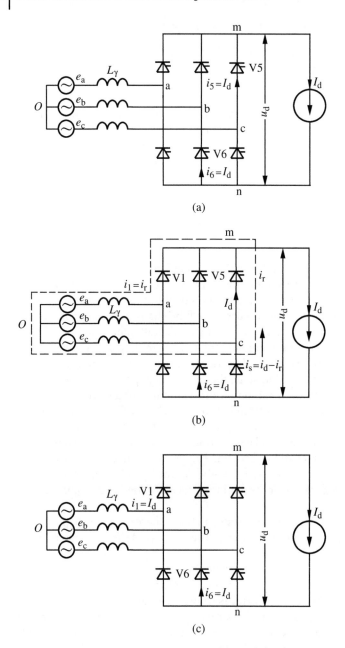

Figure 6.3 Equivalent circuit in Type I commutation state: (a) rectifying elements V5 and V6 are in the on state, (b) elements V5 and V1 are commutating, (c) elements V6 and V1 are in the on state.

When $\omega t = \alpha$ (α is the trigger control angle), rectifying elements V1 and V5 will begin to commutate, until the current flowing through V5 is zero, but the current flowing through V1 is the total current.

See Figure 6.4b for the equivalent circuit when elements V5 and V1 are commutating.

Therefore, the following equation applies to the equivalent circuit:

$$L_\gamma \frac{di_1}{dt} - L_\gamma \frac{di_5}{dt} = e_a - e_c \tag{6.3}$$

In the above loop, because $e_a > e_c$, the direction of current i_γ flows from point a to point c; therefore:

$$i_1 = i_\gamma, \quad i_5 = I_d - i_\gamma \tag{6.4}$$

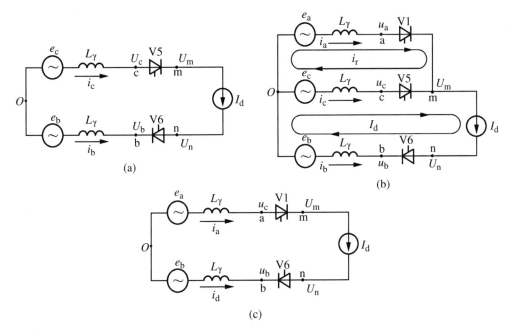

Figure 6.4 Commutation equivalent circuit diagram corresponding to Figure 6.3: (a) V5 and V6 are in the on state, (b) V5 and V1 are commutating, (c) V6 and V1 are in the on state.

where

I_d — total current of the rectifying loop.

By substituting Eq. (6.4) into Eq. (6.3), we obtain

$$L_\gamma \frac{di_\gamma}{dt} - L_\gamma \frac{d(I_d - i_\gamma)}{dt} = e_a - e_c \tag{6.5}$$

which becomes

$$2L_\gamma \frac{di_\gamma}{dt} = \sqrt{2}E \sin \omega t \tag{6.6}$$

Integrating the above equation:

$$i_\gamma = \frac{\sqrt{2}E}{2\omega L_\gamma} \cos \omega t + A = -\frac{\sqrt{2}E}{2X_\gamma} \cos \omega t + A = -I_{S2} \cos \omega t + A \tag{6.7}$$

$$I_{S2} = \frac{\sqrt{2}E}{2\omega L_\gamma} \quad X_\gamma = \omega L_\gamma$$

where

X_γ — commutating reactance of each phase between the AC power supply and the AC input of the rectifier bridge

ω — angular frequency for the fundamental wave of the AC system;

I_{S2} — the forced component amplitude of the short circuit when the AC system is two-phase shorted on the AC side of the current converter;

A — integral constant;

E — RMS value for the AC power line voltage

Since the time when $\omega t = \alpha$, the rectifying loop will transfer from the state where V5 and V6 are on to the state where another element among V5, V6, and V1 is conducting. The current will not abruptly change at the moment that V6 and V1 elements are commutating. That is $i_1 = i_\gamma = 0$; thus, from Eq. (6.7), we obtain

$$A = \frac{\sqrt{2}E}{2X_\gamma}\cos\alpha = I_{S2}\cos\alpha \tag{6.8}$$

Substituting Eq. (6.8) into Eq. (6.7) yields

$$i_1 = i_\gamma = \frac{\sqrt{2}E}{2X_\gamma}(\cos\alpha - \cos\omega t) = I_{S2}(\cos\alpha - \cos\omega t) \tag{6.9}$$

From Eq. (6.9), we see that i_γ actually denotes the short-circuit current value when the AC system is two-phase shorted at points c and after the rectifying element V1 is in the on state.

In Eq. (6.9), Item 1 refers to the free component (DC component) of the short circuit, and Item 2 refers to the forced component (power frequency AC component).

In addition:

$$i_5 = I_d - i_\gamma \tag{6.10}$$

See Figure 6.6 for the changes in commutating phase currents i_5 and i_1 in the commutation process.

It should be noted that the direction of the commutating current i_γ as specified in the above analysis is only an assumption for simplifying the analysis. As a matter of fact, the rectifying element V5 cannot flow through the reverse current; the current flowing through V5 during the commutation is the total current $I_d - i_\gamma$. At the same time, when $I_d > i_\gamma$, the direction of the total current will still conform to the unilateral conduction characteristics of the rectifying element.

From Figures (6.6) and (6.9), we see that as ωt increases, the current flowing through V1 will increase gradually, and the current flowing through V5 will decrease gradually. In the time interval corresponding to the commutation angle γ, the current i_1 will increase to I_d. Therefore, when $\omega t = \alpha + \gamma$, we can substitute $i_1 = I_d$ into Eq. (6.9), obtaining

$$\begin{aligned} i_1 = i_\gamma &= \frac{\sqrt{2}E}{2X_\gamma}[\cos\alpha - \cos(\alpha + \gamma)] \\ &= I_{S2}[\cos\alpha - \cos(\alpha + \gamma)] = I_d \end{aligned} \tag{6.11}$$

Moreover,

$$i_5 = I_d - i_\gamma = 0 \tag{6.12}$$

Due to the unilateral conduction characteristics of the rectifying element, the current flowing through i_5 will drop to zero. When $\omega t = a + \gamma$, V5 will switch off.

When the rectifying element V5 switches off, the current converter will again change from the state where the three elements V5, V6, and V1 are on to the state where V5 and V1 are on, as shown in Figures 6.1c and 6.4c. Its cathode is connected to point "a" via the conducting V1. Therefore, the cathode potential of V5 is e_a, and the anode potential is e_c. When $e_a > e_c$, V5 will withstand a reverse voltage, and V5 is reliably switched off at point D5.

The period from the moment when V1 starts conducting instantly at point T1 to the moment when V5 switches off at point D5 is called the commutation process, during which the DC current I_d flows through V5 via the "c" phase, transfers to the "a" phase, and flows through V1. At the same time, the rectifier's connection to the line potential e_{bc} via the original two conducting elements V5 and V6 changes to the connection to the line potential e_{ba} via the two conducting elements V6 and V1.

Similarly, at point C2 in Figure 6.5, when $\omega t = 60°$, V1 for the cathode assembly in the upper half-bridge of the rectifier is still conducting, but the cathode potential e_c of V2 for the anode assembled in the lower half-bridge is lower than the anode potential e_b. Because of the trigger pulse action of the same phase interval,

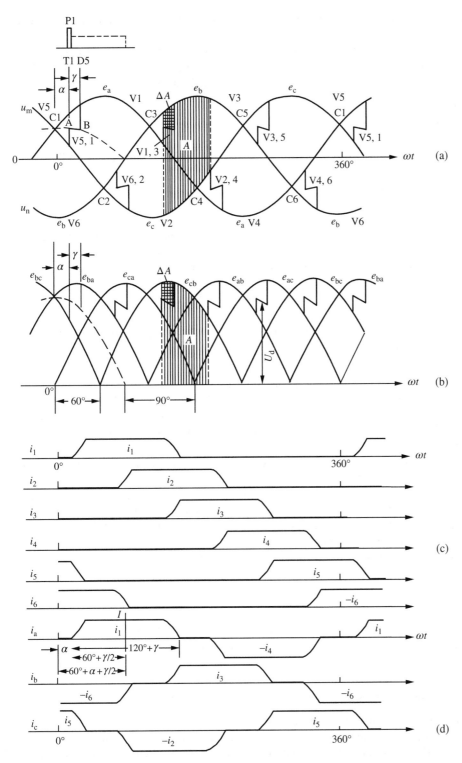

Figure 6.5 Rectifier current and voltage waveform: (a) voltage waveforms at positive and negative terminals "m," "n" of the rectified voltage to the neutral point, (b) waveform of the rectified voltage U_d, (c) current waveform of the rectifying element, (d) alternating current waveform.

V2 conducts by triggering when $\omega t = 60° + \alpha$, then starts commutating with V6, and all the other elements also commutate in turn.

In essence, the role of a controlled rectifier is equivalent to that of a set of six controllable electric switches. By the action of different triggering signals, the two phases (three phases in the case of commutation) of the power source at the three-phase AC side will be connected to the rectifier in turn, with AC converted to DC.

6.4 Commutation Angle

In the commutation process of the rectifier, the commutation angle γ is a critical parameter; from Eq. (6.11), we see that

$$\gamma = -\alpha + \cos^{-1}\left(\cos\alpha - \frac{2X_\gamma I_d}{\sqrt{2E}}\right) \tag{6.13}$$

When the parameters E, X_γ, and α are unchanged, the commutation angle γ will increase with as the rectified current I_d increases. When the commutation voltage E drops, the control angle α decreases; or when the commutating reactance X_γ and the rectified current I_d increase, the commutation angle γ will also increase accordingly.

When the commutation angles γ are varied, the number of elements that conduct while the current converter is working will also change accordingly.

Figure 6.7 shows that for the six-pulse bridge rectifier, when the commutation angle γ is different, the number of elements conducting will change at the same time.

From Figure 6.7:

(1) When $0 < \gamma < 60°$, the rectifier works in the Type I commutation state; that is, only two elements conduct at the same time in the non-commutating interval, and three elements work at the same time in the commutating interval, hence the name 2–3 mode. At the boundary state, when $\gamma = 0$, two elements conduct at the same time; however, when $\gamma = 60°$, three elements conduct at the same time, free from the non-commutating state.
(2) When $\gamma = 60°$ is unchanged, the rectifier works in the Type II commutation state, in which three elements always conduct at the same time at any time, working in the 3 mode.
(3) When $60° < \gamma < 120°$, the rectifier works in the Type III commutation state; that is, three and four elements always conduct at the same time alternately, working in the 3–4 mode. At the boundary state, when $\gamma = 60°$, three elements conduct at the same time, working in the 3 mode; however, when $\gamma = 120°$, four elements conduct at the same time, working in the 4 mode.
(4) It should be noted that the 2–3 mode is normal, but the 3–4 mode is abnormal.

6.5 Average Rectified Voltage

For the six-pulse bridge rectifier circuit, the average value of the DC voltage output by the rectifier will vary depending on the different trigger control angles α and commutation angles γ, which will be discussed separately for the following cases.

6.5.1 Zero Rectifier Control Angle α and Commutation Angle γ

See Figure 6.8 for the corresponding rectified voltage waveform.

When the control angle $\alpha = 0$, the rectifier will conduct at the natural commutation point C1, and due to the commutation angle $\gamma = 0$, the commutation occurs and is done instantly. Before the commutation point C1, elements V5 and V6 conduct, and therefore

$$U_m = U_c = e_c$$
$$U_n = U_b = e_b$$

The terminal voltage on the DC side of the rectifier:

$$U_d = U_m - U_n = e_c - e_b = e_{bc}$$

By that analogy, each element will commutate at the natural commutation point in turn, that is, the line voltage ZCP.

Since the sum of the voltages for the neutral points at the DC terminals "m," "n" for the rectifier equals the rectified voltage U_d, the corresponding waveforms are shown as the solid-line-shaded part in Figures 6.8a,b. There are six pulses in one cycle of the rectifier power supply voltage frequency. Therefore, the pulse frequency of the output rectified voltage for the three-phase bridge rectifier circuit will be $6f$, and f is the power supply voltage frequency.

Since in one cycle, the rectified voltage consists of six identical partial sine curve segments, when determining its average value, just one of the segments need be taken for calculation. It is assumed that the reference axis Y–Y of the vertical coordinates takes the zero point of the online voltage e_{ba} curve; then the line voltage e_{ba} is expressed by the equation $e_{ba} = \sqrt{2}E \cos\theta$. The rectified voltage area in this segment is be solved by using the following integral equation, with the integral range $-\frac{\pi}{6} - \frac{\pi}{6}$.

$$A_0 = \int_{-\frac{\pi}{6}}^{\frac{\pi}{6}} \sqrt{2}\, E \cos\omega t \; d\,(\omega t) = \sqrt{2}E[\sin \omega t]_{-\frac{\pi}{6}}^{\frac{\pi}{6}} = \sqrt{2}E$$

When A_0 is divided by $\frac{\pi}{3}$, the average rectified voltage can be solved for $\alpha = 0$ and $\gamma = 0$; that is,

$$U_{d0} = \frac{A_0}{\frac{\pi}{3}} = \frac{3\sqrt{2}}{\pi}E = 1.35E \tag{6.14}$$

where

E — voltage rms value for the power line.

When the load current I_d is zero, the commutation angle γ is also zero, and therefore the voltage indicated in Eq. (6.14) can also be regarded as the no-load DC voltage output value for the rectifier. See Figures 6.8c,d for the waveforms of the current flowing through each element when loaded and the AC-side current. The conductive width of each current will be an electrical angle of 120°.

6.5.2 $\alpha > 0, \gamma = 0$, for Zero Rectifier Load

In this case, see Figure 6.9 for the waveforms for u_m, u_n, and u_d. As compared to Figure 6.8, in the control angle α (C1–T1) interval, element V5 does not commutate to V1 immediately, and the DC voltage will still be determined by the line voltage e_{bc}. In the control angle α interval, a gap occurs in the DC voltage waveforms, and therefore the average rectified voltage will be less than the above U_{d0} in this case.

However, when taking the 1/6 waveform in a cycle to calculate the average rectified voltage, the upper and lower integral limits differ from the above ones. The integral range is between T1 and T2, and its area is

$$A = \int_{-\left(\frac{\pi}{6}-\alpha\right)}^{\frac{\pi}{6}+\alpha} \sqrt{2}E \cos\omega t \; d\,(\omega t) = 2\sqrt{2}E \sin\frac{\pi}{6}\cos\alpha = \sqrt{2}E \cos\alpha$$

The average rectified voltage is

$$U_d = \frac{A}{\frac{\pi}{3}} = \frac{3\sqrt{2}}{\pi}E\cos\alpha = U_{d0}\cos\alpha \tag{6.15}$$

6.5.3 $\alpha > 0, \gamma = 0$, in Controllable and Loaded State

The rectified voltage waveform is different from the above case; in the commutation angle γ interval, that is, in T1–D5 of Figure 6.5a, as elements V5 and V1 are commutating, which makes the rectifier two-phase shorted at AC terminals c and a, the line voltage e_{ca} fully drops onto the two-phase commutating reactance $2X_\gamma$, and the voltage drop of each phase X_γ is $e_{ca}/2$. Therefore, the positive potential "m" of the rectifier will be the midpoint, located at half of the sum of the two curves e_c and e_a. See Figure 6.5a for the real segment AB.

In this case, the average rectified voltage can be written as

$$U_d = \frac{1}{\frac{\pi}{3}}(A - \Delta A) = U_{d0}\cos\alpha - \Delta U \tag{6.16}$$

where

A   – all the shaded area under the e_{cb} curve in Figure 6.5b;

ΔA – voltage drop area due to commutation, as shown by the rectangular shaded area in Figure 6.5b;

ΔU – voltage drop for the average DC voltage due to commutation.

We will solve the average voltage for each part separately.

From Figures 6.5a,b, we see that the length of the vertical coordinates for the ΔA part equals half the instantaneous line voltage value between the two commutating phases (phases "a" and "b" in the figure), which is $\frac{1}{2}\sqrt{2}E\sin\omega t$, yielding the following:

$$\Delta A = \int_\alpha^{\alpha+\gamma} \frac{1}{2}\sqrt{2}E\sin\omega t\ d(\omega t) = \frac{\sqrt{2}}{2}E[\cos\alpha - \cos(\alpha+\gamma)] \tag{6.17}$$

The average commutation voltage drop:

$$\begin{aligned}\Delta U &= \frac{\Delta A}{\frac{\pi}{3}} = \frac{3\sqrt{2}}{2\pi}[\cos\alpha - \cos(\alpha+\gamma)] \\ &= \frac{U_{d0}}{2}[\cos\alpha - \cos(\alpha+\gamma)] \\ &= U_{d0}\sin\left(\alpha + \frac{\gamma}{2}\right)\sin\frac{\gamma}{2}\end{aligned} \tag{6.18}$$

From Eq. (6.11), the $[\cos\alpha - \cos(\alpha+\gamma)]$ value can be obtained; after simplification, the following can be solved by substitution into the above equation:

$$\Delta U = \frac{3\omega L_\gamma}{\pi} \times I_d = \frac{3}{\pi}X_\gamma I_d = 6fL_\gamma I_d = d_\gamma I_d \tag{6.19}$$

$$d_\gamma = \frac{3\omega L_\gamma}{\pi}$$

wherein

f – supply frequency in, Hz;

d_γ – voltage drop generated by the unit current during commutation.

d_γ – sometimes denoted by R_γ and known as the equivalent commutating resistance. Its physical meaning is the equivalent DC voltage drop corresponding to the equivalent AC voltage drop generated by the commutating current on the commutating reactance during commutation. This resistance does not consume active power.

By substituting Eqs. (6.18) and (6.19) into Eq. (6.16), various expressions can be obtained for the average rectified voltage values with the load and under controllable conditions:

$$\begin{aligned}U_d &= \frac{U_{d0}}{2}[\cos\alpha + \cos(\alpha+\gamma)] \\ &= U_{d0}\cos\left(\alpha + \frac{\gamma}{2}\right)\cos\frac{\gamma}{2} \\ &= U_{d0}\cos\alpha - \frac{3X_\gamma}{\pi}I_d \\ &= U_{d0}\cos\alpha - d_\gamma I_d\end{aligned} \tag{6.20}$$

By substituting Eq. (6.14) into Eq. (6.20), an expression can be obtained that relates the average rectified voltage value to the control angle α and the commutation angle γ:

$$U_d = \frac{U_{d0}}{2}[\cos\alpha + \cos(\alpha + \gamma)]$$
$$= \frac{3\sqrt{2}}{2\pi} E[\cos\alpha + \cos(\alpha + \gamma)] \quad (6.21)$$

6.6 Instantaneous Rectified Voltage Value

The voltage withstood by the rectifying element can be determined by the element anode and cathode voltages relative to the neutral point potential.

Taking the element V1 shown in Figure 6.10a, for instance, the voltage curve at its anode voltage u_n and cathode voltage u_m that is relative to the neutral point O will be the waveform for the voltage withstood by the rectifying element V1. In Figure 6.10a, the anode voltage of V1 is denoted with a dashed line and the cathode voltage of V1 with a solid line. The difference between their vertical coordinates is the voltage between the anode and cathode of V1.

In the C1–T1 interval before the conduction of V1, it will be in the blocking period that is under the forward voltage action. Its anode voltage will equal e_a, and the cathode voltage will equal e_c (since V5 is conducting, the potential at point "m" will be the same as that at point "c"; see Figure 6.1). The step value for the forward blocking voltage withstood by V1 will be $U_{ca} = \sqrt{2}E_{ca} \sin\alpha$; see segment D5 in Figure 6.10b. After V1 conducts at point D5, the conduction angle during the conduction period will be 120° and to point T3, V1 and V3 commutate to point D1, and the commutation is completed. After that, the anode voltage of V1 equals e_b, and as $e_b > e_a$, V1 withstands a reverse blocking voltage, and the step value for the reverse turn-off voltage at D1 is $U_{ab} = \sqrt{2}\sin(\alpha + \gamma)$. After that, see the part marked with a solid line in Figure 6.10b for the waveform for the reverse voltage withstood based on 240° from D1 → C1.

6.7 Effective Element Current Value

Figure 6.5c shows the current oscillogram of each rectifying element. See Figure 6.6 for the wave forms in the current-rising and current-falling segments of the two elements involved in the commutating during the commutation period. In the case of conduction without commutation, the current flowing through each element will equal I_d. The conduction time period in which the current flows through each element will be $120° + \gamma$.

First, the effective current value for the rectifying element is determined when $\gamma = 0$. In this case, the waveform of the current flowing through each element will be a square wave of 120° width, which is unrelated to the control angle α. From to Figure 6.8c, the valid current value of the element can be determined as

$$I_{V0} = \sqrt{\frac{1}{2\pi}\int_0^{\frac{2}{3}\pi} I_d^2 d\omega t} = \sqrt{\frac{1}{2\pi}\left(\frac{2\pi}{3}I_d^2\right)}$$
$$= \frac{1}{\sqrt{3}}I_d = 0.577 I_d \quad (6.22)$$

The RMS current value on the AC side of the rectifier bridge will consist of the currents flowing through the two elements in the same phase. Taking the "a" phase, for example, the current flowing through the a-phase secondary winding will consist of the currents for V1 and V4. As shown in Figure 6.8d, note that the current direction passing the top half of element V1 will be from the power supply to the element, and the current direction passing the bottom half of element V4 will be from the element V4 to the power supply. Therefore, the effective alternating current value will be

$$I = \sqrt{2\frac{1}{2\pi} \times \frac{3\pi}{3}I_d^2} = \sqrt{\frac{2}{3}}I_d = 0.816 I_d \quad (6.23)$$

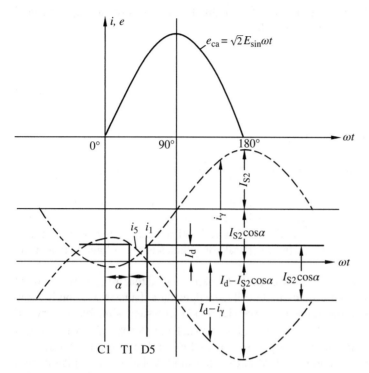

Figure 6.6 Current change in rectifier commutation process.

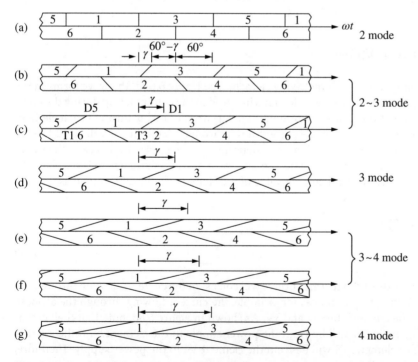

Figure 6.7 Number of elements conducting at different commutation angles: (a) $\gamma = 0$, (b) $\gamma = 20°$, (c) $\gamma = 40°$, (d) $\gamma = 60°$, (e) $\gamma = 80°$; (f) $\gamma = 100°$, (g) $\gamma = 120°$.

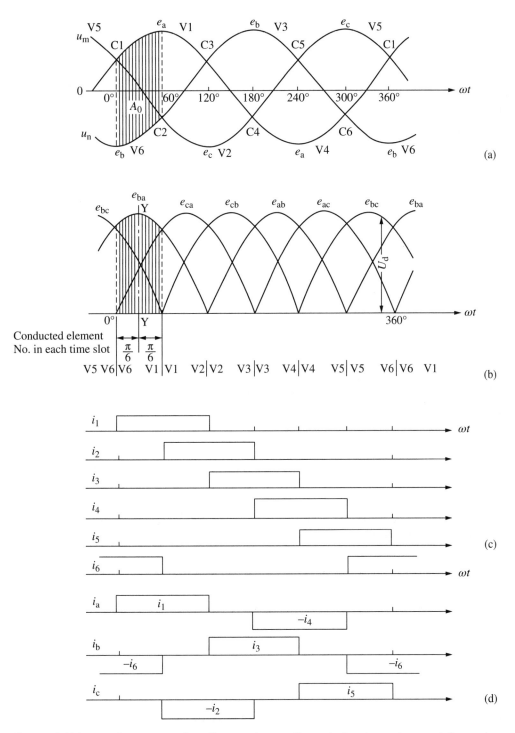

Figure 6.8 Voltage and current waveform diagram when rectifier control angle α and commutation angle γ are zero: (a) voltage oscillogram at the positive and negative terminals "m," "n" for DC output to the neutral point O, (b) DC voltage U_d oscillogram, (c) element current oscillogram, (d) rectifier AC-side current oscillogram.

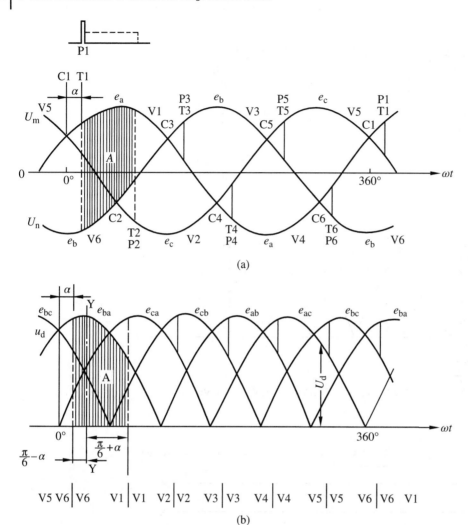

Figure 6.9 Rectified voltage oscillogram when rectifier is working at $\alpha > 0$, $\gamma = 0$: (a) voltage oscillogram at the positive and negative terminals "m" and "n" for DC output to the neutral point O, (b) DC voltage waveform.

According to a comparison between Eqs. (6.22) and (6.23): the effective current value on the AC side of the rectifier bridge will be $\sqrt{2}$ times the effective element current value.

When the commutation angle $\gamma > 0$, let us consider the waveform changes in the current-rising and current-falling segments during commutation when determining the effective element current value.

During the rise of the commutating current ($\alpha \leqslant \omega t \leqslant \alpha + \gamma$), the following can be obtained according to Eq. (6.9):

$$i_s = \frac{E}{\sqrt{2}X_\gamma}(\cos\alpha - \cos\omega t) \tag{6.24}$$

During commutation, when $\omega t = \alpha + \gamma$, the element current will equal the DC current I_d; that is,

$$I_d = \frac{E}{\sqrt{2}X_\gamma}[\cos\alpha - \cos(\alpha + \gamma)] \tag{6.25}$$

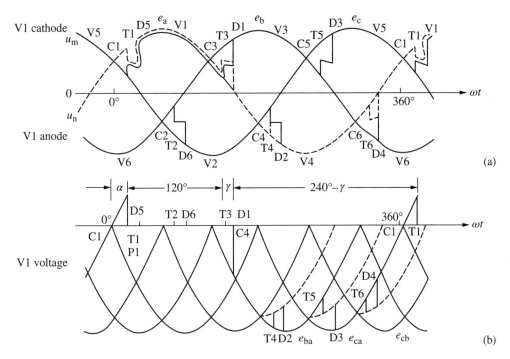

Figure 6.10 Voltage waveform of rectifier bridge element: (a) voltage at the anode and cathode of V1 relative to the neutral point, (b) V1 voltage.

According to the above i_s and I_d expressions, the expression for the commutating current-rising segment during commutation can be obtained as

$$i_s = I_d \times \frac{\cos \alpha - \cos \omega t}{\cos \alpha - \cos(\alpha + \gamma)} \tag{6.26}$$

Similarly, the element current during the falling of the commutating current ($120° + \alpha \leqslant \omega t \leqslant 120° + \alpha + \gamma$) will be

$$i_j = I_d - I_d \times \frac{\cos \alpha - \cos\left(\omega t - \frac{5}{6}\pi\right)}{\cos \alpha - \cos(\alpha + \gamma)} \tag{6.27}$$

Accordingly, integration can be performed on each time interval to determine the effective element current value as

$$I_V = \left\{ \frac{1}{2\pi} \left[\int_\alpha^{\alpha+\gamma} i_s^2 d(\omega t) + \int_{\frac{2}{3}\pi+\alpha}^{\frac{2}{3}\pi+\alpha+\gamma} i_j^2 d(\omega t) + I_d^2 \left(\frac{2\pi}{3} - \gamma \right) \right] \right\}^{\frac{1}{2}}$$

$$= I_d \frac{1}{\sqrt{3}} \sqrt{1 - 3\psi(\alpha, \gamma)}$$

$$= 0.577 I_d \sqrt{1 - 3\psi(\alpha, \gamma)} \tag{6.28}$$

where $\psi(\alpha, \gamma) = \frac{1}{2\pi} \frac{\sin\gamma[2+\cos(2\alpha+\gamma)]-\gamma[1+2\cos\alpha\cos(\alpha+\gamma)]}{[\cos\alpha-\cos(\alpha+\gamma)]^2}$

Figure 6.11 shows the characteristic relating the effective value coefficients $\sqrt{1 - 3\psi(\alpha, \gamma)}$ to α, γ.

As discussed above, since the current on the AC side consists of the currents of the two rectifying elements in the same phase, the effective current value on the AC side I will be $\sqrt{2}$ times the effective element current

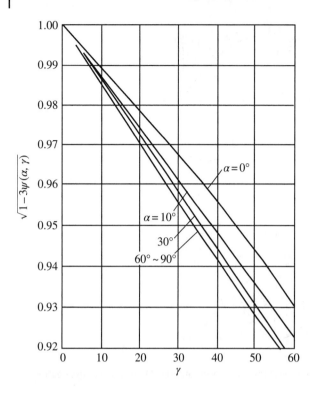

Figure 6.11 Characteristic relating effective current value coefficients $\sqrt{1 - 3\psi(\alpha, \gamma)}$ and α, γ.

value:

$$I = \sqrt{2}I_V = 0.816 I_d \sqrt{1 - 3\psi(a, \gamma)} \tag{6.29}$$

Under in regular operation, the coefficient $\sqrt{1 - 3\psi(a, \gamma)}$ will be about 0.955; thus, we obtain

$$I \approx 0.78 I_d = \frac{\sqrt{6}}{\pi} I_d$$

6.8 Fundamental Wave and Harmonic Value for Alternating Current

6.8.1 Fundamental Wave for Alternating Current

The fundamental component $I_{(1)}$ of the alternating current can be analyzed and solved using the Fourier series. If the $I_{(1)}$ commutation voltage is decomposed into active and reactive components, then the active component of the fundamental current $I_{(1)}$ will be

$$\begin{aligned}
I_{(1)p} &= I_{(1)} \cos \varphi_{(1)} \\
&= \frac{\sqrt{2}}{\pi} \left[\int_{\alpha}^{\alpha+\gamma} i_s \sin\left(\omega t + \frac{\pi}{6}\right) d(\omega t) + \int_{\alpha+\gamma}^{\frac{2}{3}\pi+\alpha} I_d \sin\left(\omega t + \frac{\pi}{6}\right) d(\omega t) \right. \\
&\quad \left. + \int_{\frac{2}{3}\pi+\alpha}^{\frac{2}{3}\pi+\alpha+\gamma} i_j \sin\left(\omega t + \frac{\pi}{6}\right) d(\omega t) \right] \\
&= \frac{\sqrt{6}}{\pi} I_d \frac{[\cos \alpha + \cos(\alpha + \gamma)]}{2}
\end{aligned} \tag{6.30}$$

where

$\varphi_{(1)}$ – fundamental power factor angle on the AC side.

Similarly, the reactive component expression for the effective fundamental current value $I_{(1)}$ can be obtained as

$$I_{(1)Q} = I_{(1)} \sin \varphi_{(1)}$$
$$= \frac{\sqrt{2}}{\pi} \left[\int_{\alpha}^{\alpha+\gamma} i_s \cos\left(\omega t + \frac{\pi}{6}\right) d(\omega t) + \int_{\alpha+\gamma}^{\frac{2\pi}{3}+\alpha} I_d \cos\left(\omega t + \frac{\pi}{6}\right) d(\omega t) \right.$$
$$\left. + \int_{\frac{2}{3}\pi+\alpha}^{\frac{2}{3}\pi+\alpha+\gamma} i_j \cos\left(\omega t + \frac{\pi}{6}\right) d(\omega t) \right]$$
$$= \frac{\sqrt{6}}{\pi} I_d \frac{\sin 2\alpha - \sin 2(\alpha+\gamma) + 2\gamma}{4[\cos\alpha - \cos(\alpha+\gamma)]} \qquad (6.31)$$

where

γ – commutation angle, rad.

The fundamental current RMS value can be solved on the basis of Eqs. (6.30) and (6.31):

$$I_{(1)} = \sqrt{I_{(1)P}^2 + I_{(1)Q}^2} = \frac{\sqrt{6}}{\pi} I_d K_{(1)} \qquad (6.32)$$

where

$$K_{(1)} = \frac{\pi \sqrt{4[\cos^2\alpha - \cos^2(\alpha+\gamma)]^2 + [\sin 2\alpha - \sin 2(\alpha+\gamma)2\gamma]^2}}{4[\cos\alpha - \cos(\alpha+\gamma)]}$$

6.8.2 AC Harmonic Value [12, 13]

In an excitation system employing the rectifier, irrespective of whether it is supplied by an AC exciter or an excitation transformer, the AC phase currents of the rectifier will also be distorted into a non-sinusoidal waveform in the presence of a higher harmonic current component. Therefore, a harmonic analysis of the non-sinusoidal current is needed to determine the supplementary loss due to the harmonic component. For the six-pulse bridge, it can be seen from the expansion principle of the Fourier series that, for an AC waveform that is point symmetric about the origin (that is, $i_{(\omega t)} = -i_{(\pi+\omega t)}$) and with a cycle of 2π, the DC component and the even harmonic do not exist in it except for the fundamental wave; it only has odd harmonics. Since the six-pulse bridge rectifier circuit has no connection at the neutral point, no harmonics that are multiples of three exist in the odd harmonic component; that is, the times of the odd harmonics it contains are 1, 5, 7, 11, 13, 17, 19, …, and so on. The Fourier series expansion equation for the non-sinusoidal current waveform $i_{(\omega t)}$ can be written as

$$i(\omega t) = \sum_{n=1}^{\infty} [a_n \cos(n\omega t) + b_n \sin(n\omega t)] \qquad (6.33)$$

$$a_n = \frac{1}{\pi} \int_0^{2\pi} i_{(\omega t)} \cos(n\omega t) d(\omega t)$$

$$b_n = \frac{1}{\pi} \int_0^{2\pi} i_{(\omega t)} \sin(n\omega t) d(\omega t)$$

$$c_n = \sqrt{a_n^2 + b_n^2}$$

$$\tan \varphi_n = \frac{a_n}{b_n}$$

where

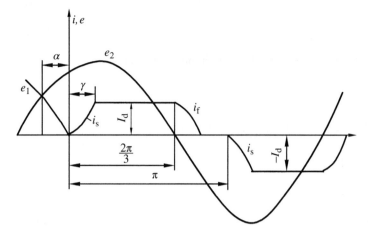

Figure 6.12 AC voltage and current oscillogram.

The calculation of the harmonic current coefficient is related to the origin of the selected coordinate system; for example, when deriving the harmonic current coefficient in Reference [8], the origin of the power supply phase voltage is taken as the reference point, that is, the origin for the e_2 voltage waveform in Figure 6.12.

When the origin of the coordinates is selected, the expression for the AC commutating current-rising segment is

$$i_s = \frac{\cos\alpha - \cos\left(\omega t - \frac{\pi}{6}\right)}{\cos\alpha - \cos(\alpha + \gamma)} \tag{6.34}$$

The expression for the current in the falling segment is

$$i_j = \frac{\cos\left(\omega t - \frac{5\pi}{6}\right) - \cos(\alpha + \gamma)}{\cos\alpha - \cos(\alpha + \gamma)} \tag{6.35}$$

Accordingly:

$$a_n = \frac{1}{\pi}\int_{\frac{\pi}{6}+\alpha}^{\frac{\pi}{6}+\alpha+\gamma} \frac{\cos\alpha - \cos\left(\omega t - \frac{\pi}{6}\right)}{\cos\alpha - \cos(\alpha+\gamma)} \times \cos(n\omega t)\mathrm{d}(\omega t)$$

$$- \int_{\frac{5}{6}\pi+\alpha}^{\frac{5}{6}\pi+\alpha+\gamma} \frac{\cos(\alpha+\gamma) - \cos\left(\omega t - \frac{5\pi}{6}\right)}{\cos\alpha - \cos(\alpha+\gamma)} \times \cos(n\omega t)\mathrm{d}(\omega t)$$

$$+ \int_{\frac{\pi}{6}+\alpha+\gamma}^{\frac{5}{6}\pi+\alpha} \cos(n\omega t)\mathrm{d}(\omega t) \tag{6.36}$$

We obtain

$$a_n = \frac{2\sin\left(n\frac{\pi}{3}\right)}{\pi n(n^2-1)[\cos\alpha - \cos(\alpha+\gamma)]}$$
$$\times \left\{ n\sin\alpha\sin\left[n\left(\alpha+\frac{\pi}{2}\right)\right] - n\sin(\alpha+\gamma)\times\sin\left[n\left(\alpha+\gamma+\frac{\pi}{2}\right)\right] \right.$$
$$\left. + \cos\alpha\cos\left[n\left(\alpha+\frac{\pi}{2}\right)\right] - \cos(\alpha+\gamma)\cos\left[n\left(\alpha+\gamma+\frac{\pi}{2}\right)\right] \right\} \tag{6.37}$$

Similarly, one can obtain

$$
\begin{aligned}
b_n = & \frac{2\sin\left(n\frac{\pi}{3}\right)}{\pi n(n^2-1)[\cos\alpha - \cos(\alpha+\gamma)]} \\
& \times \left\{ n\sin\alpha\cos\left[n\left(\alpha+\frac{\pi}{2}\right)\right] - n\sin(\alpha+\gamma)\times\cos\left[n\left(\alpha+\gamma+\frac{\pi}{2}\right)\right] \right. \\
& \left. + \cos(\alpha+\gamma)\sin\left[n\left(\alpha+\gamma+\frac{\pi}{2}\right)\right] - \cos\alpha\sin\left[n\left(\alpha+\frac{\pi}{2}\right)\right] \right\}
\end{aligned}
\tag{6.38}
$$

The harmonic amplitude and the tan of the phase angle are

$$
c_n = \sqrt{a_n^2 + b_n^2}, \quad \tan\varphi_n = \frac{b_n}{a_n} \tag{6.39}
$$

The RMS value for the harmonic component:

$$
I_n = \frac{c_n}{\sqrt{2}}\lambda I_d \tag{6.40}
$$

where

λ — connection factor, relative to the six-pulse bridge connection $\lambda = 2$.

The fundamental coefficient is

$$
\left.
\begin{aligned}
a_1 &= \frac{\sqrt{3}}{4\pi[\cos\alpha - \cos(\alpha+\gamma)]} \times [2\gamma + \sin 2\alpha - \sin 2(\alpha+\gamma)] \\
b_1 &= \frac{\sqrt{3}}{2\pi}[\cos\alpha + \cos(\alpha+\gamma)] \\
c_1 &= \sqrt{a_1^2 + b_1^2}, \tan\varphi_1 = \frac{b_1}{a_1}
\end{aligned}
\right\} \tag{6.41}
$$

The fundamental current RMS value is

$$
I_1 = \frac{c_1}{\sqrt{2}}\lambda I_d \tag{6.42}
$$

Another harmonic current component expression is given in Reference [14]. The initial point of the alternating current is taken as the origin of the coordinates. Therefore, the expression for the AC rising segment is

$$
i_s = \frac{\cos\alpha - \cos(\omega t + \alpha)}{\cos\alpha - \cos(\alpha+\gamma)} \tag{6.43}
$$

The expression for the alternating current-falling segment is

$$
i_j = \frac{\cos\left(\omega t + \alpha - \frac{2\pi}{3}\right) - \cos(\alpha+\gamma)}{\cos\alpha - \cos(\alpha+\gamma)} \tag{6.44}
$$

Accordingly:

$$
\begin{aligned}
a_n = & \frac{1}{\pi}\int_0^\gamma \frac{\cos\alpha - \cos(\omega t + \alpha)}{\cos\alpha - \cos(\alpha+\gamma)}\cos(n\omega t)\mathrm{d}(\omega t) \\
& + \int_{\frac{2}{3}\pi}^{\frac{2}{3}\pi+\gamma} \frac{\cos\left(\omega t + \alpha - \frac{2\pi}{3}\right) - \cos(\alpha+\gamma)}{\cos\alpha - \cos(\alpha+\gamma)}\cos(n\omega t)\mathrm{d}(\omega t) \\
& + \int_\gamma^{\frac{2}{3}\pi} \cos(n\omega t)\mathrm{d}(\omega t)
\end{aligned}
\tag{6.45}
$$

$$a_n = \frac{2\sin\left(n\frac{\pi}{3}\right)\sin\left(n\frac{\pi}{2}\right)}{\pi n(n^2-1)[\cos\alpha - \cos(\alpha+\gamma)]}$$
$$\times \{n\sin(\alpha+\gamma)\sin[n(\alpha+\gamma)] - n\sin\alpha\sin(n\alpha)$$
$$+ \cos(\alpha+\gamma)\cos[n(\alpha+\gamma)] - \cos\alpha\cos(n\alpha)\} \tag{6.46}$$

$$b_n = \frac{2\sin n\frac{\pi}{3}\sin\left(n\frac{\pi}{2}\right)}{\pi n(n^2-1)[\cos\alpha - \cos(\alpha+\gamma)]}$$
$$\times \{n\sin\alpha\cos(n\alpha) - n\sin(\alpha+\gamma)\cos[n(\alpha+\gamma)] + \sin[n(\alpha+\gamma)]$$
$$\times \cos(\alpha+\gamma) - \sin(n\alpha)\cos\alpha\} \tag{6.47}$$

The harmonic amplitude and the tan of the phase angle are

$$c_n = \sqrt{a_n^2 + b_n^2}, \quad \tan\varphi_n = \frac{b_n}{a_n} \tag{6.48}$$

The RMS value of the harmonic component:

$$I_n = \frac{c_n}{\sqrt{2}} \times I_d \quad (n = 1, 5, 7, 11, 13, \cdots) \tag{6.49}$$

The fundamental component coefficient:

$$\left. \begin{array}{l} a_1 = \dfrac{\sqrt{3}}{2\pi[\cos\alpha - \cos(\alpha+\gamma)]}[2\gamma + \sin 2\alpha - \sin 2(\alpha+\gamma)] \\ b_1 = \dfrac{\sqrt{3}}{\pi}[\cos\alpha + \cos(\alpha+\gamma)] \\ c_1 = \sqrt{a_1^2 + b_1^2}; \quad \tan\varphi_1 = \dfrac{a_1}{b_1} \end{array} \right\} \tag{6.50}$$

The RMS value of the fundamental component current:

$$I_1 = \frac{c_1}{\sqrt{2}} \times I_d \tag{6.51}$$

6.9 Power Factor of Rectifying Device

From the AC voltage and current oscillograms shown in Figure 6.13, when taking separately the longitudinal axes V–V and I–I representing the midlines of the positive half-waves for the voltage and current waveforms while assuming the current waveform to be approximately a trapezoidal wave, the interval between the midline I–I and the on–off point of the current waveform is $60° + \frac{\gamma}{2}$. Therefore, the included angle between the two midlines for the voltage and current is the power factor angle φ, which can be solved on the basis of Figure 6.13:

$$\varphi = \alpha + \frac{\gamma}{2} \tag{6.52}$$

Therefore, the greater the control angles α and γ during the operation of the rectifying device, the lower the power factor on the AC side of the rectifier power. Under the condition of identical DC output power, the required rectifier transformer power will also be greater. Equation (6.52) represents the approximate expression for the power factor, and the accurate calculation procedure is as follows.

We can obtain the following according to Eq. (6.20):

$$U_d = \frac{3\sqrt{2}}{\pi} E \cos\left(\alpha + \frac{\gamma}{2}\right) \cos\frac{\gamma}{2}$$

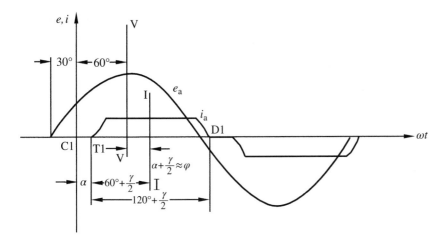

Figure 6.13 Alternating current and voltage oscillogram.

Or:

$$E = \frac{\pi U_d}{3\sqrt{2}} \times \frac{1}{\cos\left(\alpha + \frac{\gamma}{2}\right)\cos\frac{\gamma}{2}}$$

Eq. (6.29) yields

$$I = \sqrt{\frac{2}{3}}I_d \sqrt{1 - 3\psi(\alpha, \gamma)}$$

The AC input apparent power of the rectifying device:

$$\begin{aligned} S &= \sqrt{3}EI \\ &= \sqrt{3} \times \frac{\pi}{3\sqrt{2}} \frac{U_d}{\cos\left(\alpha + \frac{\gamma}{2}\right)\cos\frac{\gamma}{2}} \times \sqrt{\frac{2}{3}}I_d\sqrt{1 - 3\psi(\alpha, \gamma)} \\ &= \frac{\pi}{3} U_d I_d \frac{\sqrt{1 - 3\psi(\alpha, \gamma)}}{\cos\left(\alpha + \frac{\gamma}{2}\right)\cos\frac{\gamma}{2}} \\ &= \frac{\pi}{3} P_d \frac{\sqrt{1 - 3\psi(\alpha, \gamma)}}{\cos\left(\alpha + \frac{\gamma}{2}\right)\cos\frac{\gamma}{2}} \end{aligned} \quad (6.53)$$

Generally, the active power loss of the rectifying device is smaller, and therefore the active power P that is input from the AC side can be regarded as equal to the DC power P_d on the rectifier side, thus yielding the following:

$$P = P_d$$

The active power on AC side is given by

$$P = \sqrt{3}EI\cos\varphi$$

Therefore, one can obtain the power factor:

$$\cos\varphi = \frac{P}{\sqrt{3}EI} = \frac{P}{S} = \frac{3}{\pi} \times \frac{\cos\left(\alpha + \frac{\gamma}{2}\right)\cos\frac{\gamma}{2}}{\sqrt{1 - 3\psi(\alpha, \gamma)}} \quad (6.54)$$

Under in regular operation, the factor $\sqrt{1-3\psi(\alpha+\gamma)}$ of the rectifying device is about 0.955, which can be approximately given by

$$\cos\varphi \approx \cos\left(\alpha+\frac{\gamma}{2}\right)\cos\frac{\gamma}{2} = \frac{1}{2}[\cos\alpha+\cos(\alpha+\gamma)] \tag{6.55}$$

The power factor of the fundamental wave for the rectifying device can be obtained from Eqs. (6.30) and (6.32):

$$\cos\varphi_{(1)} = \frac{I_{(1)P}}{I_{(1)}} = \frac{\sqrt{6}}{\pi}I_d\frac{[\cos\alpha+\cos(\alpha+\gamma)]}{2} \times \frac{\pi}{\sqrt{6}I_d K_{(1)}}$$

$$= \frac{1}{K_{(1)}} \times \frac{[\cos\alpha+\cos(\alpha+\gamma)]}{2} \tag{6.56}$$

By substituting $K_{(1)}$ in Eq. (6.32) into the above equation, one can obtain:

$$\cos\varphi_{(1)} = \frac{2[\cos^2\alpha-\cos^2(\alpha+\gamma)]}{\sqrt{4[\cos^2\alpha-\cos^2(\alpha+\gamma)]^2+[\sin 2\alpha-\sin 2(\alpha+\gamma)+2\gamma]^2}} \tag{6.57}$$

A comparison of Eqs. (6.54) and (6.57) shows that the total power factor of the rectifying device differs from the fundamental power factor, which is due to the fact that the alternating current has a non-sinusoidal waveform.

According to the definition of the active power in the three-phase AC circuit, its value is the average value of the sum of the phase voltage instantaneous values multiplied by the current instantaneous values in a cycle. Therefore, the active power on the AC side is

$$P = \frac{3}{T}\int_0^T ei\,dt = \frac{3}{T}\int_0^T ei_{(1)}\,dt + \frac{3}{T}\sum_{n>1}\int_0^T ei_{(n)}\,dt$$

where

T — supply voltage cycle;

e — instantaneous value of the sinusoidal phase voltage on AC side of the rectifier, with its amplitude equal to $\sqrt{\frac{2}{3}}E$.

From the AC circuit principle, the active power generated by the fundamental current is

$$P = \frac{3}{T}\int_0^T ei_{(1)}\,dt = \sqrt{3}EI_{(1)}\cos\varphi_{(1)} \tag{6.58}$$

From Eq. (6.21), E can be written as

$$E = \frac{2\pi U_d}{3\sqrt{2}} \times \frac{1}{\cos\alpha+\cos(\alpha+\gamma)}$$

From Eq. (6.33), $I_{(1)}\cos\varphi_{(1)}$ can be written as

$$I_{(1)}\cos\varphi_{(1)} = \frac{\sqrt{6}}{\pi}I_d\frac{\cos\alpha+\cos(\alpha+\gamma)}{2}$$

By substituting the above two equations into Eq. (6.58), the active power value generated by the fundamental current is obtained as

$$P = U_d I_d$$

The average value of the product of the harmonic currents and the fundamental voltages on a circle must be zero; that is,

$$\frac{3}{T}\int_0^T ei_{(n)}\,dt = 0$$

This means only the fundamental current can supply the commutation device with active power.

The expression for the apparent power supplied by an AC power supply to the rectifying device can be written as

$$S = \sqrt{3}EI = \sqrt{3}E\sqrt{I_{(1)}^2 + \sum_{n>1} I_{(n)}^2}$$

The power of the rectifying device includes three parts: the active power, the reactive power, and the distortion power caused by the harmonic current, and the expression that relates them is

$$S^2 = P^2 + Q^2 + N^2 \tag{6.59}$$

where

P — active power;
Q — reactive power;
N — distortion power.

From Eq. (6.31), the reactive power Q of the fundamental current can be written as

$$Q = \sqrt{3}EI_{(1)}\sin\varphi_{(1)} = U_d I_d \frac{\sin 2\alpha - \sin 2(\alpha+\gamma) + 2\gamma}{2[\cos^2\alpha - \cos^2(\alpha+\gamma)]} \tag{6.60}$$

Q in the above equation is the value of the reactive power consumed in the rectification process.

The distortion power generated by the harmonic current is

$$N = \sqrt{3}E\sqrt{\sum_{n>1} I_{(n)}^2} = \sqrt{3}E\sqrt{I^2 - I_{(1)}^2} \tag{6.61}$$

Due to the presence of the distortion power, the total power factor of the rectifying device is slightly smaller than the fundamental power factor.

Table 6.1 shows the calculation formula table of the rectifying circuit in the Type I commutation state when $0 < \gamma < 60°$.

Table 6.1 Theoretical calculation formula of three-phase bridge rectifier circuit in Type I commutation state ($0 < \gamma < 60°$).

SN	Item	Calculation formula
1	No-load DC voltage when $\alpha = 0$	$U_{d0} = \frac{3\sqrt{2}}{\pi}E = 1.35E$ (wherein E – effective voltage value for power line)
2	No-load DC voltage when $\alpha \neq 0$	$U_{d0} = \frac{3\sqrt{2}}{\pi}E\cos\alpha = 1.35E\cos\alpha$
3	Direct current	$I_d = \frac{\sqrt{2}E}{2\omega L_\gamma} \times [\cos\alpha - \cos(\alpha+\gamma)] = \frac{\sqrt{2}E}{\omega L_\gamma} \times \sin\left(\alpha+\frac{\gamma}{2}\right)\sin\frac{\gamma}{2}$ $= 2I_{S2}\sin\left(\alpha+\frac{\gamma}{2}\right)\sin\frac{\gamma}{2}$
4	Voltage drop for commutating reactance	$\Delta U = d_\gamma I_d = \frac{3\omega L_\gamma}{\pi} \times I_d = \frac{3X_\gamma}{\pi} \times I_d = 6fL_\gamma I_d = \frac{U_{d0}}{2}[\cos\alpha - \cos(\alpha+\gamma)]$ $= U_{d0}\sin\times\left(\alpha+\frac{\gamma}{2}\right)\sin\frac{\gamma}{2}$
5	Commutation angle	$\gamma = -\alpha + \cos^{-1}\left[\cos\alpha - \frac{6\omega L_\gamma I_d}{\pi U_{d0}}\right]$
6	DC voltage (ignoring the resistance voltage drop caused by the commutating current in the loop)	$U_d = U_{d0}\cos\alpha - \frac{3\omega L_\gamma}{\pi} \times I_d = \frac{U_{d0}}{2}[\cos\alpha + \cos(\alpha+\gamma)]$ $= U_{d0}\cos\left(\alpha+\frac{\gamma}{2}\right)\cos\frac{\gamma}{2} = \frac{3\sqrt{2}}{\pi}E\cos\left(\alpha+\frac{\gamma}{2}\right)\cos\frac{\gamma}{2}$ $= 1.35E\cos\left(\alpha+\frac{\gamma}{2}\right)\cos\frac{\gamma}{2} \approx U_{d0}\cos\varphi$

(Continued)

Table 6.1 (Continued)

SN	Item	Calculation formula
7	Max. element voltage in the blocking period (ideal waveform)	$U_p = \dfrac{\pi}{3} U_{d0} = 1.047 U_{d0} = \sqrt{2} E$
8	Instantaneous value for the element current in the commutation period (rising front edge) (falling trailing edge)	$i_s = I_d \times \dfrac{\cos\alpha - \cos\omega t}{\cos\alpha - \cos(\alpha+\gamma)}$ In the equation, the initial phase of α is taken as the original point of $t=0$ $i_j = I_d \times \dfrac{\cos\left(\omega t - \tfrac{5}{6}\pi\right) - \cos(\alpha+\gamma)}{\cos\alpha - \cos(\alpha+\gamma)}$
9	Effective element current value	$I_V = \dfrac{I_d}{\sqrt{3}} \sqrt{1 - 3\psi(\alpha,\gamma)} = 0.577 I_d \sqrt{1 - 3\psi(\alpha,\gamma)}$ where $\psi(\alpha,\gamma) = \dfrac{1}{2\pi} \times \dfrac{\sin\gamma[2 + \cos(2\alpha+\gamma)] - \gamma[1 + 2\cos\alpha\cos(\alpha+\gamma)]}{[\cos\alpha - \cos(\alpha+\gamma)]^2}$ Under regular operation, $\sqrt{1 - 3\psi(\alpha,\gamma)} \approx 0.955$
10	Rectifying device AC	$I = \sqrt{2} I_V = \sqrt{\dfrac{2}{3}} I_d \sqrt{1 - 3\psi(\alpha,\gamma)} = 0.816 I_d \sqrt{1 - 3\psi(\alpha,\gamma)} \approx 0.78 I_d$
11	Fundamental component of the rectifying device AC	$I_{(1)} = \dfrac{\sqrt{6}}{\pi} I_d K_{(1)}$ where $K_{(1)} = \dfrac{\sqrt{4[\cos^2\alpha - \cos^2(\alpha+\gamma)]^2 + [\sin 2\alpha - \sin 2(\alpha+\gamma) + 2\gamma]^2}}{4[\cos\alpha - \cos(\alpha+\gamma)]}$
12	Apparent power of the rectifying device or the capacity of the rectifier transformer	$S = \dfrac{\pi}{3} U_d I_d \times \dfrac{\sqrt{1 - 3\psi(\alpha,\gamma)}}{\cos\left(\alpha + \tfrac{\gamma}{2}\right) \cos\tfrac{\gamma}{2}} = 2.094 P \times \dfrac{\sqrt{1 - 3\psi(\alpha,\gamma)}}{\cos\alpha + \cos(\alpha+\gamma)} = \sqrt{3} EI$
13	Active power converted in the rectifying device (excluding the device loss)	$P = P_d = U_d I_d = \sqrt{3} EI \cos\varphi = \sqrt{3} EI_{(1)} \cos\varphi_{(1)}$
14	AC fundamental reactive power	$Q = \sqrt{3} EI_{(1)} \sin\varphi_{(1)} = \dfrac{U_d I_d}{2} \times \dfrac{\sin 2\alpha - \sin 2(\alpha+\gamma) + 2\gamma}{\cos^2\alpha - \cos^2(\alpha+\gamma)}$
15	Total power factor of the rectifying device	$\cos\varphi = \dfrac{3}{\pi} \times \dfrac{\cos\left(\alpha + \tfrac{\gamma}{2}\right) \cos\tfrac{\gamma}{2}}{\sqrt{1 - 3\psi(\alpha,\gamma)}} \approx \cos\left(\alpha + \tfrac{\gamma}{2}\right) \cos\tfrac{\gamma}{2} = \tfrac{1}{2}[\cos\alpha + \cos(\alpha+\gamma)]$
16	Fundamental power factor of the rectifying device	$\cos\varphi_{(1)} = \dfrac{2[\cos^2\alpha - \cos^2(\alpha+\gamma)]}{\sqrt{4[\cos^2\alpha - \cos^2(\alpha+\gamma)]^2 + [\sin 2\alpha - \sin 2(\alpha+\gamma) + 2\gamma]^2}}$

Figure 6.14 shows the characteristic curves in the Type I commutation state for the rectified voltage U_d, the rectified current I_d, and the control angle α when $0 < \gamma < 60°$ and the supply AC voltage E is constant.

The expression for the rectified voltage is

$$U_d = \dfrac{3\sqrt{2} E}{\pi} \cos\alpha - \dfrac{3}{\pi} X_\gamma I_d (\gamma < 60°)$$

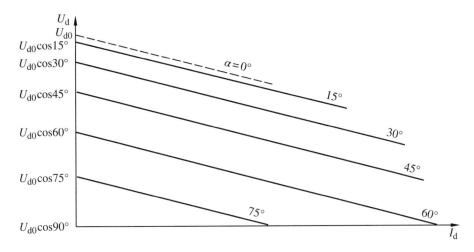

Figure 6.14 External characteristic curves of rectifier in Type I commutation state when $0 < \gamma < 60°$.

6.10 Type III Commutation State

The Type I commutation state when $0 < \gamma < 60°$ is discussed in Section III, that is, the 2–3 mode. This section will focus on Type II operating states when $\gamma = 60°$ and Type III operating states when $60° < \gamma < 120°$. To facilitate the description, we shall first discuss the Type III commutation state.

As discussed above, when $60° < \gamma < 120°$, the rectifier will work in the Type III commutation state. In this case, the rectifying element will work in the 3–4 mode; that is, there are always 3 and 4 elements conducting at the same time alternately.

It should be noted that this operating mode is irregular, and is is enabled only when the rectifier is heavily loaded, the supply voltage drops greatly on the AC side, and a short circuit occurs on the DC side.

The Type III commutation state has an obvious operation feature: a minimum fixed forced control angle α' exists in the rectifying circuit, and $\alpha' = 30°$. When the operating control angle set in the circuit is $\alpha_p \leq 30°$, the set control angle cannot play a role at all. The operating state of the rectifying circuit is then determined by the minimum fixed forced control angle α' of the circuit. Only when $\alpha_p > \alpha' = 30°$ is the operating state of the rectifying circuit determined by the set control angle.

Figure 6.15 shows the resulting rectified voltage waveform when $60° \leq \gamma < 120°$ in terms of the operating state. The number in the figure represents the corresponding number of the rectifying element.

First, it is assumed that the operating control angle $\alpha_p < 30°$, the initial state is that the rectifying elements V3 and V4 are conducting, with C5 as the natural commutation point. When supplying a trigger pulse at T5 for P5, the rectifying element V5 is conducting, and V3 and V5 are commutating. In the meantime, it is assumed that since the load current increases after commutation, the commutation angle γ increases to over 60° and it is still over 60° after being in the steady state. Therefore, in the case of 60° after P5, when P6 sends a trigger pulse, the commutation between elements V3 and V5 has not finished yet, element V6 has a trigger pulse but its cathode potential changes with the curve between T5 and D3, with its anode potential at point "n" based on the voltage e_a (because element V4 is in conducting mode, the potential at point "a" equals that at point "n"). At the same time, since e_a is less than the commutation voltage between elements V3 and V5, even when the trigger pulse is applied to element V5, it still cannot be switched to the on state, because its anode

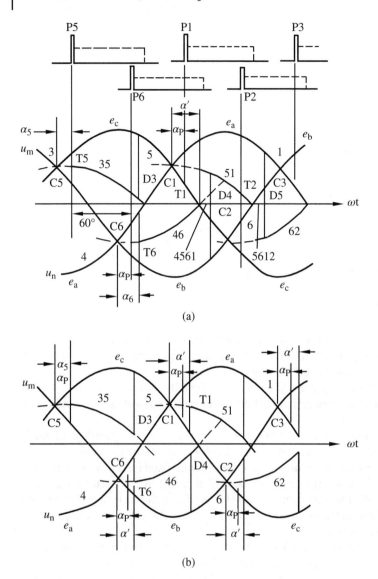

Figure 6.15 Commutation process when (a) $60° < \gamma < 120°$, (b) $\gamma = 60°$.

voltage is less than the cathode voltage. Such a state persists at point D3, where the commutation between elements V3 and V5 is completed. Then the short circuit between phases "b" and "c" disappears. The cathode potential of the element V6 is determined by the b-phase voltage e_b, which is less than the cathode voltage e_a. The element V6 is switched on if the trigger pulse is of a certain width. According to the above process, an angle will lag behind for the conduction time of the element V6 as compared to its operating control angle α_p, which is $\Delta\alpha = \alpha' - \alpha_p$.

After the element V6 is conducting at point T6, V4 and V6 will start to commutate. In this case, since the commutation angle γ is still more than 60°, a pulse is sent to the element V1 at point P1, which is at an angle of 60° after P6. The commutation between the elements V4 and V6 is not completed yet, and therefore the anode potential of V1 will change along the T1–T6 voltage curve, which is less than its cathode potential e_c (since the element V5 is conducting, the potential at point "m" will equal e_c); therefore, it cannot conduct immediately. However, after point T1 (lagging by an angle of 30° behind point C1), the anode potential of V1

Figure 6.16 Equivalent circuit of rectifying circuit when $60° < \gamma < 120°$: (a) rectifying elements V1, V5, V4, and V6 commutate at the same time, (b) elements V1 and V5 commutate while V1, V5, and V6 conduct simultaneously, (c) elements V1, V5, V2, and V6 commutate simultaneously.

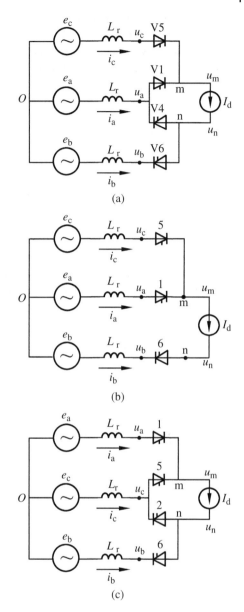

tends to rise along the dash-lined voltage curve of T6–T1, which will be conduct instantly when its cathode potential e_c is exceeded. After that, by separately commutating and conducting, the four elements including elements V1 and V5 that are with the top half as well as elements V4 and V6 that are with the bottom half, a three-phase short circuit on the AC side and a short circuit on the DC side will be formed, with a rectified voltage of zero. See Figure 6.16a for its equivalent circuit.

Then, the commutation between V4 and V6 at point D4 is completed, with V4 switched off. In the following operating state, elements V5 and V1 continue to commutate while V5, V1, and V6 conduct at the same time. See Figure 6.16b for the equivalent circuit.

Then, a trigger pulse is sent at point P2, and the element V2 is not switched on immediately; instead, begins conducting at point T2 after lagging behind point C2 at an angle of 30°. While the elements V2 and V6 commutate, V5 and V1 continue to commutate, and thus a state arises in which four elements such as V5, V1, V2, and V6 conduct at the same time, until the element V5 is switched off at point D5. See Figure 6.16c for the equivalent circuit. Then, the rectifier continues to work in the 3–4 mode, in which 3 elements and 4 elements conduct alternately.

From the above discussion, the rectifier works in the steady state of $60° < \gamma < 120°$, and if the operating control angle $\alpha_p < 30°$, the control angle will increase until $\alpha' = 30°$ when in conducting mode; if $\alpha_p > 30°$, the conducting control angle will be $\alpha = \alpha_p \geq 30°$. Obviously, the minimum control angle in the circuit will be $30°$. This phenomenon is due to the two-phase short circuit formed in the previous commutation process. In this case, the adjacent elements that enter the commutating interval later can be triggered by the reverse voltage action but cannot conduct. They cannot conduct until the voltage between the anode and cathode of the elements becomes positive.

It should be noted that: when the elements under the reverse voltage action are applied with a trigger pulse, the reverse leakage current of the corresponding elements will increase. In this case, the voltage between each element will be distributed more unevenly, which is quite unfavorable for voltage-sharing safety.

The Type III commutation process will now be analyzed quantitatively. See Figure 6.17a for the corresponding commutation oscillogram. We emphasize that in the following discussion, if it is assumed that point C1 is taken as the initial point of $\omega t = 0$, the subsequent commutation process can be divided into three time intervals, as shown in Figure 6.17.

In the T1–D4 range of the first time interval, four elements such as V4, V6, V1, and V5 conduct simultaneously, thus forming a three-phase short-circuit on the AC side. From Figure 6.16a, we see that the current i_1

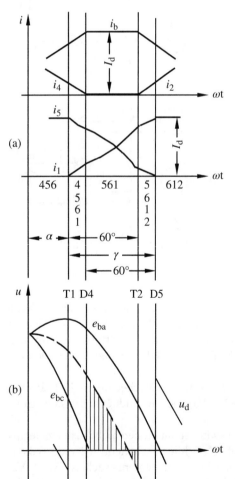

Figure 6.17 Rectifier commutating current and rectified voltage oscillogram when $60° < \gamma < 120°$: (a) commutating current oscillogram, (b) rectified voltage oscillogram.

of V1 is the c-phase short-circuit current, with the phase position inverted by 180°; using the equation that is the same as the above case in which the rectifier works in the 2–3 mode, we obtain

$$i_1 = -\sqrt{\frac{2}{3}} \times \frac{E}{X_\gamma} \cos(\omega t + 150°) + A_1 = -I_{S3} \cos(\omega t - 30°) + A_1 \tag{6.62}$$

including: $I_{S3} = \sqrt{\frac{2}{3}} \times \frac{E}{X_\gamma}$

where

I_{S3} – amplitude of the forced component for short-circuit current, when a three-phase short-circuit occurs in the AC power supply for the rectifier.

A_1 – a factor to be determined for the free component of the short-circuit current.

For point T1, we obtain the following by substituting $\omega t = \alpha$ and $i_1 = 0$ into Eq. (6.62):

$$A_1 = I_{S3} \cos(\alpha - 30°) \tag{6.63}$$

By substituting the A_1 value into Eq. (6.62), we obtain the i_1 value in the range T1~D4:

$$i_1 = I_{S3}[\cos(\alpha - 30°) - \cos(\omega t - 30°)] \tag{6.64}$$

At point D4, when $\omega t = \alpha + \gamma - 60°$, then

$$i_1 = I_{S3}[\cos(\alpha - 30°) - \cos(\alpha + \gamma - 90°)] \tag{6.65}$$

In the D4–T2 range of the second time interval, three elements such as V1, V5, and V6 conduct simultaneously, elements V1 and V5 commutate, and a short circuit occurs in phases "c" and "a" on the AC side. See Figure 6.16b for its equivalent circuit.

i_1 is expressed as

$$i_1 = -\frac{\sqrt{2}E}{2X\gamma} \cos \omega t + A_2 = -\frac{\sqrt{3}}{2} I_{S3} \cos \omega t + A_2 \tag{6.66}$$

At point D4, the i_1 value is determined on the basis of Eq. (6.65), and the value does not change abruptly; thus, we obtain

$$A_2 = I_{S3} \left[\cos(\alpha - 30°) - \cos(\alpha + \gamma - 90°) + \frac{\sqrt{3}}{2} \cos(\alpha + \gamma - 60°) \right]$$

$$= I_{S3} \left[\cos(\alpha - 30°) + \frac{1}{2} \cos(\alpha + \gamma + 30°) \right] \tag{6.67}$$

In the D4–T2 range, i_1 is expressed as

$$i_1 = I_{S3} \left[\cos(\alpha - 30°) + \frac{1}{2} \cos(\alpha + \gamma + 30°) - \frac{\sqrt{3}}{2} \cos \omega t \right] \tag{6.68}$$

At point T2, when $\omega t = \alpha + 60°$, we have

$$i_1 = I_{S3} \left[\cos(\alpha - 30°) + \frac{1}{2} \cos(\alpha + \gamma + 30°) - \frac{\sqrt{3}}{2} \cos(\alpha + 60°) \right] \tag{6.69}$$

In the T2–D5 range of the third time interval, four elements such as V1, V5, V2, and V6 conduct simultaneously; a three-phase short-circuit occurs on the AC side, with the corresponding i_1 value being obtained from Figure 6.16c; the a-phase current flows through the element V1, then:

$$i_1 = -\sqrt{\frac{2}{3}} \times \frac{E}{X_\gamma} \cos(\omega t + 30°) + A_3 = -I_{S3} \cos(\omega t + 30°) + A_3 \tag{6.70}$$

At point T2, the i_1 value when $\omega t = \alpha + 60°$ can be obtained from Equation i_1 (6.69), with the corresponding A_3 value being

$$A_3 = I_{S3}\left[\cos(\alpha - 30°) + \frac{1}{2}\cos(\alpha + \gamma + 30°) - \frac{\sqrt{3}}{2}\cos(\alpha + 60°) + \cos(\alpha + 90°)\right]$$

$$= I_{S3}\left[\frac{1}{2}\cos(\alpha - 30°) + \frac{1}{2}\cos(\alpha + \gamma + 30°)\right] \tag{6.71}$$

Therefore, in the T2–D4 range, i_1 can be expressed as

$$i_1 = I_{S3}\left[\frac{1}{2}\cos(\alpha - 30°) + \frac{1}{2}\cos(\alpha + \gamma + 30°) - \cos(\omega t + 30°)\right] \tag{6.72}$$

The commutation at point D5 is completed when $\omega t = \alpha + \gamma$:

$$i_1 = I_{S3}\left[\frac{1}{2}\cos(\alpha - 30°) + \frac{1}{2}\cos(\alpha + \gamma + 30°) - \cos(\alpha + \gamma + 30°)\right]$$

$$= \frac{1}{2}I_{S3}[\cos(\alpha - 30°) - \cos(\alpha + \gamma + 30°)]$$

$$= \frac{1}{\sqrt{3}}I_{S2}[\cos(\alpha - 30°) - \cos(\alpha + \gamma + 30°)]$$

$$= I_d \tag{6.73}$$

From Eq. (6.73), we see that when $\gamma > 60°$, the relational expression between I_d and α and $\gamma < 60°$ does not correspond to that of Eq. (6.11).

The waveform change for the rectified voltage U_d when $60° < \gamma < 120°$ will be discussed next. As shown in Figure 6.17a, in the T1–D4 range, elements V1, V5, V4, and V6 conduct simultaneously, with a three-phase short circuit forming on the AC side; therefore, $U_d = 0$.

In the D4–T2 time range, a two-phase short circuit is formed between phases "c" and "a" when elements V1, V5, and V6 conduct simultaneously, and the rectified voltage on terminal "m" of the rectified voltage is determined by the average value of the two-phase commutating voltages e_a and e_c, with the rectified voltage on terminal "n" determined by the e_c-phase voltage curve. In Figure 6.17b, the shaded part shows the waveform for the rectified voltage in the 1/6 cycle, with the remaining waveforms being identical to it. In the D4–T2 range, the algebraic sum for the rectified voltage area can be obtained according to the following equation:

$$A = \int_{\alpha+\gamma-60°}^{\alpha+60°} \frac{3}{2}\sqrt{\frac{2}{3}}E\cos\omega t\, d\omega t$$

$$= \sqrt{\frac{3}{2}}E[\cos(\alpha - 30°) + \cos(\alpha + \gamma + 30°)] \tag{6.74}$$

The average rectified voltage is

$$U_d = \frac{A}{\pi} = \frac{3\sqrt{6}}{2\pi}E[\cos(\alpha - 30°) + \cos(\alpha + \gamma + 30°)]$$

$$= \frac{\sqrt{3}}{2}U_{d0}[\cos(\alpha - 30°) + \cos(\alpha + \gamma + 30°)]$$

$$= \sqrt{3}U_{d0}\cos\left(\alpha + \frac{\gamma}{2}\right)\cos\left(\frac{\gamma}{2} + 30°\right) \tag{6.75}$$

According to Eq. (6.73):

$$I_d = \frac{1}{2}\sqrt{\frac{2}{3}} \times \frac{E}{X_\gamma}[\cos(\alpha - 30°) - \cos(\alpha + \gamma + 30°)] \tag{6.76}$$

According to Eq. (6.75):

$$U_d = \sqrt{3}U_{d0}\cos(\alpha - 30°) - 3\frac{3X_\gamma}{\pi} \times I_d$$
$$= \sqrt{3}U_{d0}\cos(\alpha - 30°) - 3d_\gamma I_d \tag{6.77}$$

In this operating state, the relationship between the current on the AC side and the element current will still be the following, taking the "a" phase as an example:

$$i_a = i_1 - i_4$$

However, in the two domains T1–D4 and T2–D5, it will differ from the expression for the 2–3 mode. For example, in the T1–D4 range:

$$i_4 = I_d - i_6, \quad i_a = i_1 - i_4 = i_1 + i_6 - I_d \tag{6.78}$$

Besides, in the T1–D4 range, the waveform of i_6 is the same as that of i_1 in the T2–D5 range, but with a leading angle of 60° in terms of the phase position; therefore, we can obtain the following by phase-shifting in Eq. (6.72):

$$i_6 = I_{S3}\left[\frac{1}{2}\cos(\alpha - 30°) + \frac{1}{2}\cos(\alpha + \gamma + 30°) - \cos(\omega t + 90)\right] \tag{6.79}$$

By substituting Eqs. (6.64), (6.73), and (6.79) into Eq. (6.78), we obtain

$$i_a = I_{S3}[\cos(\alpha - 30°) + \cos(\alpha + \gamma + 30°) - \cos(\omega t + 30°)] \tag{6.80}$$

In the other time interval, i_a will equal i_1 or $-i_4$.

6.11 Type II Commutation State

The Type III commutation state in the case of $60° < \gamma < 120°$ has been discussed in the above section, that is, the 3–4 operation mode.

When $\gamma = 60°$ is unchanged, three elements will be working simultaneously at any moment during this operating mode, which is known as the Type II commutation state of the 3 operation modes. For the calculation for such an operating state, each equation adopted for the 2–3 operation mode will still be applicable. However, note that in the Type II commutation state, the phenomenon of a forced control angle α may also exist. When the control angle $\alpha_p < 30°$ is given, the phenomenon of a real control angle $\alpha > \alpha_p$ may occur.

The operation process is explained by taking Figure 6.15b as an example. Before triggering the element V6, elements V3 and V5 will commutate, with the element V4 conducting. When the element V6 is triggered, if the control angle α_p of V6 is less than the real forced control angle $\alpha = 30°$, the element V6 will only conduct after the element V3 is switched off at point D3, and thus the real control angle will be α and more than the set α_p.

If the elements V3 and V5 commutate, the DC current increases and then remains unchanged, and the α angle increases to reduce the γ angle. If in this case, the commutation angle also achieves a steady-state value of $\gamma = 60°$, and then the real conduction angle for each element will be the same but more than α_p. See the curve after T6 in Figure 6.15b.

If I_d further increases and α_p still remains unchanged, then the real control angle α will increase. When the α angle increases to $\alpha = 30°$ with I_d, that is, when a forced control angle is achieved, $\alpha = 30°$ and $\gamma = 60°$, it is at the critical transition point of the Types II and III operating states, with the corresponding direct current expressed as

$$I_d = \frac{\sqrt{2}E}{2X_\gamma} \times [\cos\alpha - \cos(\alpha + 60°)] = \frac{\sqrt{2}E}{2X_\gamma} \times \sin(\alpha + 30°)$$
$$= \frac{1}{2} \times \sqrt{\frac{3}{2}} \times \frac{E}{X_\gamma} \tag{6.81}$$

Then, if I_d is further increased, the rectifier will transit from the Type II to the Type III operating state; that is, $\alpha = 30°$ and $\gamma > 60°$.

On the other hand, when reducing I_d and keeping α_p unchanged, α will decrease. When α is reduced to $\alpha = \alpha_p$ with I_d and $\gamma = 60°$, the corresponding direct current value will be

$$I_d = \frac{E}{\sqrt{2}X_\gamma} \times \sin(\alpha_p + 30°) \tag{6.82}$$

If I_d continues to decrease, the rectifier will transit from the Type II to the Type I operating mode, with $\alpha = \alpha_p$.

From the above, when $\alpha_p < 30°$ and the direct current varies within the range $\frac{E}{\sqrt{2}X_\gamma} \times \sin(\alpha_p + 30°) \sim \frac{1}{2} \times \sqrt{\frac{2}{3}} \times \frac{E}{X_\gamma}$, that is, when it changes between the two critical points in the Type II operating state, the corresponding control angle α will automatically increase and decrease as I_d changes, with its value being

$$\alpha = \sin^{-1}\frac{I_d X_\gamma}{\sqrt{2}E} - 30°$$

If the above case still occurs when $\alpha_p \geq 30°$, that is, α equals α_p, the direct current corresponding to $\gamma = 60°$ has only one value of $E\sin(\alpha_p + 30°)/\sqrt{2}X_\gamma$. If I_d is less than this value, the rectifier will transit to the Type I operating state. If I_d is more than this value, it will transit to the Type II operating state. For the Type II commutation state when $\gamma = 60°$, the average rectified voltage can also be obtained by using the area integral approach for the curve in Figure 6.15b. The integral range is α to $\alpha + 60°$, with its average value being

$$U_d = \frac{3\sqrt{2}}{\pi} E[\cos\alpha + \cos(\alpha + 60°)]$$

Or it can be written as

$$\frac{2\pi U_d}{3\sqrt{2}E} = 2\cos E(\alpha + 30°)\cos 30°$$

$$\frac{2\pi U_d}{3\sqrt{6}E} = \cos(\alpha + 30°) \tag{6.83}$$

When $\gamma = 60°$, according to Eq. (6.81), the expression for the direct current I_d can be

$$\frac{2I_d X_\gamma}{\sqrt{2}E} = [\cos\alpha - \cos(\alpha + 60°)]$$

$$= 2\sin(\alpha + 30°)\sin 30°$$

$$= \sin(\alpha + 30°) \tag{6.84}$$

By squaring both sides of Eqs. (6.83) and (6.84) and then adding the squared results, we obtain

$$\frac{2\pi^2}{27} \times \frac{U_d^2}{E^2} + \frac{2I_d^2 X_\gamma^2}{E^2} = 1 \tag{6.85}$$

This shows that the equation for the Type II commutation state is an elliptic equation.

6.12 External Characteristic Curve for Rectifier

In the above sections, the characteristics of the three-phase bridge rectifier circuit in Types I, II, and III operating states have been discussed. When the line voltage E of the AC power supply is constant, the relationship between the rectified voltage U_d, the rectified current I_d, and the control angle α is generally known as the external characteristic curve for the rectifier. The external characteristic family of the rectifier can be obtained in the plane taking U_d–I_d as the coordinates, which is very convenient for application. Next, we discuss the construction of the external characteristic curve of the rectifier.

According to the external characteristic expression for the rectifier, for the Type I commutation state:

$$U_d = \frac{3\sqrt{2}}{\pi} E \cos \alpha - \frac{3X_\gamma}{\pi} I_d$$

For the Type II commutation state:

$$\frac{2\pi^2}{27} \times \frac{U_d^2}{E^2} + \frac{2I_d^2 X_\gamma^2}{E^2} = 1$$

For the Type III commutation state:

$$U_d = \frac{3\sqrt{6}}{\pi} E \cos(\alpha - 30°) - \frac{9}{\pi} X_\gamma I_d$$

To facilitate application, the external characteristics of the rectifier will be expressed in per-unit values. For the rectified voltage U_d, the reference per unit will be U_{d0}; that is,

$$U_d^* = U_{d0} = \frac{3\sqrt{2}}{\pi} E$$

For the rectified current I_d, the reference per unit will be the amplitude of the two-phase short-circuit current on the AC side of the rectifier bridge; that is,

$$I_d^* = \frac{E}{\sqrt{2} X_\gamma}$$

Accordingly, Types I, II, and III commutation state equations can be converted into the per-unit expressions. When the control angle $\alpha = 0$, the individual per-unit expressions for the Types I, II, and III commutation states are

$$U_d^* = 1 - \frac{I_d^*}{2} \quad \left(0 < I_d^* \leq \frac{1}{2}\right) \tag{6.86}$$

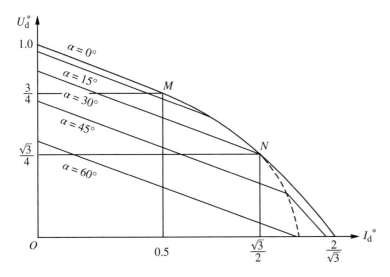

Figure 6.18 External characteristic curve for three-phase bridge rectifier circuit expressed in per-unit values.

$$\frac{4}{3}U_d^{*2} + I_d^{*2} = 1 \quad \left(\frac{1}{2} < I_d^* \leq \frac{\sqrt{3}}{2}\right) \tag{6.87}$$

$$\frac{2}{3}U_d^* + \sqrt{3}I_d^* = 2 \quad \left(\frac{\sqrt{3}}{2} < I_d^*\right) \tag{6.88}$$

See Figure 6.18 for the external characteristic curve of the rectifier as expressed by per-unit values.

6.13 Operating Principle of Three-Phase Bridge Inverter Circuit

In the three-phase full-control-bridge rectifying circuit, the excitation power is supplied by the rectified AC power supply. In the case of an inductive load, when the control angle is $60° < \alpha < 90°$, the instantaneous rectified voltage value U_d at the output of the rectifier will alternately be a positive value and a negative value. As shown in Figure 6.19, to facilitate the discussion, it is assumed that $\gamma = 0$.

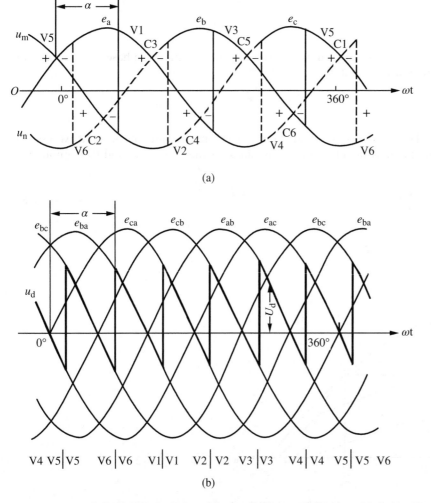

Figure 6.19 Rectified voltage waveform in the case of $60° < \alpha < 120°$ and $\gamma = 0$.

From Figure 6.19, when $\alpha > 60°$, the instantaneous rectified voltage value will be negative; when $\alpha = 90°$, the instantaneous rectified voltage value will be positive and negative equally, with the average value for the rectified voltage being zero; that is, $U_d = U_{d0} \cos 90° = 0$. When the load is pure resistance, only when the instantaneous voltage value is positive can an intermittent current flows through the load in the time interval. If the load has a greater inductance value (such as the excitation winding of the generator), the direct current will be continuous, and its average value will also be positive.

If the control angle keeps increasing until $60° < \alpha < 120°$, the negative area determined by the instantaneous value of the rectified voltage U_d will be larger than the positive area, which makes the synthesized voltage negative, with the result that the rectified voltage polarity changes inversely.

When $120° < \alpha < 180°$, as shown in Figure 6.20, all the rectified voltage areas will be negative, until the negative voltage in the case of $\alpha = 180°$ achieves the maximum $U_d = U_{d0} \cos \alpha = U_{d0} \cos 180° = -U_{d0}$.

From the above discussion, $\alpha > 90°$ will invert the rectified voltage polarity, whose direction is opposite to the direct current that is still flowing in the positive direction. In such a case, it cannot continue to supply the energy to the load; rather, it feeds back the energy stored in the load inductance to the AC side, making the rectifier enter the inverting state.

In the generator excitation loop, the inverting state is mainly used for de-excitation in the generator excitation loop. In the meantime, since the DC load side is passive, it will attenuate to zero with the generator excitation current, in which case the inversion process is completed. Such an inversion process is quite short, which is different from the steady state when the inverter works in the DC transmission line.

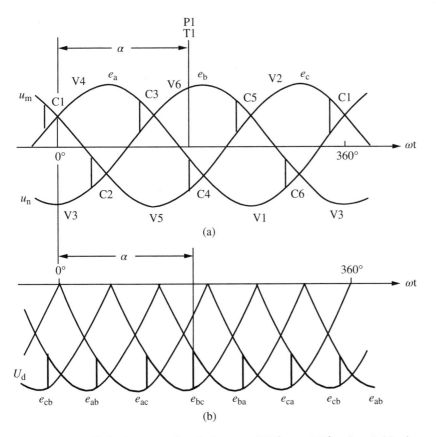

Figure 6.20 Rectified voltage waveform in the case of $120° < \alpha < 180°$ and $\gamma = 0$: (a) voltage oscillogram at the positive and negative terminals "m" and "n" of the direct output against the neutral point O, (b) the DC voltage U_d oscillogram.

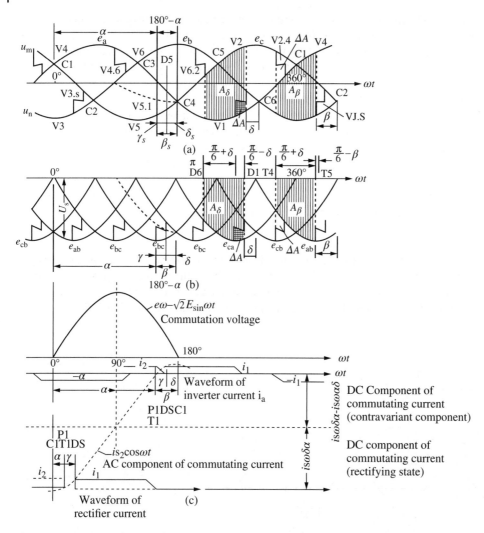

Figure 6.21 Inverter voltage and current oscillogram: (a) phase voltage oscillogram, (b) inverter voltage oscillogram, (c) current oscillogram.

To facilitate comparison with the rectifier, we shall relate the phase control angle of the inverter to β ; the relationship expression between β and α is as follows:

$$\beta = 180° - \alpha \tag{6.89}$$

where

β – inversion angle.

By taking Figure 6.21 as the example, the operation process of the inverter will be explained here. At P1, to trigger the element V1 with an angle of β_1, β will be the angle of the leading line voltage ZCP C4, the interval between C1 and C4 being 180°.

Elements V1 and V5 commutate, and the commutation is completed at D5 via the commutation angle γ_{51}. The element V5 is switched off at the δ_5 angle that corresponds to the leading line voltage ZCP C4, the element V1 fully conducts, and the relationship between the above angles expressed as

$$\beta_1 = \delta_5 + \gamma_{51} \tag{6.90}$$

6.13 Operating Principle of Three-Phase Bridge Inverter Circuit

When the operating angles of each element are equal, the following general expression is obtained:

$$\beta = \delta + \gamma \tag{6.91}$$

where

δ – leading turn-off angle.

In addition, when the inverter is working, Figures 6.21a,b show the waveforms of the voltages U_m and U_n at the DC terminals "m" and "n" of the inverter relative to the neutral point and the waveform of the DC voltage U_d. From the figure, we see that if the operating angles δ and γ of the inverter are respectively the same as the operating angles α and γ of the rectifying states, then the waveforms of U_m and U_n will fully correspond to that of U_d. The inverter voltage waveform can only be obtained when the rectified voltage waveform in Figure 6.15 is rotated by 180° in the horizontal axis ωt plane. Thus, the expression for the inverter voltage can only be solved by substituting δ with α in the formula of the rectified voltage.

When the voltage waveform between terminals "m" and "n" as shown in Figure 6.21a is constructed by taking the ωt axis as the zero point as shown in Figure 6.21b, the figure indicates the inverting negative voltage waveform between the anode and the cathode of each element.

The expression for the DC voltage of the rectifier shows that the inverter can take the phase angle δ as the limit to divide the waveform of the instantaneous DC voltage value U_d in a cycle into six identical and equal parts, with 1/6 (such as between D6–D1) taken to calculate the average inverse voltage of the inverter. We then obtain the following by ignoring the minus sign:

$$U_d = \frac{1}{\pi/3}(A_\delta - \Delta A) \tag{6.92}$$

where

A_δ – all the shaded area under the current curve of Figure 6.21b;

ΔA – loss area caused by commutation, which is marked by the rectangular shaded area.

As with Eq. (6.16) for the rectified voltage, A_δ in Eq. (6.92) can be written as

$$\frac{3}{\pi}A_\delta = U_{d0}\cos\delta \tag{6.93}$$

whereas:

$$\frac{3}{\pi}\Delta A = \frac{1}{2}U_{d0}[\cos\delta - \cos(\delta + \gamma)] = \frac{3X_\gamma}{\pi}I_d = d_\gamma I_d \tag{6.94}$$

We obtain

$$U_d = U_{d0}\cos\delta - d_\gamma I_d = \frac{1}{2}U_{d0}[\cos\delta + \cos(\delta + \gamma)] \tag{6.95}$$

When calculating the average inverter inverse voltage by taking the β phase angle as the limit, the area between T4 and T5 is taken to obtain

$$U_d = \frac{1}{\frac{\pi}{3}}(A_\beta + \Delta A) \tag{6.96}$$

where

A_β – all the shaded area under the e_{ab} current curve of Figure 6.21b;

ΔA – added area for the voltage waveform caused by commutation, with its value equaling the ΔA area in Eq. (6.92).

Therefore, we obtain

$$U_d = U_{d0}\cos\beta + d_\gamma I_d \tag{6.97}$$

From Eq. (6.94), we obtain

$$d_\gamma I_d = \frac{1}{2} U_{d0}[\cos\delta - \cos(\delta + \gamma)]$$

Substituting this into Eqs. (6.95) or (6.97) yields

$$U_d = \frac{1}{2} U_{d0}[\cos\delta + \cos(\delta + \gamma)] \tag{6.98}$$

Eqs. (6.95) and (6.97) are the expressions when the phase angles δ and β are respectively taken as the operating parameters, with identical results.

As shown in Eq. (6.98), the operating parameters are represented individually by δ and β; Eq. (6.98) becomes

$$U_d = U_{d0}\cos\left(\delta + \frac{\gamma}{2}\right)\cos\frac{\gamma}{2} = U_{d0}\cos\left(\beta - \frac{\gamma}{2}\right)\cos\frac{\gamma}{2} \tag{6.99}$$

According to the above discussion: When $\alpha > 90°$ or $\beta < 90°$, the rectifier will shift to inverter operation.

7

Excitation System for Separately Excited Static Diode Rectifier

7.1 Harmonic Analysis for Alternating Current

In the separately excited static excitation system that adopts the diode rectifier, since the exciter supplies power to the rectifier load, distortion of the phase current of the stator with a non-sinusoidal waveform occurs. Therefore, when determining the relevant parameters for the excitation system, generally a harmonic analysis of this non-sinusoidal current becomes necessary.

Therefore, if we consider only the fundamental component of alternating current, its expression is

$$i_1 = A_1 \cos\theta + B_1 \sin\theta \tag{7.1}$$

The amplitude of the fundamental current is

$$I_{m1} = \sqrt{A_1^2 + B_1^2} \tag{7.2}$$

If the initial point in the A-phase potential $e_A = \sqrt{2}E_x \sin\theta = e_x$ curve is taken as the origin of the coordinates, as shown in Figure 7.1, then the fundamental current component of the A-phase current can be written as

$$i_A = I_{m1} \sin(\theta - \varphi_1)$$

where

φ_1 — is the phase angle of the lagging power potential of the fundamental current.

The tangent value for φ_1 is

$$\tan\varphi_1 = \frac{B_1}{A_1} \tag{7.3}$$

The cosine Fourier coefficient A_1 and the sinusoidal Fourier coefficient B_1 in the AC fundamental components separately denote the active and reactive current components of the fundamental current related to the potential, which are

$$I_{am1} = |A_1|, \quad I_{rm1} |B_1| \tag{7.4}$$

If

$$A_1^* = \frac{A_1}{I_f}$$

$$B_1^* = \frac{B_1}{I_f} \tag{7.5}$$

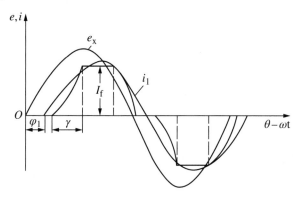

Figure 7.1 Decomposition of fundamental current component.

then

$$I_{m1} = \sqrt{A_1^2 + B_1^2} = \sqrt{A_1^{*\,2} + B_1^{*\,2}}\ I_f = C_1^*\ I_f$$

$$C_1^* = \sqrt{A_1^{*\,2} + B_1^{*\,2}} \tag{7.6}$$

where

C_1^* – the ratio between the fundamental current amplitude and the rectified current.

From Eqs. (7.5) and (7.6), we obtain

$$I_{rm1} = A_1^*\ I_f$$
$$I_{am1} = B_1^*\ I_f \tag{7.7}$$

Table 7.1 lists the Fourier coefficient values for the fundamental waves that change with the rectified current I_f and are expressed in per-unit and non-per-unit values. The relevant per-unit basic values are $I_{fb} = \frac{\sqrt{6}E_x}{2X_\gamma}$, $U_{fb} = \sqrt{6}E_x$.

Figure 7.2 shows the relationship curve, which indicates that the fundamental wave coefficients change with the rectified current.

To simplify the calculations, the relevant coefficient curves in Figure 7.2 can be linearized, so we can use the approximation $C_1^* = 1.07$. When the rectified current I_f^* changes between 0 and 1.15, its error will be no more

Table 7.1 Expressions used to calculate Fourier coefficient for fundamental current in three-phase bridge rectifier circuit.

Operating state	A_1	A_1^*	B_1	B_1^*
I	$\dfrac{3\sqrt{2}E_x}{2\pi X_\gamma}\sin^2\gamma$	$\dfrac{\sqrt{3}}{\pi}(1+\cos\gamma)$	$\dfrac{3\sqrt{2}E_x}{2\pi X_\gamma}[2\gamma - \sin 2\gamma]$	$\dfrac{\sqrt{3}}{2\pi} \times \dfrac{(2\gamma - \sin 2\gamma)}{(1-\cos\gamma)}$
II	$\dfrac{3\sqrt{2}E_x}{2\pi X_\gamma}\sin\left(\gamma + \dfrac{\pi}{3}\right) \times \sin 2\left(\alpha + \dfrac{\gamma}{2}\right)$	$\dfrac{3}{\pi}\cos\left(\alpha + \dfrac{\pi}{6}\right)$	$\dfrac{3\sqrt{2}E_x}{2\pi X_\gamma}\left[\gamma - \sin\left(\gamma + \dfrac{\pi}{3}\right)\right.$ $\left. \times \cos 2\left(\alpha + \dfrac{\gamma}{2}\right)\right]$	$\dfrac{\sqrt{3}}{2\pi} \times \dfrac{\left[2\gamma - \sqrt{3}\cos\left(2\alpha + \dfrac{\pi}{3}\right)\right]}{\sin\left(\alpha + \dfrac{\pi}{6}\right)}$
III	$\dfrac{3\sqrt{2}E_x}{2\pi X_\gamma}\sin\left(\gamma + \dfrac{\pi}{3}\right) \times \sin 2\left(\alpha + \dfrac{\gamma}{2}\right)$	$\dfrac{3}{\pi}\left[1+\cos\left(\gamma + \dfrac{\pi}{3}\right)\right]$	$\dfrac{3\sqrt{2}E_x}{2\pi X_\gamma}\left[\gamma - \sin\left(\gamma + \dfrac{\pi}{3}\right)\right.$ $\left. \times \cos 2\left(\alpha + \dfrac{\gamma}{2}\right)\right]$	$\dfrac{3}{2\pi} \times \dfrac{\left[2\gamma - \sin\left(2\gamma + \dfrac{2}{3}\pi\right)\right]}{1-\cos\left(\gamma + \dfrac{\pi}{3}\right)}$

Figure 7.2 Curve of the relationship between harmonic current coefficients A_1^*, B_1^*, C_1^* with the rectified current I_f^*.

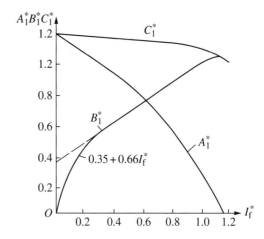

than 3%. The coefficient B_1^* can be linearized in segments:

When $0.15 \leqslant I_f^* \leqslant 0.95$, $B_1^* = 0.35 + 0.66\ I_f^*$

When $0.95 \leqslant I_f^* \leqslant 1.15$, $B_1^* = 1$ (7.8)

In addition, the phase angle E_x between the fundamental current and the potential φ_1 can be approximately written as

$$\varphi_1 = 0.69(\alpha + \gamma) \tag{7.9}$$

The effective value of the fundamental current is

$$I_1 = \frac{C_1^* I_f}{\sqrt{2}} \tag{7.10}$$

7.2 Non-distortion Sinusoidal Potential and Equivalent Commutating Reactance

7.2.1 Non-distortion Sinusoidal Potential and Equivalent Commutating Reactance

As mentioned above, since the exciter supplies power to the rectifier load, its phase current has a non-sinusoidal characteristic. Consequently, severe distortion exists in the phase voltage waveform of the exciter. Therefore, this voltage cannot be taken as the internal potential for the power supply when calculating the rectifying mode. To improve the waveform of the resultant magnetic flux of the exciter, a damping winding can be mounted, so that most of the higher harmonic in the magnetic flux would be absorbed in the damping loop. Therefore, the resultant flux linkage in the rotor loop and the sub-transient potential E'' that is in proportional to it could actually change sinusoidally, which can be approximately used as the internal potential when calculating the external characteristic of the rectifier. For a non-damping winding exciter, the higher harmonic in the resultant magnetic flux will increase, and the increase is not very obvious, because some harmonic components are compensated for by the powerful excitation winding loop. On account of the AC exciter in the non-damping winding loop, the sub-transient potential E'' can be taken as the non-distortion potential of the rectifier power.

Just as the generator power characteristic can be expressed with steady-state, transient, and sub-transient parameters, the external characteristics of the rectifier can also be expressed with different state parameters

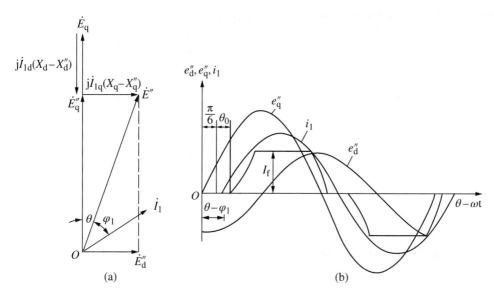

Figure 7.3 Fundamental current phasor diagram expressed with sub-transient potential: (a) phasor diagram, (b) oscillogram.

mentioned above. The Type I rectifying state external characteristic equation expressed with the sub-transient parameters I will be:

$$U_f = \frac{3\sqrt{3}}{\pi} E'' \cos(\theta_0 - \theta) - \frac{3}{\pi} I_f \left[\frac{X_d'' + X_q''}{2} - (X_q'' - X_d'') \sin\left(2\theta_0 + \frac{\pi}{6}\right) \right] \quad (7.11)$$

See Figure 7.3 for the phasor diagram corresponding to the sub-transient potential.

By this time, the analysis on the commutation process will be quite complicated, because the exciter is in a state in which the two phases are abruptly shorted during the commutation process. The commutating reactance parameters will not only be determined by the sub-transient parameters, but will also change as the rotor position changes, which is also a function of time. Therefore, it will be very complicated to represent this process with a mathematical equation.

In general, in the time interval before commutation, the phase current of the exciter will be equal to the rectified current I_f, which is constant. In the commutation process, the superimposed short-circuit current flows through the commutating phase. When leaving the commutating phase, the commutating current can be decomposed into two components: the constant rectified current I_f and the two-phase short-circuit current i_2, as shown in Figure 7.4. For the AC exciter with an asymmetric rotor magnetic pole ($X_d'' \neq X_q''$), the DC component current I_f flowing through the exciter stator winding will cause double-frequency potential in the rotor. However, a change in the two-phase short circuit is related to the exciter armature impedance and the rotor angular position, thus making the analysis more complicated. The expression for the commutating current will be

$$i = \left\{ \sqrt{3} E_m'' \left[\cos(\theta_0 - \theta) - \cos(\omega t + \theta_0 - \theta) \right] - I_f (X_q'' - X_d'') \left[\sin\left(2\omega t + 2\theta_0 + \frac{\pi}{6}\right) \right. \right.$$
$$\left. \left. - \sin\left(2\omega t + \frac{\pi}{6}\right) \right] \right\} \div \left[(X_d'' - X_q'') - (X_q'' - X_d') \cos(2\theta + 2\theta_0) \right] \quad (7.12)$$

7.2.2 Non-distortion Sinusoidal Potential Determined by Simplification

As mentioned above, the commutating current is a function of time and space. To simplify the analysis, decomposing the non-sinusoidal harmonic current and replacing the armature current during commutation

Figure 7.4 Decomposition of the commutating current into DC component I_f and two-phase short-circuit current i_2 during the commutation process.

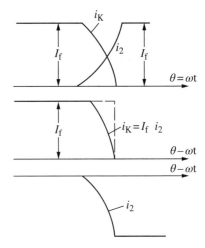

with the fundamental current can greatly simplify the analysis. To ensure that this type of replacement leaves the excitation current unchanged, an equivalent potential is used to replace the sub-transient potential E'' at the same time. According to the analysis, the potential after the commutating reactance E_x is very close to E'', enabling E_x to be utilized as the non-distortion rectifying potential. See Figure 7.5 for the corresponding phasor diagram.

The above approach was proposed by Mr. Shechtman.

By this time, the AC exciter's operation is equivalent to working in the steady state, with the exciter parameters separately expressed with respect to the direct axis and the quadrature axis of the coordinate system.

When constructing the phasor diagram shown in Figure 7.5, first one should make the fundamental current $\dot{I}_1$ lag behind $\dot{E}_x$ by the φ_1 phase angle that is after the non-distortion rectifying potential $\dot{E}_x$, and then draw the armature reaction voltage drop $j\dot{I}_{1q}(X_q - X_\gamma)$, $j\dot{I}_{1d}(X_d - X_\gamma)$ for both the direct axis and the quadrature axis the at $\dot{E}_x$ terminal, thus obtaining the interval potential $\dot{E}_q$.

7.2.3 Calculation of Equivalent Commutating Reactance [14]

For the Type I commutation state, the rectifier element will work in the 2–3 mode. In the commutation process, the two commutation voltages (such as the B and C phases) will be equivalent to the short circuit between the

Figure 7.5 AC exciter phasor diagram expressed by fundamental current and non-distortion potentials: (a) non-salient pole, (b) salient pole.

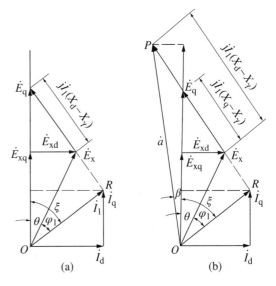

line voltages, with its short-circuit current expressed as

$$i = -\frac{\sqrt{2} \times \sqrt{3} E_x (\sin\theta - \sin\theta_0)}{(X_d'' + X_q'') - (X_d'' - X_q'')\cos 2\theta} \tag{7.13}$$

where

E_x — non-distortion sinusoidal phase potential;

X_d'', X_q'' — the sub-transient reactance for the direct axis and quadrature axis of the AC exciter;

θ_0 — included angle between the d-axis magnetic potential and the commutating phasor (such as the A phase) which is not involved at the moment of starting to commutate. See Figure 7.6 for the commutating current and magnetic potential phasor diagrams. δ is the included angle between the commutating magnetic potential and the d-axis magnetic potential.

Since the commutation from the B phase to the C phase starts from the natural commutation point, the included angle between their resultant magnetic potential $\dot{F}$ and the A-phase phasor will always be $\theta_0 = \frac{\pi}{2}$, which is unchanged in the commutation process. However, the commutating magnetic potential F changes with the value of commutating current, its direction always orthogonal to the winding axis that is not commutated, thus yielding:

$$\theta = \frac{\pi}{2} - \delta$$

Substituting the above θ value into Eq. (7.13), it can be simplified as

$$i = \frac{\sqrt{2} \times \sqrt{3}(1 - \cos\delta) E_x}{(X_d'' + X_q'') + (X_d'' - X_q'')\cos 2\delta} \tag{7.14}$$

Since the commutation process is quite short, with the winding resistance excluded, we will assume that the commutating current i does not decay. If $\omega t = \delta$, $\alpha = 0$, by substitution into Eq. (6.9), another expression for the commutating current can be obtained:

$$i = I_f = \frac{\sqrt{2} \times \sqrt{3}\,(1 - \cos\delta)E_x}{2X_\gamma} \tag{7.15}$$

A comparison of Eqs. (7.14) and (7.15) yields the following:

$$X_\gamma = \frac{1}{2}\,[(X_d'' + X_q'') + (X_d'' - X_q'')\cos 2\delta] \tag{7.16}$$

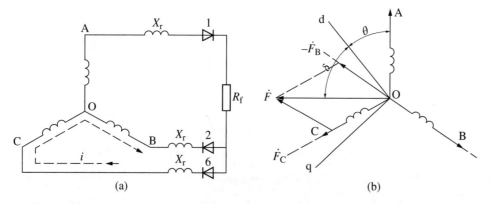

Figure 7.6 Commutating current and magnetic potential phasor diagram: (a) equivalent circuit of the commutating current, (b) spatial phasor diagram of the commutating magnetic potential.

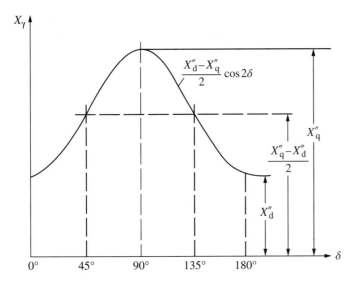

Figure 7.7 Under Type I commutation state, the changes in commutating reactance X_γ with δ when $\alpha = 0$.

A curve can be drawn on the basis of the function $X_\gamma = f(\delta)$, as shown in Figure 7.7.

When $\alpha \neq 0$, the spatial position δ of the magnetic pole during commutation will change from α to $\alpha + \gamma$, and substituting this expression into Eq. (7.16) can integrate the change in the commutation interval $\cos 2\delta$; then the average commutating reactance value during commutation is

$$\begin{aligned}
X_\gamma &= \frac{1}{\gamma} \int_\alpha^{\alpha+\gamma} f(\delta) d\delta \\
&= \frac{X_d'' + X_q''}{2} + \frac{X_d'' - X_q''}{2} \times \frac{1}{2} \times \frac{1}{\gamma} \times \int_\alpha^{\alpha+\gamma} \cos 2\delta \, d(2\delta) \\
&= \frac{X_d'' + X_q''}{2} + \frac{X_d'' - X_q''}{2} \times \frac{\sin 2(\alpha+\gamma) - \sin 2\alpha}{2\gamma} \\
&= \frac{X_d'' + X_q''}{2} + \frac{X_d'' + X_q''}{2} \times K \\
K &= \frac{\sin 2(\alpha+\gamma) - \sin 2\alpha}{2\gamma} = f(\alpha, \gamma)
\end{aligned} \qquad (7.17)$$

where

K – commutating coefficient.

According to Eq. (7.17), the calculation of the commutating reactance is complicated. We will simplify Eq. (7.17) according to the specific conditions in the actual applications, for example:

(1) For the Type I operating state, when the γ angle is smaller, if the extreme value for K is taken, one will get $K \approx 1$, in this case, $X_\gamma \approx X_d''$.
(2) For the Type II operating state, the commutating reactance X_γ is larger, with $K \approx 0$. In this case, $X_\gamma = \frac{X_d'' + X_q''}{2} = X_2$ equals the negative sequence reactance.

For the AC exciter that is provided with a damping winding and is weak in the salient pole effect, since the difference between X_d'' and X_q'' is small, the commutating reactance X_γ can be taken as

$$X_\gamma = \frac{X_d'' + X_2}{2} \qquad (7.18)$$

If the salient pole effect is strong, which is $X''_d \neq X''_q$ or $X'_d + X'_q$, the error in the X_γ value obtained by using the above approximation method will be larger. In this case, the commutating reactance is between $\frac{1}{2}(X''_d + X''_q)$ and $\sqrt{X''_d \times X''_q}$, and the commutating reactance can be calculated by using the following equations:

$$X_\gamma = \frac{\frac{1}{2}(X''_d + X''_q) + \sqrt{X''_d \times X''_q}}{2} \tag{7.19}$$

or

$$X_\gamma = \sqrt{X''_d X''_q} \tag{7.20}$$

7.3 Expression for Commutation Angle γ, Load Resistance r_f, and Commutating Reactance X_γ [15]

In general, the Type I operating state is often chosen at the time of designing the rectifying circuit, because the slope of the commutating phase voltage drop is minimal in this case. Now we will determine the expression for the relevant parameters that affect the commutating voltage drop slope and the various operating states. In the steady operating state, the rectified voltage U_f and the rectified current I_f have the following relationship:

$$U_f = I_f r_f \tag{7.21}$$

where

r_f — the field winding resistance of the generator.

When the control angle $\alpha = 0$, the expression for the rectified voltage and current in Type I operating state will be

$$\left. \begin{array}{l} U_f = \frac{3\sqrt{6}}{\pi} \times E_x - \frac{3}{\pi} I_f X_\gamma \\ I_f = \sqrt{\frac{3}{2}} \times \frac{E_x}{X_\gamma}(1 - \cos\gamma) \end{array} \right\} \tag{7.22}$$

Calculation after substituting Eq. (7.22) into Eq. (7.21) yields

$$\cos\gamma = \frac{1 - \frac{3}{\pi} \times \frac{X_\gamma}{r_f}}{1 + \frac{3}{\pi} \times \frac{X_\gamma}{r_f}} \tag{7.23}$$

In the Type I commutation state, the commutation angle is $0 \leqslant \gamma \frac{\pi}{3}$, or $\frac{1}{2} \leqslant \cos\gamma \leqslant 1$, and therefore the scope of $\frac{X_\gamma}{r_f}$ under the Type I operating state is

$$0 \leqslant \frac{X_\gamma}{r_f} \leqslant \frac{\pi}{9} \tag{7.24}$$

In the Type II commutation state, the commutation angle $\gamma = \frac{\pi}{3}$ is unchanged. However, the forced lagging control angle changes between $0 \leqslant \alpha' \leqslant \frac{\pi}{6}$. By giving $\alpha' = 0, \frac{\pi}{6}$, and substituting them into the Type II operating state, the rectified voltage and current expressions are separately as follows:

$$U_f = \frac{9\sqrt{2}E_x}{2\pi} \cos\left(\alpha' + \frac{\pi}{6}\right)$$

$$I_f = \frac{\sqrt{6}E_x}{2X_\gamma} \sin\left(\alpha' + \frac{\pi}{6}\right)$$

7.3 Expression for Commutation Angle γ, Load Resistance r_f, and Commutating Reactance $X_γ$

By combining with Eq. (7.21), the variation range of $\frac{X_γ}{r_f}$ in the Type II operating state is obtained as

$$\frac{\pi}{9} \leq \frac{X_γ}{r_f} \leq \frac{\pi}{3} \tag{7.25}$$

The separate expressions for the rectified voltage and current in the Type III operating state are

$$U_f = \frac{9\sqrt{2}E_x}{\pi}\cos\left(\alpha' - \frac{\pi}{6}\right) - \frac{9}{\pi}I_f X_γ$$

$$I_f \frac{E_x}{\sqrt{2}X_γ}\left[\cos\left(\alpha' - \frac{\pi}{6}\right) - \cos\left(\alpha' + γ + \frac{\pi}{6}\right)\right]$$

In the Type III operating state, the lagging control angle $\alpha' = \frac{\pi}{6}$ is unchanged. From this, one can obtain r_f by substituting it into the above rectified voltage and current expressions and then substituting the relevant U_f and I_f value into Eq. (7.21). After calculations, one can obtain the expression between the commutation angle γ and $\frac{X_γ}{r_f}$ in the Type III operating state:

$$\cos(γ + 2\alpha') = \frac{1 - \frac{9}{\pi} \times \frac{X_γ}{r_f}}{1 + \frac{9}{\pi} \times \frac{X_γ}{r_f}} \tag{7.26}$$

or

$$γ = \cos^{-1}\frac{1 - \frac{9}{\pi} \times \frac{X_γ}{r_f}}{1 + \frac{9}{\pi} \times \frac{X_γ}{r_f}} - 60°$$

At this time, the $\frac{X_γ}{r_f}$ ratio changes in the range

$$\frac{\pi}{3} \leq \frac{X_γ}{r_f} \leq \infty$$

See Table 7.2 for the variation ranges of the commutation angles in the various operating states. See Figure 7.8 for the changes in the corresponding curves.

Table 7.2 Expression for commutation angle $γ = f\left(\frac{X_γ}{r_f}\right)$ in three-phase bridge rectifier circuit.

Operating state	Computerized relationship	γ, α″
I	$U_f = \frac{3\sqrt{6}}{\pi}E_x - \frac{3}{\pi}I_f X_γ$ $I_f = \sqrt{\frac{3}{2}} \times \frac{E_x}{X_γ}(1 - \cos γ)$	$\cos γ = \frac{1 - \frac{3}{\pi} \times \frac{X_γ}{r_f}}{1 + \frac{3}{\pi} \times \frac{X_γ}{r_f}}$
II	$U_f = \frac{9}{2\pi}\sqrt{2E_x^2 - \frac{4}{3}(I_f X_γ)^2}$ $I_f = \sqrt{\frac{3}{2}} \times \frac{E_x}{X_γ}\sin\left(\alpha' + \frac{\pi}{6}\right)$	$γ = \frac{\pi}{3}$ unchanged $\sin\left(\alpha' + \frac{\pi}{6}\right) = \frac{1}{\sqrt{1 + \frac{\pi^2}{81} \times \left(\frac{X_γ}{r_f}\right)^2}}$
III	$U_f = \frac{9\sqrt{2}E_x}{\pi}\cos\left(\alpha' - \frac{\pi}{6}\right) - \frac{9}{\pi}I_f X_γ$ $I_f = \frac{E_x}{\sqrt{2}X_γ}\left[\cos\left(\alpha' - \frac{\pi}{6}\right) - \cos\left(\alpha' + γ + \frac{\pi}{6}\right)\right]$	$\alpha = \frac{\pi}{3}$ unchanged $\cos(γ + 2\alpha') = \frac{1 - \frac{9}{\pi} \times \frac{X_γ}{r_f}}{1 + \frac{9}{\pi} \times \frac{X_γ}{r_f}}$

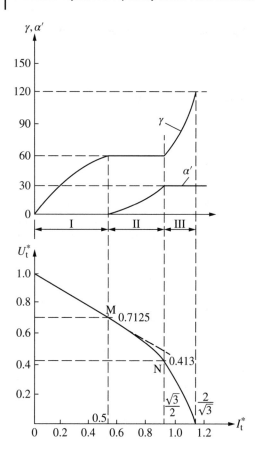

Figure 7.8 Relation curve for commutation angle $\gamma = f\left(\frac{X_\gamma}{r_f}\right)$ in three-phase bridge rectifier circuit under various operating states.

7.4 Rectified Voltage Ratio β_u and Rectified Current Ratio β_i

7.4.1 Rectified Voltage Ratio β_u

$$\beta_u = \frac{\frac{1}{T}\int_0^T u_f d\theta}{\sqrt{\frac{1}{T}\int_0^T e^2 d\theta}} = \frac{U_f}{E_x} \tag{7.27}$$

where

U_f, u_f — average value and instantaneous value for rectified voltage

E_x, e — effective value and instantaneous value for AC power supply

For the Type I operating state, the rectified voltage ratio β_u can be obtained from the corresponding equations of U_f and u_f as shown in Table 7.2 and Eq. (7.23) simultaneously:

$$\beta_u = \frac{U_f}{E_x} = \frac{3\sqrt{6}}{\pi\left(1 + \frac{3}{\pi} \times \frac{X_\gamma}{r_f}\right)} \tag{7.28}$$

Table 7.3 shows the expressions for the rectified voltage ratio β_u with $\frac{X_\gamma}{r_f}$ in the various operating states. Figure 7.9 shows the curve of the rectified voltage ratio $\beta_u = f\left(\frac{X_\gamma}{r_f}\right)$ in the three-phase bridge rectifier circuit in the various operating states.

Table 7.3 Expressions for the rectified voltage ratios $\beta_u = f\left(\frac{X_\gamma}{r_f}\right)$ in various operating states in three-phase bridge rectifier circuit.

Operating state	$\dfrac{X_\gamma}{r_f}$	Expression	β_u
I	$0 \sim \dfrac{\pi}{9}$	$U_f = \dfrac{3\sqrt{6}}{\pi}E_x - \dfrac{3}{\pi}I_f X_\gamma$ $I_f = \sqrt{\dfrac{3}{2}} \times \dfrac{E_x}{X_\gamma}(1 - \cos\gamma)$	$\dfrac{3\sqrt{6}}{\pi\left(1 + \dfrac{3}{\pi} \times \dfrac{X_\gamma}{r_f}\right)}$
II	$\dfrac{\pi}{9} \sim \dfrac{\pi}{3}$	$U_f = \dfrac{9}{2\pi}\sqrt{2E_x^2 - \dfrac{4}{3}(I_f X_\gamma)^2}$ $I_f = \sqrt{\dfrac{3}{2}} \times \dfrac{E_x}{X_\gamma}\sin\left(\alpha' + \dfrac{\pi}{6}\right)$	$\dfrac{9}{\sqrt{2\pi^2 + 54\left(\dfrac{X_\gamma}{r_f}\right)^2}}$
III	$\dfrac{\pi}{3} \sim \infty$	$U_f = \dfrac{9\sqrt{2}}{\pi}E_x \cos\left(\alpha' - \dfrac{\pi}{6}\right) - \dfrac{9}{\pi}I_f X_\gamma$ $I_f = \dfrac{E_x}{\sqrt{2}X_\gamma}\left[\cos\left(\alpha' - \dfrac{\pi}{6}\right) - \cos\left(\alpha' + \gamma + \dfrac{\pi}{6}\right)\right]$	$\dfrac{9\sqrt{2}}{\pi\left(1 + \dfrac{9}{\pi} \times \dfrac{X_\gamma}{r_f}\right)}$

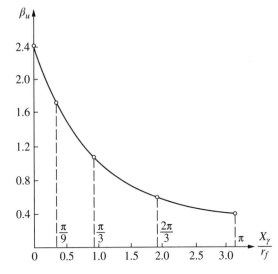

Figure 7.9 Relationship curve of rectified voltage ratio $\beta_u = f\left(\dfrac{X_\gamma}{r_f}\right)$.

7.4.2 Rectified Current Ratio β_i

The rectified current ratio is defined as

$$\beta_i = \frac{\frac{1}{T}\int_0^T i_f d\theta}{\sqrt{\frac{1}{T}\int_0^T i^2 \, d\theta}} = \frac{I_f}{I} \tag{7.29}$$

where

I_f, i_f — average value and instantaneous value for the rectified current;

I, i — average value and instantaneous value for the alternating current.

For the Type I commutation state, when calculating the rectifier current ratio, it can be assumed approximately that the changes in the commutating current are based on linear regularity in the commutation process.

In the $0 \leq \theta \leq \gamma$ interval: $i = \frac{I_f}{\gamma} = \theta$

In the $\gamma \leq \theta \leq \frac{2\pi}{3}$ interval: $i = I_f$

In the $\frac{2\pi}{3} \leq \theta \leq \frac{2\pi}{3} + \gamma$ interval: $i = \frac{I_f}{\gamma}\left(\frac{2\pi}{3} + \gamma - \theta\right)$

The phase current waveform in the negative half cycle is the same as that in the positive half cycle; therefore, the effect current value in one cycle can be obtained as

$$I = \sqrt{2\frac{1}{2\pi}\left[\int_0^\gamma \left(\frac{I_f}{\gamma}\theta\right)^2 d\theta + \int_\gamma^{\frac{2}{3}\pi} I_f^2 \, d\theta + \int_{\frac{2}{3}\pi}^{\frac{2}{3}\pi+\gamma} \left(\frac{2\pi}{3} + \gamma - \theta\right) \times \left(\frac{I_f}{\gamma}\right)^2 d\theta\right]}$$

By integrating the above equation and simplifying it, one obtains

$$I = I_f \times \frac{\sqrt{2 - \frac{\gamma}{\pi}}}{\sqrt{3}}$$

or

$$\beta_i = \frac{I_f}{I} = \frac{\sqrt{3}}{\sqrt{2 - \frac{\gamma}{\pi}}} \tag{7.30}$$

For the Types II and III commutation states, due to the presence of various forced lagging control angles α'' and various commutation angles γ, the values of the rectified current ratio β_i will differ. In this case, the expression relating the effective AC value I and the rectified current I_f of the element can be obtained from Table 6.1:

$$I = \frac{I_f}{\sqrt{3}}\sqrt{1 - 3\psi(\alpha', \gamma)} \tag{7.31}$$

$$\psi(\alpha', \gamma) = \frac{1}{2\pi} \frac{\sin\gamma[2 + \cos(2\alpha' + \gamma)] - \gamma[1 + 2\cos\alpha' + \cos(\alpha' + \gamma)]}{[\cos\alpha' - \cos(\alpha' + \gamma)]^2}$$

The waveform of the current flowing through the AC exciter in the negative half cycle is the same as that in the positive half cycle. Therefore, the expression relating the effective AC value I and the average rectified current value I_f in a cycle is

$$I = \sqrt{\frac{2}{3}I_f}\sqrt{1 - 3\psi(\alpha', \gamma)} \tag{7.32}$$

Thus, the corresponding rectified current ratio β_i is

$$\beta_i = \frac{I_f}{I} = \sqrt{\frac{3}{2}} \times \frac{1}{\sqrt{1 - 3\psi(\alpha', \gamma)}} \tag{7.33}$$

7.5 Steady-State Calculations for AC Exciter with Rectifier Load

The steady-state calculation for the AC exciter with a rectifier load is mainly used to determine the internal potential E_q.

When using the AC exciter fundamental current I_1, the commutating reactance X_γ, and the non-distortion potential E_x to calculate the external characteristic of the rectifier and the steady-state parameters related to the exciter, the following procedures are used. The effective value of the fundamental current can be determined from Eq. (7.10):

$$I_1 = \frac{C_1^*}{\sqrt{2}} \times I_f$$

From Figure 7.5, we see that the phase angle φ_1 between the fundamental current $\dot{I}$ and the potential $\dot{E}_x$ can be determined from Eq. (7.9). The commutation angle is determined by the load parameters γ_f and X_γ, with a fixed value. For different steady operating statuses, the included angle φ_1 between the exciter potential and the fundamental current also has a fixed value, but with different the phasor values. The included angle between the potential $\dot{E}_q$ and the fundamental current $\dot{I}_1$ can be obtained from Eq. (7.34):

$$\xi = \arctan \frac{E_x \sin \varphi_1 + I_1(X_q - X_\gamma)}{E_x \cos \varphi_1} \tag{7.34}$$

The included angle θ between the quadrature-axis potentials $\dot{E}_q$ and $\dot{E}_x$ is

$$\theta = \xi - \varphi_1 \tag{7.35}$$

The direct and quadrature-axis components of the fundamental current are

$$\left.\begin{aligned} I_d &= I_1 \sin \xi \\ I_q &= I_1 \cos \xi \end{aligned}\right\} \tag{7.36}$$

The internal potential of the exciter is

$$E_q = E_{xq} + I_d(X_d - X_\gamma) \tag{7.37}$$

The direct and quadrature-axis components of the sub-transient potential are

$$\left.\begin{aligned} E_q'' &= E_{xq} - I_d(X_d'' - X_\gamma) \\ E_d'' &= E_{xq} + I_q(X_q'' - X_\gamma) \end{aligned}\right\} \tag{7.38}$$

The quadrature and direct axis air-gap potential are

$$\left.\begin{aligned} E_{iq} &= E_{xq} - I_d(X_\gamma - X_\sigma) \\ E_{id} &= E_{xq} - I_q(X_\gamma - X_\sigma) \end{aligned}\right\} \tag{7.39}$$

As shown in Figure 7.5b, the internal potential phasor $\dot{E}_q$ in the salient pole exciter is very close to the phasor $\dot{a}$, that is,

$$\dot{a} = \dot{E}_x + j\dot{I}_1 (X_d - X_\gamma)$$

Since the included angle β between the phasors $\dot{a}$ and $\dot{E}_q$ is smaller, it will not cause a major error if replacing phasor $\dot{a}$ with $\dot{E}_q$. So, the potential $\dot{E}_q$ and the reactance X_d can be used to calculate the operation process of the salient pole exciter. The absolute value of the phasor $\dot{a}$ is determined by the OP side of the triangle OPR, yielding

$$a = \sqrt{(E_x \cos \varphi_1)^2 + [E_x \sin \varphi_1 + I_1(X_d - X_\gamma)]^2} \approx E_q \tag{7.40}$$

From the above equations, the steady-state process of the AC exciter supplying power to the diode rectifier can be calculated. The calculations involve two steps: the first step is to determine the operating state of the rectifier on the basis of the known load resistance r_f, exciter commutating reactance X_γ, and rectified current I_f; the second step is to determine the internal potential E_q.

Example 7.1
The calculations for the operating states of the exciter for the TBB-320-2 turbine generator.
Parameters for the generator:

7 Excitation System for Separately Excited Static Diode Rectifier

$P = 300$ MW, $\cos\varphi = 0.85$, $U_{GN} = 20$ kV, $I_{fN} = 2900$ A, $U_{fN} = 447$ V, $r_f = 0.154\Omega$. Parameters for the exciter:

$P_N = 1320$ kW, $\cos\varphi_N = 0.95$, $U_{eN} = 350$ V, $I_{eN} = 2300$ A, $f = 150$ Hz,
$X_d^* = 1.68$, $X_q^* = 1.12$, $X_\sigma^* = 0.139$, $X''^*_d = 0.288$, $X''^*_q = 0.252$.

The rectifier is wired with three-phase bridge connections, and all the reactance values are per-unit values.

Solution

The commutating reactance is obtained from Eq. (7.18) as

$$X_2^* = 0.5(X''^*_d + X''^*_q) = 0.5(0.228 + 0.252) = 0.24$$
$$X_\gamma^* = 0.5(X''^*_d + X''^*_2) = 0.5(0.228 + 0.24) = 0.234$$

$$X_b = \frac{U_{eN}}{I_{eN}} = \frac{350}{\sqrt{3} \times 2300} = 0.0885 (\Omega)$$

The commutating reactance is converted to an actual value:

$$X_\gamma = X_\gamma^* \, X_b = 0.234 \times 0.0885 = 0.0206 (\Omega)$$

where

X_b — basic reactance value.

As the ratio of the exciter commutating reactance to the excitation winding resistance is $\frac{X_\gamma}{r_f} = \frac{0.0206}{0.154} = 0.127 < \frac{\pi}{9} = 0.348$, according to Table 7.3, when the rectifier works in the Type I commutation state, the cosine of the commutation angle can be confirmed from Table 7.2 as

$$\cos\gamma = \frac{1 - \frac{3}{\pi} \times \frac{0.0206}{0.154}}{1 + \frac{3}{\pi} \times \frac{0.0206}{0.154}} = 0.774$$

$$\gamma = 39°$$

Therefore:

The commutating potential E_x determined by the external characteristics of rectification in the Type I operating state is

$$E_x = \frac{\pi}{3\sqrt{6}} \left(447 + \frac{3}{\pi} \times 0.026 \times 2900\right) = 216 (V)$$

Expressed in per-unit values:

$$E_x^* = \frac{216 \sqrt{3}}{350} = 1.065$$

The fundamental current amplitude is

$$I_{m1} = C_1^* \, I_f = 1.07 \times 2900 = 3100 (A)$$

(C_1^* can be obtained from Figure 7.2)

The effect per-unit value for the fundamental current is

$$I_1^* = \frac{3100}{\sqrt{2} \times 2300} = 0.955$$

The included angle between the fundamental current $\dot{I}_1$ and the potential $\dot{E}_x$ is

$$\varphi_1 = 0.69\gamma = 0.69 \times 39.29° = 27.11°$$

$$\tan \xi = \frac{E_x^* \sin \varphi_1 + I_1 \ (X_q - X_\gamma)}{E_x \cos \varphi_1}$$
$$= \frac{1.065 \times 0.4557 + 0.955 \times (1.12 - 0.234)}{1.065 \times 0.8917}$$
$$= 1.402$$

$$\xi = 54.5°$$
$$\theta = 54.5° - 27.11° = 27.39°$$

The direct and quadrature-axis components for the potential E_x and the armature current are

$$E_{xd}^* = E_x^* \ \sin \theta = 1.065 \times 0.460 = 0.490$$
$$E_{xq}^* = E_x^* \ \cos \theta = 1.065 \times 0.888 = 0.946$$
$$I_d^* = I_1^* \ \sin \xi = 0.955 \sin 54.5° = 0.777$$
$$I_q^* = I_1^* \ \cos \xi = 0.955 \cos 54.5° = 0.555$$

The internal potential E_q^* is obtained from Eq. (7.37):

$$E_q^* = E_{xq}^* + I_d^* \ (X_d^* - X_\gamma^*)$$
$$= 0.946 + 0.777 \times (1.68 - 0.23)$$
$$= 2.07$$

Calculating E_q based on the OP side of the triangle OPR yields

$$a^* = \sqrt{(E_x^* \ \cos \varphi_1)^2 + [E_x^* \ \sin \varphi_1 + I_1^* \ (X_d^* - X_\gamma^*)]^2}$$
$$= \sqrt{(1.065 \times 0.888)^2 + [1.065 \times 0.46 + 0.955 \times (1.68 - 0.23)]^2}$$
$$= 2.09$$

Therefore, $E_q^* \approx a^*$.

7.6 General External Characteristics of Exciter [16]

As mentioned above, the steady-state operation process of the AC exciter can be calculated by using the fixed external characteristics of the rectifier with the non-distortion commutating potential E_x. Although this method is effective, first, the non-distortion potential E_x, which changes with the operation process, must be obtained; this makes the calculation process complicated.

At the time of calculation, when considering only the fundamental current I_1 and the coefficient C_1 as constants, the exciter can use any one of the following potentials and the reactance corresponding to the potential to express the external characteristics for $U_f^* = f \ (I_f^*)$. These corresponding potentials and reactances are the non-distortion potential E_x and the commutating reactance X_γ, the air-gap potential E_i and the leakage reactance X_σ, the internal potential of the non-salient pole generator E_q and the synchronous reactance X_d, as well as the internal potential of the salient pole generator E_Q and the synchronous reactance X_q separately. Thus, the concept of the general external characteristics of the exciter can be introduced. The general external characteristics of the exciter refer to the expression relating the rectified voltage U_f and the rectified current I_f when the potential of the selected exciter is fixed. However, the basic per-unit value for the external characteristics $U_f^* = f \ (I_f^*)$ will correspond to the selected potential and reactance. When the internal potential of the selected exciter is fixed, for the non-salient pole exciter, the corresponding per-unit value will be

$$\left. \begin{array}{l} U_f^* = \dfrac{U_f}{\sqrt{6}E_q} \\ I_f^* = \dfrac{2X_d}{\sqrt{6}E_q} I_f \end{array} \right\} \tag{7.41}$$

For the salient pole exciter, the corresponding per-unit value will be

$$\left. \begin{array}{l} U_f^* = \dfrac{U_f}{\sqrt{6}E_Q} \\ I_f^* = \dfrac{2X_d}{\sqrt{6}E_Q} I_f \end{array} \right\} \quad (7.42)$$

In the steady state $U_f = I_f r_f$, by substituting into Eqs. (7.41) and (7.42) separately, for the non-salient pole exciter, calculation yields

$$\frac{U_f^*}{I_f^*} = \frac{r_f}{2X_d} \quad (7.43)$$

For the salient pole exciter, calculation yields

$$\frac{U_f^*}{I_f^*} = \frac{r_f}{2X_q} \quad (7.44)$$

Eqs. (7.43) and (7.44) indicate that in the external characteristic curve of the exciter $U_f = f\ (I_f^*)$ which is obtained when the internal potential of the exciter is fixed, the steady-state operating point of the exciter can be obtained from the point of intersection between the external characteristic curve and the DC resistance wire, which is determined by the ratio of the per-unit values of the rectified voltage to the rectified current; these can greatly simplify the calculation process.

Example 7.2
To determine the steady operating point and the internal potential E_q of the exciter from the given parameters of the generator and the exciter as mentioned in Example 7.1.

Solution
It is known that $X_q = 1.12 \times 0.0885 = 0.098\Omega$

The DC resistance wire is

$\tan \varphi = \dfrac{0.154}{0.098 \times 2} = 0.78, \quad \varphi = \arctan 0.78 = 37.95°.$

The straight line OM is led out from the coordinate origin, and the included angle between the straight and the horizontal axis coordinates is angle φ; line OM and external characteristics of the exciter intersect at M, yielding $U_f^* = 0.55, \ I_f^* = 0.71$, as shown in Figure 7.10.

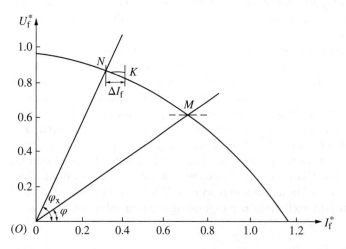

Figure 7.10 General external characteristic of three-phase bridge rectifier circuit $U_f = f(I_f^*)$.

The reactive component per-unit value for the direct axis of the exciter can be obtained from Eqs. (7.7) and (7.8):

$$I_d^* = \frac{B_1 \, I_f}{\sqrt{2} I_{eN}} = \frac{2900}{\sqrt{2} \times 2300} \times (0.35 + 0.66 \times 0.17) = 0.74$$

$$E_Q^* = \frac{U_f}{\sqrt{2} U_f^* \, U_{eN}} = \frac{447}{\sqrt{2} \times 0.55 \times 350} = 1.63$$

$$E_q^* = E_Q^* + I_d^* \, (X_d^* - X_q^*) = 1.63 + 0.74 \times 0.56 = 2.06$$

Thus, it is seen that the calculation process can be remarkably simplified by using the DC resistance wire and the external characteristics of the exciter. For estimating the transient process, at the beginning, the rectified voltage variation of the exciter can be based on the general external characteristic curve $U_f^* = f(I_f^*)$ of the exciter as shown in Figure 7.1, but at this time the per-unit value should be taken as

$$U_{fb}^* = \sqrt{6} E_x^*, \quad I_{fb}^* = \frac{\sqrt{6} E_x^*}{2 X_\gamma^*}$$

The original operating point of the exciter is determined by the intersection point N between the external characteristic curve and the corresponding DC resistance wire, with the slope of ON being

$$\tan \varphi_x = \frac{r_f^*}{2 X_\gamma^*} = \frac{0.154}{2 \times 0.0206} = 3.47, \quad \varphi_x = 75°$$

$U_f^* = 0.85$ can be obtained from the intersection point N, with

$$E_x = \frac{447}{\sqrt{6} \times 0.85} = 216 \text{ V},$$

$$E_x^* = 1.065$$

At the moment when the transient process starts, the potential E_x is unchanged. If the excitation current increment ΔI_f of the generator is known, the voltage drop of the exciter can be easily obtained. For the 300 MW turbine generator mentioned in Example 7.1, when a three-phase short-circuit abruptly occurs, the excitation current increment of the generator is $\Delta I_f = 1070$ A, which is converted into per-unit values as

$$\Delta I_f^* = \frac{\Delta I_f \times 2 X_\gamma^*}{\sqrt{6} E_x} = \frac{1070 \times 2 \times 0.0206}{\sqrt{6} \times 216} = 0.083$$

From point K shown in Figure 7.10, we obtain $\Delta I_f^* \approx 0.06$.

7.7 Transient State Process of AC Exciter with Rectifier Load

When studying the transient process of the AC exciter, which is closely related to the excitation system, the generator, and the excitation regulator, if each link of the system is described in detail, the number of orders in the system equation will increase. Therefore, we will try to simplify the study.

When studying the transient process of the AC exciter, we will generally adopt the following assumptions:

(1) The slip potential, transformer potential, and armature resistance of the AC exciter will not be considered.
(2) The forward resistance of the rectifier element will be zero, and the reverse resistance is infinite.

When adopting the above assumptions, the transient equation of the AC exciter can be written as

$$\left.\begin{array}{l} E_{iq}(s) = \psi_{id}(s) \\ E_{id}(s) = -\psi_{iq}(s) \\ U_p(s) = I_{fe}(s)r_{fe} + s\psi_{fe}(s) \\ 0 = I_{ld}(s)r_{ld} + s\psi_{ld}(s) \\ 0 = I_{ld}(s)r_{lq} + s\psi_{lq}(s) \end{array}\right\} \quad (7.45)$$

where

E_{id}, E_{iq} — direct and the quadrature-axis components of the air-gap potential;

ψ_{id}, ψ_{iq} — direct and quadrature-axis components of the air-gap flux linkage;

$r_{ld}, r_{lq}, I_{ld}, I_{lq}, \psi_{ld}, \psi_{lq}$ — direct and quadrature-axis components of the effective resistance, current, and flux linkage in the damping loop;

$r_{fe}, I_{fe}, \psi_{fe}$ — effective resistance, current, and flux linkage of the exciter excitation winding;

U_p — pilot exciter voltage.

The corresponding equivalent circuits are shown in Figure 7.11.

From the direct and quadrature axis equivalent circuits of the exciter, the following flux linkage equations are obtained:

$$\left.\begin{array}{l} \psi_{id}(s) = X_{ad}[I_d(s) + I_{ld}(s) + I_{fe}(s)] \\ \psi_{iq}(s) = X_{aq}[I_q(s) + I_{lq}(s)] \\ \psi_{fe}(s) = I_{fe}(s)X_{fe} + X_{ad}[I_{ld}(s) + I_d(s)] \\ \psi_{ld}(s) = I_{ld}(s)X_{ld} + X_{ad}[I_{fe}(s) + I_d(s)] \\ \psi_{lq}(s) = I_{lq}(s)X_{lq} + X_{aq}I_q(s) \end{array}\right\} \quad (7.46)$$

where

X_{fe}, X_{ld}, X_{lq} — the direct and quadrature-axis reactance in the excitation winding and damping winding loops

$$\left.\begin{array}{l} X_{fe} = X_{\sigma fe} + X_{ad} \\ X_{ld} = X_{\sigma ld} + X_{ad} \\ X_{lq} = X_{\sigma lq} + X_{aq} \end{array}\right\} \quad (7.47)$$

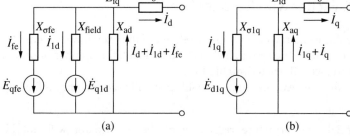

Figure 7.11 Transient state equivalent circuit of the AC exciter: (a) direct axis equivalent circuit, (b) quadrature-axis equivalent circuit.

All the above equations are related to the d, q coordinate systems, which change with time. When reducing them to the A, B, and C coordinate system that has variables, for the air-gap potential e_i, the following equation is available:

$$\left.\begin{array}{l} e_{iA} = e_{id} \cos \omega t + e_{iq} \sin \omega t \\ e_{iB} = e_{id} \cos\left(\omega t - \frac{2\pi}{3}\right) + e_{iq} \sin\left(\omega t - \frac{2\pi}{3}\right) \\ e_{iC} = e_{id} \cos\left(\omega t + \frac{2\pi}{3}\right) + e_{iq} \sin\left(\omega t + \frac{2\pi}{3}\right) \end{array}\right\} \quad (7.48)$$

For the armature current:

$$\left.\begin{array}{l} i_A = i_d \cos \omega t + i_q \sin \omega t \\ i_B = i_d \cos\left(\omega t - \frac{2\pi}{3}\right) + i_q \sin\left(\omega t - \frac{2\pi}{3}\right) \\ i_C = i_d \cos\left(\omega t + \frac{2\pi}{3}\right) + i_q \sin\left(\omega t + \frac{2\pi}{3}\right) \end{array}\right\} \quad (7.49)$$

For the rectifier power supply loop:

$$\left.\begin{array}{l} e_{iA} - \frac{di_A}{dt}L_\sigma - i_{Ar} = u_A \begin{pmatrix} i_A > 0, & u_A = u_k \\ i_A < 0, & u_A = u_i - u_a \end{pmatrix} \\ e_{iB} - \frac{di_B}{dt}L_\sigma - i_{Br} = u_B \begin{pmatrix} i_B > 0, & u_B = u_k \\ i_B < 0, & u_B = u_i = u_k - u_a \end{pmatrix} \\ e_{iC} - \frac{di_C}{dt}L_\sigma - i_{Cr} = u_C \begin{pmatrix} i_C > 0, & u_B = u_k \\ i_C < 0, & u_B = u_i = u_k - u_a \end{pmatrix} \end{array}\right\} \quad (7.50)$$

where

u_k, u_a — cathode and anode potentials which are related to the neutral point in the rectifier.

All the above equations constitute the system of equations for the transient of the AC exciter excitation system. The difficulty of solving this system of equations lies in the nonlinearity of the operating state of the rectifier and the non-sinusoidal waveform of the air-gap potential, and so on.

7.8 Simplified Transient Mathematical Model of AC Exciter with Rectifier Load

One of the approaches to simplifying the system of equations for the transient of the AC exciter is to further simplify the expressions for the rectifier. Therefore, we shall add the following assumptions:

(1) The higher harmonic component of the rectified voltage will not be considered.
(2) We shall replace the non-sinusoidal phase current of the exciter stator with the fundamental current.

As a matter of fact, in the transient process of the generator, its time constant is about a few tenths of a second or longer; however, the pulse recurrence interval of the rectified voltage waveform is less than 0.003 s, and therefore the above assumptions will be allowed. In this way, the transient process of the rectifier can be expressed with the external characteristic equations.

7.8.1 Simplified Computation Method of External Characteristics of Rectifier

Equations using the external characteristics of the rectifier to calculate the transient of the exciter can be divided into two groups: one is the system of equations using the d, q coordinates to express the transient of the AC exciter, to obtain the sub-transient potential according to the known rectified voltage and fundamental armature current components; the other group is the system of equations for the rectifier, to determine the average rectified voltage according to the known rectified current and sub-transient potential.

It is assumed that the transformer potential and the slip potential are not considered; the equations for the AC exciter armature loop will then be

$$\left.\begin{array}{l} E_{iq}(s) = \psi_{id}(s) \\ E_{id}(s) = -\psi_{iq}(s) \\ E''_d(s) = E_{id}(s) + I_q(s)(X''_d - X_\sigma) \\ E''_q(s) = E_{iq}(s) - I_d(s)(X''_d - X_\sigma) \\ E'' = \sqrt{E''^2_d + E''^2_q} \end{array}\right\} \tag{7.51}$$

The system of equations for the damping and excitation loops are the same as Eq. (7.45), and the equations for the flux linkage should be referred to Eq. (7.46). From the known sub-transient potential and rectified current, the external characteristics of the rectifier are used to determine the average rectified voltage value in a cycle. For the different operating states, the corresponding external characteristics are

$$\left.\begin{array}{ll} 0 < I_f^* \leqslant 0.5: & U_f^* = \frac{3}{\pi}\left(1 - \frac{I_f^*}{2}\right) \\ 0.5 \leqslant I_f^* \leqslant \frac{\sqrt{3}}{2}: & U_f^* = \frac{3\sqrt{3}}{2\pi}\sqrt{1 - I_f^{*2}} \\ \frac{\sqrt{3}}{2} \leqslant I_f^* \leqslant \frac{2}{\sqrt{3}}: & U_f^* = \frac{3\sqrt{3}}{\pi}\left(1 - \frac{\sqrt{3}}{2}I_f^*\right) \\ \frac{2}{\sqrt{3}} < I_f^*: & U_f^* = 0 \end{array}\right\} \tag{7.52}$$

The basic per-unit value in Eq. (7.52) is

$$U_{fb} = \sqrt{6}E'', \quad I_{fb} = \frac{\sqrt{6}E''}{2X_\gamma}$$

In the above equations:

I_f, U_f, E'', and X_γ all take the rated value of the exciter rotor as the basic value.

When calculating the fundamental component of the rotor current, Eq. (7.6) can be used, and $C_1^* = 1.05$ can be taken. The included angle between the fundamental current and the sub-transient potential is φ_1, and it can be obtained from Eq. (7.9). Besides, from Eqs. (7.34) and (7.36), if $E_x = E''$, the d-axis and q-axis components for the exciter armature current can be obtained. It can thus be seen that, when calculating the transient process of the exciter by using the external characteristics of the rectifier, although the system of equations is simplified, the nonlinear expression Eq. (7.34) and Eq. (7.36) are still employed in order to determine the d-axis and q-axis components of the stator armature current. Using the external characteristics of the exciter to calculate the transient process can further simplify the calculation.

7.8.2 Simplified Computation Method of External Characteristics of Exciter

Compared with the above-mentioned simplified computation method for the external characteristics of the rectifier, the main advantage in adopting the simplified computation method for the external characteristics of the exciter is that it can determine the d-axis and q-axis components of the armature current very easily.

In fact, in comparison with the potential $\dot{E}_q$ (or $\dot{E}_Q$) phasor, the direct axis armature current component is lagging behind by $\frac{\pi}{2}$; therefore, for the potential $\dot{E}_q$ ($\dot{E}_Q$) phasor, the direct axis armature current will be a reactive current.

On the other hand, as shown in Eq. (7.7), the fundamental component of the armature current is related to the commutation coefficient B_1. It can be seen from Eq. (7.8) that using such this expression enables us to ignore the equation for the quadrature-axis state of the exciter.

The direct axis equations for the AC exciter have the following forms:

$$\left.\begin{aligned}
E_q(s) &= X_{ad}[I_{fe}(s) + I_{1d}(s)] \\
U_p(s) &= [I_{fe}(s)r_{fe} + s\psi_{fe}(s)] \\
0 &= I_{1d}(s)r_{1d} + s\psi_{1d}(s) \\
\psi_{fe}(s) &= I_{fe}(s)X_{fe} + X_{ad}[I_{1d}(s) + I_d(s)] \\
\psi_{1d}(s) &= I_{1d}(s)X_{1d} + X_{ad}[I_{fe}(s) + I_d(s)] \\
I_d(s) &= -\frac{B_1}{\sqrt{2}}I_f(s) \\
B_1^* &= f(I_f^*)
\end{aligned}\right\} \quad (7.53)$$

If all the above components in the equation are reduced to the per-unit system which takes the rated value of the exciter rotor as the reference value, the external characteristics of the general exciter are expressed by the following system of equations:

$$\left.\begin{aligned}
0 < I_f^* \leqslant 0.5 &: \quad U_f^* = \frac{3}{\pi}\left(1 - \frac{I_f^*}{2}\right) \\
0.5 \leqslant I_f^* \leqslant \frac{\sqrt{3}}{2} &: \quad U_f^* = \frac{3\sqrt{3}}{2\pi}\sqrt{1 - I_f^{*2}} \\
\frac{\sqrt{3}}{2} \leqslant I_f^* \leqslant \frac{2}{\sqrt{3}} &: \quad U_f^* = \frac{3\sqrt{3}}{\pi}\left(1 - \frac{\sqrt{3}}{2}I_f^*\right) \\
\frac{2}{\sqrt{3}} \leqslant I_f^* &: \quad U_f^* = 0 \\
U_f^* &= \frac{U_f}{\sqrt{6E_q}} \\
I_f^* &= \frac{2X_d I_f}{\sqrt{6E_q}}
\end{aligned}\right\} \quad (7.54)$$

The salient pole exciter is expressed by the following system of equations:

$$U_f^* = \frac{U_f}{\sqrt{6E_Q}}$$

$$I_f^* = \frac{2X_q I_f}{\sqrt{6E_Q}}$$

$$E_Q = E_q - I_d(X_d - X_q)$$

Ignoring the influence of the damping loop, or replacing the open circuit time constant with an equivalent time constant whose value is equivalent to the sum $(T'_{d0} + T'_{1d0})$ of the time constants for the excitation winding and the damping winding, Eq. (7.53) can be further simplified as

$$\left.\begin{aligned}
E_q(s) &= X_{ad}I_{fe}(s) \\
U_p(s) &= I_{fe}(s)\frac{X}{T'_{d0} + T'_{1d0}} + s\psi_{fe}(s) \\
\psi_{fe}(s) &= I_{fe}(s)X_{fe} + I_d(s)X_{ad} \\
I_d(s) &= -\frac{B_1}{\sqrt{2}}I_f(s) \\
B_1^* &= 0.35 + 0.66 I_f^*
\end{aligned}\right\} \quad (7.55)$$

Therefore, these simplified systems of Eqs. (7.54) and (7.55) make it very convenient to analyze the operation process of the exciter.

It should be noted that Eq. (7.54) still includes the nonlinear factor; that is, the rectifier includes three-phase commutation states. In the transient process, the rectifier may transit from one state to another state.

If it is required to further simplify the system equation the equation can be linearized for the external characteristics of the exciter. At this time, it is assumed that the rectifier works in Type I or II commutation states under all kinds of operating states. In this case, the system of equations can be written as

$$\left.\begin{array}{l} U_p(s) = I_{fe}(s)\dfrac{X_{fe}}{T'_{do}+T'_{1do}} + s\psi_{fe}(s) \\ \psi_{fe}(s) = I_{fe}(s)X_{fe} + I_d(s)X_{ad} \\ I_d(s) = -\dfrac{1}{\sqrt{2}}B_1 I_f(s) \\ U_f(s) = k_1 E_q(s) - k_2 I_f(s)X_d \end{array}\right\} \quad (7.56)$$

where

k_1, k_2 – linearization coefficients for the external characteristics of the exciter.

7.9 Transient State Process of Excitation System in Case of Small Deviation Change in Generator Excitation Current

Any discussion of the excitation regulation performance always involves the excitation system performance when the main generator excitation current has a small deviation change.

Since the difference between the time constants for the main generator and the exciter is one order of magnitude, in the study of the forced excitation process which causes the increase in the exciter excitation current due to the terminal voltage drop of the generator, the change in the excitation current of the generator can be regarded as insignificant. Approximately, it can be assumed that for the duration of the exciter potential increase, the excitation current of the generator will be fixed; that is, I_f is constant. Similarly, in the dynamic process after the event, the exciter voltage will change from the peak value to zero (or to the negative ceiling voltage). Since the time constant for the main generator excitation loop is very large, in the transient process, the rectified current I_f can still be regarded as fixed.

If the influence of the damping loop is ignored, the system of equations for the excitation winding loop of the exciter in the transient state is as follows:

$$\left.\begin{array}{l} u_p = i_{fe} r_{fe} + \dfrac{d\psi_{fe}}{dt} \\ \psi_{fe} = i_{fe} X_{fe} + i_d X_{ad} \end{array}\right\} \quad (7.57)$$

where

ψ_{fe} – excitation winding flux linkage of the exciter.

In addition, from Eqs. (7.54) and (7.55), taking into consideration the expression that relates B_1 to I_f, we obtain

$$i_d = -\left(0.35 + 0.66 \times \dfrac{2X_d I_{f0}}{i_{fe} X_{ad} \sqrt{6}}\right)\dfrac{I_{f0}}{\sqrt{2}} \quad (7.58)$$

Eq. (7.58) indicates that when the generator excitation current I_f is fixed, by increasing the excitation current i_{fe} (or the potential E_q) of the exciter, the armature reaction current of the direct axis can be reduced. This conclusion is very important, as it indicates that under the above conditions, the increase in the excitation current of the exciter and the decrease in the armature reaction current will make the flux linkage of the exciter increase.

7.9 Transient State Process of Excitation System in Case of Small Deviation Change in Generator Excitation Current

By substituting Eq. (7.58) into (7.57) and assuming that the rectified current I_f is fixed, we obtain

$$u_p = i_{fe} r_{fe} + \frac{di_{fe}}{dt}\left(X_{fe} + \frac{0.66 X_d}{\sqrt{3}} \times \frac{I_{f0}^2}{i_{fe}^2}\right)$$

If the saturation is ignored, the excitation current can be converted into an expression for the potential:

$$i_{fe} = \frac{e_q}{X_{ad}} = \frac{(E_{q0} + \Delta e_q)}{X_{ad}}$$

Substitution into the u_p expression yields

$$u_p = i_{fe} r_{fe} + \frac{di_{fe}}{dt}\left[X_{fe} + \frac{0.66 X_d X_{ad}^2 I_{f0}^2}{\sqrt{3} E_{q0}^2 \left(1 + \frac{\Delta e_q}{E_{q0}}\right)^2}\right] \quad (7.59)$$

In this way, a first-order nonlinear differential equation can be used to express the transient characteristics of the exciter. Equation (7.59) indicates that the value of the equivalent inductance of the exciter's excitation winding, which is determined by the synthesis of the excitation current and the armature reaction demagnetization current of the exciter, will be greater than the corresponding value in the armature open circuit of the exciter's field winding.

During forced excitation, the excitation potential of the exciter will increase ($\Delta e_q > 0$) and the equivalent inductance will decrease, but the resultant value will still be greater than X_{fe}; during demagnetization ($\Delta e_q < 0$), the equivalent inductance will increase. On the basis of the above discussion, strictly speaking, the transient of the exciter cannot be expressed with a one order inertial link, because the derivate coefficient of the excitation current i_{fe} in Eq. (7.59) is the function of the excitation current. Under a small deviation, the Δe_q value is very small, thus the coefficient can be regarded as constant. In this case, Eq. (7.59) expressed by the operation format will be

$$I_{fe}(s) = \frac{U_p(s)}{r_{fe}\left(1 + s \times \frac{X'_{fe}}{r_{fe}}\right)} \quad (7.60)$$

where $X'_{fe} = X_{fe} + \frac{0.66 X_d X_{ad}}{\sqrt{3}} \times \frac{I_{f0}^2}{E_{q0}^2}$

if $\frac{X'_{fe}}{r_{fe}} = T'_{fe}$

where

T'_{fe} — transient time constant of the exciter when the generator excitation current is fixed and the excitation current of the exciter changes with a small deviation.

Obviously, the time constant for the excitation winding of the exciter connected to the rectifier load will be greater than the time constant in the exciter with open circuit armature winding, which is

$$T_{fe0} = \frac{X_{fe}}{r_{fe}} < T'_{fe} = \frac{X_{fe} + \frac{\sqrt{0.66 X_d X_{ad}^2}}{\sqrt{3}} \times \frac{I_{f0}^2}{E_{q0}^2}}{r_{fe}} \quad (7.61)$$

In fact, during the increase in the potential of the exciter, the excitation current of the generator is not fixed, but increases slowly, and the equivalent time constant will be less than the computed value from Eq. (7.61).

Note that the above conclusion contradicts the general concept of the demagnetization theory for synchronous generators. Now, we will discuss the difference between these two. In the transient theory of synchronous generators, the excitation winding reactance X_{fe} is normally solved based on the flux linkage balance equation for each winding of the generator; if the damping effect is ignored, X'_{fe} can be solved using the flux linkage equation of the excitation winding:

$$\psi_{f\Sigma} = i_f X_f + i_d X_{ad}$$

where

$\psi_{f\Sigma}$ – the excitation winding flux linkage which is co-established by the excitation current i_{fe} and the direct axis current component i_d.

When the stator winding is shorted, the direct axis current component can be determined using the following equation:

$$i_d = -\frac{e_q}{X_d} = -\frac{i_f X_{ad}}{X_d}$$

Substituting the above equation into the flux linkage equation yields

$$\psi_{f\Sigma} = i_f \left(X_f - \frac{X_{ad}^2}{X_d} \right) = i_f X'_f$$

According to the above explanations, the flux linkage, which is co-established by the excitation current and the direct axis armature demagnetization current, will equal the flux linkage generated by the single excitation winding when the excitation current flows through, and whose armature demagnetization action is considered for its reactance.

When the stator armature winding is shorted, due to the influence of the demagnetization, the excitation winding impedance X'_f will be less than the corresponding value in the stator winding open circuit, which is

$$X'_f < X_f$$

Conversely, as shown in Eq. (7.61), for the AC exciter connected to the rectifier load, its excitation winding reactance and its time constant will be greater than the corresponding value in the case of the exciter armature open circuit. This is because when increasing the excitation current of the exciter, its output rectified current will still be fixed, reducing the commutation angle, and in turn reducing the included angle between the fundamental armature current and the quadrature axis, which also reduces the direct axis armature reaction current.

Now, by using ΔX to express the influence of the change in the direct axis armature reaction on the excitation winding reactance, the following is obtained from Eq. (7.59):

$$\Delta X = \frac{0.66 X_{ad}^2 X_d}{\sqrt{3}} \times \frac{I_{f0}^2}{E_{q0}^2 \left(1 + \frac{\Delta e_q}{E_{q0}}\right)^2} \tag{7.62}$$

Now, using examples, we will discuss the influence of ΔX on the exciter time constant.

Example 7.3

A 300 MW turbine generator has a rated excitation current

I_{fN} = 2900 A, with the per-unit value on the armature side of the exciter reduced to 1.26 per unit (p.u.). Find the time constant for the exciter at this time under double-forced excitation and an exciter potential E_q = 2.062.

Solution

$$I_f^* = \frac{2X_d I_f}{\sqrt{6} E_q} = \frac{2 \times 1.68 \times 1.26}{\sqrt{6} \times 2.062} = 0.84 > 0.5$$

For an exciter working in the Type II commutation state, the corresponding equation is

$$U_f^* = \frac{3\sqrt{3}}{2\pi}\sqrt{1 - I_{f0}^{*\,2}}$$

$$\frac{U_f}{\sqrt{6}E_q} = \frac{3\sqrt{3}}{2\pi}\sqrt{1 - \frac{2X_d I_{f0}^2}{\sqrt{6}E_{q0}^2}}$$

The above equation can be written as

$$\frac{I_{f0}^2}{E_{q0}^2} = \frac{81}{2\pi^2} \times \frac{1}{r^2 + \left(\frac{3\sqrt{3}}{\pi}X_d\right)^2} = 0.358$$

To solve ΔX from Eq. (7.62), when $\Delta e_q = 0$:

$$\Delta X = \frac{0.66}{\sqrt{3}} \times 1.54^2 \times 1.68 \times 0.358 = 0.616$$

For the operating state discussed, the time constant for the exciter is

$$T'_{fe} = \frac{X_{fe} + \Delta X}{X_{fe}} \times T_{fe0} = \left(1 + \frac{\Delta X}{X_{fe}}\right)T_{fe0} = k_e T_{fe0}$$

$$X_{fe} = X_{fe0} + X_{ad} = 0.21 + 1.54 = 1.75$$

Thus, the coefficient is

$$k_e = 1 + \frac{0.616}{1.75} = 1.35$$

By adopting the double-forced excitation voltage, the internal potential E_q of the exciter changes from E_{q0} to $2E_{q0}$. The corresponding coefficient k_e changes between 1.35 and 1.15. By substituting the average value $k_e = \frac{1.35 + 1.17}{2} = 1.26$ into Eq. (7.60), we obtain

$$I_{fe}(s) = \frac{U_p(s)}{r_{fe}(1 + 1.25sT_{fe0})} \tag{7.63}$$

If the saturation of the magnetic circuit of the exciter is ignored, the expression for the exciter potential is

$$\Delta E_q(s) = \frac{U_p(s)X_{ad}}{r_{fe}(1 + 1.25sT_{fe0})} \tag{7.64}$$

Because the time constant of the generator is much larger than the time constant value of the exciter, it can be approximately assumed that when the output voltage of the exciter with a smaller time constant reaches the instantaneous peak value, the field current of the generator with a larger time constant has not changed, and still maintains the original current value; in this case, overshoot exists in the output voltage of the exciter. The following is the external characteristic equation of the exciter:

$$U_{f0} + \Delta U_f = 2.34(E_{q0} + \Delta E_q) - \frac{3}{\pi}X_d I_{fe}$$

For the rated state:

$$U_{f0} = 2.34E_{q0} - \frac{3}{\pi}X_d I_{fe}$$

The corresponding increment equation is

$$\Delta U_f = 2.34\Delta E_q$$

7.10 Influence of Diode Rectifier on Time Constant of Generator Excitation Loop

With respect to to the free component of the generator rotor current caused by the change in operating state, its decaying process is related to the parameters of the generator rotor loop (including the excitation winding and damping winding loops).

In the open circuit of the generator stator winding, the time constant of the generator excitation loop is determined by the ratio between the rotor winding reactance and its effective resistance, which is

$$T'_{d0} = \frac{X_f}{\omega r_f}$$

$$T'_{1d0} = \frac{X_{1d}}{\omega r_{1d}}$$

where

T'_{d0}, T'_{1d0} — time constants for the field winding and damping winding in the open circuit of the generator stator winding.

When the stator winding is shorted, the time constant of the excitation loop will be determined by the ratio between each reactance and its effective resistance in the excitation winding loop whose stator is shorted; that is,

$$T'_d = \frac{X'_f}{\omega r_f}$$

$$T'_{1d} = \frac{X_{1d}}{\omega r_{1d}}$$

If the generator excitation winding is powered by the AC power supply via the rectifier, then when the generator is shorted, the resulting free component current may form a closed circuit via the rectifier and the power supply winding, thus affecting the time constant value of the excitation loop. Its influence is discussed in the following.

Since the time constant of the exciter is obviously less than that of the generator excitation loop, when studying the transient process of the generator exciter winding, the time constant of the exciter can be ignored. Under this assumption, the transient process of the main generator excitation loop can be expressed by the following equation:

When $0 \leqslant I_f \leqslant \frac{0.5\sqrt{6E_q}}{2X_d}$:

$$U_f(s) = 2.34E_q(s) - I_f(s)\frac{3}{\pi}X_d = I_f(s)(r_f + sX'_f)$$

When $\frac{3\sqrt{2E_q}}{4X_d} \leqslant I_f \leqslant \frac{\sqrt{2E_q}}{X_d}$:

$$U_f(s) = \frac{9}{\pi}\sqrt{2}E_q(s) - I_f(s)\frac{9}{\pi}X_d = I_f(s)(r_f + sX'_f)$$

From the above equation, we obtain

$$2.34E_q(s) = I_f(s)\left(r_f + \frac{3}{\pi}X_d + sX'_f\right)$$

$$\frac{9\sqrt{2}}{\pi}E_q(s) = I_f(s)\left(r_f + \frac{3}{\pi}X_d + sX'_f\right)$$

Therefore, when $0 \leqslant I_f \leqslant \frac{\sqrt{6}E_q}{4X_d}$:

$$T'_f = T'_{d0} \frac{r_f}{r_f + \frac{3}{\pi}X_d} \quad (7.65)$$

When $\frac{3\sqrt{2}E_q}{4X_d} \leqslant I_f \leqslant \frac{\sqrt{2}E_q}{X_d}$:

$$T'_f = T'_{d0} \frac{r_f}{r_f + \frac{9}{\pi}X_d} \quad (7.66)$$

where

T'_f – time constant for the generator exciting winding when considering the influence of the diode rectifier.

As mentioned above, due to the influence of the diode rectifier, the equivalent time constant for the excitation loop is reduced.

7.11 Excitation Voltage Response for AC Exciter with Rectifier Load

For the AC exciter excitation system supplying power to the diode rectifier, its excitation voltage response will change with the operation mode of the main generator, which, in general, can be divided into three operating states: no-load operating state, loaded operating state, and forced operating state in the case of a three-phase short circuit at the generator end.

7.11.1 Excitation Voltage Responses of AC Exciter without Load

At this time, the AC exciter output is not connected to any load, and therefore no armature reaction exists. When the forced excitation voltage is supplied by the excitation winding side of the AC exciter, the output voltage at the AC exciter terminal will be related to the AC exciter no-load time constant. Besides, since the AC exciter has the maximum time constant value when it has no load, the excitation voltage response determined by this state is quite low.

7.11.2 Excitation Voltage Response of AC Exciter with Load

If the time constant for the AC exciter with no load is T_{d0e}, then the corresponding time constant at no load will be

$$T_e = T_{d0e} \frac{r_e^2 + (X'_d + X_e)(X_q + X_e)}{r_e^2 + (X_d + X_e)(X_q + X_e)} = KT_{d0e}$$

$$r_e = Z_e \cos \varphi_1$$
$$X_e = Z_e \sin \varphi_1 \quad (7.67)$$

where

T_{d0e} – time constant when the AC exciter has no load;

Xd, X'_d – direct axis synchronous reactance and transient reactance of the AC exciter;

r_e, X_e – equivalent active component and reactive component determined by the load power factor of the AC exciter;

Z_e – load impedance for the AC exciter, using the per-unit value to express $Z_e^* = 1$

φ_1 – power factor angle of the fundamental current.

In general, the load time constant for the exciter is about 1/2 less than the no-load time constant; that is, $K \approx 0.5$.

Since the load of the AC exciter is the excitation winding of the main generator, the time constant of the main generator is one order of magnitude greater than that of the AC exciter, which determines the characteristics of the excitation voltage response in the loaded AC exciter. Thus, in general, the transient process for forced excitation can be divided into the following two processes:

In the first phase, the excitation current of the AC exciter with the smaller time constant can reach the peak value more quickly. The AC exciter voltage can also reach the peak value at the same time. In this process, the current in the excitation winding loop of the generator with the greater time constant will still approximately remain unchanged compared to the original current, which causes overshoot in the excitation voltage. The rising segment of the corresponding excitation voltage is known as the constant-current characteristic curve segment, and is shown in Part I of Figure 7.12.

In the second phase, the generator excitation current will start to increase, due to the influence of the armature reaction, the exciter voltage will start dropping from the ceiling voltage in the constant-current characteristic curve and finally reach the steady value, as shown in the Part II of Figure 7.12. The ceiling voltage overshoot ΔU_f will change with $\frac{T_{d0}}{T_e}$, and the greater the ratio, the greater the overshoot ΔU_f.

The time duration in the first phase is shorter, which determines the time constant of the AC exciter. The time duration in the second phase is longer, which determines the time constant of the generator excitation winding.

The curvilinear coordinates of the saturation characteristics of the AC exciter can show the constant-current characteristics, as shown in the curves of Figure 7.13. In fact, in the first phase and before the AC exciter reaches the ceiling voltage, the generator excitation current still rises to some extent; therefore, the real exciter voltage variation curve is represented by curve 4, as shown in Figure 7.13. When the time constant for the generator exciting winding is smaller, the generator excitation current will rise proportionally with the excitation voltage, and the corresponding characteristics are called the constant-value resistance characteristics, as shown in curve 2 of Figure 7.13.

In the figure, 1, 2, and 3 denote the external characteristic curve of the rectifier when the generator is under no-load, load, and forced excitation states; 1', 2', 3' denote the external characteristic curve of the exciter when the generator is under no-load, load, and forced excitation states – when E_Q (or E_q) is unchanged.

As shown in Figure 7.14, the excitation voltage overshoot problem caused by the constant-current characteristic curve can also be solved by using the external characteristics of the AC exciter and the rectifier together with the graphical method. That is, curves 1, 2, and 3 refer to the external characteristics of the rectifier when the non-distortion potential E_x is fixed; on the other hand, curves 1', 2', and 3' refer to the external characteristics of the exciter based on the relation $U_f^* = f(I_f^*)$ when the internal potentials E_Q (or E_q) of the AC exciter is fixed. The operating point of the generator that is under the rated load is determined by the intersection point of the external characteristic curve 2 for the rectifier with the external characteristic curve 2' for the exciter in Figure 7.14, which is point "g," as shown in Figure 7.14. The generator excitation current is I_{fN}^*.

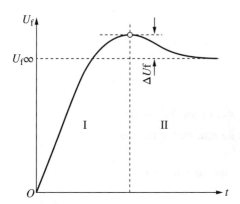

Figure 7.12 Voltage build-up curve of the AC exciter at the time of forced excitation.

Figure 7.13 Saturation characteristic curve for AC exciter:
1 – characteristic curve with no load, 2 – constant-value resistance characteristic curve with load, 3 – constant-value current characteristic curve with load, 4 – real variation characteristic curve.

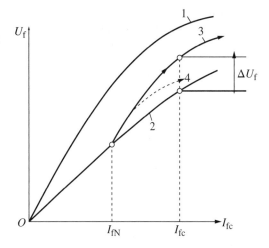

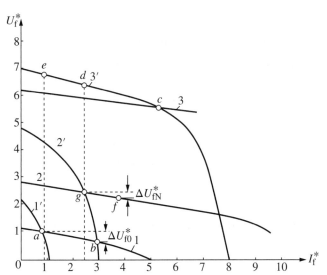

Figure 7.14 External characteristic curve of AC exciter and rectifier.

In the forced excitation, curve 3′ in Figure 7.14 shows the external characteristics of the exciter corresponding to multiples of the selected forced excitation voltage. See Curve 3′ in Figure 7.14 for the external characteristics of the corresponding rectifier. The steady-state forced excitation voltage is determined by point "c" in Figure 7.14. The constant-current characteristic curve corresponds to the fixed I_{fN}^*, which is point d in Figure 7.14. Therefore, we obtain

$$\Delta U_f = \frac{6.4 - 5.3}{5.3} \times 100\% = 20\%.$$

When the generator is under no load, the ceiling excitation voltage corresponding to the constant-current characteristic curve will be 6.8 V (see point "e" in Figure 7.14). In this case, the excitation voltage overshoot will be $\Delta U_f = \frac{6.8-5.3}{5.3} \times 100\% = 28\%$. Therefore, before the forced excitation, the lower the initial current value of the generator, the higher the exciter voltage overshoot at the end of the first phase will become.

Example 7.4
Determining the expression for the external characteristics of the AC exciter when the internal potential E_q is fixed.

Solution

From the phasor diagram for the salient pole AC exciter as shown in Figure 7.5b, we can obtain the non-distortion sinusoidal potential:

$$E_x = \sqrt{E_q^2 - [(X_d - X_\gamma)\beta_i I_f \cos \varphi_1]^2} - \beta_i I_f (X_d - X_\gamma) \sin \varphi_1 (\dot{\alpha} \approx \dot{E}_q)$$

For the Type I operating state, the external characteristic equation for the rectifier is

$$U_f = \frac{3\sqrt{6}}{\pi} \times E_x - \frac{3}{\pi} I_f X_r$$

By combining the above two equations, we obtain

$$U_f = \frac{3\sqrt{6}}{\pi} \{ \sqrt{E_q^2 - [(X_d - X_\gamma)\beta_i I_f \cos \varphi_1]^2} - \beta_i I_f (X_d - X_\gamma) \sin \varphi_1 \} - \frac{3}{\pi} I_f X_r$$

The above equation is the external characteristic expression for the exciter when the internal potential E_q of the AC exciter is fixed.

7.11.3 Field Voltage Response When Three-Phase Short-Circuit Occurs in the Generator

When a three-phase short-circuit occurs in the generator, its rotor excitation winding will generate the free component current. When this current flows through the rectifier and the AC exciter armature, the equivalent commutation voltage will increase to make the generator excitation voltage drop ΔU_f instantaneously, as shown in Figure 7.15.

The process which cause the voltage drop in the AC exciter can be understood as follows: at the moment the generator is shorted, the generator rotor current-free component flows through the AC exciter armature, which increases the commutation angle γ. However, at this time, the non-distortion commutating potential E_x of the exciter does not change, with the characteristics similar to those of the sub-transient potential E''. Therefore, the rectified voltage U_f will reduce ΔU_f on the basis of the external characteristics of the rectifier,

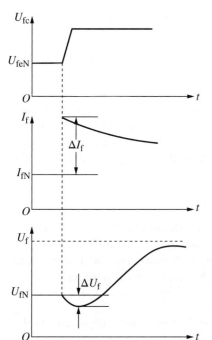

Figure 7.15 AC exciter voltage drop when three-phase short-circuit occurs in generator.

Figure 7.16 Excitation voltage response for AC exciter under various forced excitation conditions: 1 – forced excitation for generator load, 2 – forced excitation in the shorted generator, 3 – constant-current characteristics.

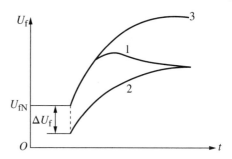

and its value can be solved from the external characteristics of the AC exciter and the rectifier in Figure 7.14. If before the forced excitation the generator is in the rated load state, the increment of the transient rotor current-free component caused by the short-circuit will be $\Delta I_f^* = 1.0$. Accordingly, the operating point will drop from point "g" in Figure 7.14 to point "f," with $\Delta U_{fN}^* = \frac{2.4-2.25}{2.25}\% = 6.6\%$. If the operating mode was "no load" before the short circuit, then $\Delta I_f^* = 2.0$. By this time, the operating point will have changed from point "a" in Figure 7.14 to point "b," $\Delta U_{f0}^* = \frac{1.0-0.8}{0.8}\% = 25\%$. Now, the rectifier has entered the Type II commutation state.

From the above, the excitation voltage response of the diode rectifier excitation system of the AC exciter will differ with various operation modes, including no-load AC exciter, loaded generator, and short-circuit failure, and so on. The excitation voltage responses of the AC exciter are shown in Figure 7.16 for different operational modes and conditions. From this figure, it is seen that the excitation voltage response will be the lowest when the generator is shorted, and the excitation voltage response will be the highest under the constant-current characteristics condition.

7.12 Short-Circuit Current Calculations for AC Exciter [17]

7.12.1 Abrupt Short Circuit on DC Side of Rectifier

In the static separately excited diode excitation system, when the rectifier output terminal is directly shorted, if the effective resistance to the rectifier is ignored, it will equal the three-phase direct short circuit at the AC exciter terminal. The sub-transient reactance will restrict the initial value of the short-circuit current. Then, with the strengthening of the armature reaction effect, the AC short-circuit current will gradually decay to a steady value. The DC component of the sub-transient short-circuit current will be determined by the time constant T_a. If the following equation is used to express the instantaneous value of the A-phase voltage:

$$u_{ea}(t) = \sqrt{2}U_e \sin(\omega t + \theta) \tag{7.68}$$

Then, when the excitation current of the exciter is constant, the instantaneous value of the three-phase short-circuit current of the no-load separately excited AC exciter can be expressed by the following equation:

$$i_{e0}(t) = \left[\frac{1}{X_d} + \left(\frac{1}{X_d'} - \frac{1}{X_d}\right)e^{-\frac{1}{T_d'}} + \left(\frac{1}{X_d''} - \frac{1}{X_d'}\right)e^{-\frac{1}{T_d''}}\right]$$
$$\times \sqrt{2}U_e \cos(\omega t + \theta) - \frac{1}{X_d''}\sqrt{2}U_e e^{-\frac{1}{T_a}}\cos\theta$$

$$T_a = \frac{X_2}{\omega R_a}$$

$$T_d'' \approx T_{d0}'' \frac{X_d''}{X_d'} \tag{7.69}$$

where

U_e — effective value of the no-load voltage of the AC exciter;
θ — voltage phase angle at the moment of short circuit;
T_a — decaying time constant of the DC component of the AC exciter short-circuit current;
T'_d — transient time constant;
T''_d — sub-transient time constant;
T''_{d0} — sub-transient time constant in the open circuit of AC exciter stator winding.

The expressions for the B-phase and C-phase short-circuit currents of the AC exciter are similar to that of the A phase; one need only replace θ in Eq. (7.69) with $\left(\theta - \frac{2\pi}{3}\right)$ and $\left(\theta - \frac{4\pi}{3}\right)$. From Eq. (7.69), it is seen that the first term in the square brackets is the steady-state current value determined by X_d, and the second and third terms in the square brackets are the AC component values decaying in accordance with the time constants T'_d and T''_d. Since T''_d and T'_d are smaller, the decay to zero will occur only in several cycles.

The second term in Eq. (7.69) is the DC component value determined when the generator air-gap flux is kept unchanged at the moment of short circuit, its decaying time constant being T_a. The DC component value is related to the phase angle of the generator voltage at the moment of short circuit. When $\theta = 0$ or π, the DC component will be maximum. θ will be replaced with the remaining two-phase DC components $\left(\theta - \frac{2\pi}{3}\right)$ and $\left(\theta - \frac{4\pi}{3}\right)$. See Figure 7.17 for the oscillogram of the relevant short-circuit current.

When a three-phase short-circuit occurs in the separate excitation AC exciter with load, its short-circuit current can be solved on the basis of the composite values of the no-load short-circuit current and the load current. The expression for the short-circuit current with load is

$$i_{eN}(t) = \left\{\left[\left(\frac{1}{X''_d} - \frac{1}{X'_d}\right)e^{-\frac{1}{T''_d}} + \left(\frac{1}{X'_d} - \frac{1}{X_d}\right)e^{-\frac{1}{T'_d}}\right]\sqrt{2}U_e\cos\delta \right.$$

$$\left. + \frac{\sqrt{2}E_q}{X_d} - \frac{\sqrt{2}U}{X''_d}e^{-\frac{1}{T_a}}\cos(\omega t + \delta)\right\} \times \cos(\omega t + \theta)$$

$$+ \left[\left(\frac{1}{X''_q} - \frac{1}{X_q}\right)e^{-\frac{1}{T''_q}}\sqrt{2}\sin\delta - \frac{\sqrt{2}U}{X''_q}e^{-\frac{1}{T_a}}\sin(\omega t + \theta)\right]\sin(\omega t + \theta) \quad (7.70)$$

where

δ — rotor power angle of the AC exciter;
θ — phase angle of the instantaneous voltage at the moment of short circuit;
E_q — internal potential of the AC exciter.

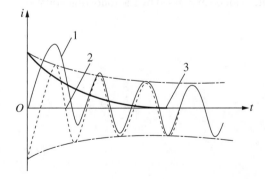

Figure 7.17 Short-circuit current oscillogram at the three-phase short circuit of the AC exciter with no load: 1 – all short-circuit current (asymmetrical component), 2 – fundamental wave curve (symmetrical component), 3 – DC component.

Figure 7.18 AC exciter phasor diagram.

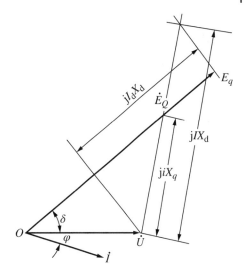

From the phasor diagram shown in Figure 7.18, we obtain

$$E_q = U \cos \delta + IX_d \sin(\varphi + \delta) \tag{7.71}$$

$$\tan \delta = \frac{IX_q \cos \varphi}{U + IX_q \sin \varphi} \tag{7.72}$$

From Eq. (7.70), we obtain the fundamental component of the short-circuit current:

$$i_{eN}(t) = \left\{ \left[\left(\frac{1}{X_d''} - \frac{1}{X_d'} \right) e^{-\frac{t}{T_d''}} \right. \right.$$

$$\left. + \left(\frac{1}{X_d'} - \frac{1}{X_d} \right) e^{-\frac{t}{T_d'}} \right] \sqrt{2} U_e \cos \delta + \frac{\sqrt{2} E_q}{X_d} \right\} \cos(\omega t + \theta)$$

$$+ \left(\frac{1}{X_q''} - \frac{1}{X_q'} \right) e^{-\frac{t}{T_q''}} \sqrt{2} U \sin \delta \sin(\omega t + \theta) \tag{7.73}$$

When $\theta = 0$, which is equivalent to no load, if a no-load short-circuit occurs, the short-circuit current fundamental components of Eqs. (7.73) and (7.69) will be the same.

Example 7.5
Calculating the short-circuit current of the 100 kVA analog AC generator. The relevant parameters are

$$U = \frac{225}{\sqrt{3}} V, \quad X_d'' = 0.0393 \Omega, \quad X' = 0.1529 \Omega, \quad X_d = 0.478 \Omega,$$

$$X_q'' = 0.01023 \Omega, \quad X_q = 0.2481 \Omega, \quad T_d'' = 0.018 s,$$

$$T_d' = 0.285 s, \quad T_q'' = 0.223 S, \quad \cos \varphi = 0.8$$

Solution
See Figure 7.19 for the corresponding computed results.
I_{syN} – short-circuit current of rated load (symmetrical envelope);
I_{sy0} – no-load short-circuit current (symmetrical envelope);
From Figure 7.19, it can be seen that at any time the difference ΔI_{sy} between the no-load and load short-circuit current envelopes will be nearly the same as the $\Delta I_{sy(0)}$ value when $t = 0$. From Eq. (7.73), the

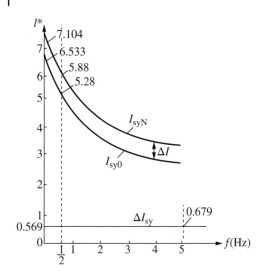

Figure 7.19 Decay curve for the symmetrical component envelope of the short-circuit current.

effective value of the symmetrical AC component for the short-circuit current when $t = 0$ can be obtained. From Eq. (7.71), we obtain

$$I_{syN(0)} = \left[\frac{1}{X_d''}U_e \cos\delta + \frac{IX_d \sin(\varphi+\delta)}{X_d}\right]\cos\theta$$
$$+ \left[\left(\frac{1}{X_q''} - \frac{1}{X_q}\right)U_e \sin\delta\right]\sin\theta \tag{7.74}$$

If it is assumed that the short-circuit initial phase angle $\theta = 0$, then the above equation can be simplified as

$$I_{syN(0)} = \frac{1}{X_q''}U_e \cos\delta + I\sin(\varphi+\delta) \tag{7.75}$$

Besides, in Eq. (7.69), if $t = 0$ and $\theta = 0$, the effective value of the symmetrical AC component for the no-load short-circuit current can be obtained:

$$I_{sy0(0)} = \frac{\sqrt{2}U_e}{X_d''} \tag{7.76}$$

From Eqs. (7.75) and (7.76), the value of the difference is obtained when $t = 0$:

$$\Delta I_{sy(0)} = \frac{U_e}{X_d''}(\cos\delta - 1) + I\sin(\varphi+\delta) \tag{7.77}$$

When δ is smaller, compared with the second terms, the first term is smaller and can be ignored. However, δ in the second term can also be ignored. Therefore, we obtain

$$\Delta I_{sy(0)} \approx I\sin\varphi \tag{7.78}$$

For this example, if $I = 1$p.u., $\sin\varphi = 0.569$, then we obtain $\Delta I_{sy(0)} = 0.569$, as shown in Figure 7.19. Accordingly, the short-circuit current of the separate excitation AC exciter operating at the rated power factor and the rated load can be expressed as

$$i_{eN}(t) = \left[\frac{1}{X_d} + \left(\frac{1}{X_d'} - \frac{1}{X_d}\right)e^{-\frac{t}{T_d'}} + \left(\frac{1}{X_d''} - \frac{1}{X_d'}\right)e^{-\frac{t}{T_d''}}\right]\sqrt{2}U_e \cos(\omega t + \theta)$$
$$- \frac{1}{X_d''}\sqrt{2}U_e e^{-\frac{t}{T_a}}\cos\theta + \sqrt{2}I\sin\varphi \tag{7.79}$$

Example 7.6
In the separate excitation static diode excitation system of a 200 MW turbine generator, to determine the no-load short-circuit current curve of the AC exciter when the rectifier output end has a direct short-circuit.

It is known that
$X_d = 2.31$, $T_a = 0.024$s; $X'_d = 0.21$, $T'_d = 0.064$s; $X''_d = 0.12$, $T''_d = 0.008$s; $f = 100$ Hz, $U_e = 2$ a multiple of the forced ceiling voltage

Solution
During calculation, by taking the maximum short-circuit current when the short-circuit instantaneous voltage phase angle difference $\theta = 0$, the following can be obtained by substituting the relevant parameters into Eq. (7.69):

$$i_a(t) = \left[\frac{1}{2.31} + \left(\frac{1}{0.12} - \frac{1}{2.31}\right) \times e^{-\frac{t}{0.064}} + \left(\frac{1}{0.12} - \frac{1}{0.21}\right) \times e^{-\frac{t}{0.008}}\right]$$
$$\times 2\sqrt{2}\cos\omega t - \frac{2}{0.12} \times \sqrt{2}e^{-\frac{t}{0.0242}}$$

Since the AC exciter frequency f is 100 Hz and the cycle T is 0.01 s, during calculation, the time interval is taken as $\Delta t = \frac{T}{2} = 0.005$s. The computed results are shown in Table 7.4.

See Figure 7.20 for the corresponding $i_a(t) = f(t)$ curve.

When using Eq. (7.69) to express the short-circuit current values for the B and C phases, if $\theta_b = \theta - \frac{2\pi}{3}$, $\theta_c = \theta + \frac{2\pi}{3}$, the short-circuit current values of the B and C phases can be obtained. By connecting each amplitude point in the short-circuit current $i_a(t)$ curve for each phase (such as the A phase), the envelope of $i_a(t)$ can be obtained. By converting this curve into the actual values and multiplying them by the current transformation ratio, the ampere-second characteristics of the rectifier can be obtained. These characteristics can be used as the basis for selecting the number of parallel circuits for the fast-acting fuse and the rectifier.

7.12.2 Determination of Transient Rotor Current-Free Component in Case of Abrupt Generator Terminal Three-Phase Short Circuit

When the generator is shorted, the transient rotor current-free component will flow through the rectifier; from this, the ampere-second characteristics of the rectifier can be determined. When a three-phase short circuit abruptly occurs in the generator end in the load state, the transient rotor current increment is

$$\Delta I_{fd} = \left[\left(\frac{X_{11d}X_{ad} - X_{f1d}X_{a1d}}{X_{11d}X_{ffd} - X_{f1d}^2} \times \frac{1}{X''_d} - \frac{X_{ad}}{X_{ffd}} \times \frac{1}{X'_d}\right)e^{-\frac{t}{T''_d}} + \frac{X_{ad}}{X_{ffd}} \times \frac{1}{X_d}e^{-\frac{t}{T'_d}}\right]$$
$$\times U\cos\delta - \frac{X_{11d}X_{ad} - X_{f1d}X_{a1d}}{X_{11d}X_{ffd} - X_{f1d}^2} \times \frac{U}{X''_d}e^{-\frac{t}{T_a}}\cos(\omega t + \delta) \quad (7.80)$$

However,

$$X_{11d} = X_{1ds} + X_{ad}$$
$$X_{1ds} = \frac{1}{\frac{1}{X''_d - X_s} - \frac{1}{X_{ad}} - \frac{1}{X_f}}$$
$$X_{f1d} = X_{a1d} = X_{ad}$$
$$X_{ffd} = X_f + X_{ad}$$

Table 7.4 Ampere-second characteristics when DC short-circuit occurring to AC exciter via rectifier.

$t \times 10^{-2}$ (s)	0	0.5	1	1.5	2	2.5	3	3.5	4	4.5	5	5.5
$i_a(t)$	0	−35.63	−3.73	−21.95	−28.95	−14.36	−1.89	−8.22	−0.866	−6.79	−0.186	−4.9

$t \times 10^{-2}$ (s)	6	6.5	7	7.5	8	8.5	9	9.5	10	10.5	11	
$i_a(t)$	0.28	−365	0.596	−2.81	0.8	−2.28	0.95	−1.95	1.04	−1.67	1.09	

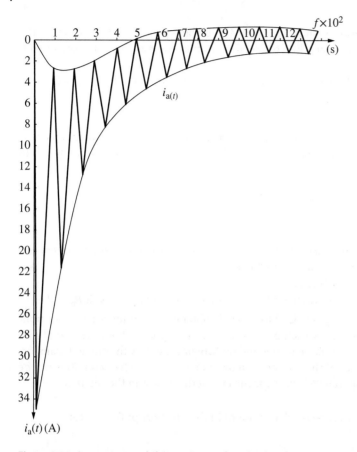

Figure 7.20 Ampere-second characteristics when DC short-circuit occurs in AC exciter via rectifier.

If the transient rotor current-free component is converted into an actual value, then $\Delta i_{f1} = \Delta I_{fd} i_{f\delta}$, and here, $i_{f\delta}$ is the excitation current value corresponding to the air-gap potential $E = X_{ad}$ as obtained from the straight-line segment of the generator no-load characteristic curve, with the corresponding total current of

$$i_{f1} = \Delta i_{f1} + I_{fN} \tag{7.81}$$

In addition, when a three-phase short-circuit abruptly occurs in the main generator end, the excitation current increment provided by the AC exciter is

$$\Delta i_{f2} = (I_{fc} - I_{fN}) \left[1 - \frac{(T'_d - T'_{1d})T'_d}{(T'_d - T''_d)(T'_d - T_e)} e^{-\frac{t}{T'_d}} \right.$$
$$\left. - \frac{(T''_d - T'_{1d})T''_d}{(T''_d - T'_d)(T''_d - T_e)} e^{-\frac{t}{T''_d}} \right.$$
$$\left. - \frac{(T_e - T'_{1d})T_e}{(T_e - T'_d)(T_e - T''_d)} e^{-\frac{t}{T_e}} \right] \tag{7.82}$$

where

I_{fc} — forced field current value corresponding to the multiples of the forced field voltage when the rotor current is stable;

T_e — time constant value of the AC exciter with the load.

Figure 7.21 Transient rotor current change curve when three-phase short circuit occurs in generator end and forced excitation occurs in AC exciter.

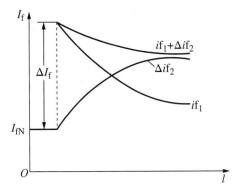

When a three-phase short-circuit occurs in the generator end and the AC exciter provides forced excitation, the total current flowing through the rectifier will be

$$\sum I_f = i_{f1} + \Delta i_{f2} \tag{7.83}$$

Accordingly, the ampere-second load characteristics of the rectifier can be determined; see Figure 7.21 for the corresponding resultant rotor current change curve.

7.13 Calculations for AC Rated Parameters and Forced Excitation Parameters

For the static diode rectifier excitation system with separate excitation, the parameters for the AC main and pilot exciters can be determined according to the following procedures.

7.13.1 Determination of Distortion Sinusoidal Potential E_x and Rated Power P_N

From the rated excitation voltage U_{fN}, the forced excitation voltage U_{fc}, and the forced excitation current I_{fC} of the generator, expressions for the rectified current and voltage in the Type I operating state are obtained:

$$I_f = \frac{\sqrt{6}E_x}{2X_\gamma} \times (1 - \cos\gamma)$$

$$U_f = \frac{3\sqrt{6}E_x}{\pi} - \frac{3}{\pi}I_f X_\gamma$$

The expression for the phase current I_e of the AC exciter is

$$I_e = \sqrt{\frac{2 - \frac{\gamma}{\pi}}{3}} \times I_f \approx \sqrt{\frac{2}{3}}I_f$$

By substituting the above equation into the expression for the excitation current I_f, calculations yield

$$1 - \cos\gamma = \frac{X_\gamma}{\frac{E_x}{I_e}} = X_\gamma^* \tag{7.84}$$

To make the rectifier work in the Type I commutation state, the γ angle should be less than $\frac{\pi}{3}$. We can use the approximation $\gamma_N = \frac{\pi}{6}$ and substitute $\cos\gamma_N = \cos\frac{\pi}{6} = \frac{\sqrt{2}}{2}$ into Eq. (7.84), obtaining

$$X_\gamma^* = 1 - \cos\gamma = 1 - \frac{\sqrt{3}}{2} = 0.134$$

Thus, we can take $X_\gamma^* = 0.15 \sim 0.2$.

By selecting X_γ^* and $\cos\gamma_N$ as known quantities, E_{xN} can be determined from the expression for the external characteristics of the rectifier.

$$E_{xN} = \frac{2\pi}{3\sqrt{6}} \times \frac{U_{fN}}{(1 + \cos\gamma_N)}$$

The rated phase current of the AC exciter is

$$I_{eN} = I_{fN}\sqrt{\frac{2 - \frac{\gamma}{\pi}}{3}}$$

The rated power factor is

$$\cos\varphi_N = \frac{3}{\pi} \times \sqrt{\frac{2}{2 - \frac{\gamma}{\pi}}} \times \cos^2\frac{\gamma}{2}$$

The rated power of the AC exciter is

$$P_N = 3E_{xN}I_{eN}\cos\varphi_N$$

Under forced excitation, the forced excitation voltage U_{fc} and the forced excitation current I_{fc} of the generator are known, and from $I_{fc} = \frac{\sqrt{6}E_{xc}}{2X_\gamma}(1 - \cos\gamma_c)$ and $U_{fc} = \frac{3\sqrt{6}E_{xc}}{\pi} - \frac{3}{\pi}I_{fc}X_\gamma$, we can obtain E_{xc} and γ_c simultaneously. Similarly, the AC exciter power P_c under forced excitation can be obtained.

7.13.2 Determination of Rated Parameters for AC Pilot Exciter

When adopting the three-phase half-controlled bridge rectifying loop, the external characteristic expression for the corresponding rectifier is

$$U_{fe} = \frac{3\sqrt{6}}{2\pi} \times E_x(1 + \cos\alpha) - \frac{3}{\pi}I_{fe}X_\gamma \qquad (7.85)$$

where

U_{fe}, I_{fe} — current and voltage in the excitation loop of the AC exciter supplied by the AC pilot exciter.

Under forced excitation, if the control angle for the cathode assembly $\alpha = 0$ (or taking α as $10°-20°$), the following is obtained:

$$U_{fec} = \frac{3\sqrt{6}}{\pi}E_x - \frac{3}{\pi}I_{fec}X_\gamma \qquad (7.86)$$

The terminal voltage of the pilot exciter working under self-excitation and a constant voltage is constant, and thus $E_{xc} = E_{xN}$. The terminal voltage of the permanent-magnet pilot exciter is determined by the regulating characteristics of the pilot exciter. As discussed above, by giving X_γ and getting $\cos\gamma_c$, we can obtain E_{xN} from the equation:

$$E_{xN} = E_{xc} = \frac{2\pi}{3\sqrt{6}} \times \frac{U_{fN}}{1 + \cos\gamma_N}$$

When the generator is in the rated no-load state, the control angle α in the half-controlled bridge loop can be obtained from the following equations:

$$U_{fe0} = \frac{\sqrt{6}E_{xN}}{2\pi}(1 + \cos\alpha_0) - \frac{3}{\pi}I_{fe0}X_\gamma$$

$$U_{feN} = \frac{\sqrt{6}E_{xN}}{2\pi}(1 + \cos\alpha_N) - \frac{3}{\pi}I_{feN}X_\gamma$$

The rated current of the AC pilot exciter is

$$I_{PN} = I_{feN}\sqrt{\frac{2 - \frac{\gamma_N}{\pi}}{3}}$$

The forced excitation current is

$$I_{Pc} = I_{fec}\sqrt{\frac{2 - \frac{\gamma_c}{\pi}}{3}}$$

and

$$\cos\gamma_N = 1 - \frac{2I_{feN}X_\gamma}{\sqrt{6}E_{xN}}$$

$$\cos\gamma_c = 1 - \frac{2I_{fec}X_\gamma}{\sqrt{6}E_{xc}}$$

The rated power factor is

$$\cos\varphi_N = \frac{3}{\pi} \times \sqrt{\frac{2}{2 - \frac{\gamma_N}{\pi}}} \times \cos^2\frac{\gamma_N}{2}$$

The rated power is

$$P_N = 3E_{xN}I_{PN}\cos\varphi_N$$

The forced excitation power is

$$P_c = 3E_{xc}I_{Pc}\cos\varphi_c$$

8

Brushless Excitation System

8.1 Evolution of Brushless Excitation System

The research results show that when the excitation current of the generator is greater than 8000 A, it is difficult to manufacture the slip ring of the corresponding capacity given the slip ring material, cooling condition, carbon brush current equalizing, and other factors. For this, it is appropriate to use a brushless excitation system for large-capacity turbogenerators.

8.1.1 Diode Brushless Excitation System

An important achievement the United States–based Westinghouse has made in excitation system research and development is the introduction of a brushless excitation system for large-capacity steam turbines in the early 1960s. The prototype delivered a power of 180 kW.

On the basis of the excitation voltage response ratio, the brushless excitation systems developed by Westinghouse fall into the following categories:

(1) Brushless excitation systems with the standard excitation voltage response ratio $R \leqslant 05$,
(2) Brushless excitation systems with a high response ratio $R \geqslant 20$,
(3) Brushless excitation systems with a high initial response ratio (HIR systems).

For HIR systems, the output voltage can reach 95% of the ceiling voltage within 0.1 s.

According to incomplete statistics, Westinghouse supplied brushless excitation systems for 24 units of 220–600 MW steam turbines during 1968–1973, and delivered another 60 units during 1975–1981. So far, it has supplied brushless excitation systems for a total of 500 steam turbines.

The first HIR system (1720 kW, 500 V), designed in July 1977, was supplied to Hunter No. 1 unit (496 MVA, 24 kV, 3600 r min^{-1}) of the United States–based Utah Power and Light.

China imports the 600 MW steam turbine prototype from Westinghouse, also a HIR system, which can supply excitation power up to 3250 kW.

The two 900 MW steam turbine brushless excitation systems deployed at Dayawan Nuclear Power Plant in Guangdong, China, are both developed and supplied by the United Kingdom–based GEC.

The brushless excitation systems developed by the United Kingdom–based Parsons fall into two categories: diode-based brushless excitation systems and SCR (silicon controlled rectifier)-element-based brushless controlled excitation systems.

In the brushless excitation system, three-phase bridge wiring is generally used for the rectifier. In the ideal case, this rectifier circuit delivers the highest energy conversion ratio coefficient. To deliver every 1 kW DC output power, the rectifier circuit requires an AC power capacity of only 1.05 kVA.

The brushless exciter is connected to the rotating diode rectifier in two ways, as shown in Figure 8.1.

In Figure 8.1a, the exciter's field windings in parallel are connected to parallel rotating diodes. The advantage of this wiring is ease of cable connection. But due to the power supply to parallel diode rectifiers, the load current distribution of the diodes in each parallel branch is dependent on the positive potential drop.

Design and Application of Modern Synchronous Generator Excitation Systems, First Edition. Jicheng Li.
© 2019 China Electric Power Press. Published 2019 by John Wiley & Sons Singapore Pte. Ltd.

8 Brushless Excitation System

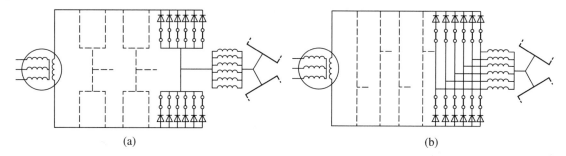

Figure 8.1 Connection of rotating diode rectifier and armature: (a) the exciter's field windings in parallel are connected to parallel rotating diodes, (b) each branch armature winding is directly connected to the independent branch diode rectifier.

In Figure 8.1b, each branch armature winding is directly connected to the independent branch diode rectifier. There is no problem concerning unevenly distributed current for each branch diode. So, it is not necessary to take the current-equalizing impact into consideration. At the same rated rectified current, the capacity of the rectifying element in Figure 8.1b can be at least 15% lower than that in Figure 8.1a. However, since the exciter field windings need to be led out via multiple branches, the base size of the exciter will be expanded by about 10% accordingly.

The Soviet Union has done a lot of research on the brushless excitation system as well and has made remarkable progress.

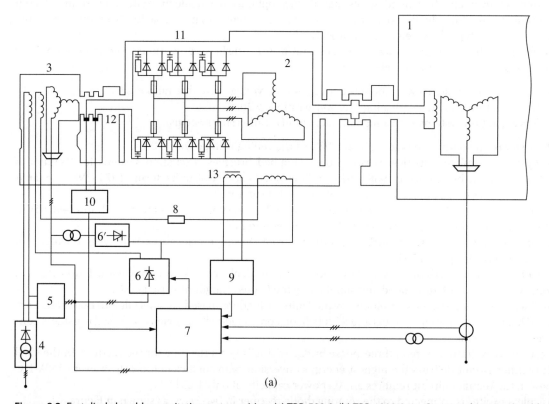

Figure 8.2 Fast diode brushless excitation system wiring: (a) TBB-500-2, (b) TBB-1200-2. 1 – Steam turbine; 2 – Brushless exciter; 3 – Pilot exciter; 4 – Excitation starter; 5 – Pilot exciter voltage regulator; 6 – Automatic and manual circuit SCR rectifier; 7 – Automatic excitation regulator; 8 – Additional resistor; 9 – Non-contact detection and measurement circuit; 10 – Protection and measurement circuit; 11 – Rotating rectifier; 12 – Test slip ring; 13 – Rotor current sensor.

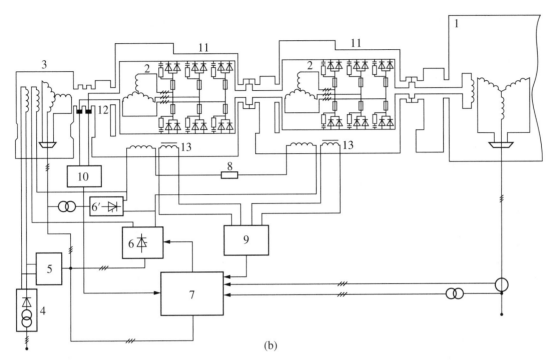

(b)

Figure 8.2 (Continued)

Russia has developed two types of AC exciters for large steam turbines: (i) three-phase AC exciter with a sinusoidal emf waveform in rotating armature and (ii) multi-phase AC exciter with a trapezoidal emf waveform. Such a design can further improve capacity utilization of AC exciters.

Russia has a proven track record in developing diode brushless excitation systems for large-capacity steam turbines. Figures 8.2a,b show the brushless excitation systems for 500 MW and 1200 MW steam turbines manufactured by Russia, respectively. The difference is that two exciters are deployed in the TBB-1200-2 1200 MW steam turbine. Each exciter independently supplies power to its own rotating rectifier, and the two rectifiers are connected in parallel on the DC side. In addition, the field windings of the two AC rectifiers are connected in series, where the same excitation current flows through them. In order to protect the generator rotor winding from the abnormal overvoltage caused in the operation of the breakdown, a slip ring (12) is deployed on the generator shaft for the connection of the protection unit. The non-contact detection and measurement circuit (9) is powered by the rotor current sensor (13).

The Russian diode brushless excitation system for the 1200 MW steam turbine delivers a rated field voltage of 530 V, a rated field current of 7,640 A, and a ceiling voltage factor of 2. Given the rated excitation power up to 4050 kW and the force excitation power of approximately four times the rated excitation power, if supplied by a set of AC exciter and rotating rectifier, their size needs to be significantly expanded, which is hardly ever possible due to the mechanical strength limit. To address the problem, Russian engineers deployed two identical sets of brushless exciters and rectifier rings on the same steam turbine shaft. Each setting of the exciter delivers a capacity equal to half the rated excitation power. The two sets of rectifier rings are connected in parallel at the DC output terminal, while the DC field windings of the brushless AC exciter are connected in series and the resulting excitation current is adjusted by the automatic excitation regulator. The rotating rectifier is wired like a three-phase bridge. Each bridge arm consists of 12 parallel diodes with a rated capacity of 500 A, 2000 V. Every pair of diodes is connected to a 750 A, 1300 V fuse for protection. The wiring is as shown in Figure 8.2b.

In addition to the above countries that manufacture diode brushless excitation steam turbines, the Sweden-based ASEA, Swiss-based ABB, Germany-based KWU (now Siemens), France-based Alstom,

Japan-based Mitsubishi, Belgium-based ACEC, and Italy-based Marelli all produce large-capacity brushless excitation steam turbines.

8.1.2 SCR Brushless Excitation System

In the brushless excitation system, the replacement of the rotating diode rectifier with the SCR will significantly improve the speed of excitation system control. However, it is technically challenging since in the SCR brushless excitation system, the trigger pulse on the static side needs to be supplied to the rotating SCR; that is, a control mechanism will have to be set up between the static part and the rotating part. The United Kingdom–based Parsons installs a small-capacity rotating armature control exciter with the same number of poles as the main exciter on the shaft of the brushless main exciter. The end of each phase of the armature thereof is connected to the control pole of the rotating SCR rectifier to deliver a trigger pulse synchronized with the main exciter armature voltage waveform. The phase shift between the voltage of the control exciter and the voltage of the main exciter is dependent on the physical angle between the fixed magnetic systems thereof. The change in the angle will then lead to a change in the firing angle of the SCR rectifier.

In fact, the control exciter is provisioned with a field winding on both the direct axis and the quadrature axis. The two field windings introduce a control current that is proportional to automatic control and makes the synthetic magnetomotive force thereof a constant. The change in the ratio of the two control portions can lead to a change in the synthetic magnetomotive force in the spatial position, which will in turn lead to a change in the SCR rectifier control firing angle. In this way, the voltage of the main exciter can be subject to control from the positive voltage maximum to the negative voltage maximum, with a control time less than 0.01 s. The AC exciter provisioned with the SCR rectifier rather than the diode is bigger in size, since it works at a constant voltage at the time.

For SCR brushless excitation systems designed on the basis of the above theory, the following aspects have to be considered:

(1) The rate of rise of the forward current through the SCR must not exceed the specified value, otherwise damage will be done due to localized heating of SCR elements.
(2) The rate of rise of the forward voltage must not exceed the specified value, otherwise false conduction may be triggered before the pulse is applied on the control pole of the element.
(3) The control pole current must rapidly be applied to all parallel SCR elements that will conduct at the same time.
(4) Effective ways must be identified to guarantee even distribution of the output current between parallel SCR elements with different forward voltage drop and conduction characteristics.
(5) Inverter subversion of the rectifier must be prevented. In case that the rectifier works at a control angle close to 180°, the fault where a specific bridge arm maintains conduction may occur.
(6) The control system must feature anti-jamming capability.

The United Kingdom–based Parsons has even developed a pilot SCR brushless excitation system on the basis of the above theory to validate the design. The prototype 500 kVA AC exciter, although applicable for a 60 MW, 3000 r min^{-1} steam turbine in output, is also applicable for a 660 MW steam turbine brushless excitation system in terms of the rectifier ring, armature diameter, current load, and SCR element selection.

Parsons has also introduced the rotating pulse transformer trigger mode, which features pulse amplification on the shaft.

The provisioning of the SCR brushless excitation system on the steam turbine not only improves the speed of the control system, but also enables inverter-based de-excitation.

Russia has also achieved remarkable results in developing SCR brushless excitation systems for large-capacity steam turbines. Regarding the transmission of trigger pulse signals from the static side to the rotating SCR rectifier side, a simple and reliable non-contact control system is used to transmit the control pulse with the rotating pulse transformer mounted on the same shaft as the rotating SCR element. The magnetic circuit of the stator and the rotor of the pulse transformer are annular, with a small air gap between

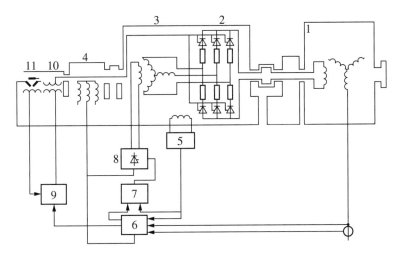

Figure 8.3 Wiring diagram of the Russian SCR brushless excitation system for 300 MW steam turbine. 1 – Steam turbine; 2 – Rotating SCR rectifier; 3 – AC exciter; 4 – Permanent magnet pilot exciter; 5 – Non-contact fuse detection and excitation current measurement device; 6 – Steam turbine automatic voltage regulator; 7 – AC exciter automatic voltage regulator; 8 – SCR rectifier; 9 – Pulse forming unit; 10 – Dynamic pulse transformer; 11 – Control pulse synchronization unit.

the stator and the rotor. The primary and secondary windings of the transformer are accordingly placed in the slot of the stator and the rotor. On the basis of these studies, a prototype SCR brushless excitation system for 300 MW steam turbines may be produced. The basic wiring of the excitation system is as shown in Figure 8.3. The rotating SCR rectifier (2) is structured on the basis of a three-phase bridge circuit. Each bridge arm consists of six parallel rotating SCR elements (500 A, 2000 V) and six fuses. The exciter armature winding consists of six parallel branches that are respectively led out. The control system consists of the pulse forming unit, the synchronization unit (9 and 11), and the dynamic control pulse transformer (10).

The studies conducted on the prototype show that when the pulse transformer secondary winding control pulse leading time is 70 μs, the SCR elements respectively connected to the AC exciter field winding parallel branches successfully conduct. The change in the trigger pulse duration within 300–900 μs is sufficient for the SCR excitation system with the steam turbine field winding as the load.

8.2 Technical Specifications for Brushless Excitation System

The difference between the AC exciter in the brushless excitation system and the ordinary synchronous generator lies in the fact that in addition to the parameter requirements, the ceiling value, response ratio, and other special performance requirements will be considered. In the design, the coordination with the main exciter circuit will also be enabled, and the compliant technical values can thus be calculated.

The technical specification for the brushless excitation system defines the specifications for the AC exciter and rotating rectifier [18].

8.2.1 Ratings

The ratings of the brushless excitation system include the voltage, current, and power. Their values are determined on the basis of the requirement of the generator excitation circuit. The ratings are represented by DC values.

$$\left. \begin{array}{l} \text{Rated current } I_{EN} = K_1 I_{fN} \\ \text{Rated voltage } U_{EN} = K_2 U_{fN} = K_Z I_{fN} R_{fN} \\ \text{Rated power } P_{EN} = I_{EN} V_{EN} \times 10^{-3} \end{array} \right\} \quad (8.1)$$

Table 8.1 Exciter system margins K_1, K_2.

Manufacturer	Hitachi	Toshiba	Mitsubishi	Fuji	Meidensha
Thermal power	1.05	1.05–1.1	Approximately 1.1	1.1	–
Hydropower	1.1	1.1–1.2	Approximately 1.1	1.15	1.1

where

K_1 – Current margin

K_2 – Voltage margin

I_{fN} – Excitation current at rated load

The voltage margin and current margin are decided by the manufacturer. Table 8.1 gives the brushless exciter system margins defined by Japanese manufacturers.

From Table 8.1, the value of K_1, K_2 is 1.05–1.2. For the margin setting, the design error and the excitation power change as a result of the change in the operation mode are taken into account.

8.2.2 AC Exciter

The AC rating of the AC exciter is decided on the basis of the DC rating of the brushless excitation system. The initial calculation is as follows:

(1) Rated line voltage of AC exciter U_{ac}:

$$U_{ac} = \frac{\pi}{3\sqrt{2}}(U_{EN} + U_\gamma) \tag{8.2}$$

where

U_{EN} – Rated DC voltage of brushless exciter

U_γ – Commutation reactance voltage drop

(2) Rated current of AC exciter (I_{ac}):

$$I_{ac} = \sqrt{\frac{2}{3}} \times I_{EN} \sqrt{1 - 3\psi_\gamma} \tag{8.3}$$

where

I_{EN} – Rated DC current of brushless exciter

$$\psi_\gamma = \frac{(2 + \cos\gamma)\sin\gamma - \gamma(1 + 2\cos\gamma)}{2\pi(1 - \cos\gamma)^2} \tag{8.4}$$

where

γ – Commutation angle of rectifier (when $\gamma = 40°$, $\psi_\gamma = 0.03$)

The AC voltage U_{ac} is equal to the voltage rectification ratio coefficient K_V multiplied by the rated DC voltage U_{EN}, that is,

$$U_{ac} = K_V U_{EN} \tag{8.5}$$

For an AC exciter with damper winding: $K_V = 0.80$–0.95.

For an AC exciter without damper winding: $K_V = 0.80\text{--}1.10$.

The AC current I_{ac} is equal to the current rectification ratio coefficient K_I multiplied by the rated DC current I_{EN}, that is,

$$I_{ac} = K_I I_{EN} \tag{8.6}$$

(3) Rated capacity of AC exciter P_{ac}:

$$P_{ac} = \sqrt{3} U_{ac} I_{ac} \times 10^{-3}$$

8.2.3 AC Pilot Exciter

The capacity of the pilot exciter is selected according to the following principles:

(1) The rated capacity of the pilot exciter should be greater than the rated excitation parameter of the generator.
(2) The output ceiling voltage and current of the excitation system are regarded as the maximum output capacity of the pilot exciter.

Table 4.2 shows a design example of the main excitation circuit of the brushless exciter.

8.2.4 Rotating Rectifier

(1) *The number of parallel branches of rotating rectifier n_p.* Currently, the rotating rectifier of the diode brushless excitation system usually features a three-phase bridge rectifier circuit. In terms of the combination of rectifier bridge arms, the wiring of one element for each branch and multiple branches in parallel applies in most cases, and the rectifier elements are connected to fuse in series so as to disconnect a faulty rectifying element.

Given the unbalance rate of current between parallel rectifier elements, a certain margin will be guaranteed in the determination of the number of parallel branches n_p. So

$$n_p \geq \frac{(1+K_u)\,I_{EN}}{3i_{av}} + K_{HM} \tag{8.7}$$

where

K_u – Current unbalance coefficient, usually set to 0.1–0.25;
i_{av} – Average current allowed by rectifier element in three-phase bridge rectifier circuit, A;
K_{HM} – Parallel margin degree.

(2) *Overvoltage protection unit.* Overvoltage protection of the rotating rectifier is enabled for two purposes:
  1) Peak overvoltage as a result of commutation. In this case, the capacitive elements connected in parallel at the ends of the rectifier element absorb the surge overvoltage.
  2) Hosts are not put into operation at the same time. The external overvoltage caused by the grid is applied on both ends of the rotating rectifier. Generally, the overvoltage is absorbed by parallel resistors at the field winding, which will then increase the capacity of the exciter.

The resistance of parallel resistors is 20–50 times that of the generator field winding.

8.3 Composition of Brushless Excitation System

8.3.1 Wiring

Figure 8.4 shows wiring examples of brushless excitations systems for large-capacity steam turbines. Table 8.2 lists the elements of brushless excitation systems.

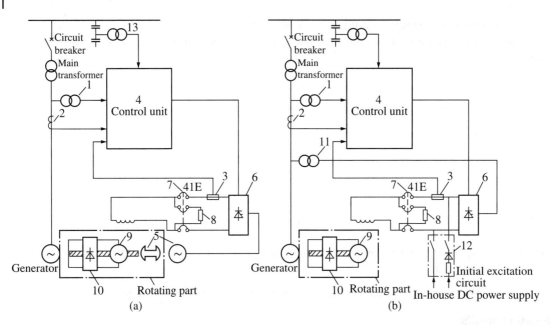

Figure 8.4 Wiring examples of brushless excitation system: (a) separately excited brushless excitation system, (b) self-excited brushless excitation system. 1 – Voltage transformer; 2 – Current transformer; 3 – Shunt; 4 – Control unit; 5 – Pilot exciter; 6 – SCR rectifier; 7 – Magnetic field breaker (magnetic field switching); 8 – De-excitation resistor; 9 – AC exciter; 10 – Rotating rectifier; 11 – Excitation transformer; 12 – Initial excitation circuit.

In Figure 8.4a, the armature (9) and rectifier (10) of the AC exciter are both placed on the rotating shaft to directly supply field current to the generator. The field winding of the AC exciter is placed on the static side. The excitation of the AC exciter is supplied by the permanent magnet pilot exciter (5) via the SCR rectifier. Since its excitation is supplied by the pilot exciter, the AC exciter is known as a separately excited system.

In the brushless excitation system shown in Figure 8.4b, the SCR rectifier (6) is powered by the excitation transformer (11) on the synchronous generator side, which makes it a self-excited system. Meanwhile, before the self-excited system works successfully and the generator sets up the voltage, the initial excitation will be enabled by the in-house DC power supply.

8.3.2 Composition of AC Excitation Unit

Broadly speaking, the AC excitation unit consists of a rotating armature AC main exciter, rotating rectifier, and permanent magnet pilot exciter. The features of the AC main exciter and rotating rectifier are described below.

8.3.2.1 AC Main Exciter

In the brushless excitation system, due to the requirements on control, the design of the AC main exciter is based on the following considerations:

(1) *Time constant* T_{d0e}. Since in the brushless excitation system the automatic excitation regulator works on the SCR rectifier connected to the excitation circuit of the AC main exciter, the excitation control signal will not work on the generator's field-winding side until it goes through the AC exciter inertia element with a high-rotor-winding time constant. Lowering the time constant T_{d0e} of the AC exciter is an effective way to improve the excitation system's voltage response ratio. T_{d0e} can be represented as

$$T_{d0e} = \frac{SK_d}{2\pi \ f(X_d - X_d'') \ P_{EX}} \tag{8.8}$$

Table 8.2 Components of brushless excitation system.

S/N in Figure 8.12	Designation	Description	Applicable wiring	
			Separately excited	Self-excited
1	Voltage transformer TV	Generator voltage detection	○	○
2	Current transformer TA	Generator current detection	○	○
3	Shunt SH	Generator excitation current detection	○	○
4	Control unit	Supply SCR rectifier trigger voltage signals and regulate voltage and current at the generator side	○	○
5	Pilot exciter PEX	Usually the permanent magnet type, as AC power supply of SCR rectifier and control unit	○	–
6	SCR rectifier	Convert AC of PEX or exciter transformer into DC according to signals of the control unit	○	○
7	Field breaker or field discharge switch 41E	Connect or disconnect the excitation circuit of ACEX	○	○
8	De-excitation resistor R_D	Enable the excitation current of ACEX into a circuit when 41E is disabled	○	○
9	AC exciter ACEX	The rotating armature type AC exciter; the armature generates voltage when the current flows through the excitation winding	○	○
10	Rotating rectifier	The rectifier is placed on the shaft of the synchronous generator to rectify the AC output of the AC exciter into DC	○	○
11	Excitation transformer	Power supply of the SCR rectifier and control unit	–	○
12	Initial excitation circuit	Initial excitation is enabled by the in-house DC power supply so that the generator can produce voltage	–	○
13	Voltage divider PD	Voltage detection at the power transmitter end (for PSVR)[a]	○	–

a) PSVR, power system voltage regulator, is only applicable to large-capacity units. No application example is available in China yet.

where

K_d – Damping coefficient, usually set to 1.0–1.1;
S – Rated apparent power of AC exciter, kVA;
P_{EX} – Excitation power of AC exciter at no-load rated voltage, kW;
f – Frequency of AC exciter, Hz.

According to Eq. (8.8), the increase in the frequency of the AC exciter f can decrease the time constant of exciter T'_{d0}. The frequency can be increased by increasing the number of poles of the AC exciter. However, considering the various possibilities and economic efficiency in terms of structural design, for steam turbines, it is advisable to set the frequency of the AC exciter (f) to 100–200 Hz. A further increase in f will no longer lead to a significant decrease in the time constant of the exciter (T'_{d0}).

(2) *Commutation reactance X_γ*. For the AC main exciter in a brushless excitation system, the commutation reactance X_γ is determined as described in Section 8.2 by using the damping winding and salient pole effect.

(3) *Wiring of armature winding of AC main exciter*. To use the AC exciter as a low-voltage and high-current excitation power supply, a number of rectifier elements will be connected in parallel on each bridge arm branch. Two methods are available to address, as shown in Figure 8.5.

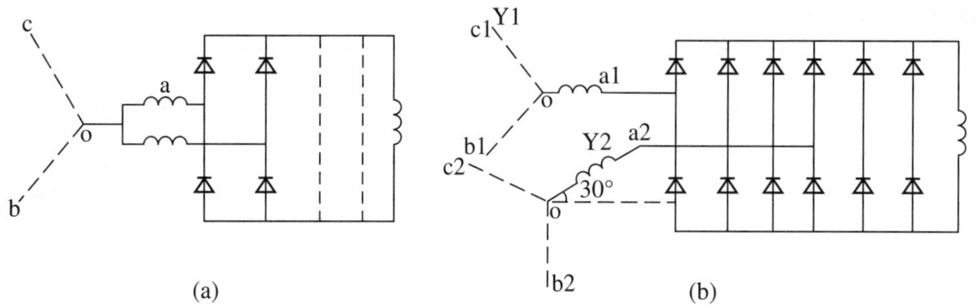

Figure 8.5 Wiring of armature winding of AC main exciter: (a) multidrop winding supply, (b) dual Y30° phase difference six-phase winding supply.

In Figure 8.5a, the number of parallel elements is reduced by a multidrop winding supply. In Figure 8.5b, the wiring of 6-phase dual-γ and 30° difference apply, which can not only reduce the number of parallel elements on each branch, but also improve the ripple coefficient and increase the average value of the rectified voltage.

8.3.2.2 Rotating Rectifier

To fix the rotating three-phase bridge rectifier on the rectifier ring, the cathode rectifier elements of the upper half are usually subject to normal sintering; that is, cathode elements are fixed on the positive rectifier ring.

The anode elements of the lower half are subject to reverse sintering so as to fix the common anode elements on the negative rectifier ring.

To decide on the rated current of the diode element, it is necessary to consider the aperiodic component transient current induced in the rotor winding when the main generator encounters a short-circuit. The selection of the reverse repetitive peak voltage rating is made on the basis of the peak value of the maximum rotor induction voltage when the generator operates asynchronously.

8.4 Voltage Response Characteristics of AC Exciter

The evaluation of large signal transient response characteristics of a brushless excitation system involves the excitation voltage response characteristics of the excitation system. For a brushless excitation system, the excitation voltage response characteristics are closely related to the features of the AC exciter. This section describes the excitation voltage response characteristics of the AC exciter in different working states.

The AC exciter voltage response shall cover the following three scenarios:

1. *No-load excitation voltage response.* When the output open circuit of the AC exciter is not connected to any load, the armature reaction voltage drop will not exist in its armature circuit. When the force excitation voltage is supplied at the field-winding side of the AC exciter, the change rate of the output voltage of the AC exciter is related to its no-load time constant. Since the AC exciter delivers the maximum time constant in the no-load case, the field excitation voltage response ratio thus determined will be relatively low.
2. *Load excitation voltage response.* When the time constant of the AC exciter in the no-load case is set to T_{d0e}, the time constant in the load case will be

$$T_e = T_{d0e} \times \frac{r_e^2 + (X_d' + X_e)(X_q + X_e)}{r_e^2 + (X_d + X_e)(X_q + X_e)} = KT_{d0e} \qquad (8.9)$$

$$r_e = Z_e \cos\phi_1$$
$$X_e = Z_e \sin\phi_1$$

where

- T'_{d0e} — No-load time constant of AC exciter;
- X_d, X'_d — Direct axis synchronous and transient reactance of AC exciter;
- r_e, X_e — Equivalent active and reactive components determined by the load power factor of the AC exciter;
- Z_e — Load impedance of the AC exciter, represented by p.u. $Z_e^* = 1$;
- φ_1 — Fundamental current power factor angle.

Typically, the time constant of the exciter in the load case is 1/2 lower than that in the no-load case; that is, $K \approx 0.5$.

Since the load of the AC exciter is the field winding of the main generator, the time constant of the load is greater than that of the AC exciter by one order of magnitude. This difference determines the voltage response characteristics of the AC exciter in the load case, which divides the force excitation transient process into the following two phases.

In the first phase, for the AC exciter with a lower time constant, the excitation current will reach the ceiling value more quickly. The AC exciter voltage will reach the ceiling value at the same time. During the process, the current in the field-winding circuit of the generator with a higher time constant approximately maintains the original current so that the excitation voltage is subject to overshoot. The corresponding excitation voltage rising segment is referred to as the constant current characteristics curve segment, as shown in Part I in Figure 8.6.

In the second phase, the generator excitation current starts to rise. Due to the armature reaction, the exciter voltage starts to decline from the maximum ceiling voltage of the constant current characteristics curve to the stable value, as shown in Part II in Figure 8.6. The ceiling voltage overshoot ΔU_f varies by $\frac{T'_{d0}}{T_e}$. The higher the ratio, the higher is ΔU_f.

The first phase does not last long, being dependent on the time constant of the AC exciter; the second phase lasts long, being dependent on the time constant of the generator field winding, as shown in Curve 3 of Figure 8.7, the constant current characteristics indicated in the AC exciter saturation characteristics curvilinear coordinates. In fact, in the first phase before the AC exciter voltage reaches the ceiling value, the excitation current of the generator will still rise to a certain extent, so the actual exciter voltage variation curve is as shown in Curve 4 of Figure 8.7. When the time constant of the generator field winding is relatively low, the excitation current of the generator rises in proportion to the excitation voltage. The corresponding characteristics are known as constant resistance characteristics, as shown in Curve 2 of Figure 8.7.

Figure 8.6 Voltage response curve of AC exciter in the case of force excitation.

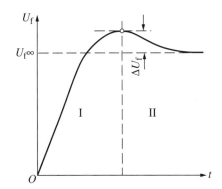

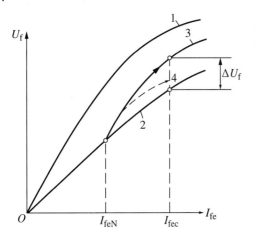

Figure 8.7 Saturation characteristic curve of AC exciter. 1 – No-load characteristics; 2 – Load constant resistance characteristics; 3 – Load constant current characteristics; 4 – Actual change characteristics.

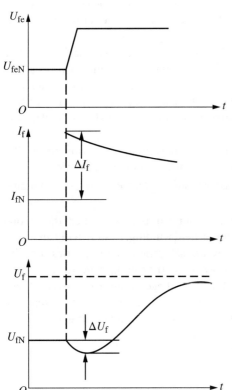

Figure 8.8 Drop of the AC exciter voltage when the generator is subject to three-phase short circuit.

8.4.1 Excitation Voltage Response when Generator Is Subject to Three-Phase Short Circuit

When the generator is subject to a three-phase short circuit, a free-component current will be induced in its rotor excitation winding. When this current flows through the rectifier and AC exciter armature, the equivalent commutation voltage increases to enable instantaneous drop of the generator excitation voltage ΔU_f, as shown in Figure 8.8. The excitation voltage response of the AC exciter drops.

In summary, in the diode brushless excitation system, the excitation voltage response characteristics of the AC exciter vary greatly by the operating mode. As shown in Figure 8.9, the excitation voltage response when the generator is short-circuited is the lowest, while that under the condition of constant current characteristics is the highest.

Figure 8.9 Excitation voltage response of the AC exciter under different force excitation conditions. 1 – Generator load force excitation; 2 – Generator short-circuit fault force excitation; 3 – Constant current characteristics.

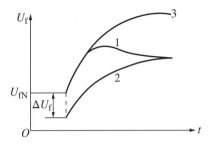

Figure 8.10 Excitation voltage response curve of the AC exciter.

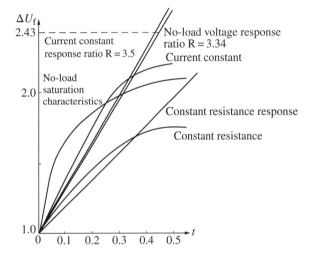

Manufacturers can hardly measure the excitation voltage response of the AC exciter with the operating conditions of the generator. Instead, they measure it with the no-load of the AC exciter or the equivalent resistance to the rotor winding during the design and delivery test. Figure 8.10 shows the excitation voltage response curve of the AC exciter of a China-made 200 MW steam turbine. In Figure 8.10, within 0.5 s, when the output voltage of the AC exciter rises, its output current remains at the rated value.

8.5 Control Characteristics of Brushless Excitation System

In the diode brushless excitation system, since excitation control acts on the field-winding side of the AC exciter, the regulation of generator excitation can only be enabled through the AC exciter inertia element with a high time constant. To improve the speed of excitation regulation, the time constant must be compensated for. Meanwhile, compensation measures vary depending on the disturbance signals. The details are described in the following.

8.5.1 Time Constant Compensation in Case of Small Deviation Signals

Under the action of small deviation signals, all elements of the system still stay within the linear unsaturated working range. The slight incremental change in the voltage and load induced by small deviation signals under the condition of static steadiness falls under the operating state.

Under the condition of small deviation signals, to minimize the impact of the AC exciter inertia time constant and improve the excitation regulation speed of the diode brushless excitation system, it is advisable to connect the proportional negative feedback element in parallel in the AC exciter first-order inertia element system to improve the response of the excitation system.

228 | 8 Brushless Excitation System

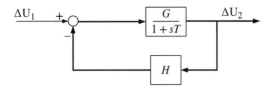

Figure 8.11 Proportional negative feedback compensation of first-order inertia element.

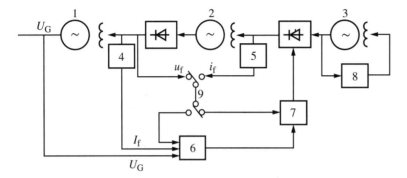

Figure 8.12 Block diagram of diode brushless excitation system. 1 – Generator; 2 – Exciter; 3 – Pilot exciter; 4 – Current converter; 5 – Voltage converter; 6 – Automatic excitation regulator; 7 – SCR rectifier trigger circuit; 8 – Automatic voltage regulator; 9 – Hard negative feedback circuit changeover switch.

In the primary inertia element with time constant T and gain of G as shown in Figure 8.11, if the proportional negative feedback amount H is imported at the output thereof, then the original transfer function $W(s) = \frac{G}{1+sT}$ will become

$$W_1(s) = \frac{\frac{G}{1+sT}}{1+\frac{GH}{1+sT}} = \frac{1}{1+GH} \times \frac{G}{1+\frac{sT}{1+GH}} \tag{8.10}$$

The equivalent gain after synthesis will be

$$G_1 = \frac{G}{1+GH} \tag{8.11}$$

The equivalent time constant will be

$$T_1 = \frac{T}{1+GH} \tag{8.12}$$

As shown in Eqs. (8.11) and (8.12), the introduction of proportional negative feedback can make the system's gain and time constant drop to $\frac{1}{1+GH}$. The decline in the gain can be compensated through the pre-placed amplification element, while the decline in the time constant can help improve the speed of the system.

Similarly, the AC exciter time constant T_e of the diode brushless excitation system shown in Figure 8.12 can also be compensated for by means of proportional negative feedback.

When the proportional negative feedback element is connected in parallel to both ends of the AC exciter with time constant T_e, the synthesis transfer function $W_1(s)$ of the AC exciter can be calculated from the following equation:

$$W_1(s) = \frac{K_C}{1+K_f K_0} \times \frac{K_e}{1+\frac{sT_e}{1+K_f K_0}} \tag{8.13}$$

where

K_C – Total amplification factor of all elements that are not included in the feedback circuit;

K_f – Negative feedback coefficient;

K_e – Amplification coefficient of AC exciter;

T_e – Time constant of AC exciter;

K_0 – Total amplification factor of all elements that are included in the proportional negative feedback element.

According to Eq. (8.13), when the proportional negative feedback element is connected in parallel, the exciter remains an aperiodic element, but the equivalent time constant and amplification factor both drop. Further, when the feedback coefficient K_f becomes high enough, the equivalent exciter time constant can drop to be low enough. The proportional negative feedback element can consist of elements in series of one or several excitation systems. In this case, the regulation circuit time constant included by feedback will drop to

$$T = \frac{T_e}{1 + K_f K_0} \tag{8.14}$$

where

K_0 – Total amplification coefficient of all elements included in the proportional negative feedback element.

Obviously, when K_0 is high enough, T_e can be significantly lowered as long as a low K_f value applies. $(1 + K_f K_0)$ in the above equation is defined as the feedback coefficient. The corresponding amplification coefficient is

$$K = \frac{K_e K_0}{1 + K_f K_0} \tag{8.15}$$

The above discussion shows that the adoption of the proportional negative feedback correction element can lower the time constant of the exciter, while the reduction in the total amplification coefficient can be compensated for by improving the amplification coefficient of some other elements that are not included in feedback.

In the brushless excitation system, proportional negative feedback signals can be obtained from the generator excitation voltage U_f or self-exciter excitation current I_{fe}.

Under the condition of minor deviation linearization, the change in feedback deviation will lead to proportional incremental changes in all elements. In fact, since the exciter and SCR control circuit are subject to the impact of nonlinear factors such as saturation, the output voltage of the exciter does not grow in a linearly proportional manner, but varies exponentially by the time constant T, which drops due to saturation.

This leads to the following inequation:

$$K_S(U_{AVR} - K_f U_f) \leqslant U_{fe,max} \tag{8.16}$$

where

K_S – SCR rectifier amplification coefficient;

U_{AVR} – Excitation regulator output voltage of feedback summing point;

$U_{fe \cdot max}$ – Maximum output voltage of SCR rectifier.

According to the above inequation, the case where the excitation regulator output voltage and feedback voltage synthesis input signals are still lower than the maximum voltage of the exciter after being amplified via the SCR rectifier indicates that all elements included in proportional negative feedback are not saturated.

8 Brushless Excitation System

The feedback effect can lower the time constant of the regulation system and make the exciter voltage vary exponentially along the saturation curve by the lowered time constant T. If some elements included therein are already saturated, then

$$K_S \left(U_{AVR} - K_f U_f\right) \geqslant U_{fe,max} \tag{8.17}$$

The excitation voltage will vary exponentially along the saturation curve with the natural time constant T_e. In this case, the feedback element will not lower the time constant of the regulation system.

8.5.2 Time Constant Compensation in Case of Large Disturbance Signal

Under the action of large disturbance signals, the excitation system elements present a nonlinear saturation state. The method of lowering the time constant of the AC exciter based on proportional negative feedback theory does not work anymore. In this case, it is necessary to look for other solutions, which are described in the following.

8.5.2.1 Connect Additional Resistors in Series

In the case of a large disturbance, the most simple and effective way to lower the time constant of the AC exciter is to connect additional resistors in series in the field-winding circuit thereof, which will also increase the capacity of the pilot exciter. If the externally connected additional resistance is K times the field-winding resistance, then the time constant T_e of the exciter will drop to $\frac{1}{(1+K)}$. If the voltage of the exciter field winding remains unchanged, the pilot exciter power after additional resistors are connected in series will rise to $(1 + K)$ times, that is,

$$P = I_{fe}^2 \left(r_{fe} + \Delta R\right) = I_{fe}^2 \, r_{fe} \, (1 + K) = P_0 \, (1 + K) \tag{8.18}$$

where, P, P_0 represent the pilot exciter power with/without additional resistance, respectively.

For example, for the exciter with $T_e = 1.7$ s, if T_e drops to 0.17 s, then the power of the pilot exciter will rise 10 times.

8.5.2.2 Increasing the Excitation Voltage Multiple of the Exciter

Under the condition of the same exciter time constant T_e, if different multiples of the force excitation voltage are applied to the exciter field winding, then the exciter will deliver different growth rates of the excitation voltage. The higher the force excitation voltage multiple, the faster the excitation voltage grows, as shown in Figure 8.13. When the excitation voltage is applied to the field winding $U_{fec} > U_{feN}$, under the condition of

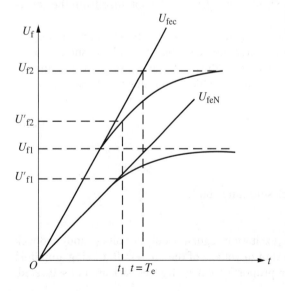

Figure 8.13 Impact of the exciter field voltage U_{fe} on excitation voltage response.

the same time t_1, $U'_{f2} > U'_{f1}$; that is, the higher U_{fec} is, the higher the initial growth rate of the exciter output voltage U_f. For a high-initial-response brushless excitation system, $\frac{U_{fec}}{U_{feN}}$ is usually set to a value greater than 50. In practice, it is not required that the exciter voltage reach the level of the force excitation voltage U_{fec}. The purpose to adopt a high value of the excitation voltage U_{fec} is to improve the initial excitation voltage response speed. For this reason, when the exciter voltage reaches the required ceiling voltage multiple, it is necessary to impose restrictions on the excitation voltage immediately.

8.5.2.3 Comparison of the Above Two Methods

In the exciter field-winding circuit, the growth rate of the excitation voltage can be improved by either connecting additional resistors in series or by improving the force excitation voltage of the exciter field winding, whose effect, however, is different in magnitude.

First, let us discuss the change in the excitation current during the force excitation process. Set the force excitation voltage applied to the exciter field winding to U_{fec}. When the exciter armature is open-circuited, the excitation current i_{fe} will vary exponentially as follows:

$$i_{fe} = I_{fes} \left(1 + e^{-\frac{1}{T_{e0}}}\right) \tag{8.19}$$

where

I_{fes} — The excitation current steady-state value

The excitation current growth rate can be calculated from Eq. (8.19):

$$\frac{di_{fe}}{dt} = \frac{I_{fes}}{T_{e0}} \times e^{-\frac{1}{T_{e0}}}$$

Given that $I_{fes} = \frac{U_{fe0}}{R_{fe}}$, $T_{e0} = \frac{L_{fe}}{R_{fe}}$, the above equation can be rewritten as

$$\frac{di_{fe}}{dt} = \frac{U_{fe0}}{R_{fe}} \times \frac{R_{fe}}{L_{fe}} \times e^{-\frac{t}{T_{e0}}} = \frac{U_{fe0}}{L_{fe}} \times e^{-\frac{t}{T_{e0}}} \tag{8.20}$$

where

R_{fe}, L_{fe} — Total resistance and inductance of the exciter field circuit.

At the moment $t = 0$, the excitation current delivers the maximum growth rate, which is independent of the excitation circuit resistance value and is represented by Eq. (8.21):

$$\left(\frac{di_{fe}}{dt}\right)_{t=0} \quad \frac{U_{fe0}}{L_{fe}} \tag{8.21}$$

For the case that additional resistors are connected, the exciter time constant T_e in the Eq. (8.20) is lowered, but $e^{-\frac{1}{T_{e0}}}$ is also lowered, which leads to a reduced change rate of the excitation current. Therefore, under the same external force excitation voltage U_{fe0}, the connection of additional resistors will lead to a reduced change rate of the excitation current. This conclusion is applicable to both states with and without load. When the generator is subject to a three-phase short circuit, representing the impact of additional resistors on the change in the excitation current can be complex. Figure 8.14 demonstrates the impact of increasing the multiple of the force excitation voltage of the pilot exciter or connecting additional resistors in series on the excitation voltage response ratio when the generator operates in different working states. The results are calculated utilizing computer simulation.

From Figure 8.14, under the same excitation response ratio, the connection of additional resistors can lead to a significantly increased pilot exciter capacity.

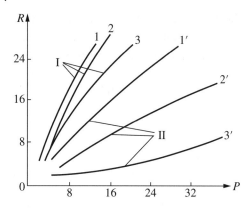

Figure 8.14 Mapping of the excitation response ratio R and the pilot exciter capacity P_{EX} in different generator working modes. I – Increasing the force excitation voltage multiple; II – Connecting additional resistors; 1, 2, 3 – mapping in the case of no-load, load, and three-phase short circuit, respectively, when the force excitation voltage multiple is increased; 1′,2′,3′ – mapping in the case of no-load, load and three-phase short circuit, respectively, when additional resistors are connected.

Table 8.3 Some mathematical models for excitation systems.

Excitation system	IEEE model in 1968	IEEE model in 1981	Excitation system	IEEE model in 1968	IEEE model in 1981
DC exciter	I	DC-1, DC-2	Static self-excitation system	1S	ST1
Brushless excitation system	I	AC1	Static self compound excitation system	III	ST2
High initial brushless excitation system	I	AC2	High initial brushless excitation system		

8.6 Mathematical Models for Brushless Excitation System

Addressing the need for excitation and power system research, mathematical models for excitation systems are usually set up for the study of the role of relevant parameters. Models cannot represent actual elements in all aspects, but can depict their input, output, and internal characteristics. Table 8.3 lists some mathematical models for excitation systems.

In 1967, the United States–based Institute of Electrical and Electronics Engineers (IEEE) introduced some mathematical models for excitation systems. In most cases, the application of these models can deliver satisfactory results. However, in recent years, with the development of new excitation systems, the legacy models can hardly represent certain characteristics in an accurate manner. For this reason, IEEE introduced new mathematical models in 1981.

It should be noted that the new excitation system models introduced by IEEE in 1981 are applicable to the case where the frequency deviation stays within ±5% of the ratings and the oscillation frequency is not greater than 3 Hz. So, it is not appropriate to analyze the sub-synchronous resonance phenomenon and the excitation system's action on torsional vibration torque with these models. Instead, the analysis will be conducted on the basis of more detailed mathematical models that cover low time constants.

Moreover, although these models cannot serve as the main basis for checking the performance of excitation systems, the results calculated with the given models are sufficiently aligned with the measured results.

Furthermore, in 1988, IEC also introduced similar optimized mathematical models for excitation systems in Part II Excitation System of Rotating Electrical Machines.

Further, organizations such as the International Electrotechnical Commission have done some research on excitation system models, and some well-known international motor manufacturers such as the United States–based Westinghouse and GE, Switzerland-based ABB, Sweden-based ASEA, and United Kingdom–based R-R and GEC have all introduced dedicated excitation system models for the purpose of design and research.

8.6 Mathematical Models for Brushless Excitation System

China's Excitation Society has also introduced excitation system models for the purpose of power system stability research and testing.

This section will describe the IEEE mathematical models that the United States–based Westinghouse has introduced for the design of brushless excitation systems.

8.6.1 Determination of Saturation Coefficient

For the creation of mathematical models for excitation systems, saturation of the exciter characteristic curve will first be considered. To enable quantification, the expression for the saturation coefficient is introduced, which is defined as

$$S_E = \frac{A - B}{B} \tag{8.22}$$

where A and B are coefficients determined by the no-load and constant resistive load characteristic curve of the AC exciter, as shown in Figure 8.15.

For the calculation of mathematical models for excitation systems, the saturation coefficient will be represented by mathematical expressions, which can take many forms. One of the mathematical expressions is as follows:

$$S_E = C_1 e^{C_2 E_{FD}} \tag{8.23}$$

where C_1 and C_2 are undetermined coefficients. Generally, for the determination of the saturation coefficient, the exciter operating point change will be considered. It is usually determined by the maximum force excitation voltage $E_{FD,max}$ and $0.75 E_{FD,max}$ and then validated by $E_{FD,N}$. Then Eq. (8.23) applies. For the saturation characteristic curve of the brushless exciter shown in Figure 8.16, for the case that the maximum force excitation voltage $E_{FD,max} = 605$ V,

$$S_{E,max} = \frac{A - B}{B} = \frac{320 - 170}{170} = 0.882$$

$$0.75 E_{FD,max} = 0.75 \times 605 = 454 \quad (V)$$

$$S_{E,0.75\,max} = \frac{210 - 125}{125} = 0.68$$

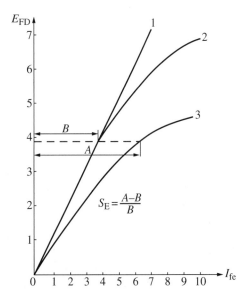

Figure 8.15 Determination of saturation coefficient. 1 – Air-gap line; 2 – No-load saturation characteristic curve; 3 – Constant resistive load characteristic curve.

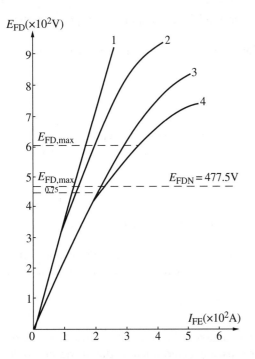

Figure 8.16 Saturation characteristic curve of the 1695 kW brushless exciter. 1 – Air-gap line; 2 – No-load saturation characteristic; 3 and 4 – Constant current/constant resistive load characteristic.

From Eq. (8.23), the saturation coefficient is as follows:

$$S_E = C_1 e^{C_2 E_{FD}}$$

Taking natural logarithms on both sides:

$$\ln S_E = \ln C_1 + C_2 E_{FD}$$

C_1 and C_2 can be calculated from S_E and E_{FD}. For the sake of convenience, E_{FD} will sometimes be translated into p.u., for which the generator's no-load excitation voltage $E_{FD0} = I_{FD0} \times R_{f75°C} = 1094 \times 0.1392 = 145(\text{V})$ applies. So

$$\ln 0.882 = \ln C_1 + C_2 \times \frac{605}{145}$$

$$\ln 0.68 = \ln C_1 + C_2 \times \frac{454}{145}$$

$C_1 = 0.31$; $C_2 = 0.25$.

8.6.2 I-Type Model

8.6.2.1 Block Diagram of I-Type Model

The block diagram of the I-type excitation system model is as shown in Figure 8.17. U_T denotes that the generator-side voltage signal, after comparison with the reference voltage at the summing point, forms a voltage deviation, which after being amplified by the amplifier and subjected to the amplitude limit, is supplied to the exciter as the excitation voltage. The excitation system stabilizer functions as a stabilization circuit to maintain a high static gain for the excitation system and make the system's effective gain drop in the transient state.

In the model, all parameters are represented in p.u. For the generator's terminal voltage base value, the rating U_{Tb} applies; for the excitation voltage base value E_{FDb}, the excitation voltage against the unit rated voltage generated on the extension of the exciter no-load characteristics curve segment (i.e., the air-gap line) applies; for the excitation voltage U_{Rb}, the output voltage of the excitation regulator against the unit rated excitation

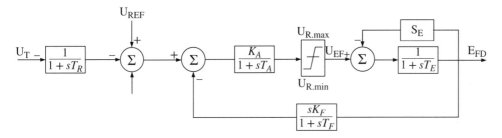

Figure 8.17 Block diagram of I-type mathematical model.

voltage on the extension of the exciter no-load characteristic curve straight segment (i.e., the air-gap line) applies. In the derivation model, all parameters are represented in p.u.

8.6.2.2 Derivation of I-Type Model

Assume the equivalent circuit of the AC exciter is as shown in Figure 8.18, and the saturation characteristic curve as shown in Figure 8.19.

From Figures 8.18 and 8.19, the excitation circuit equation can be obtained:

$$U_\mathrm{R} = U_\mathrm{fe} = I_\mathrm{fe} R_\mathrm{fe} + L_\mathrm{fe}\frac{\mathrm{d}I_\mathrm{fe}}{\mathrm{d}t} \tag{8.24}$$

$$I_\mathrm{fe} = \frac{E_\mathrm{FD}}{R_\mathrm{g}} + \Delta I_\mathrm{fe} \tag{8.25}$$

$$R_\mathrm{g} = \frac{E_\mathrm{FD}}{I_\mathrm{fe}}$$

$$\Delta I_\mathrm{fe} = S_\mathrm{e} E_\mathrm{FD} \tag{8.26}$$

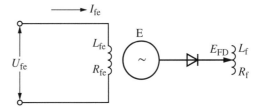

Figure 8.18 Equivalent circuit of AC exciter.

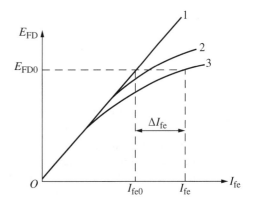

Figure 8.19 Saturation characteristic curve of exciter. 1 – Air-gap line; 2 – No-load saturation characteristic curve; 3 – Constant resistive load characteristic curve.

where

R_g — Slope of the straight segment of the exciter no-load characteristic curve;
ΔI_{fe} — Excitation current increment caused by saturation;
S_e — Saturation ratio coefficient, A/V.

Substituting the value of S_e into Eq. (8.25) yields

$$I_{fe} = \frac{E_{FD}}{R_g} + S_e E_{FD} \tag{8.27}$$

In Eq. (8.27), all parameters are represented in p.u. For the excitation current base value I_{feb} of the exciter, the excitation current against the unit rated excitation voltage on the straight extension of the no-load characteristic curve of the exciter (i.e., the air-gap line) applies. Eqs. (8.26) and (8.24) can thereby be written as

$$I_{fe}^* = E_{FD}^* + S_e E_{FD}^* R_{gb} \tag{8.28}$$

$$U_{fe}^* = I_{fe}^* + \frac{L_{fe}}{R_{fed}} \times \frac{dI_{fe}^*}{dt} \tag{8.29}$$

If the saturation coefficient is represented by p.u., then

$$S_E^* = \frac{\Delta I_{fe}^*}{E_{FD}^*} = S_e R_{gb} \tag{8.30}$$

Substituting Eq. (8.28) into Eq. (8.29) yields $\frac{dI_{fe}^*}{dt} = \frac{dI_{fe}^*}{dE_{FD}^*} \times \frac{dE_{FD}^*}{dt}$. Then Eq. (8.29) can be rewritten as

$$U_{fe}^* = E_{FD}^* + S_e E_{FD}^* R_{gb} + T_e \times \frac{dI_{fe}^*}{dE_{FD}^*} \times \frac{dE_{FD}^*}{dt} \tag{8.31}$$

where

T_e — The exciter time constant that varies with the exciter's degree of saturation $\frac{dI_{fe}^*}{dE_{FD}^*}$.

If

$$T_e = \frac{T_E}{\frac{dI_{fe}^*}{dE_{FD}^*}} \tag{8.32}$$

where

T_E is the exciter unsaturated time constant,

substituting Eq. (8.30) into Eq. (8.31) yields

$$U_{fe}^* = (1 + S_E^*)\ E_{FD}^* + T_E \times \frac{dE_{FD}^*}{dt} \tag{8.33}$$

In Eq. (8.33), the exciter time constant T_E is set to an unsaturated value, while the impact in the case of saturation is represented by S_E^*. $U_{FE}^* = U_{fe}^*$, From Eq. (8.33), we obtain the equivalent block diagram of the AC exciter, as shown in Figure 8.20.

In addition, according to the regulation element proportional negative feedback operation rules, Figure 8.20 can also be represented by the block diagram in Figure 8.21. To simplify the expression, the p.u. symbols of all parameters will be omitted in the following discussion.

Figure 8.20 Equivalent block diagram of I-type AC exciter.

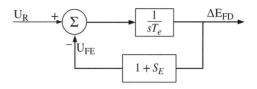

Figure 8.21 Another representation of the equivalent block diagram of I-type AC exciter.

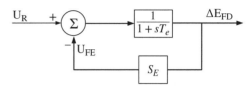

Figure 8.22 AC exciter linearization equivalent block diagram.

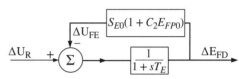

8.6.2.3 Linearization of Small Deviation Signal AC Exciter Model

To study the working state of the excitation system under small signal conditions such as dynamic stability, it is necessary to linearize elements that represent the saturation impact on the exciter so as to obtain the relevant incremental equation. For the saturation coefficient S_E branch shown in Figure 8.20, if represented by p.u., the following equation can be obtained by definition:

$$I_{fe} = U_{FE} = S_E E_{FD} = C_1 e^{C_2 E_{FD}} E_{FD}$$

Under the effect of small signal, the linearization incremental equation is as follows:

$$\Delta U_{FE} = \Delta E_{FD} \, (C_1 e^{C_2 E_{FD0}}) + C_1 E_{FD0} C_2 e^{C_2 E_{FD0}} \, \Delta E_{FD}$$
$$= \Delta E_{FD} C_1 e^{C_2 E_{FD0}} \, (1 + C_2 E_{FD0})$$

That is,

$$\frac{\Delta U_{FE}}{\Delta E_{FD}} = S_{E0} \, (1 + C_2 E_{FD0}) \tag{8.34}$$

The AC exciter linearization equivalent block diagram under the condition of small signal can be obtained from the above equation, as shown in Figure 8.22.

Under the effect of small deviation signal, given the impact of the saturation factor, the equivalent AC exciter time constant will drop. An example is given below.

Example 8.1 According to the AC exciter characteristic curve shown in Figure 8.16, the linearization AC exciter transfer function expression under the effect of small deviation signal is determined. The saturation constants have been obtained: $C_1 = 0.31$, $C_2 = 0.25$, $E_{FDN} = 477.5\,\text{V} = 3.13$ p.u., and $T_E = 2.49\,\text{s}$.

First, determine the branch saturation coefficient shown in Figure 8.22:

$$S_{E0} = C_1 e^{C_2 E_{FD0}} = 0.31 \; e^{0.25 \times 3.13} = 0.678$$
$$S_{E0} \, (1 + C_2 E_{FD0}) = 0.678 \, \times \, (1 + 0.25 \times 3.13) = 1.208$$

Substitute the value of this branch saturation coefficient S_{ED} into the linearization AC exciter equivalent block diagram shown in Figure 8.21, and then simplify the original model with the following equation:

$$\frac{\Delta E_{FD}}{\Delta U_R} = \frac{\dfrac{1}{1+sT_E}}{1 + \dfrac{S_{E0}\,(1+C_2 E_{FD0})}{1+sT_E}} = \frac{\dfrac{1}{1+2.49s}}{1 + \dfrac{1.208}{1+2.49s}} = \frac{0.452}{1+1.13s}$$

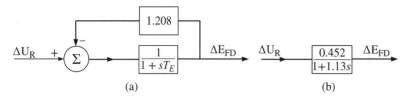

Figure 8.23 AC exciter block diagram where saturation is considered.

The simplified AC exciter block diagram is shown in Figure 8.23b. This figure shows that due to saturation, the AC exciter time constant can be reduced by half at the rated operating point from $T_E = 2.49\,\text{s}$ to $1.13\,\text{s}$. But at the same time, the gain of the AC exciter also drops from 1 to 0.452, which can be compensated for by the gain of other elements.

8.6.3 AC-I Model

8.6.3.1 The Block Diagram

For brushless excitation systems with a low excitation voltage response ratio, for example, $R = 0.5\,\text{s}^{-1}$, the I-type model applies, because in the case of force excitation the AC exciter voltage basically varies according to the constant resistive load characteristics curve, that is, $E_{FD}^* = I_{FD}^*$. In the I-type model, the saturation coefficient is determined by the constant resistive load saturation characteristic curve. In fact, the saturation coefficient S_E at any instant takes not only the saturation factor but also the combined effect of the exciter armature reaction and the rectifier commutation voltage drop into consideration.

For high-response-ratio excitation systems (e.g., $R = 2.0\,\text{s}^{-1}$), it is not appropriate to apply the I-type excitation system models. The reason is that, although the operating point of the exciter in the steady state is located on the constant resistance load saturation characteristic curve, in the case of force excitation, since the generator field winding features a high time constant, when the excitation voltage rises, the generator excitation current I_{FD} remains approximately constant. The transient operating point is close to the constant current saturation characteristic curve, that is, in the transient state, $E_{FD}^* \neq I_{FD}^*$.

On the basis of the above, IEEE introduced the new AC1 model for high-response-ratio brushless excitation systems, which is different from the I-type model since it takes saturation, the exciter armature reaction, rectifier commutation voltage drop, and other factors into consideration. The saturation coefficient S_E is obtained based on the exciter no-load saturation characteristic curve and takes only the saturation factor into consideration. The exciter armature reaction and commutation voltage drop are respectively represented by K_D and K_C. The block diagram of the AC1 model is shown in Figure 8.24, and the corresponding AC exciter block diagram is shown in Figure 8.25. This model considers the exciter's load effect more accurately. The determination of the saturation coefficient S_E, exciter armature reaction, and rectifier commutation voltage drop coefficients K_D and K_C will be discussed below.

8.6.3.2 Saturation Coefficient S_E

In the AC1 model, the saturation coefficient S_E is obtained on the basis of the exciter no-load saturation characteristic curve, as shown in Figure 8.26.

8.6.3.3 Armature Reaction Coefficient K_D

The armature reaction of the AC exciter induced by the load current I_{FD} is represented by the armature reaction coefficient K_D. The value is set to the function of the direct axis synchronous reactance of the exciter X_d.

The coefficient K_D can be determined from Eq. (8.35):

$$K_D = (1 + S_{ED}) \left(\frac{U_E - E_{FD}}{E_{FD}} - \frac{K_C}{\sqrt{3}} \right) \tag{8.35}$$

8.6 Mathematical Models for Brushless Excitation System

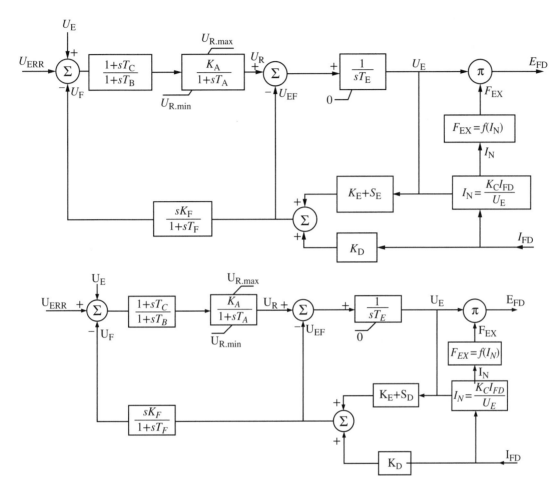

Figure 8.24 AC1 model block diagram.

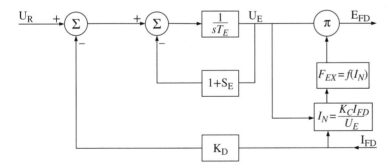

Figure 8.25 AC exciter block diagram of AC1 model.

where

U_E – The voltage on the exciter no-load voltage curve against the excitation voltage E_{FD} on the constant resistance line;

E_{FD} – The generator excitation voltage decided by the constant resistance line operating point;

K_C – The rectifier commutation voltage drop coefficient.

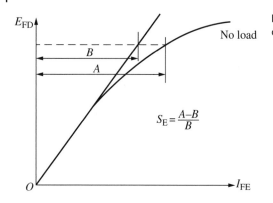

Figure 8.26 Determination of AC exciter saturation coefficient: (a) the original model, (b) the simplified model.

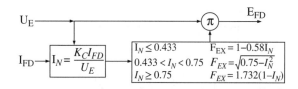

Figure 8.27 Determination of the rectifier coefficient F_{EX}.

For the I-type commutation state, the coefficient K_C is calculated as follows:

$$K_C = \frac{3\sqrt{3}}{\pi} \times \frac{X_{CE}}{R_{FDb}} \tag{8.36}$$

$$X_{CE} = \frac{X_d'' + X_2}{2} \tag{8.37}$$

where

X_{CE} – The equivalent commutation reactance of the exciter;
X_d'' – The direct axis sub-transient reactance of the exciter;
X_2 – The negative sequence reactance of the exciter.

In addition, for the convenience of calculation, the concept of the rectifier coefficient is introduced into the AC1 model. For example, for the I-type commutation state, the rectifier coefficient is as follows:

$$F_{EX} = 1 - \frac{K_C I_{FD}}{\sqrt{3} U_E} \tag{8.38}$$

$$F_{EX} = 1 - \frac{3}{\pi}\left(\frac{X_{CE} I_{FD}}{R_{FDb} U_E}\right) \tag{8.39}$$

The rectifier coefficient F_{EX} of the other working states is shown in Figure 8.27. The excitation voltage of a given working state: $E_{FD} = F_{EX} U_E$. The rectifier external characteristic curve represented by $\frac{E_{FD}}{U_E}$ and $I_N = \frac{K_C I_{FD}}{U_E}$ as the p.u. base value is shown in Figure 8.28.

Figure 8.29 indicates the physical significance of the saturation coefficient S_{E0}, armature reaction coefficient K_D and the commutation voltage drop coefficient K_c. In Figure 8.29, the exciter voltage drop, $\Delta U_E = I_{FD0}\left(K_D + \frac{K_C}{\sqrt{3}}\right)$, represents the sum of the armature reaction and the commutation voltage drop, while curve segment b'c' represents the incremental excitation current component determined by the saturation coefficient S_{E0}. Furthermore, it should be noted that for the study of the stability of large signal power systems, the AC exciter block diagram indicated in the AC1 model is also applicable.

8.6 Mathematical Models for Brushless Excitation System | 241

Figure 8.28 Rectifier external characteristic curve $\frac{E_{FD}}{U_E} = f\left(\frac{K_C I_{FD}}{U_E}\right)$.

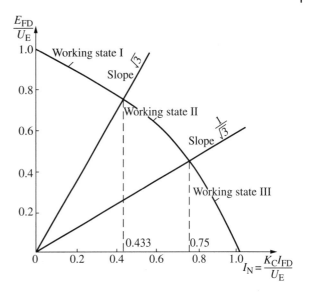

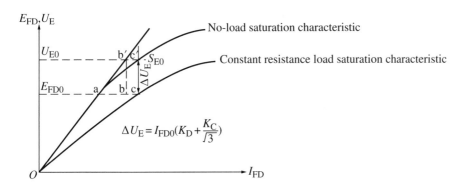

Figure 8.29 Determination of S_E, K_D, and K_C by the exciter no-load and constant resistance load saturation characteristic curve.

Example 8.2 Determine the AC1 model parameters of the following brushless excitation system. Known: under the condition E_{FDN} (the rated excitation voltage of the exciter) = 477.5 V and I_{FDN} (the rated excitation current) = 5898 A, R_{FDB} (the excitation field-winding resistance) = 0.0772 Ω at 75 °C and S (the AC exciter capacity) = 3,661 kVA. The rated terminal voltage and current are 417 V and 5074 A respectively. X_d'' (the direct axis sub-transient reactance of the exciter) = 22.11%; X_2 (negative sequence reactance) = 23.34%; L_{FE} (the exciter field-winding inductance) = 0.08 H and R_{FE} (the exciter field-winding resistance) = 0.0321 Ω.

When the generator force excitation voltage is 800 V (according to the constant resistance load saturation characteristic), the exciter field current I_{FEC} is 750 A, and the pilot exciter side voltage is 265 V. The additional resistance externally connected to the exciter field circuit R_{EXT} is 0.32 Ω. The exciter characteristic curve is as shown in Figure 8.30.

The calculation procedures of the relevant parameters are as follows:

1. Calculation of the rectifier commutation voltage drop coefficient K_C

   First, determine the exciter's p.u. base value impedance:

   $$Z_{EN} = \frac{417}{\sqrt{3} \times 5074} = 0.0474 \ (\Omega)$$

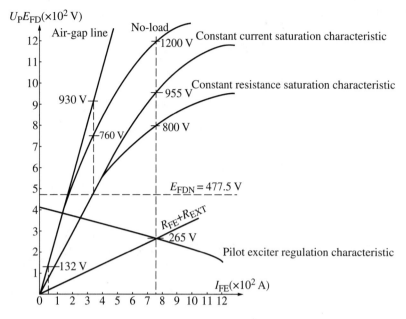

Figure 8.30 AC exciter saturation characteristic and pilot exciter external characteristic curve.

The exciter commutation reactance p.u.:

$$X_{CE}^* = \frac{X_{d''}^* + X_2^*}{2} = \frac{22.11\% + 23.34\%}{2} = 22.73\%$$

Rms: $X_{CE} = X_{CE}^* Z_{EN} = 0.2273 \times 0.0474 = 0.01$ (Ω)

The commutation voltage drop coefficient represented by the generator base value:

$$K_C = \frac{3\sqrt{3} X_{CE}}{\pi R_{FDB}} = \frac{3\sqrt{3} \times 0.01}{\pi \times 0.0772} = 0.214$$

2. Calculation of the armature reaction coefficient K_D
   From Eq. (8.35), we obtain

$$K_D = (1 + S_{E0}) \left(\frac{U_E - E_{FD}}{E_{FD}} - \frac{K_C}{\sqrt{3}} \right) = 1.028 \times \left(\frac{760 - 477.5}{477.5} - \frac{0.214}{\sqrt{3}} \right) = 0.48$$

S_{E0} is obtained from the exciter air-gap line and the no-load saturation curve. In the case of $E_{FD} = 477.5$ V, S_{E0} is obtained from the curve in Figure 8.16:

$$S_{E0} = \frac{180 - 175}{175} = 0.028$$

In the above calculation, the coefficients K_C and K_D are calculated under the assumption that the rectifier is working in the I-type commutation state. So, it is necessary to validate the calculation results. First, validate the commutation voltage drop coefficient K_C. For the I-type commutation state, the terminal voltage of the AC exciter can be written as

$$U_E = E_{FD} + \frac{3}{\pi} I_{FD} X_{CE} \tag{8.40}$$

Substitute the original calculation point of $E_{FD} = 477.5$ V into Eq. (8.40):

$$U_E = 477.5 + 0.955 \times 5898 \times 0.01 = 533.8 \quad (V)$$

$$\frac{E_{FD}}{U_E} = \frac{477.5}{533.8} = 0.8945$$

From the rectifier external characteristic curve in Figure 8.28, we obtain

$$I_N = \frac{K_C I_{FD}}{U_E} = 0.18$$

Thus:

$$K_C = \frac{I_N U_F}{I_{FD}} = \frac{0.18 \times \frac{533.8}{132}}{\frac{5898}{1709}} = 0.21 \approx 0.214$$

(similar to the calculated value).

In the above equation, the generator excitation current base value I_{FDb} is set to the excitation current that is needed to generate the unit rated voltage on the generator no-load saturation characteristic air-gap line; that is, $I_{FDb} = 1709$ A.

Now validate the armature reaction coefficient K_D. From Figure 8.30:

$$\Delta U_E = 760 - 477.5 = 282.5 \quad (V)$$

Substitute the relevant values into $\Delta U_E = I_{FD0}\left(K_D + \frac{K_C}{\sqrt{3}}\right)$

$$\frac{282.5}{132} = \frac{5898}{1709} \times \left(K_D + \frac{0.214}{\sqrt{3}}\right)$$

Thus, $K_D = 0.496$. The result indicates that K_D calculated from Eq. (8.35) is close to the actual value.

The compensated exciter time constant:

$$T_E = \frac{L_{FE}}{R_{RFE} + R_{EXT}} = \frac{0.08}{0.0321 + 0.32} = 0.22(s)$$

The uncompensated exciter time constant: $T_E = 2.49$ s.

The exciter voltage differential negative feedback circuit feedback factor typical value: $K_F = 0.03$, $T_F = 1.0$ s.

8.7 AC2 Model

In the high-initial-response brushless excitation system, the pilot exciter provides a high multiple of the force excitation voltage to get HIR. When the voltage of the exciter reaches the ceiling value required, the exciter field current limiting circuit imposes limits instantaneously to prevent severe over-current in the excitation winding. Moreover, to lower the exciter's equivalent time constant under the condition of small deviation, an additional time constant compensation circuit is enabled by the excitation voltage proportional negative feedback circuit. Generally, the circuit can compensate for the exciter time constant by an order of magnitude. The AC2 model is designed to represent the high-initial-response brushless excitation system. The resulting block diagram is shown in Figure 8.31. The AC2 model has the following features:

(1) The determination of K_C, K_D, S_E is the same as in the AC-I model.
(2) The set multiple of the exciter field current limit U_{LR} is dependent on the multiple of the force excitation ceiling voltage multiple.

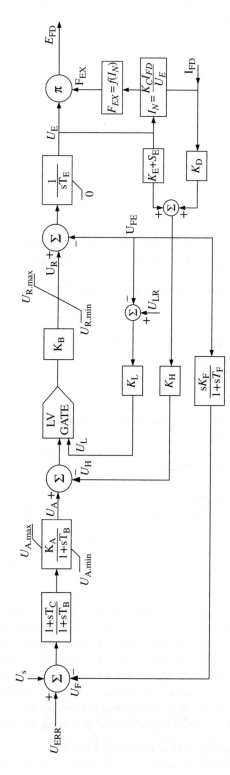

Figure 8.31 Block diagram of the AC2 model.

(3) The total gain of the excitation regulator consists of K_A and kb. kb represents the gain of the power amplifier in the excitation regulator.
(4) To deliver a high initial effect, the maximum output voltage of the excitation regulator $U_{R,\cdot max}$ is rather high, usually over 50 times the p.u.
(5) The exciter field current limit coefficient, if too low, will lead to low sensitivity, and if too high, will lead to an unstable regulation system. General, the typical value is set to $K_B K_L \geqslant 100$.
(6) The proportional negative feedback coefficient K_H is dependent on the magnitude of the compensation of the exciter time constant. General, the total feedback coefficient is set to $1 + K_B K_H \geqslant 100$.

Example 8.3 Determine the parameters of the AC2 model from Figure 8.30 and the exciter data in 8.2. In this case, assume the additional resistance of the exciter field-winding circuit $R_{EXT} = 0$. Relevant p.u. already known: $I_{FDb} = 1709 A$, $E_{FDb} = 132 V$, $R_{FDb} = 0.0772 \Omega$, $I_{FEb} = 50 A$, and $U_{FEb} = 1.57 V$. The calculation of S_{E0}, K_C, and K_D is the same as in the previous instance. The p.u.: $S_{E0} = 0.028$; $K_C = 0.214$, and $K_D = 0.48$.

Get the exciter time constant:

$$T_E = \frac{0.08}{0.0321} = 2.49(s)$$

(1) Calculate the exciter field current limit multiple and the force excitation saturation coefficient. In the case of force excitation, the excitation current is 750 A. The excitation current limit multiple:

$$U_{LR,max} = \frac{750}{50} = 15$$

In the case of force excitation, the voltage on the exciter no-load saturation characteristic curve is 1200 V. The p.u. value:

$$U^*_{E,max} = \frac{1200}{132} = 9.09$$

The saturation coefficient: $E_{FD} = 477.5 V$, $E_{FD} = 800 V$.

$$S_{EN} = \frac{180 - 175}{175} = 0.028$$

$$S_{E,max} = \frac{385 - 285}{385} = 0.259$$

(2) Calculate the voltage and current of the pilot exciter. In case of field forcing, the excitation current supplied by the exciter, $I_{FE} = 750 A$, converted into the AC value:

$$I_P = \frac{750}{1.2} = 615 \quad (A)$$

From Figure 8.30, the voltage regulation characteristic of the pilot exciter, the DC voltage value is obtained: $U_{P\cdot max} = 270 V$.

Converting into the AC value:

$$U_{P,max} = \frac{270}{1.17} = 230 \quad (V)$$

(3) Determine the maximum/minimum output voltage multiple of the excitation regulator:

$$U_{R,max} = \frac{U_{P,max}}{U_{FE}} = \frac{270}{50 \times 0.0321} = 171.97$$

$$U_{R,min} = -0.9 U_{R,max} = -154.8$$

(4) Determine the maximum/minimum output voltage multiple of the first-stage amplifier. The maximum output voltage of the first-stage amplifier: $\pm 12 V$. So,

$$U_{A,max} = \frac{12}{1.57} = 7.64 \quad (V)$$

$$U_{A,min} = -U_{A,max} = -7.64(V)$$

(5) Calculate the amplifier coefficient of the amplifier K_B. In the nominal case, $I_{FEB} = 350$ A.
The AC value:

$$I_{FEB2} = \frac{350}{1.2} = 291 \text{ (A)}$$

From Figure 8.30, the voltage of the pilot exciter is obtained:

$$U_{PN} = U_{FEN} = 335.5 \text{ (V)}$$

In the case of force excitation, the exciter field current: $I_{FEC} = 750$ A; the pilot exciter voltage is therefore $U_{FEC} = 265$ V.

When the input voltage of the SCR rectifier phase shifter in the excitation regulator is set to 5 V, the amplification factor of the power amplifier will be

$$K_B = \frac{265}{5} = 53$$

In the AC2 model, the typical value: $K_L K_B \geq 100$, Now, $K_L kb = 150$, giving

$$K_L = \frac{150}{K_B} = \frac{150}{53} = 2.83$$

(6) Calculate the feedback coefficient K_H. Set the total feedback amplification coefficient: $1 + K_H K_B = 10$. We obtain

$$K_H = \frac{9}{K_B} = \frac{9}{53} = 0.17$$

(7) Calculate the first-stage amplification coefficient K_A. Since the two ends of the forward-path transfer function K_B are connected to the negative feedback path K_H, the amplification coefficient of the synthetic circuit is as follows:

$$K_{BCL} = \frac{53}{1 + 53 \times 0.17} = 5.25$$

From the typical value $K_A K_{BCL} = 200$, we obtain

$$K_A = \frac{200}{5.25} = 38.$$

8.8 Generator Excitation Parameter Detection and Fault Alarm

8.8.1 Determination of Excitation Voltage and Current of Rotor Excitation Winding

In the brushless excitation system, due to the elimination of the slip ring, it is necessary to take special steps to measure the voltage and current in the rotor excitation circuit [19].

Generally, an additional small measurement slip ring is employed to measure the rotor excitation voltage. For the measurement of the rotor excitation current, a simple method is a combination of test and calculation. The calculation basis is that, in the normal rated operating range, the pilot generator delivers an extremely low saturation condition. Under the condition that the generator field-winding resistance is set to a specific value, the excitation current of the generator is linearly proportional to that of the exciter.

Furthermore, for steam turbines with brushless excitation systems, the external DC power supply and analog test slip ring are used to measure the generator's no-load and short-circuit characteristics in the delivery test. The excitation current measured at any instant is accurate and can be used for the calculation of the generator's rotor excitation current in different operating modes. Thus, the equation is as follows:

$$I_{fN} = \sqrt{I_{f1}^2 + K_{f2}^2 + 2K_0 I_{f1} I_{f2} \sin \phi} \tag{8.41}$$

where

I_{f1} – The excitation current when the generator no-load voltage is the sum of the rated voltage U_{GN} and the rotor armature resistance voltage drop $r_a I_{GN}$;

I_{f2} – The excitation current when the generator stator current is set to the rating in the generator short-circuit characteristics test;

K_0 – The saturation coefficient;

ϕ – The power factor angle.

The values according to the Japan-based JEC-114-1979 standard are as shown in Table 8.4.

For the saturation coefficient of the power factor that is not indicated in Table 8.4, the K_0 that is close to the power factor should be selected. Moreover, under the condition of zero power factor, if the saturation coefficient $K_0 = 1 + \sigma$ is below 1.25, then $K_0 = 1.25$ will apply. σ represents the saturation coefficient that is determined when the voltage is $1.2U_{GN}$ in the generator no-load saturation curve. The resulting value is calculated as follows:

$$\sigma = \frac{c_1 c}{bc_1} \tag{8.42}$$

$c_1 c$ and bc_1 can be obtained from Figure 8.32.

The generator rotor excitation current of the brushless excitation system can also be determined as follows. As mentioned earlier, the generator's no-load and short-circuit excitation currents I_{f1} and I_{f2} are measured with the analog slip ring in the delivery test, and the measurement results are accurate. Moreover, generally, the AC main exciter for a brushless excitation system is unsaturated in the rated operation state and saturated in the force excitation state. So, the excitation current I_{fe} and output current I_f (the generator excitation current) of the AC exciter are linearly proportional. From this, when the excitation current in the case of the generator no-load rated voltage is I_{f1}, the measured excitation current of the AC exciter is I_{fe1}. Likewise, the excitation current of the generator is obtained from the three-phase short-circuit characteristics, and the corresponding excitation current of the AC excitation is I_{f2}. The rated excitation current of the AC exciter in the

Table 8.4 The saturation coefficient K_0.

$\cos\phi$	1.0	0.95	0.9	0.85	0.8	0
Salient pole machine	1.0	1.1	1.15	1.2	1.25	$1+\sigma$
Non-salient pole machine	1.0	1.0	1.05	1.1	1.15	$1+\sigma$

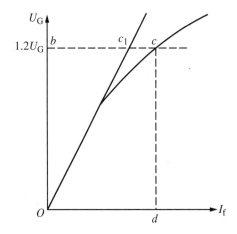

Figure 8.32 Determination of the generator's saturation coefficient σ. O_b – Voltage = $1.2U_{GN}$; O_{c1} – Air-gap line; O_c – No-load saturation characteristic curve; O_d – No-load excitation current.

rated operation state according to Eq. (8.41) is as follows:

$$I_{feN} = \sqrt{I_{fe1}^2 + K_0^2 I_{fe2}^2 + 2K_0 I_{fe1} I_{fe2} \sin\phi} \qquad (8.43)$$

where

I_{fe1}, I_{fe2} — Excitation current that maps to I_{f1}, I_{f2};
K_0 — Saturation coefficient of AC exciter;
ϕ — Power factor angle of AC exciter.

After I_{feN} is calculated from Eq. (8.43), the generator's rated excitation current I_{fN} can be calculated according to the following equation:

$$I_{fN} = K I_{feN} \times \frac{I_{fe2}}{I_{f2}} \qquad (8.44)$$

where

I_{fe2} — The AC exciter excitation current against the generator excitation current in the generator rated three-phase short-circuit characteristics test;
K — Temperature correction coefficient that represents the difference between the field-winding temperatures in the generator three-phase short-circuit characteristics test and during the actual nominal load operation.

The expression for K:

$$K = \frac{\sqrt{r^2 + 1 + 2r\sin\phi}}{\sqrt{(r \times \rho)^2 + 1 + 2r\rho\sin\phi}}$$

$$\sin\phi = \sqrt{1 - \cos^2\phi} \qquad (8.45)$$

where

r — The design value of the short-circuit ratio of AC exciter;
ρ — The ratio of the field-winding resistance in the test state to that in the case of rated load;
$\cos\phi$ — The design value of the rated power factor of the AC exciter.

8.8.2 Protection, Fault Monitoring, and Alarm of Brushless Exciter

In the brushless excitation system, the possible faults are usually of the following types.

8.8.2.1 Fault Types

(1) *Rectifier element failure.* A rectifier element failure usually leads to short circuit of elements. For the three-phase bridge rectifier circuit, if any bridge arm consists of only one element in series and multiple elements in parallel, then when the short-circuit fault occurs in an element, an interphase fault will be caused. In this case, the fault must be eliminated. The adoption of rapid fuse protection is one of the effective measures. For large-capacity steam turbines, it is necessary to increase the rectifier element's parallel capacity margin at the same time.
(2) *Field ground fault.* In the brushless excitation system, due to the elimination of the carbon brush, the excitation circuit grounding failure rate is generally low. Moreover, since one-point grounding of the excitation circuit will not cause equipment damage, grounding protection is not provided for small generators in most cases, and magnetism loss protection is adopted as the final protection measure. For large-capacity

steam turbines, an alarm will be generated when a ground fault occurs, and the operating personnel will then take appropriate actions.

In regard to the generator excitation circuit over-current fault, the automatic excitation regulator protection limit circuit will take care of it.

8.8.2.2 Fault Monitoring

(1) *Rectifier element fault.* The faults of a rotating diode are mostly of two types: elements are short- or open-circuited. When such a fault occurs, the armature reaction magnetic field generated by the AC exciter armature winding is different from the normal case, and the AC voltage induced in the AC exciter field winding is superposed on the DC excitation current component. In this case, an additional winding can be employed on the AC exciter's field pole or between the field poles to measure the signal voltage waveform for the determination of the fault form. Figure 8.33 gives the waveforms in different fault states. When an arm is open-circuited, the rectifier elements commute four times per cycle. One commutation is conducted between cathode elements, so the armature current change rate is the highest, and the electric potential induced in the measuring winding is high, as shown in Figure 8.33b.

When an arm is short-circuited as shown in Figure 8.33c, the a-phase cathode element tube is short-circuited, and short-circuit current flows from b-phase and c-phase cathode elements to the a-phase fault element. The current varies with the armature current base wave frequency, so the same potential as the AC exciter base wave frequency is induced in the measuring winding. The resulting waveform is shown in Figure 8.33d.

An example of the rectifier element fault detection circuit is shown in Figure 8.34.

For rectifier element fault detection, the fault detection circuit shown in Figure 8.35 also applies. When a fault occurs in the generator excitation circuit rectifier element, an abnormal voltage can be induced in the exciter field winding.

In the circuit, the shunt is connected in series to the AC exciter's excitation circuit. When a fault occurs in the rectifier element, the armature reaction magnetic field immediately induces a harmonic potential in the exciter field winding. An AC component related to the fault state is superposed in the field winding (the frequency depends on the element open- or short-circuit fault).

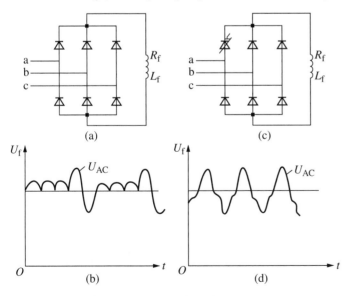

Figure 8.33 Signal waveform in measuring winding when a fault occurs in the rotating diode: (a) rectifier circuit when an arm is disconnected, (b) fault signal waveform when an arm is disconnected, (c) rectifier circuit when an arm is short-circuited, (d) fault signal waveform when an arm is short-circuited.

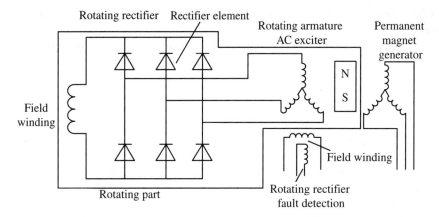

Figure 8.34 Detection of rectifier component fault circuit by additional winding.

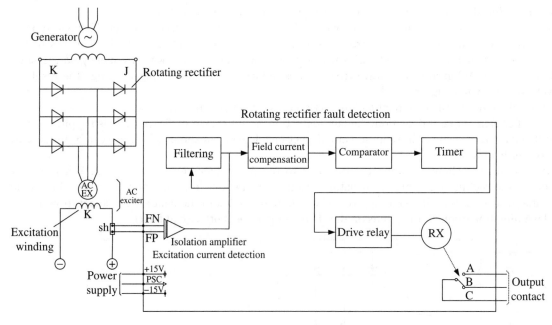

Figure 8.35 Use of the current shunt to detect the rectifier component fault circuit.

The fault signal is amplified via the shunt (SH) and isolation amplifier, operated via the filter, excitation current compensator, comparator and timer, and then output by the output relay for the purpose of warning.

(2) *Field-winding ground fault*. In the brushless excitation system, the generator field-winding ground fault can be detected as follows:
  1) Rely on the measuring slip ring to connect the carbon brush on the positive end of the field winding via the current limiting resistor to the negative end of the external rectifier current. Once a ground fault occurs and the DC voltage at the grounding point is greater than the set DC voltage, grounding current will flow through the ground detection relay to generate a warning signal. The detection circuit is shown in Figure 8.36.
  2) The rotating transformer and SCR switch grounding test circuit are shown in Figure 8.37.
     In this circuit, the reference potentiometers R_1 and R_2 are connected between one end of the generator field winding and the shaft for voltage division. The MΩ component is powered by the rotating

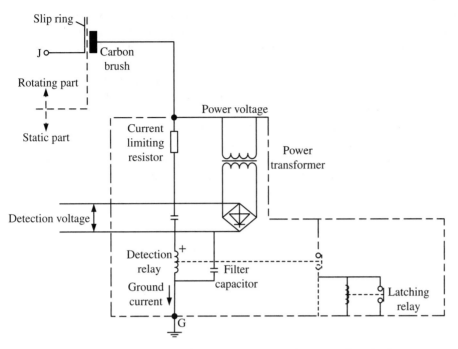

Figure 8.36 Use the slip ring to detect the rotor excitation winding grounding circuit.

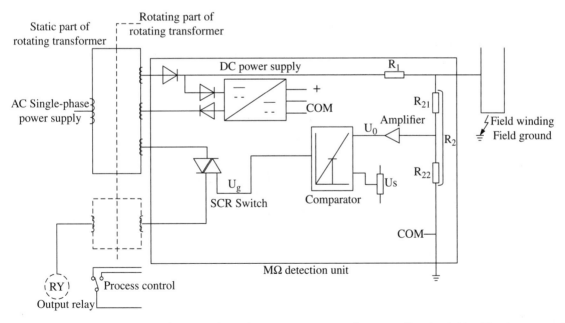

Figure 8.37 Use the rotating transformer to detect the rotor excitation winding grounding circuit. *Note*: The above ground detection circuits are designed to measure the insulation resistance to ground of a specific point of the field winding. The rated voltage in the circuits represents the set value of the insulation resistance to the ground of the field winding. When the insulation resistance to the ground of a specific point of the field winding is lower than the set value, an alarm will be generated. For this reason, it is more appropriate to call the above circuits the field-winding insulation resistance to ground detectors.

secondary winding of the rotating transformer. When a ground fault occurs and the divided voltage of the reference potentiometers is greater than the set voltage of the comparator, the AC SCR switch enables conduction of the operation output relay, and the alarm is generated by the corresponding contact.

3) The AC exciter is supplied with additional winding to power the auxiliary rectifier. The positive terminal of the rectifier is connected to the detection coil on the rotating side via the resistor. Once a ground fault occurs between the host excitation windings JK, the ground point and the negative terminal ground point of the rectifier will constitute a pathway. The ground fault signal is generated by the detection coil on the fixed side. The circuit is shown in Figure 8.38.

(3) *Generator excitation over-current fault.* When a fault occurs in the generator excitation circuit, the over-excitation limit of the AC exciter field current is usually employed to render indirect protection.

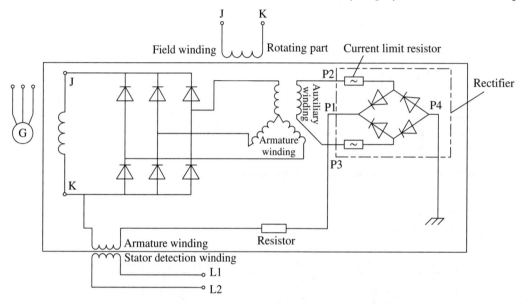

Figure 8.38 Use of the additional winding of AC exciter to detect the grounding fault circuit of rotor winding.

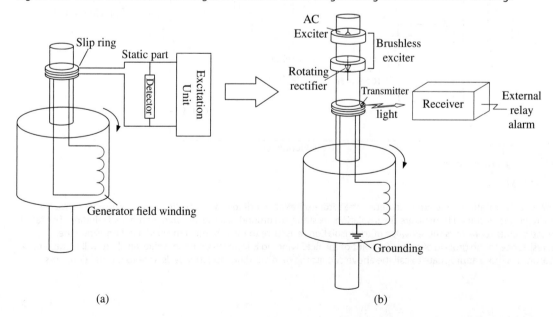

Figure 8.39 Light pulse signal ground detection circuit: (a) the traditional method, (b) the light pulse non-contact method.

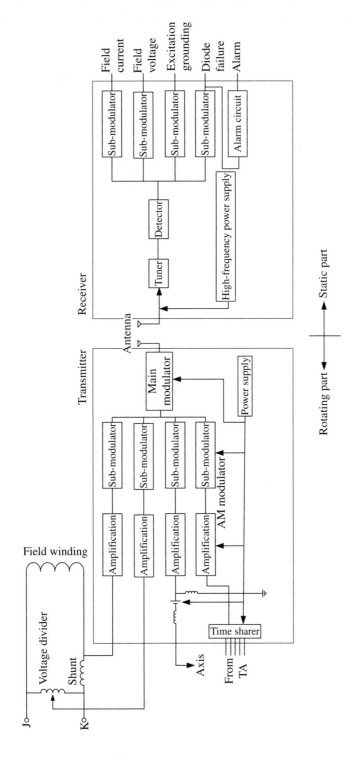

Figure 8.40 FM modulation signal transmitting and receiving system.

8.8.2.3 Telemetry Detection Technology

In recent years, thanks to the advances in telemetry, the technologies that measure brushless excitation system faults with light pulse or radio transmitting and receiving signals instead of an electric brush and slip ring are widely used.

Figure 8.39 gives an example where a light impulse detects the generator field-winding ground fault in a brushless excitation system.

For the light pulse non-contact scheme shown in Figure 8.39b, if the excitation winding is subject to one-point grounding and trace current flows through the light emitting diode, then the detector at the rotating side will detect fault signals, which will be received by the detector at the static side. Then a trip or alarm signal will be generated.

Figure 8.40 depicts the fault detection system where FM modulation transmits signals.

In the circuit, output signals of the excitation system correlated measurements are received by the transmitter in the rotating part, converted into trace voltage signals, subject to FM modulation, output in the form of transmit radio waves, and received by the receiving antenna for related treatment.

9

Separately Excited SCR Excitation System

9.1 Overview

In a separately excited silicon-controlled rectifier (SCR) excitation system, the excitation of the generator is supplied by a coaxial self-excited constant voltage AC exciter (also called an auxiliary generator) via an SCR. In the case of a given load, the terminal voltage of the auxiliary generator can be approximately constant. Its own excitation is supplied by the self-excitation circuit connected to the auxiliary generator terminal.

In a separately excited SCR excitation system, changes in the generator excitation voltage are achieved by virtue of changes in the control angle of the SCR connected to the main excitation winding circuit of the generator. Since the excitation voltage regulation acts on the SCR trigger circuit, whose time constant is only a few microseconds, rather than on the excitation winding head of the auxiliary generator, the system has a high field flashing voltage response ratio. The diagram of the working principle of a separately excited SCR excitation system is shown in Figure 9.1.

9.2 Characteristics of Separately Excited SCR Excitation System

9.2.1 Characteristics of Hydro-generator SCR Excitation System

As can be seen from Figure 9.1, the excitation of the generator is supplied by an auxiliary generator coaxial with the main generator via an SCR. The control signal of the automatic excitation regulator (5) of the generator is taken from the voltage and current at the generator terminal. The output of the excitation regulator is connected to the phase shifter (4) to adjust the control angle α in order to change the excitation voltage of the generator. After the contactor in parallel with the discharger acts, the discharger (7) as an overvoltage protection device of the rotor excitation winding can short it out. Similarly, the nonlinear resistor (6) is provided in order to limit the rotor overvoltage. In the separately excited SCR excitation system, in addition to the de-excitation resistance, the SCR can be used for inversion de-excitation.

The excitation of the auxiliary generator (2) is composed of the transformer (8) and the SCR (9). The automatic voltage regulator (11) of the auxiliary generator acts on the phase shifter (10) to change the control angle of the SCR and keep its terminal voltage at a constant value. The field flashing of the auxiliary generator is supplied by the power supply (12). In the following, examples are given to introduce design characteristics of some separately excited SCR excitation systems for large hydro-generators. For hydro-generators, the world's largest separately excited SCR excitation system is the one installed on a 500 MW hydro-generator at the Hydro Power Station of Krasnoyarsk in the Soviet Union which was put into operation in 1976. The auxiliary generator has high- and low-voltage AC armature windings, with a capacity of 7650 kW, rated voltage of 1460 V, rated current of 3020 A, and power factor of 0.4. The SCR adopts a three-phase bridge connection, with a rated voltage of 1900 V, rated current of 7400 A, and an output excitation power of 13 600 kW in the forced state.

Figure 9.2 shows the structure of the auxiliary generator for the corresponding excitation system. The SCR excitation system for the hydro-generator is shown in Figure 9.3.

9 Separately Excited SCR Excitation System

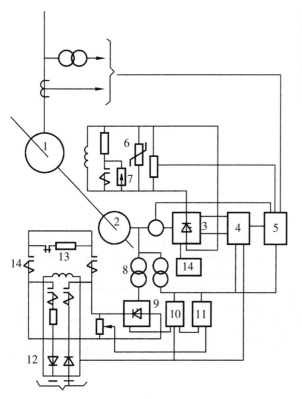

Figure 9.1 Diagram of working principle of separately excited SCR excitation system: 1 – generator; 2 – auxiliary generator; 3 – main SCR; 4 – trigger phase shifter; 5 – automatic excitation regulator; 6 – nonlinear resistance; 7 – discharger; 8 – transformer; 9 – SCR; 10 – phase shifter; 11 – automatic voltage regulator; 12 – field flashing source; 13 – de-excitation resistance; 14 – field circuit breaker.

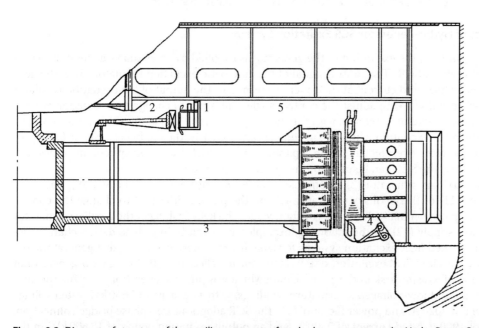

Figure 9.2 Diagram of structure of the auxiliary generator for a hydro-generator at the Hydro Power Station of Krasnoyarsk: 1 and 2 – auxiliary generator's rotor and stator, respectively; 3 and 4 – main generator's rotor and stator, respectively; 5 – upper bracket of hydro-generator.

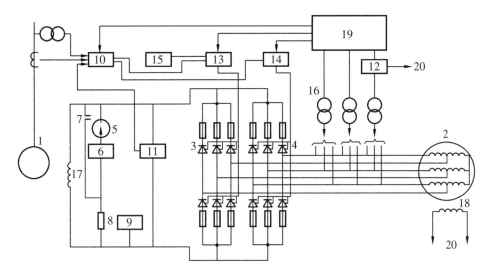

Figure 9.3 Diagram of connection of separately excited SCR excitation system: 1 – main generator; 2 – auxiliary generator; 3 – SCR set for normal working; 4 – SCR set for forced excitation; 5 – discharger; 6 – current relay acting on contactor operating coil; 7 – contactor; 8 – de-excitation resistance; 9 – insulation monitoring relay; 10 – generator excitation regulator; 11 – feedback correction element; 12 – auxiliary generator voltage regulator; 13 and 14 – phase shifters; 15 – manual control circuit; 16 – transformer for self-use; 17 and 18 – excitation winding of main and auxiliary generators respectively; 19 – auxiliary generator power supply; 20 – auxiliary generator field circuit.

In a separately excited SCR excitation system, the following factors should be taken into account when selecting SCR series components:

(1) The commutating voltage in the forced excitation state.
(2) Possible operating overvoltage of the auxiliary generator.
(3) Aging of rectifying elements.
(4) Unevenness in the voltage distribution of series elements.
(5) Overvoltage and discharger operation voltage during non-synchronous operation.

In the circuit, each arm consists of three 2000 V components in series and ten 320 A components in parallel. When the rectifier is short-circuited, it is protected by the fast-acting fuse. The overvoltage caused by commutation of the rectifier is limited by the RC circuit.

In the normal working state, the phase control angle of the SCR set for normal working $\alpha_N = 34°$, while that of the SCR set for forced excitation $\alpha_f = 131°$; and the commutation angle $\gamma_f = 8°$. This is the incomplete commutation state. The generator excitation current in the rated state $I_{fN} = 3700$ A, while the current supplied by the set for forced excitation $I_{ff} = 400$ A. In the forced excitation state, $\alpha_f = 0$; the current supplied by the set for forced excitation $I_{ff} = 2I_{fN} = 7400$ A; and the commutation angle $\gamma_f = 40°$. In the forced excitation state, I_{ff} is 7400 A, $\alpha_f = 75°$, $\alpha_N = 16°$, $\gamma_{fN} = 37°$, and $\gamma_{Nf} = 9°$. At this point, γ_{fN} is the commutation angle at the time of the commutation from the set for forced excitation to the set for normal working, while γ_{Nf} is the commutation angle at the time of the commutation from the set for normal working to the set for forced excitation.

Under the above working conditions, the rectified current of the set for forced excitation $I_{ff} = 5200$ A, while that of the set for normal working $I_{fN} = 2200$ A. Thus, the requirements for the forced excitation voltage multiple $K_U = 4$ and forced excitation current multiple $K_I = 2$ are met.

When a rectifier is used for inversion de-excitation, $\alpha_f = 130°$, and the corresponding negative rectified voltage is 1580 V. When the stator winding is three-phase short-circuited, $I_{ff} = 2I_{fN}$, and the corresponding de-excitation duration is 1.1 s. When the stator winding is open-circuited, the excitation current has the same value, and the de-excitation duration is about 2 s.

In the excitation system, with the help of an automatic excitation regulator, the following functions can be achieved:

(1) Achieve normal and forced excitation limits and limiting forced excitation.
(2) Limiting the excitation output to a value corresponding to the rated power and $\cos\varphi = 1$ when the trigger control system is deactivated and two of the fuses in one of the arms of the rectifier are overheated.
(3) Switching on of the contactor and the de-excitation resistance in the event of the generator's loss of field.
(4) Load reduction and shutdown of the unit when the rectifier cooling water is interrupted.

9.2.2 Characteristics of SCR Excitation System for Steam Turbine Generator

The typical connection of a separately excited SCR excitation system for large steam turbines generator is shown in Figure 9.4.

The excitation of the generator is supplied by auxiliary generator 2, which is coaxial with the main generator, via SCR 3. The SCR is three-phase bridge-connected, with DC outputs in parallel. The rectifier is connected to the generator with a bus that is about 8 m long. It is connected directly to the input end of the middle SCR set. At the point of connection, there are two branches connected to the other two rectifier sets with buses that are respectively 2.4 and 2.7 m long. Each rectifier set has a separate trigger device. The excitation of the generator is regulated by the automatic excitation regulator. When the excitation regulator is deactivated, the

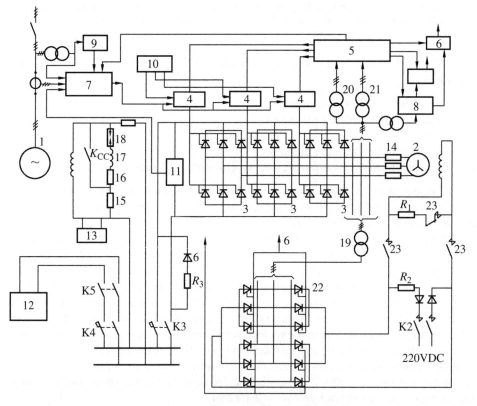

Figure 9.4 Separately excited SCR excitation system for 800 MW steam turbine: 1 – generator; 2 – auxiliary generator; 3 and 22 – main SCR and auxiliary generator SCR respectively; 4 – main SCR trigger; 5 – auxiliary generator power supply; 6 – auxiliary generator self-excitation circuit SCR trigger; 7 and 8 – excitation regulator of generator and that of auxiliary generator, respectively; 9 – difference adjustment circuit; 10 – manual control circuit; 11 – rotor voltage soft negative feedback circuit; 12 – standby exciter; 13 – rotor current monitoring circuit; 14 – current-sharing reactor; 15 – self-synchronizing resistance; 16 – current limiting resistor; 17 – current relay; 18 – discharger; 19 – self-excited transformer; 20 and 21 – exciter regulator power transformer for generator and that for auxiliary generator, respectively; 23 – de-excitation switch.

control angle of the SCR is changed with manual control circuit 10. The auxiliary generator is self-excited by rectifier transformer 19 at its end via SCR 22 to maintain the self-excitation constant voltage characteristics. Rectifier 22 is in parallel on the DC side, but trigger device 6 is dedicated. Automatic voltage regulator 8 of the auxiliary generator is proportional, regulating the excitation on the basis of voltage deviation.

In the excitation system, the generator's and the auxiliary generator's excitation regulator and SCR trigger circuits are powered by the self-use transformer connected to the auxiliary generator terminal. The field flashing of the auxiliary generator is achieved via the battery pack and the additional resistor R_2 circuit.

In all operating modes, the steam turbine's excitation weakening and de-excitation are achieved with the SCR working in the inversion mode. In the no-load state, the time constant of the steam turbine is about 7.9 s. In the event of a fault, the time for the inversion de-excitation to reduce the rotor current to zero is about 0.7 s. The remaining 7.2 s is primarily determined by the attenuation of the free component of the rotor current in the rotor core and the damping winding of the steam turbine. When the rotor current reaches zero, the generator terminal voltage is close to 57%. That is, the de-excitation circuit absorbs only 43% of the magnetic field energy. As many studies have pointed out, further acceleration of the rotor current attenuation process has little effect on de-excitation. If necessary, the negative excitation current should be provided by the excitation system to accelerate the de-excitation process. Compared with a hydro-generator, the steam turbine plays an insignificant role in de-excitation.

In the development of large-capacity separately excited SCR excitation systems, parallel current sharing of SCRs is an important topic. Test results show that direct parallelization on the SCR AC side will significantly degrade the current-sharing characteristics. The waveform of each rectifier set is shown in Figure 9.5. For the 800 MW steam turbine, the no-load excitation current is 1000 A, and the current nonuniformity between the three rectifier sets in Phase C reaches a factor of 2 or greater. When the excitation current increases to 2000 A, the nonuniformity factor drops to 1.7. To improve the current-sharing characteristics, the current-sharing reactors X_T are added to the circuit. The primary has $W_I = 1$ turn, and the secondary has $W_{II} = 200$ turns. The current-sharing reactors are connected in series to each phase of the SCR, causing an additional reactance of about 0.025 Ω. The optimum parameters of the current-sharing reactors can be determined with the SCR DC-side short-circuit conditions. For the system, the optimal connection of the current-sharing reactors is shown in Figure 9.6. The resistance R in parallel with the current-sharing reactors is 46–48.6 Ω. When these measures are taken, the current nonuniformity between the rectifier sets is smaller than 15%, even in the forced excitation state, where the excitation current can go up to 7000 A.

Besides, since all the SCR sets in the field circuit are powered by the auxiliary generator at full voltage, when the generator is unloaded, the rectifiers are in a deeply controlled state; the control angle is large; and the rectified voltage shows a high ripple component. If a transition to the standby exciter system occurs in this state, the ripple current will reach thousands of amperes or even damage the SCRs. To solve this problem,

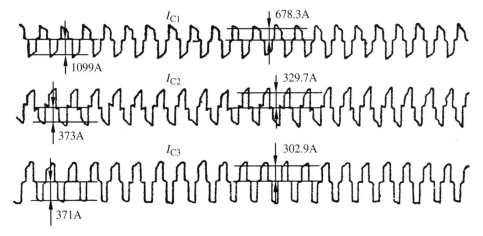

Figure 9.5 AC current distribution of C-phase SCR when 800 MW steam turbine is unloaded.

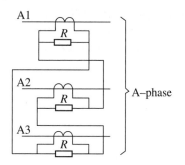

Figure 9.6 Connection of current-sharing reactors.

Table 9.1 Key parameters of auxiliary generators for 500 MW hydro-generators and 800 MW steam turbines.

Parameter	500 MW hydro-generator	800 MW steam turbine	Parameter	500 MW hydro-generator	800 MW steam turbine
Nominal power (kVA)	7650	5700	Rated excitation current (A)	586	195
Power factor	0.4	0.5	Vertical axis synchronous reactance X_d^*	0.526	1.36
Rated voltage (V)	1460	940	Horizontal axis synchronous reactance X_q^*	0.356	1.36
Rated current (A)	3020	3500	Vertical axis transient synchronous reactance $X_d'^*$	0.172	0.127
Rated excitation voltage (V)	191	107	$X_q'^*$	0.356	0.127
$X_d''^*$	0.107	0.078	$T_{d2}'^*$	–	0.99[a]
$X_q''^*$	0.108	0.086	$T_{d3}'^*$	–	1.14[a]
T_{d0}^* (s)	2.61	6.5	$T_d'^*$	0.271	0.08[a]
T_{d1}^*	0.853[a]	0.61[a]			

a) The time constant of the vertical axis damping circuit when the excitation winding and damping system circuits are closed, in s.

some of Soviet power stations made SCRs complete the transition to the inversion state in advance before switching to the standby excitation unit, to reduce the rotor current shock in the transfer process.

In the separately excited SCR excitation system, one of the causes for damage to the rectifiers is excessive growth of currents of the SCR elements at the time of turn-on. The current-sharing reactors connected in series to the circuit can help limit the current growth.

Besides, in order to improve the reliability of the operation in the inversion state, when the SCR current is interrupted, a small current path can be maintained, and R-V and diode circuits can be connected in parallel on the DC side. The R-V should be 70 Ω, and the diode should allow long-term flow of 5 A.

The key parameters are listed in Table 9.1 for the purpose of comparison of the design characteristics of auxiliary generators for hydro-generators and steam turbines.

9.3 Influence of Harmonic Current Load on Electromagnetic Characteristics of Auxiliary Generator

9.3.1 Harmonic Current mmf

9.3.1.1 Harmonic Current Value

In a separately excited SCR field circuit, a three-phase bridge rectifier circuit is usually adopted. The corresponding relationship between the AC/DC voltage and the current is shown in Figure 9.7.

Figure 9.7 Three-phase bridge rectifier circuit voltage and current waveforms: (a) AC voltage, (b) DC voltage, (c) AC current, (d) three-phase bridge circuit.

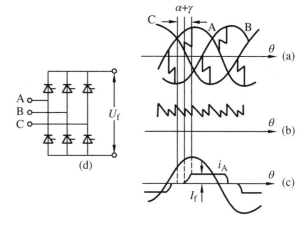

Table 9.2 Harmonic current component values [control angle $\alpha = 0$].

Harmonic current	5	7	11	13	Remark
Harmonic component I_n (%)	20	14.3	9.1	7.7	Commutation angle $\gamma = 0$
Harmonic component I_n (%)	14	7.2	2	1.5	Commutation voltage drop = 12%

As can be seen from Figure 9.7, the waveform of the AC current is basically a trapezoidal wave and therefore contains a harmonic component. Besides, since the phase of the stator current lags behind the stator voltage, the stator current can be divided into active and reactive components.

The value of the harmonic current can be determined with Eq. (9.1):

$$I_n = \frac{K_n}{n} \times I_1 = \frac{K_n}{Kp \pm 1} \times I_1 (K = 1, 2, 3 \cdots) \quad (9.1)$$

where

n — order of harmonics, $n = 5, 7, 11, 13, \ldots$;

K_n — coefficient determined by the control angle and commutation angle;

p — rectifier phase number (it equals 6 for a three-phase bridge rectifier circuit).

When $K_n = 1$, the order of harmonics is $n = Kp \pm 1$, and $K = 1, 2, 3, \ldots$, Thus, the corresponding order of harmonics is 5, 7, 11, 13, 17, 19 …. The relevant harmonic values are shown in Table 9.2.

As can be seen from the table, when the commutation reactance increases, the harmonic current component values decrease.

9.3.1.2 mmf Generated by Harmonic Current

When the harmonic current flows through the three-phase windings of the synchronous motor, a harmonic current rotating magnetic field will be generated. The expressions for the currents flowing through the windings in the phases are as follows:

$$i_A = \sqrt{2}[I_1 \sin(\omega t + \varphi_1) + I_5 \sin(5\omega t + \varphi_5) + I_7 \sin(7\omega t + \varphi_7) + \cdots] \quad (9.2)$$

$$i_B = \sqrt{2}\left\{I_1 \sin\left(\omega t + \frac{2\pi}{3} + \varphi_1\right) + I_5 \sin\left[5\left(\omega t + \frac{2\pi}{3}\right) + \varphi_5\right] + I_7 \sin\left[7\left(\omega t + \frac{2\pi}{3}\pi\right) + \varphi_7\right] + \cdots\right\} \quad (9.3)$$

$$i_C = \sqrt{2}\left\{I_1 \sin\left(\omega t + \frac{2}{3}\pi + \varphi_1\right) + I_5 \sin\left[5\left(\omega t + \frac{4\pi}{3}\right) + \varphi_5\right]\right.$$
$$\left. + I_7 \sin\left[7\left(\omega t + \frac{4\pi}{3}\right) + \varphi_7\right] + \cdots\right\} \tag{9.4}$$

Thus, the mmf generated in the Phase A winding can be obtained:

$$M_A = A_0 i_A \left[K_{W1} \cos\left(\frac{\pi}{\tau}x\right) - \frac{K_{W3}}{3} \times \cos\left(3 \times \frac{\pi}{\tau}x\right) + \frac{K_{W3}}{5} \times \cos\left(5 \times \frac{\pi}{\tau}x\right)\right.$$
$$\left. - \frac{K_{W7}}{7} \times \cos\left(7 \times \frac{\pi}{\tau}x\right) + \cdots + \frac{K_{Wm}}{m} + \cos\left(m \times \frac{\pi}{\tau}x\right) + \cdots\right]$$

$$A_0 = \frac{3\sqrt{2}qN}{\pi} \tag{9.5}$$

where

- A_0 — constant;
- q — number of slots per phase per pole;
- N — number of conductors per slot;
- τ — polar distance;
- K_W — winding factor.

Similarly, the mmf in Phase B and that in Phase C can be obtained.

The total resultant mmf of the fundamental current three-phase windings can be obtained as follows:

$$M_s(\omega t) = M_A(\omega t) + M_B(\omega t) + M_C(\omega t)$$
$$= A_0 I_1 \left[K_{w1} \sin\left(\omega t - \frac{\pi}{\tau}x\right) + \frac{K_{w5}}{5} \times \sin\left(\omega t - 5 \times \frac{\pi}{\tau}x\right)\right.$$
$$\left. - \frac{K_{w7}}{7} \times \sin\left(\omega t - 7 \times \frac{\pi}{\tau}x\right) + \cdots\right] \tag{9.6}$$

The total resultant mmf of the fifth harmonic current three-phase windings can be obtained as follows:

$$M_s(5\omega t) = A_0 I_5 \left[K_{w1} \sin\left(5\omega t - \frac{\pi}{\tau}x\right) + \frac{K_{w5}}{5} \times \sin\left(5\omega t - 5\frac{\pi}{\tau}x\right)\right.$$
$$\left. - \frac{K_{w7}}{7} \times \sin\left(5\omega t - 7 \times \frac{\pi}{\tau}x\right) + \cdots\right] \tag{9.7}$$

The total resultant mmf of the seventh harmonic current three-phase windings can be obtained as follows:

$$M_s(7\omega t) = A_0 I_7 \left[K_{w1} \sin\left(7\omega t - \frac{\pi}{\tau}x\right) + \frac{K_{w5}}{5} \times \sin\left(7\omega t - 5\frac{\pi}{\tau}x\right)\right.$$
$$\left. - \frac{K_{w7}}{7} \times \sin\left(7\omega t - 7 \times \frac{\pi}{\tau}x\right) + \cdots\right] \tag{9.8}$$

In the general bracket of formula Eqs. (9.6)–(9.8), the first term represents the spatial fundamental magnetic field, and the subsequent terms represents the spatial harmonic magnetic filed terms of order. The higher the number m of the winding coefficient K_{Wm}, the smaller its value. Therefore, $K_{Wm} < K_{W1}$, and it can be ignored compared with the value of the first item. Thus, the total mmf equals the sum of the first terms of Eqs. (9.6)–(9.8). That is,

$$\Sigma M = M_s(\omega t) + M_s(5\omega t) + M_s(7\omega t) + \cdots$$
$$= A_0 K_{W1} \left[I_1 \sin\left(wt - \frac{\pi}{\tau}x\right) + I_5 \sin\left(5\omega t - \frac{\pi}{\tau}x\right) + I_{1\sin}\left(7\omega t - \frac{\pi}{2}x\right) + \cdots\right] \tag{9.9}$$

In Eq. (9.9), the first term, (ωt), represents the rotating mmf generated by the fundamental current. The second item, $M_S(5\omega t)$, represents the mmf generated by I_5. Its amplitude is expressed as I_5/I_1 (per-unit value).

Table 9.3 Number of harmonics present on stator side and rotor side of auxiliary generator.

Number of harmonics	Stator harmonic		Rotor induced voltage harmonics		Magnetic field direction
	6-phase	12-phase	6-phase	12-phase	
1	1 (100)	1	–	–	Forward
5	5 (20)	–	6	–	Reverse
7	7 (14.3)	–	6	–	Forward
11	11 (9.1)	11 (9.1)	12	12	Reverse
13	13 (7.7)	13 (7.7)	12	12	Forward
17	17 (5.9)	–	18	–	Reverse
19	19 (5.3)	–	18	–	Forward
23	23 (4.4)	23 (4.4)	24	24	Reverse
25	25 (4.0)	25 (4.0)	24	24	Forward

Its length is the same as the fundamental length. It rotates in the direction opposite to the fundamental field at a speed five times the fundamental speed. This mmf is called the time harmonic mmf.

The third $M_S(7\omega t)$ is the mmf generated by the seventh harmonic current I_7. Its amplitude is I_7/I_1. Its length is the same as the fundamental length. It rotates in the fundamental direction at a speed seven times the fundamental speed.

For a rotor rotating at the fundamental speed, usually the negative-sequence $M_S(5\omega t)$ mmf speed is $5+1=6$ times the fundamental speed, while the positive-sequence $M_S(7\omega t)$ mmf speed is $7-1=6$ times the fundamental speed. So, for the rotor circuit, both mmfs will generate a voltage at a frequency six times the fundamental frequency.

Similarly, for the mmf generated by high-order harmonics such as 11th or 13th harmonic, since the 11th harmonic is negative-sequence and 13th harmonic is positive-sequence, both induce a voltage at a frequency 12 times the fundamental frequency for the rotor circuit. This is true of the cases when n is any other value. Thus, for the rotor side, the expression for the mmf coupled to it is as follows:

$$M_r = A_0 K_{W1} \left[I_1 \sin\left(\omega t - \frac{\pi}{\tau}x\right) + I_5 \sin\left(6\omega t \frac{\pi}{\tau}x\right) + I_7 \sin\left(6\omega t - \frac{\pi}{\tau}x\right) \right.$$
$$\left. + I_{11} \sin\left(12\omega t + \frac{\pi}{\tau}x\right) + I_{13} \sin\left(12\omega t - \frac{\pi}{\tau}x\right) + \cdots \right] \tag{9.10}$$

The number of harmonics present on the stator side and the rotor side is shown in Table 9.3. The values shown in parentheses in the table represent the percentage of the harmonic amplitude when $K_n = 1$.

9.3.2 Influence of Harmonic mmf on Voltage Waveform of Generator without Damping Winding

For a salient pole rotor generator without damping winding, the harmonic mmf has a significant effect on the distortion of the generator waveform. As mentioned earlier, for five or seven harmonics, an AC i_{fac} at a frequency six times the fundamental frequency will be induced in the vertical axis winding. If the number of turns of the excitation windings per pole is W_p, the resulting alternating magnetic field whose mmf is $i_{fac} W_p$ acts on the stator winding side. As discussed above, the magnetic field will generate a rotating mmf whose positive sequence is $(6+1)\omega = 7\omega$ and negative sequence is $(6-1)\omega = 5\omega$ on the stator side.

$$M_r(5\omega t) = -\frac{D}{2}\sin\left(5\omega t + \frac{\pi}{\tau}x + \xi\right)\Big|$$
$$M_r(7\omega t) = -\frac{D}{2}\sin\left(7\omega t + \frac{\pi}{\tau}x + \xi\right)\Big| \tag{9.11}$$

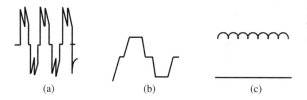

Figure 9.8 Voltage and current waveforms of generator without damping winding (rectified load): (a) terminal voltage, (b) AC, (c) excitation current.

Table 9.4 Calculated and measured values of harmonic current.

Number of harmonics	Parameter value		
	Terminal voltage (%)		Load current (%)
	Measured value	Calculated value	Measured value
5	12.6	13.1	9.2
7	9	8.9	2.5

where

D – function depending on A_0, W_p, ω, and τ;

ξ – function of A_0 and ψ.

Thus, the sum of M_S (5 ωt) and M_S (7 ωt) generated by the harmonic current in the stator winding and M_r (5 ωt) and M_r (7 ωt) generated on the rotor side will decide the final resultant mmf in the air gap. Thus, a back emf at a frequency five or seven times the fundamental frequency in the stator winding will be generated. This is the main reason for the distortion of the auxiliary generator voltage waveform. The test waveforms are shown in Figure 9.8. The calculated and measured values of the harmonic current are shown in Table 9.4.

As can be seen from Table 9.4, the harmonic current $I_7 = 2.5\%$. Its value is much smaller than the value of I_5 (9.2%), but the value of E_7 in the voltage waveform has reached 9.0%. Thus, its influence should not be ignored.

9.3.3 Influence of Harmonic mmf on Voltage Waveform of Generator with Damping Winding

The situation will be different when the generator rotor surface is equipped with damping winding. If the rotor stays still, for the fifth or seventh harmonic current, a rotating magnetic field will be generated at a speed six times its own speed. But the mmf generated by the damping winding determined by $s = 6$ does not completely offset the stator harmonic mmf due to the presence of the leakage reactance X_2 (s) and the resistance R_2 (s) in the damping winding. Figure 9.9 shows the corresponding vertical and horizontal axis equivalent circuit diagrams of the auxiliary generator.

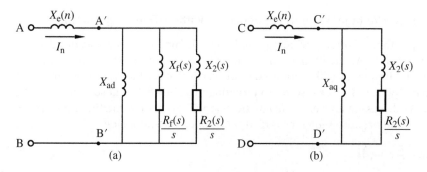

Figure 9.9 Equivalent circuit diagram for generator with damping winding: (a) vertical axis circuit, (b) horizontal axis circuit.

Figure 9.10 Voltage and current waveforms for auxiliary generator with damping winding when it supplies power to rectified load: (a) terminal voltage, (b) AC, (c) excitation current, (d) damping current.

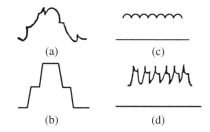

The harmonic component ΔU_N generated by the nth harmonic current in the output voltage waveform is as follows:

$$\Delta U_N = \frac{I_n}{2}(X_{AB} + X_{CD}) \tag{9.12}$$

where

X_{AB} and X_{CD} — total damping reactance from AB and CD, respectively.

The mmf of the time harmonics is alternately coupled to the vertical and horizontal axis circuits shown in Figure 9.9 and causes induced currents in the circuits. In every 1/6 cycle of the power frequency, the rectifying elements are alternately commutated once. As the change occurs in an instant, it can be considered that the excitation flux of the auxiliary generator is constant. Half the sum of X_d'' and X_q'' can be approximately taken as the commutation reactance. The emf after the commutation reactance can be taken as the non-distorted emf. Besides, although the rotor slip varies in a range where s is 1, 6, 12, 18, and 24, the rotor circuit impedances at the ends (A′B′ and C′D′) in Figure 9.9 are almost unchanged. Therefore, the commutation reactance can still be obtained from X_d'' and X_q'' measured when $s = 1$. For the auxiliary generator with damping winding, the voltage and current waveforms are shown in Figure 9.10.

As can be seen from Figure 9.10, the terminal voltage waveform is significantly improved compared with the case where the generator has no damping winding. The current flowing through the damping winding is primarily the sixth harmonic current. Although the current flowing through the excitation winding is the sixth harmonic current as well, its value has become very small.

It should be noted that for a solid-core magnetic pole generator, the rotor core surface is the equivalent of an infinite number of damping windings which cause eddy current damping.

9.3.4 Damping Winding Loss and Allowable Value

When current flows through the damping winding, an additional loss I^2R is generated. For general synchronous generators, the allowable value of the negative-sequence current flowing through the damping winding is shown in Table 9.5.

For salient pole generators, I_2 is 10%–12%. For non-salient pole generators, $I_2 \approx 8\%$. According to the standard identified at the International Council on Large Electric Systems (CIGRE), if a negative-sequence current twice the value specified in the table continuously flows through a general motor, varying degrees of damage to the motor will be caused.

For general synchronous generators, only when an unbalanced load is connected, a negative-sequence current at a frequency twice the fundamental frequency may be generated, and the damping winding will withstand an additional loss at a frequency twice the frequency when $s = 2$.

The situation will vary for a generator that supplies power to the rectifier load. In that case, induced currents in the rotor circuit include not only currents at a frequency twice the fundamental frequency but also currents at frequencies 6 and 12 times the fundamental frequency. For calculation of the loss of the damping winding, a simpler method is to convert the harmonic influence into an "equivalent negative-sequence current" and determine the allowable load under the same criteria as the average negative-sequence current. The equivalent negative-sequence current can be obtained with the following procedure.

Table 9.5 Allowable value of negative-sequence current.

IEC34–1 (1969)	VDE0530 (1972)	CIGRE11 (1972)
100 MW or smaller salient pole generators: 12%		Salient pole generators: 10%
Steam turbine: 8%; temperature rise of each part had better not exceed 5 °C		
100 MW or greater ones: Agreed by manufacturer and user		Steam turbines (indirect cooling): 10%
		Steam turbines (direct cooling):
		$S < 960$ MVA: 8%
		$960 < S < 1200$ MVA: 6%
		$1200 < S < 1500$ MVA: 5%

It is assumed that f_n is the cycle of n harmonics; R_0 is the DC resistance; and I_n is the nth harmonic current. The damping winding equivalent resistance R_n at f_n can be obtained as follows:

$$R_n = \sqrt{\frac{f_n}{f_1}} R_0 \tag{9.13}$$

The harmonic current loss can be obtained as follows:

$$P_n = \Sigma K \sqrt{\frac{f_n}{f_1}} I_n^2 \tag{9.14}$$

The negative-sequence current loss can be obtained as follows:

$$P_2 = K \sqrt{\frac{f_2}{f_1}} I_2^2 \tag{9.15}$$

Equation (9.14) is made equal to Eq. (9.15), and $f_n = nf_1$ and $f_2 = 2f_1$ are substituted into it. The equivalent negative-sequence current I_{2eq} can be obtained:

$$I_{2eq} = \sqrt{\sum_n \left(\sqrt[4]{\frac{n}{2}} I_n \right)^2} \tag{9.16}$$

where

n – harmonic order ($n = 5, 7, 11, 13, 17, 19\ldots$; for a three-phase bridge circuit).

If the value shown in Table 9.2 is taken as an example, $I_5 = 14\%$ and $I_7 = 7.2\%$. Thus, $I_{2eq} = 28\%$ can be obtained. Obviously, the proportions of I_5 and I_7 are small, but the equivalent negative-sequence current $I_{2eq} = 28\%$ is large enough. Therefore, in the design of an auxiliary generator that supplies power to the rectifier load, the damping circuit resistance should not increase with the frequency, and there should be sufficient heat capacity to absorb the negative-sequence current. Special attention should be paid to the conductive and electrical contact sections.

9.3.5 Stator Winding and Core Loss

9.3.5.1 Stator Winding Loss

When the harmonic current flows through the stator winding, an additional loss will be generated. The relationship between the total current I_g and each harmonic current is as follows:

$$I_g^2 = I_1^2 + I_5^2 + I_7^2 + I_{11}^2 + \cdots \tag{9.17}$$

When the harmonic current flows through the winding, the change in the winding resistance can generally be obtained from Eq. (9.18):

$$\frac{R_n}{R_0} = (1 + An^2) \tag{9.18}$$

where

R_0 – DC resistance;

A – coefficient.

Thus, the expression for the winding loss can be written as follows:

$$P_c = I_g^2 R_{eq} = \sum_1^\infty I_n^2 R_n = \sum_1^\infty I_n^2 R_0(1 + An^2) \tag{9.19}$$

Equation (9.19) shows that due to the skin effect of the harmonic current, the value of the AC resistance increases and is directly proportional to the square of the harmonic order. Usually, only the fundamental loss $I_1^2 R$ is taken account of in the design. When the generator supplies power to the rectifier load, the loss added due to the harmonic current is as follows:

$$P_c = I_1^2 R_1 = \sum_1^\infty I_n R_0 (1 + An^2) \tag{9.20}$$

The coefficient A is related to the winding wire thickness d. If the current density is kept constant, $A \propto d^2$. So, a thinner wire is favorable as it can reduce the AC resistance. If the allowable negative-sequence current I_2 shown in Table 9.5 is 10%, the resulting temperature rise will not exceed 5 °C. If the additional iron loss is taken into account, the temperature rises by 6%–7%, which is allowable.

9.3.5.2 Stator Core Loss

Generally, the iron loss of a synchronous motor is obtained in the no-load state. The loss of the tooth and that of the yoke are calculated on the basis of the respective magnetic field distribution waveforms. These magnetic fields include only the spatial harmonic component. In the case where power is supplied to the rectifier load, the aforementioned time harmonic magnetic field exists as well. However, at no load, the time harmonic magnetic field does not exist. Therefore, the time harmonic magnetic field loss cannot be determined through the no-load test. It is usually estimated. The corresponding iron loss is as follows:

$$P_F = \sum_n K_F(\sigma_H f_H + \sigma_w \Delta^2 f_n^2) B_n^2 \tag{9.21}$$

where

σ_H – hysteresis loss coefficient;

σ_W – eddy current loss coefficient;

Δ – core thickness;

B_n – sum of harmonic magnetic fields in the air gap.

The iron loss increment is related to the design load. It is generally 10%–20% of the no-load iron loss.

9.3.6 Armature Reaction and Power Factor

Under the condition that power is supplied to the rectifier load, the armature reaction and the power factor of a synchronous generator will be different from those of a general load generator. Usually, the armature reaction involves only the fundamental component. That is, under the same effective current condition, the fundamental armature reaction of a synchronous generator with rectifier load is smaller and is related to the phase number. For example, when $p = 6$, the fundamental current is 95.5%. When $p = 12$, it decreases to 57.1%.

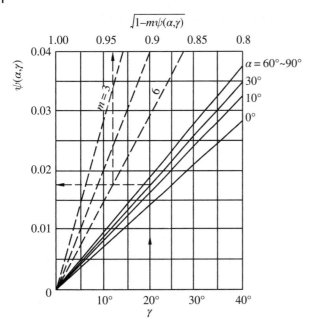

Figure 9.11 $\psi\,(\alpha,\gamma) = f\,(\alpha,\gamma)$ characteristic curve.

When the commutation angle γ and the control angle α are taken into account, the relationship between the effective value of the stator current and the rectified current is as follows:

$$I_{\text{eff}} = \sqrt{\frac{2}{3}} I_d \sqrt{1 - 3\psi(a,\gamma)} \tag{9.22}$$

The power factor under the rectifier load condition can generally be expressed as the ratio of the active power to the apparent power:

$$\cos\varphi = \frac{U_d I_d}{3 U_1 I_{\text{eff}}} = \frac{3}{\pi} \times \frac{\cos\frac{\gamma}{2}\cos\left(\alpha + \frac{\gamma}{2}\right)}{\sqrt{1 - m\psi(\alpha,\gamma)}} \tag{9.23}$$

$\psi\,(\alpha,\gamma)$ in Eqs. (9.22) and (9.23) is a coefficient determined by α and γ, which is obtained from Figure 9.11. For a three-phase bridge circuit, m = 3.

9.4 Parameterization of Separately Excited SCR Excitation System

9.4.1 Auxiliary Generator

In a separately excited SCR excitation system, the excitation power is taken from the auxiliary generator terminal and the working state is self-excitation at a constant voltage, hence the following are the performance characteristics:

(1) The excitation power is taken from the generator shaft end and is independent of the main generator's mode of operation. Thus, this is a good operation mode.
(2) The separately excited SCR excitation system has an inherently high field flashing system performance.

(3) The excitation power reliability depends to a large extent on the reliability of the self-excitation circuit of the auxiliary generator. Many countries focus on simplicity and operational reliability in the design of the self-excitation circuit.
(4) For an auxiliary generator with the SCR as load, the distortion of its terminal voltage waveform is more serious than that when power is supplied to an uncontrolled diode rectifier. Therefore, the auxiliary generator must be added with a damping winding to improve the voltage waveform and reduce local overheating of the core caused by harmonic losses.
(5) The added damping winding will increase the time constant of the auxiliary generator, but this is insignificant for a separately excited SCR excitation system, because the auxiliary generator is working in the self-excitation constant voltage state, and the response speed of the excitation system depends only on the time lag of the SCR trigger circuit.
(6) To ensure that the rectifier works in the Ith commutation state, the commutation reactance value should be limited to the corresponding value.

An auxiliary generator acting as a power supply for a separately excited SCR excitation system can be parameterized in accordance with the following procedure.

9.4.1.1 Preselected Commutation Reactance Value X_γ

The rectifier's commutation state depends primarily on the commutation reactance value. For a non-salient pole generator:

$$X_\gamma = \frac{X_d'' + X_2}{2} (X_d'' \approx X_q'') \tag{9.24}$$

For a salient pole generator:

$$X_\gamma = \frac{\frac{1}{2}(X_d'' + X_q'') + X_d'' X_q''}{2} (X_d'' \neq X_q'') \tag{9.25}$$

In the preliminary design, the per-unit value $X_\gamma^* = 0.15$, and

$$X_\gamma^* = \frac{\frac{X_\gamma}{U_{eN}}}{I_{eN}} \tag{9.26}$$

where

U_{eN} and I_{eN} — rated phase voltage and phase current of auxiliary generator, respectively;
$X\gamma$ — commutation reactance, in Ω.

As can be seen from Eqs. (9.6)–(9.11),

$$\cos\alpha - \cos(\alpha + \gamma) = \sqrt{\frac{2}{3}} \times \frac{I_f X_\gamma}{E_x} \tag{9.27}$$

In the forced excitation state, $\alpha \approx 0$ and $I_{fc} = K_1 I_{fN}$, I_{fc}, I_{fN} are respectively the excitation current in the forced excitation state and that in the rated state. K_1 is the strong excitation current multiple: $KI = \frac{I_{fc}}{I_{fN}}$.

$I_{fN} = \sqrt{\frac{3}{2}} I_{eN}$ is substituted into Eq. (9.27). Through simplification, the expression between the cosine commutation angle and the commutation reactance in the forced excitation state can be obtained:

$$\cos\gamma = 1 - K_1 K_\gamma$$

9.4.1.2 Calculated emf and Actual AC Stator Current of Auxiliary Generator

The non-distorted emf E_x after the commutation reactance acts as the calculated emf of the rectification external characteristic, and the stator fundamental current I_1 replaces the actual AC stator current.

9.4.1.3 Rated Phase Current of Auxiliary Generator

The rated phase current of an auxiliary generator consists of two parts: the excitation current supplied to the generator I_{fN} and the rated self-excitation current I_{feN}. The ACs corresponding to the two parts are respectively I_{e1N} and I_{e2N}. For the sum of the two, when a margin is considered:

$$I_{eN} = K_1(I_{e1N} + I_{e2N}) \tag{9.28}$$

where

K_1 – current margin coefficient $K_1 = 1.1\text{–}1.15$.

Equation (9.28) can also be written as follows:

$$I_{eN} = K_1 \left(\sqrt{\frac{2}{3}} I_{fN} + \sqrt{\frac{2}{3}} I_{feN} \right)$$

9.4.1.4 Rated Phase Voltage of Auxiliary Generator

When the auxiliary generator works in the self-excitation constant voltage state, the terminal voltage in the forced excitation state is the same as the rated value. The rectified voltage and current output in the forced excitation state are respectively U_{fc} and I_{fc}. The external expression of the three-phase bridge rectifier is as follows:

$$U_{fe} = \frac{3}{\pi} \times \sqrt{6} E_x \cos\alpha - \frac{3}{\pi} \times I_{fc} X_\gamma$$

The following can be obtained:

$$U_N = E_x = K_2 \times \frac{U_{fc} \frac{3}{\pi} \times I_{fc} X_\gamma}{\frac{3}{\pi} \sqrt{6}} \tag{9.29}$$

where

K_2 – voltage margin coefficient K_2 is $1.1\text{–}1.15$.

9.4.1.5 Capacity of Auxiliary Generator

The capacity of an auxiliary generator can be obtained from the following equation:

$$S = 3U_N I_N \tag{9.30}$$

9.4.2 Calculation of SCR Control Angle in Different Operating States

9.4.2.1 No-Load Operating Status

With the equation

$$U_{f0} = \frac{3}{\pi} \times \sqrt{6} U_{eN} \cos\alpha_0 - \frac{3}{\pi} \times I_{f0} X_\gamma \tag{9.31}$$

the following can be obtained:

$$\cos\alpha_0 = \frac{U_{f0} + \frac{3}{\pi} \times I_{f0} X_\gamma}{\frac{3}{\pi} \times \sqrt{6} U_{eN}} \tag{9.32}$$

9.4.2.2 Rated Operating Status
With the equation

$$U_{fN} = \frac{3}{\pi} \times \sqrt{6} U_{eN} \cos \alpha_N - \frac{3}{\pi} \times I_{fN} X_\gamma \qquad (9.33)$$

the following can be obtained:

$$\cos \alpha_N = \frac{U_{fN} + \frac{3}{\pi} \times I_{fN} X_\gamma}{\frac{3}{\pi} \times \sqrt{6} U_{eN}} \qquad (9.34)$$

9.4.2.3 Forced Excitation Status
It is assumed that the control angle in the forced excitation state is limited to α_f. With the equation

$$U_{fc} = \frac{3}{\pi} \times \sqrt{6} U_{eN} \cos \alpha_f - \frac{3}{\pi} \times I_{fc} X_\gamma$$

the following can be obtained:

$$\cos \alpha_c = \frac{U_{fc} + \frac{3}{\pi} \times I_{fc} X_\gamma}{\frac{3}{\pi} \times \sqrt{6} U_{eN}} \qquad (9.35)$$

When the control angle $\alpha_f = 0$, the maximum forced excitation voltage value can be obtained.

9.4.3 Current-Sharing Reactor

In high-power SCRs, a current-sharing reactor is usually connected to the circuit in order to improve the dynamic and static steady flow characteristics of the SCRs connected in parallel. For the purpose of simplicity, a two-branch example is taken to discuss this point. The result is also applicable to multi-branch cases. The corresponding circuit is shown in Figure 9.12. It is assumed that SCR 1 is turned on, and then the back emf e_1 applied to both ends of the inductance L_1 will remain until SCR 2 is turned on. Thus, current sharing can be achieved, and the current rise rate $\frac{di_1}{dt}$ of SCR 1 can be limited in the allowable range.

The following can be obtained under the above conditions:

$$U_e = i_1 r_1 + L_1 \frac{di_1}{dt} \qquad (9.36)$$

where

U_e — AC phase voltage;

L_1 and r_1 — inductance and resistance of air reactor of the first branch, respectively.

Figure 9.12 Connection diagram of current-sharing reactor.

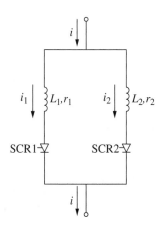

Before SCR 2 is turned on, all the series elements of the SCR 1 branch are subjected to the back emf $e_1 = -L_1 \frac{di_1}{dt}$. The general solution to Eq. (9.36) is as follows:

$$i_1 = \frac{U_e}{r_1} \times \left(1 - e^{-\frac{t}{T_1}}\right) \tag{9.37}$$

$$T_1 = \frac{L_1}{r_1} \tag{9.38}$$

The time constant of the current equalizer should meet the following condition, so that the back emf e_1 can remain until SCR2 is turned on:

$$T_1 > \Delta t = t_{SCR1} - t_{SCR2} \tag{9.39}$$

where

t_{SCR1}, t_{SCR2} — corresponding on-time of SCR 1 and SCR 2, respectively.

With Eq. (9.39), the following can be obtained:

$$L_1 \ldots r_1 \Delta t \tag{9.40}$$

In addition, the relational expression of the inductance L_1 can be obtained from Eqs. (9.37) and (9.38), so that the current rise rate $\frac{di_1}{dt}$ can be limited:

$$\frac{di_1}{dt} = \frac{U_e}{r_1 T_1} \times e^{-\frac{t}{T_1}} = \frac{U_e}{L_1} \times e^{-\frac{t}{T_1}}$$

At the instant of turning on, when $t = 0$:

$$L_1 \ldots \frac{U_e}{\frac{di_1}{dt}} \tag{9.41}$$

Equation (9.41) shows that when the allowable current rise rate is constant, the higher the applied voltage U_e, the greater the value of the current limiting inductance.

9.5 Separately Excited SCR Excitation System with High-/Low-Voltage Bridge Rectifier

When an excitation system is required to provide a high ceiling for the excitation voltage multiple, it is usually adopts a single three-phase fully controlled rectifier circuit. In order to meet the all requirements of no-load, rated and forced excitation operations, the SCR must be in a deeply controlled state with a large control angle α, so that there is sufficient excitation reserve when it works in the forced excitation state ($\alpha = 0$). This setting not only reduces the power factor of the power supply but also seriously distorts the rectified voltage waveform, resulting in a significant increase in the loss of the auxiliary generator. For the purpose of improvement, a three-phase high-/low-voltage bridge rectifier circuit can be adopted, as shown in Figure 9.13.

In this circuit, two rectifier sets in parallel – one for normal working (SCRN) and the other for forced excitation (SCRf) – are adopted. The rectifiers are connected in parallel on the DC side and powered respectively by the high- and low-voltage windings on the AC side. Under normal circumstances, the low-voltage sectional winding emf e_N supplies power to the SCR element of the set for normal working (SCRN), providing the rated excitation. The SCR element of the set for high-voltage forced excitation (SCRf) supplied by the winding emfs $e_N + e_f$ of the two sections (the set for normal working and the set for forced excitation) is in the deeply controlled state since the control angle α_f is large. In each power cycle, only a small part of the excitation is supplied. In the forced excitation state, the SCR element (SCRf) is fully open and provides

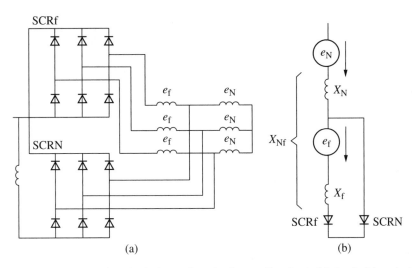

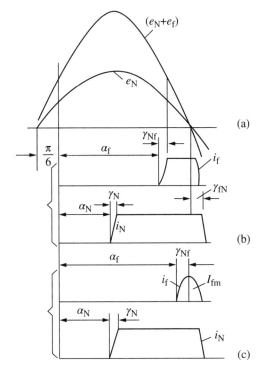

Figure 9.13 Three-phase high-/low-voltage bridge rectifier circuit: (a) circuit; (b) equivalent commutation circuit for high-/low-voltage rectifier circuit.

Figure 9.14 Voltage and current waveforms of three-phase high-/low-voltage bridge rectifier circuit: (a) supply voltage, (b) phase current in complete commutation state, (c) phase current in incomplete commutation state.

forced excitation. At this point, the cathode potential of the SCR element of the set for low-voltage normal working (SCRN) is higher than the anode potential and is latched. The voltage and current waveforms of the three-phase high-/low-voltage bridge rectifier circuit are shown in Figure 9.14. In Figure 9.14b, γ_{Nf} and γ_{fN} are respectively the commutation angle from the set for high-voltage forced excitation to the set for low-voltage normal working and that from the set for low-voltage normal working to the set for high-voltage forced excitation.

In the high/low-voltage bridge rectifier circuit, the control angles α_f and α_N of the SCR elements (SCRf and SCRN) of the set for forced excitation and the set for normal working are respectively set and powered by different sectional windings. Thus, there are several commutation states:

(1) The rectifying element's commutation state between the phases of the set for low-voltage normal working (corresponding to the no-load excitation state).
(2) The rectifying element's commutation from the set for low-voltage normal working to the set for high-voltage forced excitation in the same phase (rated working state) or the opposite state.
(3) The rectifying element's commutation (the forced excitation state) state between the phases in the set for high-voltage forced excitation.

For the commutation between the phases of the set for low-voltage normal working, the commutation reactance is as follows:

$$X_N = \sqrt{\frac{3}{2}} \times \frac{E_N}{I_f}[\cos \alpha_N - \cos(\alpha_N + \gamma_N)] \tag{9.42}$$

For the commutation between the set for low-voltage normal working and the set for high-voltage forced excitation in the same phase, the commutation reactance is as follows:

$$X_f = \sqrt{2} \times \frac{E_f}{I_f} \times \left[\cos\left(\alpha_f + \frac{\pi}{6}\right) - \cos\left(\alpha_f + \gamma_{Nf} + \frac{\pi}{6}\right)\right] \tag{9.43}$$

For the commutation between phases of the set for high-voltage forced excitation, the commutation reactance is as follows:

$$X_{Nf} = \sqrt{\frac{3}{2}} \times \frac{E_N + E_f}{I_f} \times [\cos \alpha_f - \cos(\alpha_f + \gamma_f)] \tag{9.44}$$

In the three equations:

X_N — winding commutation reactance of the set for low-voltage normal working;

X_f — commutation reactance between windings of the set for normal working and the set for forced excitation in the same phase;

X_{Nf} — commutation reactance of the set for high-voltage forced excitation;

γ_N, γ_f — commutation angle of the set for normal working and that of the set for forced excitation, respectively;

γ_{Nf} — commutation angle between set for normal working and the set for forced excitation in the same phase;

α_N, α_f — control angle of the set for normal working and that of the set for forced excitation, respectively;

E_N — effective value of supply voltage of the set for low-voltage normal working;

E_f — effective value of supply voltage of the set for high-voltage forced excitation;

I_f — excitation current of generator.

X_N equals the sum of the super-transient reactance and the negative-sequence reactance for the auxiliary generator power supply and equals the winding leakage resistance for the transformer power supply. X_f equals the leakage resistance at the time of the sudden single-phase short circuit of the forced excitation sectional winding for the auxiliary generator power supply and equals the leakage resistance at the time of the short circuit of the forced excitation sectional winding for the transformer power supply. X_{Nf} is calculated in the same way as X_N. However, for the winding section, the sum of the windings in the two sections (the set for normal working and the set for forced excitation) is taken.

In this case, it should be noted that the commutation between the set for normal working and the set for forced excitation in the same phase occurs under the condition of low emf [because the value of the control angle α_f of the rectifying element of the set for forced excitation], which significantly lengthens the commutation process. If the commutation does not end when the power supply emf equals zero, as shown in Figure 9.14c, and the set for forced excitation commutates to the set for normal working after the power supply voltage crosses zero, the state is called incomplete commutation. If the commutation ends before the voltage crosses zero, the state is called complete commutation, as shown in Figure 9.14b. The commutation current of the set for forced excitation can be obtained from Eq. (9.43):

$$i_f = \sqrt{2} \times \frac{E_f}{X_f} \times \left[\cos\left(\alpha_f + \frac{\pi}{6}\right) - \cos\left(\alpha_f + \gamma_{Nf} + \frac{\pi}{6}\right)\right]$$

If it is assumed that the commutation ends exactly when the voltage crosses zero, the following equation is obtained:

$$\alpha_f + \gamma_{Nf} + \frac{\pi}{6} = \pi$$

If $i_f = I_f$ is substituted into the equation, the expression of the control angle α_{f1} corresponding to the condition can be obtained:

$$\cos\left(\alpha_{f1} + \frac{\pi}{6}\right) = \frac{X_f I_f}{\sqrt{2} E_f} - 1 \tag{9.45}$$

Obviously, when $\alpha_f < \alpha_{f1}$, the commutation will end before the voltage crosses zero. At this point, part of the rectified voltage will be provided after the SCR element of the set for forced excitation is turned on, and the rectified current will flow in the winding of the set for forced excitation. So, when the set for normal working works in the rated state, the rectified voltage outputted by the high-/low-voltage bridge rectifier circuit is as follows:

$$U_{fN} = \Delta U_{fN} + \Delta U_{fc} \tag{9.46}$$

where

ΔU_{fN} and ΔU_{fc} — rectified voltage outputted by the set for normal working and that by the set for forced excitation, respectively

That can be expressed as the following equation:

$$\Delta U_{fN} = \frac{3\sqrt{6}}{2\pi} \times E_N[\cos\alpha_N + \cos(\alpha_N + \gamma_N)]$$

$$\Delta U_{fN} = \frac{3}{\pi} \int_{\frac{\pi}{6}+\alpha_f+\gamma_{Nf}}^{\pi} E_{fm} \sin\theta d\theta = \frac{3\sqrt{2}}{\pi} \times E_f \left[1 + \cos\left(\alpha_f + \gamma_{Nf} + \frac{\pi}{6}\right)\right]$$

For the rectified voltage in the forced excitation state, if the initial state is rated, the initial value is as follows:

$$U_{fc} = \frac{3\sqrt{6}}{\pi} \times (E_N + E_f) - \frac{3}{\pi} X_{Nf} I_{fN} \tag{9.47}$$

In Eq. (9.44), $\alpha_f = 0$. The expression of the commutation angle in the forced excitation state can be obtained.

$$\cos\gamma_f = 1 - \sqrt{\frac{2}{3}} \times \frac{X_{Nf} I_{fN}}{E_f + E_N} \tag{9.48}$$

In addition, in the high-/low-voltage bridge rectifier circuit, there are certain requirements for the selection of the control angle of the set for forced excitation α_f. Its value should not be greater than the limiting value α_{f1}, as shown in Figure 9.15.

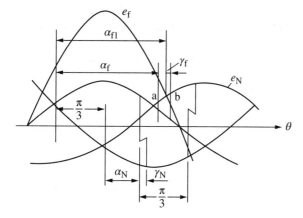

Figure 9.15 Determination of control angle limit value of the set for forced excitation α_{f1}.

The smaller value of the control angle of the set is assumed for normal working α_N. For example, it is turned on at point "b." If the selected value of the control angle of the set for forced excitation $\alpha_f \geq \alpha_{f1}$, the SCR element of the set for high-voltage forced excitation has a pulse input at point b. But it cannot be turned on since its cathode potential is lower than the corresponding value of the set for normal working. The α_{f1} limiting value can be determined in accordance with the following procedure. It is assumed that $E_f = 2E_N$. Thus, the following equation can be obtained:.

$$2\sqrt{2}E_N \sin\left(\alpha_{f1} + \frac{\pi}{6}\right) = \sqrt{2}E_N \sin \alpha_{f1}$$

The solution is as follows.

$$\alpha_{f1} = 127°$$

9.6 Parameterization of High-/Low-Voltage Bridge Rectifier

The connection diagram of a separately excited SCR excitation system with a high-/low-voltage bridge rectifier circuit is shown in Figures 9.11–9.16.

The stator winding of auxiliary generator 2 has taps, which supply power respectively to the high-/low-voltage bridge rectifier circuit's set for forced excitation and the set for normal working.

The following symbols are used in the calculation:

$i_1, \ldots, i_6$ — current of the set for normal working;
$i'_1, \ldots, i'_6$ — current of the set for forced excitation;
$u_1 \ldots u_6$ — counter voltage of the set for normal working;
$u'_1, \ldots, u'_6$ — counter voltage of the set for forced excitation;
L_f, r_f — inductance and resistance of excitation winding of synchronous generator, respectively.

For a high-/low-voltage bridge rectifier circuit, in the normal working state, the generator excitation is primarily powered by the SCR set for normal working, and the set for forced excitation is in a deeply controlled state. For forced excitation, the SCR set for forced excitation is fully open and supplied forced excitation, and the set for normal working is latched.

For an auxiliary generator whose stator winding has taps, the reactance of the forced excitation winding is denoted by X_f, the reactance of the normal winding by X_N, and the total reactance of all the windings $(X_{Nf} > X_N + X_f)$ by X_{Nf}. For the forced excitation and inversion working states, since only the set for forced excitation works, the formula obtained with one three-phase bridge rectifier set is still applicable for the purpose of the calculation of the state.

Figure 9.16 Connection diagram of separately excited SCR excitation system with high-/low-voltage bridge rectifier circuit: 1 – generator; 2 – auxiliary generator; 3 – SCR set for normal working; 4 – SCR set for forced excitation.

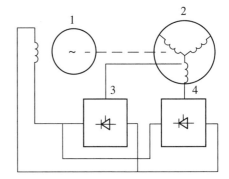

Figure 9.17 Commutation of SCR of forced excitation.

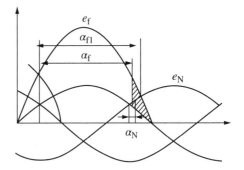

In the normal working state, there are two rectifier sets working simultaneously. It is assumed that the control angle of the set for forced excitation is α_f. When the commutation current commutates from the set for low-voltage normal working to the set for high-voltage forced excitation in the same phase, it will occur under the condition of a low emf e_f value, as shown in the shaded area in Figure 9.17. Since the voltage difference is low, the commutation will be lengthened and may not end when the emf e_f reaches zero. When the voltage value crosses zero and becomes negative, the rectifier of the forced excitation will commutate to the set for normal working in the direction opposite to the original.

It is assumed that the commutation process from the set for forced excitation to the set for normal working ends at the point where the phase adjacent to the set for normal working begins working. Under this condition, at the forward and reserve commutation intervals of the set for forced excitation and the set for normal working, the forced excitation winding of the power supply is shunted by its own rectifier. Therefore, under the condition that the power supply emf is stable and the reactance X_f is large enough, the rectified voltages of the two sets are the same as the voltage in only one set.

So, under the above complete commutation conditions, the calculation formula for the three-phase bridge rectifier with only one winding supplying power is still applicable. In the calculation, E_N is taken as the emf and X_N as the commutation reactance.

The following is an analysis of the commutation current change in the case of a non-complete commutation. From the equivalent circuit shown in Figure 9.18, the following commutation equation can be obtained:

$$X_f \times \frac{di'_1}{dt} = E_{mf} \sin \omega t \tag{9.49}$$

Thus, the following equation can be obtained:

$$i_1 = -\frac{E_{mf}}{X_f} \times \cos \omega t + C \tag{9.50}$$

where

E_{mf} – maximum value of voltage of set for high-voltage forced excitation.

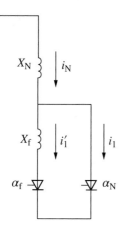

Figure 9.18 Equivalent commutation circuit of high-/low-voltage rectifier in the case of non-complete commutation.

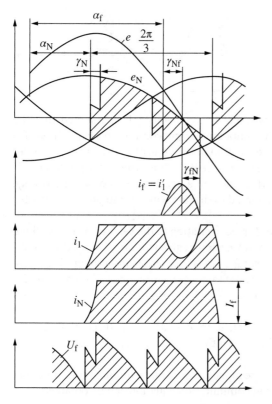

Figure 9.19 Commutation current change curves for sets for high-/low-voltage purposes.

Under the initial conditions, when $\omega t = \frac{6}{\pi} + \alpha_f$ and the current of the set for forced excitation $i_1 = 0$, the constant C is obtained. Then, the following equation can be obtained:

$$\left.\begin{array}{l} i_f = i'_1 = \frac{E_{mf}}{X_f} \times \left[\cos\left(\frac{\pi}{6} + \alpha_f\right) - \cos \omega t\right] \\ i_1 = I_f - i'_1 \end{array}\right\} \quad (9.51)$$

From Eq. (9.51), the commutation current change curves for the set for forced excitation and the set for normal working can be drawn, as shown in Figure 9.19.

9.6 Parameterization of High-/Low-Voltage Bridge Rectifier

The angle at the time of a commutation from the set for normal working to the set for forced excitation is γ_{Nf}, and the angle at the time of an inverse commutation is γ_{fN}. The two angles are equal at the time of a complete commutation.

As can be seen from Figure 9.18, the current i_N in the normal winding is the sum of the currents in the high- and low-voltage windings. That is, $i_N = i_1 + i'_1$. The rectified voltage U_f is obtained from the oblique part of the supply voltage curve (see Figure 9.19).

If $i_f = I_f$ when $\frac{\pi}{6} + a_f + \gamma_{Nf} = \pi$, it is a complete commutation. With Eq. (9.51), the condition for the complete commutation can be obtained:

$$\cos\left(\frac{\pi}{6} + \alpha_{f1}\right) = \frac{X_f I_f}{E_{mf}} - 1 \tag{9.52}$$

where

α_{f1} — limiting value of control angle of rectifier in forced excitation.

With Eq. (9.51), the average of the currents flowing through the forced excitation winding can be obtained:

$$I_{fc} = 2 \times \frac{3}{\pi} \times \frac{E_{mf}}{X_f} \times \int_0^{\gamma_{Nf}} \left[\cos\left(\frac{\pi}{6} + \alpha_f\right) - \cos\left(\omega t + \frac{\pi}{6} + \alpha_f\right)\right] d\omega t$$

Since $a_f = \frac{5\pi}{6} - \gamma_{Nf}$, the following can be obtained:

$$I_{fc} = \frac{6}{\pi} \times \frac{E_{mf}}{X_f} \times (\sin \gamma_{Nf} - \gamma_{Nf} \cos \gamma_{Nf})$$

Similarly, with Eq. (9.51), the amplitude of the currents flowing through the forced excitation winding can be obtained:

$$I_{mf} = \frac{E_{mf}}{X_f} \times (1 - \cos \gamma_{Nf}) \tag{9.53}$$

The fundamental current flowing through the winding of the set for normal working and its phase angle can be obtained from the following equation:

$$a_1 = \frac{\sqrt{3}}{2\pi[\cos \alpha_N - \cos(\alpha_N + \gamma_N)]} \times [2\gamma_N + \sin 2\alpha_N - \sin 2(\alpha_N + \gamma_N)]$$

$$b_1 = \frac{\sqrt{3}}{\pi} \times [\cos \alpha_N + \cos(\alpha_N + \gamma_N)]$$

$$c_1 = \sqrt{a_1^2 + b_1^2}$$

$$\tan \varphi_{1N} = \frac{a_1}{b_2}$$

The fundamental component of the current flowing through the forced excitation winding is determined as follows. The range of the current flowing through the winding is as follows:

$$\pi - r_{Nf} \leqslant \omega t \leqslant \pi + r_N$$

With the simultaneous Eqs. (9.51) and (9.53), the current flowing through the forced excitation winding during the commutation can be obtained:

$$i_f = i'_1 = -\frac{I_{mf}}{1 - \cos \gamma_{Nf}}(\cos \gamma_{Nf} + \cos \omega t)$$

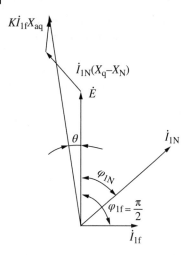

Figure 9.20 Voltage phasor diagram of auxiliary generator with two rectifier sets.

The cosine term Fourier series coefficient is as follows:

$$A_1 = \frac{2}{\pi} \times \int_{\pi-\gamma_{Nf}}^{\pi+\gamma_{Nf}} i'_1 \cos \omega t \, d\omega t = \frac{2}{\pi} \times \frac{I_{mf}}{1 - \cos \gamma_{Nf}} \times \int_{\pi-\gamma_{Nf}}^{\pi+\gamma_{Nf}} (\cos \gamma_{Nf} + \cos \omega t) \cos \omega t \, d\omega t$$

$$= \frac{2 I_{mf}}{\pi(1 - \cos \gamma_{Nf})} \left(\gamma_{Nf} - \frac{1}{2} \sin \gamma_{Nf} \right)$$

Since the current waveform symmetrical longitudinal axis is an even function, the sine term coefficient $B_1 = 0$. Thus, the following can be obtained:

$$I_{1f} = \frac{A_1}{\sqrt{2}}$$

$$\varphi_{1f} = \frac{\pi}{2}$$

The voltage phasor diagram for an auxiliary generator with two rectifier sets is shown in Figure 9.20.

If the turns ratio of all the windings to the normal winding is expressed as K, the emf built by the mmf in the forced excitation winding in the normal winding equals $K I_{1f} X_{aq}$. X_{aq} is the armature reaction reactance of the normal winding on the horizontal axis.

As mentioned earlier, there are two working states (complete commutation and incomplete commutation) for commutations between the set for normal working and the set for forced excitation. However, since higher currents flow through the forced excitation winding in the case of a complete commutation, the loss increases. So, it is appropriate to adopt incomplete commutation in the design.

For the incomplete commutation state, the formula for the three-phase bridge rectifier circuit obtained with one rectifier set is still applicable, but E_{mN}, X_N, α_N, and γ_N should be taken as calculated values. Besides, E_{mN} is the maximum voltage of the low-voltage normal set.

In order to calculate the voltages and currents of the two rectifier sets in the rated, forced excitation, and inversion states, the reactance of the auxiliary generator should be determined first:

$$X_N = \frac{X_\gamma}{K_E}$$

$$K_E = \frac{\frac{K^2 + K}{2} + \frac{1}{\frac{1/K^2 + 1/K}{2}}}{2}$$

$$X_\gamma = \frac{(X''_d + X_2)}{2} \tag{9.54}$$

When the two rectifier sets work simultaneously, the forced excitation winding is short-circuited. The result is equivalent to a single-phase short circuit. Thus:

$$X_f = \frac{X_d'' + X_2 + X_0}{3K_E'} \tag{9.55}$$

where K_E' is obtained from Eq. (9.54), but K should be substituted with $\frac{K}{K-1}$.

The voltage, current, and capacity of the auxiliary generator can be obtained by referring to the calculation method for an auxiliary generator supplying power to one rectifier set.

It should be noted that in order to reduce the capacity of the auxiliary generator, the cross-sectional area q_f of the forced excitation winding is usually smaller than the cross-sectional area q_N of the working winding. At this point, the power of the auxiliary generator is as follows:

$$S_N = 3 \times \frac{1 + \frac{W_f}{W_N} \times \frac{q_f}{q_N}}{1 + \frac{W_f}{W_N}} \times UI \tag{9.56}$$

It is assumed that $\frac{Wf}{WN} = 2$ and $\frac{q_f}{q_n} = \frac{1}{2}$. When they are substituted into Eq. (9.56), the following is obtained:

$$S_N = 2UI$$

That is, the power of the auxiliary power supplying power to two rectifier sets is 2/3 of the capacity of the auxiliary power supplying power to one rectifier set.

9.7 Transient Process of Separately Excited SCR Excitation System

The calculation of the transient process of a separately excited SCR excitation system is complex because of the presence of an auxiliary generator with rectified load in the circuit.

Usually, in the calculation of the process, the transient processes at the time of forced excitation and de-excitation are most important. At the time of a forced excitation, the excitation regulator moves the control angle of the SCR to $\alpha_c \approx 0$. Through a 0.01–0.02 s delay, the excitation output voltage reaches the ceiling. Under the action of the voltage, the generator excitation current reaches a value twice the rated value. If the excitation system has a higher forced excitation voltage multiplier (K_U is 3–4), the rotor current limiting unit in the field regulator will limit the rotor current to a value twice the rated value.

The complexity of the calculation of the transient process of the auxiliary generator is primarily caused by the influence of the armature reaction on the vertical axis magnetic potential. However, in practical cases, the abovementioned armature reaction is largely compensated by the free component in the excitation winding and the action of the excitation regulator of the auxiliary generator. Thus, it can be approximately considered that the horizontal axis transient emf E_q' is approximately invariant. Since the non-distorted emf E_x after the commutation reactance X_γ is close to the E_q' value, the procedure will be greatly simplified if the non-distorted emf (Ex) is taken as the rectifier power supply AV power supply emf for the calculation of the transient process.

Thus, the following generator field circuit equation can be obtained for the transient process:

$$L_f \frac{di_f}{dt} + \left(\frac{3}{\pi} \times X_\gamma + r_f\right) i_f = \frac{3\sqrt{3}}{\pi} \times E_{mx} \cos\alpha - 2\Delta U \tag{9.57}$$

where

E_{mx} — maximum value of non-distorted emf of auxiliary generator;
L_f and r_f — inductance and resistance of rotor excitation winding of generator.

The time constant of the field circuit can be obtained from Eq. (9.57):

$$T'_{d0} = \frac{L_f}{r_f + \frac{3}{\pi} \times X_\gamma} \tag{9.58}$$

If the saturation of the generator excitation winding is taken into account, the relationship curve between the instantaneous value of the inductance L_f and the current should be determined. At this point, Eq. (9.57) can be written as follows:

$$L_f \frac{di_f}{dt} = \frac{3\sqrt{3}}{\pi} \times E_{mx} \cos\alpha - \left(r_f + \frac{3}{\pi} \times X_\gamma\right) i_f - 2\Delta U \tag{9.59}$$

It is appropriate to solve Eq. (9.59) with the numerical method. Meanwhile, the influence of the stator and damping windings on the equivalent excitation winding should also be taken into account. For example, when the generator stator winding is short-circuited, the value of the equivalent excitation winding is minimal.

Besides, in a more accurate calculation of the transient process and steady state of the rectifier circuit, the resistance of the connecting wires in the rectifier AC and DC circuits should be taken into account. The resistance in the DC circuit can be attached to the resistance of the excitation winding and the cable on the AC side to the phase resistance of the auxiliary generator.

The steady-state equation of the rectifier circuit when the resistance is taken into account is as follows:

$$U_f = \frac{3\sqrt{3}}{\pi} \times E_{mx} \cos\alpha - \left[r_f + \frac{3}{\pi} \times X_\gamma + \left(2 - \frac{3\gamma}{2\pi}\right) r\right] i_f - 2\Delta U$$

where

r – phase resistance including resistance of connecting cables;

r_f – resistance of excitation winding including resistance of cable on DC side.

In the transient state, when the resistance is taken into account and the voltage drop of the rectifier is ignored, Eqs. (9.57) and (9.59) can be written as follows:

$$L_f \frac{di_f}{dt} + \left[r_f + \frac{3}{\pi} \times X_\gamma + \left(2 - \frac{3\gamma}{2\pi}\right) r\right] i_f = \frac{3\sqrt{3}}{\pi} \times E_{mx} \cos\alpha \tag{9.60}$$

$$T'_{d0} = \frac{L_f}{r_f + \frac{3}{\pi} \times X_\gamma + \left(2 - \frac{3\gamma}{2\pi}\right) r} \tag{9.61}$$

$$L_f \frac{di_f}{dt} + \left[r_f + \frac{3}{\pi} \times X_\gamma + \left(2 - \frac{3\gamma}{2\pi}\right) r\right] i_f = \frac{3\sqrt{3}}{\pi} \times E_{mx} \cos\alpha \tag{9.62}$$

Equations (9.57)–(9.62) can be used to calculate the transient process of forced excitation of a hydro-generator. This is applicable to both small deviations and large disturbance signals.

If the original state is rated and the control angle is α_N, when the control angle undergoes a transition from α_N to $\alpha_c \approx 0$, the expression of the generator rotor current is as follows:

$$I_f = I_{fN} + \frac{\frac{3\sqrt{3}}{\pi} \times E_{mx}(1 - \cos\alpha_N)}{r_r + \frac{3}{\pi} X_\gamma} \times \left(1 - e^{-\frac{t}{T'_{d0}}}\right) \tag{9.63}$$

The time constant of the excitation winding T'_{d0} corresponds to the working state of the generator. Its saturation can be determined with the calculation method or the graphical method.

For example, the hydro-generator works under the no-load condition. The influence of the damping winding circuit is not taken into account. $L_f = 0.8$ H, $r = 0.222\,\Omega$, $X_\gamma = 0.05\,\Omega$, $r = 0.005\,\Omega$, and $\gamma = 40°$. The time constant of the excitation winding can be obtained as follows:

$$T'_{do} = \frac{L_f}{r_f + \frac{3}{\pi} \times X_\gamma + \left(2 - \frac{3\gamma}{2\pi}\right)r}$$

$$= \frac{0.8}{0.222 + \frac{3}{\pi} \times 0.05 + \left(2 - \frac{3 \times 40}{360}\right) \times 0.005}$$

$$= 2.88 (s)$$

When $X_\gamma = 0$ and $r = 0$,

$$T'_{do} = \frac{L_f}{r_f} = \frac{0.8}{0.222} = 3.77 (s)$$

When $X_\gamma \neq 0$ and $r = 0$,

$$T'_{do} = \frac{L_f}{r_f + \frac{3}{\pi} \times X_\gamma} = \frac{0.8}{0.222 + \frac{3}{\pi} \times 0.05} = 2.96 (s)$$

The T'_{do} value obtained shows that the influence of the resistance and reactance in the rectifier circuit on the time constant value is great.

If the influence of the generator damping winding and the magnetic circuit saturation are not considered, the rotor current expression when the generator is unloaded can be obtained from Eq. (9.60):

$$I_f = I_{f0}\, e^{-\frac{t}{T'_{do}}} + \frac{\frac{3\sqrt{3}}{\pi} \times E_{mx} \cos\alpha}{r_f + \frac{3}{\pi} \times X_\gamma + \left(2 - \frac{3\gamma}{2\pi}\right)r} \times \left(1 - e^{-\frac{t}{T'_{do}}}\right) \quad (9.64)$$

where

I_{f0} — initial value of excitation current of generator in initial state.

Its value is as follows:

$$I_{f0} = \frac{\frac{3\sqrt{3}}{\pi} \times E_{mx} \cos\alpha_0}{r_f + \frac{3}{\pi} \times X_\gamma + \left(2 - \frac{3\gamma}{2\pi}\right)\gamma}$$

where

α_0 — control angle at beginning of forced excitation.

The inversion de-excitation of a separately excited SCR excitation system is discussed below. When the rectifier works in the inversion state, the magnetic field energy stored in the DC load can be fed to the AC side to change the polarity of the rectified voltage and realize the inversion de-excitation. The corresponding connection is shown in Figure 9.21.

In the normal working, the rectifier circuit supplies the generator excitation, as shown in Figure 9.21a. In the inversion de-excitation, the control angle α is increased to $\frac{\pi}{2} \sim \pi$, so that the rectifier is in the inversion state outputting the negative rectified voltage. The magnetic field energy stored in the generator excitation winding is gradually fed to the stator side of the auxiliary generator. Since the excitation winding of the synchronous generator acting as a load is passive, the inversion process will end with the attenuation of the magnetic

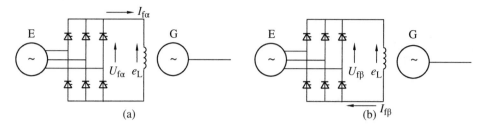

Figure 9.21 Inversion de-excitation of separately excited SCR excitation system: (a) $0 \leq \alpha \frac{\pi}{2}$ in rectification state, (b) $\frac{\pi}{2} \leq \alpha \leq \pi$ in inversion state.

field energy stored and the reduction of the inversion current. With Figure 9.21b, the generator field circuit equation at the time of the inversion can be obtained:

$$e_L - U_{f\beta} = r_f i_{f\beta}$$

$$-U_{f\beta} = -e_L + r_f i_{f\beta} = L_f \frac{di_{f\beta}}{dt} + r_f i_{f\beta} \tag{9.65}$$

where

e_L – generator excitation winding induced emf;

$i_{f\beta}$ – instantaneous value of inversion current.

With Eq. (9.65), the following can be obtained:

$$i_{f\beta} = C\, e^{-\frac{r_f}{L_f} \times t} - \frac{U_{f\beta}}{r_f} \tag{9.66}$$

When $t = 0$ and $I_{f\beta} = I_{f0}$, the coefficient $C = I_{f0} + \frac{U_{f\beta}}{r_f}$. When it is substituted into Eq. (9.66), the following can be obtained:

$$I_{f\beta} = \left[I_{d0} + \frac{U_{f\beta}}{r_f} \right] e^{-\frac{r_f}{L_f} \times t} - \frac{U_{f\beta}}{r_f} \tag{9.67}$$

When the inversion de-excitation ends, $I_{f\beta} = 0$. With Eq. (9.67), the inversion de-excitation time constant can be obtained as follows:

$$t_m = \frac{L_f}{r_f} \times \ln \frac{U_{f\beta 0} + U_{f\beta}}{U_{f\beta}} \tag{9.68}$$

where

$U_{f\beta 0}$ – rectified voltage at time of start of inversion;

$U_{f\beta}$ – inverting rectified voltage.

As can be seen from Eq. (9.68), the higher the inversion voltage $U_{f\beta}$, the smaller the initial value $U_{f\beta 0}$, and the shorter the de-excitation time constant (t_m).

When $U_{f\beta} = 5U_{f\beta 0}$ is substituted into Eq. (9.68), the following can be obtained:

$$t_m = 0.182 T'_{d0}$$

$$T'_{d0} = \frac{L_f}{r_f}$$

where

T'_{d0} – time constant of excitation winding when generator is unloaded.

The above value is close to the ideal no-load de-excitation time $t_m = 0.167\, T'_{d0}$.

10

Static Self-Excitation System

10.1 Overview

In China, different periods have witnessed different bases for selection of large synchronous generator excitation modes. In the early 1950s, China-made steam turbines featured a capacity of 25–50 MW and a coaxial DC exciter. Such DC exciters mostly adopted a self-shunt excitation connection, without a pilot exciter. This excitation mode had been applied even to steam turbines with a capacity of 100–125 MW. However, due to the influence of multiple factors such as speed, mechanical strength, and voltage between commutator segments on exciter capacity, there was the problem of limiting capacity. For a steam turbine with a speed of 3000 r min^{-1}, the theoretical analysis shows that the limit capacity of the coaxial DC exciter was 600 kW. Since the excitation power of a 1000 kW or greater full-speed steam turbine exceeded 600 kW and a coaxial DC exciter could not be used, there was the need to develop new excitation systems. In this period, Harbin electrical machinery factory (HEC) developed the first separately excited static diode rectifier excitation system with an inductive high-frequency pilot exciter and an AC main exciter, which is conventionally called the three-generator excitation system. The frequency of the pilot exciter is 500 Hz and that of the AC main exciter is 100 Hz. The purpose of the high frequency of the main exciter is to reduce the time constant of the excitation system to improve the speed of the excitation voltage response ratio of the excitation system.

The three-generator excitation mode has been used to date as the typical excitation mode for 100, 125, 200, and 300 MW steam turbines. In the late 1970s, due to the need of the reform and opening up, China embarked on comprehensive technical communication and cooperation with famous foreign motor manufacturers in the motor manufacturing industry. In the early 1980s, China introduced 300 and 600 MW steam turbines with a brushless excitation system from United States–based Westinghouse.

The static self-excitation system is called a high-initial-response excitation system (HIR system) since its excitation voltage response time $t < 0.1$ s. Most of the current China-made 300 and 600 MW steam turbines adopt a three-generator brush excitation system with a high excitation voltage response ratio ($R \geqslant 2.0$ times per second). A brushless excitation system with a high-initial-voltage response ratio is not provided unless specifically required.

It is well known that, in a brushless excitation system with a main exciter and a pilot exciter, the excitation control action point is located at the input end of the excitation winding of the AC main exciter as the AC synchronous generator of the main exciter. In essence, it is a small steam turbine with a capacity of thousands of kilowatts and a rotor time constant that is generally up to 2 to 3 s. In order to overcome the influence of this inertial element to achieve a high-initial-voltage response ratio, there is the need to take such measures as excitation current feedback compensation and increasing the forced excitation voltage ratio of the excitation winding of the exciter to achieve speed. Moreover, the excitation regulation circuit should be specially designed with necessary overcurrent limit protection measures to prevent generator overexcitation and saturation caused by the loss of control of regulation. Three-generator excitation systems (including brushless excitation systems) adopted by large steam turbines play a strong supporting role in ensuring and improving the stability of the power system.

Design and Application of Modern Synchronous Generator Excitation Systems, First Edition. Jicheng Li.
© 2019 China Electric Power Press. Published 2019 by John Wiley & Sons Singapore Pte. Ltd.

In the early 1980s, some countries such as the United States (GE) and Sweden (ASEA) were promoting static self-excitation systems in large steam turbines, while China was initiating the application of static self-excitation systems with hydro-turbines.

Static excitation systems refer to thyristor excitation systems with an excitation transformer connected to the generator terminal as an excitation power source. The remarkable performance characteristic of this excitation mode is the high field flashing voltage response speed, which makes it easy to achieve a high-initial-voltage response performance. Most static excitation systems for steam turbines at home and abroad adopted the self-compound excitation mode at the early stage. The main difference between a self-compound excitation system and a self-excitation system is that the former is powered by the voltage source (the excitation transformer) and the current source (the series transformer) in the form of phase compensation. Thus, in the case of a short-circuit fault at the generator terminal or the near end, the series transformer provides an additional excitation source, so that the thyristor rectifier voltage can be maintained at a high value.

However, series transformers bring considerable problems in applications, such as large layout space and a high-capacity current busbar and cable, which increase investment and maintenance. Due to the presence of series transformers, a larger series reactance degrades the commutation working condition of the thyristor and causes overvoltage, compromising the operational safety of the rotor excitation winding at the instant when the thyristor element is turned on or off.

Due to these problems as well as the influence of such factors as improper system design and unsound operation and maintenance, at the early stage of the application of static self-compound excitation systems to hydro-turbines in China, accidents seriously affecting normal operation of generators were seen frequently.

Therefore, in terms of excitation modes, China tends to simplify self-compound excitation to self-shunt excitation, in a bid to further improve operational reliability and reduce equipment costs.

In the mid-1980s, China vigorously explored and studied the application of self-shunt excitation systems to large steam turbines with a focus on simple self-excitation systems' ability to maintain sufficient excitation capacity in the case of near-end short circuit. Conventionally, the conditions for a three-phase short circuit on the high-voltage side of a generator are taken as the basis for a comparison between an AC self-excitation system and an AC separate excitation system in regard to the ability to maintain forced excitation capacity. In the case of a closed busbar, the possibility of ta hree-phase short circuit at the generator terminal is virtually eliminated. So, the high-voltage side of the step-up transformer can be single-phase grounded as a basis for the comparison. In this case, the advantage of the self-excitation system is more obvious.

In recent years, China and Switzerland have extensively studied the application of self-excitation systems and relay protection in combination. It is generally believed that, in the event of a generator failure, the generator excitation winding has a large time constant and the excitation current attenuates with the time constant. Therefore, the reliability of the relay protection action of the first-order instantaneous operation in milliseconds can be ensured. For backup protection when the set time is smaller than 0.5 s, relay protection is feasible. However, for backup protection when the set time is greater than 0.5 s, a delay low-voltage protection relay with "memory" started with overcurrent is taken to ensure its reliable action. If the voltage has not been restored within the set time, the voltage relay trips after a time delay to improve the reliability of the backup protection action.

It should be noted that a self-excitation system's ability to maintain generator voltage is considerably related to unit operation mode in addition to the fault point. Figure 10.1 shows the current attenuation curves under various short-circuit conditions Swedish ASEA provided for the 36 MW generator at the Bapanxia Hydro Power Station.

As can be seen from Figure 10.1, for an asymmetric short circuit in any form, the self-excitation system can maintain a high forced excitation capacity. Only when the generator terminal is short-circuited in three phases and the first and second states of the main protection fail to operate is the self-excitation system's ability to maintain the generator terminal voltage significantly reduced. However, as mentioned above, when an enclosed busbar is adopted, the possibility of generator terminal three-phase short circuit is virtually eliminated. For any other asymmetric short circuit, the self-excitation system has good characteristics.

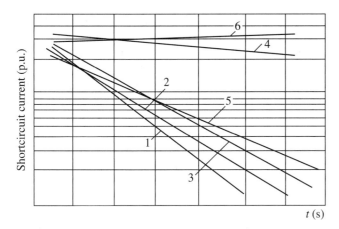

Figure 10.1 Generator short circuit stator current attenuation curve: 1 – three-phase short circuit at generator terminal; 2 – three-phase short circuit on high-voltage side of transformer and stand-alone; 3 – two-phase short circuit at generator terminal; 4 – two-phase short circuit on high-voltage side of transformer and stand-alone; 5 – three-phase short circuit on high-voltage side of transformer (three generators in operation); 6 – two-phase short circuit on high-voltage side of transformer (three generators in operation).

Self-excitation systems have been successfully applied to hydro-turbines, which has created favorable conditions for their popularization in and application to steam turbines.

In 1990, Hebei Institute of Technology (the predecessor of Hebei University of Technology), Northeast Electric Power Test and Research Institute, and Qinghe Power Station in Liaoning, which were in close cooperation, successfully developed China's first self-excitation system for a Soviet Union–made 200 MW steam turbine at Qinghe Power Station. In 1997, Hebei University of Technology successfully applied multi-microcomputer self-excitation systems to another two 200 MW steam turbines.

At present, China has seen extensive application of self-excitation systems to steam turbines. The 600 MW steam turbines co-produced by DFEM and Japan's Hitachi use self-excitation systems as a typical excitation mode as well.

In the introduced generators, two 350 MW steam turbines and two China-made ones at Huaneng Shang'an Power Station Phase I and II adopt US GE's Generrex and UK R-R's three-mode redundant (TMR) self-excitation systems, respectively. A 300 MW steam turbine at Tianjin Dagang Power Station and a 350 MW one at Jiangsu Ligang Power Station selected self-excitation systems supplied by Italia. A 125 MW steam turbine at Hebei Douhe Power Station chose a self-excitation system imported from Japan. A Westinghouse 600 MW steam turbine put into operation in 1998 at Yangzhou No. 2 Power Station and a 740 MW one at Zhuhai Power Station also applies self-excitation systems. In recent years, self-excitation systems have also been applied in some nuclear power stations. For example, a static self-excitation system with a brushless exciter has been used at Tianwan Nuclear Power Station.

In short, self-excitation systems characterized by simple connection, easy maintenance, and inherent high initial fast response have fully demonstrated their superior performance in applications.

Moreover, self-excitation systems boast significantly better reliability than AC exciter systems since they eliminate rotating parts. In terms of plant investment, due to the smaller floor space, they can reduce power station construction costs. In addition, they can shorten the shaft system and play an important role in improvement of generator torsional vibration conditions and reduction of torsional vibration order. Nonetheless, there are some noteworthy issues in the application of self-excitation systems to large steam turbines.

For example, there is still a lack of in-depth research on a self-excitation system's own stability in the event of large system voltage fluctuations caused by a serious system fault, given the fact that the excitation output of a self-excitation system is constrained by the generator terminal voltage. If the problem is not handled properly, with the self-excitation system, the generator's loss of field will further aggravate the system voltage fluctuations, which will eventually lead to a voltage collapse and an extensive power outage.

When the generator is unloaded and the excitation regulator is out of control, a faulty forced excitation will be caused, which is very harmful. When the generator terminal voltage varies greatly, a problem of coordination between the low excitation limit of the excitation regulator and the generator's field loss protection occurs. The normal operating conditions should ensure that the low excitation limit acts before the field loss protection.

In addition, the possibility of field loss caused by a system fault should also be taken into account. A generator's loss of field depends on such factors as short-circuit fault duration, system voltage recovery capacity, and generator excitation attenuation time constant.

For large steam turbines with a self-excitation system, the existence of shaft voltage has always been one of the operational issues. Since a self-excitation system with a thyristor provides a considerable high-frequency harmonic voltage source for the generator excitation winding circuit, axis voltage has become a more prominent and complex issue and the focus of international attention. Besides, the coupling of the high-frequency harmonic voltage source and the capacitor brings the generator rotor circuit insulation resistance against ground (shaft) down, to which full attention should be paid in operation.

10.2 Characteristics of Static Self-Excitation System

10.2.1 Main Circuit Connection

The typical connection mode for the main circuit of a static self-excitation system is to connect the excitation transformer to the outlet end of the synchronous generator. This connection mode is simple, enabling high reliability of the excitation source. The excitation transformer in the self-excitation system can also be connected to the system side of the main circuit breaker, but this mode is rarely adopted since it is greatly affected by system voltage fluctuations. The excitation transformer can also be connected to the station service power bus or used as an excitation source which supplies power to the rectifier through the intermediate part of the station transformer, but this connection mode is inadvisable if the synchronous generator excitation system requires high reliability of power supply.

A self-excitation system's excitation transformer is typically not provided with an automatic switch. The high-voltage side can be added with a high-voltage fuse. A high-capacity oil-cooled excitation transformer should be provided with gas protection. The connection on the high-voltage side of the transformer must be included in the differential protection range of the generator.

The excitation transformer winding connection set is generally Yy0. Yd11 is adopted if the secondary side current is high. Generally, for a three-phase rectifier circuit, the voltage calculation formula is calculated using the secondary winding star connection method. When a triangular connection is adopted on the secondary side of the excitation transformer, special attention should be paid to the calculation.

When the excitation transformer is installed outdoors, due to the voltage drop, the feeder between the secondary side of the transformer and the rectifier bridge should not be too long, which must be considered especially if the excitation current is high. Besides, single-core armored cable is inadvisable, but rubber cable is applicable. The reason is that when single-core armored cable is powered by AC, high voltage and considerable current will be induced in steel armor, which will cause interference in the communication cable.

The high-power rectifier in a self-excitation system adopts a three-phase bridge connection. The advantage of this connection mode is that the voltage the thyristor component can withstand is low, and the transformer capacity utilization is high. Three-phase bridge circuits can be half-controlled or fully controlled. Both modes have the same ability to enhance excitation. However, in excitation suppression, a half-controlled bridge can only control the excitation voltage to zero, while a fully controlled bridge can produce a negative excitation voltage and make the excitation current quickly drop to zero in inversion operation. Since a fully controlled bridge has these advantages in terms of control, self-excitation systems in China mostly adopt the three-phase fully controlled bridge. In other countries, many self-excitation systems for hydro-turbines use the fully controlled circuit, while some self-excitation systems for steam turbines apply the half-controlled circuit. In particular, United States–based GE uses the half-controlled circuit as a typical mode.

It should be emphasized that for a conventionally connected self-excitation system with a set of power rectifier bridges, which is required to provide a high ceiling of the excitation voltage multiple, the silicon controlled rectifier (SCR) must be in a deeply controlled state to satisfy the forced excitation capacity reserves under the long-term rated operating conditions, in order to meet the no-load, rated, and forced excitation requirements. This will cause a serious distortion of the waveform of the rectified voltage, which is very unfavorable for the excitation system's operation safety and reliability. The disadvantages are reflected in the following aspects:

(1) As the SCR is in a deeply controlled state, the rectified voltage waveform distortion causes a high spike overvoltage, which sometimes compromises the safety of the rotor circuit insulation.
(2) The increase in the harmonic current of the rectified current increases the loss of the excitation transformer and its temperature rise.
(3) The power factor of the excitation transformer decreases, so that the equipment capacity increases.
(4) The thyristor component for the deeply controlled state withstands a higher commutation overvoltage.

In view of this, the Soviet Union once developed a high-/low-voltage bridge rectifier circuit powered by two sets of excitation transformer windings, which could take into account the no-load, rated, and forced excitation requirements.

10.2.2 Field Flashing Circuit

For a self-excitation system with an excitation transformer connected to the generator terminal, when the generator starts and the speed is close to the rated value, the generator terminal voltage is a residual voltage, whose value is generally low (1%–2% of the rated voltage). At this point, the trigger circuit in the excitation regulator cannot work, because the synchronous voltage is too low. The thyristor is not triggered and cannot output an excitation current to enable the generator to build up a voltage. Therefore, measures must be taken. First, a field flashing should be supplied to the generator, so that the generator can gradually reach the voltage for stable operation. This process is called field flashing.

If the excitation transformer is not connected to the generator terminal but is connected to the system side of the generator's main circuit breaker or the station service power bus, there is no need to additionally consider field flashing measures. Although the generator terminal voltage is a residual voltage, the AC side of the rectifier bridge is a full voltage. So, the rectifier bridge and the excitation regulator can work properly. Moreover, as the regulator is in the forced excitation state (because the generator terminal voltage is low at the beginning of the build-up), when an AC power supply is added, an excitation current can be outputted to enable the generator to build up a voltage.

When the excitation transformer is connected to the generator terminal, there are two field flashing measures. The first measure is called separate field flashing. That is, a field flashing source and a field flashing circuit are provided separately to supply field flashing. The other measure is called residual voltage field flashing. That is, the residual voltage generated by the remanence of the generator is used to supply field flashing.

When the separate field flashing supplies the field flashing current, the generator voltage will gradually increase. When the excitation regulator can work (for example, the generator voltage reaches 10% of the generator rated voltage), the field flashing circuit will be automatically disconnected and enter the self-excitation state.

In order to quickly build up the generator voltage, the field flashing source capacity and voltage should be selected to enable the build-up of 50%–70% of the rated voltage in the short term. Thus, the selected battery voltage is about 1/4 of the rated excitation voltage. The field flashing voltage should not be too high. The generator terminal voltage it builds up should not exceed the lower limit of the generator voltage set by the excitation regulator.

The working principle and voltage selection principle are the same as above for the field flashing circuit rectified by the station service power.

If a separate field flashing is adopted, some equipment should be added. If a battery is used as a field flashing source, the burden on the battery of the power station will increase. If the station service AC power is used as a field flashing source, the generator will fail in field flashing when the station service power disappears.

When the residual voltage of the generator is high, the residual voltage instead of a separate power supply can be considered for the field flashing. There are two measures. First, although the regulator cannot work at the time of the field flashing, technical measures can be taken to temporarily turn on the thyristor branch in the rectifier bridge to form an uncontrollable rectifier device to provide the field flashing current. This is similar to a self-shunt excited DC generator's self-excitation process of building up a voltage with the residual voltage. Second, measures are taken for the synchronization circuit in the regulator to make working at the residual voltage equivalent to working at the rated voltage. Besides, the AC side of the regulated power supply of the regulator is taken from another independent AC power supply. Thus, the excitation regulator can be immediately engaged in work at the time of the field flashing to control the thyristor turn-on and supply the field flashing current until the generator voltage continues to rise to the set voltage of the regulator.

10.2.3 Stable Operating Point of Static Self-Excitation System

The generator voltage build-up process for a static self-excitation system whose excitation power is taken from the generator terminal, similar to that for a self-shunt excitation DC exciter, is subject to the problem of a stable operating point. In a self-excitation system, the stable operating point depends on the intersection of the generator no-load (or load) characteristic curve and the rectifier external characteristic curve. The external characteristics of a three-phase bridge rectifier can be written as follows:

$$U_f = \frac{3\sqrt{6}}{\pi} \times \frac{U_G}{K_T} \times \cos\alpha - \frac{3}{\pi} \times I_f X_\gamma - 2\Delta U \tag{10.1}$$

where

K_T — transformer ratio of rectifier transformer.

$$U_f = I_f r_f \tag{10.2}$$

where

r_f — excitation winding resistance.

When Eq. (10.2) is substituted into Eq. (10.1), the following is obtained:

$$I_f = \frac{\frac{3\sqrt{6}}{\pi} \times \frac{U_G}{K_T} \times \cos\alpha - 2\Delta U}{r_f + \frac{3}{\pi} \times X_\gamma} \tag{10.3}$$

In Eq. (10.3), when the $\cos\alpha$ value is kept constant and a set of U_G values is given, the points can be connected to obtain the characteristic curve $I_f = f(U_G^*) \cos\alpha = C$, as shown in Figure 10.2. For example, when $\cos\alpha = 70°$ is constant, the intersection A of the rectifier external characteristic curve and the generator no-load characteristic curve is just the stable operating point of the static self-excitation system.

As can be seen from Eq. (10.3), in the $U_G^* - I_f^*$ coordinate system, the intersections of the rectifier external characteristic curve and the horizontal and vertical axes are respectively as follows:

$$\left. \begin{array}{l} I_{f0} = -\dfrac{2\Delta U}{r_f + \dfrac{3}{\pi} \times X_\gamma} \\ U_{G0} = \dfrac{2\Delta U}{\dfrac{3\sqrt{6}}{\pi} \times \cos\alpha} \times K_T \end{array} \right\} \tag{10.4}$$

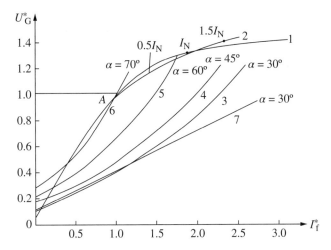

Figure 10.2 Operating characteristic curve of static self-excitation system when generator is unloaded: 1 – generator no-load characteristic curve; 2 – generator load characteristic curve; 3–6 – external characteristics of rectifier at different control angles; 7 – external characteristics when temperature is factored in.

In fact, since the total voltage drop per arm of the SCR $2\Delta U$ is relatively small, the rectifier external characteristic curve is virtually a straight line across the origin of the coordinates. Since the residual voltage of the generator is low and the starting part of the no-load characteristic curve is nonlinearly bent, the straight line tangent of the no-load characteristic curve also passes through the origin of coordinate. Thus, there is no stable intersection of the static self-excitation system and the rectifier external characteristic curve. A stable operating point can be obtained only from the saturation section.

It should be noted that when the variation of the rotor excitation winding temperature is factored in, the rectifier external characteristic curve will not be a straight line, as shown in Figure 10.2 (Curve 3). This is favorable for the operational stability of the static self-excitation system.

When the generator is loaded, the determination of the stable operating point of the excitation system is virtually the same as that in the case where the generator is unloaded, though the load characteristic curve is applied to replace the no-load characteristic curve, as shown in Figure 10.2 (Curve 2). The so-called generator load characteristic curve refers to the curve of the relationship between the generator stator voltage U_G and the excitation current I_f when the generator's voltage and power factor are constant. At this point, the intersection of the rectifier external characteristic curve and the generator load characteristic curve is just the stable operating point of the excitation system.

10.2.4 Short-Circuit Current of Self-Excited Generator

For the sake of comparison, what is first discussed is the characteristics of the variation of the short-circuit current when the synchronous generator adopts a separate AC excitation system or a static self-excitation system and a short-circuit fault occurs. On the basis of the characteristics of the connection mode of the excitation system and the principle of electrical machinery, the following conclusions can be drawn:

(1) For both excitation modes, the generator short-circuit current super-transient components are equal, because the super-transient component is determined by the generator damping winding parameters, independent of the excitation mode.
(2) For both excitation systems, the initial short-circuit current transient components are equal, because the initial value is determined by the principle of conservation of the excitation winding flux linkage, independent of the excitation system connection mode.
(3) For both excitation systems, the short-circuit current transient component attenuation time constants are different. Moreover, in the case of a near-end three-phase short circuit, the self-excited generator's

short-circuit current will attenuate to zero, without a stable value, while the separate excitation system's generator short-circuit current has a certain stable value, which is related to the excitation current provided by the excitation system.

What is discussed below is the relevant expressions of the various variables when the self-excited generator encounters a three-phase short circuit.

10.2.4.1 Relevant Expressions in Case of Three-Phase Short Circuit

It is assumed that the generator undergoes a three-phase short circuit under the following conditions:

(1) The generator is at the no-load rated voltage.
(2) The short-circuit point is from the generator terminal via the external resistance X_e, as shown in Figure 10.3.
(3) The aperiodic component in the stator current is omitted.
(4) The super-transient component is considered separately and not reckoned in the short-circuit current.

If the external reactance X_c is reckoned in the generator stator leakage reactance, the following can be obtained:

$$X_{dc} = X_d + X_c$$
$$X'_{de} = X'_d + X_e$$

When the generator terminal undergoes a three-phase short circuit, the terminal voltage is as follows: $U_G = i_d X_e$

The stator winding longitudinal axis flux linkage is as follows:

$$\psi_d = i_f X_{ad} - i_d X_{de} = 0$$

With the above formula, the following equation can be obtained:

$$i_d = \frac{X_{ad}}{X_{de}} \times i_f \tag{10.5}$$

where

X_{ad} — vertical-axis armature reaction reactance between generator rotor and stator windings.

As can be seen from Eq. (10.5), $i_d \propto i_f$.

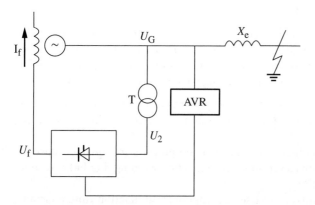

Figure 10.3 Connection diagram for main system of elf-excited generator.

If the rectifier bridge tube voltage drop and commutation voltage drop are omitted, the rectified voltage ratio can be written as follows:

$$\beta_u = \frac{U_f}{U_2} = \frac{U_f}{\frac{U_G}{K_T}} = \frac{K_T U_f}{U_G} \tag{10.6}$$

If the thyristor control angle α is not equal to zero, the following is obtained:

$$U_{f0} = \frac{\beta_u}{K_T} \times U_{G0} \cos \alpha_0 \tag{10.7}$$

where

- U_{f0} — generator no-load excitation voltage;
- U_{G0} — generator no-load rated voltage;
- α_0 — control angle of rectifier bridge when generator is unloaded at rated voltage.

In the event of a three-phase short circuit, Eq. (10.7) can be written as follows:

$$U_{fc} = \frac{\beta_u}{K_T} \times U_{Gc} \cos \alpha_c \tag{10.8}$$

where

- α_c — control angle in case of forced excitation.

If Eq. (10.8) is reduced to the per-unit value, the following can be obtained.

$$U_{fc}^* = U_G^* \times \frac{\cos \alpha_c}{\cos \alpha_0} = K_\alpha U_G^*$$

$$K_\alpha = \frac{\cos \alpha_c}{\cos \alpha_0}$$

$$U_G^* = \frac{U_{GC}}{U_{G0}} \tag{10.9}$$

where,

- K_α — forced excitation ceiling voltage coefficient;
- U_G^* — generator terminal voltage per-unit value in case of short circuit.

For a three-phase half-controlled bridge rectifier circuit, when the generator voltage is rated:

$$U_{f0} = \frac{\beta_u}{K_T} \times U_{G0} \times \frac{1 + \cos \alpha_0}{2} \tag{10.10}$$

When the generator is short-circuited:

$$U_{fc} = \frac{\beta_u}{K_T} \times U_{GC} \times \frac{1 + \cos \alpha_c}{2} \tag{10.11}$$

It can be expressed as a per-unit value as follows:

$$U_{fc}^* = \frac{1 + \cos \alpha_C}{1 + \cos \alpha_0} \times U_G^* = K_a U_G^* \tag{10.12}$$

As can be seen from the above equation, in the event of a three-phase short circuit of the self-excited generator, under the assumed conditions, $U_f \propto U_G$, $i_d \propto i_f$.

10.2.4.2 Transient Characteristics of Three-Phase Short Circuit

In the three-phase short-circuit transient process, the expression of the per-unit value of the generator excitation voltage is as follows:

$$U_f^* = K_\alpha U_G^* = K_\alpha i_d^* \quad X_e = K_\alpha \times \frac{X_e}{X_{de}} \times E_q^*$$

$$i_d^* = \frac{1}{X_{de}} \times E_q^* \tag{10.13}$$

where $E_{q\infty}^* = U_f^*$.

Since

$$E_q' = E_q \times \frac{X_d' + X_e}{X_d + X_e} = \frac{X_{de}'}{X_{de}} \times E_q \tag{10.14}$$

when the electromagnetic transient equation for a synchronous generator is taken into account,

$$E_{q\infty} = E_q + T_{d0}' + \frac{dE_q'}{dt} \tag{10.15}$$

If Eqs. (10.13) and (10.14) are substituted into Eq. (10.15), the following can be obtained:

$$E_q + T_{d0}' \times \frac{X_{de}'}{X_{de}} \times \frac{dE_q}{dt} = K_\alpha \times \frac{X_e}{X_{de}} \times E_q \tag{10.16}$$

or it can be written as follows:

$$\left(1 + S \times \frac{X_{de}'}{X_{de}} \times T_{d0}'\right) E_q = K_\alpha \times \frac{X_e}{X_{de}} \times E_q$$

$$S = \frac{d}{dt} \tag{10.17}$$

where

S — differential operation symbol.

With Eq. (10.17), the following can be obtained:

$$E_q = E_{q(0+)} \; e^{\frac{-t}{T_{dk}}}$$

$$E_{q(0+)} = \frac{X_{de}}{X_{de}'} \times E_q' \tag{10.18}$$

where

$E_{q(0+)}$ — E_q value at instant after short circuit.

The equivalent time constant in the event of a short circuit of the self-excited generator is as follows:

$$T_{dk} = \frac{T_{d0}' \times \frac{X_{de}'}{X_{de}}}{1 - K_\alpha \times \frac{X_e}{X_{de}}} \tag{10.19}$$

When a conventional excitation generator is short-circuited, the excitation circuit time constant is as follows:

$$T_d' = T_{d0}' \times \frac{X_{de}'}{X_{de}}$$

Obviously, $T_{dk} > X_d'$.

Figure 10.4 Curve $E_{q(t)}$, $i'_{d'(t)}$, $U_{G(t)}$ when generator is short-circuited.

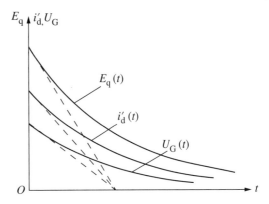

As can be seen from Eq. (10.19), a short circuit of the self-excited generator is the equivalent of the introduction of a negative resistance into the rotor excitation circuit. Its per-unit value is $K_\alpha \times \frac{X_e}{X_{de}}$. Thus, the excitation circuit equivalent resistance decreases. So, the equivalent time constant T_{dk} increases.

Figure 10.4 shows Curve $E_q = f(t)$ in Eq. (10.18) as well as Curve $i'_d = f(t)$ and $U_G = f(t)$. The three show similar variation rules and the same time constant T_{dk}. On the basis of Eq. (10.18), the transient component i'_d and the terminal voltage U_G of the short-circuit current can be expressed as follows:

$$i'_d = i'_{d(0+)} e^{-\frac{t}{T_{dk}}} \tag{10.20}$$

$$U_G = U_{G(0+)} e^{-\frac{t}{T_{dk}}} \tag{10.21}$$

If $U_G = U_G^* = 1$ p.u., before the short circuit, the i'_d value at the instant after the short circuit is as follows:

$$i'_{d(0+)} = \frac{1}{X'_{de}}$$

The instantaneous value of U_G at the instant after the short circuit is as follows:

$$U_{G(0+)} = i'_{d(0+)} X_e$$

The transient component i_d of the short-circuit current of a self-excited generator should be the sum of i'_d obtained in Eq. (10.20) and the super-transient component i''_d:

$$i_d = \left(\frac{1}{X''_{de}} - \frac{1}{X''_d}\right) e^{-\frac{t}{T''_d}} + \frac{1}{X'_{de}} e^{-\frac{t}{T_{dk}}} \tag{10.22}$$

Under the same conditions, the short-circuit current i_d of a separately excited generator is as follows:

$$i_d \left(\frac{1}{X''_{de}} - \frac{1}{X'_{de}}\right) e^{-\frac{t}{T''_d}} + \left(\frac{1}{X'_{de}} - \frac{1}{X_{de}}\right) e^{-\frac{t}{T'_d}} + \frac{1}{X_{de}} \tag{10.23}$$

The corresponding short-circuit current variation curves are shown in Figure 10.5.

In Eq. (10.23), the last term $\frac{1}{X_{de}} = I_\infty$, is expressed as the steady-state short-circuit current value in the case of constant excitation. If there is excitation regulation, a short-circuit current component which corresponds to the regulator and increases with time should be attached, so that the short-circuit current can achieve a new high steady-state value.

In the event of the near-end three-phase short circuit, as can be seen from the comparison between Eqs. (10.22) and (10.23), the short-circuit current of the separately excited generator with or without excitation regulation eventually tends to a steady-state value, while that of the self-excited generator eventually tends to zero. Besides, the short-circuit current transient component attenuation time constant of the separately excited generator is T'_d, while that of the self-excited generator is T_{dk}.

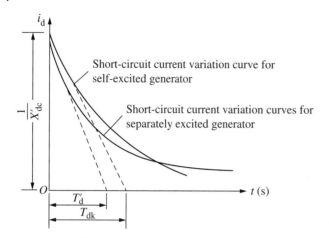

Figure 10.5 Short-circuit current variation curves for self-excited and separately excited generator, respectively.

As can be seen from Figure 10.5, both excitation modes show the same short-circuit current variation in the initial short super-transient process. In the transient process, since $T_{dk} > T'_d$, in the initial period of time, the short-circuit current attenuation of the self-excited generator is slower than that of the separately excited generator. However, since the short-circuit current of the separately excited generator eventually tends to a steady-state value and that of the self-excited generator eventually tends to zero, the short-circuit current attenuation of the self-excitation generator appears to be faster in the later period of time.

The above conclusion is based on a sudden three-phase short-circuit at the no-load rated voltage of the synchronous generator. If the three-phase short circuit occurs when the generator is loaded, the conclusion holds as well. Nevertheless, the stator current $i_{d(0+)}$ and the excitation current $i_{f(0+)}$ are different. That is, the current component caused by the load before the short circuit must be taken into account. If the generator terminal has a regional load which needs to be considered in the calculation, it can be replaced by a constant impedance. In the calculation of the time constant T_{dk}, the external reactance X_e can be connected in parallel with the regional load constant impedance. Thus, the calculated T_{dk} is less than that when there is no regional load, and the attenuation of the short-circuit current to zero is faster, which is detrimental to the self-excited generator.

10.2.5 Critical External Reactance

In the analysis of the transient process of the three-phase short-circuit current of a self-excited generator, if accurate calculation of the equivalent time constant T_{dk} is required, the commutation voltage drop of the rectifier bridge can be considered and the time constant T'_{d0} of the rotor excitation winding should be corrected. The commutation voltage drop is the equivalent of a positive equivalent resistance R_r introduced into the rotor excitation circuit, and its value is proportional to the leakage resistance of the excitation transformer.

For a three-phase bridge rectifier circuit, the equivalent resistance R_r is as follows:

$$R_r = \frac{3}{\pi} \times X_r$$

R_r is about 5%–10% of the generator excitation winding resistance. Thus, the corrected T'_{d0} value should be as follows:

$$\frac{T'_{d0}}{1.05 \sim 1.1} = (0.91 \sim 0.95)\, T'_{d0}$$

So, when the commutation voltage drop is considered, the equivalent time constant of Eq. (10.19) is rewritten as follows:

$$T_{dk} = (0.91 \sim 0.95) \times \frac{T'_{d0} \times \frac{X'_{de}}{X_{de}}}{1 - K_\alpha \times \frac{X_e}{X_{de}}} = (0.91 \sim 0.95) \times \frac{T'_d}{1 - K_\alpha \times \frac{X_e}{X_{de}}} \tag{10.24}$$

The equivalent time constant is related to the external reactance X_e. If X_e increases, T_{dk} will increase.

When the denominator of Eq. (10.24) is zero, that is, $\left(1 - K_\alpha \times \frac{X_e}{X_{de}}\right) = 0$, T_{dk} tends to infinity and the short-circuit current is not attenuated. Thus, the critical value of the external reactance can be obtained as follows:

$$K_\alpha X_{ec} = X_{de} = X_d + X_{ec}$$

The following can be obtained:

$$X_{ec} = \frac{X_d}{K_\alpha - 1} \tag{10.25}$$

where

X_{ec} — critical external reactance.

When $T_{dk} = \infty$, the transient component of the short-circuit current is not attenuated and the initial value is maintained constant.

The following can be seen from Eq. (10.5):

(1) If the short-circuit point occurs within the critical external reactance (i.e., $X_e < X_{ec}$), the transient component of the short-circuit current is attenuated.
(2) If the short-circuit point occurs at the boundary of the critical external reactance (i.e., $X_e \approx X_{ec}$), $T_{dk} \approx \infty$, and the transient component of the short-circuit current is not attenuated.
(3) If the short-circuit point occurs outside the critical external reactance (i.e., $X_e > X_{ec}$), since $K_\alpha = \frac{\cos \alpha_c}{\cos \alpha_0} > 1$, the equivalent time constant T_{dk}, obtained by Eq. (10.19), is negative. In the transient process, the short-circuit current and the excitation current are not attenuated but rise with time until the maximum excitation current limiting device acts. Or the excitation regulator works to increase the control angle α of the thyristor to reduce the excitation.
(4) If the limit reactance X_{ec} calculated with Eq. (10.25) is small, the range of the short-circuit current attenuation is small. Once the limit reactance exceeds the X_{ec} value, the short-circuit current will remain constant or rise with time. This is very beneficial for the operational stability of the self-excitation system.

10.2.6 Influence of Short-Circuit Mode on Short-Circuit Current

Figure 10.6 shows the three-phase short-circuit current variation curve for a 1000 MW self-excited steam turbine.

The following can be seen from Figure 10.6:

(1) If the short circuit is at Point "a" on the high-voltage side and the forced excitation multiple K_α increases from 1.6 to 2 as the ceiling voltage coefficient K_α increases, the equivalent time constant T_{dk} also increases. Although the short-circuit current attenuation is slower, the effect is not obvious.
(2) The further the short-circuit point is away from the generator terminal, the smaller the initial value of the short-circuit current will be, and the slower the short-circuit current attenuation will be.
(3) If the short-circuit point is at the generator terminal and $X_e = 0$, $T_{dk} = T'_d$, the short-circuit current will be attenuated to zero by T'_d, which is very detrimental for the operation of the self-excitation system, as shown by Curve "a" in Figure 10.7.

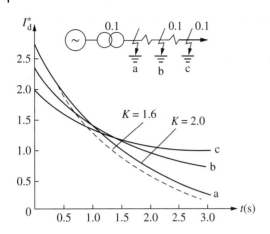

Figure 10.6 Three-phase short-circuit current variation curve for 1000 MW self-excited steam turbine: a – three-phase short circuit on high-voltage side; b – three-phase short circuit at point 50 km away; c – three-phase short circuit at point 100 km away.

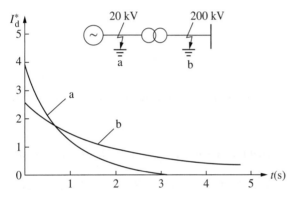

Figure 10.7 Three-phase short-circuit current variation curve for 450 MW self-excited steam turbine. a – short circuit at generator terminal; b – short circuit on busbar on high-voltage side.

Asymmetric short circuit is discussed in the following. Grid short-circuit faults are usually asymmetric. For a separately excited generator (single-phase or two-phase), asymmetric short circuit is more favorable than three-phase symmetrical short circuit. The reason is that in the case of an asymmetrical short circuit, the generator terminal voltage is higher, the equivalent time constant T_{dk} is greater, and the short-circuit current and the excitation current are attenuated more slowly and may even rise.

When the power system is asymmetrically short-circuited, the generator terminal voltage is not symmetrical. With the symmetric component method, the voltage is decomposed into positive-sequence voltage, negative-sequence voltage, and zero-sequence voltage. The positive-sequence voltage decreases gradually from the generator terminal to the short-circuit point, while the negative-sequence voltage decreases gradually from the short-circuit point to the generator terminal. The connection on the low-voltage side of a step-up transformer is generally star-shaped, without a zero sequence. So, only the positive-sequence voltage component and the smaller negative-sequence voltage component exist at the generator terminal. If the negative-sequence voltage component is omitted, only the positive-sequence voltage component exists at the generator terminal. Besides, the positive-sequence component of the short-circuit current corresponds to the DC current in the excitation winding, and the negative-sequence current generates a 100 Hz AC component in the excitation winding. So, the calculation of the excitation system and the analysis of the short-circuit process will be the same as for the case of a three-phase short circuit described above, though the short-circuit equivalent time constant T_{dk} needs to be corrected.

In the event of an asymmetric short circuit, the positive-sequence component (I_{A1}) of the short-circuit current (Phase A) can be calculated with Eq. (10.26):

$$\dot{I}_{A1}^{(n)} = \frac{\dot{E}_{A\Sigma}}{X_{A\Sigma} + X_{\Delta}^{(n)}} \tag{10.26}$$

Table 10.1 $X_\Delta^{(n)}$ value in various asymmetrical short-circuit modes.

Mode	Symbol	Additional reactance ($X_\Delta^{(n)}$)	Mode	Symbol	Additional reactance ($X_\Delta^{(n)}$)
Single-phase ground short circuit	$X_\Delta^{(1)}$	$X_{2\Sigma}+X_{0\Sigma}$	Two-phase ground short circuit	$X_\Delta^{(1.1)}$	$X_{2\Sigma}//X_{0\Sigma}$
Two-phase short circuit	$X_\Delta^{(2)}$	$X_{2\Sigma}$			

In the equation, the superscript at the upper-right part (n) represents the short-circuit mode. $X_\Delta^{(n)}$ is called the additional reactance (i.e., the portion beyond the positive-sequence reactance), which varies with the short-circuit mode, as shown in Table 10.1.

Eq. (10.26) shows that in the case of an asymmetric short circuit, the calculation of the positive-sequence component of the short-circuit current needs only an additional reactance $X_\Delta^{(n)}$ connected in each phase. So, the asymmetric short-circuit calculation can be converted to a symmetrical three-phase short-circuit calculation. In this case, only a correction of the equivalent time constant of the positive-sequence current in accordance with Eq. (10.27) is needed:

$$X_{dk}^{(n)} = (0.91 \sim 0.95) \times \frac{T'_{d0} \times \frac{X'_{de}+X_\Delta^{(n)}}{X_{de}+X_\Delta^{(n)}}}{1 - K_\alpha \times \frac{X_e+X_\Delta^{(n)}}{X_{de}+X_\Delta^{(n)}}}$$

$$= (0.91 \sim 0.95) \times \frac{T_d^{(n)}}{1 - K_\alpha \times \frac{X_e+X_\Delta^{(n)}}{X_{de}+X_\Delta^{(n)}}}$$

$$T_d^h = T'_{d0} \times \frac{X'_{de} + X_\Delta^{(n)}}{X_{de} + X_\Delta^{(n)}} \tag{10.27}$$

If $1 - K_\alpha \times \frac{X_e+X_\Delta^{(n)}}{X_{de}+X_\Delta^{(n)}} = 0$, the critical external reactance in the case of an asymmetric short circuit can be obtained as follows:

$$X_{ec}^{(n)} = \frac{X_d}{K_\alpha - 1} - X_\Delta^{(n)} \tag{10.28}$$

As can be seen from the comparison between Eqs. (10.28) and (10.25), the critical external reactance in the case of an asymmetric short circuit is smaller than that in the case of a three-phase short circuit. If an asymmetrical short circuit occurs at the near end of the step-up transformer, the excitation current will not be attenuated but will instead rise. Figure 10.8 shows the excitation current variation curves of a self-excited generator in various short-circuit modes. As can be seen from the figure, in the event of an asymmetrical short circuit beyond the critical external reactance, the excitation current rises.

10.2.7 Terminal Voltage Recovery of Self-Excited Generator after Short-Circuit Fault is Cleared

The generator voltage recovery process after the short circuit is cleared is described below, with a 600 MW self-excited generator as an example.

In Figure 10.9, Curve "a" represents the case of a fault clearing 1.5 s after the short circuit, where it takes about 0.7 s for the generator terminal voltage U_G to recover to the rated value. Curve "b" represents the case of a fault clearing 2 s after the short circuit, where the recovery time is about 1 s. Curve "c" represents the case of a fault clearing 3 s after the short circuit, where the recovery time is about 1.5 s. The clearing time t_c and

10 Static Self-Excitation System

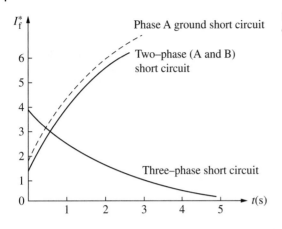

Figure 10.8 Variation of short-circuit current of self-excited generator in different short-circuit modes.

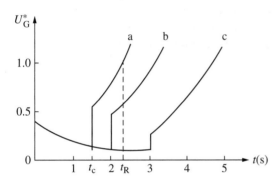

Figure 10.9 Curve of voltage recovery after short-circuit clearing: a – clearing 1.5 s after short circuit; b – clearing 2.0 s after short circuit; c – clearing 3.0 s after short circuit.

the voltage recovery time t_R are related as follows:

$$t_R \approx \frac{-T_{\Sigma 0}}{T_{dk}} \times t_c = \frac{X_{de}}{X'_{de}} \times \frac{\left(1 - K_\alpha \times \frac{X_e}{X_{de}}\right) t_c}{(K_\alpha - 1)}$$

$$T_{\Sigma 0} \approx \frac{T'_{d0}}{1 - K_\alpha} \tag{10.29}$$

where

$T_{\Sigma 0}$ – equivalent time constant of voltage recovery after clearing of short circuit;
T_{dk} – equivalent time constant in case of short circuit at external reactance X_e.

As can be seen from Eq. (10.29), the longer the short-circuit clearing time, the longer the voltage recovery time; that is, $t_R \propto t_c$. Figure 10.10 shows the result of a test on a simulation generator. For a self-excited generator, the voltage recovery after the short-circuit clearing is relatively fast, especially in the case of a quick clearing of the short circuit.

10.2.8 Inversion De-excitation of Self Thyristor Excitation System

The inversion de-excitation process of a self thyristor excitation system is virtually similar to that of a separate thyristor excitation system. Since the DC load side is passive, the inversion process is instantaneous and non-sustainable.

Figure 10.10 Curve of relationship between clearing time t_c and recovery time t_R.

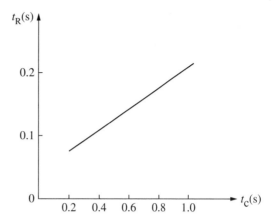

For the generator inversion de-excitation time, the corresponding value of the separate thyristor excitation system is as shown in Eq. (9.68):

$$t_m = \frac{L_f}{r_f} \times \ln \frac{U_{f\beta 0} + U_{f\beta}}{U_{f\beta}}$$

$$T = \frac{L_f}{r_f}$$

where

$U_{f\beta 0}$ – voltage at beginning of inversion [function of control angle α];
$U_{f\beta}$ – inversion voltage [function of inversion angle β];
$\frac{L_f}{r_f}$ – time constant of excitation winding depending on generator's working state.

Thus, the above equation can be rewritten as follows:

$$t_m = T \ln \frac{\cos \alpha + \cos \beta}{\cos \beta} \tag{10.30}$$

In the process of inversion de-excitation of a self thyristor excitation system, the amplitude of the inversion voltage decreases with the attenuation of the generator terminal voltage, but it is constant in a separately excited system. Figure 10.11 shows a comparison of the inversion process between a self thyristor excitation system and a separate one.

Thus, in the determination of the de-excitation time of a self thyristor excitation system, Eq. (10.30) can still be used. In this case, only the initial amplitude of the inversion voltage $U_{f\beta}$ should be multiplied by the exponential attenuation factor $e^{-\frac{1}{T_c}}$ to take into account the attenuation of the amplitude. Thus, the de-excitation time expression for the self-excitation system can be written as follows:

$$t_m = T \ln \frac{\cos \alpha + \cos \beta e^{-\frac{1}{T_c}}}{\cos \beta e^{-\frac{1}{T_c}}} \tag{10.31}$$

where

T_c – inversion voltage attenuation time constant of self-excitation system.

Typically, the generator has a certain remanence voltage, which is about 5%. For the corresponding demagnetization time, $t_m = 3T_c$ is taken. Then, $e^{-\frac{3T_c}{T_c}} = e^{-3} \approx 0.05$.

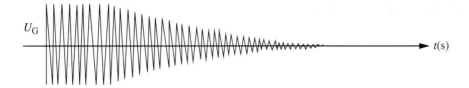

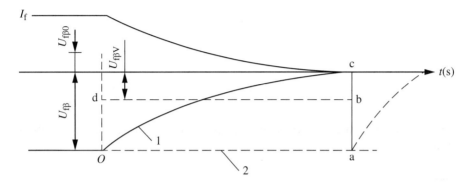

Figure 10.11 Comparison of inversion process between self and separate thyristor excitation systems: 1 – variation of inversion voltage of self-excitation system, whose amplitude attenuates; 2 – variation of inversion voltage of separate excitation system, whose amplitude is constant.

For ease of engineering calculations, Eq. (10.31) can be simplified. The steps include the following: (i) the inversion voltage curve that is attenuated exponentially is converted to a square wave voltage whose equivalent amplitude is the same; (ii) the areas of the two are made equal; and (iii) the equivalent constant amplitude voltage $U_{f\beta V}$ is obtained. The area enclosed by the curve is calculated as follows:

$$S_{oac} = U_{f\beta} \int_0^{3T_c} e^{-\frac{1}{T_c}} \, dt = U_{f\beta} T_c \left(e^{-\frac{1}{T_c}} \right)_0^{3T_c} = 0.95 T_c U_{f\beta} \tag{10.32}$$

If the equivalent square wave voltage curve oabd is equal to the above area, the average voltage amplitude can be obtained as follows:

$$U_{f\beta V} = \frac{0.95 T_c}{3 T_c} \times U_{f\beta} = 0.318 U_{f\beta} \tag{10.33}$$

That is, the average constant amplitude is 0.318 times the initial amplitude. Thus, the expression of the de-excitation time expressed as the curve of the square wave voltage whose equivalent amplitude is the same can be written as follows:

$$t_m = T \ln \frac{\cos \alpha + 0.318 \cos \beta}{0.318 \cos \beta} \tag{10.34}$$

It should be noted that, when different residual voltage values are taken, the average coefficient values are different.

The determination of the de-excitation time is described with an example. It is assumed that the generator is operating without load, $\alpha = 78.5°$, and $\cos\alpha = 0.2$. In the case of an inversion, $\beta = 37°$, $\cos\beta = 0.8$, and $T = T_{d0} = 6.04\,\text{s}$. When they are substituted into Eq. (10.34), the following is obtained:

$$t_m = 6.014 \ln \frac{0.2 + 0.318 \times 0.8}{0.318 \times 0.8} = 3.5(s)$$

10.2.9 Excitation Transformer Protection Scheme [20]

In static self-excitation systems, power stations employ a variety of generator and excitation transformer protection schemes, as shown in Figure 10.12.

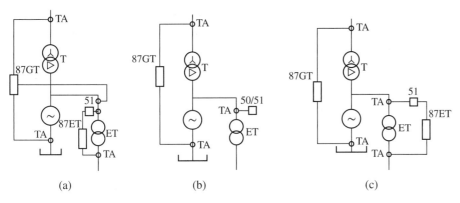

Figure 10.12 Excitation transformer protection schemes: (a) generator and excitation transformer are provided with differential protection; (b) excitation transformer is within the range of differential protection of generator; (c) excitation transformer deprived of separate differential protection of high-current transformer itself on high-voltage side. 87GT – differential protection of generator-transformer set; 51 – AC time-limited overcurrent protection relay; 87ET – differential protection of excitation transformer; 50 – quick-break overcurrent protection; TA – current transformer.

In Figure 10.12a, the generator and the excitation transformer are provided with differential protection. The two sets of differential protection have their own protection range. The excitation transformer is provided with the same protection scheme as that of the station transformer. In the scheme, the biggest difficulty is the installation of a set of current transformers with the same ratio of transformation as the generator outlet circuit on the high-voltage side of the excitation transformer. This current transformer has a high ratio of transformation (25 000/5 for the current transformer of a 600 MW generator). Thus, it is costly and hard to lay out and should be placed in the enclosed busbar on the high-voltage side of the transformer.

In Figure 10.12b, the excitation transformer is placed within the range of the differential protection of the generator-transformer set, and a set of quick-break overcurrent protection is equipped separately (50/51 in the figure). However, in most cases, the protection sensitivity is not adequate. In some short-circuit cases, the excitation transformer has no quick-acting protection, which is clearly unacceptable. So, this protection scheme is not perfect.

Compared to the protection scheme shown in Figure 10.12a, the protection scheme shown in Figure 10.12c omits the set of high-current current transformers on the high-voltage side of the excitation transformer, which solves the problem of difficulty in installing the current transformer. Besides, the excitation transformer itself is provided with differential protection, which can ensure adequate sensitivity in the event of an excitation transformer fault.

The focuses of the above three protection schemes are as follows:

(1) Whether the set of current transformers on the high-voltage side of the excitation transformer which is used for the large differential motion of the generator-transformer set can be omitted
(2) Whether the excitation transformer itself needs to be provided with differential protection.

At a large power station, in the event of a fault on the high-voltage side of the excitation transformer or the high-voltage station transformer, since the short-circuit current is high, the high-voltage side of the excitation transformer and that of the station transformer are generally not equipped with circuit breakers (for a small generator, the high- and low-voltage sides of the high-voltage station transformer are equipped with circuit breakers). In the event of a high-voltage station transformer fault, its own differential protection action trips not only the circuit breaker on the high-voltage side of the generator-transformer set but also the circuit breaker on the low-voltage side of the station transformer through the protection outlet. The standby power supply is automatically switched on. The standby transformer drives the station service load, so that the generator can continue to operate or halt safely.

The low-voltage side of the excitation transformer is connected to the SCR, without a circuit breaker. So, in the event of an excitation transformer or thyristor silicon rectifier cabinet fault, the object of the protection action is the generator-transformer set's protection outlet, which is the same as the object of the outlet action of the differential protection action of the generator-transformer set. This generic characteristic does not involve the selection of an excitation transformer protection scheme.

The following is a discussion on several important characteristics that influence the selection of an excitation transformer protection scheme and a further analysis of excitation transformer protection connection modes based on an example as well as a verification of the feasibility of the scheme shown in Figure 10.12c, where the high-current transformer on the high-voltage side is omitted and the excitation transformer is separately provided with differential protection.

Known raw data: The generator's rated capacity $S_{GN} = 728$ MVA, rated voltage $U_{GN} = 22$ kV, and super-transient reactance X_d'' 0.19. The main transformer's rated capacity $S_{TN} = 750$ MVA and short-circuit impedance $U_{TK} = 12\%$. The excitation transformer's rated capacity $S_{EN} = 7100$ kVA and short-circuit impedance $U_{EK} = 8\%$. The base capacity $S_b = 100$ MVA. The per-unit value of the impedance of each device can be determined as follows:

$$X_d^{*''} = \frac{X_d''\%}{100} \times \frac{S_b}{S_{GN}} = \frac{19}{100} \times \frac{100}{728} = 0.026 \text{ (p.u.)}$$

$$X_{TK}^* = \frac{U_{TK}}{100} \times \frac{S_b}{S_{TN}} = \frac{12}{100} \times \frac{100}{750} = 0.016 \text{ (p.u.)}$$

$$X_{EK}^* = \frac{U_{EK}}{100} \times \frac{S_b}{S_{EN}} = \frac{8}{100} \times \frac{100}{7.1} = 1.126 \text{ (p.u.)}$$

The connection of the generator-transformer set and the excitation transformer is shown in Figure 10.13.

(1) The high-voltage side of the excitation transformer does not have to be equipped with a current transformer for differential protection of generator-transformer set. The state is normal operation.

The excitation transformer's rated capacity is 7100 kVA. Its primary rated current is as follows:

$$I_{1NE} = \frac{7100}{\sqrt{3} \times 22} = 186 \text{ (A)}$$

Its secondary rated current is as follows:

$$I_{2NE} = \frac{186}{25000/5} = 0.037 \text{ (A)}$$

where

25 000/5 — ratio of transformation of current transformer for large differential protection.

In the case of normal operation, there is an unbalanced current that is no higher than 0.037 A flowing in the differential circuit of the generator-transformer set.

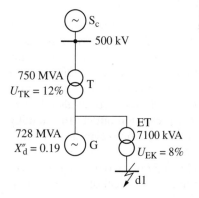

Figure 10.13 Connection diagram for generator-transformer set and excitation transformer.

As can be seen from the relay protection setting calculation, generally the generator-transformer set differential protection setting is (0.2–0.5) I_{2NT} (I_{2NT} is the main transformer's rated current).

The main transformer's primary rated current is as follows:

$$I_{1NT} = \frac{750000}{\sqrt{3} \times 22} = 19681 \text{ (A)}$$

The main transformer's secondary rated current is as follows:

$$I_{2NT} = \frac{19681}{25000/5} = 3.936 \text{ (A)}$$

It is assumed that the protection setting is $0.2I_{2NT}$, and so the following can be obtained:

$$0.2 \times 3.936 = 0.78 \text{ (A)}$$

$$0.78 \text{ A} \gg 0.037 \text{ A}$$

It can be seen that there is a difference of almost one order of magnitude between the two values. So, in the case of normal operation, the excitation transformer's load current will not cause a faulty action of the generator-transformer set differential protection.

Unlike an excitation transformer, a high-voltage station transformer features high capacity and a high unbalanced current generated by the load current. It is known that the rated capacity of a high-voltage station transformer is 40 MVA. Then its primary rated current is as follows:

$$I_{1NS} = \frac{40000 \times 2}{\sqrt{3} \times 22} = 2099.4 \text{ (A)}$$

Its secondary rated current is as follows:

$$I_{2NS} = \frac{2099.4}{25000/5} = 0.42 \text{ (A)}$$

In the generator-transformer set differential circuit, the unbalanced current generated by the load of the station transformer is 0.42 A. Compared to 0.78 A, $I_{2NS} = 0.42$ A is on the same order of magnitude. In some cases, it may cause a faulty action of the generator-transformer set differential protection. So, for the sake of reliability, a current transformer with a large ratio of transformation for generator-transformer set differential protection must be installed on the high-voltage side of the station transformer.

(2) For one generator category, the excitation transformer's capacity is much smaller than that of the high-voltage station transformer, and its impedance is much larger than that of the latter. So, if the excitation transformer is protected with the generator-transformer set differential protection, the sensitivity is not adequate. In other words, in the event of an excitation transformer fault, the generator-transformer set differential protection may not act. But the generator-transformer set differential protection can act in the event of a fault of the bus on the low-voltage side of the high-voltage station transformer. If the high-voltage side of the station transformer is not equipped with a high-current transformer, in the event of a fault of the bus on the low-voltage side of the station transformer, the generator-transformer set differential protection will act, resulting in an override trip and expanding the fault coverage. So, the high-voltage side of the high-voltage station transformer must be equipped with a high-current transformer for complete generator-transformer set differential protection.

(3) For the same converted capacity, the impedance of the excitation transformer is much greater than that of the generator and that of the main transformer. So, even if the excitation transformer is included in the generator-transformer set differential protection range (the high-voltage side is not equipped with a high-current transformer), the generator-transformer set differential protection does not act in most cases. This will be described with an example. The following calculation can be used for the description.

When the low-voltage side of the excitation transformer is short-circuited at Point d1 in Figure 10.13, the following two cases arise:

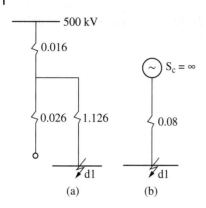

Figure 10.14 Equivalent short-circuit impedance diagram: (a) main transformer's high-voltage circuit breaker is broken; (b) main transformer's high-voltage circuit breaker is closed.

(1) The calculation is based on the minimum operation mode (the circuit breaker on the high-voltage side of the main transformer is not closed). The impedance diagram is shown in Figure 10.14a. The impedance is calculated as follows:

$$\sum X^* = 1.126 + 0.026 = 1.152 \text{ (p.u.)}$$

$$\sum X_b^* = \sum X \times \frac{S_{GN}}{S_b} = 1.152 \times \frac{728}{100} = 8.386 \text{ (p.u.)}$$

The per-unit value of the short circuit current is as follows:

$$I_b^* = \frac{1}{8.386} = 0.119 \text{ (p.u.)}$$

The short-circuit current supplied by the generator is as follows:

$$I'' = I_b \times I_{1NG} = 0.119 \times \frac{72800}{\sqrt{3} \times 22} = 2278 \text{ (A)}$$

The short-circuit current converted to the secondary side is $2278/(25\,000/5) = 0.45$ A.

(2) The calculation is based on the maximum operation mode (the system is calculated based on infinity). The impedance diagram is shown in Figure 10.14b.

The short-circuit current supplied by the generator and the system is as follows:

$$I'' = I_b \times I_{1NE} = \frac{1}{0.08} \times \frac{7100}{\sqrt{3} \times 22} = 2329 \text{ (A)}$$

The short-circuit current converted to the secondary side is $2329/(25\,000/5) = 0.466$ A.

It is known from the foregoing calculation that the main transformer's primary rated current is 19.681 A and the secondary rated current 3.936 A.

So, in the maximum and minimum operation modes, the variation range of the corresponding short circuit current is $\frac{0.066}{3.936} \sim \frac{0.45}{3.936}$ of the rated current, that is, $(0.118–0.114)\, I_{2N}$.

The set value of the generator-transformer set differential protection action is generally $(0.2–0.5)\, I_{2N}$. When the minimum set value of the action $0.2\, I_{2N}$ is taken and compared to the above short-circuit current variation range, the short-circuit current is less than the protection action value. So, in the event of a short circuit at Point d1, the generator-transformer set differential protection will not act.

On the basis of data analysis results and the foregoing conclusions, the author of Ref. [20] believes that the connection scheme shown in Figure 10.2c is feasible for excitation transformers and it boasts such advantages as economy, rationality, and high reliability.

10.3 Shaft Voltage of Static Self-Excitation System [21, 22]

10.3.1 Overview

The existence of shaft voltage has been one of the issues of concern for the operation of large steam turbines. In recent years, with the wide application of static self-excited systems, the research on high-frequency shaft voltage of generators has attracted more and more attention.

Serious bearing or steam turbine front gear box electric corrosion caused by shaft voltage is seen from time to time on generators that adopt the static self-excitation system. For example, at a Chinese power station, the DC exciter of a 25 MW steam turbine was replaced by a static self-excitation system. Soon after, the operator found that what was between the generator shaft and the bearing block was live. Later, the operator heard an abnormal sound in the steam turbine front gear box, resulting in a strong vibration. That triggered the action of the emergency governor and eventually led to a stoppage. An inspection revealed that the reduction gear was severely damaged. After the gear was replaced, a similar accident occurred again.

Then, the shaft voltage and the shaft current were measured, with their values are shown in Table 10.2.

The cause for the electric corrosion and damage of the gear was deduced. The equivalent capacitance formed when the pulsating rectified voltage outputted by the static self-excitation system acted on the generator's rotor excitation winding and passed between the excitation winding and the rotor and between the bearing oil film, the gear oil film, and the main circuit component, generating a loop current. That is, the shaft current destroyed the lubricating oil film and caused electric corrosion of the various parts on the bearing bush surface and the gear contact surface.

In view of that, the plant undertook improvement measures. It eliminated the star-shaped ground RC protection circuit connected to the AC side of the SCR to cut off possible shaft current. Moreover, it installed a ground brush on the generator shaft, which can only reduce the shaft-to-ground potential but cannot completely prevent the generation of shaft current.

10.3.2 Source and Protection of Shaft Voltage

Steam turbines with static excitation systems generate four forms of shaft voltage. Table 10.3 summarizes the sources of these voltages. The first two forms are inductive, their magnitude usually being much less than 10 V. However, if the induced voltage circuit goes through a low-resistance closed branch, a high current will pass between the bearings and the auxiliary equipment, resulting in serious damage. The last two forms appear between the shaft and the ground. If the shaft has no contact with the ground, a 100 V or greater voltage will be generated easily. The shaft voltage covers a wide frequency band, from DC to 500 kHz. The meaning of the letter symbols in Table 10.3 is shown in Figure 10.15.

Practice has shown that the insulation of all the bearings and the shaft seal at the generator excitation end can effectively prevent the generation of an induced shaft voltage and the resulting high ring current in the generator. Meanwhile, in order to determine the working state of the various parts, generally two separate insulating layers are used for the bearing and the shaft seal insulation.

Table 10.2 Determination of shaft voltage and shaft current of generator.

Measured value of shaft voltage	Shaft voltage at both ends of generator (V)	Shaft-to-ground voltage (V)		Shaft-to-ground current (A)	
		Shaft-to-ground voltage	No. 4 bearing block-to-ground voltage	No. 3 bearing block	No. 4 bearing block
The first measurement	0.5	185	90	0.32	0.35
The second measurement	2.0	170	95	0.226	0.225

Table 10.3 Types and handling of rotating motor shaft voltage.

Source	Resulting shaft voltage type	Outcome
(a) Magnetic circuit asymmetry • Stator lamination joint • Rotor eccentricity • Rotor or stator sag causing various magnetic fluxes	A variational flux link chain induces the shaft voltage through the rotating shaft – block – bearing circuit	The induced voltage will cause a high current in any low-impedance circuit, resulting in the corresponding damage
(b) Axial flux • Remanence • Rotor eccentricity • Rotor winding symmetry	The rotating magnetic flux induces the unipolar voltage in the bearing and rotor parts	The induced voltage will cause a high current and corresponding damage in the bearings and the shaft seal
(c) Electrostatic charge • Steam flushing steam turbine blades	The shaft capacitance related to the grounding condition is charged and generates a static emf due to the internal insulation	If the oil film is broken down by the voltage between the shaft and the bearing (the ground), there will be a discharge, generating spots and damaging the bearing and shaft seal surfaces
(d) External voltage acting on rotor winding • Static excitation device • Voltage source or rotor winding insulation asymmetry • Active rotor winding protection device	Through the power supply and the winding as well as the insulation capacitance and resistance related to the grounding conditions, the external voltage makes the shaft generate an emf	

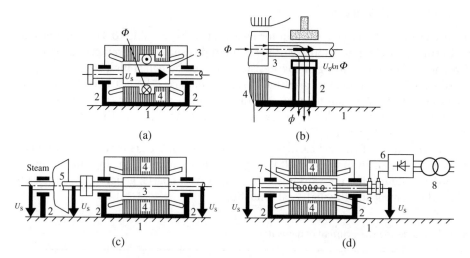

Figure 10.15 Generation of shaft voltage and circuit for large steam turbine: (a) shaft voltage generated by asymmetry of magnetic circuit; (b) shaft voltage generated by axial flux; (c) shaft voltage generated by electrostatic charge; (d) shaft voltage generated by external voltage acting on rotor winding. 1 – block; 2 – bearing; 3 – rotating shaft; 4 – stator; 5 – steam turbine; 6 – rectifier; 7 – rotor winding; 8 – transformer.

When the shaft voltage discharges on the oil film between the shaft and the bearing bush, a shaft-to-ground voltage whose amplitude is 20 V or greater will be generated. This will give rise to a shaft current and cause continuous bearing corrosion.

In order to prevent the shaft voltage from giving rise to a ring current, generally the shaft is grounded with a grounding carbon brush. However, a lot of experience in operation has shown that it is very difficult to achieve good and reliable grounding at the generator terminal. The reason is that the contact resistance is affected by the high surface speed of the rotating shaft, and the surface speed of the shaft can reach 100 m s^{-1}, which is almost twice the allowable value specified by the manufacturer. Moreover, the oil mist contamination causes poor contact with the grounding brush.

Besides, the static self-excitation system provides a new bypass for the shaft current. The periodic pulsating current up to $10 \text{ A } \mu\text{s}^{-1}$ caused by the thyristor during the commutation can be coupled to the rotating shaft through a rotor winding-to-ground capacitance.

Oil mist and high speed will cause scratches on the contact surface between the carbon brush and the rotating shaft. Thus, a sharp-angled waveform will be formed. These scratches may also degrade the electric contact. Even with regular maintenance including periodic grinding of the brush and the shaft, the grounding carbon brush in this position is unreliable.

Despite the small shaft diameter and low surface speed at the free end of the high-pressure cylinder rotor of the steam turbine, the grounding carbon brush installed here cannot eliminate the high-frequency current pulse during the commutation. This is caused by the frequency characteristics of the shaft impedance. Although the shaft inductance is small, large impedance to high frequencies (up to 500 kHz) is exhibited.

From the above, shaft voltage prevention measures available at present include insulation of all the bearings and the shaft seal at the generator exciter end, which can effectively prevent the induced shaft voltage from generating a ring current, and regular maintenance of the grounding carbon brush in any form at the steam turbine end. So, no reliable shaft current protection measure can be provided.

10.3.3 Shaft Grounding System for Large Steam Turbine Generator

A new shaft grounding system must meet the following requirements:

(1) Ability to decrease shaft-to-ground voltage to less than 20 V in any mode of operation;
(2) Compatibility with existing devices;
(3) Ease of maintenance and good accessibility;
(4) Possibility of monitoring of operation of grounding device.

The bases for these requirements are the following:

(1) Surveys have confirmed that shaft voltages whose amplitude is less than 20 V will not cause electric corrosion of steam turbine bearings.
(2) Any new grounding system must be compatible with existing shaft protection devices and monitoring device and easy to install on steam turbines in various forms.
(3) Any shaft grounding device must be easy to access, observe, and maintain without harm to operators or radioactive radiation during maintenance at nuclear power stations.
(4) Monitoring of grounding device primarily refers to monitoring of electric contact and local damage of grounding carbon brush.

10.3.4 New Shaft Grounding Device

When the carbon brush is operating at a surface speed much lower than the maximum set by the manufacturer, its contact performance is greatly enhanced. Given this, only two positions along the shaft of a large steam turbine can accommodate the grounding carbon brush.

The first position is the free end of the high-pressure cylinder of the steam turbine. But this mode cannot protect the high-frequency shaft voltage pulse from the static self-excitation system.

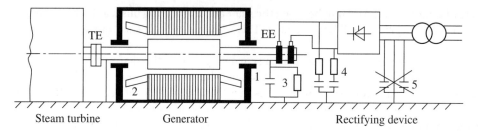

Figure 10.16 Shaft voltage protection connection for static self-excitation system: 1 – shaft seal; 2 – grounding carbon brush at steam turbine end; 3 – passive RC protection; 4 – thyristor RC ground protection; 5 – conventional RC protection for excitation transformer.

The second position is the generator excitation end. The carbon brush is insulated, bypassing the bearings. If there is any unexpected shaft-to-ground contact at the steam turbine end, the shaft voltage induced by the generator shaft will cause a high ring current.

Figure 10.16 shows that a steam turbine has a conventional grounding carbon brush at the steam turbine end and a grounding carbon brush at the excitation end grounded through a new set of passive RC circuits. Here, the resistance value is about 500 Ω, which is high enough to limit the current to a few milliamperes (which is harmless) and low enough to prevent the build-up of a DC emf. The parallel capacitor is about 10 μF, which is valid for all shaft voltages produced by the protective static excitation system.

A fuse added between the carbon brush and the RC circuit can prevent a high ring current flowing through the shaft in the event of an accident, such as a ring current caused by damage to Capacitor C.

The new RC grounding device added at the excitation end can be used together with any existing grounding carbon brush at the steam turbine end. This new grounding system also applies to the existing rotor ground fault monitoring device. The monitoring device itself is a shaft voltage source. An alarm will be issued in the event of bad grounding.

Figure 10.17 shows the shaft voltage variation measured when the grounding carbon brush is in different locations.

Figures 10.17a,b show respectively the shaft-to-ground voltage at the steam turbine end and that at the excitation end measured when all the grounding carbon brushes are removed. A voltage curve can be understood as a superposition of the various variables, including a DC component (the electrostatic effect) of about 45 V, a 50 Hz component (a rotor eccentricity induced shaft voltage) of about 2.5 V, and a 150 Hz rectangular wave component (a common mode voltage of the static excitation system) of about 15 V.

Figure 10.17d shows the high-frequency shaft voltage during the thyristor commutation, which exceeds 15 V.

Figures 10.17b,e show the shaft voltage measured in the case of only a conventional grounding carbon brush at the steam turbine end. The shaft voltage at the steam turbine end is almost zero. At the excitation end, there is still an induced low-frequency voltage of a few volts and a peak voltage generated by the static excitation whose sum is for all practical purposes no greater than 20 V. If the current grounding carbon brush at the steam turbine end can operate reliably, it can effectively reduce the shaft voltage.

Figures 10.17c,f show the test results obtained for the case with no grounding carbon brush at the steam turbine end but only a new RC grounding device at the excitation end. Due to the voltage drop of Resistor R, a DC voltage component appears in the shaft voltage. The role of the smaller induced voltage component can be seen from the shaft voltage at the steam turbine end. The shaft voltage spikes at both ends of the generator are virtually eliminated. All the shaft voltages are reduced to below the safety limit (20 V) extremely reliably over a long period of time.

A grounding system with a conventional grounding carbon brush at the steam turbine end and an RC grounding device at the excitation end is also measured. The measurement results confirm that all the shaft voltages are significantly reduced and there is no adverse effect between the two sets of grounding devices.

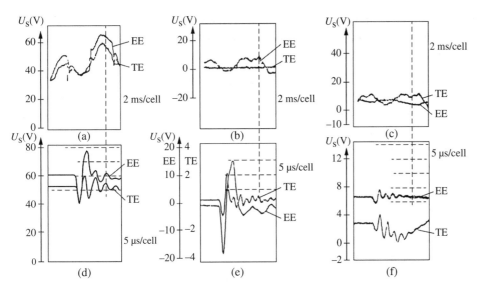

Figure 10.17 Shaft-to-ground voltage for 1200 MVA steam turbine: (a) without grounding carbon brush (low-frequency); (b) with grounding carbon brush at steam turbine end (low-frequency); (c) grounded only via RC impedance at excitation end (low-frequency); (d) without grounding carbon brush (high-frequency); (e) with grounding carbon brush at steam turbine end (high-frequency); (f) grounded only via RC impedance at excitation end (high frequency). EE – excitation end; TE – steam turbine end.

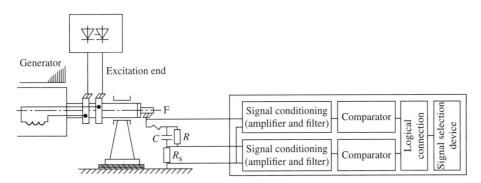

Figure 10.18 Shaft grounding monitoring system: R_s – Shunt resistor for current measurement.

This new shaft grounding system provides the possibility of installing a new grounding monitoring system. Figure 10.18 is a block diagram for such a monitoring system. Its design basis is that the contact of the grounding carbon brush and the working state of the ground circuit can be identified through a measurement of the characteristic frequency component of the voltage and current of the circuit.

10.4 Coordination between Low Excitation Restriction and Loss-of-Excitation Protection [23]

For a synchronous generator, loss of excitation means that its excitation current is lower than the minimum excitation current value required to ensure its stable operation under certain power conditions. When the low excitation limit cannot stop the excitation current from continuously decreasing, the generator's operating point will exceed the stability limit, and then the generator will absorb reactive power from the system. Thus, there will be a slip and even loss of synchronism.

The general setting principle for an excitation system is that, in the above case, the low excitation limit acts first. If the limit fails, the loss-of-excitation protection acts through a set delay, to identify the cause for the loss-of-excitation fault and apply a trip based on the control logic.

The impedance criterion is the loss-of-excitation protection action criterion widely used at present. The criterion has been widely used in the field of loss-of-excitation protection, but we cannot intuitively decide if the coordination between the low excitation and loss-of-excitation protection characteristics is appropriate. The reason is that a low excitation limit is expressed in the P-Q plane coordinate system, but loss-of-excitation protection is expressed in the R-X impedance plane. The two cannot be directly compared in terms of their parameters, and a conversion is needed.

In view of this, Siemens once introduced a new loss-of-excitation protection action principle expression, that transforms the generator operating limit diagram into a coordinate plane represented by admittance and plots the generator's low excitation limit and loss-of-excitation protection characteristics in the same plane, so as to assess the rationality of the parameter setting of the two protection systems.

10.4.1 Generator Operating Limit Diagram

For ease of discussion, the following basic electric variables are defined as follows:

Symbol convention: output power (P, $Q > 0$) is positive.
Apparent power: S, in VA (kVA, MVA).
Active power: P, in W (kW, MW).
Reactive power: Q, in var. (kvar, Mvar).

If per-unit values are adopted, the corresponding calculations should be based on the nameplate data of the generator such as the rated apparent power S_N, rated voltage U_N, and rated current I_N.

On the basis of the definition of the Cartesian coordinate system (x-axis = real part and y-axis = imaginary part), the first quadrant is the operating range of the generator $P > 0$, $Q > 0$. Under the condition of under-excitation, the generator is operating in the second quadrant $P > 0$, $Q < 0$. The corresponding operating limit diagram of the generator is shown in Figure 10.19.

With the active power and reactive power expressions, the expression for the static stability limit can be obtained. For a salient pole generator, since its vertical-axis reactance parameter and horizontal-axis reactance parameter are different, that is, X_d X_q, a circle with a diameter of $U \frac{X_d - X_q}{X_d X_q}$ is formed. This circle represents the steady-state power that the generator can generate when no excitation is applied ($E = 0$):

$$P = \frac{EU}{X_d} \sin \delta + \frac{U^2}{2} \frac{X_d - X_q}{X_d X_q} \sin 2\delta \qquad (10.35)$$

$$Q = \frac{EU}{X_d} \cos \delta - \frac{U^2}{X_d} \left(1 + \frac{X_d - X_q}{X_q} \sin 2\delta\right) \qquad (10.36)$$

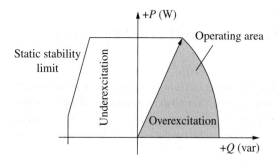

Figure 10.19 Generator operating limit diagram.

where

- E – stator winding emf;
- U – generator terminal voltage;
- X_d – vertical-axis synchronous reactance;
- X_q – horizontal-axis synchronous reactance;
- δ – rotor power angle.

For a non-salient pole generator, since the vertical-axis synchronous reactance X_d and the horizontal-axis synchronous reactance X_q are approximately equal, Eqs. (10.35) and (10.36) can be simplified. The theoretical static stability limit power angle $\delta = 90°$. The corresponding stability limit can also be obtained through the vertical-axis synchronous reactance X_d. For a salient pole steam turbine, at this point, the stability limit value is related to the vertical-axis synchronous reactance X_d, the horizontal-axis synchronous reactance X_q, the excitation system, and the generator terminal voltage. The theoretical limit on Axis Q depends on the horizontal-axis synchronous reactance X_q. The allowable power angle δ is less than 90°. Figures 10.20 and 10.21 respectively show the operating limit diagrams for non-salient and salient generators represented through the phasor diagrams of the current and voltage.

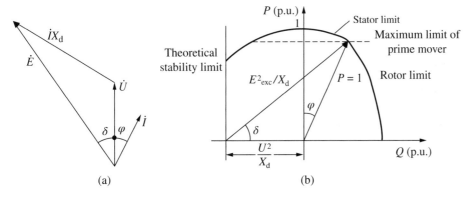

Figure 10.20 Phasor diagram and operating limit diagram for non-salient pile generator ($X_d = X_q$): (a) voltage phasor diagram; (b) operating limit diagram. E_{exc} – stator winding emf; I – stator current

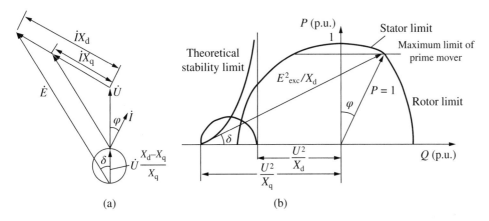

Figure 10.21 Phasor diagram and operating limit diagram for salient pole generator ($X_d \neq X_q$): (a) voltage phasor diagram; (b) operating limit diagram.

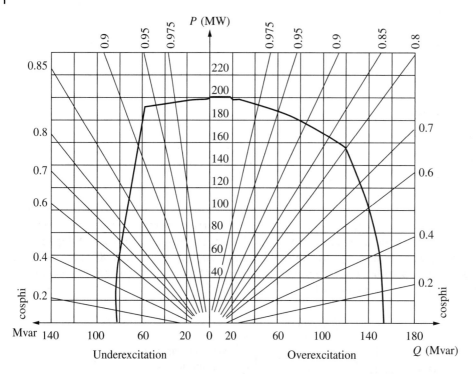

Figure 10.22 Operating limit diagram for non-salient pole generator.

As can be seen from Figures 10.20 and 10.21, the operating range of the generator is restricted by the prime mover input power and the temperature rise of the excitation winding in the overexcitation area and by the prime mover input power, the stator end temperature rise, and the stability limit in the under-excitation area.

When the generator is connected to the grid, its actual stability limit will change. In this case, the contact reactance between the generator and the system (e.g., the reactance of the step-up transformer) as well as the appropriate safety margin should be taken into account. So, the actual allowable stability limit of the generator will be less than the theoretical value. Generally, the stability limit diagram given by the generator factory is different from the operating limit diagram.

Figures 10.22 and 10.23 show the operating limit diagrams for non-salient and salient pole generators, respectively.

For a non-salient generator, taking into consideration the contact reactance between the generator and the system, the actual stability limit tilts to the right side of the coordinate axis compared to the theoretical stability limit. But for a salient pole generator, the theoretical stability limit diagram shifts to the right of the coordinate axis and is approximately symmetrical about the center of the circle.

The operating limit diagram applies to a generator under the condition of rated voltage (U_N) and rated current (I_N). In actual operation, the voltage cannot be guaranteed to be constant. The influence of the voltage variation on the operating limit of the generator can be calculated by the following formula for Q.

If no excitation is applied, the stator winding emf $E = 0$. The maximum reactive power absorbed by the generator $Q = -U^2/X_d$. When the voltage varies by 10%, the reactive power varies as follows:

When $U = 0.9$,

$$Q = -\frac{U^2}{X_d} = -\frac{0.9^2}{X_d} = -\frac{0.81}{X_d}$$

When $U = 1.1$,

$$Q = -\frac{U^2}{X_d} = -\frac{1.1^2}{X_d} = -\frac{1.21}{X_d}$$

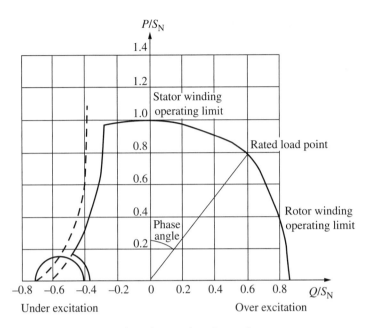

Figure 10.23 Operating limit diagram for salient pole generator.

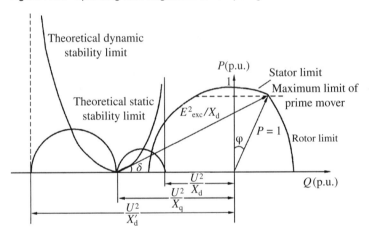

Figure 10.24 Dynamic stability limit (salient pole generator).

Compared to that at the rated voltage, the stability limit of the generator at a low voltage shifts to the right, which further restrains the reactive power of the system that can be absorbed by the generator. This reactive power is proportional to the square of the voltage. In the case of an overvoltage, the stability limit of the generator shifts to the left. So, there is no need to pay attention to it.

The foregoing discussion on static stability primarily involves the influence of small interferences occurring in the system on the static stability limit of the generator. If there is a sudden large interference in the system's load or operation mode, the parameters of the generator will make corresponding transient changes and responses. This will involve the dynamic stability of the generator. In order to simplify the discussion, the transient variables (X'_d, X'_q, and E'') can replace the corresponding variables in Eqs. (10.35) and (10.36). Figure 10.24 shows the basic results of the dynamic stability. In this process, it is assumed that the transient vertical-axis reactance of the generator is equal to the horizontal-axis synchronous reactance; that is, $X'_d = X_q$. As can be seen from Figure 10.24, in the dynamic process, even if the power angle is greater than 90°, the generator can remain stable. A similar conclusion can be drawn if a non-salient pole generator is analyzed. In this

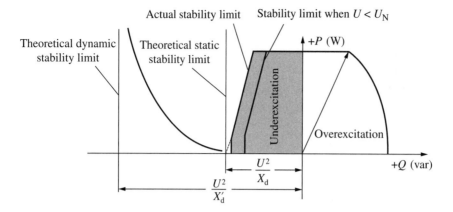

Figure 10.25 Actual stability limit of generator.

case, the dynamic stability limit of the generator depends on the transient vertical-axis reactance. In actual operation, the power angle is greater than 90°, generally 110°–120°.

The characteristics of the various stability limits for the generator in the loss-of-excitation state are described in Figure 10.25:

(1) The actual stability limit of the generator in operation is on the right side of the theoretical stability limit, generally determined by the operating limit diagram of the generator.
(2) If the generator terminal voltage $U < U_N$, the stability limit will shift to the right.
(3) Taking into consideration the dynamic working conditions in operation, the dynamic stability limit of the generator is introduced. If the operating point exceeds the dynamic stability limit, the generator must be disengaged from the system immediately, otherwise there may be a pole slip and loss of synchronism.

10.4.2 Admittance Measurement Principle and Derivation

On the basis of the definition of complex power, the apparent power of the generator can be expressed as follows:

$$\dot{S} = P + jQ \tag{10.37}$$

or it can be written as follows:

$$\dot{S} = \dot{U}\hat{I} \tag{10.38}$$

where

$\hat{I}$ — conjugate value of Phasor I.

In rectangular coordinates, the phasor voltages U, I, and $\hat{I}$ can be written as follows, respectively,

$$\dot{U} = U_a + jU_b \tag{10.39}$$
$$\dot{I} = I_a + jI_b \tag{10.40}$$
$$\hat{I} = I_a + jI_b \tag{10.41}$$

On the basis of the definition of the complex power, Eq. (10.38) can be written as

$$\dot{S} = \dot{U}\hat{I} = (U_a + jU_b)(I_a - jI_b)$$
$$= (U_aI_a + U_bI_b) + j(U_bI_a - U_aI_b) \tag{10.42}$$

10.4 Coordination between Low Excitation Restriction and Loss-of-Excitation Protection

When $P = U_a I_a + U_b I_b$, $Q = U_b I_a - U_a I_b$ and they are substituted into Eq. (10.42), the following is obtained:

$$\dot{S} + P + jQ \tag{10.43}$$

It can be concluded that Eq. (10.42) is equivalent to Eq. (10.37).
The conjugate root ($\hat{S}$) of the complex power ($\dot{S}$) is discussed below.

$$\hat{S} = \hat{U}\dot{I} = P - jQ = (U_a - jU_b)(I_a + jI_b)$$
$$= (U_a I_a + U_b I_b) - j(U_b I_a - U_a I_b) \tag{10.44}$$

Similarly, when $P = U_a I_a + U_b I_b$, $Q = U_b I_a - U_a I_b$ and they are substituted into Eq. (10.44), the following is obtained:

$$\hat{S} = P - jQ \tag{10.45}$$

The expression for the complex admittance is deduced below.
From the definition of the electric engineering admittance:

$$\dot{Y} = \frac{\dot{I}}{\dot{U}} = \frac{\dot{I}\hat{U}}{\dot{U}\hat{U}} = \frac{\hat{S}}{U^2} = \frac{P - jQ}{U^2} = \frac{P}{U^2} - j\frac{Q}{U^2} \tag{10.46}$$

$$\dot{Y} = G + jB \tag{10.47}$$

where

Y – admittance;
G – conductance (real part of admittance);
B – susceptance (imaginary part of admittance).

With Eq. (10.46) and (10.47), the following can be obtained:

$$\left. \begin{array}{l} G = \dfrac{P}{U^2} \\ B = -\dfrac{Q}{U^2} \end{array} \right\} \tag{10.48}$$

Thus, the generator operating limit diagram expressed in the admittance plane can be obtained only when the P and Q values in the generator operating limit diagram are divided by the square of the voltage, and the symbol of the parameter of the imaginary part is reversed. When the generator terminal voltage $U = U_N = 1$, the per-unit parameters in the operating limit diagram are equivalent to those in the admittance diagram (see Figure 10.26). So, the set parameter of the loss-of-excitation protection can be obtained directly from the generator operating limit diagram.

In the generator operating limit diagram expressed as the admittance parameter, the per-unit value of each parameter is as follows:

$$\frac{1}{X_{d,pu}} = \frac{1}{X_d} \times \frac{U_N}{\sqrt{3}I_N}$$

$$G_{pu} = \frac{P/S_N}{(U/U_N)^2}$$

$$B_{pu} = \frac{Q/S_N}{(U/U_N)^2}$$

The loss-of-excitation protection measurement algorithm is based on Eqs. (10.37), (10.38), and (10.42). The protection device obtains the phasors of the three-phase voltage to ground and the three-phase current through instantaneous value sampling, and then the current and voltage positive-sequence components are

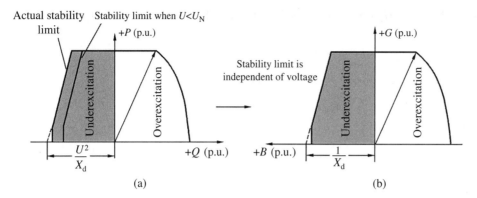

Figure 10.26 Operating limit diagram and admittance diagram for generator: (a) P-Q plane; (b) G-B plane.

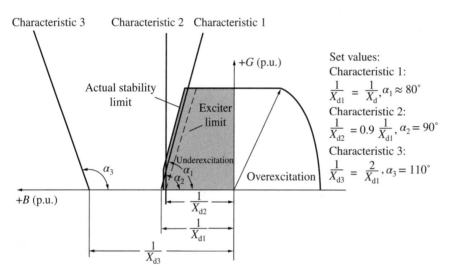

Figure 10.27 Loss-of-excitation protection characteristic curves of admittance characteristics (non-salient pole generator). (For salient pole generator, set value of Characteristic 1 approximates $1/X_d + 1/2(1/X_q - 1/X_d)$; set value of Characteristic 2 approximates X_d; and $\alpha_2 = 100°$).

calculated with the phasor data. The generator's active power and reactive power are calculated with the voltage and current positive-sequence components as defined in Eqs. (10.36) and (10.38). Then, they are divided by the square of the positive-sequence voltage U_1 in accordance with Eq. (10.42). Thus, the power plane can be converted to the admittance plane.

The characteristic parameters required for the loss-of-excitation protection can be directly obtained from Figure 10.25. The static stability limit given in the figure, generally expressed as two straight lines, must be monitored. Besides, there is a threshold for the dynamic stability limit. In the deeply underexcited operating area (the left side of Characteristic 3 in Figure 10.27), it is no longer possible for the generator to enter synchronization. So, an immediate trip is required at this point. This is different from the case where the static stability limit (Characteristic 1 and 2 in Figure 10.27) is exceeded. In the latter, as long as the excitation voltage increases, it is entirely possible to pull the generator back to synchronous operation. So, the decrease in the excitation voltage U_{ex} can be taken as an auxiliary criterion for the loss-of-excitation protection. This auxiliary criterion controls the trip time for Characteristics 1 and 2. In this way, after static stable working conditions for the generator are restored, in the event of a transient ride-through caused by the dynamic pulse signal's entry into the static stability limit, any fault action of the loss-of-excitation protection can be effectively prevented.

The settings of Characteristics 1, 2, and 3 are as follows.
For a non-salient pole generator:
Characteristic 1:

$$\lambda_1 = \frac{1}{X_{d1}} = 1.05 \frac{1}{X_d}; \alpha_1 = 60° \sim 80°$$

Characteristic 2:

$$\lambda_2 = 0.9\lambda_1; \alpha_2 = 90°$$

Characteristic 3:

$$\lambda_3 = \frac{1}{X_d} \sim \frac{1}{X'_d}$$

where, λ_3 must be greater than 1; and $\alpha_3 = 80°–110°$ (generally 110° can be taken).
For a salient pole generator:
Characteristic 1:

$$\lambda_1 \approx 1.05 \left[\frac{1}{X_d} + 0.5 \left(\frac{1}{X_d} - \frac{1}{X_q} \right) \right], \quad \alpha_1 = 60° \sim 80°$$

Characteristic 2:

$$\lambda_2 = 0.9\lambda_1; \quad \alpha_2 = 90°$$

Characteristic 3:

$$\lambda_3 = \frac{1}{X_d} \sim \frac{1}{X'_d}$$

where λ_3 must be greater than 1; and $\alpha_3 = 80°$ to $-110°$ (generally 110° can be taken).

For the settings of these loss-of-excitation protection curves, the datum points and the inclination angles of the straight lines can be determined only on the basis of the intersections of the setting characteristic lines and Axis B. The set loss-of-excitation protection characteristic curves should be as close as possible to the given generator stability limit characteristic curve. Then, the characteristics of the excitation circuit control device are considered.

The actions characteristics of loss-of-excitation protection can be seen from Figure 10.27:

(1) Beyond Characteristics 1 and 2. The excitation voltage monitor is not activated. In this case, an alarm must be issued. A long delay (10 s or so) should be set for the trip order.
(2) Beyond Characteristics 1 and 2. The excitation voltage monitor is activated. In this case, a short delay (0.5–1.5 s) should be set for the trip order.
(3) Beyond Characteristic 3. In this case, a short delay (shorter than 0.3 s) or no delay can be set for the trip order.

10.4.3 Comparison of Impedance and Admittance Measurement

In order to obtain the valued parameter of the loss-of-excitation protection impedance principle, the generator operating limit diagram must be transformed (mapped) to the impedance plane. The reciprocal of the trajectory is adopted for the mathematical transformation. But the result of this transformation cannot be intuitively compared with the operating limit diagram. The reason is that, according to the trajectory theory, when a straight line that does not pass through the origin passes through the reciprocal and transforms itself into a circle, it is tangent to the origin. When the theoretical stability limit is transformed to the impedance plane, a circle is obtained. The transformation mechanism is shown in Figure 10.28. All the points on the left side of the stability limit in the operating limit diagram are located in the semicircle (the shaded area) in the impedance plane after being transformed.

10 Static Self-Excitation System

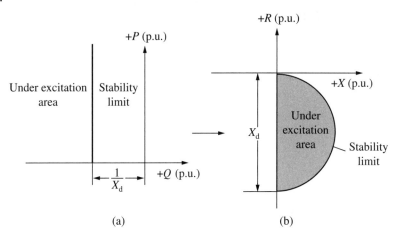

Figure 10.28 Transformation to impedance plane. (a) P-Q plane; (b) R-X plane.

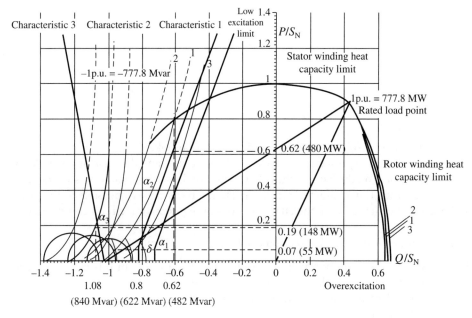

Figure 10.29 Coordination between loss-of-excitation protection and low excitation limit for VGS 777.8 MW hydro-turbine at Three Gorges Hydro Power Station: Characteristic 1 – static stability limit, $\alpha_1 = 70°$; Characteristic 2 – theoretical static stability limit, $\alpha_2 = 90°$; Characteristic 3 – dynamic stability limit, $\alpha_3 = 100°$. 1 – $U_G = 1.0$, $I_G = 1.0$, $\cos\varphi = 0.9$; 2 – $U_G = 1.05$, $I_G = 0.95$, $\cos\varphi = 0.9$; 3 – $U_G = 0.95$, $I_G = 1.05$, $\cos\varphi = 0.9$; rated capacity: $S_N = 777.8$MVA, $U_{GN} = 20$ kV, $\cos\varphi_N = 0.9$, $N = 75$ r min^{-1}.

The above paragraphs describe the setting of the admittance loss-of-excitation protection and the principle of its coordination with the generator operating limit diagram and underscore that the loss-of-excitation protection characteristics set in accordance with the admittance principle can be very close to the generator operating limit diagram and the results of the setting characteristics can be intuitively assessed between the two. As an example, the setting of the loss-of-excitation protection and the low excitation limit of a VGS 777.8 MW hydro-turbine at Three Gorges Hydro Power Station is shown in Figure 10.29.

Characteristic 1: static stability limit, $\alpha_1 = 70°$, $U_G = 1.0$, $I_G = 1.0$, $\cos\varphi = 0.9$.
Characteristic 2: theoretical static stability limit, $\alpha_2 = 90°$, $U_G = 1.05$, $I_G = 0.95$, $\cos\varphi = 0.9$.
Characteristic 3: dynamic stability limit, $\alpha_3 = 100°$, $U_G = 0.95$, $I_G = 1.05$, $\cos\varphi = 0.9$.
Rated capacity: $S_N = 777.8$MVA, $U_{GN} = 20$ kV, $\cos\varphi_N = 0.9$, $N = 75$ r min^{-1}.

10.5 Electric Braking of Steam Turbine [24]

In recent years, an array of 700 MW large hydro-turbines have been put into operation at such hydropower stations as Three Gorges, Longtan, Laxiwa, and Xiaowan. These hydro-turbines have extremely great rotor rotational inertia. For example, the rotational inertia GD^2 of the generator at Three Gorges Hydro Power Station is $GD^2 \geqslant 450,000 \text{t} \cdot \text{m}^2$. Therefore, braking has become even more important. Moreover, since hydro-turbines generally play such roles as frequency modulator, peak shaver, and standby for emergency in grids, they start and stop frequently. The conventional mechanical braking has far from been able to meet the requirements of the current control and operation modes. So, large hydro-turbines' need for electric braking has become more prominent. At present, all the hydro-turbines at Three Gorges Hydro Power Station have been equipped with an electric braking system.

The conventional electric braking technology virtually uses the diode rectifier to provide generator excitation current. This mode is now rarely used.

The flexible electric braking which is widely used at present uses the regulator and the SCR of the generator excitation system to provide a controllable braking excitation current at the time of the braking. It also regulates the braking current on the basis of the water head, the speed, and the water leakage of the guide blade at the time of the braking to change the braking characteristics, so that the entire control process can be smooth and capable of being regulated.

10.5.1 Selection of Braking System

When a vertical shaft hydro-turbine stops, the small amount of leakage in the hydro-turbine and the mechanical shaft inertia will cause a residual rotation torque. If the hydro-turbine is not equipped with a braking system and it stops only by virtue of the wind resistance and the bearing friction, the process is long.

Generally, when the steam turbine is operating under the following conditions, a fast braking mode is required:

(1) It starts and stops many times within 24 h (e.g., the turbine at a pumped-storage power station or peak shaving power station).
(2) It has reliable bearings but has no automatic hydraulic control system, so that the bearing system is likely to operate under non-optimal lubrication conditions.

There are two braking modes, including mechanical braking and electric braking.

10.5.1.1 Mechanical Braking
Mechanical braking is based on a symmetrical circular brake, which directly acts on the rotor or a separate brake ring. Wind resistance and bearing friction can also play the role of natural brake at high speed. In order to avoid too large braking equipment, mechanical braking is generally applied when the speed decreases to 30%–25% of the rated speed. When no fast braking is needed, mechanical braking starts generally at 10% of the rated speed, so that brake shoe saving can be achieved. Brake cylinder control and drive can be pneumatic or hydraulic. This system acts on a brake shoe made of a non-metallic material with a high coefficient of friction, creating a high coefficient of friction on the rotor or the brake ring. The brake surface requirements depend on the pressure of the braking system, the inertia of the mechanical part relative to the shaft, the maximum temperature rise allowed on the rotor or the brake ring, and the required braking time.

10.5.1.2 Electric Braking
After the stop operation is executed, the braking of the generator is achieved with a method in which the stator winding three-phase short circuit is constantly excited in the rotor excitation winding to generate an electric braking torque.

Another possible electric braking method uses a two-phase short circuit. In this case, the reverse magnetic field additional loss is conducive to the increase of the braking torque. The advantages of this method are as follows:

(1) In order to generate the same braking torque as the three-phase short circuit, the electric braking system is required to provide a certain value excitation current.
(2) A higher additional loss can appear at 2/3 of the copper loss. The braking time is short. The additional loss depends on the characteristics of the generator. It can reach 10 times the additional loss caused by the three-phase short circuit. In this method, it should be noted that there will be a great swing torque in the braking process, which will generate a strong stress on the shaft and other connections.

There are other electric braking methods:

(1) A resistance is connected to the outside of the stator.
(2) A DC current is provided for the stator winding.
(3) A static frequency converter (SFC) is connected to the stator.

The three methods are generally used for pumped-storage power stations for high-priority rapid starts and stops.

For a hydropower station that requires its generators to undergo 2–3 starts and stops within 24 h, the best braking method is the stator winding three-phase short circuit. The result is that both mechanical stress and stator winding overheating are small.

In the realization of electric braking, since a large generator generally has a static self-excitation system powered by the generator terminal, the method used by the rotor winding to provide an excitation current must be studied. Figures 10.30 and 10.31 provide two possible methods. The method shown in Figure 10.30 uses a low- or medium-voltage power supply to supply power to the thyristor rectifier bridge of the excitation system. A three-phase change-over switch can be provided in the excitation cabinet. The excitation current of the system can be controlled with the current control mode of the automatic voltage regulator (AVR). The circuit can also be used for commissioning and testing, such as the setting of generator characteristic determination or relay protection. One power transformer can be shared by two adjacent generators, since generally only one of them brakes at a time.

Figure 10.31 shows another method that uses a field flashing circuit to provide the excitation current required for the braking process. In this method, the DC current in the braking process depends on the voltage of the AC auxiliary power supply and the temperature of the rotor winding.

Generally, for electric braking, the excitation current provided is equal to the excitation current when the generator is unloaded.

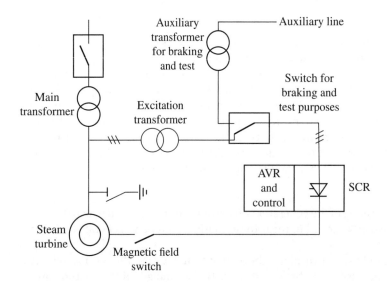

Figure 10.30 Electric braking circuit with SCR.

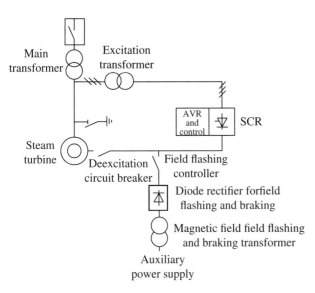

Figure 10.31 Field flashing circuit used for electric braking.

10.5.2 Basic Expression for Electric Braking

When the generator is disengaged, the stator winding is short-circuited in three phases and a constant DC current is input into the rotor winding. At this point, a short-circuit current will be generated in the stator winding. The current generates a copper damage braking torque in the stator winding, causing the generator's retarding braking and stop. It should be emphasized that, in the electric braking process, the short-circuit current (I_k) in the stator winding is a constant value which does not change with a decrease in the speed of the generator. It can be determined from the theory of synchronous motors that the excitation current flowing through the rotor winding is relatively constant in the electric braking process. So, it can be considered that the fundamental flux per pole (Φ) is also constant.

The stator-winding-induced internal emf and the vertical-axis synchronous reactance are as follows, respectively,

$$E = 4.44\ fW\Phi K_f = \frac{4.44}{60}PW\Phi K_f n = K_E n \tag{10.49}$$

$$X_d = \omega L = 2\pi f L = \frac{2\pi P n}{60}L = K_d n \tag{10.50}$$

where

- E – stator-winding-induced internal emf;
- K_f – waveform factor;
- n – speed of generator;
- P – number of pole pairs of generator;
- K_E – voltage coefficient, $K_E = \frac{4.44}{60}PW\Phi K_f$;
- X_d – vertical-axis synchronous reactance;
- K_d – armature reaction coefficient, $K_d = \frac{2\pi PL}{60}$.

The short-circuit current I_k is as follows:

$$I_k = \frac{E}{\sqrt{X_d^2 + R^2}} = \frac{K_E n}{\sqrt{(K_d n)^2 + R^2}} \tag{10.51}$$

With Eq. (10.51), when the stator winding resistance R is neglected, the following can be obtained:

$$I_k = \frac{K_E n}{K_d n} = \frac{K_E}{K_d} = \text{constant} \tag{10.52}$$

As can be seen from Eq. (10.52), in the braking process, theoretically the stator short-circuit current (I_k) is constant, which does not change with the speed (n).

The following is the expression for the electric braking torque (M_E):

$$M_E = \frac{P_E}{\omega} = \frac{3I_k^2 R}{2\pi f} \tag{10.53}$$

Substituting Equation (10.51) into Eq. (10.53) and since $f = \frac{nP}{60}$, the following can be obtained:

$$M_E = \frac{3I_k^2 R}{2\pi f} = K_m \frac{n}{K_d^2 n^2 + R^2} \tag{10.54}$$

$$K_m = \frac{90 R K_E}{\pi P}$$

In order to obtain the speed at which the maximum braking torque appears, the speed (n) in Eq. (10.54) can be differentiated, and the derivative can be made equal to zero as follows:

$$n = \frac{R}{K_d} \tag{10.55}$$

Equation (10.55) shows that the maximum electromagnetic torque appears when $n = \frac{R}{K_d}$. When the speed is zero, the electromagnetic torque is zero as well. When the generator applies electric braking, the core magnetic flux is not saturated. Meanwhile, the fact that the braking time is long, much greater than the generator's transient and super-transient time constants, is considered. Its working characteristics are similar to those of a double-winding transformer in the steady state. The main reaction involves copper loss instead of iron loss of the excitation reactance. This is similar to the short-circuit rising current test characteristics of a normal generator.

In the electric braking system, the following conclusions can be drawn from Eq. (10.54). Since I_k and R are constant, the electric braking torque increases with the decrease in the generator speed. It can be seen that the characteristics of electric braking have a unique effect on a low-speed stop. The electromagnetic torque is directly proportional to the square of the stator short-circuit current and inversely proportional to the generator speed. So, increasing the stator short-circuit current is very effective in reducing downtime. The key to the stop process is the steepness of the decrease of the generator speed in the low-speed area. As long as the electric braking current is equal to or even greater than the rated current of the stator, electric braking can achieve a satisfactory speed decrease rate in the low-speed area.

10.5.3 Selection of Braking Transformer

In the design of an electric braking system, an important issue that should be considered is the need to apply a braking transformer, though the required DC excitation current can be obtained through the SCR with the 380 V station service power in the absence of a braking transformer. The reason is that, once the thyristor is out of control, an overcurrent that compromises the generator safety will be generated in the stator winding circuit.

The rated line voltage on the secondary side of the braking transformer is calculated as follows:

$$\frac{3\sqrt{2}}{\pi} U_2 \cos \alpha_{\min} = K_i I_f R_f + \Sigma \Delta U \tag{10.56}$$

where

- U_2 — rated line voltage on secondary side of braking transformer;
- α_{min} — minimum control angle outputted by excitation regulator (10°–15°);
- K_i — short-circuit current overload factor;
- I_f — rotor current value when stator winding short-circuit current is rated;
- R_f — rotor winding resistance;
- $\sum\Delta U$ — forward voltage drop, lateral voltage drop, line resistance voltage drop, and slip ring carbon brush voltage drop of elements including two arms of thyristor conduction (4 V in calculations).

Firstly, the short-circuit current overload factor is determined. China's regulations on technical conditions for hydro-turbines stipulate that a hydro-turbine shall be able to withstand a short-time overcurrent 1.5 times the rated current at the rated load for 2 min without damage. Generally applied in the case of a stator short circuit, electric braking generally lasts less than 10 min. In accordance with the stator heat capacity curve, theoretically, in the case of stator electric braking, a stator current 1.0–1.3 times the rated current is rational, and K_i is 1.0–1.3 correspondingly. At present, there is no relevant standard for electric braking time. So, rated current is generally applied in projects.

Second, the rated excitation voltage/current can be considered as the rotor winding resistance. The line current on the secondary side of the braking transformer is obtained as follows:

$$I_2 = 1.1 \times \sqrt{\frac{2}{3}} I_k \tag{10.57}$$

The rated capacity of the braking transformer is calculated as follows:

$$S = \sqrt{3} U_2 I_2 \tag{10.58}$$

Since the entire braking process takes less than the 10 min short-time duty, 1/2 of the capacity can be taken. The design capacity is 1/2S. The curve provided by the transformer manufacturer can be referred to. As can be seen from the capacity characteristic curve provided by the transformer manufacturer, in the case of zero initial load, the transformer can be overloaded at twice the rated capacity for a short term of 20 min. A general hydropower station has multiple generators, which vary in downtime. Under this condition, sharing by two to three generators' of one braking transformer can be considered.

10.5.4 Key Points of Design of Electric Braking Circuit

Electric braking lasts long, generally a few minutes, much longer than the motor transient process. So, in the case of variable speed, the stator and rotor currents still maintain a constant proportional relation. But in the start and stop process, they no longer have a linear proportional relation.

It has been demonstrated that electric braking by itself can pull the speed to 0.5%–1% of the rated speed point to enable stop, without the need for mechanical braking. However, since mechanical braking acts as a standby when electric braking fails, it is necessary to design a hybrid (electric and mechanical) braking mode. Air braking can be utilized from time to time to test the working state of mechanical braking.

In the electric braking process, the rotor constant excitation current control method is feasible to keep the stator current constant.

In the design of a braking transformer, it is necessary to keep the stator current smaller than the upper limit of the specified stator heat capacity in consideration of the loss of control.

10.6 Electric Braking Application Example at Pumped-Storage Power Station [25]

The pumped-storage units at Tongbai Pumped Storage Power Station are taken as an example of the application of electric braking. The power station has four 300 MW vertical shaft single-stage mixed flow reversible pump turbine-generator motor sets with a total installed capacity of 1200 MW.

The generator motor-transformer combination features unit connection, which is equipped with a generator circuit breaker and reversing disconnector. On the 500 kV side, two sets of generator-transformer unit connections are connected into a combined unit to form a two-inlet and two-outlet internal bridge connection.

Playing the roles of peak and valley load regulator, frequency and phase modulator, and standby for emergency, the power station, connected to Zhuji Substation via a two-circuit 500 kV line, accesses the East China power system at 500 kV voltage. The electric braking connection of the power station is shown in Figure 10.32.

10.6.1 Composition of Electric Braking System

An electric braking system consists of a stator three-phase short-circuit switch, a logic control device, and a braking excitation device. The logic control device and the braking excitation device are provided by the excitation system. In the event of electric braking, the excitation source is taken from the excitation transformer. The basic connection diagram is shown in Figure 10.32. The relevant parameters are shown in Table 10.4–10.6.

10.6.2 Control of Electric Braking System

10.6.2.1 Setting of Control Logic of Monitoring System

When a generator stop or condition transformation order is sent, the monitoring system judges that the generator is in transition from power generation or pumping to the idle state (a special transition process for pumped-storage units) and the generator is not operating in the line charging mode before the transformation. When the generator speed decreases to 50% of the rated speed, it checks if the guide blades are in the closed position and if there is no protection latching and electric braking put-in signal. Then, it sends an electric braking put-in order.

The protection latching and electric braking put-in signal are generated in the generator protection action electric stop, the main transformer protection action stop, the main transformer protection action adjacent generator stop, the excitation system trip, or an SFC trip during the generator SFC drag.

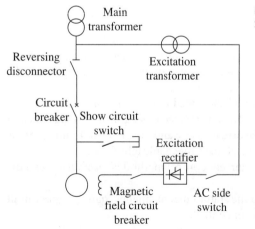

Figure 10.32 Electric braking connection diagram of Tongbai Pumped Storage Power Station.

Table 10.4 Key parameters of generator motor.

Parameter	Working condition of generator	Working condition of pump
Rated power	334 MVA	336 MW
Rated power factor	0.9	0.975
Rated voltage	18 kV	
Rated current	10 713.1 A	
Rated speed	300 r min^{-1}	
Rated excitation voltage	258 V DC	
Rated excitation current	1 773 A DC	
Rated no-load excitation voltage	104 V DC	
Rated no-load excitation current	471 A DC	
Rotational inertia	2 775 000 kg m^2	

Table 10.5 Key parameters of excitation transformer.

Rated capacity	3 × 510 kVA	Rated current	53.69/1757.2 A
Rated voltage	18 000/550 V	Connection set	Yd5

Table 10.6 Key parameters of short-circuit switch.

Model	SDCEM Type SB250
Rated voltage	24 kV
Rated current	12 500 A (for 10 min)
Breaking/closing time	≤10 s.
Operating mechanism	Type MP180; three-phase linkage

10.6.2.2 Excitation System Electric Braking Execution Procedure

After the electric braking put-in order sent by the monitoring system is received, the excitation initiates the program operation. The short-circuit switch, the AC-side switch, and the magnetic field circuit breaker are closed, and the excitation is put in.

When the speed decreases to 1% of the rated speed, the excitation stops the program operation. The magnetic circuit breaker, the AC-side switch, and the branch switch are opened, and the excitation terminates.

When electric braking is put into operation, the excitation regulator is operating in the excitation current regulation mode. The actual excitation current can only be regulated within the allowable range set in the debugging. In fact, the excitation current is set in a way where the braking current is 1.1 times the rated stator current in the event of electric braking.

In the event of an electric braking time-out, the excitation system will send a braking time-out trip signal, and the electric braking execution program will terminate.

10.6.2.3 Short-Circuit Switch Hard Wiring Latching

When the short-circuit switch opening/closing operation is executed, in order to ensure safety, the hard wiring circuit is provided with the corresponding latching. That is, the opening/closing operation is allowed only when all the criteria, including opened generator circuit breaker, zero generator terminal voltage, and opened magnetic field circuit breaker, are met.

10.6.2.4 Protection Latching at Time of Electric Braking Put-In

When the electric braking is put in, the setting of the short-circuit point and the electric variable changes in the braking process may cause faulty actions by the generation protection system. So, the relevant protections should be latched when electric braking is put in. At Tongbai Pumped Storage Power Station, the relevant protections that are latched when electric braking is put in include negative-sequence current protection, loss-of-excitation protection, reverse power protection, low-impedance protection, stator ground protection, large differential protection, and low-frequency overcurrent protection.

Taking into consideration the fact that electric braking put-in is not allowed in the event of an electric braking equipment failure or a generator internal electrical failure, Tongbai Pumped Storage Power Station adopts a hybrid braking mode for normal stop. In this mode, electric braking and mechanical braking are used in combination. When the generator speed decreases to 50% of the rated speed, the electric braking system is put into operation. When the speed decreases to 5% of the rated speed, the mechanical braking system is put into operation.

Pump storage units' transformation from power generation and pumping to stop is long due great inertia of rotation, which will delay their transition to another working condition or entry into standby state. Electric braking can effectively shorten generators' deceleration time, thus meeting the requirements of quick condition switch of generators.

11

Automatic Excitation Regulator

11.1 Overview

Since the 1960s, analog excitation regulators have been dominant in applications, their functions by and large being able to meet large synchronous generators' requirements for excitation control.

However, analog excitation regulators have many shortcomings. In particular, they face many difficulties in the realization of self-test and hardware modification. So, it is necessary to provide a variety of special functional components to meet different control requirements.

The dilemma did not come to an end until the mid-1980s. With the rapid development of digital microprocessor technology, conventional excitation regulators based on analog technology have been undergoing a gradual transformation to digitization.

As microprocessor technology has been widely used in all industrial fields, many functions which were realized with a lot of hardware in the past can be integrated in one chip. This microprocessor-based device has undergone great improvements in operational speed and functionality.

Meanwhile, as electronics manufacturers from various countries have reached agreements on regulations on standard buses that have already been open (e.g., VME bus, multi-purpose bus, STD bus, and PCI bus), international standards for the fields have been formed. This has further promoted the possibility of electronics manufacturers' development of highly integrated microprocessing card integrating multiple components (e.g., memory, timer, serial/parallel/Ethernet ports, and fieldbus management) and various input/output I/O interfaces compliant with international standards.

Moreover, great improvements have been seen in the field of software. With professional software companies' support and investment, a strong development environment has been created. This environment includes debugger, software analyzer, I/O interface, graphics, and databases, making program development easier and less dependent on hardware.

In this context, projects have been increasingly tending to implement modern excitation system control and protection functions with digital electronic technology. It should be emphasized that such digital excitation systems or automatic excitation regulators are not just digital changes in analog devices but also providers of more sophisticated and complex control functions. Besides, the application of digital control in excitation systems is not a vision proposed today; it can be traced back to some dissertations in the late 1970s. In recent years, with the popularization of digital technology and the rapid development of digital control technology, it has been technically possible to implement digital control excitation systems. In addition, excellent cost performance and high reliability have laid a favorable foundation for digital control excitation systems. Table 11.1 shows a comparison of features between digital and analog automatic voltage regulators (AVRs).

Design and Application of Modern Synchronous Generator Excitation Systems, First Edition. Jicheng Li.
© 2019 China Electric Power Press. Published 2019 by John Wiley & Sons Singapore Pte. Ltd.

Table 11.1 Comparison of features between digital and analog AVRs.

Performance	Digital	Analog	Remark
Composition	Hardware system consisting of hardware with microprocessor as core and power component. Measurement and operation digitally controlled by software	Consisting of IC analog control component and power component	
Functionality	High functionality realized through software extension. AVR functions integrated or decentralized	Multi-functionality realized through hardware addition. AVR features and other functions decentralized	Other functions, such as electronic speed control and program control
Reliability	Characteristics of digital operation loop durable. Easy to achieve multiple control and high reliability (able to detect faults and automatic switch with self-diagnosis function). Multiplicity avoiding function abortion except for anomaly of whole system	Necessary to periodically check characteristics of test operation component. Multiplicity subject to complex control loop (fault self-diagnosis complex). Aborting corresponding functions of fault component in operation loop	
Maintainability	Easy setting of tracking and self-diagnosis functions at specific points	Necessary to test characteristics of each component to determine operating points	Data held in memory and recalled in case of failure of digital AVR tracking function
Operability	Easy to improve control function through software's processing of set values	Hard improvement in control function, requiring changes in hardware and set values	

11.2 Theoretical Basis of Digital Control [25]

11.2.1 Digital Discrete Technology

When digital control technology is used to control a continuous system, the information of the continuously controlled system must firstly be sampled at discrete instantaneous equal intervals. So, the control signal is a discontinuous constant, and it only changes at the discrete point of time. Thus, the discrete model equivalent to the continuous system can be obtained. This method is called the discrete similarity method. The discrete similarity method is different from the classical numerical integration method applied in continuous systems, that is, the numerical integration solving method that approximately describes an integrand with an interpolation polynomial.

The discrete similarity method can be used to simulate the continuous system. Finally, the difference equation describing the discrete controller can be derived. Thus, computer programming can be achieved. The solution obtained from the difference equation is the sampled value of the motion of the discrete controller. Although it is different from the sampled value of the motion of the continuous system controller, the solution of the difference equation can be considered to be the sampled value of the continuous system, as long as the discrete system and the continuous system are similar. If the discrete similarity method of the applied continuous system is correct, the solution obtained from the difference equation listed in the discrete system will be able to simulate the motion of the continuous system.

In discrete system study, the Z-transform method can be used for a mathematical description of the sampling process.

The simulation process of a digital continuous system is as follows: (1) the system is discretized with a sampler and a holder; (2) the pulse function and the difference equation of the system are obtained with the Z-transform method; and (3) the difference equation is solved, and digital control is applied.

11.2.2 Z-Transform

The role of the Z-transform in a discrete system is very similar to that of the Laplace transform in a continuous system. The continuous signals should be discretely sampled, so that the continuous system can be discretized.

First, the role of the sampler S and that of the holder H in the discrete system are discussed. S closes every T seconds to let the input signal pass and converts the continuous signals u(t) into a series of pulse signals $u^*(t)$ occurring at the instant of the sampling 0, T, $2T$, …. T is the sampling period. The output of the sampler can be expressed as follows:

$$U^*(t) = \delta(t)u(t)$$

where

$\delta(t)$ – a series of unit pulses.

Most of the time functions discussed here are equal to zero when $t < 0$. So,

$$\delta_T = \sum_{k=0}^{\infty} \delta(t - kT)$$

$$u^*(t) = \sum_{k=0}^{\infty} u(t)\delta(t - kT)$$

$$u^*(t) = \sum_{k=0}^{\infty} u(kT)\delta(t - kT)$$

The holder converts the sampled signal into continuous signals $u_h(t)$. These continuous signals approximately reproduce the signals acting on the sampler. The simplest holder shown in Figure 11.1 converts the sampled signals into a signal that remain constant between two consecutive sampling instants. The expression is as follows:

$$u_h(kT + t) = u(kT) \quad (0 < t < T) \tag{11.1}$$

The Z-transform is discussed below.

The Laplace transform of the sampling sequence $u^*(t)$ is as follows:

$$u^*(s) = L[u^*(t)] = \sum_{k=0}^{\infty} u(kT)e^{-kT}$$

It is assumed that $e^{Ts} = z$ or $s = T^{-1} \ln z$. $u^*(s)$ is written as $u(z)$. The following can be obtained:

$$u(z) = u^*(s) = u^*(T^{-1} \ln z) = \sum_{k=0}^{\infty} u(kT)z^{-k}$$

So, $u(z)$ is called the Z-transform of $u^*(t)$ and is expressed as $Z[u^*(t)]$. Since only the signal value at the sampling instant is considered in the Z-transform, the Z-transform of $u(t)$ has the same result as that of $u^*(t)$. That is,

$$Z[u(t)] = Z[u^*(t)] = u(z) = \sum_{k=0}^{\infty} u(kT)z^{-k} \tag{11.2}$$

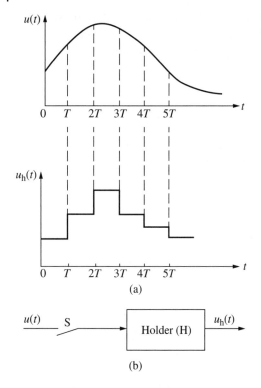

Figure 11.1 Functional processes of sampler and holder: (a) sampling discretization, (b) block diagram of sampling and holding.

For example, the Z-transform of the unit step function $1(t)$ is as follows:

$$Z[1(t)] = \sum_{k=0}^{\infty} l(kT)z^{-k} = 1 + z^{-1} + z^{-2} + \cdots = \frac{z}{z-1} \tag{11.3}$$

Equation (11.3) is the summation of the series.

11.2.3 Discretization of Continuous System

The steps to discretize a continuous system with a transfer function $G(s)$ are as follows:

(1) First, the transfer function $G(s)$ of the continuous system is given, as shown in Figure 11.2a. It is assumed that the input signal is $u(t)$. When this signal is discretized and sampled, a sampling switch S1 can be provided at the system input as shown in Figure 11.2b. The smaller the sampling period T, the closer the discrete system will be to the continuous system.
(2) The continuous signal $u(t)$ is converted into a series of discrete signals $u(kT)$ whose period is T, which can usually be abbreviated as $u(k)$, after passing through the sampling switch. To ensure model equivalence, system equivalence should be required first. So, $u(k)$ cannot directly enter the original continuous system. A holder must be added, so that the sampled value can be restored to the continuous signal. Here, the concern is the mathematical model of signal reproduction, rather than the reproduction of the original signal.
(3) A sampling switch S2 should be added at the output end of the system, as shown in Figure 11.2c. It should be synchronized with the sampling switch S1 at the input end. Then, $y(t)$ is changed to $y(k)$.
(4) The Z-transform is taken for the discrete input and output signals u(k) and y(k), respectively. Thus, $u(z)$ and $y(z)$ are obtained. Meanwhile, $G(z) = \frac{y(z)}{u(z)}$. $G(z)$ is a discrete model equivalent to the original system.

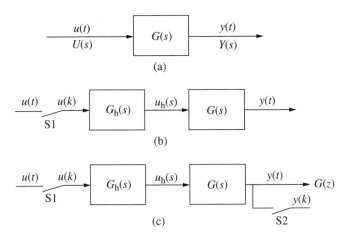

Figure 11.2 Discretization of continuous system: (a) block diagram of continuous system; (b) discretization of input signal and signal holding; (c) discretization of output signal.

A Z-transform is taken for $G(z)$. Thus, a difference equation for operations on a digital computer can be obtained. From the Z-transform principle, for the discrete similarity model shown in Figure 11.2, the Z-transform transfer function $G(z)$ is as follows:

$$G(z) = Z[G_h(s)G(s)] \tag{11.4}$$

The right-hand side of Eq. (11.4) shows that the Z-transform is taken for the functions $G_h(s)$ and $G(s)$. A appropriate sampling period T should be chosen in the Z-transform, so that the function $G(z)$ can accurately express $G(s)$. The shorter the period, the closer the discrete signal will be to the continuous value. Meanwhile, the sampling period T is required to satisfy the required value of the sampling theorem, so that the discrete signal can hold all the information of the continuous signal. That is,

$$T \leqslant \frac{\pi}{\omega_m} \tag{11.5}$$

where

ω_m — maximum angular frequency of the continuous system.

In addition, an appropriate holder should be chosen, so that the recovered discrete signal is as close as possible to the original signal.

11.2.4 Transfer Function of Holder

The following are characteristics of several commonly used holders:

(1) Zero-order holder. The mathematical expression of a zero-order holder is as follows:

$$u_h(t) = u(kT) \quad [kT \leqslant t < (k+1)T]$$

The characteristics of the zero-order holder are shown in Figure 11.3.

As can be seen from Figure 11.3, the zero-order holder can keep $u(kT)$ constant for the time interval of $KT \sim (K+1)T$. If $u(kT)$ is regarded as an ideal unit step pulse input with a constant amplitude of the zero-order holder, its output is a square wave with an amplitude of $u(kT)$ for a duration of T. Its form is shown in Figure 11.4.

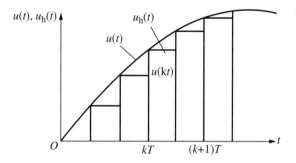

Figure 11.3 Characteristics of zero-order holder.

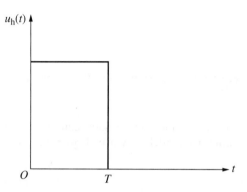

Figure 11.4 Pulse transition function of zero-order holder.

The expression of the pulse transition function of the zero-order holder is

$$g_h(t) = \begin{cases} 1 & (1 \leq t \leq T) \\ 0 & (t < 0) \end{cases}$$

(11.6)

With the Laplace transform, the transfer function of the holder can be obtained as

$$L[g_h(t)] = L[g_1(t)] + L[g_2(t)]$$

where

$$g_1(t) = \begin{cases} 1 & (t \geq 0) \\ 0 & (t < 0) \end{cases}$$

(11.7)

$$g_2(t) = \begin{cases} 1 & (t \geq T) \\ 0 & (0 \leq t \leq T) \end{cases}$$

(11.8)

With the Laplace transform, the following equation can be obtained:

$$L[g_1(t)] = \frac{1}{s}$$

$$L[g_2(t)] = \frac{e^{-sT}}{s}$$

$$L[g_h(t)] = \frac{1}{s} - \frac{e^{-sT}}{s} = \frac{1 - e^{-sT}}{s} = G_h(s)$$

(11.9)

As can be seen from Figure 11.3, the signal $u_h(t)$ recovered by the zero-order holder is significantly different from the original signal $u(t)$. Obviously, when T is greater, the difference is greater. However, when $u(t)$ is a unit pulse signal, the signal recovered by the zero-order holder will be distortion-less, which is the characteristic of the zero-order holder.

(2) **First-order holder.** The mathematical expression of a first-order holder is as follows.

$$u_h(t) = u(kT) + \frac{u(kT) - u[(k-1)T]}{T} \times (t - kT) \tag{11.10}$$

The characteristics of the first-order holder are shown in Figure 11.5. With a similar analytical method, the transfer function of the first-order holder can be obtained as follows:

$$G_h(s) = T(1 + sT)\left(\frac{1 - e^{-sT}}{sT}\right)^2 = \frac{1 + sT}{T} \times \left(\frac{1 - e^{-sT}}{s}\right)^2 \tag{11.11}$$

As can be seen from Figure 11.5, the signal $u_h(t)$ recovered by the first-order holder is very different from the original signal $u(t)$. However, if the waveform of the input signal $u(t)$ is a constant-slope ramp signal, the signal recovered by the first-order holder will not be distorted.

(3) **Triangle holder.** The mathematical expression of a triangular holder is

$$u_h(t) = u(kT) + \frac{u[(k + 1 + T] - u(kT)}{T} \times (t - kT)[kT \leq t < (k+1)T] \tag{11.12}$$

The triangular holder's characteristics are shown in Figure 11.6, and its pulse transition function is shown in Figure 11.7. The holder is called a triangular holder because its pulse transition function is triangular. Its transfer function is as follows:

$$G_h(s) = \frac{1 - e^{-sT}}{sT^2} \times e^{sT} \tag{11.13}$$

As can be seen from Figure 11.6, the output signal $u_h(t)$ of the triangular holder is very close to the original signal $u(t)$, and it can recover the original signal well. However, it can be seen from Eq. (11.12) that KT and $(K + 1)T$ must be known to calculate $u_h(t)$. But this condition cannot be met in some cases. In view of this, the predictor–corrector method is adopted. That is, $u_h(K + 1)T$ is predicted with any other holder, a zero-order holder, for example and then the obtained value is substituted into Eq. (11.13), so that a more accurate $u_h(t)$ can be obtained.

Figure 11.5 Characteristics of first-order holder.

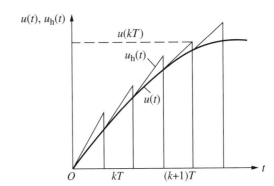

Figure 11.6 Characteristics of triangular holder.

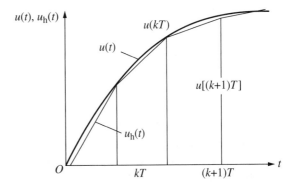

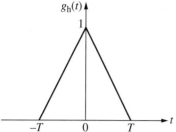

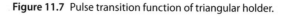

Figure 11.7 Pulse transition function of triangular holder.

11.2.5 Discrete Similarity Model and Bilinear Transformation

(1) Discrete similarity model of integration element. $G(s) = s^{-1}$ of the integration element is discretely sampled and can form a discrete similarity model after passing through the holder:

$$G(z) = Z[G_h(s)G(s)] = \frac{Y(z)}{u(z)} \tag{11.14}$$

where

$G_h(s)$ — transfer function of holder.

If a zero-order holder is used, Eq. (11.14) can be written as follows:

$$G(z) = Z\left[\frac{1-e^{-sT}}{s} \times \frac{1}{s}\right] = (1-z^{-1})Z \times \frac{1}{s^2} = (1-z^{-1}) \times \frac{zT}{(z-1)^2} = \frac{T}{z-1}$$

Through an inverse transformation of $G(z)$, the following difference equation can be obtained:

$$y_{k+1} = y_k + Tu_k \tag{11.15}$$

This difference equation is the same as the numerical calculation formula of the Euler method, because $t_{k+1} - t_k = T$ and the right-hand function of the differential equation of the integration element is u, that is, $f(y_k, t_k) = u_k$.

If a triangular holder is used, Eq. (11.14) can be written as follows:

$$G(z) = Z\left[\frac{(1-e^{-sT})}{sT^2}e^{sT} \times \frac{1}{s}\right] = \frac{T(z+1)}{2(z-1)} \tag{11.16}$$

Through an inverse Z-transform, the following difference equation can be obtained:

$$y_{k+1} = y_k + \frac{T}{2}(u_{k+1} + u_k) \tag{11.17}$$

Equation (11.17) is the numerical integration formula of the implicit trapezoidal method.

(2) Bilinear transformation. For a continuous system expressed as a transfer function, based on the transformation formula of the triangular holder of the integration element described in the above section, s in the transfer function can be replaced by the right-hand term of Eq. (11.16). That is,

$$\frac{1}{s} \rightarrow \frac{T(z+1)}{s(z-1)}$$

Or

$$s \rightarrow \frac{2(z-1)}{T(z+1)} = \frac{2(1-z^{-1})}{T(1+z^{-1})} \tag{11.18}$$

Thus, the transfer function expressed as z^{-1} can be obtained directly. This transformation is called the bilinear transformation.

Since $f(z)z^{-k}$ can be written directly as f_{k-1}, the inverse Z-transform can be implemented directly. After the difference equation is listed and this transformation is implemented, the output of the discrete similarity mode of the control system depends only on the sampled values of several current and previous input variables and the sampled values of several previous output variables. These variables are all known, without the need for prediction–correction calculations. Moreover, the equation listing and solving are quick and easy, which is the reason for the wide application to fast closed-loop control.

(3) Characteristics of the discrete similarity method. In many cases, the difference equation derived from the discrete similarity method is the same as the calculation formula of the numerical integration method. But the discrete similarity method has a clear physical meaning. The discrete similarity method can conduct a phase shift through the transfer function of the integrator and control the numerical accuracy and stability through the amplitude adjustment, while the numerical integration method does not feature frequency domain or time domain adjustability. For many systems, the computational errors of the discrete similarity method originate from the sampling switch and the holder, and they can be analyzed in the frequency domain. The errors are expressed in the frequency domain as phase shifts (delays) and amplitude attenuations (distortions). Generally, the longer the sampling period is, the greater the delays and the distortions. These delays and distortions can be corrected through compensation.

11.3 Digital Sampling and Signal Conversion

The operating parameters related to the generator state variables should be taken in the excitation regulator as feedback variables for operations, so that generator excitation regulation, control, and limiting functions can be achieved. There are two ways to process these feedback variables: analog sampling and AC sampling.

For analog variable sampling, generally an analog variable transmitter is used as a measuring element. The output of the analog variable transmitter is a DC voltage proportional to the input variable, which passes through the A/D converter interface circuit for computer sampling. This method is easy to implement and can guarantee measurement accuracy. So, it was used for microcomputer excitation in the early stages.

When the transmitter converts AC to DC, a filter circuit is generally needed to ensure accuracy. From the point of view of improving the response speed of the excitation regulator, the filter time constant of the transmitter should be minimized. The relevant standard specifies that the time constant shall not exceed 50 ms. A high-frequency active filter can easily meet this requirement, since its time constant is only 7–10 ms. The analog variable transmitter has such shortcomings as complex circuit hardware and the need for regulation and maintenance. This section focuses on the functionality of the AC sampling circuit.

11.3.1 AC Sampling

AC sampling corresponds to analog variable sampling. Through the AC interface, the generator voltage and the current transformer secondary voltage and current signals are converted into an AC voltage which is proportional to the original signal quantitatively but lower in amplitude. Thus, the computer can implement sampling processing, and the relevant generator voltage U_G, current I_G, and active power P, Q can be obtained through operations. AC sampling technology is the key technology of microcomputer excitation and one of the markers of the digitization depth of exciter devices.

There are two AC interface categories: voltage interface and current interface. Both are front analog channels consisting of a signal amplitude conversion device, isolation shielding, and an analog low-pass filter, as shown in Figure 11.8.

It should be noted that, in the design of an AC sampling circuit, due to the existence of the low-pass filter element, there is a phase shift between the input and output voltage signals, which affects the active and reactive power measurement accuracy. So, the AC voltage interface and the AC current interface must have the same phase shift, supplemented by software phase compensation. Besides, the A/D converter's acquisition in the multiway switch order affects the calculation of the phase between the waveforms and the active and reactive power. So, a sample holder is provided before the A/D conversion, so that the associated variables can be held simultaneously at the instant before the sampling. An AC interface can only be used to sample and calculate the AC variables. For the DC rotor current measurement, a DC transmitter is generally used.

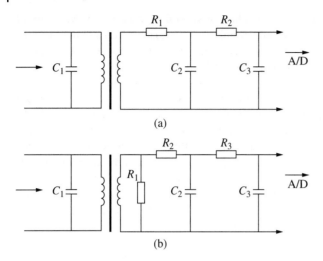

Figure 11.8 AC interface circuit.

11.3.2 Fourier Algorithm for AC Sampling

In the case of AC sampling, no transmitter is used. The measured AC signal should be arithmetically processed before it can become a control or display value. The Fourier algorithm is generally used for AC sampling. This algorithm is described as follows.

For the periodic function $u(t)$ whose period is T, that is, the signal $u(t) = u(t + T)$ can be evenly sampled at equal intervals of N points in one cycle, and its RMS, average, and power can be obtained. In the Fourier algorithm, the real and imaginary part values of each harmonic should be calculated first. Thus, the RMS of each harmonic can be determined. At the number of sampling points $N = 12$, the expression of the Fourier algorithm is as follows.

The real part of the voltage:

$$\text{Re}U = \frac{1}{6}\left[h_0 - h_6 + \frac{\sqrt{3}}{2}(h_1 + h_{11} - h_5 - h_7) + \frac{1}{2}(h_2 + h_{10} - h_4 - h_8)\right] \tag{11.19}$$

The imaginary part of the voltage:

$$\text{Im}U = \frac{1}{6}\left[h_3 - h_9 + \frac{\sqrt{3}}{2}(h_3 + h_4 - h_8 - h_{10}) + \frac{1}{2}(h_1 + h_5 - h_7 - h_{11})\right] \tag{11.20}$$

where

$h_0, h_1, \ldots h_{11}$ — sampled value at each point.

The computational formulas for the current real part $\text{Re}I$ and imaginary part $\text{Im}I$ are the same as in Eqs. (11.19) and (11.20), although the h's represent the sampled values of the current. The fundamental peaks of the voltage and the current are respectively as follows:

$$U = \sqrt{\text{Re}U^2 + \text{Im}U^2} \tag{11.21}$$

$$I = \sqrt{\text{Re}I^2 + \text{Im}I^2} \tag{11.22}$$

The power values are as follows:

$$\tilde{S} = \dot{U}\dot{I} = P + jQ = \frac{1}{2}[(\text{Re}U\ \text{Re}I + \text{Im}U\ \text{Im}I) + j(\text{Im}U\ \text{Re}I - \text{Re}U\ \text{Im}I)] \tag{11.23}$$

$$P = \frac{1}{2}(\text{Re}U \ \text{Re}I + \text{Im}U \ \text{Im}I) \tag{11.24}$$

$$Q = \frac{1}{2}(\text{Im}U \ \text{Re}I - \text{Re}U \ \text{Im}I) \tag{11.25}$$

The physical meaning of the Fourier 12-point algorithm is described in Figure 11.9.

Note that in the Fourier 12-point algorithm, the phasors of the variables are projected on 6 Cartesian coordinates at intervals of 30° and the projections of the 12 variables projected on the real or imaginary axis are superimposed. Thus, the real and imaginary parts of the original phasors can be calculated. Through this decomposition, 2nd, 3rd, 4th, and 5th harmonics and DC component can also be filtered out. So, this is a commonly used algorithm.

11.3.3 Three-Phase One-Point Algorithm

The three-phase one-point algorithm is characterized by simple and easy calculation, but it requires three-phase symmetry; otherwise, serious calculation errors may occur.

The formula of the three-phase one-point algorithm is as follows:

$$U = \frac{1}{\sqrt{3}}\sqrt{U_{ab}^2 + U_{bc}^2 + U_{ca}^2} \tag{11.26}$$

where

U_{ab}, U_{bc}, U_{ca} — sampled values of line voltage (instantaneous values).

$$I = \frac{1}{\sqrt{3}}\sqrt{I_a^2 + I_b^2 + I_c^2} \tag{11.27}$$

where

I_a, I_b, I_c — sampled values of phase current (instantaneous values).

$$P = \frac{1}{9}[U_{ab}(I_a - I_b) + U_{bc}(I_b - I_c) + U_{ca}(I_c - I_a)] \tag{11.28}$$

$$Q = \frac{1}{3\sqrt{3}}(U_{ab}I_c + U_{bc}I_a + U_{ca}I_b) \tag{11.29}$$

Figure 11.9 12-point Fourier algorithm graphics.

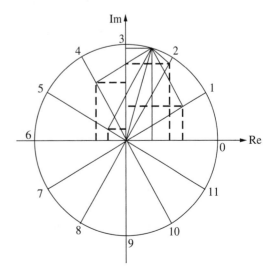

11.3.4 Speed Measurement Algorithm

In a digital excitation regulator, some elements require that the speed should act as the regulatory signal. For example, in a power system stabilizer (PSS), ω acts as the regulatory signal

In the digital speed measurement method, the frequency f is measured. The basic method of frequency measurement is period measurement. That is, the period T of each cycle of the AC voltage and the crystal oscillator frequency f_0 in the microcomputer are measured. After an appropriate frequency division, this can be used as the counting frequency f_c. The corresponding pulse train is Φ. A pulse whose period is $\frac{1}{f_c}$ of Φ is used as the standard timing unit of the measurement period T. It is assumed that the measured width of T is equivalent to m standard chronograph pulses. Then

$$T = \frac{1}{f_c} \times m \tag{11.30}$$

Thus, the measured frequency can be obtained as follows:

$$f = \frac{f_c}{m} \tag{11.31}$$

The angular frequency is

$$\omega = 2\pi f$$

If the AC voltage signal of the frequency measurement is from the generator stator voltage, the measured ω is the angular frequency of the generator voltage. If the AC voltage signal of the frequency measurement is from the AC tachogenerator on the rotating shaft of the generator, the measured ω is the angular velocity of the generator.

11.4 Control Operation

Control operation is the core of a microcomputer excitation regulator. With the support of the microcomputer hardware, the following operations are implemented with the application software:

(1) Data acquisition, timing sampling, and operation, for checking the correctness of the measurement data, scale conversion, and selection of display.
(2) Regulation algorithm, for calculations under the regulation law.
(3) Control output, for conversion of results of the regulation algorithm and clamped output as well as thyristor control through phase shift trigger element.
(4) Others, such as entry of setting values, modification of parameters, change of mode of operation, sound-light alarm, and implementation of other functions.

The key regulation algorithms are described separately below.

Proportional–integral–differential (PID) control is a correction method for design based on the frequency domain method of classical control theory. This design method can be used to improve the static and dynamic performance of the generator voltage.

1. Differential equation expression of PID regulation
   For an analog excitation regulator, the PID regulation law can be expressed as the following differential equation:

$$\begin{aligned} Y(t) &= K_p e(t) + K_i \int_0^t e(t)dt + K_d \times \frac{de(t)}{dt} \\ &= K_p \left[e(t) + \frac{1}{T_i} \int_0^t e(t)dt + T_d \times \frac{de(t)}{dt} \right] \end{aligned} \tag{11.32}$$

$$e(t) = U_g - U_c$$

where

- $Y(t)$ — control output;
- $e(t)$ — generator terminal voltage deviation signal;
- U_g — given voltage;
- U_c — measured voltage, proportional to generator terminal voltage;
- K_p — scale factor, used to improve response speed of control system to reduce static deviation;
- T_i — integral time constant, used to eliminate static errors;
- T_d — derivative time constant, used to improve dynamic performance of system.

For computer control, Eq. (11.32) must be discretized, and a difference equation is used in place of the differential equation. The trapezoidal integral is used to approximate the integral, and a backward difference is used to approximate the differential. The PID digital control algorithm can be obtained as follows:

$$Y(k) = K_p \left\{ e(k) + \frac{T}{T_i} \sum_{j=1}^{k} e(j) + \frac{T_d}{T}[e(k) - e(k-1)] \right\}$$

$$e(k) = U_g - U_c(k) \tag{11.33}$$

where

- T — sampling period;
- $e(k)$ — generator terminal voltage deviation in kth sampling.

Equation (11.33) is the PID position control algorithm of digital control, and has some drawbacks. For example, each output is related to all the past sampled values, which increases memory usage. Moreover, any wrong calculation will lead to a great cumulative error of $Y(k)$, thus affecting operational safety. So, an incremental algorithm is generally adopted for PID regulation.

When k in Eq. (11.33) is replaced by $k-1$, the following equation can be obtained:

$$Y(k-1) = K_p \left\{ e(k-1) + \frac{T}{T_i} \sum_{j=1}^{k-1} e(j) + \frac{T_d}{T}[e(k-1) - e(k-2)] \right\} \tag{11.34}$$

When Eq. (11.34) is subtracted from Eq. (11.34), the following incremental equation is obtained:

$$\Delta Y(k) = Y(k) - Y(k-1)$$
$$= K_p \left\{ [e(k) - e(k-1)] + \frac{T}{T_i} \times e(k) + \frac{T_d}{T}[e(k) - 2e(k-1) + e(k-2)] \right\}$$

So,

$$Y(k) = Y(k-1) + K_p[e(k) - e(k-1)] + K_i e(k) + K_d[e(k) - 2e(k-1) + e(k-2)]$$

$$K_i = K_p \times \frac{T}{T_i}$$

$$K_d = K_p \times \frac{T_d}{T} \tag{11.35}$$

where

- K_i — integral coefficient;
- K_d — differential coefficient.

The key parameters that need to be determined for PID regulation include K_p, K_i, and K_d. With the critical ratio setting method, the parameters can be determined easily.

11 Automatic Excitation Regulator

When the critical oscillation period T_u and the corresponding proportional coefficient K_u are measured, the following can be obtained:

$$K_p = 0.6K_u, \; T_i = 0.5T_u, \; K_i = K_p \times \frac{T}{T_i}, \; T_d = 0.125T_u, \; K_d = K_p \times \frac{T_d}{T}$$

In Eq. (11.35), if $K_d = 0$, it is a PI regulation algorithm. If $K_i = 0$, it is a proportional regulation algorithm. With the algorithm formula, the corresponding programming is not difficult.

It should be noted that the expression of the transfer function of the PID filter in the frequency domain is as follows:

$$G_c(s) = K_p \times \frac{(1+sT_{C2})(1+sT_{C1})}{(1+sT_{B2})(1+sT_{B1})} \tag{11.36}$$

The amplitude-frequency characteristics expressed as Eq. (11.36) are shown in Figure 11.10.

In Figure 11.10, K_R represents the DC gain, which is used to determine the regulator's regulation accuracy. Through the integral section determined by the integral bandwidth control time constant T_{B1} and the integral time constant T_{C1}, it appears as a proportional gain K_P of a transient gain turn-down in the intermediate frequency area, so as to improve the transient stability of the system. Through the differential section determined by the differential time constant T_{C2} and the differential bandwidth control time constant T_{B2}, it appears as a high-frequency gain K_∞ of a differential gain suppression in the high-frequency area, so as to prevent high-frequency spurious signals' interference with the differentiation element. In an excitation system with an exciter, the differential section is primarily used to compensate for the influence of the exciter lag on the gain and the phase margin to improve the stability of the regulation system. In the absence of differential regulation (or lead-lag correction of transient gain turn-up), the soft feedback of the rotor voltage or the soft or hard feedback of the exciter excitation current can be applied for the same purpose. K_P is related to the large signal regulation performance. When the generator terminal voltage decreases to 80% of the rated value, the minimum value of K_P should be such that it does not limit the top value of the forced excitation in the frequency band.

In a microcomputer excitation digital control system, the aforementioned discretization method and bilinear transformation should be used to convert Eq. (11.36) into a difference equation. The following can be used in Eq. (11.36), so that the operation can be simplified:

$$a = \frac{2T_{C1}}{T}, \quad b = \frac{2T_{C2}}{T}, \quad c = \frac{aK_R}{K_p}, \quad d = \frac{bK_p}{K_\infty}$$

The difference equation of the lead-lag correction element (1) is as follows:

$$Y_a(k) = \frac{(b+1)e(k) - (b-1)e(k-1) + (d-1)Y_a(k-1)}{d+1} \tag{11.37}$$

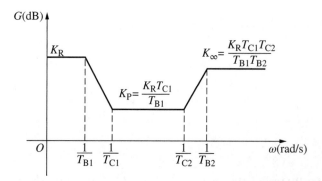

Figure 11.10 Amplitude-frequency characteristics of standard PID filter.

The difference equation of the lead-lag correction element (2) is

$$Y_b(k) = \frac{K_R(a+1)Y_a(k) - (a-1)Y_a(k-1) + (c-1)Y_b(k-1)}{c+1} \tag{11.38}$$

where,

T — sampling period;

$e(k)$ — generator terminal voltage deviation in k^{th} sample, $e(k) = U_r - U_G(k)$;

$Y_b(k)$ — PID output control voltage, acting on phase shift trigger element.

$Y_b(k)$ should be added with an appropriate clipping element before being used for the phase shift trigger.

2. Linear optimal excitation control (LOEC) algorithm

When the grid structure is weak, the transmission power limit is primarily constrained by the static stability limit. The problem of small interference is prominent, and there are low-power oscillations of various frequencies. Linear optimal control is a power system small interference stability control theory and method that is designed to suppress low-power oscillations of various frequencies. It takes advantage of the constant power network structure and parameters. The dynamic model of the load group can be replaced by a constant impedance characteristic. The three assumptions can be processed linearly around a certain equilibrium point, so that the problem is reduced to optimal control of a multi-input-multi-output (MIMO) linear system. That is,

$$\dot{X} = AX + BU \tag{11.39}$$

where

$\dot{X}$ — differential component of X;

A — state coefficient matrix;

B — control coefficient matrix;

X — n-dimensional state vector;

U — r-dimensional control vector.

In the design of the controller of the above system, a quadratic performance index is chosen.
With

$$J = \frac{1}{T}\int_0^\infty (X^T Q X + U^T R U)dt \tag{11.40}$$

the optimal control solution can be obtained as follows:

$$U = -R^{-1}B^T P X = -KX \tag{11.41}$$

where

P — solution of the algebraic Riccati equation.

As can be seen from the above solution, the solution to optimal control is the constant feedback of the full state variable of the system. However, since the power system is highly decentralized geographically, real-time state information transmission between generators is extremely difficult. So, it is almost impossible to achieve small interference stability optimal control of the power system. So, in the early stage, designs were generally based on the single generator on infinite system model. When the generator rotor excitation circuit is powered by a power thyristor, the third-order equation is used. When the generator rotor excitation circuit is powered by a thyristor and an excitation regulator controls the AC exciter circuit thyristor, the fourth-order equation is used.

For the third-order state equation, generally $X = [\Delta P \Delta \omega \Delta U]^T$. The quadratic performance index is given for the solving of the problem of optimization. The feedback gain matrix can be obtained as follows:

$$K = [K_P K_\omega K_U] \tag{11.42}$$

The control increment is

$$U = -KX = -[K_P \Delta P + K_\omega \Delta \omega + K_U \Delta U] \tag{11.43}$$

where

ΔP – difference between benchmark P_b and measured value P of active power;

$\Delta \omega$ – difference between dynamic benchmark ω_b and measured value ω of speed;

ΔU – difference between given value U_g and measured value U_c of generator terminal voltage.

Since the linear model used is derived under the condition of linearization of the operating points, the state equilibrium point of the system is the dynamic benchmark of the state variable. When the system is operating, the state variable always fluctuates near its equilibrium point. It can be considered that the state variable's fluctuation center is just its equilibrium point, that is, the dynamic benchmark to be obtained. The fluctuation center can be obtained with software from several sampled values.

Some of the elements in the state matrix A vary with the operating points of the generator. When the microcomputer excitation optimal control is realized, the optimal feedback gain matrix K at each operating point (P, Q) should be calculated in advance on the basis of the given parameters and stored in tabular form in the memory of the microcomputer after being converted. When the excitation regulator is operating, the microcomputer conducts timing sampling. ΔP, $\Delta \omega$, and ΔU are calculated. The corresponding K_P, K_ω, and K_U are identified. Thus, U is obtained. $Y = U_0 + U$ is outputted. Since U is an increment, it is necessary to add the base value Y_0 to get the full output. There are several ways to calculate U_0. A simple method is to take $U_0(k)$ as the arithmetic mean value for the first N times. The recursive formula is as follows:

$$Y_0(k) = \frac{1}{N} \sum_{i=1}^{N} Y_0(k-i) \tag{11.44}$$

Another method is

$$Y_0(k) = Y_0(k-1) + K_I e(k) \tag{11.45}$$

where

$K_I e(k)$ – integral term;

K_I – integral coefficient.

3. Nonlinear optimal excitation control (NOEC)

At present, great progress has been made in improving the stability and suppressing low-frequency oscillations under small interference conditions in China's power systems. The main problem affecting grid operation safety has shifted from small interference to large disturbance transient stability. So, it is vital to develop an optimal excitation controller directly based on the multi-generator system accurate nonlinear model. Since the 1970s, nonlinear controllers based on differential geometry theory have been applied in many areas, and a sound theoretical system has been developed. Differential geometry is a feedback linearization method. Under its basic principle, the necessary coordinate transformation is conducted for the mathematical model of the controlled system with the differential geometric mathematical method. Thus, a nonlinear feedback is obtained. Under the action of this nonlinear feedback, the original nonlinear system is mapped to an integral precise linear system. Through the coordinate transformation, the control characteristics of the system remain unchanged.

The basic idea of the nonlinear control strategy is to assume that the characteristics of the nonlinear power system as follows:

$$\begin{cases} \dot{x} = f(x) + \sum_{i=1}^{m} g_i(x)u_i \\ y_i = h_i(x) \quad (i = 1, 2, 3, \cdots, m) \end{cases} \quad (11.46)$$

Under the principle of differential geometry, a set of nonlinear coordinate transformation and control transform is obtained as follows:

$$\begin{cases} Z = T(x) \\ V = \alpha(x) + \beta(x)u \end{cases} \quad (11.47)$$

The nonlinear expression of the power system in Eq. (11.46) is precisely linearized into the following linear control system:

$$\begin{cases} \dot{Z} = AZ + BV \\ Y = CZ \end{cases} \quad (11.48)$$

Through the coordinate transformation and control transformation in the above equations, control of the nonlinear system can be converted to control of the precise linear system. The control optimization strategy is as follows:

$$U = \frac{-KT(x) + -\alpha(x)}{\beta(x)} \quad (11.49)$$

In the application to an excitation control system, through a certain approximation, the nonlinear optimal control algorithm can be obtained as follows:

$$U_c = \frac{HT'_{d0}}{\omega_0 I_q}\left(K_1 \Delta \delta + K_2 \Delta_\omega - \frac{\omega_0}{H} \times K_3 \Delta P_e\right) - \frac{T'_{d0} E'_q}{I_q} \times \dot{I}_q + E_q + \frac{HT'_{d0}}{\omega_0 I_q} \times K_v \Delta V_t \quad (11.50)$$

11.5 Per-Unit Value Setting

In a digital control excitation system, the per-unit value setting is the same as the conventional setting. That is, per-unit value is defined as the ratio of the actual value of a physical variable to the selected base value in the same unit:

$$\text{Per-unit value (p.u.)} = \frac{\text{actual value}}{\text{base value in same unit}}$$

Base values are selected arbitrarily. However, for convenience, rated values are taken as base values in an excitation system. Application of base values to operations in a digital excitation regulator can make the physical variables such as voltage and current frequency vary in the vicinity of 1.0, thus avoiding larger errors caused by calculations with too large and small numerical values. The commonly used base values are as follows:

Base value of generator stator voltage $\quad U_b = U_{GN}$
Base value of stator current $\quad I_b = I_{GN}$
Base value of impedance $\quad Z_b = U_b/I_b$
Base value of power $\quad S_b = S_N$
Base value of stator angular frequency $\quad \omega_b = \omega_N$
Base value of reference angular velocity $\quad \Omega_b = \omega_b/P$
Base value of torque $\quad M_b = S_b/\Omega_b$

The per-unit base value of an excitation current is the excitation current value at the point of the extension line of the air gap line of the generator no-load characteristic curve corresponding to the generator voltage. The excitation current per-unit base value is I_{fb}. The corresponding excitation voltage base value is I'_{fb}, and the excitation winding resistance base value is U_{fb}/I_{fb}.

11.6 Digital Phase Shift Trigger

11.6.1 Digital Phase Shift Trigger

Generally, for a conventional analog regulator, the linear phase shift trigger circuit at the moment of the triggering is determined on the basis of the comparison between the synchronous sawtooth wave and the control voltage, and the cosine phase shift trigger circuit at the moment of the trigger is determined on the basis of the comparison between the cosine synchronous voltage and the control voltage. However, in the linear phase shift circuit, the so-called linearity means that silicon-controlled rectifier (SCR) control angle (α) is linearly proportional to the control voltage. The control voltage is not linearly proportional to the rectifier voltage $u_d = u \cos_\alpha$. In the cosine phase shift circuit, since the control angle is determined by the intersection of the control voltage curve and the cosine voltage curve, the control voltage is linearly proportional to the rectifier voltage.

In a digital excitation regulator, the control voltage of the phase shift trigger circuit is a digital signal. For a trigger completed by software, the trigger time can be expressed as the angle or time.

By hardware type, the digital phase shift can be divided into two types: the counter-comparator method and the counter direct method.

By the location of hardware in a microcomputer excitation regulator, the digital phase shift can be divided into two types: the bond-free method (some of the hardware in the digital phase shift element is implanted into a single chip microcomputer or a chip) and the peripheral circuit method (composition by peripheral I/O devices). As mentioned earlier, by control voltage expression, the digital phase shift can be divided into two modes: linear phase shift and cosine phase shift.

It should be noted that calculations are only allowed in systems with a floating-point processor because cosine and arc cosine functions are transcendental functions. In systems without a floating-point processor, direct calculations are not allowed, though approximate operations can be conducted with the look-up table or the Taylor series expansion method.

Here are the basic composition principles of several digital phase shift forms:

1. Counter-comparator method

    The principle of the counter-comparator method: A cycle counter is used as a timing standard. The zero crossing point of the synchronous voltage is associated with the timing standard. The trigger angle is converted into the sum of the synchronous voltage zero crossing point time and the time corresponding to the trigger angle, and fed into the latch. The comparator compares the timing standard with the latch. When the two are equal, a trigger pulse is outputted. After the time required for the pulse width elapses, the trigger pulse disappears. Only one cycle counter and one comparator are needed, while one or more latches are generally needed.

    Digital components usually only accept digital signals, such as level, level variation, and level duration. A synchronous signal generally taken from the power supply of the three-phase fully controlled bridge is isolated by the transformer before it becomes a weak signal. So, an analog synchronous signal must be converted to a digital signal, so that it can be recognized by the computer system. A zero crossing turnover voltage comparator can be used for such a conversion. The synchronous signal is converted into a square

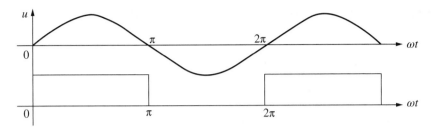

Figure 11.11 Zero crossing conversion of synchronous signal.

wave signal. The positive zero crossing of the synchronous signal is converted into the rising edge of the square wave signal, while the negative zero is converted into the falling edge of the square wave signal, as shown in Figure 11.11. In the figure, it is assumed that the zero crossing point of the synchronous signal is the moment when the control angle is zero, which can be achieved through an analog conversion circuit. For a digital microcomputer excitation regulator phase shift trigger circuit, when the counter-comparator method is adopted, the internal devices of an 8096 or 80C196 single chip microcomputer can be used to achieve a bond-free digital phase shift trigger.

2. Counter direct method

For the abovementioned counter-comparator digital phase shift circuit, the cycle count is needed. Meanwhile, the up count is generally used. The function is achieved with a sawtooth wave trigger level-comparator analog circuit. Both a circuit inside the microcomputer chip and one outside the microcomputer chip need a counter. Such a circuit appears to be quite complicated.

The advantage of the counter direct method is the use of a down counter. Moreover, a CPLD can also achieve a bond-free phase shift trigger. The counter is triggered by an external synchronous signal. When a jump occurs, the initial count value is counted. When it is reduced to zero, a trigger pulse is automatically outputted. The simplest circuit can complete the generation of the prototype pulse with only one 8253 chip. It should be noted that the synchronous signal frequency should be acquired as a timing benchmark because the trigger angle of the counter direct method is expressed as a time interval. In addition to an 8253 counter, which can achieve a digital phase shift, a programmable system device like the PSD5 ×× can achieve a bond-free counter direct digital phase shift.

11.6.2 Characteristics of Digital Phase Shift

A digital phase shift circuit is characterized by high integration, phase shift angle symmetry, and good accuracy and accuracy of the angle and width of each phase trigger pulse.

It should be noted that the digital phase shift circuit also has the general characteristics of digitization. For example, digitization is discrete, and there is a time delay in the discrete process. In an analog phase shift circuit, the trigger angle is continuous without delay variation. For a digital phase shift circuit, the trigger angle depends on the microcomputer's operation speed, which is subject to certain restrictions in the application. For example, in a three-machine excitation system with a pilot exciter, the supply frequency of the AC pilot exciter is 500 Hz. At this frequency, a new trigger angle cannot be calculated under each 60° electrical angle condition with the operation speed of any current microcomputer, which to some extent affects the dynamic response speed.

Moreover, for the counter-comparator method, such problems as the impossibility of using a three-phase synchronous signal and the failure to achieve an accurate phase shift in the case of three-phase synchronous voltage asymmetry need to be solved.

11.7 External Characteristics of Three-Phase Fully Controlled Bridge Rectifier Circuit

11.7.1 Three-Phase Fully Controlled Bridge Rectifier Circuit

The power output stage of a static excitation system is generally the voltage source – a three-phase fully controlled rectifier bridge, as shown in Figure 11.12. The excitation regulator is used to control the control angle of the rectifier bridge. When the control angle is in the range 0°–90°, the rectifier bridge is in the rectifying state, where the average value of the output voltage is positive. When the control angle is (90°–180°) – γ (γ is the commutation angle), the rectifier bridge is in the inversion state, where the average value of the output voltage is negative. In general, the control angle is maintained at about 70°, and the control angle is approximately 0° in the case of forced excitation. The inversion state can be used for generator de-excitation. For example, the control angle is fixed at 150°, so that the energy stored in the rotor excitation winding can be returned to the rectifier power supply through the rectifier bridge.

In the three-phase bridge fully controlled rectifier circuit, there is a commutation voltage drop. When the control angle is constant, the output voltage decreases as the output current (or the load) increases. Within the range of application of an excitation system, the three-phase fully controlled rectifier circuit generally works in State I. The external characteristic expression is Eq. (11.51), and the per-unit external expression is Eq. (11.52).

$$U_f = \frac{3\sqrt{2}}{\pi} U_E \cos\alpha - \frac{3}{\pi} I_f X_\gamma \tag{11.51}$$

$$U_f^* = U_E^* \cos\alpha - \frac{3}{\pi} I_f^* \times \frac{X_\gamma}{R_{fB}} \tag{11.52}$$

Figure 11.13 shows the external characteristics of the three-phase fully controlled bridge rectifier circuit expressed as per-unit values.

11.7.2 Linear Phase Shift Element

The phase shift phase element where the phase shift control voltage and the control angle are linearly relative is called the phase shift phase element. The corresponding phase shift characteristics are shown in Figure 11.14.

In Figure 11.14, the operating points of the straight line ("a" in the figure) are determined by the maximum control voltage $u_{R,max}$, the minimum control voltage $u_{R,min}$, the maximum control angle α_{max}, and the minimum

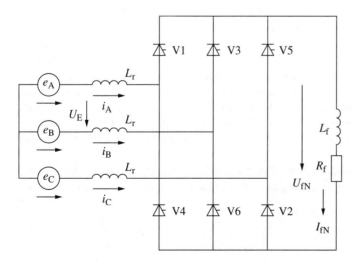

Figure 11.12 Three-phase fully controlled bridge rectifier circuit.

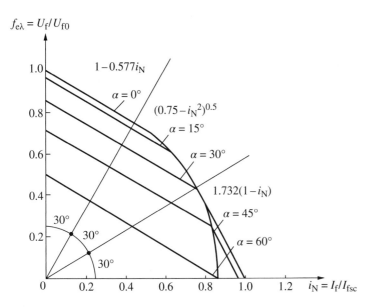

Figure 11.13 External characteristics of three-phase fully controlled bridge rectifier circuit.

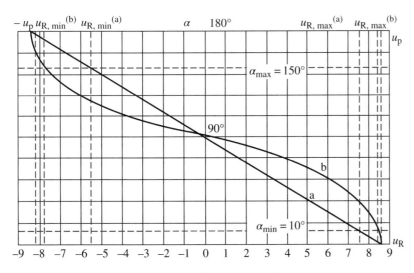

Figure 11.14 Phase shift characteristics of three-phase fully control bridge.

control angle α_{min}. In general, $\alpha_{max} = 150°$ and $\alpha_{min} = 10°$, while $u_{R,max}$ and $u_{R,min}$ depend on the excitation system's excitation voltage range. u_p is the no-load voltage outputted by the rectifier bridge. Besides, in the case of a variation in the ratio of the rectifier power supply voltage to the rated value, it is assumed that the power supply variation coefficient is K_u, and the characteristics of the phase shift element can be expressed as the following equations:

$$\alpha = -\frac{\pi}{2u_p} \times u_R + \frac{\pi}{2} \quad (\alpha_{min} \leq \alpha \leq \alpha_{max}) \tag{11.53}$$

$$u_f = K_u u_R \cos\alpha - \frac{3}{\pi} i_f \times \frac{X_\gamma}{R_{fb}} \tag{11.54}$$

$$u_{R,min} = u_p \cos\alpha_{max}$$
$$u_{R,max} = u_p \cos\alpha_{min} \tag{11.55}$$

11.7.3 Cosine Phase Shift Element

When there is a cosine relation between the control voltage and the control angle outputted by the control system, it is the cosine phase shift element, as shown in Figure 11.14 (Curve "b"). The functions of the cosine phase shift element and the corresponding relations are as follows:

$$\alpha = \arccos \frac{u_R}{u_p} \quad (\alpha_{min} \leq \alpha \leq \alpha_{max}) \tag{11.56}$$

$$u_f = K_u u_R - \frac{3}{\pi} i_f \times \frac{X_\gamma}{R_{fb}}$$

$$u_{R,min} = u_p \cos \alpha_{max}$$

$$u_{R,max} = u_p \cos \alpha_{min}$$

The cosine phase shift can produce a linear relationship between the output voltage of the rectifier bridge and the control voltage.

For a three-phase fully controlled bridge rectifier circuit, the cosine phase shift function is expressed as follows:

$$\alpha = \arccos \frac{u_R + \frac{3}{\pi} i_f \times \frac{X_\gamma}{R_{fb}}}{K_u u_p} \quad (\alpha_{min} \leq \alpha \leq \alpha_{max}) \tag{11.57}$$

$$u_f = u_R \tag{11.58}$$

To simplify the expression, in Eq. (11.57), the rectifier bridge load factor is chosen as

$$K_c = \frac{3 X_\gamma}{\pi R_{fb}}$$

It is substituted into Eq. (11.55). Then, the following can be obtained:

$$\left. \begin{array}{l} u_{R,min} = K_u u_p \cos \alpha_{max} - K_c i_f \\ u_{R,max} = K_u u_p \cos \alpha_{min} - K_c i_f \end{array} \right\} \tag{11.59}$$

11.7.4 Mathematical Model of Three-Phase Fully Controlled Rectifier

In order to decrease the simulation time, the one-cycle average model is generally used in place of the instantaneous value model for the rectifier circuit in a digital excitation system. That is, it is considered that the calculated trigger angle α can immediately get the response of the average value of the excitation voltage, as shown in Eqs. (11.54) and (11.55). For different phase shift elements, the models shown in Figure 11.15–11.17 can be obtained. If the phase shift element is stored in the simulation system, it can be combined with the voltage source–three-phase fully controlled bridge into a unit scale element, so that the model can be simplified.

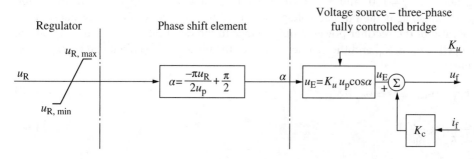

Figure 11.15 Voltage source – three-phase fully controlled bridge rectifier circuit and linear phase shift element characteristics.

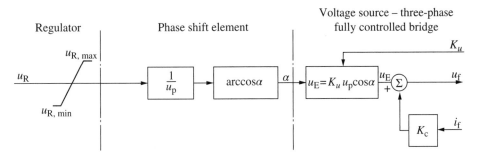

Figure 11.16 Voltage source – three-phase bridge full controlled rectifier circuit and linear cosine phase shift.

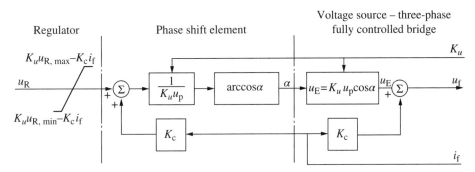

Figure 11.17 Voltage source – three-phase bridge full controlled rectifier circuit and linear phase shift with compensation.

11.8 Characteristics of Digital Excitation Systems

11.8.1 Overview

The block diagram of a typical digital excitation system is shown in Figure 11.18.

In Figure 11.18, the dotted lines are controlled with digital technology. Specifically, control is achieved with one or more microprocessors. The D/A or A/D interface unit is used to convert analog variables to digital variables or digital variables to analog variables.

The reference input of the digital excitation is typically a number stored in the microprocessor's random access memory (RAM). The reference input is compared with the feedback variable representing the generator output voltage, so that the voltage deviation can be obtained. The generator output variable should be converted to a variable that can be accepted by the computer via a scaling circuit and then be converted to a digital variable via the A/D converter. The comparison between the given reference value and the feedback variable is done via a mathematical program in the microprocessor. The comparison result acts on the control logic stored in the microprocessor to complete the desired control algorithm.

Conventional control algorithms include the PID algorithm. In addition, new theoretical control algorithms such as linear or nonlinear control, fuzzy logic, and adaptive control can also be used. The control program output converted into an analog signal is used to drive the power amplifier unit in the excitation system. The analog signal outputted can be obtained via the D/A converter. Also, a pulse can be directly sent by the microprocessor o the power amplifier unit. The generator's excitation voltage and current are supplied by the power amplifier unit.

In a digital excitation system, complex-control-equation operations can be performed. In an analog excitation system, the additional unit power is supplied by a separate device. But in a digital excitation system, it is only part of the normal control algorithm, and the put-in and cut-out regulation is very smooth.

In a digital excitation system, the simplest communication is achieved via a local keyboard and the display. The complex communications include local serial communication, remote serial communication,

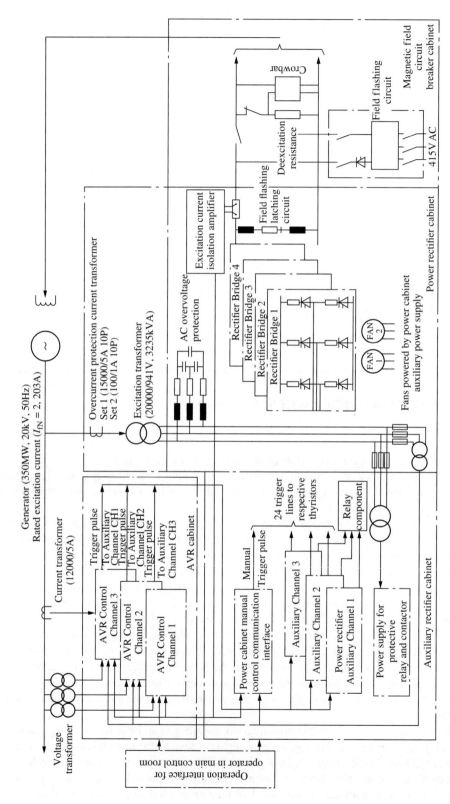

Figure 11.18 Block diagram of typical digital excitation system.

modulation-demodulation communication, and a regional network. The communication data include input values, output values, set values, internal signals, limit values, control relay status, and fault conditions. The communication data can be exchanged with information on the other controllers in the system, such as the governor or the monitor. Also, the operator can directly control the variables to shift the system to a specific operating state.

The reference-dependent adjustments in a digital excitation system such as the generator voltage setting, manual excitation setting, and reactive power and power factor setting are all digital. These digital variables are inputted into the microprocessor's remote increase/decrease node or adjusted directly via the power station's computer control network.

Besides, in a digital system, various parameters including parameters inputted into the excitation system and its internal parameters can be recorded, sent to external data loggers via the D/A converter, or loop-recorded inside the computer. The data record should have the filter and reset functions.

In terms of instrumentation, most analog systems require added instrumentation for measurement of various parameters in system, while digital systems can display these parameters without the need for added instrumentation such as a transmitter. Moreover, digital systems can send these data to the upper computer of a power station, eliminating the need for measuring instruments and wire connections.

Compared to analog systems, digital systems can provide more information and control functions which are easy to achieve. Digital systems generally have the self-test function, can identify system internal failures in advance, and solve them in a safe and orderly way. Moreover, digital systems feature easy parameterization, thus significantly reducing field debugging time and workload.

11.8.2 Characteristics of Typical Digital Excitation System

Figure 11.19 shows a three-channel three-mode redundant (TMR) digital microcomputer excitation system for large steam turbines developed by the United Kingdom's Rolls-Royce. The following is a brief introduction to the characteristics of this system.

In the excitation system, a TMR controller is adopted. The AVR consists of three identical and independent control channels. Each control channel can be automatically or manually regulated. In each control channel of the TMR structure, the protection stop function can be easily achieved. Meanwhile the "majority voting" mode is used to process the stop signal, which not only ensures generator safety but also improves system availability. The TMR controller follows the working principle of "majority voting." If the three AVR channels are all working properly, the main regulatory signal will be outputted in accordance with the "majority voting" (two out of three) control program.

When one channel fails, an alarm signal will be sent, and the mid-value of the other two signals will be taken as a regulatory signal. If two channels fail, the AVR will be switched to the manual control channel.

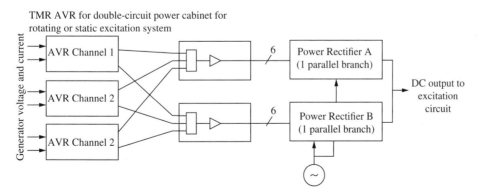

Figure 11.19 Principle diagram for TMR digital microcomputer excitation system.

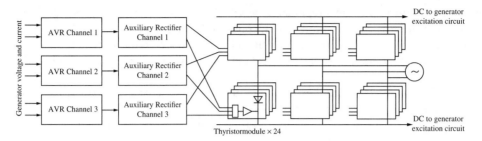

Figure 11.20 Working principle diagram of AVR control channel of TMR microcomputer excitation system.

In this system, the TMR controller hardware is the DSP produced by Texas Instruments, which is a 32-bit high-speed digital signal microprocessor. Moreover, an input/output circuit specially designed for this application is provided. The DSP module inputs the AC sampling signals from the generator VT and AT circuits. The operation program is executed via the adapter. The control algorithm executes an operation every 3.3 ms, to ensure that the excitation system has a good response speed. Figure 11.20 shows a schematic diagram where the AVR control signal is outputted to the thyristor component end, and the final output trigger signal is determined through "majority voting" (two out of three).

The excitation system's power rectifier cabinet also adopts a redundant structure. In this design, there are four rectifier bridge branches. When one branch quits, the other three can still handle the rated load and twice the forced excitation.

Besides, in the rectifier, the auxiliary cabinet also has a completely independent manual control channel, which is put in automatically as a standby channel when two of the AVR's three manual channels fail simultaneously.

11.8.3 Operation Modes of Digital Excitation System

1. AVR automatic mode
   The basic function of the automatic mode is to compare the generator actual terminal voltage with the automatic given value and use that to control the generator terminal voltage. The comparison (between the generator actual terminal voltage and the automatic given value) error signal is outputted to the PID controller. The output of the PID controller is modulated into a trigger pulse, which is sent to the power rectifier.
   The accuracy of the AVR's regulation from idling to the continuous maximum rated voltage is ±0.5%.
2. AVR manual mode
   The basic function of the manual mode is to compare the generator actual excitation current with the manual given value and use that to control the generator excitation current. The comparison (between the generator actual excitation current and the manual given value) error signal is sent to the PI controller. The output of the PI controller is modulated into a trigger pulse signal, which is supplied to the power rectifier. This mode can ensure that the generator excitation current is constant.
   The switch between the two modes is undisturbed (unless the excitation in the case of a switch from the automatic mode to the manual mode is lower than the manual constraint limit line). This switch can be done by the operator or is done automatically in the event of a subsystem failure.
3. Rectifier device manual mode
   A rectifier device manual regulation control channel (PI current controller) is provided between the AVR and the rectifier device. When there is an AVR automatic channel failure or the AVR cannot be put in temporarily during debugging, the control channel can provide manual excitation control.

4. Mode selection
   The system has three control operation modes:
   (1) AVR automatic operation mode.
   (2) AVR manual operation mode, which is automatically put in when two VTs fail or is manually put in during generator carbon brush maintenance.
   (3) Rectifier device manual operation mode, which is automatically put in when two AVR control channels fail or is applied during system debugging.
5. Operating modes
   System control can be operated locally or remotely via the buttons of the main control room or DCS microcomputer terminals.

11.8.4 Functions of Digital Excitation System

In a modern intelligent power system, power impact has been more frequent with the increase in the proportion of various types of environmentally friendly power generation (wind and PV power). Meanwhile, the capacity of a single conventional thermal power unit or nuclear power unit has exceeded 1000 MW. So, the requirements for an excitation system's performance, especially in terms of safety and reliability, have been higher. In addition, the excitation regulator's limiting function and relay protection coordination have attracted much attention. For example, DL/T 843 – 2010 5.18 provides that an AVR shall have additional functional units whose actions are in keeping with the characteristics of the generator-transformer set and the power system, such as an overexcitation limit (including the top current instantaneous limit and overexcitation inverse time limit), under-excitation limit, V/Hz limit, and PSS. The AVR's limiting characteristics and setting values should ensure that the generator-transformer group performs in accordance with its potential short-time load capacity and be matched with the relevant generator-transformer set and excitation transformer relay protection. The limiting process should be fast and smooth. After the internal protection of the excitation regulator acts, there will be a switch to the standby. Besides, the limiting and protection functions should feature the same configuration type, coordinated set values, and the same action principle.

1. Reference given value control
   In both the automatic control mode and the manual control mode, the operator can increase or decrease the excitation value by adjusting the given reference value of the excitation regulator.
   Figure 11.21 shows the excitation regulator's excitation increase/decrease function.
   The automatic excitation regulator's excitation increase/decrease order is externally regulated. The requirement of excitation increase/decrease is met through a change in the given reference value. The magnitude of the change in the reference value is determined by the excitation increase/decrease rate. The automatic tracking circuit tracks the reference value adjustment to ensure that there is no disturbance switch in the event of a fault.
2. Generator operating capacity curve
   The operating capacity curve of the generator expresses the relationship between the active power and the reactive power of the generator under the rated working conditions as well as the range that can ensure long-term safe operation. Figure 11.22 shows the operating capacity curve of a typical salient pole synchronous generator.
   In Figure 11.22, for the rotor current limit curve, the circle's center is at $1/x_q$ on Axis $-Q$ ($1/x_d$ on Axis $-Q$ for a non-salient pole synchronous generator), and its radius is I_{fn}. The limit value is determined by the heat capacity of the generator excitation winding. The angle δ between the phasor I_{fn} and Axis $-Q$ is equal to the rotor power angle. The stator current limit Curve 2 is determined by the stator rated current. The circle's center is at the origin O of the coordinates, and its radius is I_{Gn}. The angle φ between

356 | 11 Automatic Excitation Regulator

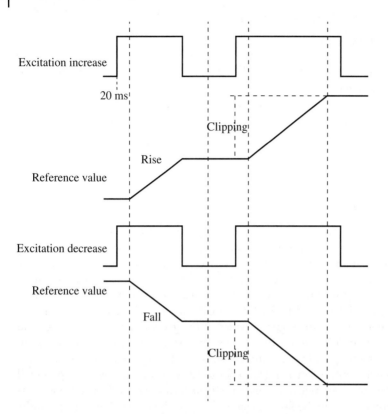

Figure 11.21 Reference value adjustment of excitation regulator.

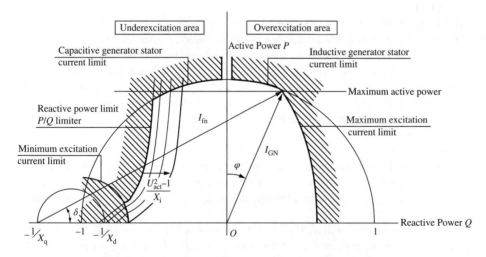

Figure 11.22 Operating capacity curve of typical salient pole synchronous generator.

the radius phasor and Axis P is equal to the generator power factor angle. The active power limit value (3) is determined by the generator rated active power. The leading phase capacity limit curve (4) is determined by such factors as the generator static stability, the heat generation at the generator stator end, and the minimum excitation current limit. The minimum excitation current limit value (5) depends on the generator static stability limit. Generally, the minimum excitation current $I_{fmin} = 0.1\, I_{fn}$ is taken.

Within the normal range of the generator operating capacity curve, the operating point excitation is regulated d by the excitation regulator, so that the generator can operate within the allowable range of the power factor. The input variable of the PID controller in the excitation regulator is the generator voltage deviation signal, and it is the main control signal. If the generator voltage deviation main ring closed-circuit control signal cannot meet the operational requirements for some reasons, the limiter will be activated under the preset rule. In this case, the deviation signal outputted by the auxiliary ring composed of the limiter will, in place of the voltage deviation main signal, regulate the excitation.

3. Reactive power limiter

    In the automatic operation mode, if the generator is based on the leading power factor (i.e., in the leading phase operation) and the common operating points may compromise the static stability of the synchronous generator, the excitation system output will increase to or exceed the corresponding value of the "automatic reactive power limit line." Meanwhile, an alarm will be issued and excitation decrease prohibited.

    In manual control mode, if the AVR automatic channel controller is operating normally, the manual given value will be prohibited from decreasing to below the "manual reactive power limit." If the active power increase causes the manual constraint limit to increase, the manual given value will also increase.

    For the calculation of the "manual constraint limit line" in an automatic controller, if the automatic controller is abnormal when the manual controller is selected, the constraint limit line value before the abnormality should be chosen.

    In the manual control mode, if the reactive or active power measurement is wrong, the operator can arbitrarily increase or decrease the manual given value to ensure an adequate stability margin.

    If the manual given value decreases to the manual constraint limit value, a signal alarm will occur.

    In the reactive power limiting function, there are two limiting modes: reactive power low excitation limit and reactive power overexcitation limit.

    Here, it should be emphasized that the depth of the leading phase is closely related to the generator terminal voltage value. The higher the generator terminal voltage, the deeper the leading phase, and vice versa.

4. V/Hz overexcitation limiter

    When the generator is operating without load, the V/Hz overexcitation limiter can ensure that the ratio of the generator voltage to the frequency does not exceed the preset limit value.

    When the generator is loaded, the limiter can ensure that the generator voltage does not exceed the rated value.

    The limiter action limiting mode is definite or inverse time lag. The characteristic curves of the V/Hz overexcitation limiter are shown in Figure 11.23.

    Section 6.5.11 of the power industry standard DL/T 843 – 2010 *Specification for Excitation System for Large Turbine Generators* provides that an AVR's V/Hz limiting characteristics shall be matched with the overexcitation characteristics of the generator and the main transformer. An AVR's V/Hz limiting shall have definite and inverse time lag characteristics. The excitation regulation for the dynamic process of the generator shall not be affected by the V/Hz ratio limiting unit. The inverse time lag characteristics had better adopt multi-point expression in non-functional form and should be coordinated with the definite and inverse time lag characteristics of the overexcitation protection. Table 11.2 shows the coordination between the set values of the V/Hz overexcitation protection action of a large generator and the set values of the V/Hz overexcitation limiter action of the excitation regulator.

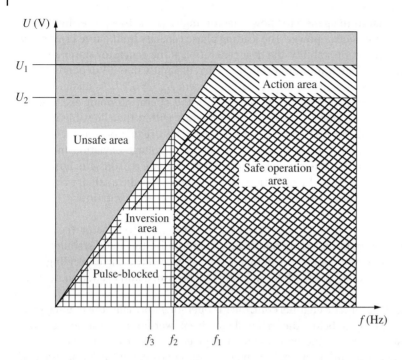

Figure 11.23 Characteristic curves of V/Hz overexcitation limiter.

Table 11.2 Coordination between set values of V/Hz overexcitation protection of generator and set values of V/Hz overexcitation limiting of excitation regulator.

A. Set values of V/Hz overexcitation protection of generator						
V/Hz multiple	1.10	1.14	1.18	1.20	1.26	1.30
Action time (s)	80	16	6	2	0.2	0
B. Set values of V/Hz overexcitation limiting of excitation regulator						
V/Hz multiple	1.08	1.12	1.16	1.20	1.24	1.28
Duration (s)	40	8.8	3.0	1.0	0.1	0

Figure 11.24 shows the generator V/Hz overexcitation characteristic curves provided by HEC, DFEM, and Alstom. Figure 11.25 shows the main transformer V/Hz overexcitation characteristic curves provided by different manufacturers.

5. Maximum excitation current limiter

   The functions of a maximum excitation current limiter are as follows:
   (1) It can adopt the superposition control mode, as it instantaneously quits through an instantaneous action to quickly limit the excitation maximum output current in the event of a direct short circuit at the excitation system output end.
   (2) It can limit any excitation current that exceeds the ceiling current in the event of a faulty forced excitation.
   (3) It can effectively limit the short-circuit current resulting from a short circuit in the rectifier bridge AC input and DC output circuit or between the rotor excitation winding turns.

   The functional characteristic curves of the maximum excitation current limiter are show in Figure 11.26.

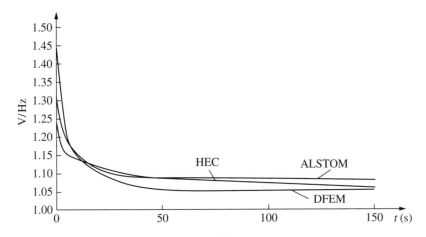

Figure 11.24 Steam turbine V/Hz overexcitation characteristic curves.

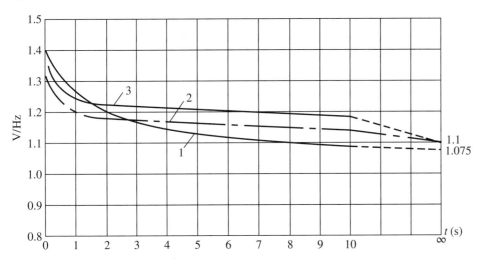

Figure 11.25 Main transformer V/Hz overexcitation characteristic curves provided by different manufacturers.

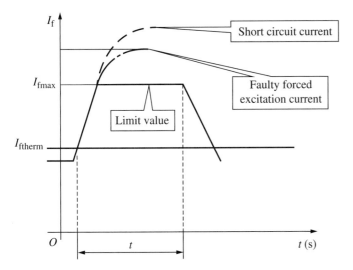

Figure 11.26 Functional characteristics of maximum excitation current limiter.

11 Automatic Excitation Regulator

6. **Generator rotor current limiter**
   In the automatic operation mode, the system provides three levels of rotor excitation current limit:
   (1) Instantaneous current limit;
   (2) Given current limit level with preset time limit function;
   (3) $I^2 t$ over temperature protection system in compliance with the heat capacity requirements set out in ANSI C50.13 1989.

   If the current value (including the delay value) or the $I^2 t$ value exceeds the preset value, the current limiter acts and the limit value drops to the current limit given level. When the excitation current drops to below an allowable continuous value, the current limit level will rise.

   In the manual mode, the operator directly controls the operating current, but the maximum current that can operate is limited to the preset level of the system parameter.

   Figure 11.27 shows the heat capacity model of the generator rotor excitation winding. The model expresses the heat balance between heating and cooling. The limiting takes temperature limit exceeding as a standard and takes into account heating and cooling delay. In order to prevent stator or rotor excitation winding overload caused by repeated forced excitation, when the cooling time is insufficient, the allowable overcurrent time is correspondingly shortened.

   Under the basic principle of current thermal effect and the provisions on generator rotor overload in DL/T 684-2012 *Guide of Calculating Settings of Relay Protection for Large Generator and Transformer*, the generator rotor winding thermal limit calculation formula is as follows:

   $$t = \frac{C}{I_{fn}^2 - 1}$$

   where

   C – heating coefficient $C = 33.75$.

   Tables 11.3 and 11.4 respectively show examples of the coordination between the set values of the overcurrent tolerance of the generator rotor excitation winding and the set values of the forced excitation inverse time lag limit of the excitation regulator

7. **Generator stator current limiter**
   The stator current limiter will instantaneously act when the generator is operating at the leading power factor but delay in action when the generator is operating at the lagging power factor. The stator current limiter characteristic curve is shown in Figure 11.28.

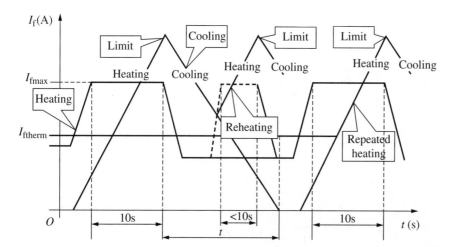

Figure 11.27 Heat capacity model of generator rotor windings.

Table 11.3 Overcurrent multiple tolerance of generator rotor excitation (based on GB/T 7064, C = 33.75).

Current multiple	Duration (s)
2.09	10
1.46	30
1.25	60
1.13	120

Table 11.4 Set values of forced excitation inverse time lag of excitation regulator (C should not exceed 30).

Current multiple	Duration (s)
2.09	8.9
1.46	26.5
1.22	53
113	108

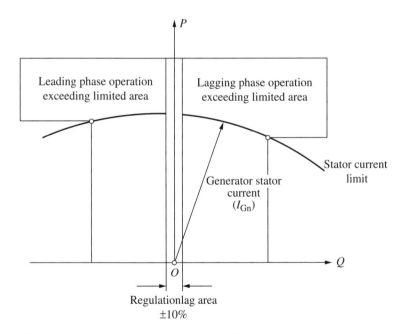

Figure 11.28 Generator stator current limiter characteristic curve.

Function description for a stator current limiter:
(1) For the excitation system, the stator current limiter can only limit the reactive component of the generator stator current but cannot limit the active component of the stator current.
(2) When the generator current goes beyond the limited area, the limiter will act to increase the excitation current if the generator is operating in the leading phase and will act to decrease the excitation current if the generator is operating in the lagging phase. In short, the role of the stator current limiter is to

regulate the reactive component of the generator stator current, to limit the generator current to the limited area.

(3) A regulation lag area should be provided in the vicinity of the power factor $\cos\varphi = 1$ ($\pm 10\%$). When the generator stator current is in this area, the limiter will not participate in the regulation or limitation.

(4) The purpose of the regulation lag area: The limiter's participation in the regulation in this area will cause frequent fluctuations in the generator stator current, which will affect the operational stability of the generator. When the generator current goes beyond this area, the regulator will implement the regulation, limiting the generator stator current to the limited area by decreasing the active component of the generator current.

(5) The basis for the stator current limit is the heat capacity of the generator stator windings. When the generator stator current is lower than the generator stator current limit setpoint, the generator stator windings are in a dissipate heat process. The regulator performs a reverse integration, until the stator heat accumulation is 0. When the generator stator current is greater than the generator stator current limit setpoint, the regulator begins to make an integral computation of the generator stator heat accumulation. When this value reaches the allowable heat capacity value of the generator stator (a certain safety margin is taken into account), the regulator will decrease the generator stator current to a value smaller than the generator stator current limit setpoint, so that the generator rotor can be in a cooling state and generator operation safety can be ensured.

(6) The stator current limit can also be achieved with the power factor constant operation mode, in which the allowable reactive power outputted by the generator is determined on the basis of the allowable active power outputted by the generator.

In accordance with GB/T 7064, the corresponding expression of the overcurrent tolerance of the generator stator current windings is as follows

$$t = \frac{C}{I_{Gn}^2 - 1}$$

where

C − heating coefficient (37.5).

Tables 11.5 and 11.6 respectively show examples of the coordination between the set values of the heat resistance of the generator stator windings and the set values of the stator current limit of the excitation regulator.

8. Excitation regulator superposition control mode

Generally, in a generator excitation regulation system, the deviation of the generator terminal voltage from the given reference value is the regulation target.

Another mode is superposition control, in which the system voltage on the high-voltage side of the generator main transformer provides an additional signal which acts on the given reference value integration point of the excitation regulator. The regulator formed under this principle can also be called the power system voltage regulator (PSVR). Since the system voltage on the high-voltage side of the main

Table 11.5 Generator stator over current multiple ($C = 37.5$).

Current multiple	Duration (s)
2.0	12.5
1.8	16.74
1.5	30
1.15	113

Table 11.6 Set values of generator stator current limiter (C should not exceed 33.75).

Current multiple	Duration (s)
2.0	11.25
1.8	15.0
1.5	2.7
1.15	104.7

transformer is regulated, the AVR control mode that keeps the generator voltage constant can further improve the ultimate output reactive capacity of the generator. A PSVR simplified model is shown in Figure 11.29.

In the system voltage superposition control mode, when the generator is unloaded, the generator voltage tracks the system voltage changes and can replace the synchronizing voltage regulator to accelerate the generator's grid connection and reduce the reactive power impact at the time of grid connection.

After the generator is connected to the grid, depending on the system voltage changes, it automatically regulates the generator reactive power and maintains the stability of the generator voltage, thus further improving system operation stability.

9. ST5B model

In the international standard IEEE 421-5 2005, the model of the voltage source self-shunt excitation system ST5B is provided. The block diagram of the transfer function of the model is shown in Figure 11.30.

The ST5B model has the following features:

(1) A series PID correction element is connected into the AVR main ring.
(2) The high- and low-flux gates are respectively introduced into the limiting function in the limiting circuit as required by logic control.
(3) An independent series PID correction element is introduced into the limiting auxiliary element.

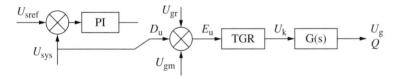

Figure 11.29 Simplified PSVR superposition control model: U_{sref} - reference voltage; U_{sys} - synchronous signal voltage; U_{gr} - generator voltage.

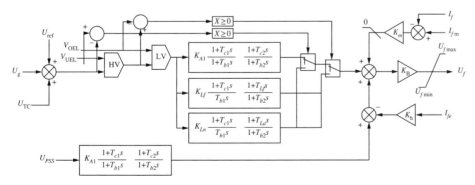

Figure 11.30 ST5B model.

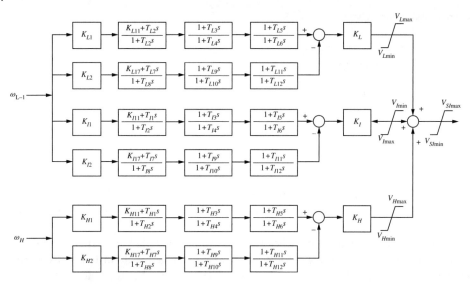

Figure 11.31 PSS-4B model.

10. PSS-4B model

    The PSS-4B model complies with the international standard IEEE 421-5 2005. It features three-band frequency full-range optimal compensation, with the optimal damping characteristics lying within the range 0.1–1.0 Hz. In addition, it can effectively suppress reactive power reverse regulation. Thus, it is adaptable to various modes of contact with grid.

    Figure 11.31 shows the PSS-4B model.

12

Excitation Transformer

12.1 Overview

In recent years, with the continuous growth of capacity of hydropower and thermal power units, in terms of choice of excitation mode, there has been a transition from conventional rotary AC exciters to static self-excitation systems. So, as one of the important components of a static self-excitation system, the excitation transformer has attracted the attention of manufacturers and operation departments of power stations.

By insulation mode, the following are the excitation transformer types used in excitation systems for large hydropower and thermal power units:

(1) Resin cast dry-type transformer with epoxy resin as insulating material
(2) Alkali-free glass-fiber-wound impregnated dry-type transformer
(3) MORA dry-type transformer
(4) Nomex® dry-type transformer

In these transformer insulation modes, resin cast insulation and wound dry insulation are used more extensively at present.

(1) For a dry-type transformer with cast insulation, the wire-wound windings (including the high-voltage winding and the low-capacity low-voltage winding) are based on the vacuum casting process. The foil winding is primarily used for the high-capacity low-voltage winding.

    The vacuum casting process eliminates the possibility of bubbles inside the transformer to reduce the partial discharge of the transformer windings, thus ensuring the service life of the transformer. Meanwhile, the casting process makes the windings a steel body with strong ability to withstand short circuit. Conductive materials of windings of cast dry-type power transformers are primarily copper and aluminum.

(2) For an alkali-free glass-fiber-wound impregnated transformer, the low-voltage winding is generally foil, while the high-voltage winding is wrapped around the winding machine, with an epoxy glass cloth roller as its internal mold. The wire and the glass fiber are wrapped simultaneously, and then the fiber impregnated with resin is compound-wound around the already wound wire via the resin tank. After the whole winding is completed, it is heated in an oven for curing, so that it can be made a whole. The biggest advantage of the wrap-around structure is that no special mold or casting equipment is needed. However, since the resin is added under conventional conditions rather than being cast under vacuum conditions, there is inevitably air in it, which can easily cause partial discharge and thus reduce its operational reliability. Moreover, wrap-around transformers require many labor hours. In summary, a wound transformer is more costly than a general cast transformer.

(3) The MORA dry-type transformer is a new dry-type transformer developed by the German transformer manufacturer MORA in the recent two decades under the premise of adaptability to the new concept of environmental protection and improvement of flame retardance with new processes, technologies, and materials.

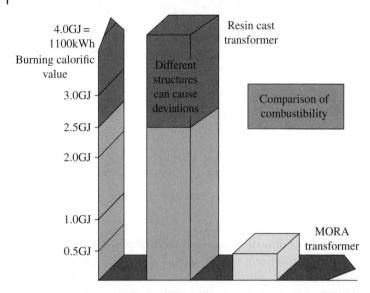

Figure 12.1 Comparison of combustibility between MORA dry-type transformer and resin cast dry-type transformer.

The MORA transformer's high-voltage winding is edgewise wound layer by layer around the well-insulated ceramic bracket. There are cooling ducts in the high- and low-voltage windings and vertically and horizontally between the windings. So, the transformer has good short-term overload and resistance to short-circuit stress.

For the MORA transformer, the winding material is impregnated with insulating paint and then dried in a vacuum state. The process is simple, easy, and low-cost.

The MORA transformer winding is made of glass fiber or Nomex insulating material. Its allowable temperature is −50 °C to 155 °C, in Insulation Class F or H.

It should be emphasized that, in order to improve its flame retardance, the MORA transformer as a whole adopts a very small amount of combustible insulating material. The combustible material accounts for only about 1% of its total weight. Figure 12.1 shows a comparison of combustibility between the MORA transformer and a resin cast transformer. It can be seen from Figure 12.1 that the MORA dry-type transformer has an obvious advantage in flame retardance compared to the resin cast dry-type transformer. For excitation devices applied to underground plants as well as excitation devices where the excitation transformer and power rectifier cabinet are configured together, such a transformer with low combustibility and zero smoke or toxic gas release at the time of combustion appears to be more important for improvement of the operational safety of power stations.

Besides, as an environmentally friendly product, the MORA transformer can be disassembled after it loses efficacy, and its winding copper can be recycled; it is not easy to achieve this with the resin cast dry-type transformer.

(4) Another open impregnated dry-type transformer is an American transformer with Nomex insulating material, which differs from the German dry-type transformer as shown in Table 12.1.

The Nomex and MORA open impregnated dry-type transformers have the following similarities in insulation and process:

1) Foil low-voltage coil or layered low-voltage coil with multiple paralleled conductors
2) Fully paralleled (open) high-voltage winding
3) Windings based on the impregnation process
4) Class C or H insulation system

Table 12.1 Comparison of insulation mode and process between two open impregnated dry-type transformers.

Nomex (American)	MORA (German)
Turn insulation: high-temperature-resistant Nomex insulation paper	Turn insulation: glass fiber
Cushion block: Nomex cardboard or comb stay	Cushion block: ceramic chip
Impregnation process: vacuum pressure impregnation (VPI)	Impregnation process: vacuum impregnation (VI)

In addition, compared to the resin cast dry-type transformer, the open impregnated dry-type transformer boasts a simpler manufacturing process similar to that of the oil-immersed transformer, lower manufacturing cost, and better fire proofing and flame retardance. Moreover, for the Nomex impregnated dry-type transformer, an encapsulated structure that can further improve its fouling resistance and moisture protection has been developed.

It should be emphasized that, with technical progress, the requirements for transformer insulation have been increasing. In particular, for excitation transformers applied to underground plants, the insulation should carefully chosen.

In the 1950s, transformers generally used mineral oil as a cooling medium, but mineral oil is associated with great fire safety risk. In the 1960s, resin cast dry-type transformers were developed by Germany. Since then, they have almost become the recognized alternative to mineral-oil-cooled transformers. Over the nearly half a century of application of resin cast transformers, it has been noted that such transformers boost performance reliability on the whole. However, failures were seen with resin cast dry-type transformers made by both Chinese and other manufacturers. For example, a large hydropower station experienced an inter-turn short circuit on an excitation transformer with a single-phase capacity of 3000 kVA, which expanded into an inter-layer short circuit. Moreover, due to the poor flame retardance, there was an open flame. Even after the relay protection action cut off the electrical circuit, the open flame continued and emitted toxic gases, endangering the personal safety of firefighters.

So, exploring and developing impregnated transformers using a new insulating liquid to bring about improvements in transformers' fireproofing, environmental friendliness, and operational reliability has once again become a topic of concern in engineering.

At present, such ester-based synthetic ester transformers have made remarkable progress and are widely used in fields requiring high environmental friendliness and fireproofing such as high-rise public buildings, factories, high-risk sites as well as renewable energy such as wind energy (e.g., offshore wind power) and tidal energy. Besides, fire is a risk for activities in offshore environments. Since ester insulating liquids are biodegradable, once they are leaked, they can be dissolved in water, without side effects on aquatic life. With their ignition point up to 322 °C, they feature good flame retardance. So, the MIDEL-7131 synthetic ester insulating liquid has been widely used in the offshore oilfield industry.

So, for devices installed in underground plants of hydropower stations such as main transformer sand excitation transformers, synthetic ester insulating liquids as a cooling medium are a good choice in terms of fireproofing and environmental friendliness.

12.2 Structural Characteristics of Resin Cast Dry-Type Excitation Transformer

In this section, a brief description of the performance characteristics of resin cast dry-type field transformers, which are widely used at present, is presented.

12.2.1 Core

The material of the core of a resin cast dry-type transformer is high-quality cold-rolled grain-oriented silicon steel sheet, of a 45° fully inclined joint structure. The core column is banded with insulating tape, with its surface sealed with special resin for moisture and rust protection.

12.2.2 Winding Insulation Structure

Currently, there are two resin cast dry-type transformer winding processes: casting and winding.

12.2.3 Winding Material

The winding material can be copper wire, copper foil, aluminum wire, or aluminum foil. However, for the sake of design, some non-Chinese manufacturers promote a certain winding material. For example, Germany's Siemens recommended an epoxy cast dry-type transformer with aluminum foil windings in its bidding document for excitation systems for the generators of Three Gorges Hydro Power Station to take the thermal expansion coefficient into account. Since the thermal expansion coefficient of copper is 17 and that of resin is 28–30, aluminum windings can withstand smaller internal stress, and the capacitance distribution between the aluminum foil windings can make the potential distribution more even.

Moreover, Siemens believes that it is easier to achieve temperature rise reduction for aluminum foil windings than for copper windings.

Since different winding materials have different thermal expansion coefficients, composite epoxy resins, for example, quartz powder epoxy resin or glass fiber epoxy resin, should be doped.

Table 12.2 lists the performance parameters of the insulating materials.

It can be seen from Table 12.2 that the linear expansion coefficient of the epoxy resin and glass fiber composite material is close to that of copper. So, windings made of the glass fiber and epoxy resin composite insulating material and copper have significant impact resistance, resistance to temperature changes, and crack resistance.

12.2.4 Heat Dissipation and Cooling

There are two cooling modes for dry-type transformers: air natural circulation and air forced circulation, respectively, represented by AN and AF (A: air; N: natural circulation; F: forced circulation). The cooling mode of products with air cooling system can be represented by AN/AF.

In the AN mode, under normal working conditions, the transformer can continuously output 100% of the rated capacity. In the AF mode, varying degrees of overload operation are allowed.

The allowable overload multiple in the AF mode is applicable to various emergency overload or intermittent overload operations. In this case, the large load loss and short-circuit impedance increases will inevitably

Table 12.2 Performance parameters of resin composite materials.

Performance Name	Dielectric strength (kV mm^{-1})	Impact strength (kJ m^{-2})	Bending strength (MPa)	Linear expansion coefficient ($\times 10^{-6}$/°C)
Pure epoxy	16–19	18–25	130–150	60–70
Quartz powder epoxy	18–20	10–12	110–120	28–35
Glass fiber epoxy	18–20	40–60	420–480	16–20
Copper	—	—	—	17
Aluminum	—	—	—	24

lead to an increase in the temperature rise, transformer overheating, and acceleration of insulating material aging and failure. So, continuous overload operation in the AF mode is not recommended. For an excitation transformer, in capacity selection, the generator's forced excitation multiple is taken into account, and a certain margin is set aside. So, for an epoxy cast dry-type excitation transformer, the addition of an AF cooling device may not be considered. However, in a place with poor ventilation, the addition of an AF cooling device must be considered.

12.3 Application Characteristics of Resin Cast Dry-Type Excitation Transformer

Since an excitation transformer supplies DC power to the synchronous generator's excitation winding via a (controllable or non-control) rectifier, it in essence is a non-standard rectifier transformer, different from a power transformer. So, the following special operating statuses should be taken into account in the design:

(1) Since the harmonic current in the high-/low-voltage windings of the excitation transformer generates an additional loss, the design capacity of the excitation transformer should be higher than the rated capacity determined by the fundamental current.
(2) The harmonic current determined by the grid and transformer impedances causes excitation transformer core excitation voltage waveform distortion (harmonic voltage), which may lead to core saturation. So, the design size of the core should be properly amplified to prevent saturation.
(3) The harmonic current generates a harmonic voltage drop in the corresponding harmonic impedance. This harmonic voltage will increase the no-load loss and transformer noise. The harmonic current will produce an increase in the load loss and an unevenly distributed eddy current loss, thus causing local overheating.
(4) There is a transient commutation overvoltage for the excitation transformer.

In order to avoid the induced overvoltage caused by the capacitive coupling between the high- and low-voltage windings, a ground shield is generally provided between the windings.

12.4 Specification for Resin Cast Dry-Type Excitation Transformer

12.4.1 Rated Capacity of Excitation Transformer

In addition to supplying the maximum continuous load excitation capacity to the generator, the excitation transformer provides the instantaneous output capacity of the forced excitation voltage and current in the case of forced excitation. By definition, the working capacity of the excitation transformer is as follows:

$$S_N = U_{1N} I_{1N} = U_{2N} I_{2N}$$

where

U_{1N} — primary rated voltage of excitation transformer, equal to generator terminal rated voltage;

I_{1N} — primary rated current of excitation transformer;

U_{2N} — secondary rated voltage of excitation transformer, which should meet the requirement of provision of maximum forced excitation voltage;

I_{2N} — secondary rated current of excitation transformer, which should meet the requirement of provision of forced excitation current. It can be determined by excitation current of maximum continuous load of generator, since forced excitation current duration is generally 10–20 s.

Besides, since the excitation transformer takes the rectifier as a load, the harmonic current component increases the harmonic loss. In view of this, in the determination of the rated capacity of the excitation transformer, it is necessary to taken into account the harmonic current additional loss factor, appropriately increase the secondary rated current value, and take this current as the temperature rise test current. This is the method that is widely used to determine the rated capacity of the excitation transformer.

12.4.2 Connection Set

The Chinese standard provides five three-phase double-winding power transformer connection sets: Yyn0, Yd11, YNd11, YNy0, and Yy0. In terms of excitation transformer connection set selection, China follows the power transformer standard, generally choosing Yd11. The reason for this is that when the star connection is adopted on the primary side of the excitation transformer, the primary winding phase voltage is only $1/\sqrt{3}$ of the line voltage, reducing the withstand voltage level of the primary winding. The secondary winding triangular connection can provide a branch for the third harmonic short-circuit current to offset the third harmonic flux and improve the secondary phase voltage waveform.

It should be emphasized that for a static self-excitation system of a hydro-turbine, for ease of connection of the excitation transformer and the generator's closed bus and prevention of interphase short circuit, a three single-phase transformer bank composed of three single-phase transformers is generally adopted for the excitation transformer. Since the magnetic circuits of the cores of the three phases are independent of each other, the main magnetic flux and the third harmonic flux are closed along the same magnetic circuit. Since the magnetic resistance of the magnetic circuit of each phase is small, the third harmonic magnetic flux is large. The third harmonic emf caused by the third harmonic is high, with its amplitude up to 45%–60% of the fundamental amplitude. This will distort the phase emf waveform, increase the phase voltage, and compromise the transformer's operational safety. So, for an excitation transformer consisting of a three-phase transformer set, a triangular connection must be adopted for the primary or secondary winding, so that the third harmonic current flowing through the triangularly connected winding can be used to offset the third harmonic flux.

Besides, when a Yd connection, for example, Yd11, is adopted for the excitation transformer, the secondary phase voltage lags the primary phase voltage by 30° in phase. So, this factor should be taken into account in the selection of the synchronous transformer connection of the thyristor trigger circuit.

In terms of the excitation transformer connection set selection, in some cases the primary and second connection sets are the same, so that the phase coordination of the trigger synchronous transformer can be simplified. For example, in the static self-excitation system of a 600 MW steam turbine generator produced by Hitachi, the connection set Dd0 is adopted for the excitation transformer. Hitachi believes that with the advancement of insulation technology, the withstand voltage level of the triangular connection on the primary side is safe. However, during the excitation equipment debugging, an excitation transformer winding breakdown occurred.

In order to improve the waveform of an excitation transformer with a thyristor rectifier load, Yd is generally adopted for the excitation transformer. The connection set affects the excitation regulator's phase coordination.

It should be emphasized that the excitation transformer connection set also affects the winding structure. For example, foil structure is ideal for the low-voltage-side winding of the excitation transformer as it can maintain the reactance balance and improve the sudden short-circuit strain capacity. When its capacity is above 1250 kVA, the transformer winding reactance height is generally 950–1000 mm. If a star connection is adopted for the low-voltage side, the secondary phase current will be $\sqrt{3}$ times that in the case of a triangular connection. The thickness of the foil winding also increases when the reactance height is kept constant. When its capacity is between 1250 and 2500 kVA, the secondary winding copper foil thickness is 1–1.6 mm. The increase in the copper foil thickness not only increases the eddy current loss (up to 14% of the total copper consumption) but also makes the en-winding extremely difficult. So, for a large excitation transformer, a triangular connection is a better choice for the secondary low-voltage winding.

12.4.3 Insulation Class and Temperature Rise

Currently, the insulation class of excitation transformers supplied from outside to China is generally F or F/H, while that of China-made ones is F or F/B. The temperature rise of the insulation class is based on the international standard, as shown in Table 12.3.

Table 12.3 Temperature rise limit for transformer.

Insulation class	Insulation system temperature (°C)	Winding hot spot temperature (°C)		Rated current winding average temperature rise (K) ($\Delta\theta_{WT}$)
		Rated value (θ_c)	Maximum allowable value (θ_{cc})	
A	105	95	140	60
E	120	110	155	75
B	130	120	165	80
F	155	145	190	100
H	180	175	220	125
C	220	210	250	150

In a transformer, the components involving temperature rise include the windings, the core, and the additional fasteners.

It should be emphasized that when determining the temperature rise limit of excitation transformers, some users often request a reduction of one class in temperature rise based on the insulation class. For example, for the excitation transformers of Venezuela's Guri-2 Power Station, the insulation class is F, and the allowable temperature rise 100 K. But the user requests a temperature rise assessment based on Class B. That is, the maximum temperature rise is 80 K. The actual temperature is the sum of the temperature rise and the ambient temperature. The ambient temperature is set at 40 °C.

The connection lines outside the core or windings are generally not subject to the temperature rise limit. But it is required that the corresponding temperature should not be such that it damages the core or the other parts or the adjacent materials. For an epoxy cast dry-type excitation transformer, since the magnetic density is low, the core temperature rise is low and does not damage the core or the other parts or the adjacent materials.

12.4.4 Impedance Voltage

The transformer short-circuit impedance voltage is an important parameter, which affects the rectifier's phase commutation working state and the transformer's short-circuit current value.

In the event of a generator magnetic field short circuit or a collector ring flashover and a rectifier bridge arm short circuit, the short-circuit current will be determined by the transformer's impedance voltage.

The short-circuit impedance voltage of a large-capacity excitation transformer is significantly higher than that of an ordinary power transformer. For example, non-Chinese manufacturers' bidding documents for excitation systems of the generators of Three Gorges Hydro Power Station show that the excitation transformer short-circuit impedance is generally 6%–12%, though some manufacturers' recommendations are up to 16%.

It should be noted that in the selection of the impedance voltage of an excitation transformer, various factors such as the short-circuit current limit, rectifier commutation working state, and function of de-excitation mode should be taken into account.

For example, when the AC de-excitation mode is adopted, in the event of a direct shot circuit on the DC side of the rectifier, the short-circuit current can be limited by means of the overcurrent trip protection function, without needing to depend only on the short-circuit impedance of the excitation transformer.

Besides, if the excitation transformer's short-circuit impedance is too large, under the forced excitation conditions, the decrease in the secondary voltage of the excitation transformer and the increase in the excitation current may cause a transition of the working point of the external characteristic of the rectifier to the Commutation State II, resulting in a significant decrease in the excitation output voltage.

12.4.5 Short-Time Current Overload Capacity

The transformer's short-time current overload capacity is affected by many factors such as structure, material, operating conditions, and initial load. Various excitation transformer types have different overload capacity curves. Figure 12.2 shows the overload capacity curves of the SCB9 resin cast dry-type transformer.

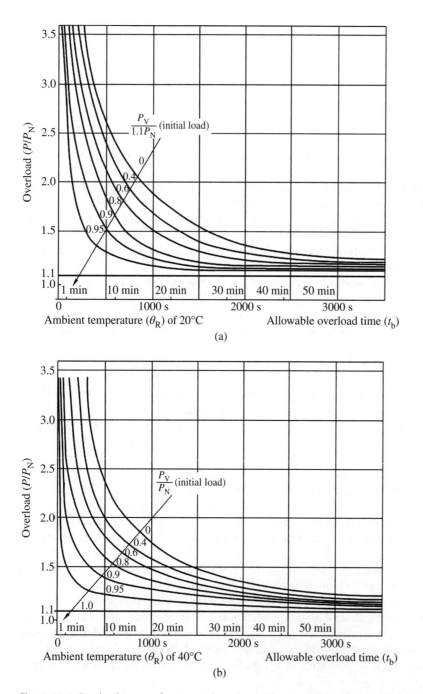

Figure 12.2 Overload curves of resin cast dry-type transformer (I): (a) ambient temperature $\theta_R = 20\,°C$; overload curves of resin cast dry-type transformer (II), (b) ambient temperature $\theta_R = 40\,°C$. P_N – rated capacity; P_V – initial load; P – overload capacity.

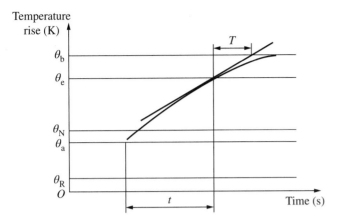

Figure 12.3 Variation of temperature rise over rated capacity in a short time.

It should be noted that an excitation transformer is also subject to voltage overload capacity requirements, because voltage overload will lead to increases in core loss and excitation current.

For example, in an actual application, after an overhaul or start of the generator, if the excitation transformer in the self-excitation system works, in the generator inter-turn withstand voltage test, the excitation transformer will withstand a voltage 1.15–1.30 times the rated voltage. Attention should be paid to this when determining the excitation transformer specifications.

The constraint on an excitation transformer's short-time current overload capacity is the winding's maximum temperature rise. For example, a Class-F insulation system's winding temperature is 155 °C and its winding maximum temperature rise is 100 K, while a Class B insulation system's winding temperature is 130 °C and its winding maximum temperature rise is 80 K. This provision is consistent with Table 12.3 and the international standard.

Under the condition of short-time current overload, the variation law of the average temperature rise θ_e of the excitation transformer windings can be expressed as follows [27]:

$$\theta_e = \theta_a + (\theta_b - \theta_a)(1 - e^{-t/T}) \tag{12.1}$$

where

$$\theta_a = \theta_N \times (S_a/S_N)^{1.6} \quad (K) \tag{12.2}$$
$$\theta_b = \theta_N \times (S_b/S_N)^{1.6} \quad (K) \tag{12.3}$$
$$t = T \ln \frac{\theta_b - \theta_a}{\theta_b - \theta_e} \tag{12.4}$$

where

- θ_a — initial steady-state temperature rise of windings K;
- θ_b — short-time overload steady-state temperature rise of windings K;
- T — heating time constant of windings;
- t — short-time over nameplate rated capacity s;
- θ_N — rated temperature rise K;
- S_N — rated capacity of excitation transformer kVA;
- S_a — initial capacity before overload kVA;
- S_b — capacity over rated value in a short time kVA.

The physical meaning of Eq. (12.1) is depicted in Figure 12.3, where θ_R represents the ambient temperature, which can be considered to be 20 or 40 °C.

As can be seen from Eq. (12.1), the winding average temperature rise θ_e caused by the short-time overload is related to parameters such as θ_a, θ_b, T, and t.

For copper windings, the heating time constant $\mathbb{T}$ can be approximately expressed as follows:

$$T = 165 \times \frac{\theta_b}{J_b^2} \qquad (12.5)$$

where

J_b — winding current density corresponding to short-time overload A mm^{-2}.

In Eq. (12.5), the coefficient (165) depends on the copper conductor's density and resistivity as well as the specific heat capacity of the conductor and the resin insulating material.

In $J_b = \left(\frac{S_b}{S_N}\right) J_N$, J_N is the winding rated current density. If $\frac{S_b}{S_N} = K_b$, the following can be obtained:

$$J_b = K_b J_N$$

$J_b = K_b J_N$ and Eq. (12.3) are substituted into Eq. (12.5), yielding Eq. (12.6):

$$T = 165 \times \frac{\theta_N \times K_b^{1.6}}{(K_b J_N)^2} \qquad (12.6)$$

As can be seen from Eq. (12.6), decreasing the winding rated current density J_N can increase the winding heating time constant. When the overload time $t = (3-4)\,T$, the winding heating is close to the short-time overload steady-state temperature rise θ_b.

Example 12.1

For an excitation transformer, the rated capacity is 2450 kVA, primary voltage 15.75 kV, secondary voltage 0.6 kV, and impedance voltage $U_X^* = 6\%$. The insulation class is F. The high-voltage winding rated temperature rise $\theta_{N1} = 80.93$ K. In terms of the low-voltage winding rated temperature rise, the first layer is 80.41 K, the second layer 790.22 K, and the third layer 80.93 K. The low-voltage winding rated average temperature rise $\theta_{N2} = 80.2$ K. In terms of the winding current density, $J_{N1} = 1.618$ A mm^{-2} and $J_{N2} = 1.964$ A mm^{-2}. How should the low-voltage winding's temperature rise and temperature rise time be calculated when the generator undergoes 20 s of forced excitation and the forced excitation current reaches twice the rated value?

Solution

It is known that the rated temperature rise $\theta_{N2} = 80.2$ K ≈ 81 K. With Eq. (12.2), the winding initial steady temperature rise can be calculated.

$$\theta_a = 81 \times \left(\frac{2450}{2450}\right)^{1.6} = 81(\text{K})$$

When the generator undergoes forced excitation and the forced excitation current reaches twice the rated value, the steady-state temperature rise can be calculated with Eq. (12.3).

$$\theta_b = 81 \times \left(\frac{2 \times 2450}{2450}\right)^{1.6} = 245.5 \ (\text{K})$$

When the forced excitation current reaches twice the rated value, the low-voltage winding's current density is as follows:

$$J_{b2} = \frac{2 \times 2450}{2450} \times 1.964 = 3.928 \ (\text{A/mm}^2)$$

With Eq. (12.5), the heating time constant can be obtained.

$$T = 165 \times \frac{245.5}{3.928^2} = 2622(\text{s}) = 43.7(\text{min})$$

With Eq. (12.1), the low-voltage's winding average temperature rise after 20 s of forced excitation can be obtained.

$$\theta_e = 81 + (245.5 - 81) \times (1 - e^{20/2622}) = 82.3 \text{ (K)}$$

There is an increase of only 1.3 K over the rated average temperature rise.

According to Germany's industry standard VDE0532, $\theta_C \leq 120$ K ensures conformance. If this average temperature rise is reached, the temperature rise time can be obtained from Eq. (12.4):

$$t = 2622 \times \ln \frac{245.5 - 81}{245.5 - 120} = 710 \text{ (s)} = 11.8 \text{ (min)}$$

So, the low-temperature winding's temperature rise is 82.3 K, and the temperature rise time is 11.8 min.

12.4.6 Resistance to Sudden Short-Circuit Current

12.4.6.1 Calculation of Sudden Short-Circuit Current

The sudden short-circuit current generated in the event of a direct short circuit at the rectifier output end is the most serious. When the power supply internal resistance and the circuit impedance are ignored, the short-circuit RMS current expressed as a per-unit value can be written as follows:

$$I_d^* = \frac{1}{Z_d} \tag{12.7}$$

where

Z_d — short-circuit impedance of excitation transformer expressed as per-unit value.

12.4.6.2 Thermal Stability

In the event of a sudden short-circuit fault, the heat generated by the high-value short-circuit current in a short time can be approximately considered to be absorbed by the windings. It cannot be dissipated into the surrounding media in a short time. The temperature rise of the windings depends on the magnitude and duration of the short-circuit current. According to the applicable standard, the duration for verification of a transformer's thermal stability when it withstands a short circuit current is 2 s. The maximum average temperature $\theta_2 \leq 350\,°\text{C}$.

The expression for the windings' achievement of the maximum average temperature is as follows:

$$\theta_1 = \theta_0 + aJ_D^2 t \times 10^{-3} \tag{12.8}$$

where

θ_0 — initial temperature of windings before short circuit;

J_D — current density of winding in short-circuit steady state A mm^{-2}.

When the thermal stability is calculated, its maximum value can be considered to be

$$\theta_1 = \theta_0 + \theta_t = 40 + 100 = 140(°\text{C})$$

For copper windings, the coefficient a is as follows:

$$a = 0.01976 \times [235 + (\theta_2 + \theta_1)/2]$$

When $\theta_2 = 350\,°\text{C}$ and $\theta_1 = 140\,°\text{C}$ are substituted into this equation. $a = 9.48$ is obtained.

The original data in Example 12.1 are taken. The corresponding short-circuit current density can be obtained:

$$J_D = J_D^* \times J_{N2} = \frac{1}{0.06} \times 1.964 = 32.8 (\text{A/mm}^2)$$

With Eq. (12.8), the following can be obtained:

$$\theta_1 = 140 + 9.48 \times (32.8)^2 \times 2 \times 10^{-3} = 141.4(°C)$$

Since $\theta_1 \leq 350\,°C$, conformance is ensured.

It can be seen that the temperature rise generated by the sudden short-circuit current within 2 s is only 21.4 K higher than the specified steady-state temperature rise 120 K, much smaller than the average maximum temperature limit 350°C.

12.4.6.3 Dynamic Stability under Action of Sudden Short-Circuit Current

An excitation transformer's dynamic stability when it withstands a short-circuit current should be assessed through a test. The structure of the excitation transformer affects its dynamic stability. For example, for a copper foil low-voltage winding, in terms of the axial direction, each winding has only one turn. So, the axial force generated in the case of a sudden short circuit is much lower than that for a winding with multiple parallel strands.

12.4.7 Overcurrent Protection

A short circuit on the DC side (e.g., a collector ring short circuit) is equivalent to a short circuit of the secondary winding of the excitation transformer. In the event of such a fault, the impedance voltage can be increased to limit the short-circuit current. Other protection methods can also be utilized. For example, a quick fuse can be adopted on the high-voltage side of the excitation transformer to improve the short-circuit capacity the rectifier can withstand. Or a choke reactance can be connected in series to the DC side of the rectifier to limit the short-circuit current's growth rate on the time integral. Or a fast overcurrent detection relay can be adopted on the DC side to latch the silicon-controlled rectifier (SCR) that is turned on when the short-circuit current reaches the limit value.

In addition to the above protection methods, in a modern self-excitation system, an AC transformer and an overcurrent protection relay that is started in timing or inverse time lag mode are generally connected to the secondary winding of the excitation transformer to send the trip order of the generator's main circuit breaker and magnetic circuit breaker when the specified value is reached.

12.4.8 Overvoltage Suppression

In addition to the impacts of external overvoltage, operating overvoltage, and fault overvoltage, there may be also impacts of commutation overvoltage, de-excitation overvoltage, transformer leakage inductance, and oscillation overvoltage composed of distributed capacitance on an excitation transformer in operation. At the initial stage, these overvoltages will generate a high potential gradient at the end of the excitation transformer and cause spike voltage oscillations during the subsequent absorption. So, they must be suppressed.

12.4.8.1 Closing Surge Overvoltage

Since there is parasitic capacitance between the windings of the excitation transformer, power supply put-in or cut-out, and atmospheric overvoltage, may generate an AC overvoltage in the transformer. In particular, when the excitation transformer is not powered by the generator terminal voltage but is powered by the station service power or the main transformer of the generator, special attention should be paid to this aspect.

For an overcurrent caused by closing, it is assumed that C_{12} is the parasitic capacitance between the primary and secondary windings and C_{20} is the capacitance between the secondary winding and the core. At the instant of the primary closing of the excitation transformer, the charging circuit composed of C_{12} and C_{20} causes a displacement charge. The overvoltage transmitted from the primary side to the secondary side can be obtained as follows:

$$u_2 = \frac{C_{12}}{C_{12} + C_{20}} \times u_1 \qquad (12.9)$$

The higher the capacity of the transformer, the higher the primary voltage u_1 will be, and the more serious the overvoltage caused by the closing will be. So, effective protection measures must be taken.

One of the measures to limit such an operating overvoltage is to add a metal electrostatic shield layer between the primary and secondary windings. The shield is connected to the grounded core. Thus, the capacitance C_{12} between the primary and secondary windings is made close to zero. When there is an external overvoltage intrusion, it is shorted out by the shield capacitance, so that it will no longer harm the secondary winding.

Another measure to suppress an overvoltage is to connect a ground capacitance C_{02} to the secondary winding end. When it is determined that the overvoltage multiple is K_{u2}, the value of C_{02} can be obtained from the following equation:

$$C_{02} = \frac{C_{12}u_1 - Ku_2(C_{12} + C_{20})}{Ku_2} \tag{12.10}$$

It should be noted that when the shield layer or the ground capacitance is connected to the excitation transformer, its ability to withstand the full-wave impact test voltage is affected, and the value of the ground capacitance current increases. Careful attention must be paid to this.

12.4.8.2 Opening Overvoltage

An overvoltage caused by the opening of the transformer is much higher than that caused by its closing. When the transformer circuit is broken, the magnetic energy stored in the circuit is released. If there is no absorption device, the magnetic energy will be converted into arc energy, and an overvoltage will be generated. The measure to suppress such an overvoltage is to feed the energy stored in the inductor into an overvoltage absorption R-C network in parallel with the secondary winding of the transformer.

12.4.8.3 Atmospheric Overvoltage

Compared to the operating overvoltage multiple, the atmospheric overvoltage multiple is higher. But an atmospheric overvoltage's action time is short, generally only dozens of microseconds. Devices like a lightning arrester can be used for protection against such an overvoltage.

12.4.9 AC R-C Protection

12.4.9.1 Role of R-C Protection

Since a rectifier load is connected to the excitation transformer, the thyristor of each arm of the rectifier bridge undergoes commutation turn-on or turn-off in turn in operation. When the thyristor is in the off state, the magnetic field energy stored on the secondary winding side of the excitation transformer in the corresponding phase will cause a transient overvoltage that will endanger device safety during its release. So, In addition to R-C dampers in parallel at both ends of the thyristor, an R-C protection device for suppression of the AC overvoltage should be connected to the secondary winding side of the excitation transformer. The typical connection mode is shown in Figure 12.4. Here, a triangular or star connection can be adopted for the transformer R-C protection. The voltage at both ends of the capacitor, which is connected to the secondary winding side of the excitation transformer, cannot change suddenly, thus effectively inhibiting the spike overvoltage caused by commutation of the rectifier. The capacitor connected in series with the capacitor is used to consume the energy during the energy conversion and suppress the oscillations in the LC circuit.

The role of the other set of nonlinear resistance protection (the selenium pile or varistor) in Figure 12.4 will be described later.

12.4.9.2 Energy Conversion Caused by Commutation Voltage Effect

The R-C protection is connected in parallel on the secondary winding side of the excitation transformer to absorb the magnetic field energy released by the turn-off phase. The amplitude of this energy is also related to the control angle of the thyristor. The analysis shows that the waveform distortion is the most serious when the thyristor control angle $\alpha = 90°$.

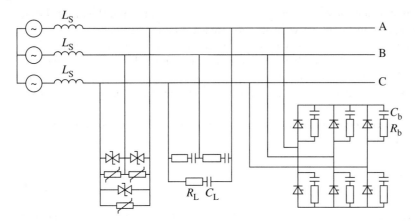

Figure 12.4 Connection of R-C protection.

According to the calculation formula recommended by Westinghouse, the conversion energy absorbed by the R-C components in parallel at both ends of the thyristor per period is as follows:

$$E_b = 3.5 C_b U_L^2 \tag{12.11}$$

where

C_b — capacitance in parallel with thyristor F;

U_L — RMS of line voltage inputted on AC side of three-phase bridge rectifier V.

For the R-C protection dampers connected to the secondary winding side of the excitation transformer, the conversion energy absorbed per period is as follows:

$$E_L = 6.0 C_L U_L^2 \tag{12.12}$$

where

C_L — capacitance in dampers F;

U_L — RMS of line voltage secondary-side winding of excitation transformer V.

12.4.9.3 Energy Conversion Caused by Reverse Current Recovery Effect

A small quantity of current carriers is accumulated in the thyristor in the turned-on state. When the element withstands a reverse voltage, the reverse recovery charge amount Q of the original accumulated current carriers flow out of the element and form the reverse recovery currents i_Q. When the recovery currents are quickly combined and turned off, a high overvoltage will be caused in the leakage reactance of the excitation transformer, as shown in Figure 12.5.

In Figure 12.5a, E_{Rm} represents the secondary winding line voltage peak of the excitation transformer and L_S the leakage inductance of the transformer in each phase. In Figure 12.5b, I_{rr} represents the recovery current amplitude and t_{rr} the reverse recovery current turn-off time. The energy generated by i_R will be consumed on the transformer R-C dampers R_L and C_L and the thyristor R-C dampers R_b and C_b.

Within the reverse recovery time t_{rr}, a reverse recovery current I_{rr} whose peak current is E_{Rm} and leakage inductance is $2L_S$ flows through. Its value can be determined with the equivalent circuit shown in Figure 12.5a:

$$I_{rr} = t_{rr} \times \frac{E_{Rm}}{2L_S}$$

$$E_{Rm} = \sqrt{2} U_L \tag{12.13}$$

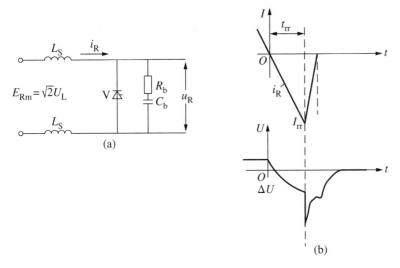

Figure 12.5 Overvoltage caused by recovery current i_Q effect: (a) equivalent circuit, (b) thyristor current and voltage waveforms.

where

E_{Rm} – line voltage peak of secondary winding of excitation transformer V;
L_S – leakage inductance per phase of excitation transformer H;
t_{rr} – reverse recovery current turn-off time µs.

12.4.9.4 Calculation of Magnetic Energy Stored

For the thyristor component, the magnetic energy stored in the excitation transformer leakage inductance per period after the commutation turn-off is calculated as follows:

$$W = \frac{1}{2} 2L_S I_{rr}^2 = L_S I_{rr}^2 \tag{12.14}$$

Since the power frequency is 50 Hz, the total magnetic field energy stored by the excitation transformer per second is as follows:

$$\Sigma W = f L_S I_{rr}^2 \tag{12.15}$$

where

f – power frequency Hz.

In the commutation turn-off process, the total energy stored by the field transformer is absorbed by the R-C dampers connected to the thyristor and the secondary side of the excitation transformer due to voltage effect and current effect, respectively.

Example 12.2

It is known that an excitation transformer's capacity $S = 5000$ kVA, secondary line voltage $U_L = 1000$ V, thyristor reverse recovery time $t_{rr} = 25\,\mu s$, thyristor damper resistance $R_b = 25\,\Omega$, and capacitance $C_b = 0.25\,\mu F$. A triangular connection is adopted for the excitation transformer secondary winding side dampers. There are six branches in parallel. Each branch's resistance $R_L = 10\,\Omega$ and capacitance $C_L = 1\,\mu F$. For the three-phase bridge rectifier, there are four branches are in parallel. The transformer's impedance voltage $U_k = 6\%$.

Solution

The leakage reactance L_S of the excitation transformer can be obtained from the following equation:

$$L_S = U_k U_L^2 / 2 \times 1000 \times \pi f S \qquad (12.16)$$

where

S – capacity of excitation transformer kVA.

The following can be obtained:

$$L_S = \frac{0.06 \times 1000^2}{314 \times 5000 \times 1000} = 38 \text{ (mH)}$$

With Eq. (12.13), the following can be obtained:

$$I_{rr} = t_{rr} \times \frac{U_m}{2L_S} = \frac{25 \times 10^{-6} \times 1000 \times \sqrt{2}}{2 \times 38 \times 10^{-6}} = 465 \text{ (A)}$$

The magnetic energy stored by the leakage reactance of the excitation transformer per second is calculated as follows:

$$W = 50 \times 38 \times 10^{-6} \times 465^2 = 411 \text{ (W)}$$

When Eq. (12.11) is multiplied by f, the energy absorbed by the capacitance C_b of the thyristor dampers per second due to the voltage effect can be obtained:

$$E_b = 3.5 f C_b U_L^2 = 3.5 \times 50 \times 0.25 \times 10^{-6} \times 1000^2 = 43.75 \text{ (W)}$$

When Eq. (12.12) is multiplied by f, the energy absorbed by the capacitance C_L of the line dampers per second due to the commutation voltage effect can be obtained:

$$E_L = 6.0 C_L U_L^2 f = 6 \times 50 \times 1.0 \times 10^{-6} \times 1000^2 = 300 \text{ (W)}$$

According to the calculation data of similar projects by Westinghouse, the current-effect energy consumed by the resistive element is about 25% for the thyristor dampers and about 30% for the line dampers.

Figure 12.6 shows the curves of the variations of the currents absorbed by the line R-C dampers, the thyristor R-C dampers, and the nonlinear resistance limiter over time when the thyristor is turned off.

Due to the charging of the capacitor, after the transient response time $t_{rr} = 25 \text{ μs}$, the line and thyristor R-C dampers absorb the magnetic energy in the excitation transformer circuit. Under the selected parameters, the current absorbed by the thyristor dampers is approximately 20% of that absorbed by the line R-C dampers.

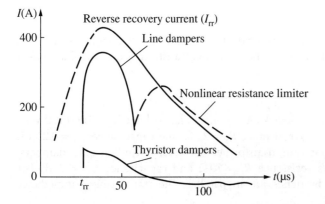

Figure 12.6 Variations of currents absorbed by line and thyristor R-C dampers and nonlinear resistor resistance limiter.

Due to the two R-C damper circuits, the reverse recovery current attenuation di/dt slows down, thus reducing the circuit overvoltage caused. Since the di/dt value is low, the nonlinear resistance limiter is turned on about 55 μs later.

As can be seen from the above discussion, in the power rectifier circuit powered by the excitation transformer, in the selection of overvoltage suppression damper parameters, not only the energy of the commutation voltage effect but also the energy of the reverse current recovery effect should be taken into account. Otherwise, the R-C elements absorbing the energy may be burned, which may cause further failure of the power rectifier. Multiple accidents of this kind have been seen in China. So, lessons should be learned from them.

12.4.10 Test Voltage

For a transformer operating at the rated voltage, the voltage amplitude is definite. However, an operation or atmospheric overvoltage intrusion will generate an abnormal overvoltage, which may damage the transformer insulation. So, a transformer should undergo the high-voltage withstand voltage test before leaving the factory. There are two test forms: (1) short-term power frequency RMS and (2) full-wave peak impact test voltage, as shown in Table 12.4.

The waveform of the full-wave impact voltage test is determined by the characteristics of the simulated atmospheric overvoltage. Figure 12.7 shows the standard waveform of the full-wave impact withstand voltage test of the transformer. The part where the immaturevoltage rises from zero to the maximum value, which is called wave front, is only 1.5 μs. The descending part is called the wave tail. The maximum peak voltage is $(12\sim17)\,U_m$. The whole process is called full wave.

Table 12.4 Power frequency and full-wave impact test voltages.

Rated voltage (kV)	Power frequency test voltage RMS (kV)	Full-wave impact voltage peak (kV)	
		I	II
<1	3	—	—
3	10	20	40
6	20	40	60
10	28	60	75
15	38	75	95
20	50	95	125
35	70	145	170

Figure 12.7 Standard waveform of full-wave impact withstand voltage test of transformer.

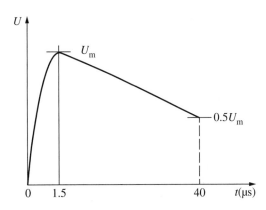

12.4.11 Noise

12.4.11.1 Overview

The noise sources of the transformer include the vibration of the core, the change in the magnetic flux in the core, and the expansion and contraction of the hysteresis loop [28]. Among these, the expansion and contraction of the hysteresis loop is the main cause of vibration. The vibration frequency consists of a fundamental frequency that is twice the supply frequency and the additional frequency of the higher harmonics.

There is an electromagnetic attraction at the silicon steel sheet joints and between the laminations due to the magnetic leakage, which causes core vibration. With the continuous improvement in the core lamination mode, the core vibration caused by the electromagnetic attraction at the joints and between the laminations is much smaller than that caused by the magnetostriction. So, this noise is generally negligible.

The leakage generated by the winding load current will cause vibration of the windings. When the rated working magnetic flux density of a dry-type transformer is within the range 1.5–1.6 T, this vibration is small compared to the core vibration caused by the magnetostriction.

Usually installed indoors, epoxy dry-type transformers have to satisfy certain noise-level requirements from the environmental perspective.

12.4.11.2 Sound Wave, Sound Pressure, Sound Intensity, and Sound Intensity Level

In order to describe the propagation characteristics of noise in a medium, two physical variables in acoustics, including sound pressure and sound intensity, are usually used. Besides, since the propagation vibration direction of a noise wave caused by a noise source is parallel to the propagation direction of the wave, the noise wave is a longitudinal wave.

When the frequency of the longitudinal wave is 20–20 000 Hz, it is audible humans. The study on noise is primarily limited to this frequency range.

The difference between the pressure intensity in the presence of acoustic wave propagation and that in the absence of acoustic wave propagation is called sound pressure.

The sound intensity is defined as energy density of a sound wave, that is, the energy passing through an area perpendicular to the acoustic wave propagation direction per unit area per unit time. The higher the noise frequency, the greater the resulting sound intensity and sound pressure.

At 1000 Hz, the maximum and minimum sound intensity normal people can hear are 1 and 10^{-12} W m^{-2}, respectively. Represented by I_0, the minimum sound intensity is generally used as the standard for the measurement of sound intensity. Since orders of magnitudes of sound intensity vary widely, the logarithmic scale, for example, the decibel (dB) is generally taken as a measure of the sound intensity level. The expression for the sound intensity level is as follows:

$$L_I = \lg(I/I_0) \tag{12.17}$$

12.4.11.3 Noise of Epoxy Dry-Type Excitation Transformer

A dry-type transformer's noise originates from the magnetostriction of the core. So, the windings around the transformer core naturally play the role of sound insulation. In order to enhance the noise absorption, generally a vibration damping device needs to be added to the appropriate position of the transformer. Part of the noise of the transformer is propagated by the air medium and the remaining part is absorbed by the device carrier.

Figure 12.8 shows the product noise data values given by Germany's GEAFOL. Figure 12.9 shows the noise increase caused by multiple noise sources when multiple transformers are operating.

12.4.11.4 Noise Reduction Measures

For the excitation transformer body, the following noise reduction measures can be taken:

(1) Improving the core material: Since the magnetostriction coefficient of the silicon steel sheet used in the core directly affects the noise intensity of the transformer body, reducing the silicon steel sheet magnetostriction is the most fundamental and effective way to reduce the transformer noise. A high-quality

Figure 12.8 Noise levels of dry-type transformers: 1 – noise value allowed by German DIN42540 conventional dry-type transformer standard, 2 – noise value allowed by German DIN42523 epoxy dry-type transformer standard, 3 – value guaranteed by GEAFOL epoxy dry-type transformer.

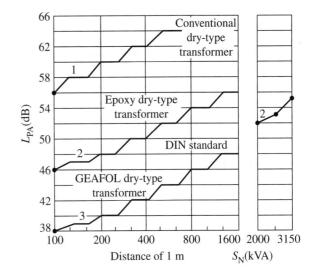

Figure 12.9 Noise increase caused by multiple noise sources.

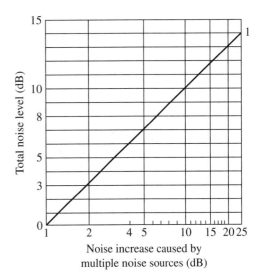

silicon steel sheet features good grain orientation integrity and has an insulating coating on its surface, which provides stress and insulation resistance to reduce the magnetostriction. Generally, a silicon steel sheet's silicon content is 2%–3%. Experimental results show that its magnetostriction is approximately zero when its silicon content is 6.5%. Only when its silicon content exceeds 3.5% will it become very brittle and hard to process and inapplicable to transformers. For cores, the good characteristics of the oriented silicon steel sheet should be utilized as much as possible. High-quality high-permeability Hi-B silicon steel sheet can reduce magnetostriction. An oblique or step joint can further reduce local magnetostriction.

(2) Optimizing the geometric dimensioning of the core: The noise generated during the core excitation is related to such factors as geometric dimensioning, structure, weight, joint mode, and lap joint area. The unevenness of the core magnetostriction is closely related to its geometric dimensioning. Under the premise that the mechanical strength of the core is met, the minimum lap joint area should be selected to reduce the noise of the core. The core's noise is closely related to its clamping force. The optimal core clamping force is 0.08–0.12 MPa. Figure 12.10 shows the curve of the relationship between the core's magnetic flux density and magnetostriction rate.

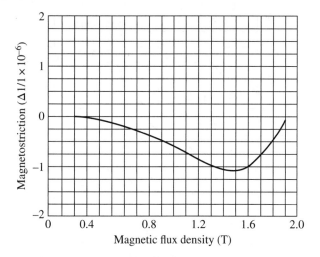

Figure 12.10 Curve of relationship between core's magnetic flux density and magnetostriction rate.

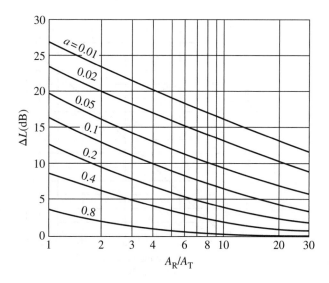

Figure 12.11 Relation between absorption coefficient (a) of wall and top and $\alpha\ A_R/A_T$.

(3) Reducing the vibration noise of the accessories: Accessories such as cooling fans will increase the transformer noise with the core vibration. When there is a strict requirement to reduce the transformer operating noise, the AN mode should be designed as far as possible. When the AF mode is a must, in addition to a solid fixing method, fans with less operating noise should be adopted.

(4) Isolating the transformer noise: In order to further control the transformer noise, isolation measures can be implemented after the on-site installation is completed. Common isolation measures include setting up a sound insulation screen, building a noise barrier, and providing a transformer room. The noise control mode depends on the actual need.

In the equipment installation process, cavity resonance of the structure and the transformer should be avoided. For example, the cavity under the installed base of the transformer will make the noise greater.

(5) Improving the structural arrangement of the transformer installation: The noise of a transformer installed indoors propagated through the air will increase due to wall reflections. The increase is related to the ratio of the surface area A_T of the transformer to the total area A_R of the room as well as the acoustic absorption coefficient α. Figure 12.11 shows the relationship between the α of the wall and the area ratio A_R/A_T of the wall and the top.

Table 12.5 lists the reference values of the surface area A_T of transformers with different capacity values. Table 12.6 shows the acoustic absorption coefficients of different materials.

In terms of air noise absorption, as can be seen from Figure 12.11, slag wool can greatly increase the absorption coefficient a and significantly reduce the noise level.

It should be noted that the data in Table 12.6 given by GEAFOL are obtained at the frequency of $f = 125$ Hz. In Germany, the standard frequency is 60 Hz, and the fundamental frequency of the transformer noise is twice the power frequency, that is, 120 Hz. In Chinas, the figures are 50 and 100 Hz, respectively. So, the absorption coefficient will vary. A noise propagated externally can be reduced after being absorbed by the noise barrier. For example, a 12 cm-thick brick wall can reduce a noise by 35 dB and a 24-cm-thick one can reduce it by 39 dB.

As the distance increases, the noise decreases, as shown in Figure 12.12.

12.4.11.5 Structure Carrier Noise Reduction Design

The noise mentioned above is emitted by the transformer and propagated through the air medium. The reduction of another noise due to the transformer structure is discussed below. Under normal circumstances, a

Table 12.5 Reference values of surface area A_T of transformers with different capacity values.

S (kVA)	A_T (m²)
100	4.5
250	5.5
400	6.1
630	8.2
1000	11.4
1600	13.5

Table 12.6 Acoustic absorption coefficient of different materials at frequency of 125 Hz.

Material	Absorption coefficient a
Non-plastered brick wall	0.024
Plastered brick wall	0.024
Concrete	0.01
3 cm glass wool board applied on hard brick	0.22
Smooth plate surface covered with 4 cm slag wool	0.74

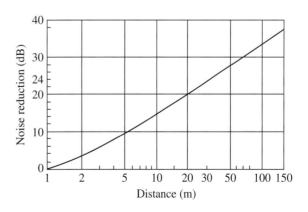

Figure 12.12 Curve of relation between distance and noise reduction.

transformer can be installed in a basement or a special small room to reduce the noise propagation level. Of course, other elastic vibration damping measures such as special elastic vibration damping support points or steel springs may be employed. As mentioned above, the core is a carrier of the transformer noise. The windings primarily play a dominant role in noise insulation. So, silicone rubber buffer positioning vibration damping fixed windings are generally used for epoxy dry-type transformers. After the vibration damping, the windings' inherent natural vibration frequency is related to the silicone rubber's static compression.

According to the analysis, the inherent natural vibration frequency f_0 of the external elastic vibration damping mass system should be less than 1/3 of the fundamental frequency of the vibration source. For a 50 Hz power frequency system, the fundamental frequency of noise is 100 Hz. So, the natural vibration frequency f_0 of the elastic vibration damping mass system should be below 33 Hz. For a China-made SCL transformer, when the static compression of silicone rubber is 2–4 mm, the windings' inherent natural vibration frequency can be reduced to 8–11 Hz.

Here is an example of calculation of noise reduction.

Example 12.3
The GEAFOL 1000 kVA epoxy dry-type transformer's body weight is 2630 kg. It is installed at the four support points. Each support point bears 657 kg or 6570 N. The transformer is mounted on a fixed foundation of the basement. The static compression of the damping material is 0.25 cm. The elastic modulus of the material is $C_D = \frac{6570}{0.25} = 26.300 \text{N/cm}$. Identify the material that can reduce the noise.

Solution
A material whose coefficient $C = 23.400 \text{ N cm}^{-1}$ and allowable static long-term heavy load is 8500 N is selected. The actual compression is as follows:

$$\Delta h = \frac{6.570}{23.400} = 0.28 \text{ (cm)}$$

This can meet the design requirements.

If the dry-type transformer is installed outside the basement and is supported by the ground-based foundation, the elastic vibration damping will be stronger, with elastic compression values of up to 5 mm.

Example 12.4
Two GEAFOL 630 kVA dry-type transformers are installed in the basement. It is assumed that elastic vibration damping measures have been taken to reduce the transformer structure carrier noise. Noise is propagated through the air medium via the top of the basement to Room A, as shown in Figure 12.13. Identify the total noise level in Room A.

Solution
On the basis of the size shown in Figure 12.13, the total surface area of the basement can be obtained:

$$A_R = 208 \text{ (m}^2\text{)}$$

When $S = 630$ kVA, the surface area of the transformer $A_T = 8.2 \text{ m}^2$, which can be found in Table 12.5. The noise of the structural carrier can be found in Figure 12.8 as follows:

630 kVA 52 dB;
Tolerance: 3 dB;
Total: 55 dB.

Up and down outside the basement, reflection will increase the noise value. So, the area ratio can be calculated first:

$$\frac{A_R}{A_T} = \frac{208 \text{ m}^2}{8.2 \text{ m}^2} = 25.4$$

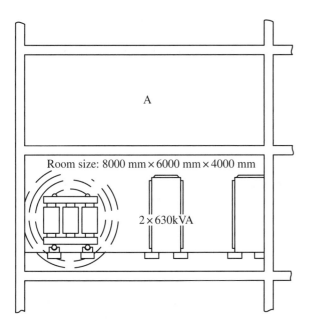

Figure 12.13 Dry-type transformer installed outside basement.

When the wall material is concrete, the acoustic absorption coefficient $a = 0.01$, which can be found in Table 12.6. From Figure 12.11, it is seen that the noise increase caused by reflection is +12 dB, and the allowable error is +3 dB. So, the total noise generated by the structure carrier can be calculated as follows:

$$L_n = 55 + 12 + 3 = 70 \text{ (dB)}$$

The basement top material is 24-cm-thick concrete, and the noise can be reduced by 39 dB when it propagates through the top. So, the noise intensity in Room A will be 70−39 = 31 dB.

The area ratio of Room A is

$$\frac{A_R}{A_F} = \frac{208 \text{ m}^2}{48 \text{ m}^2} = 4.3$$

When an acoustic absorption material, for example, a blanket and a thin protective shield, is adopted, a = 0.6, and the noise increase caused by the indoor reflection is +3 dB (see Figure 12.11). So, the total noise level in Room A is 31 + 3 = 34 dB.

12.4.12 Electrostatic Shield

The DL/T650 − 1998 *Specification for Potential Source Static Exciter Systems for Large Turbine Generators* requires that an electrostatic shield be installed between the low- and high-voltage windings of an excitation transformer. In order to achieve a good shielding effect and prevent the generation of suspended potential in the alternating magnetic field, the electrostatic shield must be well grounded. The roles of the electrostatic shield are as follows:

(1) *Overvoltage suppression.* The electrostatic shield makes the parasitic capacitance between the primary and secondary windings close to zero, thus suppressing the overvoltage transfer.
(2) *Suppression of the transmission of high-order harmonics.* The electrostatic shield suppresses the transmission of high-order harmonics generated by the rectifier circuit to the power supply side.
(3) *Static noise reduction.* The electrostatic shield provided between the high- and low-voltage windings help reduce the electrostatic noise caused by harmonic factors.

12.4.13 Operational Environmental Impact

For epoxy cast dry-type transformers, the following are the assessment criteria involving environmental factors.

12.4.13.1 Environmental Class
E0 – indoor installation, no condensation, and negligible pollution
E1 – occasional condensation and slight pollution
E2 – heavy pollution and frequent condensation

12.4.13.2 Climate Class
C1 – minimum operating temperature of −5 °C and storage and transport temperature up to −25 °C
C2 – minimum operating temperature of −25 °C and storage and transport temperature of −25 °C

12.4.13.3 Fire Class
F0 – No special flammability limit measures
F1 – All materials free of halogen, little smoke or heat generated in burning, and self-extinguishable flame

12.4.13.4 Operating Environment
A transformer's insulation is affected by environmental factors such as the ambient temperature and altitude. Normally, the rated withstand voltage of the transformer is applicable to devices operating under the following conditions:

(1) The ambient air temperature does not exceed 40 °C.
(2) The altitude of the installation site does not exceed 1000 m.

According to GB311.1 – 1997 *Insulation Co-ordination for High Voltage Transmission and Distribution Equipment*, for the test voltage of the external insulation of any equipment whose ambient temperature is above 40 °C in the dry state, the specified rated withstand voltage shall be multiplied by the temperature correction factor K_t:

$$K_t = 1 + 0.0033\,(T - 40)$$

where

T – ambient temperature °C.

For the external insulation of any equipment or a dry-type transformer at an altitude of more than 1000 m but not more than 4000 m, the insulation strength should be reduced by about 1% for every increase of 100 m in altitude. For a test on a site at an altitude of no more than 1000 m, the rated withstand voltage specified in GB311.1 should be multiplied by the altitude correction factor K_a for the test voltage:

$$K_a = \frac{1}{1.1 - H \times 10^{-1}} \quad (12.18)$$

where

H – altitude of equipment installation site in m.

With the increase in altitude, the air density will be reduced, which will affect the equipment's heat dissipation. So, the altitude of a transformer's operating environment will have an impact on its temperature rise. For a transformer operating at an altitude of more than 1000 m but tested at a normal altitude, its temperature rise limit should be corrected accordingly. In terms of altitude, for the portion beyond 1000, 500 m is taken as a level, and the temperature rise decreases by 2.5% (for self-cooled) or 5% (air-cooled). But the case where the temperature decreases 5 K or more with every increase of 1000 m in altitude should be considered as well. In that case, it can be considered that when the transformer is in the high-altitude operation, the increased transformer rise caused by the poor heat dissipation has been compensated for by the decrease in the ambient temperature. In a test at the normal altitude, the temperature rise limit will not be corrected.

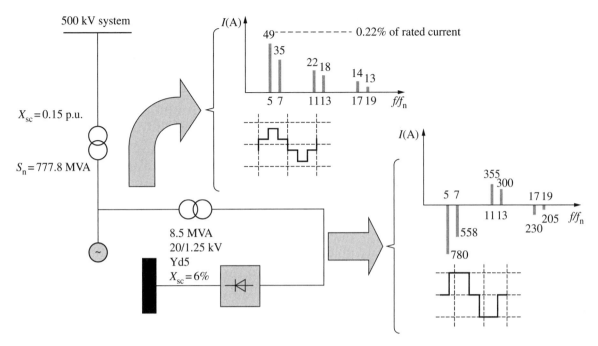

Figure 12.14 Elementary wiring diagram of self-excitation system.

12.5 Harmonic Current Analysis

12.5.1 Overview

For an excitation transformer applied to a static self-excitation system, since power is supplied to the generator's excitation winding via the three-phase bridge rectifier, the current flowing through the excitation transformer windings is a non-sinusoidal current component containing high-order harmonics.

For a three-phase bridge rectifier circuit, according to Fourier series analysis, harmonic currents except fundamental currents do not contain even harmonics as well as the third harmonic or harmonics in orders which are multiples of three. So, the orders of the harmonic currents are 5, 7, 11, 13, 17, 19, and 23.

Figure 12.14 shows the elementary wiring diagram of the self-excitation system and the proportion of each harmonic current on the primary and secondary sides of the transformer. The excitation transformer features a capacity of 8.5 MVA, 20/1.25 kV, Yd5, and short-circuit impedance $X_{sc} = 6\%$.

12.5.1.1 Harmonic Current Analysis

Harmonic currents can cause different types of additional losses, which should be given special consideration in the design of an excitation transformer.

For harmonic current analysis and calculation, the following methods are generally adopted in engineering:

(1) *Fourier series expansion method.* For a three-phase bridge rectifier circuit, as is known from the Fourier series expansion principle, for an AC current waveform with a period of 2π symmetrical to the horizontal axis (i.e., $i_{(\omega t)} = -i_{(\pi + \omega t)}$), there is no even harmonic or DC component. Apart from the fundamental harmonic, there are only odd harmonics. Besides, since there is no neutral point wiring in the three-phase bridge rectifier circuit, in the odd harmonic component, there is no third harmonic or a harmonic in an order that is a multiple of three. That is, the orders n of the harmonics it contains are respectively 1, 5, 7, 11, 13, 17, 19 …

With Eqs. (6.37) and (6.38), the harmonic coefficient is calculated as follows:

$$a_n = \frac{2\sin n\frac{\pi}{3}}{\pi n(n^2-1)[\cos\alpha - \cos(\alpha+\gamma)]} \times \left[n\sin\alpha \sin n\left(\alpha+\frac{\pi}{2}\right) - n\sin(\alpha+\gamma)\sin n\left(\alpha+\gamma+\frac{\pi}{2}\right)\right]$$
$$+ \left[\cos\alpha \cos n\left(\alpha+\frac{\pi}{2}\right) - \cos(\alpha+\gamma)\cos n\left(\alpha+\gamma+\frac{\pi}{2}\right)\right] \tag{12.19}$$

$$b_n = \frac{2\sin n\frac{\pi}{3}}{\pi n(n^2-1)[\cos\alpha - \cos(\alpha+\gamma)]} \times \left[n\sin\alpha \sin n\left(\alpha+\frac{\pi}{2}\right) - n\sin(\alpha+\gamma)\cos n\left(\alpha+\gamma+\frac{\pi}{2}\right)\right]$$
$$+ \left[\cos(\alpha+\gamma)\sin n\left(\alpha+\gamma+\frac{\pi}{2}\right) - \cos\alpha \sin n\left(\alpha+\frac{\pi}{2}\right)\right] \tag{12.20}$$

The harmonic's amplitude and phase angle tangent are respectively as follows:

$$c_n = \sqrt{a_n^2 + a_b^2}$$
$$\tan\varphi_n = \frac{a_n}{b_n} \tag{12.21}$$

The harmonic component RMS is

$$I_n = \frac{C_n}{\sqrt{2}}\lambda I_f \tag{12.22}$$

where

λ — wiring coefficient $\lambda = 2$.

For the fundamental harmonic, the harmonic coefficient is as follows:

$$a_1 = \frac{\sqrt{3}}{4\pi[\cos\alpha - \cos(\alpha+\gamma)]} \times [2\gamma + \sin 2\alpha - \sin 2(\alpha+\gamma)] \tag{12.23}$$

$$b_1 = \frac{\sqrt{3}}{2\pi}[\cos\alpha + \cos(\alpha+\gamma)]$$

$$c_1 = \sqrt{a_1^2 + b_1^2} \tag{12.24}$$

$$\tan\varphi_1 = \frac{a_1}{b_1} \tag{12.25}$$

The fundamental current RMS is

$$I_1 = \frac{c_1}{\sqrt{2}}\lambda I_f \tag{12.26}$$

The corresponding data are substituted into the equation. Each harmonic component RMS I_1, I_5, I_7, I_9, ⋯ can be obtained in turn. Then, the harmonic additional loss can be calculated, and the equivalent excitation transformer secondary current rated value can be obtained.

The harmonic current value can be determined on the basis of the excitation data of the ABB 740 MVA hydro-turbine of Three Gorges Hydro Power Station. It is known that the generator's rated excitation current $I_f = 4158$ A. Through conversion, the rated AC current value of the secondary side of the excitation transformer is $I_{2N} = 0.816 \times 4158 = 3230$ A. The thyristor rated control angle $\alpha_N = 73.5°$. The rated commutation angle $\gamma_N = 23°$. With Eqs. (12.23)–(12.26), the following can be obtained:

$$a_1 = 0.548, \quad b_1 = 0.047, \quad c_1 = 0.5495$$

Thus, the value of the fundamental current can be obtained: $I_1 = \frac{0.5495}{\sqrt{2}} \times 2 \times 4158 = 3230$ (A)

Table 12.7 Excitation transformer secondary winding side harmonic current values.

Harmonic order (h)	Each harmonic current value (A)	Proportion in fundamental harmonic (%)
1	3241.32	100
5	645.10	19.9
7	458.53	14.2
11	287.51	8.9
13	240.8	7.4
17	179.67	5.5
19	158.35	4.9
23	126.24	3.9
25	113.78	3.5

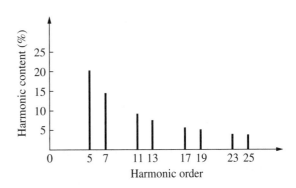

Figure 12.15 Harmonic amplitude spectrum.

The other harmonic current values are calculated through the Fourier series expansion in Eqs. (12.19)–(12.22). The results are shown in Table 12.7.

The harmonic amplitude spectrum is shown in Figure 12.15.

(2) *Equivalent coefficient method.* The following engineering expression can be used to determine the harmonic current component value:

$$I_h\% = \frac{K \times 100}{h - 5/h} \quad (12.27)$$

where

$I_h\%$ – hth harmonic current with rated fundamental current as basic value %;

K – constant 0.61 for diode system and 0.75 for thyristor system;

h – harmonic current order 5, 7, 11, 13, 17, 19, 23, and 25, respectively.

For example, for $h = 5$, $K = 0.75$ is substituted into the above equation. The following can be obtained:

$$I_5\% = \frac{0.75 \times 100}{5 - 5/5} = 18.75\%$$

12.5.1.2 Influence of Harmonic Current

The stray loss caused by harmonic currents usually increases the total loss of an excitation transformer by about 8%. The influence of harmonic currents can be reduced by applying copper or aluminum foil in place of

Table 12.8 Provisions on harmonic current limits in IEC 146.

Harmonic order	With fundamental amplitude as benchmark (p.u)
Fundamental	1.0
5	0.2
7	0.1486
11	0.0909
13	0.0769
17	0.0588
19	0.0526
23	0.0435
25	0.0400

solid conductor and by increasing of the distance between the high- and low-voltage windings in the design of the excitation transformer as well as of the special winding duct.

For a transformer that supplies power to the rectifier load, generally harmonic current value limits are specified. For example, IEC146 specifies harmonic current value limits. The corresponding values are shown in Table 12.8.

12.5.1.3 Harmonic Loss Calculation [29]

The harmonic loss calculation for excitation transformers is presented with a 740 MVA ABB hydro-turbine of Three Gorges Hydro Power Station as an example. The RMS values of the various harmonic currents are shown in Table 12.3.

IEC 61378-1:2011 *Converter transformers – Part 1: Transformers for Industrial Applications* specifies that losses caused by non-sinusoidal harmonic currents include the following:

Total load loss = resistance loss (fundamental and harmonic) + winding eddy current loss (fundamental and harmonic) + connecting line loss (fundamental and harmonic) + stray loss (fundamental and harmonic)

The increase factor of each harmonic loss is calculated as follows:

(1) *Resistance loss increase factor*. The resistance loss increase factor is calculated as follows:

$$F_R = \Sigma(I_h/I_1)^2 = 1 + 0.199^2 + 0.142^2 + 0.089^2 + 0.074^2 + 0.055^2 + 0.049^2 \\ + 0.039^2 + 0.035^2 = 1.081$$

Eddy current loss increase factor $x = 2$. The eddy current loss increase factor can be calculated as follows:

$$F_W = \Sigma[(I_h/I_1)^2 h^x] = \Sigma[(I_h/I_1)^2 h^2] \\ = (1 \times 1)^2 + (0.199 \times 5)^2 + (0.142 \times 7)^2 + (0.089 \times 11)^2 + (0.074 \times 13)^2 \\ + (0.055 \times 17)^2 + (0.049 \times 19)^2 + (0.039 \times 23)^2 \\ + (0.035 \times 25)^2 = 7.173$$

(2) *Stray loss increase factor*. With the calculation formula of the stray loss increase factor provided in IEC 61378, $x = 0.8$, the stray loss increase factor can be calculated as follows:

$$F_{ce} = \Sigma[(I_h/I_1)^2 h^x] = \Sigma[(I_h/I_1)^2 h^{0.8}] \\ = 1^2 \times 1^{0.8} + 0.199^2 \times 5^{0.8} + 0.142^2 \times 7^{0.8} + 0.089^2 \times 11^{0.8} + 0.074^2 \times 13^{0.8} \\ + 0.055^2 \times 17^{0.8} + 0.049^2 \times 19^{0.8} + 0.039^2 \times 23^{0.8} + 0.035^2 \times 25^{0.8} = 1.425$$

The load loss due to the fundamental current is determined below.

(1) Fundamental resistance loss (measured at 120 °C). $P_{R1} = 15.240$ W.
(2) Fundamental eddy current loss. $P_{W1} = 585$ W (calculated).
(3) Stray loss. The total loss at the fundamental current (measured at 120 °C). $P_{N1} = 17.535$ W.

On the basis of these measured fundamental losses and the eddy current loss found, the sum of the losses generated by the fundamental current on the connecting line and the structural members can be obtained:

$$P_{ce} = P_{N1} - P_{R1} - P_{W1} = 17535 - 15240 - 585 = 1710 \text{ (W)}$$

The total loss caused by the harmonic currents is calculated as follows:

$$\begin{aligned} P_N &= \sum (I_h/I_1)^2 P_{R1} + F_W P_{W1} + F_{ce} P_{ce} \\ &= 1.081 \times 15240 + 7.173 \times 585 + 1.425 \times 1710 \\ &= 23113 \text{ (W)} \end{aligned}$$

Thus, the equivalent fundamental power frequency temperature rise test current can be determined as follows:

$$I = \sqrt{P_N/P_{N1}} I_1 = \sqrt{23113/17535} I_1 = 1.15 I_1$$

That is, the sum of the rated load loss and the harmonic load loss is taken as the excitation transformer test loss. That is, the excitation transformer temperature rise test is conducted under the condition of 1.15 times the fundamental rated current.

12.5.2 Excitation Transformer Capacity Calculation Checking Example: 740 MVA ABB Hydro-Turbine of Three Gorges Hydro Power Station

The excitation transformer capacity checking calculation is presented with a 740 MVA ABB hydro-turbine of Three Gorges Hydro Power Station as an example.

(1) The secondary rated voltage U_{2N} of the excitation transformer is determined.

The secondary-side rated voltage of the excitation transformer is a very important parameter, because parameters such as the excitation voltage response ratio of the self-shunt excitation system and the ceiling voltage multiple are directly related to it.

The ceiling voltage of the self-shunt excitation system is generally no less than 1.8 times the rated excitation voltage for a large steam turbine and no less than twice the rated excitation voltage for a large hydro-turbine. Moreover, when the hydro-turbine terminal positive-sequence voltage is 80% of the rated value, the ceiling voltage should be guaranteed. For the hydro-turbine of Three Gorges Hydro Power Station, in view of the power station's role in supporting the stability of the power system, the ceiling voltage is increased to 2.5 times the excitation voltage in the case of the maximum capacity of the generator (840 MVA). If the secondary-side rated voltage value is inappropriately selected or calculated, the excitation voltage may be below the required value.

It is known that the generator's rated excitation voltage $U_{fN} = 475.9$ V and rated excitation current $I_{fN} = 4185$ A. When the generator voltage is $0.8 U_{GN}$, the excitation system is required to provide a $2.5 U_{fN}$ forced excitation voltage.

It is assumed that

a. The total voltage drop of the rectifier cabinet and the external connecting cable is 10 V;
b. The transformer short-circuit impedance voltage drop is 8%;
c. The rectifier cabinet minimum control angle α_{min} at the time of forced excitation is 10°;
d. The forced excitation current is limited based on $2I_{fN}$.

With the three-phase bridge fully controlled rectifier bridge calculation formula, the following formula can be written in the forced excitation state:

$$U_{fc} = 2.5 U_{fN} = \frac{3\sqrt{2}}{\pi} 0.8 U_2 \cos \alpha_{min} - \frac{3}{\pi} I_{fc} X_r - 2\Delta U$$

For an estimated engineering calculation, it is assumed that the commutation voltage drops is 10% of the forced excitation voltage U_{fc} in the forced excitation state.

It is known that $U_{fc} = 2.5 \times 475.9 = 1189.75\,V$ and $0.1U_{fc} = 118.975$. When they are substituted into the above formula, $1189.75 = 1.064\,U_2 - 118.975 - 10$ can be obtained. $U_{2N} = 1239\,V$ can be obtained. $U_{2N} = 1250\,V$ is taken.

(2) The secondary rated current I_{2N} of the excitation transformer is determined.

On the basis of the generator's maximum excitation current $I_{fmax} = 4800\,A$, the secondary-side rated AC current of the excitation transformer can be determined as follows:

$$I_{2N} = 0.816 \times 4800 = 3917\ (A)$$

(3) The rated capacity S_N of the excitation transformer is determined.

The rated capacity of the excitation transformer is determined as follows:

$$S_N = \sqrt{3}U_{2N}I_{2N} = \sqrt{3} \times 1250 \times 3917/1000 = 8480\ (kVA)$$

12.5.2.1 Specification for Technical Parameters of Excitation Transformer

Table 12.9 shows the specification for the technical parameters of the excitation transformers of the hydro-turbines of Three Gorges Hydro Power Station.

Table 12.9 Specification for technical parameters of excitation transformers of hydro-turbines of Three Gorges Hydro Power Station.

Parameter	Value		UOM	Remark
	VGS	ABB		
Rated capacity	3 × 2200	3 × 2925	kVA	
Primary side voltage	$20/\sqrt{3} \pm 5\%$	$20/\sqrt{3} \pm 5\%$	kV	
Secondary-side voltage	1024	1243	V	
Insulation class	H	H		
Withstand voltage level:				1 min power frequency
Primary side	50	50	kV	RMS
Secondary side[a]	5	5		RMS
Impact	125	125		Peak
Temperature rise	80	80	K	
Short-circuit impedance	6%	8%		
No-load loss	3100	3900	W	
Loss at maximum load	16 400	17 600	W	
No-load current	<0.5%I_{FN}	<0.5%I_{FN}		I_{FN} is rated current of generator
Connection mode	Y/D-11	Y/D-11		
Partial discharge level	<10	<10	PC	
Cooling mode	AN	AN		
Noise	60	60	dB	Three-phase
Ingress protection	IP20	IP20		
Dimensions	2700 × 1600 × 2700	2800 × 1600 × 2800	mm × mm × mm	Long × wide × high Single-phase transformer
Mass	7000	9000	kg	

a) Under the regulation on insulation level of dry-type transformers, the secondary rated voltage of the excitation transformers of Three Gorges Hydro Power Station is 1243 V, and the rated short-time power frequency withstand voltage corresponding to the voltage level 3 kV should be 10 kV.

13

Power Rectifier

13.1 Specification and Essential Parameters for Thyristor Rectifier Elements

13.1.1 Overview

With the rapid development of the power system and the growth of per-unit capacity, a significant increase has been seen in the excitation power required by large synchronous generators. For example, the rated excitation power of the generators of Three Gorges Hydro Power Station is larger than 2000 kW.

In this context, the power rectifier design philosophy has changed. In the early 1980s, due to the limitations constraining the development of power electronics technology, multi-circuit series and parallel structures were generally adopted for power rectifiers of excitation systems. For example, at Gezhouba Power Station, the structure of "three in series and five in parallel" is adopted for the power rectifier bridge elements of a 12.5 MW hydro-turbine, and one power cabinet has up to 90 elements. This not only complicates the power cabinet structure but also makes installation and replacement difficult. Moreover, the discreteness of the elements makes the current and voltage sharing problems more prominent.

In recent years, China has made significant advances in high-power high-parameter thyristor rectifier element development. The country has been able to mass-produce Φ100 thyristor rectifiers featuring KPX-3000 A/5500 V, the parameters reaching advanced international levels.

Improvements in thyristor rectifier element parameters can greatly simplify power rectifier structure. In thyristor rectifier element radiator development, considerable progress has been made in recent years. In addition to conventional aluminum and copper radiators, a new heat pipe exchanger has been applied. This heat pipe exchanger that dissipates heat through liquid circulation evaporation and condensation features much better cooling performance than solid good conductor radiators.

New progress has been made in power rectifier AC- and DC-side overvoltage protection as well.

13.1.2 Specification for Thyristor Rectifier Elements

The specification for a high-power thyristor is presented with the power rectifier applied to the THYRIPOL static self-excitation system of a 700 MW hydro-turbine of Three Gorges Hydro Power Station as an example. The corresponding parameters are shown in Table 13.1.

13.1.3 Operating Characteristics of Thyristor

The turn-on conditions of thyristor elements include a forward voltage applied between the anode and the cathode, a forward control voltage applied by the control electrode to the cathode, and the play-out of the control electrode upon the trigger of the thyristors (i.e., irreversible controllability).

The relationship between the thyristors' inter-electrode voltage and current is called their volt-ampere characteristic, as shown in Figure 13.1.

The anode volt-ampere characteristic can be divided into two regions: the forward characteristic region and the reverse characteristic region. The forward characteristics can be divided into a forward blocking state and

Table 13.1 Specification for power thyristor of THYRIPOL static self-excitation system of Three Gorges Hydro Power Station (thyristor model is EUPEC-T1401 N).

SN	Parameter	Working conditions	Symbol	Value (Typical)	Value (Maximum)	UOM
1	Forward off-state repeatable peak voltage	$T_{vj} = -40\,°C \sim T_{vjmax}$	U_{DRM}		4800 5000 5200	V
2	RMS forward current		I_{TRMSM}		3690	A
3	Forward average current	$T_c = 85\,°C, f = 50\,Hz$ $T_c = 60\,°C, f = 50\,Hz$	I_{TAVM}		1890 2520	A
4	Forward inrush current	$T_{vj} = 25\,°C, T_p = 10\,ms$ $T_{vj} = T_{vjmax}, T_p = 10\,ms$	I_{TSM}		31 28.5	kA
5	I²t value	$T_{vj} = 25\,°C, T_p = 10\,ms$ $T_{vj} = T_{vjmax}, T_p = 10\,ms$	$I^2 t$		4.8×10^6 4.0×10^6	A²·s
6	On-state current rise rate	DIN IEC 747-6 $f = 50\,Hz,$ $U_D = 0.67 U_{DRM},$ $I_{GM} = 3\,A, di_g/dt = 6\,A\,\mu s^{-1}$	di/dt		150	A μs⁻¹
7	Off-state voltage rise rate	$T_{vj} = T_{vjmax},$ $U_D = 0.67 U_{DRM}$	du/dt		2000	V μs⁻¹
8	On-state voltage drop	$T_{vj} = T_{vjmax}, I_T = 2000\,A$	U_T	1.57	1.70	V
9	Resistance slope	$T_{vj} = T_{vjmax}, 1500\,A/4500\,A$	U_{TO} I_T	0.88 0.34	0.92 0.37	V mΩ
10	On-state pressure drop (calculation formula: $U_T = A + B \times I_T \times C \times \ln(I_T + 1) + D \times \sqrt{I_T}$)	$T_{vj} = T_{vjmax}$	A B C D	0.497 0.00013 7 −0.0127 0.02	0.539 0.00019 3 0.00534 0.0164	
11	Gate trigger current	$T_{vj} = 25\,°C, U_D = 6\,V$	I_{GT}	350		mA
12	Gate trigger voltage	$T_{vj} = 25\,°C, U_D = 6\,V$	U_{GT}		2.5	V
13	Non-trigger gate voltage	$T_{vj} = 25\,°C, U_D = 0.5\,U_{DRM}$	U_{GD}	0.4		V
14	Non-trigger gate current	$T_{vj} = T_{vjmax}, U_D = 6\,V$ $T_{vj} = T_{vjmax}, U_D = 0.5\,U_{DRM}$	I_{GD}	20 10		mA
15	Holding current	$T_{vj} = 25\,°C,$ $U_D = 12\,V, R_A = 4.7\,\Omega$	I_H	350		mA
16	Latching current[a]	$T_{vj} = 25\,°C, U_D = 12\,V,$ $R_{GK} \geq 10\,\Omega, I_{GM} = 3\,A,$ $di_g/dt = 6\,A\,\mu s^{-1}, T_g = 20\,\mu s$	I_L		3	A
17	Forward and reverse off-state current	$T_{vj} = T_{vjmax}, U_D = U_{DRM}$ $U_R = U_{RPM}$	I_D, I_R		100	mA
18	Gated delay time	DIN IEC 747-6 $T_{vj} = T_{vjmax}, T_{TM} = I_{TAVM},$ $U_{RM} = 100\,V,$ $U_{DM} = 0.67 U_{DRM},$ $du/dt = 20\,V\,\mu s^{-1},$ $-di/dt = 10\,A\,\mu s^{-1}$	T_{gd}	2		μs

(continued)

Table 13.1 (Continued)

SN	Parameter	Working conditions	Symbol	Value Typical	Value Maximum	UOM
19	Current turn-off delay time	$T_{vj} = T_{vjmax}$, $T_{TM} = I_{TAVM}$, $U_{RM} = 100$ V, $U_{DM} = 0.67 U_{DRM}$, $du/dt = 20$ V μs^{-1}, $-di/dt = 10$ A μs^{-1}	T_q	400		μs
20	Reverse recoverable peak current	$T_{vj} = T_{vjmax}$, $T_{TM} = 2000$ A, $di/dt = 10$ A μs^{-1}, $U_R = 0.5 U_{RRM}$, $U_{RM} = 0.8 U_{RRM}$	I_{RM}		350	A
21	Reverse recovery charge	$T_{vj} = T_{vjmax}$, $T_{TM} = 2000$ A, $di/dt = 10$ A μs^{-1}, $U_R = 0.5 U_{RRM}$, $U_{RM} = 0.8 U_{RRM}$	Q_r		18	mAs
22	Thermal resistance loss of radiators in pairs		R_{thJC}	0.0086 0.0080		kW
23	Radiator thermal resistance loss		R_{thCH}	0.0025		kW
24	Maximum junction temperature		T_{vjmax}		125	°C
25	Operating ambient temperature		T_{cop}	$-40 \sim +125$		°C
26	Storage temperature		T_{stg}	$-40 \sim +150$		°C

a) Latching current I_L is defined as the minimum main current that can continue to maintain conduction at the instant when the thyristor is transformed from the off state to the on state and the trigger pulse is removed.

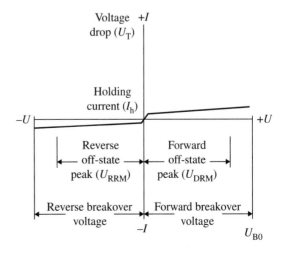

Figure 13.1 Volt-ampere characteristic of thyristors.

a forward conduction state. The forward blocking state shows different branches with different gate currents I_g. For example, when $I_g = 0$, as the forward anode voltage U_{ak} increases, there is only a very small forward leakage current within a very wide range, and the characteristic curve is very close to the horizontal axis and parallel to the horizontal axis, since Junction J2 is in the inverse voltage state and the thyristors are in the off state. When U_{ak} increases to the forward-breakover voltage U_{B0} and the thyristors suddenly turn from being blocked to being conductive or reach the reverse off-state peak voltage U_{RRM}, there will be a dramatic increase in the reverse leakage current, an increase in the thyristor reverse leakage current, and thyristor damage caused by the reverse breakdown.

The key parameters of a thyristor are now described.

13.1.3.1 Nominal Voltage

1) Off-state repetitive peak voltage U_{DRM}: This voltage refers to the forward peak voltage that is allowed to be repetitively applied to the element when the gate is open-circuited and the element junction temperature is rated. The repetitive peak voltage refers to the operating overvoltage. What corresponds to it is the off-state non-repetitive peak voltage U_{DSM}. "Non-repetitive" indicates that this voltage should not be applied repetitively over a long period of time. A non-repetitive peak voltage is usually caused by external factors, such as lightning and a circuit break. U_{DSM} should be smaller than the forward-breakover voltage U_{B0}. The margin is determined by the manufacturer. As specified, the off-state repetitive peak voltage U_{DRM} is 90% of the off-state non-repetitive peak voltage U_{DSM}.
2) Reverse repetitive peak voltage U_{RRM}: U_{RRM} refers to the reverse peak voltage that is allowed to be repetitively applied to the element when the gate is open-circuited and the junction temperature is rated. As specified, the reverse repetitive peak voltage (U_{RRM}) is 90% of the reverse non-repetitive peak voltage U_{RSM}.
3) Rated voltage U_{TN}: U_{TN} is generally U_{DRM} or U_{RRM}, whichever is smaller, which is rounded to a specified voltage class that is no larger than the value of the rated voltage of the thyristor. The voltage class should not be selected arbitrarily. Within the range below 1000 V, every 100 is taken as a class. Within a range of 1000–3000 V, every 200 V is taken as a class. In the selection of a thyristor, its rated voltage should be 2–3 times the normal working voltage peak, so that it can withstand the operating overvoltage.

13.1.3.2 Rated Current

In terms of the rated current $I_{T(AV)}$ of a thyristor (element-rated average on-state current), at the ambient temperature of +40 °C, under the specified cooling conditions, as the thyristor element undergoes a single-phase power frequency sinusoidal half-wave current of the resistive load, when the junction temperature does not exceed the rated junction temperature and remains stable, the average current in a period is called the rated average on-state current, and its integral value is taken as the rated current of the element. The rated current of the thyristor is specified by the maximum average on-state current under certain conditions, because conventional thyristors are used more for controllable rectifiers, and their rectified output current performance is usually measured with the average current. However, as specified, treating the average current as the rated current does not necessarily guarantee the safety of a thyristor in use, because the heating of the thyristor core depends on the root mean square (RMS) current flowing through the thyristor. In order to ensure the operational safety of a thyristor from the perspective of heating, the RMS current must be obtained from the rated average current.

The average on-state current flowing through the thyristor element is determined first. As mentioned above, the on-state current average $I_{T(AV)}$ of the element is obtained on the basis of the average of the sinusoidal half-wave current in one period. The applicable calculation equation is identified below.

On the basis of the definition, it is assumed that the average of the sinusoidal half-wave current whose peak is I_m in one period is $I_{T(AV)}$. The following equation can be obtained:

$$I_{T(AV)} = \frac{1}{2\pi} \int_0^\pi I_m \sin \omega t \, d\omega t = \frac{I_m}{\pi} \tag{13.1}$$

The RMS sinusoidal half-wave current in one period is as follows:

$$I_T = \sqrt{\frac{1}{2\pi} \int_0^\pi (I_m \sin t)^2 \, d\omega t} \tag{13.2}$$

On the basis of Eqs. (13.1) and (13.2), the following equation can be obtained:

$$I_m = \pi I_{T(AV)} = 2I_{T(RMS)}$$

Or the following equation can be obtained:

$$I_{T(RMS)} = 1.57 \, I_{T(AV)} \tag{13.3}$$

The relationship between the peak I_m, the RMS $I_{T(RMS)}$, and the average $I_{T(AV)}$ is shown in Figure 13.2.

In practical applications, the current flowing through each thyristor rectifier element in the bridge arm is not a sinusoidal half-wave. When the commutation angle is ignored, the current waveform flowing through the element can be approximated as a square wave with an amplitude of I_d and a width of 120°. In this case, it is the rated excitation current. Similarly, the expression relating the average $I_{T1(AV)}$ and the RMS $I_{T1(RMS)}$ of the current flowing through each element in one period can be obtained.

This calculation applies only to a three-phase fully controlled bridge circuit under the assumption that the reactance on the DC side is infinite. In this case,

$$I_{T1(AV)} = \frac{1}{2\pi} \int_0^{\frac{2}{3}\pi} I_d \, d\omega t = \frac{I_d}{3} \tag{13.4}$$

$$I_{T1(RMS)} = \sqrt{\frac{1}{2\pi} \int_0^{\frac{2}{3}\pi} I_d^2 \, d\omega t} = \frac{I_d}{\sqrt{3}} = 0.577 I_d \tag{13.5}$$

I_d, $I_{T1(AV)}$, and $I_{T1(RMS)}$ are related as shown in Figure 13.3.

It is assumed that the RMS $I_{T1(RMS)}$ is equal to $I_{T(RMS)}$ in the case of the above two conductive waveforms, when the expression relating the actual load average current I_d and the element-defined rated current average $I_{T1(AV)}$ can be obtained. When Eq. (13.3) is made equal to Eq. (13.5), the following equation can be obtained:

$$I_{T1(RMS)} = I_{T(RMS)}.$$

Or $1.57 I_{T(AV)} = 0.577 I_d$; that is,

$$I_{T(AV)} = \frac{0.577}{1.57} I_d = 0.36 \, I_d \tag{13.6}$$

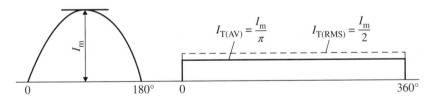

Figure 13.2 Relationship between I_m, $I_{T(RMS)}$, and $I_{T(AV)}$.

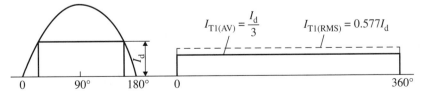

Figure 13.3 Relationship between I_d, $I_{T1(RMS)}$, and $I_{T1(AV)}$.

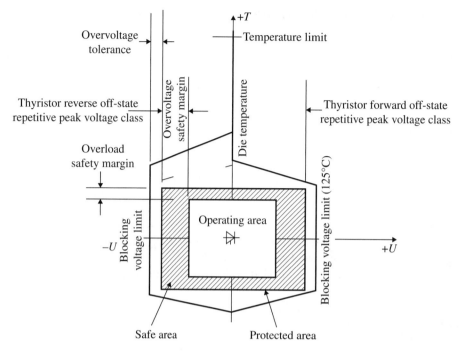

Figure 13.4 Safe operating limit range of thyristor.

Equation (13.6) indicates that for a three-phase bridge rectifier circuit with a rated excitation current of I_d, when the average current $I_{T(AV)}$ flowing through each element in the arm is selected, its value is $0.367I_d$.

It should be noted that in the selection of the rated current of an element, the forced excitation current should prevail, and an appropriate margin should be factored in. In general, for a thyristor element, in addition to its normal rated value, its working range in extreme working conditions should be given as a safety margin and protection set value reference. The corresponding working range is shown in Figure 13.4.

13.2 Parameterization of Power Rectifier

13.2.1 Thyristor Element Parameters

Power rectifier parameterization is presented with an ABB 5STP 26N6500Φ100 thyristor as an example. Table 13.2 shows the essential parameters of the 5STP 26N6500 thyristor.

13.2.2 Calculation of Reverse Repetitive Peak Voltage (U_{RRM})

As required by the original data of the hydro-turbine, it is known that the RMS generator excitation winding test voltage is 5090 V and the peak value is 7197 V. Under the applicable industry standard, the instantaneous value of the overvoltage at both ends of the excitation winding should be no larger than 70% of the maximum test voltage $7197 \times 0.7 = 5038$ V. The U_{RRM} of the selected thyristor should be above this value.

In addition, as required by the bidding document for excitation systems for the hydro-turbines of Three Gorges Hydro Power Station, the reverse repetitive peak voltage U_{RRM} should be no smaller than 2.75 times the maximum peak voltage of the secondary side of the excitation transformer. That is, U_{RRM} should be larger than $1378 \times \sqrt{2} \times 2.75 = 5358$ V.

When U_{RRM} of the selected thyristor is 6500 V, the above requirements can be met.

Table 13.2 Essential parameters of ABB 5STP 26N6500 thyristor.

Mode: 5STP 26N6500	Diameter: Φ100
Repetitive peak voltage U_{DRM}/U_{RRM}	6500 V
Average on-state current $I_{T(AV)}$	2810 A
Threshold voltage $U_{T(T0)}$	1.12 V
Slope resistance r_T	0.29 mΩ
Thermal resistance $R_{thjc} + R_{thch}$	$0.0057 + 0.001$ °C W^{-1}
du/dt	2000 V μs^{-1}
di/dt	1000 A μs^{-1}
Maximum allowable junction temperature T_{jmax}	125 °C
Gate trigger voltage (U_{GT})	2.6 V
Gate trigger current (I_{GT})	400 mA
I^2t	10125 kA^2·s ($T = 10$ ms)
	10375 kA^2·s ($T = 8.3$ ms)
On-state inrush current I_{TSM}	45 000 A ($T = 10$ ms)
	5 000 A ($T = 8.3$ ms)

13.2.3 Calculation of Thyristor Element Loss

As required by the excitation date of the hydro-turbine, the output capacity of a single rectifier bridge is 2500 A. Thus, the loss of a single thyristor element can be calculated. The calculation procedure is as follows.

1. On-state loss P_T
   There are two ways to calculate the on-state loss. Generally, the $P_T/I_{T(AV)}$ characteristic curve provided by the manufacturer is adopted for the purpose. $P_T = f[I_{T(AV)}]$ of the selected element is shown in Figure 13.5. The relationship between the thyristor element's actual average $I_{T(AV)}$ and the rectifier bridge output DC current I_f is as follows:

   $$I_{T(AV)} = I_f/3 = 833.3 (A)$$

   From Figure 13.5, the corresponding on-state loss P_T is 1700 W.
2. Turn-on loss P_{ON}
   A value of $di/dt = 10$ A μs^{-1} is taken. From Curve $[W_{ON} = f(I_{T(AV)})]$ in Figure 13.6, when $I_{T(AV)} = 833.3$ A, $W_{ON} = 0.7$ W. So, $P_{ON} = 50$ Hz × 0.7 W = 35 W.
3. Turn-off loss P_{OFF}
   Values of $di/dt = 5$ A μs^{-1} and $U_0 = 1.414 \times 1378 = 1948$ V are taken. From Curve $W_{OFF} = f(U_0)$ in Figure 13.7, $W_{OFF} = 10.5$ W. So, $P_{OFF} = 50$ Hz × 10.5 W = 525 W. The total loss of a single thyristor element is
   $$P_{TOT} = P_T + P_{ON} + P_{OFF} = 1700 + 35 + 525 = 2260 \text{ (W)}$$

13.2.4 Selection of Radiator

A panel double-sided radiator is adopted. Fom Curve $T_{CASE}/I_{T(AV)}$ in Figure 13.8, when $I_{T(AV)} = 833.3$ A, the case temperature T_{CASE} is about 113 °C.

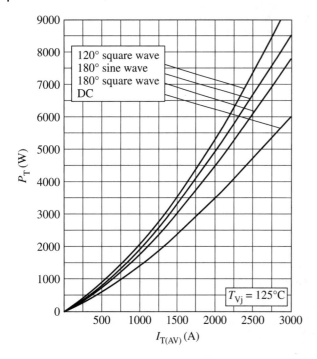

Figure 13.5 Thyristor on-state loss $P_T = f(I_{T(AV)})$ curve.

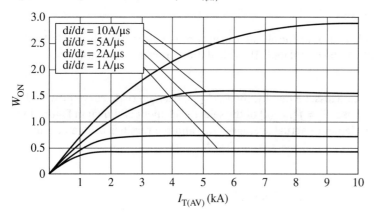

Figure 13.6 Thyristor turn-on loss curve [$W_{ON} = f(I_{T(AV)})$].

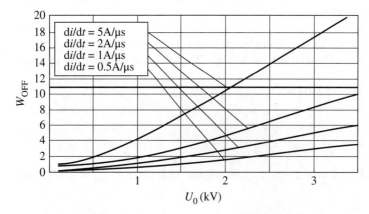

Figure 13.7 Thyristor turn-off loss curve [$W_{OFF} = f(U_0)$].

Figure 13.8 Thyristor radiator case temperature curve $[T_{CASE} = f(I_{T(AV)})]$.

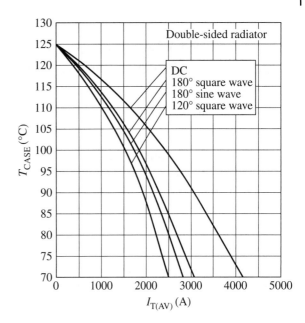

The maximum thermal resistance of the radiator can be calculated with the following formula:

$$R_{thHA(max)} = \frac{T_{C(max)} - P_{TOT}R_{thCH} - T_A}{P_{TOT}}$$

where

$T_{C(max)}$ – design maximum junction temperature of thyristor (110 °C);
P_{TOT} – total loss of single thyristor;
R_{thCH} – thyristor surface thermal resistance;
T_A – ambient temperature 40 °C.

$R_{thHA(max)} = 30 \text{ K kW}^{-1}$ can be obtained.

According to the above calculation, the thermal resistance of the selected double-sided press-fit air-cooling radiator is less than 30 K kW^{-1} when the wind speed is 6 m s^{-1}.

13.2.5 Selection of Fan Model

13.2.5.1 Calculation of Ventilation Quantity

The ventilation quantity is calculated as follows:

$$Q_f = \frac{Q}{C_p \gamma \Delta t} \tag{13.7}$$

where

Q_f – ventilation quantity m^3 hr^{-1};
Q – total loss power kcal hr^{-1};
C_p – air specific heat kcal (kg·K)$^{-1}$ 0.24;
γ – air specific gravity kg m^{-3} 1.128 kg m^{-3} at 40 °C;
Δt – air inlet and outlet temperature difference K.

$$\Delta t = \frac{2.24 I_{T(AV)} U_{T(AV)} \times 860}{864 \gamma \frac{A}{2} v} 10^3 (°C) \tag{13.8}$$

where

$I_{T(AV)}$ – actual forward average current of thyristor (calculated to be 833.3 A);

$U_{T(AV)}$ – forward voltage drop of thyristor 1.12 V;

A – cross-sectional area of air duct 18 410 mm²;

v – inlet air velocity 6 m s⁻¹;

γ – air specific gravity 1.128.

When they are substituted into Eq. (13.8), $\Delta t = 16.7\,°C$ can be obtained.

Through the thyristor output capacity calculation, a single thyristor's heating power can be obtained as 2260 W and the total heating power as Q = 2260 × 6 = 13 560 W. Through the conversion 1 kcal hr⁻¹ = 1.163 W, Q = 11 660 kcal hr⁻¹ can be obtained.

When they are substituted into Eq. (13.7), $Q_f = 2580\,m^3\,hr^{-1}$ can be obtained.

13.2.5.2 Selection of Fans

On the basis of the above calculation, two centrifugal fans can be selected, one operating and the other serving as a standby. The motor power is 1.33 kW. With one fan operating, the air flow is 2600 m³ hr⁻¹.

13.2.6 Selection of Quick Fuse Parameters

Fast fuses are connected in series to the rectifier bridge arms. In the event of any abnormal working condition, the fast fuses can protect the thyristor elements from damage. In addition, in the event of a rectifier bridge arm failure (e.g., a thyristor breakdown), the fast fuses can quickly clear the fault to prevent the operation of the entire device being impacted.

Key parameters for fast fuses include

1. *Rated current*. The rated current of the fast fuse is expressed by the RMS current value. The rated current refers to the current that is allowed to pass through under AN (A: air; N: natural circulation) conditions. For fast fuses connected in series to the bridge arm branches, the RMS current flowing through the thyristors and the fast fuses is 0.577 times the DC output current. So, when the output per rectifier is 2500 A, the RMS rated current of the quick fuses is $I_R = 2500 \times 0.577 = 1442$ (A).
2. *Breaking capacity*. The strength of the quick fuse case determines to a large extent the capacity of breaking the maximum fault current, which is called the "rated breaking capacity" in the IEC standard. Moreover, the shape of the metal fuse pieces inside the quick fuses, the filler's metal vapor adsorption and heat, and the fuse links' electrodynamic force affect the breaking capacity. Insufficient breaking capacity will lead to fast blow, continuous arc burning and explosion, and even AC/AC short circuit in serious cases. So, the rated breaking capacity is an important safety indicator. When choosing the breaking capacity, the arc voltage peak at the instant of the breaking (called the transient recovery voltage in the standard) should also be considered, and it should be lower than the maximum voltage the thyristor elements can withstand. According to the calculation, the short-circuit current in the event of a generator rotor slip ring short circuit or a direct short circuit on the DC side of the rectifier bridge is about 70 kA, which acts as the breaking capacity current of the fuses.
3. *Rated voltage*. The rated voltage of the quick fuses should be higher than the RMS voltage on the AC side of the rectifier bridge.
4. *Heat capacity I^2t*. When the current carrying capacity of the fast fuses can meet the system's short-circuit current requirements, the fast fuses can break the fault current in the event of a short-circuit fault. But the identification of the fast fuses' ability to protect the thyristor elements in series must be based on an analysis of I^2t of both the elements and the fuses. Only when the I^2t of the fast fuses is smaller than that of the thyristor elements can the fast fuses protect the thyristor elements. In the event of a short-circuit fault, I^2t is divided into two stages: fore-arc I^2t and blow I^2t.

The time interval when the melt metal turns from solid to liquid is called the fore-arc time. I^2t at this stage is just fore-arc I^2t. Experience has shown that a higher fore-arc I^2t rise rate is more conducive to reliable breaking of the fast fuses. So, in applications, low multiple overloads of the fast fuses should be avoided. That is, the possible short-circuit current in the circuit should be larger than five times the rated current of the fast fuses.

The arc ignites when the melt metal turns into vapor and then burns and quenches. I^2t at this stage is just the blow I^2t. In applications, the rated voltage of the quick fuses is always higher than the service voltage. The actual blow I^2t should be corrected to the actual voltage/rated voltage $\times I^2t$. The fast fuses' integrated I^2t is the sum of the fore-arc I^2t and the corrected blow I^2t.

I^2t is an important indicator in the selection of fast fuses. The identification of the fast fuses' ability to protect the thyristor elements in series must be based on an analysis of the I^2t of both the fast fuses and the thyristor elements. Only when the fast fuses' I^2t is smaller than the thyristor elements' I^2t can the fast fuses play a protective role.

Under this principle, a 170M7255 fast fuse produced by Danmark's Bussmann is taken as an example. Its rated voltage is 1500 V and rated current is 2100 A. The coordinative relationship between the fast fuse and the thyristor is shown in Figure 13.9.

5. *Temperature rise and power consumption.* The loss of the quick fuses is calculated as follows:

$$\Delta W = mI_t[1 + \alpha(t - t_0)]R/n_b$$

where

- m — number of rectifier bridge arms (six for three-phase rectifier bridge);
- I_t — RMS current passing through quick fuses;
- t — 120 °C in case of air-cooling;
- t_0 — may be 20 °C in calculation;
- α — resistance temperature correction factor 0.0035/°C;
- R — cold resistance of quick fuses;
- n_b — number of branches in parallel per arm (generally $n_b 1$ in excitation systems).

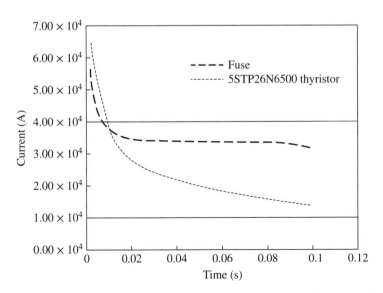

Figure 13.9 Ampere-second characteristic coordination between fast fuse and thyristor.

The fast fuses blow based on the loss or heat. In normal operation, the temperature rise should be minimized.

The fast fuses' power consumption greatly depends on their cold resistance. The cold resistance is an important indicator of fast fuses. A low cold resistance is conducive to reduction of their temperature rise, because their current carrying capacity is primarily limited by their temperature rise. Connection of joints of fast fuses affects their temperature rise as well. As required, temperature rise at quick fuse joints should not affect the operation of adjacent devices. Tests proved that fast fuses can operate for a long time when their temperature rise is below 80 K. Products based on a stable manufacturing process can still operate for a long term when their temperature rise is 100 K. A temperature rise of 120 K is the critical point of the current carrying capacity. Fast fuses cannot operate for a long time if the temperature rise is 140 K. In practical applications, if the quick fuses are placed in the cooling system of the thyristor elements, their temperature rise can be more effectively reduced.

13.2.7 Calculation of Rectifier Loss

The total loss of a single rectifier is calculated as follows:

$$P = 6P_{VTHN} + 6P_{VSN} + 6P_{VFUN} + P_{VCN} + P_{VFAN} + P_{VOVP}$$

where

P_{VTHN} — on-state loss of each thyristor;
P_{VSN} — thyristor switching loss;
P_{VFUN} — fast fuse loss;
P_{VCN} — copper bar loss;
P_{VFAN} — fan loss;
P_{VOVP} — overvoltage protection circuit loss.

It is assumed that the excitation device's output current is 4708 A, evenly distributed by four power cabinets. The loss is calculated based on a per-rectifier output current of 1177 A.

(1) The on-state loss of each thyristor is calculated as follows:

$$P_{VTHN} = 1/3 I_{dN}(U_{T0} + I_{dN}^2 r_T)$$

When the threshold voltage U_{T0} is 1.12 V and the slope resistance r_T is 0.29 mΩ,

$$P_{VTHN} = 1/3 \times 1177\ (1.12 + 1177 \times 0.00029) = 573\ (W)$$

(2) The thyristor switching loss is calculated as follows:

$$P_{VSN} = 35 + 525 = 560\ (W)$$

(3) The copper bar loss is calculated as follows (a copper bar resistance of 0.1 mΩ is taken):

$$P_{VCN} = I_{dN}^2 R_{equ} = 1177^2 \times 0.0001 = 221\ (W)$$

(4) The fan loss is calculated as follows:

$$P_{VFAN} = 1330\ (W)$$

(5) The fast fuse loss is calculated as follows:

$$P_{VFUN} = (1.732 I_{dN}/3 I_{FUN})^{2.4} \times P_{VFUN}$$

When $I_{FUN} = 2100$ A, $P_{VFUN} = 180$ W. $P_{VFUN} = 13$ W can be obtained through the calculation.

(6) The overvoltage protection unit loss is calculated as follows:

$$P_{\text{VOVP}} = \frac{(1.35U_{\text{IN}})^2}{R_1} = 2307\ (\text{W})\ (1378\text{V}/1500\Omega)$$

The total loss of a single rectifier is calculated as follows:

$$P = 6 \times (573 + 560) + 221 + 1330 + 6 \times 13 + 2307 = 10734\ (\text{W})$$

The total loss of the four rectifiers: $10\,734 \times 4 = 42\,936$ W.
If the power of the fans is neglected, the total loss of the four rectifiers is 37.6 kW.

13.3 Cooling of Large-Capacity Power Rectifier

The cooling of a power rectifier is done primarily by the radiator. Using the basic principles of heat transfer theory, a heat flow path must designed for the radiator, whose thermal resistance must be as low as possible for the thyristor elements to dissipate the heat generated by their power consumption as fast as possible, so that the junction temperature inside the thyristor core is always kept within the allowable junction temperature range and normal operation is guaranteed.

In radiator selection, structural design and the cooling medium should be taken into account. In radiator structure selection, factors such as energy consumption, the volume and weight of auxiliary equipment, complexity, ease of operation, reliability, availability, and maintainability should be taken into account. The cooling medium selection, electrical insulation, chemical stability, material corrosion, environmental impact, and flammability as well as degree of commercialization should be taken into account. Currently, there are three commonly used cooling media: air, oil, and water. The heat transfer coefficients, which decisively influence the heat dissipation effect, are 35, 350, and 3500 W $(\text{m}^2\cdot\text{K})^{-1}$, respectively, which are described below. Commonly used air-cooling modes include AN and AF (A: air; F: forced circulation).

13.3.1 AN Radiator

In the so-called AN mode, heat dissipates through natural convection and radiation. This mode is has low efficiency, but it features a simple structure, zero noise, and easy maintenance. In particular, it boasts high reliability since it does not have a rotating part. It is applicable to power devices whose rated current is smaller than 200 A and simple high-current rectifier devices.

13.3.2 AF Radiator

AF radiators are primarily used for devices with a rated current of 200–2000 A. AF radiators feature high heat dissipation efficiency, with a heat transfer coefficient 2–4 times that of AN radiators. But the AF mode requires fans, which are associated with weaknesses such as high noise, dust inhalation, relatively low reliability, and hard maintenance. The standard value of the cooling air velocity inside an AF device is 6 m s^{-1}. When the air velocity is smaller than this value, the thermal resistance corresponding to the actual air velocity should be selected on the basis of the information provided by the manufacturer.

The quality characteristics of radiator material have a significant effect on the heat dissipation efficiency. Red copper's thermal conductivity is twice that of industrial pure aluminum. At the same heat dissipation efficiency, the size of a red copper radiator is $\frac{1}{3} - \frac{1}{2}$ that of an aluminum radiator. Moreover, red copper has an advantage in salt dew resistance. However, due to the large proportion of copper and high cost, red copper radiators are less frequently employed.

13.3.3 Heat Pipe Radiator [30]

"Heat pipe" is a heat transfer element invented by G. M. Grover from Los Alamos National Lab in 1963. Taking full advantage of the heat transfer principle and the fast heat transfer properties of the refrigerating medium, a heat pipe quickly transfers heat from heat sources to the outside. Its thermal conductivity is higher than all known metals. In the past, heat pipe technology was widely used in the space and military industries. However, since it was introduced into the radiator manufacturing industry, it has triggered changes in radiator design philosophy and helped the industry become independent of the single heat dissipation mode, where a better heat dissipation effect is achieved solely with high-air-flow fans. With heat pipe technology, a satisfactory cooling effect can be achieved even with low-speed and low-air-flow fans. An appropriate design can even achieve the full AN mode and eliminate the need for AF fans, thus effectively solving the noise and dust problems in the AF mode and setting a new milestone in the application of radiators.

China has embarked on research on the heat transfer performance of heat pipes since the 1970s and the application of heat pipes to electronic device cooling and spacecraft. With the continuous advances in the science and technology, especially since the 1980s, the country has been expanding its heat pipe research and application areas and has even been applying heat pipes to high-power electronic devices.

13.3.3.1 Working Principle of Heat Pipe

Thermodynamically, heat absorption and heat dissipation are relative. Where there is a temperature difference, there is heat transfer from the high temperature to the low temperature. There are three heat transfer modes, including radiation, convection, and conduction. Among these, heat conduction is the fastest. The application of heat pipes is based on the fast conduction of heat achieved through a vapor–liquid phase transition of the working fluid.

A typical heat pipe consists of a case, a wick, and an end cap. When a heat pipe is manufactured, it is vacuumized. When a negative pressure of $1.3 \times (10^{-1}$–$10^{-4})$ Pa is reached, it is filled with an appropriate amount of working fluid, so that the capillary porous material of the wick that is closed to its inner wall is filled with liquid and sealed. As shown in Figure 13.10, one end of the heat pipe is the evaporation section (heating section) and the other is the condensation section (cooling section). An insulation section can be arranged between the two sections depending on the application. When one end of the heat pipe is heated, the liquid in the capillary wick evaporates and vaporizes. The vapor under the action of the pressure difference

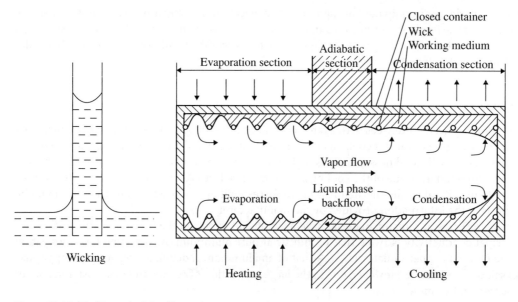

Figure 13.10 Working principle of heat pipe.

releases heat to the other end and condenses into a liquid. The liquid flows back to the evaporation section along the porous material on account of the wicking. As that cycle repeats, the heat is transferred from one end of the heat pipe to the other end. Physically, the heat pipe's heat transfer process involves the following six interrelated steps:

(1) The heat is transferred from the heat source through the heat pipe wall and the wick filled with working liquid to the liquid–vapor interface.
(2) The liquid evaporates on the liquid–vapor interface in the evaporation section.
(3) The vapor in the vapor chamber flows from the evaporation section to the condensation section.
(4) The vapor condenses on the vapor–liquid interface in the condensation section.
(5) The heat is transferred from the vapor–liquid interface through the wick, the liquid, and the pipe wall to the cold source.
(6) The condensed working fluid flows back to the evaporation section due to the wicking in the wick.

13.3.3.2 Loop Heat Pipe (LHP) Radiator

The LHP is a heat transfer device invented by Professor Yu. F. Maidanik, a Russian scientist. It uses the wicking in the evaporator to drive the circuit to operate and utilizes evaporation and condensation of the working medium to transfer heat. Thus, it can transfer a large quantity of heat at a small temperature difference and over long distances. It is an efficient two-phase heat transfer device.

An LHP consists of five parts: evaporator, evaporation section, condensation section, backflow section, and compensation chamber, as shown in Figure 13.11. There is a wick structure inside the evaporator. There are many vapor channels on the inner wall of the evaporator or the wick structure, as shown in Section A-A in Figure 13.11. The basic working principle is as follows: The wick structure itself can suck the liquid upward to be filled with the working medium liquid. When the evaporator is heated, the wick structure is heated as well. The liquid in the wick structure evaporates into a gas and passes through the vapor channel along the evaporation section into the condensation section while drawing the heat. In the condensation section, the gas condenses into a liquid, releasing latent heat. By virtue of the wicking of the wick structure, the liquids flows back into the compensation chamber along the backflow section and reaches the mix structure. Thus, a working medium circulation and heat transfer process is formed. The main role of the compensation chamber is to accommodate the liquid in the evaporation section and the condensation section at the time of the start-up and prevent the evaporator from drying up owing to delayed liquid backflow during operation.

LHP has the following advantages:

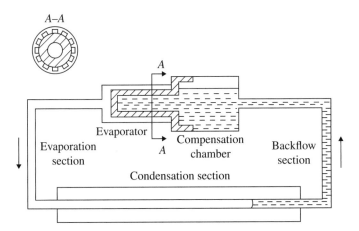

Figure 13.11 Schematic diagram of LHP.

(1) It is an efficient two-phase heat transfer device, which uses the wicking in the evaporator to drive the circuit to operate and uses evaporation and condensation of the working medium to transfer heat, without the need for external power.
(2) It features good isothermal performance, and long-distance heat transfer and change of heat transfer direction capabilities.
(3) The vapor–liquid channel separation design makes the flow direction of the medium in the two phases in the pipe consistent, accelerates the speed of movement of the medium, avoids the mutual interference between the vapor and the liquid, greatly reduces the thermal resistance, and improves the heat transfer efficiency.
(4) It allows more flexible applications since different structures can be selected depending on the needs.

Due to its good heat transfer performance, LHP technology has matured and is widely applied to aerospace. So, LHP radiators for high-power thyristor rectifiers have made the fanless AN mode for high-power thyristor rectifiers possible.

13.3.3.3 Selection of Power Radiator

In the selection of a power radiator, the power device structure is the key consideration. When a power device is equipped with a radiator, the heat dissipation path will change. The thermal resistance R_{Tj} remains unchanged. The heat of the device is on the one hand transferred through the case directly to the surroundings at a thermal resistance of R_{Tp} and on the other hand is transferred to the radiator at a thermal resistance of R_{Tc} and then dissipated by the radiator into the surrounding space at a thermal resistance of R_{Tf}. The thermal resistance network is shown in Figure 13.12. When the heat dissipation of the case itself to the surrounding environment can be ignored, that is, $R_{Tp} \geq R_{Tc} + R_{Tf}$, Figure 13.12 can be simplified to Figure 13.13. In the thermoelectric simulation method, the power consumption is simulated as a current, the temperature difference as a voltage, and the thermal resistances as resistances. Thus, the thermal resistances can be calculated:

$$R_T = R_{Tj} + R_{Tc} + R_{Tf} = \frac{T_j - T_a}{P_c} \tag{13.9}$$

$$R_{Tj} = \frac{T_j - T_c}{P_c} \tag{13.10}$$

$$R_{Tc} = \frac{T_c - T_f}{P_c} \tag{13.11}$$

$$R_{Tf} = \frac{T_f - T_a}{P_c} \tag{13.12}$$

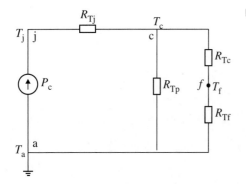

Figure 13.12 Thermal resistance network diagram for radiator installation.

Figure 13.13 Simplified thermal resistance network diagram.

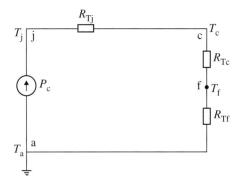

where

T_j — junction temperature of semiconductor device;
T_c — temperature of semiconductor device case;
T_f — temperature at maximum temperature point of radiator;
T_a — ambient temperature;
T_{Tj} — total thermal resistance of system;
P_c — dissipation power of semiconductor device.

As can be seen from Eq. (13.9), to improve the heat dissipated by a power device through the radiator, each of the thermal resistances should be minimized. Since the power device's internal thermal resistance R_{Tj} determined by the power device process is constant, effective measures to reduce the thermal resistance of the interface and the radiator should be considered. The interface thermal resistance R_{Tc} is the thermal resistance generated by the contact stress between the device and the radiator, and is affected by many factors. The radiator's thermal resistance R_{Tf} is the main basis for the radiator selection.

For the purpose of this chapter, the thermal resistance of the LHP radiators applied to high-power thyristor rectifiers is 0.057 K W^{-1}.

13.3.3.4 Design of High-Power Heat Pipe Rectifier

Radiators for high-power thyristor rectifiers include aluminum radiator, copper radiator, and heat pipe radiator. Aluminum radiator technology has a long history. An aluminum radiator's minimum thermal resistance can reach 0.035–0.04 °C W^{-1} (air velocity is higher than 5 m s^{-1} for the AF mode). In the absence of air, an aluminum radiator's minimum thermal resistance can reach 0.15–0.25 °C W^{-1}. The copper radiator is another kind of efficient radiator. Since copper's thermal conductivity is better than that of aluminum, a copper radiator's thermal resistance can reach 0.03 °C W^{-1}, which is better than that of an aluminum radiator under the same AF conditions. At the same air velocity and the same thermal resistance, a copper radiator's volume is smaller than that of an aluminum radiator. So, copper radiators are more suitable for applications with tight space constraints. Heat pipe radiators integrate the advantages of aluminum and copper radiators. Under AN conditions, a heat pipe radiator's thermal resistance can reach a minimum of 0.04 °C W^{-1}. So, the long-term stable output of a heat pipe radiator rectifier can reach 1500 A in the absence of air. In the case of a slight air flow (less than 3 m s^{-1}), a heat pipe radiator's thermal resistance is equal to that of an AF copper radiator, and the output of a heat pipe radiator rectifier can reach 2000 A.

An ABB 5STP52U5200 thyristor imported with original packaging is selected. Its average on-state current is 4120 A (70 °C), and its reverse withstand voltage is 4400 V. The maximum temperature rise of the thyristor radiator at which the long-term stable output of the heat pipe radiator rectifier can reach 1500 A in the absence of air is calculated.

The thyristor loss calculation formula is as follows:

$$P_{TOT} = P_T + P_{GM} \tag{13.13}$$

where

P_{TOT} – average loss of thyristor elements W;
P_T – on-state loss of thyristor elements W;
P_{GM} – switching loss of thyristor elements 20 W.

When the per-rectifier output is 1500 A, the current flowing through a single thyristor is a 120° square wave. The corresponding current average is as follows:

$$I_{T(AV)} = 1500/3 = 500 (A)$$

From the thyristor on-state loss characteristic curves in Figure 13.14, the on-state loss of the thyristor elements is 650 W. On the basis of Eq. (13.13), the total heat value of the elements of a single thyristor can be calculated:

$$P_1 = 650 + 20 = 670 \, (W)$$

From the allowable thyristor core temperature diagram in Figure 13.15, the allowable thyristor temperature is about 107 °C. The calculation is based on an ambient temperature of 40 °C. If a radiator is designed on the basis of this dissipation power, the required radiator resistance is as follows:

$$R_{Tf} = \frac{T_f - T_a}{P_c} = \frac{107 - 40}{670} = 0.1 \, (°C/W)$$

When an LHP radiator is selected, the thermal resistance R_{Tf} is 0.057 °C W^{-1} under the AN conditions, which can meet the design requirements. When the rectifier output is 1500 A in the absence of air, the average current per thyristor is calculated as follows:

$$I_{T(AV)} = 1500/3 = 500 \, (A)$$

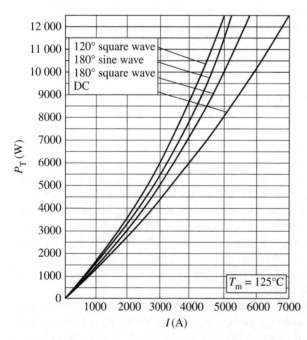

Figure 13.14 Thyristor on-state loss curves corresponding to different average on-state currents.

Figure 13.15 Allowable thyristor case temperature curves corresponding to different average on-state currents.

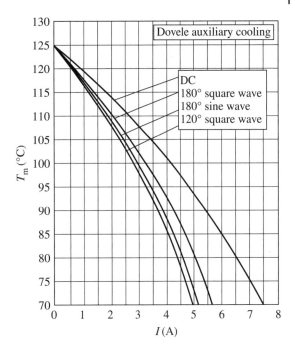

From the thyristor on-state loss characteristic curves in Figure 13.14, the on-state loss of the thyristor elements is 650 W. From Eq. (13.13), the total heat value of the elements of per thyristor can be calculated:

$$P_1 = 650 + 20 = 670 \,(\text{W})$$

So, the thyristor radiator temperature rise is as follows:

$$\Delta T = R_{\text{Tf}} P_1 = 670 \times 0.057 = 38.2 \,(°\text{C})$$

It should be noted that China has developed heat pipe high-power thyristor rectifiers with a per-rectifier capacity of 2500 A. For example, the STR-RG system high-power heat pipe thyristor rectifier developed by Nengda General Electric Co. Ltd. has achieved a per-rectifier rated capacity of 2500 A. In addition, the STR-RG series features an LHP radiator structure that can achieve gas–liquid separation, thus solving the problem of gas–liquid combination and mutual interference in conventional heat pipe structures and further improving the heat dissipation efficiency, which is the innovation of the design.

13.4 Current Sharing of Power Rectifier [31]

13.4.1 Background

In the excitation system of a large synchronous generator, there should be a good current sharing coefficient between the multiple high-power rectifiers operating in parallel, so that the equipment's capacity can be fully and effectively utilized. So, the applicable industry standards all stipulate that power rectifiers' current sharing coefficient shall be no smaller than 0.85. There are two technologies that can meet the current sharing coefficient requirement: conventional current sharing and digital current sharing.

Conventional current sharing measures include matching based on thyristor parameters such as the peak average on-state voltage drop, matching based on the rectifiers' AC/DC inlet and outlet lines, connecting a hollow current sharing reactor to the rectifier bridge arms, and making the cables of the branches from the secondary side of the excitation transformer to the power rectifier inlet line end equal in length. These measures can improve the flow sharing function to a certain extent, but also increase the difficulty of system design and engineering construction.

Another technology for improvement of current sharing is digital current sharing. The factors affecting current sharing are analyzed in detail, and conventional and digital current sharing technologies are compared in Reference [31]. This section cites the applicable conclusions without presenting many mathematical derivations.

13.4.2 Factors Influencing Effect of Current Sharing

13.4.2.1 AC-Side Inlet Line

The thyristors' on-state resistance (R_{1T}) is many times larger than the resistance $R_{1a} + R_{1d}$ of the AC side and DC side of the rectifier bridges – generally reaching $(R_{1T}/R_{1a} + R_{1d}) \approx 15$. When the on-state current of the thyristors selected by the rectifier bridges is large (e.g., larger than 3000 A), the influence of the length of the AC-side inlet line on the current sharing will be obvious since the thyristors' on-state resistance decreases (e.g., $R_{1T}/(R_{1a} + R_{1d})$ is 4–5). In order to reduce the influence of the AC inlet line on current sharing, the thyristors' on-state current margin should be a smaller value under the rectifier bridge output requirement.

When a cable is adopted for the AC inlet line, on the one hand, since the equivalent time constant $\left(\frac{L1a+L2a}{R1a+R2a}\right)$ becomes larger, the commutation process between the two commutation elements will affect the current distribution. On the other hand, since a cable's resistivity is much larger than that of a copper bar and the difference between the AC-side cable's resistance and the thyristors' on-state resistance is small, the influence of the AC-side cable on current sharing is significant. In a three-machine excitation system, a cable is generally adopted for connection between cabinets. The time constant $\left(\frac{L1a+L2a}{R1a+R2a}\right)$ is large, and the rectifier bridges' AC power supply frequency is 350–500 Hz. In that case, the commutation process has a decisive influence on the current distribution. So, the difference in AC-side cable length between the thyristor rectifier bridges has a significant influence on current sharing.

For bridges interconnected with a copper bar on their AC side, since the copper bar shows a significant resistance (i.e., the resistance is much larger than the inductance), the objective of current sharing can be achieved by changing the resistance of the copper bar. For example, grooving or punching the copper bar for resistance improvement is a method of current sharing. In particular, changing the contact resistance of the copper bar with this method can more significantly improve current sharing. Changing the inductance to improve current sharing is more difficult to implement, because the series inductance required is large, and therefore the time constant will increase. Since the current distribution is primarily determined by the commutation process, current sharing can be achieved only when the reactance is increased to the same magnitude as the thyristor on-state resistance. Westinghouse and GE apply current sharing reactors added to the AC side. Current sharing reactors have large size and complex reactance installation and copper bar arrangements. An iron core reactance can be adopted to reduce the current sharing reactance volume, but at the expense of additional heat caused by hysteresis loss.

13.4.2.2 DC-Side Outlet Line

When the difference between the AC-side parameters and the thyristor parameters is ignored, the influence of the DC-side impedance on current sharing can be obtained. It should be noted that the impedance of the DC output side is common to the AC three-phase input side. So, in the commutation process, the DC bus current change process is more complex than the AC-side single-phase impedance change process. But the influence of the DC side is less than that of the AC side. So, the AC-side impedance matching should be given priority in current sharing.

13.4.2.3 Commutation Process

If the inlet line is completely symmetrical, the current distribution is independent of the commutation since the current is evenly distributed. If the communication inlet line is not symmetrical, the impedance of Bridge 1 and that of Bridge 2 are $|Z_1| = |R_{1a} + j\omega L_{1a}|$ and $|Z_2| = R_{2a} + j\omega L_{2a}|$, respectively, and it is assumed that

$|Z_1| > |Z_2|$. In general, the current of the rectifier bridge with the smaller AC-side impedance is larger. Thus, as the excitation current increases, the commutation angle will increase. Thus, the current of the branch with the larger current will decrease accordingly, and the difference between the current of Bridge 1 and that of Bridge 2 will automatically balance.

13.4.2.4 Thyristor Parameters

In the study on current sharing distribution, the thyristor model is simplified with a voltage source plus an internal resistance. The thyristors from trigger to full turn-on should meet the threshold voltage $U_{T(TO)}$ requirements. So, the turn-on of the thyristors with a large $U_{T(TO)}$ will lag behind the turn-on of those with a small $U_{T(TO)}$. After the thyristors reach $U_{T(TO)}$, the thyristor voltage drop will decrease to U_T. The thyristor voltage drop calculation formula is as follows:

$$U_T = A + Bi_T + C \ln(i_T + 1) + D \sqrt{i_T}$$

The above equation can be first-order approximately equivalent to $U_T = A' + B'i_T$, where B′ is the reduced slope resistance (30% greater than that obtained from the slope data provided by the thyristor manufacturer). It is assumed that $U_{T(TO)1}$ of Thyristor T12 is greater than $U_{T(TO)2}$ of Thyristor T22. To turn T22 on, the following must be met:

$$U_{T1} \approx U_{T(OT)1} + R_{T1}\ i_T > U_{T(TO)2}$$

On the whole, in the rectifier bridge thyristor selection, the thyristors on the same arm should be arranged under the principle of threshold voltage similarity. The thyristor on-state resistance difference can be balanced on the basis of the AC/DC side inlet/outlet line length. For example, using an AC cable as an inlet/outlet line is just a method to balance the thyristor on-state resistance difference. It can be considered that when the current of the thyristor turned on first rises to a large value, those turned on later will primarily rely on the thyristor voltage drop difference for the current transfer after the commutation is completed. The threshold voltage drop difference has an important influence on the thyristor current distribution. So, for thyristor rectifier bridges interconnected with copper bars on their AC side, priority should be given to selection of the thyristor threshold voltage drop.

13.4.3 Digital Current Sharing

Under the principle of digital current sharing (or intelligent current sharing), the thyristor trigger pulse outputted by a digital automatic voltage regulator (AVR) is processed and its turn-on time is regulated, so that the purpose of current sharing can be achieved. There are many digital current sharing methods. One of them is to measure the RMS value or average of each current and compare the current of each thyristor with the evenly distributed value of the rotor current. The trigger pulse of the thyristor whose current is greater than the evenly distributed value of the rotor current will be appropriately delayed to achieve an even current distribution. Digital current sharing technology is more widely used in ABB excitation devices.

13.4.4 Comparison Between Digital and Conventional Current Sharing

Compared to conventional current sharing, digital current sharing has many advantages, such as (1) simple regulation, clear objective, and online implementation; (2) lower requirements for thyristor device parameter consistency, which reduces device cost and difficulty of selection; (3) strong adaptability to AC/DC inlet/outlet line impedance changes, low on-site construction requirements, and easy rectifier position arrangement; (4) independence of operating conditions from the excitation system (e.g., commutation); (5) accurate reflection of status of each thyristor and actual current sharing through measurement of current of each thyristor; and (6) online fault diagnosis and fault current limiting of power amplifier of excitation system, which is helpful for improving the mean time between failures (MTBF) of the excitation device. But digital current sharing also has some prominent weaknesses. For example, since a regulation circuit is connected to the pulse circuit, the system reliability is reduced. In the working area, an excessive pulse delay angle is bound to increase the intensity of the current surge of the thyristor that is turned on first. This is disadvantageous for an excitation system with multiple bridges in parallel, because the thyristor that is turned on first needs to withstand an impact current that is instantaneously 3–4 times higher than the normal current. So, the trigger pulse delay angle should be limited to the specified range.

It should be emphasized that excitation rectifier systems can generally meet the industry standard through conventional current sharing. However, when current sharing is judged through the meter reading, a serious imbalance between the thyristors may be hidden, because a shunt through the DC side cannot accurately reflect the current waveform of each thyristor or the current of the bridge where it is located (when there is a difference between the DC positive and negative arm currents). Conventional current sharing generally involves requirements for the rectifier inlet and outlet lines and even requires the addition of a special inlet/outlet line cabinet. Grooving and punching are economical measures but not applicable in all situations, since they lead to hard manufacturing and processing, especially inconvenient on-site rectifier bridge maintenance.

13.4.5 Conclusions

For conventional current sharing, priority should be given to the selection of the thyristor threshold voltage drop. The second consideration should be AC inlet line and DC outlet line regulation as well as AC and DC copper bar grooving or punching. And the last consideration is adoption of a current sharing reactance. Besides, in the selection of thyristor specifications, under the premise of avoiding any influence on the rectifier output, priority should be given to thyristors with a small on-state current, so that the thyristor parameters can take precedence in the current sharing influence to ensure the thyristors' current sharing. The experimental and theoretical analysis shows that for digital current sharing, the current sharing coefficient can be regulated over a wide range through regulation of the delay angle of the trigger pulse during the commutation, so that current balance of each thyristor can be truly controlled. For an excitation system with a large commutation angle under the rated operating conditions, as long as the no-load current sharing coefficient reaches 0.9 or more, the required current sharing coefficient of all operating points of the rectifier bridges can be ensured.

Compared to a system with conventional current sharing, a system with digital current sharing requires that on the one hand priority should be given to conventional current sharing measures and on the other hand the reliability of the added digital circuit and operational logic should be ensured. For a system that relies solely on conventional current sharing measures, the current of each thyristor should be measured at the time of the commissioning to ensure that the ammeter can truly reflect the current of each thyristor, or thyristor current monitoring device should be installed to monitor the current distribution. It should be noted that judging the effect of conventional current sharing measures with an ammeter may lead to inaccurate results. A digital current sharing system can achieve true intelligent current load distribution and help improve the reliability of the excitation system.

13.5 Protection of Power Rectifier [32, 33]

In the static self-excitation system of a large hydro-turbine, the overvoltage and overcurrent withstand capacity of the thyristor elements in the power rectifiers is poor. Thus, even an instantaneous overvoltage or overcurrent may damage them. Moreover, the forward voltage rise rate and current rise rate withstand capacity of the thyristor elements is limited. Thus, the out-of-limit forward voltage rise rate or current rise rate may damage them as well.

When a forward/reverse instantaneous peak voltage applied to the thyristor elements exceeds the specified value, it is referred to as an overvoltage. If the instantaneous value of the applied forward voltage exceeds the off-state non-repetitive peak voltage of the thyristor elements and reaches the forward-breakover voltage or the voltage value does not reach the forward-breakover voltage but the voltage rise rate is large and exceeds the allowable blocking voltage critical rise rate, the power components may be damaged. If the instantaneous

13.5 Protection of Power Rectifier

reverse peak voltage applied to the thyristor elements exceeds its non-repetitive reverse peak voltage and reaches the reverse breakdown voltage, reverse breakdown damage of the power components may occur.

In order to ensure safe and reliable operation of the excitation system and extend the life of the power components, in addition to improving the quality of the power components and choosing correct parameters in the design as well as setting aside a certain margin of safety, limiting their overvoltage, overcurrent, voltage rise rate, and current rise rate to the allowable ranges is a must.

From the above considerations, a high-power rectifier should generally have an anode overvoltage protection device attached at the inlet line end of the AC side, an R-C protector added at both ends of the thyristors, and an overvoltage absorption device added at the output end of the DC side.

13.5.1 Power Rectifier Anode Overvoltage Suppressor

In order to suppress the spike overvoltage caused by the thyristors in the commutation process, spike overvoltage suppressors in different forms are connected to the AC input end of the high-power rectifier to preprocess the AC-side overvoltage.

Figure 13.16 shows the connection of a typical AC spike overvoltage suppressor used by GE for the AC side of the power rectifiers in the EX2100 excitation system for 600–1000 MW large steam turbines. The specific

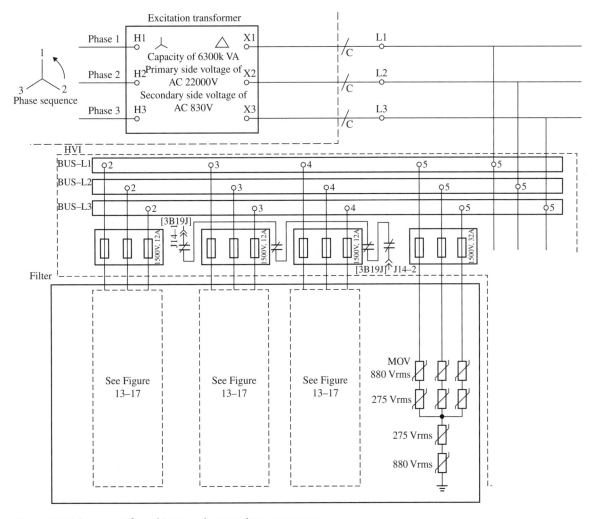

Figure 13.16 Power rectifier cabinet anode overvoltage suppressor.

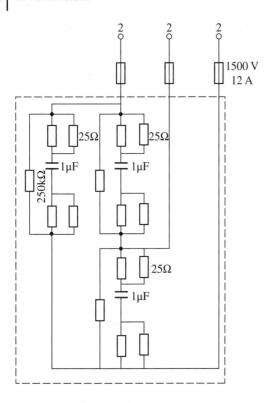

Figure 13.17 Anode overvoltage suppressor (filter).

connection of the spike overvoltage R-C protector is shown in Figure 13.17. After the spike overvoltage is filtered, its amplitude can be reduced to below 1.5 times the rated value. So, the effectiveness is very evident. In addition, in order to further limit the amplitude of the spike overvoltage, metal oxide varistor (MOV) nonlinear resistance overvoltage protection with a higher set value is provided. The MOV's interphase or ground action is set at 2310 V.

It should be emphasized that the role of the spike overvoltage R-C protector is to slow down the overvoltage rise rate. Different spike overvoltages can result in different overvoltage limits, while the MOV nonlinear resistance limit is constant. This physical process difference must be identified.

For the Unitrol-5000 excitation regulator manufactured by ABB, a spike overvoltage suppressor is used for the AC inlet end of the power rectifier. It is a DC blocking filter. After the three-phase bridge rectifier connected to the secondary side of the excitation transformer undergoes rectification, the DC voltage is applied to the R-C suppressor circuit. The advantage of this circuit is that when the capacitor releases energy, the discharge current will not be applied to the thyristor circuit in the on state due to the blocking of the diode rectifiers. The connection is shown in Figures 13.18, where the DC blocker's overvoltage protection function is presented with a switching transformer as an example.

For the THYRIPOL excitation system produced by Siemens used for the 700 MW hydro-turbines of Three Gorges Hydro Power Station, no spike overvoltage suppressor is connected to the AC inlet line end in the power rectifiers. So, the spike overvoltage is directly applied to the power rectifier thyristor circuit, seriously endangering the operation safety of the power rectifiers.

Figure 13.19 shows the anode voltage waveform of the power rectifiers of No. 7700 MW hydro-turbine of Three Gorges Hydro Power Station. In the figure, the voltage proportionality coefficient is 206, and the rated voltage of the secondary side of the excitation transformer is 1024 V.

As can be seen from Figure 13.19, the secondary rated spike voltage of the excitation transformer is as follows:

$$U_{2\max} = 7.33 \times 206 = 1509.98 \text{ (V)}$$

13.5 Protection of Power Rectifier | 419

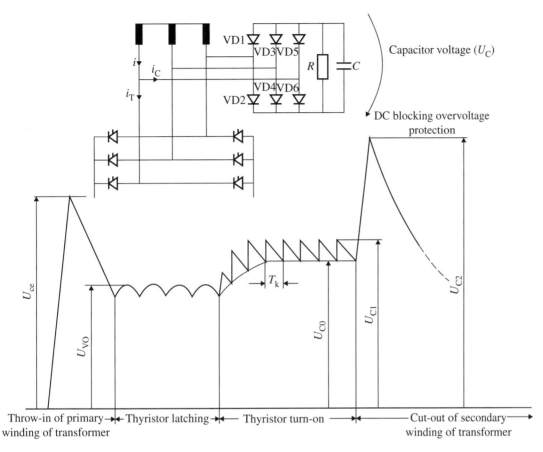

Figure 13.18 ABB DC blocking overvoltage absorber.

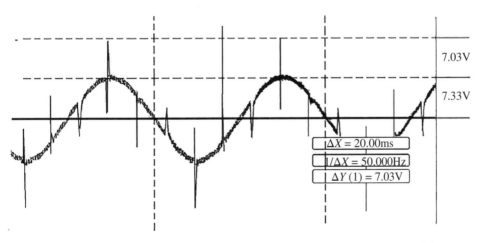

Figure 13.19 Anode voltage waveform of power rectifiers of No. 7700 MW hydro-turbine of Three Gorges Hydro Power Station (with no anode overvoltage suppressor connected).

The spike overvoltage caused by the commutation is as follows:

$$\Delta U_{2\max} = 7.03 \times 206 = 1448.18 \text{ (V)}$$

The overvoltage multiple is as follows:

$$K_V = \frac{U_{2\max} + \Delta U_{2\max}}{U_{2\max}} = \frac{1509.98 + 1448.18}{1509.98} = 1.959$$

To improve the working state at the anode spike overvoltage, Three Gorges Hydro Power Station has installed the spike overvoltage suppressor shown in Figure 13.20 on the secondary side of the excitation transformer, that is, the AC inlet line side of the power rectifiers.

The specifications of the spike overvoltage suppressor components are as follows:

Resistor R: 40 Ω/500 W, high-power non-inductive resistance.
Capacitor C: 4 μF/AC4000 V, German-made capacitor.
High-voltage cable: AGG-4.0/5000 V, resistance to high temperatures up to 180 °C.
Fuse: FWJ-35 A/1000 V, high-voltage low-current fuse.

The anode voltage waveform after the spike overvoltage suppressor is connected is shown in Figure 13.21. The oscillogram voltage rated proportionality coefficient $K = 206$.

When the anode overvoltage suppressor is connected, the following can be seen from Figure 13.21:

$$U_{2\max} = 7.33 \times 206 = 1509.98 \text{ (V)}$$

The spike overvoltage is as follows:

$$\Delta U_{2\max} = 5.47 \times 206 = 1126.82 \text{ (V)}$$

The overvoltage multiple is as follows:

$$K_V = \frac{U_{2\max} + \Delta U_{2\max}}{U_{2\max}} = \frac{1509.98 + 1126.82}{1509.98} = 1.746 < 1.959$$

The effectiveness of spike overvoltage suppression has been improved.

Besides, in order to highlight the role of the spike overvoltage suppressor, Figure 13.22 compares the overvoltage rise process with the suppressor with that without the suppressor. As can be seen from Figure 13.22,

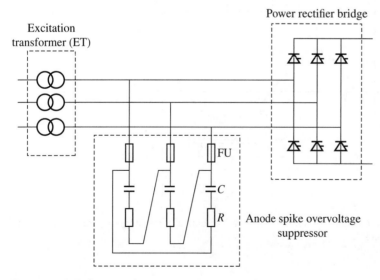

Figure 13.20 Spike overvoltage suppressor.

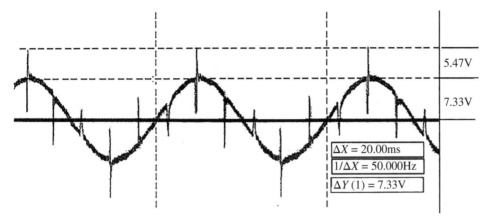

Figure 13.21 Anode voltage waveform of power rectifiers of No. 7 hydro-turbine of Three Gorges Hydro Power Station (with anode overvoltage suppressor connected).

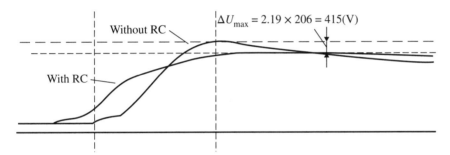

Figure 13.22 Comparison of anode overvoltage rise process with spike overvoltage suppressor and that without spike overvoltage suppressor on power rectifiers on No. 12 hydro-turbine of Three Gorges Hydro Power Station.

with the spike overvoltage suppressor connected, the overvoltage rise process becomes smooth and the overvoltage value decreases, thus reducing the impact on the power rectifier bridge thyristor elements.

13.5.2 R-C Protection of Power Rectifier Thyristor Elements

13.5.2.1 Working Principle of R-C Protection

As can be seen from Figure 13.23, before Point C3, Elements V1 and V2 are turned on and Currents i_1 and i_2 flow through them. At time t_1, Element V3 receives the pulse signal.

When Point C3 is crossed, $e_b > e_a$. The cathode of Element V1 withstands a reverse voltage. V1 and V3 begin commutation, until V1 is cut off and V3 is completely turned on. At time t_2, i_1 drops to zero while i_3 rises to the full current I_f. But at this time the commutation does not end. So, there are still a small number of carriers at Junction PN of Element V1, and a short time interval (generally only a few or tens of microseconds) is required for the discharge. So, Element V1 remains reversely turned on, until the reverse current reaches the maximum value $-i_{V1}$. A smaller number of carriers accumulated are quickly recombined. Element V1 is completely cut off, with its current reaching zero. Meanwhile, the forward current in Element V2 generates a sudden drop to the full current I_f. Since both i_1 and i_2 flow through the secondary-winding circuit of the excitation transformer with the leakage reactance (X) and the di/dt current variation rate is extremely large at the short time interval t_3 and t_4, at the instant when the reverse current in Element V1 is suddenly cut off and in an open-circuit state, an extremely high transient voltage, which is just the so-called spike overvoltage, will be induced in the AC circuit with the leakage reactance X_L, endangering the operational safety of the rectifier elements.

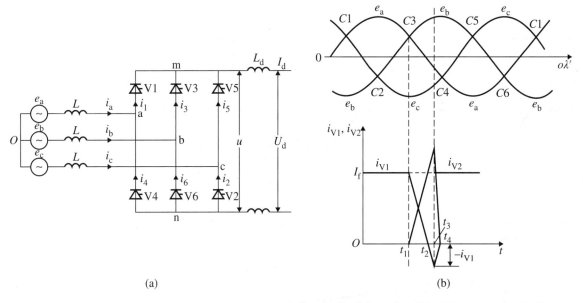

Figure 13.23 Schematic diagram of commutation overvoltage of power rectifier: (a) three-phase bridge rectifier circuit, (b) commutation overvoltage ($\alpha = 0$).

In order to reduce the commutation spike overvoltage, a discharge branch should be provided for the reverse recovery current when the elements show a reverse blocking characteristic. This branch is generally composed of R-C elements and connected to both ends of the elements to slow down the reverse current attenuation rate and limit the resulting spike voltage. This R-C protection circuit is also known as a commutation snubber.

When the energy absorbed by the commutation snubber is too large, it will be difficult to install an R-C element having an excessively large capacity in the thyristor unit. In this case, two commutation overvoltage snubber sets can be added respectively. The set added at both ends of the element is known as the element snubber. The other set, which is added on the secondary-winding side of the excitation transformer, is called the circuit snubber, that is, the aforementioned anode overvoltage suppressor. Parameter selection for commutation snubbers is a complex topic. Many power rectifiers for power stations have commutation snubbers; however, it is noteworthy that improper parameter selection for commutation snubbers cannot prevent accidents.

The principle of parameter selection for a commutation snubber is that the capacity of the R-C snubber branch should be enough to absorb all of the magnetic field energy released in the commutation process without damage. Besides, the commutation overvoltage amplitude should be limited to the allowable range.

13.5.2.2 Principle for Parameter Selection for Commutation Snubber

The parameter selection for an R-C snubber is described with the three-phase bridge rectifier circuit shown in Figure 13.23 as an example. During the commutation of V1 and V2, the equivalent circuit can be represented by the equivalent circuit shown in Figure 13.24. Since the commutation process is conducted under the action of the line voltage, the commutation circuit contains a commutation inductance of 2 L.

In Figure 13.24, $E = \sqrt{6}E_S \sin(\alpha_0 + \gamma_0)$. E_S is the RMS phase voltage. R and C are the resistance and capacitance of the snubber, respectively. L is the transformer's leakage reactance per phase. α_0 and γ_0 are the element's control angle and commutation angle at the instant of the end. At the instant when V1 in Figure 13.23 shows a blocking characteristic after the commutation and $t = 0$, Switch S is equivalently broken. The current in the circuit reaches the maximum value of the reverse recovery current $i(t)|_{t=0} = I_0 = I_{RM}$. I_{RM} can be obtained from the element specification. For the hydro-turbines of Three Gorges Hydro Power Station, the T1401 N 2000 A/5200 V elements are used for the power rectifier thyristors. The recovery charge $Q_{rr} = 18$ mA·s can be

Figure 13.24 Equivalent commutation circuit.

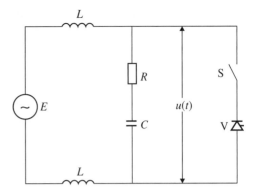

Figure 13.25 Recovery charge $Q_{rr} = f(di/dt)$ ($t_{vj} = 125\,°C$, $I_{TM} = 2000\,A$, $U_R = 0.5U_{RRM}$, and $U_R = 0.8U_{RRM}$).

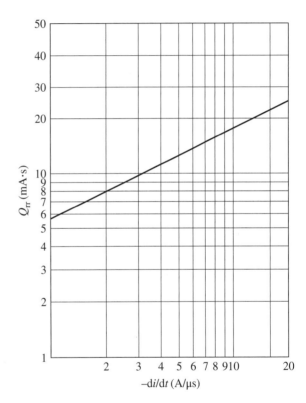

obtained from the specification. $di/dt = 10$ (A μs^{-1}) can be obtained from Figure 13.25 and $I_{RM} = 250$ A from Figure 13.26.

Since an R-C snubber branch is connected to both ends of the element, the magnetic energy stored in the secondary winding of the excitation transformer will be absorbed through the R-C branch after the commutation. Thus, the circuit voltage equation can be obtained:

$$E = L\,\frac{di\,(t)}{dt} + Ri\,(t) + \frac{1}{C}\int_0^t i(t)\,dt \tag{13.14}$$

The following equations can be obtained:

$$i(t) = \frac{(E - I_0 R)}{L\omega}e^{-\alpha t}\sin\omega t + \frac{\alpha I_0}{\omega}e^{-\alpha t}\sin\omega t + I_0 e^{-\alpha t}\cos\omega t \tag{13.15}$$

$$u(t) = E - (E - I_0R)\left(\cos \omega\, t - \frac{\alpha}{\omega} \sin \omega\, t\right)e^{-\alpha t} + \frac{I_0}{\omega C}e^{-\alpha t} \sin \omega t \tag{13.16}$$

$$\frac{du(t)}{dt} = (E - I_0R)\left(2\alpha \cos \omega\, t + \frac{\omega^2 - \alpha^2}{\omega} \sin \omega\, t\right)e^{-\alpha t} + \frac{I_0}{C}\left(\cos \omega\, t - \frac{\alpha}{\omega} \sin \omega\, t\right)e^{-\alpha t} \tag{13.17}$$

$$\alpha = \frac{R}{2L}$$

$$\omega = \sqrt{\omega_0^2 - \alpha^2} = \omega_0 \sqrt{1 - \xi^2}$$

$$\omega_0 = \frac{1}{\sqrt{LC}}$$

$$\xi = \frac{\alpha}{\omega_0} = \frac{R}{\sqrt[2]{L/C}}$$

where

- α – attenuation coefficient;
- ω – damped oscillation frequency;
- ω_0 – undamped oscillation frequency;
- ξ – damping coefficient.

From Eq. (13.16), we see that in the energy absorption process, the voltage $u(t)$ appearing at both ends of the element has the damped oscillation characteristic. The oscillation frequency is closely related to circuit parameters such as L, R, and C. Besides, at the instant when V1 is blocked and $t = 0$, the following equation can be obtained from Eq. (13.16):

$$u(t)\big|_{t=0} = I_0R = I_{RM}R \tag{13.18}$$

The voltage variation rate when $t = 0$ can be obtained from Eq. (13.17):

$$\frac{du(t)}{dt}\bigg|_{t=0} = \frac{(E - I_0R)R}{L} + \frac{I_0}{C} \tag{13.19}$$

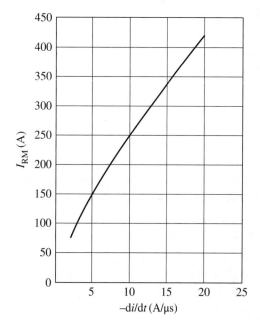

Figure 13.26 Recovery current $I_{RM} = f(di/dt)$ ($t_{vj} = 125\,°C$, $I_{TM} = 2000\,A$, $U_R = 0\,0.5\,U_{RRM}$, and $U_{RM} = 0.8\,U_{RRM}$).

If the R-C snubber parameters are selected under the condition that $\frac{du(t)}{dt}\big|_{t=0} \leqslant 0$, or the maximum spike overvoltage value appears at the instant when $t = 0$ in the whole damping process, the following formula can be obtained:

$$\frac{(E - I_0R)R}{L} + \frac{I_0}{C} \leqslant 0$$

Or it can be written as follows:

$$R \geqslant \frac{E}{2I_0} + \sqrt{\left(\frac{E}{2I_0}\right)^2 + \frac{L}{C}} \tag{13.20}$$

The variation of the transient voltage waveform at both ends of the element is shown in Figure 13.27a. In this case, the initial voltage variation rate can be reduced. However, since the resistance Ri is too large and gives $u(t)_{t=0} = I_0R$ have a high initial value, the voltage will be suddenly applied to both ends of the element and generate a $\frac{du}{dt}$ value that compromises the element's safety and causes a false trigger of the element.

So, in the actual R-C snubber parameter selection, no declining variation rate after $\frac{du}{dt}\big|_{t=0}$ is expected. In the period of time when $t > 0$, there is still a rising variation rate for $\frac{du}{dt}$. Only when $t = t_1$ does the maximum voltage u_m appear. The corresponding voltage variation characteristic curve is shown in Figure 13.27b.

In Eq. (13.19), if $\frac{du(t)}{dt}\big|_{t=0} > 0$, the following formula can be obtained:

$$R < \frac{E}{2I_0} + \sqrt{\left(\frac{E}{2I_0}\right)^2 + \frac{L}{C}} \tag{13.21}$$

With Formula (13.17), when $\frac{du(t)}{dt} = 0$, the maximum transient voltage u_M and the corresponding time t_1 can be obtained as follows:

$$t_1 = \frac{\pi - \beta}{\omega} \tag{13.22}$$

$$u_M = u(t_1) = E - (E - I_0R)\left[\cos(\pi - \beta) - \frac{\alpha}{\omega}\sin(\pi - \beta)\right]e^{-\alpha t_1} + \frac{I_0}{\omega C}e^{-\alpha t_1}\sin(\pi - \beta) \tag{13.23}$$

$$\beta = \arctan\frac{(E - I_0R)2\alpha + \frac{I_0}{C}}{(E - I_0R)\frac{\omega^2 - \alpha^2}{\omega} - \frac{\alpha I_0}{\omega C}} \tag{13.24}$$

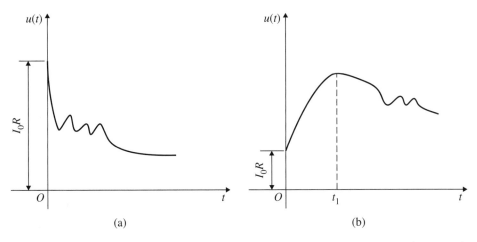

Figure 13.27 Transient voltage [u(t)] variation curves for R-C snubber and both ends: (a) $\frac{du}{dt} \leq 0$; (b) $\frac{du}{dt} > 0$.

13.5.2.3 Energy Conversion of Commutation Process

During the commutation of the power rectifier elements, since magnetic field energy is stored in the secondary winding of the excitation transformer, the stored energy will be released into the R-C snubber circuit after the commutation. The energy stored and that released should be balanced. To this end, the R-C snubber elements should have sufficient capacity reserves. Otherwise they will be burned out due to a lack of element capacity, which will lead to rectifier interphase short circuits.

According to the Westinghouse's design guidelines, the energy absorbed by the R-C snubber involves two parts:

(1) Energy conversion caused by the voltage commutation effect: Its magnitude is related to the control angle α and the commutation angle γ of the thyristor elements. On the basis of the formula recommended by Westinghouse, the energy that should be absorbed by the RC elements per second for the R-C protection in parallel at both ends of the thyristor elements is as follows:

$$W_u = 3.5Cf(\sqrt{3}E_S)^2 \tag{13.25}$$

where

f – power frequency;

E_S – RMS power supply phase voltage.

The voltage commutation effect energy expression given by Canada's CGE is as follows:

$$W_u = 1.75Cf(\sqrt{3}E_S)^2[\sin^2\alpha + \sin^2(\alpha + \gamma)] \tag{13.26}$$

where

α and γ – respectively control angle and commutation angle in given working state.

When $\alpha = 90°$ and $\gamma = 0$, Eq. (13.26) can be written as follows:

$$W_u = 1.75Cf(\sqrt{3}E_S)^2(1+1) = 3.5Cf(\sqrt{3}E_S)^2$$

This is the same as Eq. (13.25).

If a circuit R-C snubber is connected to the secondary-winding side of the transformer, the voltage effect expression is as follows:

$$W_{uL} = 6.0Cf(\sqrt{3}E_S)^2 \tag{13.27}$$

The voltage effect energy conversion primarily appears as the conversion between the magnetic energy stored by the excitation transformer and the energy absorbed by the snubber capacitance.

(2) Energy conversion caused by reverse current effect: This part is the energy consumed by the reverse recovery current I_{RM} in the snubber resistance element. The energy stored by the reverse recovery peak current I_{RM} in the secondary-side winding of the excitation transformer is as follows:

$$W_I = \frac{1}{2}(2L)\,I_{RM}^2 = LI_{RM}^2 \tag{13.28}$$

13.5.2.4 Commutation Snubber Parameter Calculation Program

(1) The selected snubber R-C element should ensure that the commutation spike overvoltage does not exceed the allowable value under various working conditions. In view of the voltage specifications of the thyristor elements, generally the safety factor is appropriate when it is (2.5–3.0) $\sqrt{6}E_S$. $\sqrt{6}E_S$ is the peak voltage of the secondary rated line voltage of the excitation transformer. The safety factor of the hydro-turbines of Three Gorges Hydro Power Station is 2.75.

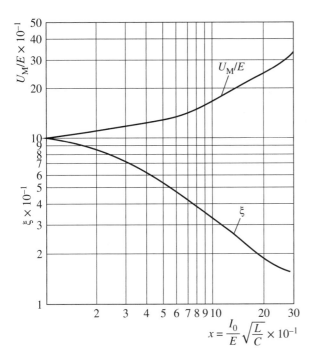

Figure 13.28 R-C snubber coefficient optimization curves.

(2) On the basis of the selected U_M/E, the damping coefficient ξ and the initial power supply coefficient χ are obtained from the coefficient optimization curves in Figure 13.28:

$$\chi = \frac{I_0}{E}\sqrt{L/C} \tag{13.29}$$

(3) The snubber capacitance C is calculated as follows:

$$C = L\frac{I_0^2}{(E\chi)^2} \tag{13.30}$$

(4) The snubber resistance R is calculated as follows:

$$R = 2\xi\sqrt{L/C}$$

(5) The total loss of the voltage effect and current effect that should be absorbed by the snubber is calculated as follows:

$$W_\Sigma \geqslant W_u + W_I = 1.75Cf(\sqrt{3}E_S)^2[\sin^2\alpha + \sin^2(\alpha+\gamma)] + LI_{RM}^2 f$$

(6) After the snubber parameters are determined, a check should be conducted on the basis of Eqs. (13.22) and (13.23), so that it can be determined that U_M meets the selected $\frac{U_M}{E}$.

13.5.2.5 R-C Snubber Parameter Verification

The R-C snubber parameters are verified with the ABB power rectifiers of Three Gorges Hydro Power Station as an example.

(1) The R-C snubber connection is shown in Figure 13.29. The parameters are as follows:
   1) $R_1 = R_2 = R_3 = R_4$, 220 Ω, 300 W;
   2) The other similar elements of L_1 are 50 µH;
   3) $C_1 = 1\,\mu F$, 2100 V AC;
   4) The other similar elements of R_5 are 680 kΩ.

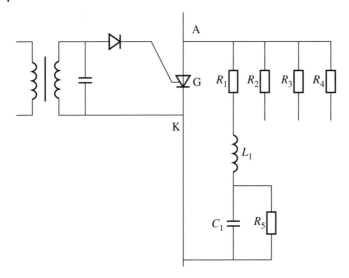

Figure 13.29 Power rectifier cabinet R-C snubber connection diagram.

(2) The capacity absorbed by the R-C snubber is calculated:

$$W_\Sigma = 1.75 C f E^2 [\sin^2\alpha + \sin^2(\alpha + \gamma)] + L I_{RM}^2 f \qquad (13.31)$$

When $\alpha = 90°$ and $\gamma = 0$ are substituted into Eq. (13.31), the following equation can be obtained:

$$W_\Sigma = 3.5 C f E^2 + L I_{RM}^2 f \qquad (13.32)$$

The phase leakage reactance of the excitation transformer is calculated as follows:

$$X = u_{k\%} \frac{U_{2N}}{I_{2N}} \qquad (13.33)$$

where

$u_{k\%}$ — short-circuit impedance of excitation transformer (8% for the ABB units);

U_{2N} — secondary rated voltage of excitation transformer;

I_{2N} — secondary phase current of excitation transformer.

$$U_{2N} = \frac{1243}{\sqrt{3}} = 718 \text{(V)}$$

$$I_{2N} = \frac{S_N}{3 U_{2N}} = \frac{3 \times 2925 \times 10^3}{3 \times 718} = 4076 \text{(A)}$$

The leakage reactance per phase is obtained from the above data:

$$X = 0.08 \times \frac{718}{4076} = 0.0141 \ (\Omega)$$

The leakage inductance per phase is calculated as follows:

$$L = \frac{0.0141}{2\pi f} = 4.49 \times 10^{-5} \ \text{(H)}$$

The excitation transformer lead cable is about 30 m. For 3–10 kV single-core cables, their reactance is $0.18 \, \Omega \, \text{km}^{-1}$. When the cable is 30 m, the additional lead reactance is calculated as follows:

$$X_C = 0.03 \times 0.18 = 0.0054 \ (\Omega)$$

The cable inductance is calculated as follows:

$$L_C = \frac{0.0054}{2\pi f} = 1.72 \times 10^{-5} \text{ (H)}$$

The total inductance of the AC lead of the excitation transformer is calculated as follows:

$$L_t = L + L_C = 6.21 \times 10^{-5} \text{ (H)}$$

It is known that $I_{RM} = 250$ A. When it is substituted into Eq. (13.32), the following can be obtained:

$$W_\Sigma = 3.5 \times 1 \times 10^{-6} \times 50 \times 1243^2 + 6.21 \times 10^{-5} \times 250^2 \times 50$$
$$= 270 + 194$$
$$= 464 \text{ (W)}$$

Since 464 W < 1200 W, it is safe.

The overvoltage multiple verification steps are as follows.

First, from Eq. (13.29), the initial current coefficient is obtained:

$$\chi = \frac{I_0}{E}\sqrt{\frac{L_t}{C}}$$

In the equation, $I_0 = I_{RM} = 250$ A. $E = \sqrt{6}E_S \sin(\alpha_0 + \gamma_0)$. When $\alpha_0 + \gamma_0$ is made equal to 90°, $E = \sqrt{6} \times 718 = 1757$ (V).

The following can be obtained:

$$\chi = \frac{250}{1757}\sqrt{\frac{6.21 \times 10^{-5}}{1 \times 10^{-6}}} = 1.12$$

From Figure 13.28, $\frac{U_M}{E} = 1.8$ times, or the spike overvoltage

$$U_M = 1.8E = 1.8 \times \sqrt{6} \times 718 = 3167.9 \text{ (V)}$$

13.6 Thyristor Damage and Failure [34]

Despite the great improvements in working reliability, performance, and quality of power electronic devices, damage to rectifier thyristors in static excitation systems caused by improper design or application is seen from time to time.

If a failed element is replaced but the root cause for its failure is not analyzed, the potential risk cannot be eliminated. Reference [34] analyses the causes for thyristor damage in terms of design and application respectively in simple language using a number of specific examples and draws useful conclusions. This section will introduce the arguments presented in the reference.

13.6.1 Factors Affecting Thyristor Operation Safety in Lectotype Design

Several key parameters that affect thyristor operation safety are as follows:

(1) Rated voltage: Generally, the manufacturer takes the off-state repetitive peak voltage U_{DRM} or the reverse repetitive peak voltage U_{RRM}, whichever is smaller, as the rated voltage of a thyristor.

A thyristor's rated voltage reflects its voltage withstand capacity. An inappropriate rated voltage can easily cause thyristor overvoltage breakdown. Generally, in thyristor lectotype design, the following are some considerations. The first consideration is the anode voltage of the excitation system (e.g., the secondary-side voltage of the excitation transformer, whose maximum value should be taken in calculations). The second consideration is the overvoltage impact coefficient. It should be 1.5, because many power stations require 1.3 times the rated voltage or more for no-load tests after generator overhaul which directly engage the generator terminal voltage and excitation system, while some power stations require 1.5 for generator overvoltage protection. The third consideration is that generally the voltage rise factor is 1.1 and the voltage margin coefficient is 2–3. Besides, the thyristor's off-state repetitive voltage U_{DRM} and reverse repetitive voltage U_{RRM} are considered on the basis of a repetition frequency of 50 Hz. So, in the lectotype design of some generators with high overspeed protection setpoint, the voltage margin coefficient should be appropriately large. This is ignored by many excitation manufacturers. Since most hydropower stations' overspeed protection may be set at 1.4 (i.e., the frequency can reach 70 Hz) and load rejection causes a voltage rise, these factors should be taken into account in thyristor rated voltage selection.

(2) Average on-state current: The rated current is a very important parameter of a thyristor. The average on-state current $I_{T(AV)}$ is measured at an ambient temperature of 40 °C and a 50 Hz (power frequency) sine half-wave current under specified cooling conditions. So, the service environment and cooling conditions should be taken into account in thyristor lectotype design.

(3) On-state current critical rise rate: The on-state current critical rise rate di/dt is an important parameter of a thyristor as well. A di/dt value that exceeds the specified value may cause breakdown of the thyristor elements.

(4) Off-state voltage critical rise rate: A parameter corresponding to the on-state current critical rate di/dt is the off-state voltage critical rise rate du/dt, which refers to the minimum voltage rise rate at which the thyristor shifts from cutoff to turn-on at the rated junction temperature when the control electrode is in the broken circuit state. If a large-amplitude spike voltage with a sharp leading edge is generated in the system, the thyristor in the cutoff state may be turned on faultily. Such a fault is hard to identify. So, in thyristor lectotype design, du/dt should be as large as possible.

13.6.2 Thyristor Damage and Failure Analysis and Identification

13.6.2.1 Thyristor Overvoltage Breakdown

Despite some overvoltage suppression measures such as a spike overvoltage suppressor on the AC side of the excitation system, thyristor overvoltage breakdown which damages the thyristor is seen from time to time. It should be noted that the cause of such a failure is often hard to identify. An element breakdown is followed by an AC-side interphase short circuit and overcurrent, which is usually misjudged as an excitation system regulator failure. In the case of an overvoltage breakdown, there are often multiple clear breakdown points on the chip. Thyristor overvoltage breakdown examples are shown in Figures 13.30 and 13.31.

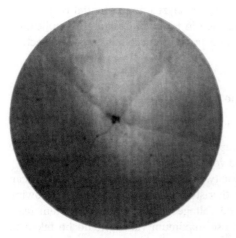

Figure 13.30 Thyristor with large overvoltage breakdown points.

Figure 13.31 Thyristor with small overvoltage breakdown points.

13.6.2.2 Thyristor Overcurrent

There are many causes of thyristor overcurrent, such as an inadequate $I_{T(AV)}$ margin identified in the lectotype design, short-out of the rotor circuit, and the fast fuses' failure to play a protective role because of some discrete parameters. Besides, in the event of a faulty forced excitation, which is generally caused by a regulator failure, there is often a loss of control and long-time high-current operation, which eventually causes thyristor overcurrent and large burnt area. A damaged thyristor is shown in Figure 13.32.

An example is a thyristor burnout by an overcurrent caused by a short-circuit of the rotor slip ring circuit. The damaged thyristor is shown in Figure 13.33.

13.6.2.3 Too Large On-State Current Critical Rise Rate (di/dt)

A large power loss will be generated at the instant when the thyristor is turned on. Since the current spreading speed is limited at the time of the turn-on, this loss is always concentrated in the cathode region near the control electrode. If the thyristor's allowable di/dt is low, the current will cause local overheating of the cathode region in the vicinity of the control electrode, resulting in permanent damage to the control electrode. In particular, the problem of di/dt is more prominent for a high-current thyristor. In fact, even if the damage is caused by an excessive di/dt, it appears as an A-K inter-electrode breakdown. The cause will be mistaken for a large-area chip burnout caused by the overcurrent. However, as can be seen from Figure 13.34, on a thyristor damaged by an excessive di/dt, there is obvious damage in the vicinity of the control electrode.

Figure 13.32 Thyristor burned out by long-time rotor overcurrent caused by faulty forced excitation.

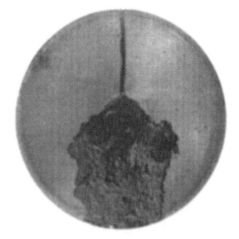

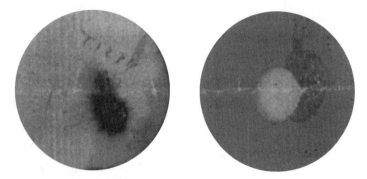

Figure 13.33 Thyristor burned out by overcurrent caused by short-circuit of generator rotor slip ring circuit.

Figure 13.34 Obvious burnout in the vicinity of control electrode caused by excessive di/dt.

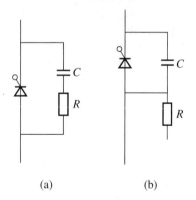

Figure 13.35 Connection of thyristor R-C snubber: (a) correct connection, (b) wrong connection.

Figure 13.35 shows another example of thyristor burnout as a result of an excessive di/dt caused by a wrong connection after an R-C snubber replacement.

Obviously, in Figure 13.34b, where the R-C snubber connection is wrong, the direct rapid discharge of Capacitor C at the time of the thyristor turn-on causes the thyristor burnout.

In summary, there are many possible causes for thyristor damage. But in all of these cases, thyristor damage appears as a "breakdown," and the outcome is large-area chip burnout caused by a high short-circuit current. Such an outcome is often mistaken for overcurrent damage. So, when a thyristor is damaged, various symptoms should be carefully analyzed and inspected, so that the true cause can be identified and similar incidents can be avoided. It should be noted that the damaged thyristor elements in the pictures provided here include China-made and foreign products, covering such sizes as 50, 75, and 100 mm.

13.7 Capacity of Power Rectifiers Operating in Parallel

For large-capacity power rectifiers used in excitation systems of large hydro-turbines and steam turbines, the number of power rectifiers operating in parallel should be redundant under the operation safety requirements. In bidding documents, the specification for power rectifiers is typically described as follows.

A three-phase fully controlled bridge connection should be adopted for thyristor rectifiers. The following requirements should be met:

(1) The generator's requirements for the excitation system under various working conditions (including forced excitation) should be met. The number of rectifier branches in parallel should be no greater than five (depending on the generator's excitation power and the thyristor size), and the number of elements in each branch should be one.
(2) The redundancy of the number of branches in parallel in the thyristor rectifier bridges should be considered under the N-1 principle, to satisfy all the functions including forced excitation in the event of the failure of one bridge and meet the requirements of all the operation modes except forced excitation in the event of the failure of two bridges.
(3) The branches in parallel should ensure that the current sharing coefficient is not less than 0.9.
(4) At the rated load operating temperature, the reverse repetitive peak voltage withstood by the thyristors should be no less than 2.75 times the maximum peak voltage of the excitation transformer.

13.7.1 Verification and Calculation of Thyristor Power Cabinet Capacity

13.7.1.1 Calculation of Power Rectifier Load

It is assumed that the T1451N thyristor manufactured by EUPEC is used. The relevant parameters are shown in Table 13.3.

It is known that the generator's rated excitation current I_{fN} is 4128 A. Under the technical requirements set out in the bidding document:

(1) When five rectifiers are in parallel, each rectifier has one rectifier bridge set. The number of elements in series or parallel in each arm is one. Long-term operation at 1.1 times the rated current is required. The output per rectifier is calculated as follows:

$$I_{d5} = \frac{I_{fN} \times 1.1}{5K_I} = \frac{4128 \times 1.1}{5 \times 0.95} = 956 \text{ (A)}$$

(K_I is the current sharing coefficient)

Table 13.3 Key technical parameters of EUPEC T1451 N thyristor.

Mode	T1451 N
Manufacturer	EUPEC
U_{DRM}–U_{RRM} (V)	4800–5200
U_T (V)	1.57
U_{T0} (V)	0.88
$I_{T(AV)}$ (A/°C)	1690/85 2300/60
du/dt (V μs^{-1})	2000
di/dt (A μs^{-1})	300
r_T (mΩ)	0.345
$I^2 t$ (kA^2·s)	9250/125 °C
U_{gt} (V)	2.5
I_{gt} (mA)	350

(2) Under the N-1 principle, one rectifier quits operation and four operate in parallel. The output per rectifier is calculated as follows:

$$I_{d4} = \frac{I_{fN} \times 1.1}{4K_I} = \frac{4128 \times 1.1}{4 \times 0.95} = 1195 \text{ (A)}$$

(3) Under the N-2 principle, two rectifiers quit operation and three operate in parallel. The output per rectifier is calculated as follows:

$$I_{d3} = \frac{I_{fN} \times 1.1}{3 \times 0.95} = 1593 \text{ (A)}$$

(4) Under the N-1 principle, four rectifiers operate in parallel at the required forced excitation current that is twice the rated current for no less than 20s. The output per rectifier is calculated as follows:

$$I_{d4f} = \frac{I_{fN} \times 2}{4 \times K_I} = \frac{4128 \times 2}{4 \times 0.95} = 2173 \text{ (A)}$$

From the above calculations, it is seen that the output current per rectifier is the highest in Case (4). So, the thyristor average on-state current should be determined in this state.

13.7.1.2 Calculation of Power Rectifier Thyristor Element Loss

The average power consumption per thyristor is calculated as follows:

$$P_{AV} = U_{T0} I_{T(AV)} + k I_{T(AV)}^2 r_T \tag{13.34}$$

where

P_{AV} — average loss of thyristor;
U_{T0} — threshold voltage of thyristor 0.88 V;
r_T — slope resistance of thyristor 0.345 mΩ;
$I_{T(AV)}$ — average current flowing through thyristor;
k — waveform coefficient (3 in three-phase fully controlled bridge rectification calculation).

(1) When the output per rectifier is 956 A, the thyristor actual on-state current average $I_{T(AV)} = 0.367 \times 956 = 350.85$ (A), and

$$P_{AV1} = 0.88 \times 0.367 \times 956 + 3 \times (0.367 \times 956)^2 \times 0.345 \times 10^{-3} = 436.1 \text{ (W)}$$

(2) When the output per rectifier is 1195 A

$$P_{AV1} = 0.88 \times 0.367 \times 1195 + 3 \times (0.367 \times 1195)^2 \times 0.345 \times 10^{-3} = 585.0 \text{ (W)}$$

(3) When the output per rectifier is 1593 A

$$P_{AV1} = 0.88 \times 0.367 \times 1593 + 3 \times (0.367 \times 1593)^2 \times 0.345 \times 10^{-3} = 868.3 \text{ (W)}$$

(4) When the output per rectifier is 2173 A

$$P_{AV3} = 0.88 \times 0.367 \times 2173 + 3 \times (0.367 \times 2173)^2 \times 0.345 \times 10^{-3} = 1360.1 \text{ (W)}$$

13.7.1.3 Calculation of Thyristor Junction Temperature and Radiator Temperature Rise

The thyristor element junction temperature is calculated as follows:

$$T_{jmax} = \Delta T_j + \Delta T_s + T_a$$
$$\Delta T_j = P_{av}(R_{jc} + R_{cs})$$
$$\Delta T_s = P_{av} R_{sa}$$
$$\Delta T = 125 - T_{jmax}$$

where

T_{jmax} — thyristor element allowable maximum junction temperature (125 °C);

R_{jc} — thyristor element junction case thermal resistance (0.0087 K W^{-1} for double-sided heat dissipation);

R_{cs} — case-radiator contact resistance (0.0025 K W^{-1});

R_{sa} — radiator thermal resistance (0.03 K W^{-1} when cooling air velocity is 6 m s^{-1});

T_a — ambient temperature (50 °C as required by bidding document);

ΔT_s — radiator temperature rise;

ΔT_j — thyristor element junction temperature rise;

ΔT — safety margin.

(1) When five rectifiers operate in parallel at a current that is 1.1 times the rated current for a long time, the output per rectifier is 956 A, and the steady-state temperature rises are calculated as follows:

Radiator temperature rise:

$$\Delta T_s = 436.1 \times 0.03 = 13.1 \, (°C)$$

Thyristor element junction temperature rise:

$$\Delta T_j = 436.1 \times 0.011 = 4.8 \, (°C)$$

Thyristor element junction temperature:

$$T_j = \Delta T_j + \Delta T_s + T_a = 4.8 + 13.1 + 50 = 67.9 \, (°C)$$

Safety margin:

$$\Delta T = 125.0 - 67.9 = 57.1 \, (°C)$$

(2) When four rectifiers operate in parallel at a current that is 1.1 times the rated current for a long time, the output per rectifier is 1195 A, and the steady-state temperature rises are calculated as follows.

Radiator temperature rise:

$$\Delta T_s = 585 \times 0.03 = 17.6 \, (°C)$$

Thyristor element junction temperature rise:

$$\Delta T_j = 585 \times 0.011 = 6.4 \, (°C)$$

Thyristor element junction temperature:

$$T_j = \Delta T_j + \Delta T_s + T_a = 17.6 + 6.4 + 50 = 74 \, (°C)$$

Safety margin:

$$\Delta T = 125.0 - 74 = 51 \, (°C)$$

(3) When three rectifiers operate in parallel at a current that is 1.1 times the rated current for a long time, the output per rectifier is 1593 A, and the steady-state temperature rises are calculated as follows.

Radiator temperature rise:

$$\Delta T_s = 868.3 \times 0.03 = 26.0 \, (°C)$$

Thyristor element junction temperature rise:

$$\Delta T_j = 868.3 \times 0.011 = 9.6 \, (°C)$$

Thyristor element junction temperature:

$$T_j = \Delta T_j + \Delta T_s + T_a = 26 + 9.6 + 50 = 85.6 \,(°C)$$

Safety margin:

$$\Delta T = 125.0 - 85.6 = 39.4 \,(°C)$$

(4) When four rectifiers operate in parallel at a forced excitation current that is twice the rated current for 20 s, the output per rectifier is 2173 A, and the temperature rise is calculated as follows. This is the highest working condition requirement. When the four rectifiers operate stably in parallel at a current that is 1.1 times the rated current, a forced excitation current that is twice the rated current occurs. The heat increment generated by the thyristor within 20 s is calculated as follows:

$$Q = (P_{AV4} - P_{AV2}) \times t = (1360 - 585) \times 20 = 15502 \,(J)$$

It is known that the radiator's mass is 10 kg, and its specific heat capacity is 903 J (kg·K)$^{-1}$. The internal conduction of heat is not considered. All heat is absorbed by the radiator. The temperature rise increment of the radiator is calculated as follows:

$$\Delta T_{sa} = 15502/(10 \times 903) = 1.72 \,(°C)$$

So, when the four rectifiers operate in parallel, the loss increment generated by the 20 s forced excitation produces a radiator temperature rise of less than 2 °C if the internal conduction of heat is not considered. Radiator temperature rise:

$$\Delta T_s = 17.6 + 1.72 = 19.3 \,(°C)$$

Thyristor element junction temperature rise:

$$\Delta T_j = 1360.1 \times 0.011 = 15.0 \,(°C)$$

Thyristor element junction temperature:

$$T_j = \Delta T_j + \Delta T_s + T_a = 19.3 + 15.0 + 50 = 84.3 \,(°C)$$

Safety margin:

$$\Delta T = 125.0 - 84.3 = 40.7 \,(°C)$$

The thyristor junction temperature and radiator temperature rise calculation results are listed in Table 13.4. As can be seen from the above calculations and analysis, the requirement for radiator temperature rise no greater than 40 °C set out in the specification is met, and there is a large safety margin for the thyristor maximum junction temperature. So, the requirements set out in the bidding document are fully met.

Table 13.4 Thyristor junction temperature and radiator temperature rise calculation results.

Operating conditions	Thyristor element temperature rise	Radiator temperature rise	Ambient temperature	Temperature rise margin
Five rectifiers in parallel, 1.1 times the rated current, long term	4.8	13.1	50	57.1
Four rectifiers in parallel, 1.1 times the rated current, long term	6.4	17.6	50	51
Three rectifiers in parallel, 1.1 times the rated current, long term	9.6	26.0	50	39.4
our rectifiers in parallel, twice the rated current, 20 s	15.0	19.3	50	40.7

13.7.2 Influence of Altitude on Output Capacity of Power Rectifier

Generally, when an excitation device is actually installed at an altitude of greater than 1000 m, due to the rectifier element heat dissipation degradation caused by the thin air, it is necessary to consider decreasing the output capacity of the power rectifiers. The decreasing coefficient calculation formula varies with manufacturers. For example, for ABB, the power rectifier output current decreasing coefficient is calculated as follows:

$$K_u = 1 - 85.7 \times 10^{-6} \times (H - 1000)$$

where

K_u – current decreasing coefficient;
H – actual installation altitude in m.

Generally, the altitude benchmark is 1000 m. Each increase of 1000 m in altitude causes a decrease of about 8% in power rectifier current capacity.

13.8 Uncertainty of Parallel Operation of Double-Bridge Power Rectifiers

13.8.1 Background

When a thyristor element is short-circuited due to damage caused by an overvoltage or for any other reason, in terms of probability, the double-bridge parallel scheme may not be able to completely ensure that the quick fuses act in accordance with the correct protection scheme. The specific action result often depends on the ampere-second characteristic of the fuses, leading to great uncertainty.

The most serious possibility is that both rectifier bridge sets quit operation when the fast fuse of the fault-free rectifier bridge arm corresponding to the faulty bridge arm blows but the fast fuse of the faulty arm does not act. This goes against the purpose of double-bridge parallel operation: improving the reliability and redundancy of the system.

13.8.2 Analysis of Uncertainty of Protection Action of Fast Fuses in Double-Bridge Parallel Scheme

From Figs. 13.36 and 13.37, respectively, we see that in the event of a thyristor short-circuit fault, the three-bridge parallel scheme can make the fast fuse of the faulty branch correctly blow but the two-bridge parallel scheme fails to do that.

In Figure 13.36, it is assumed that the thyristor of Arm T+ undergoes a short-circuit fault. In the phase sequence, the arm that should be triggered for turn-on next is R+. When Arm R+ of the three bridges is turned on, Phase R and T are short-circuited. The short-circuit current I_{CC} depends on the leakage reactance of the excitation transformer. All the I_{CC} flows through the faulty T+. In the faulty R+ of Bridge A, the short-circuit current is supplied by three rectifiers. In this current distribution scheme, the short-circuit current I_{CC} flows through the faulty T+ to make the fuse blow. $1/3 I_{CC}$ flows through each of the normal arms of R+ of Bridge B and C. So, the heating power of each normal arm is only about 1/9 of that of the faulty T+, and the normal arms will not blow.

After the fuse of Arm T+, the alarm contact gives a signal. After the monitoring logic gets the alarm, it can latch the trigger pulse of the faulty bridge in accordance with the specified logic and issue an alarm signal.

In Figure 13.37, assume that only the two bridges (A and B) are operating in parallel. In the event of a short-circuit fault of the thyristor of Arm T+ of Bridge A, the arm that should be triggered for turn-on next is R+. When Arm R+ of the two bridges is turned on, Phases R and T are short-circuited. All the I_{CC} flows through the faulty T+ of Bridge A. The short-circuit current in Arm R+ of the faulty phase is supplied by the two rectifiers (A and B). Thus, the heating power of the normal R+ of Rectifier B is about 1/4 of that of the

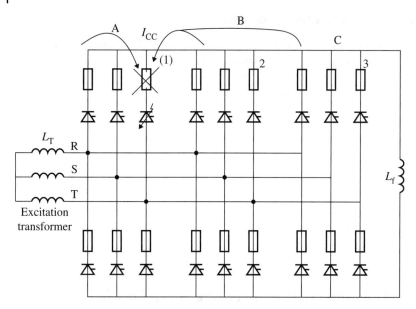

Figure 13.36 Fast fuse blow in three-bridge parallel scheme.

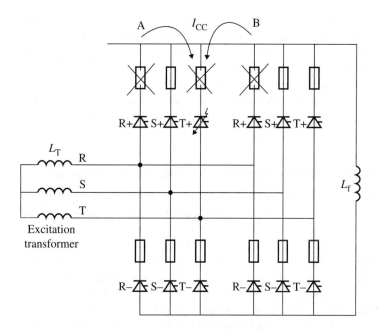

Figure 13.37 Fast fuse blow in two-bridge parallel scheme.

faulty T+, exceeding the ampere-second limit of the fuse. The normal R+ of Rectifier B may be blow together with the faulty T+ (the blow sequence may vary). After the monitoring logic gets the alarm, it will latch the trigger pulse of all the rectifier bridges that give the blow signal in accordance with the specified logic. There will be an excitation trip and redundancy failure.

An alternative to the double-bridge parallel scheme is the cold standby redundancy scheme. In the normal case, the trigger pulse is only sent to one of the rectifier bridges. Only when this rectifier bridge fails will there be a switch to another rectifier bridge.

When the thyristor of Arm T+ of the in-service bridge is short-circuited, as has been noted, Arms T+ and R+ of the in-service bridge quickly blow because of the same short-circuit current since the standby bridge is not turned on. The fault point is cut off. The in-service bridge is cut off. An alarm signal is issued. The standby bridge is immediately put into operation. This design philosophy has been accepted by GE and ABB.

13.9 Five-Pole Disconnector of Power Rectifier

Each power rectifier of the GE EX2100 excitation system is equipped with a five-pole disconnector (AC three-phase inlet line and DC two-pole outlet line). In the event of a failure of any of the thyristor rectifier elements (including thyristor elements, cooling fans, and fast fuses), the AVR will generate a trigger pulse signal to latch the faulty rectifier, so that it has no DC output. In the case, the five-pole disconnector can be manually broken, so that the faulty rectifier can be completely and isolated cut off from the AC and DC main circuit, and the faulty element can then be safely replaced. The five-pole disconnector makes it easy to replace a faulty rectifier element online without the need for stopping operations.

The five-pole disconnector is used for the GE EX2100 excitation system. The most prominent characteristic of the design philosophy is to increase the redundancy of the control system to ensure high reliability of the excitation system in operation.

With the five-pole disconnector that enables online maintenance and faulty element replacement, the N-1 redundant system can be used for the power rectifiers. That is, when one rectifier quits operation, all generator operation modes including forced excitation conditions can still be guaranteed.

It should be particularly emphasized that GE does not adopt the N-2 redundant system. That is, when two rectifiers quit operation, redundant modes including forced excitation conditions can still be guaranteed. The main design philosophy is as follows:

(1) GE believes that too many redundant rectifier connections will reduce the equipment operation reliability.
(2) The online maintenance concept can achieve the maximum operation reliability with a smaller number of rectifiers. For example, when one rectifier quits operation due to failure, with the five-pole disconnector in the rectifier, the three-phase AC input and DC positive and negative output of the power rectifier can be completely disconnected from the other rectifiers in operation simultaneously. This makes it easy to replace the faulty element and put the repaired rectifier into operation safely in a short time. So, the redundancy of the power rectifier cabinets can be restored to the N-1 level again. As can be seen, the requirements of online maintenance can be met only with the five-pole disconnector.

The US standard for large thermal power units stipulates that excitation systems shall adopt five-pole disconnectors to meet the requirements of online maintenance. In China, the 50-plus sets of GE EX2100 excitation system that have been ordered for 300–600 MW large thermal power units all provide five-pole disconnectors.

It should be noted that the GE EX2100 third-generation digital excitation represents a more sophisticated five-pole disconnector operation and structural design. For example, in the previous generations, the door of the rectifier needs to be opened during operation when the five-pole disconnector is operated, while in the third generation, one can safely operate the five-pole disconnector throw-in and cut-out without needing to open the door of the rectifier during operation.

Besides, for the operation of a five-pole disconnector, there is a series of stringent requirements for the software process design. For example, for the breaking of the five-pole disconnector, the pulse of the power rectifier must be cut off first. For the closing of the five-pole disconnector, the pulse can be provided only after the disconnector is closed.

For power rectifiers that provide a rectifier redundancy of N-1 or N-2 and do not provide a five-pole disconnector, although the forced or rated excitation state can still be guaranteed when one or two rectifiers quit operation, the AC input ends and DC output ends of all the rectifiers that quit operation are live (otherwise the pulse is cut off), which does not allow online maintenance. This obviously differs from the GE EX2100 excitation system with the five-pole disconnector that allows online maintenance.

For ABB, depending on user needs, a five-pole disconnector can be provided to allow online maintenance to further improve the MTBF of the excitation system.

However, ABB does not recommend the five-pole disconnector for static self-excitation systems of large steam turbines. The reasons are as follows:

(1) In the event of online maintenance enabled with a five-pole disconnector, when the to-be-replaced thyristor element fails to withstand the additional AC voltage, a high-value short-circuit current will flow through the five-pole disconnector.
(2) When a five-pole disconnector is applied, a number of technical safety measures must be taken to ensure the safety of maintenance personnel.
(3) Some power station users do not allow online maintenance in static self-excitation systems.
(4) All the static self-excitation systems ABB has provided for large steam turbines except a very few nuclear power units have not undergone online maintenance.

14

De-excitation and Rotor Overvoltage Protection of Synchronous Generator

14.1 Overview

The role of a de-excitation system is to quickly cut off the excitation of the generator and rapidly consume the magnetic field energy stored in the excitation winding in the energy consumption elements of the de-excitation circuit in the event of an accident, for example, a short circuit or grounding, inside or outside the generator.

Figure 14.1 shows the schematic diagram of a de-excitation system. In the figure, *a*, *b*, *c*, and *d* respectively represent a generator terminal short circuit, stator winding grounding, rotor slip ring short circuit, and rectifier failure. In all of these cases, it is required that the excitation source should quickly cut off to de-excite the generator.

It should be noted that when the generator-transformer set connection is adopted, the generator also needs to be de-excited quickly in event of a fault on the wire from the outside of the generator to the transformer or the main circuit breaker.

When the generator stator windings are grounded, a ground fault current will be generated. If the generator neutral point is grounded via a high resistance, the insulation of a stator bar will be broken down. If the fault current is small, the core damage will not be too serious. If the fault current is large, it will cause serious burnout of the copper or iron cores, in addition to breakdown of the bar insulation. This case requires at least replacement of the damaged insulation and even partial repair of the generator stator core. So, some manufacturers believe that generators should not feature a field circuit breaker. But more manufacturers believe that a simple, fast de-excitation device should be applied.

More manufacturers believe that a simple and effective fast de-excitation device is necessary since minor damage may lead to greater burnout in some cases where no fast de-excitation device is applied.

As mentioned above, the main requirement for the generator de-excitation system is the ability to reliably and quickly consume the magnetic field energy stored in the generator. The simplest de-excitation method is to disconnect the generator excitation winding from the power supply. However, this will cause a high overvoltage at both ends of the excitation winding, endangering the safety of the generator insulation. So, the excitation winding must be connected to a closed circuit that can consume the magnetic field energy.

Countries have different de-excitation mode design philosophies. For example, there are two different de-excitation principles: energy consumption de-excitation and energy transfer de-excitation. The working principle of an energy consumption de-excitation device is to consume the magnetic energy in the field circuit breaker. When the main contact of the field circuit breaker is broken, the magnetic field energy stored in the generator excitation circuit forms an arc. The arc burns in the combustion chamber to convert the electric energy into heat energy until it is quenched. The DM-2 arc chute, the most widely used de-excitation device in China, is an example.

What corresponds to the above-mentioned de-excitation mode is energy transfer de-excitation. In this de-excitation mode, the magnetic field energy is not consumed by the field circuit breaker but is transferred by the field circuit breaker to the linear or nonlinear resistance energy consumption elements. The working principle of an energy transfer de-excitation device is that an overvoltage is generated after the main contact is broken so that the nonlinear de-excitation resistance is in parallel with the generator excitation winding

Design and Application of Modern Synchronous Generator Excitation Systems, First Edition. Jicheng Li.
© 2019 China Electric Power Press. Published 2019 by John Wiley & Sons Singapore Pte. Ltd.

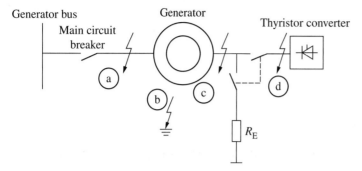

Figure 14.1 Schematic diagram of generator de-excitation system.

turned on, and then this resistance consumes the magnetic field energy of the generator. If the de-excitation resistance is linear, the de-excitation process is even simpler, requiring closing the arcing contact in advance and then opening the main contact.

As is well known, when a fully controlled rectifier bridge circuit is applied for the generator, the inversion mode can also be used to feed the generator's magnetic field energy back to its stator side. When the AC power supply is constant (e.g., in the AC exciter mode), this inversion de-excitation effect is significant. But for a self-excitation system, as the de-excitation accelerates, the generator voltage decreases and the reverse voltage acting on the inversion circuit also decreases, which will affect the attenuation of the inversion de-excitation. The inversion de-excitation can reduce the burden on the mechanical field circuit breaker to a certain extent.

There are significant differences between the de-excitation mode for hydro-turbines and that for steam turbines.

For a hydro-turbine, since the damping of the rotor itself is small, at the time of the de-excitation, the magnetic field energy in the excitation circuit is mostly absorbed by the de-excitation device. So, the field circuit breaker capacity is generally large.

For a generator with a DC exciter, the field circuit breaker is generally located in the main excitation circuit, and linear resistance de-excitation is adopted.

For a hydro-turbine with an AC exciter or self-excitation thyristor excitation system, generally nonlinear resistance de-excitation is adopted in combination with inversion de-excitation. In the normal operation mode, reversion de-excitation generally acts as the main de-excitation mode. In the case of an accident, tripping of the field circuit breaker is the main de-excitation mode, supplemented by inversion de-excitation.

For a steam turbine, since the damping of the rotor itself is strong, the damping winding time constant determined by the full inductance and the resistance of the damping winding is much greater than the super-transient time constant T_d'' determined by the ratio of the leakage inductance and the resistance of the damping winding. So, a fast de-excitation system can only accelerate the attenuation of the rotor excitation current in the vertical axis excitation winding circuit but cannot make the energy stored in the generator rotor and the horizontal axis damping winding quickly disappear to achieve fast de-excitation. So, simplified de-excitation is generally adopted for large steam turbines.

(1) For a brushless excitation system, since a field circuit breaker cannot be connected to the generator excitation winding circuit, de-excitation can be carried out only on the excitation-winding side of the AC exciter. Yet the generator excitation circuit conducts natural de-excitation via the rotating rectifiers on the basis of the corresponding generator time constant. For example, in the early 1980s, this de-excitation mode was applied to the China-made 300 and 600 MW steam turbines based on technologies introduced from the United States.

(2) In countries other than China, for a static rectifier excitation system with AC main and auxiliary exciters, a field circuit breaker provided in the AC main exciter excitation circuit is adopted as a typical de-excitation method. In China, a two- or three-break field circuit breaker and a linear de-excitation resistor provided in the generator main excitation circuit are adopted as the main de-excitation method.

(3) In countries other than China, the following de-excitation methods are adopted for a static self-excitation system:
   1) Switching off the neutral point of the thyristor rectifier AC-side power supply: An example is the Generrex-power potential source (PPS) excitation system manufactured by GE. The neutral point of the excitation transformer powered by the P winding bar is switched off, so that the thyristor rectifiers lose AC power. Thus, de-excitation can be achieved.
   2) In a 600 MW steam turbine, a field circuit breaker and a linear de-excitation resistor are provided in the generator main excitation winding circuit depending on user needs. For example, Japan's Hitachi has chosen such de-excitation systems.

Sweden's ASEA and Switzerland's ABB applies nonlinear de-excitation resistance.

It should be noted that, before the tripping of the field circuit breaker in the de-excitation system described above, the automatic voltage regulator (AVR) will issue an order to change the thyristor rectifiers to the inversion state. The field circuit breaker trips after a certain delay.

In general, for a steam turbine with an AC exciter, de-excitation is carried out on the excitation-winding side of the AC exciter. For a steam turbine with a static self-excitation system, de-excitation is achieved with the field circuit breaker and the de-excitation resistor provided in the generator main excitation circuit.

14.2 Evaluation of Performance of De-excitation System

Generally, a de-excitation system for a large synchronous generator should meet the following basic requirements:

(1) After the de-excitation device acts, the final residual voltage of the generator should be lower than the value at which the short-circuit point arc can be maintained.
(2) In the de-excitation process, the de-excitation reverse voltage that the generator rotor excitation winding withstands should not exceed the specified value.
(3) The de-excitation time should be as short as possible.

Here, the definition of the de-excitation time t_m should be emphasized. The test shows that when the generator stator voltage drops below 500 V, the arc caused by an internal fault of the generator will be quenched naturally. That is, a stator voltage less than 500 V is not enough to continue to maintain the arc burning at the faulty point. At this point, it can be considered that the de-excitation process has ended. The time from the action of the de-excitation device to the end of the de-excitation process is called the de-excitation time.

For a large synchronous generator with a rated voltage of 15 000–20 000 V, the remanence voltage is about 300 V. So, when the stator voltage drops to 500 V, the voltage established by the generator rotor excitation current will be 200 V, accounting for about 1% of the generator rated voltage. From the definition of the de-excitation time, when the rotor current drops to the excitation current corresponding to the stator voltage of 200 V, it can be considered that the de-excitation process has ended.

Since the generator voltage and the excitation current are directly proportional at this point, the time required for the rotor current's decrease from the no-load rated excitation current to 1/100 of this value is the de-excitation time. So, the de-excitation time defined on the basis of the generator voltage and that defined on the basis of the excitation current are consistent in meaning.

In order to compare different de-excitation methods in performance, there should first be comparison criteria. At present, the following ways to evaluate de-excitation performance are available.

14.2.1 Equivalent Generator Time Constant Method

This criterion is a method for evaluation of the performance of a de-excitation system with the generator no-load time constant T'_{d0} as a basic value. The de-excitation time expression is as follows:

$$t_m = K_m T'_{d0} \tag{14.1}$$

where

K_m — de-excitation $K_m < 1$.

T'_{d0} — generator no-load time constant.

The above de-excitation time t_m is defined as the time required from the action of the de-excitation device to the generator excitation current's drop to the specified value close to zero.

In order to compare different de-excitation methods in performance, the ideal de-excitation conditions are first discussed.

In the so-called ideal de-excitation, the rotor current attenuates linearly. In this case, the de-excitation time is the shortest.

The ideal de-excitation conditions are discussed with the de-excitation circuit shown in Figure 14.1 as an example. It is assumed that R_E in the circuit is a nonlinear resistor and $R_E \gg R_f$. R_f is the excitation winding resistance. The total resistance of the de-excitation circuit $\Sigma R = R + R_f$.

At the time of the de-excitation, the main contact is opened and the arc contact is closed. The de-excitation circuit equation can be obtained:

$$L_f \frac{di_f}{dt} + R_{i_f} = 0 \tag{14.2}$$

When the generator excitation current i_f decreases linearly, the de-excitation time is the shortest. The expression meeting the conditions is as follows:

$$\frac{di_f}{dt} = C \tag{14.3}$$

At the time of the de-excitation, the overvoltage generated at both ends of the excitation winding is not allowed to exceed the allowable value U_{fm}. This is taken as an additional constraint. When the relation between Eqs. (14.2) and (14.3) is taken into account, the following equation is obtained:

$$L_f \frac{di_f}{dt} = -Ri_f \leq U_{fm} = C \tag{14.4}$$

where C is a constant. Thus, the expression meeting the ideal de-excitation conditions can be obtained:

$$R = \frac{U_{fm}}{i_f} \tag{14.5}$$

In order to make $i_f R = U_{fm} = C$ in the de-excitation process, Resistor R should have nonlinear characteristics that vary with the current i_f.

The following formula can be obtained from Eq. (14.4):

$$\frac{di_f}{dt} = -\frac{U_{fm}}{L_f}$$

The above formula is integrated, and the initial current is equal to I_{f0}. The following equation is then obtained:

$$i_f = I_{f0} - \frac{U_{fm}}{L_f} t \tag{14.6}$$

If the influence of the saturation on the inductor L_f is neglected, Eq. (14.6) shows that the current will attenuate linearly. When $t = t_m$, the current reaches zero. Thus, the de-excitation time can be obtained:

$$t_m = I_{f0} \frac{L_f}{U_{fm}}$$

$$I_{f0} = \frac{U_{fm}}{R_0 + R_f} \tag{14.7}$$

where

I_{f0} — initial value of field current;

R_0 — initial resistance of nonlinear resistor.

The expression for I_{f0} is substituted into Eq. (14.7). Another expression for the de-excitation time is thus obtained:

$$t_m = \frac{U_{fm}}{R_0 + R_f} \times \frac{L_f}{U_{fm}} = \frac{1}{K+1} \times \frac{L_f}{R_f} \tag{14.8}$$

$$\frac{L_f}{R_f} = T'_{d0}$$

$$K = \frac{U_{fm}}{U_{f0}} = \frac{R_0}{R_f}$$

where

K — overvoltage multiple;

T'_{d0} — generator no-load time constant.

When $K = 5$:

$$t_m = \frac{T'_{d0}}{6} = 0.167 T'_{d0} \tag{14.9}$$

Equation (14.9) shows that the excitation time is $0.167 T'_{d0}$ under the ideal de-excitation conditions with a given overvoltage factor $K = 5$ which makes the excitation current vary linearly.

The excitation voltage and current versus time curves under the ideal de-excitation conditions, that is, the de-excitation attenuation characteristics of the ideal de-excitation system, are shown in Figure 14.2.

Figure 14.2 De-excitation attenuation characteristics of ideal de-excitation system.

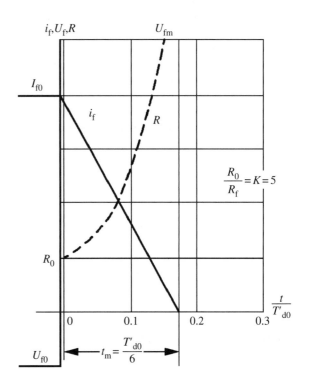

14.2.2 Valid De-excitation Time Method

The valid de-excitation time method is the evaluation criteria proposed by Sweden's ASEA. The valid de-excitation time is defined as the time determined when the time integral of the generator magnetic excitation current is divided by the field flashing current in the de-excitation process. Its expression is as follows:

$$T_1 = \frac{1}{I_{f0}} \int_0^\infty i\,dt \qquad (14.10)$$

where

T_1 – valid de-excitation time.

Equation (14.10) shows that the valid de-excitation time (T_1) is directly proportional to the energy consumed by the de-excitation system under the condition of constant arc voltage.

The valid de-excitation time can also be determined by the time integral of the square of the current, as shown in Eq. (14.11):

$$T_2 = \frac{1}{I_{f0}^2} \int_0^\infty i^2\,dt \qquad (14.11)$$

The physical meaning of this expression is that the valid de-excitation time (T_2) is directly proportional to the energy consumed by a constant resistor.

14.2.3 Generator Voltage-Based De-excitation Time Determination Method

The de-excitation of de-excitation systems used currently effectively acts only on the generator's vertical axis magnetic field system. However, in fact, the generator has a horizontal axis flux component that is free from the influence of the de-excitation effect, in addition to the vertical axis of flux. Table 14.1 shows the variation of the flux component of a 400 MVA steam turbine with the mode of operation.

As can be seen from Table 14.1, the horizontal axis flux component can vary over a wide range depending on the mode of operation. For example, when the power factor $\cos\phi = 1.0$, the horizontal axis flux component Φ_q is 87%, much larger than the vertical axis flux component Φ_d (48%). So, it is not appropriate to evaluate the performance of a de-excitation system on the basis of only the de-excitation effectiveness of the vertical axis flux with the on-load time constant of the generator as a basic value. Besides, the damping winding system of the generator also stores a certain amount of energy that is free from the influence of the de-excitation effect, which is also a factor that should not be neglected for de-excitation.

Austrian scientists proposed an evaluation method to determine the valid de-excitation time on the basis of the actual final de-excitation effectiveness, that is, the attenuation of the generator voltage over time. This can take full account of the influence of various factors on de-excitation performance and reflect the actual de-excitation effectiveness. From this definition, the corresponding expression is as follows:

$$T = \frac{1}{U_0} \int_0^\infty u\,dt \qquad (14.12)$$

Table 14.1 Variation of flux component of 400 MVA steam turbine with mode of operation.

Mode of operation	Unloaded	$\cos\phi = 0.75$	$\cos\phi = 1.0$
Stator current (%)	0	100	75
Flux $\Phi = \sqrt{\Phi_d^2 + \Phi_q^2}$ (%)	100	100	100
Φ_d (%)	100	82	48
Φ_q (%)	0	57	87

where

T – valid de-excitation time determined by time integral of generator voltage;

U_0 – initial voltage of generator.

14.3 De-excitation System Classification

14.3.1 Linear Resistor De-excitation System

14.3.1.1 Expression for Linear De-excitation Time

For a so-called linear resistor de-excitation system, the U-I characteristic of the selected de-excitation resistor conforms to the linear relation of Ohm's law.

In general, a linear de-excitation resistor made of high-temperature-resistant metal material is a reliable conventional de-excitation solution that has been widely used.

One of the advantages of a linear resistor de-excitation system is that as long as the dissipation temperature rise generated when the linear resistor absorbs the de-excitation energy does not exceed the allowable value in the de-excitation process, the linear resistor will restore to the normal state once the de-excitation process is over. There is no life loss, aging, or U-I characteristic curve variation in use.

Generally, a linear resistance is 2–3 times the thermal resistance value of the generator excitation winding. In the de-excitation process, the rotor excitation winding generates an overvoltage. Due to the constraint of the rotor excitation winding insulation strength, the rotor overvoltage multiple is generally no greater than 4–5 times the generator's rated excitation voltage.

Figure 14.3 shows the connection of a linear resistor de-excitation system. In the de-excitation process, the rotor voltage equation is

$$L_f \frac{di_f}{dt} + i_f(R + R_f) = 0 \tag{14.13}$$

From Eq. (14.13), the expression for the rotor excitation current is obtained as

$$i_f = \frac{U_{f0}}{R_f} e^{-\frac{R+R_f}{L_f}t} = I_{f0} e^{-\frac{R+R_f}{L_f}t} \tag{14.14}$$

If $I_{f0}R = U_{fm}$,
the following equation can be obtained:

$$\frac{U_{fm}}{U_{f0}} = \frac{I_{f0}R}{I_{f0}R_f} = \frac{R}{R_f} = K$$

Figure 14.3 Connection of linear resistor de-excitation system.

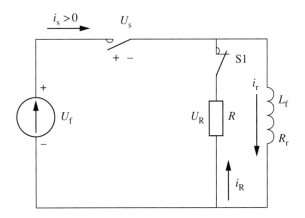

where

K – rotor excitation winding overvoltage multiple.

Equation (14.14) can be rewritten as follows:

$$\frac{I_{f0}}{i_f} = e^{\frac{R+R_f}{L_f}t}$$

The natural logarithm of both sides of the equation is taken:

$$\ln \frac{I_{f0}}{i_f} = \frac{R+R_f}{L_f}t$$

If it is considered that the de-excitation is over when $if = \frac{I_{f0}}{N}$ (where $N = 100$), the corresponding de-excitation time is

$$t_m = \frac{L_f}{R+R_f} \ln 100$$

When $T'_{d0} = \frac{L_f}{R_f}$ and $K = \frac{R}{R_f}$ are taken into account and substituted into the above equation, the following equation is obtained:

$$t_m = \frac{T'_{d0}}{1+K} \times 4.6 \tag{14.15}$$

When $K = 5$, the single-step linear resistor de-excitation time is

$$t_m = 0.767 T'_{d0} \tag{14.16}$$

14.3.1.2 Linear Resistor De-excitation Commutation Conditions

In a linear resistor de-excitation system, the greater the de-excitation resistance, the faster the de-excitation and the higher the overvoltage multiple the rotor excitation winding withstanding capacity. Conventionally, the de-excitation resistance is 3–5 times the thermal resistance of the excitation winding. Thus, in the forced excitation state, it is assumed that the forced excitation current is twice the rated excitation current. In the forced excitation state, during the de-excitation, the excitation winding will withstand 6–10 times the rated excitation voltage.

Besides, the commutation conditions should also be considered at the time of the de-excitation. This is illustrated with the de-excitation circuit shown in Figure 14.4.

In the case of a normal de-excitation, K2 is closed while K1 is broken. The de-excitation waveform at the time of a correct commutation is shown in Figure 14.5.

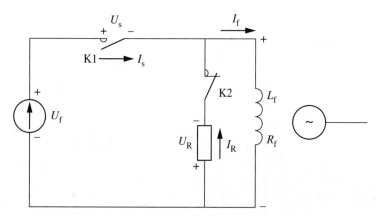

Figure 14.4 Linear resistor de-excitation system.

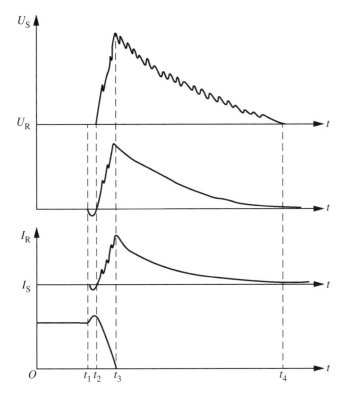

Figure 14.5 Linear resistor de-excitation waveform.

As can be seen from Figure 14.5, the operation state is normal before t_1. At t_1, K2 is closed, and Resistor R is turned on. The follow current of the power supply voltage U_f makes an additional current flow through the Resistor R circuit. Meanwhile, the current I_S flowing through the circuit breaker increases the same additional current. At t_2, after K1 is segmented, an arc is generated. As the arc elongates, the arc voltage U_S rises. Meanwhile, due to the breaking of the circuit breaker, the rotor current shows a downward trend and thus causes a reverse potential maintaining the rotor current. Its polarity is shown in Figure 14.5 (U_R).

Under the action of U_R, the current I_R flowing through Resistor R also rises. Since $I_f = I_R + I_S$, the increase in I_R will certainly decrease I_S. This process is called commutation. It lasts until the excitation current I_f is completely transferred to the de-excitation resistor. At t_3, when the commutation ends, $I_S = 0$. Then, U_S, U_R, and I_R all become attenuated exponentially to zero, and the de-excitation ends. Generally, t_1-t_2 is a few milliseconds, t_2-t_3 tens of milliseconds, and t_3-t_4 a few seconds.

It should be noted in the commutation process that the t_2-t_3 commutation is the field circuit breaker's arc burning and energy absorption process. The longer the process, the more serious the arc contact and arc cover burnout. So, a shorter commutation process is better.

The maximum value of the break voltage of the circuit breaker is determined by its structure and the characteristics of its arc cover. It is one of the key technical parameters of the field circuit breaker. Its value is related to the de-excitation commutation process. Under Kirchhoff laws, the following equation can be obtained:

$$U_R + U_f - U_s = 0$$

which can be written as follows:

$$U_R = U_s - U_f \tag{14.17}$$

In the de-excitation process, the maximum value of U_R is $U_{Rmax} = I_{Rmax}R$.

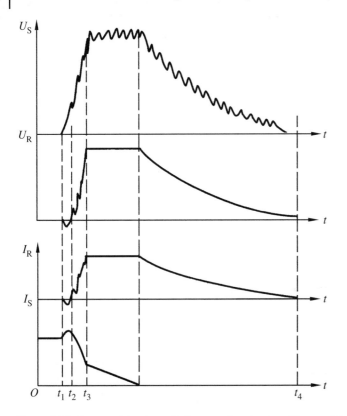

Figure 14.6 Linear resistor de-excitation waveform in case of incomplete commutation.

If the value of U_R is too large, Eq. (14.17) will become the following inequality:

$$U_{R\max} > U_s - U_f \tag{14.18}$$

As a result, the excitation current I_f cannot be completely commutated to Resistor R for energy consumption, and part of the current flows to the K1 contact. This will extend the commutation time and aggravate the burnout of the main contact of the circuit breaker.

Figure 14.6 shows the de-excitation waveform of the incomplete commutation when the de-excitation resistor R is too large.

So, the necessary condition that ensures proper commutation of the field circuit breaker is as follows:

$$U_{Sm} \geqslant U_{fm} + U_{Rm} \tag{14.19}$$

That is, the maximum break arc voltage of the circuit breaker must be greater than the sum of the maximum supply peak voltage and the maximum voltage drop of the de-excitation resistance at the time of the forced excitation.

As mentioned above, a single-step linear resistor de-excitation system is constrained by the overvoltage multiple at the time of the de-excitation in terms of the rapidity of the de-excitation and represents longer de-excitation time. To reduce the de-excitation time constant, a stepped linear resistor de-excitation system can be used to accelerate the rapidity of the de-excitation. The connection is shown in Figure 14.7.

14.3.1.3 Stepped Linear Resistor De-excitation System

In a two-step linear resistor de-excitation system, the first-step linear resistor R_1 and the second-step linear resistor R_2 are chosen under the constraint that the voltage generated by the rotor current $i_f(t)$ across the de-excitation resistor R_1 or $(R_1 + R_2)$ does not exceed the allowable voltage of the rotor windings in the de-excitation process. Generally, the overvoltage or de-excitation resistor multiplier $K = 5$ [35].

Figure 14.7 Stepped linear resistor de-excitation system.

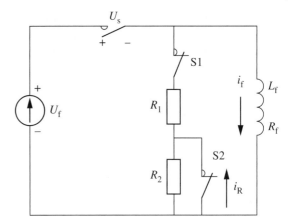

At the initial de-excitation stage, when the main contact of the field breaker is broken, the de-excitation current flows through the first-step de-excitation resistor R_1. When the rotor current drops from I_{f0} to AI_{f0} ($A < 1$), the contact S2 is broken, and the second-step de-excitation resistor R_2 is put in to accelerate the de-excitation process.

Due to the constraint of the rotor overvoltage multiple, when the second-step de-excitation resistor is put in, the de-excitation resistance increases from the first-step resistor R_1 to the sum of the first- and second-step resistors ($R_1 + R_2$). Obviously, at this point, the constraint that the overvoltage multiple should not exceed the initial overvoltage multiple is still necessary. Thus, the expression for the de-excitation time of the two-step linear resistor de-excitation system can be listed as follows:

$$t_m = \frac{T'_{d0}}{1+K_1} \ln \frac{I_{f0}}{AI_{f0}} + \frac{T'_{d0}}{1+K_2} \ln \frac{AI_{f0}}{I_{f0}/100} = \frac{T'_{d0}}{1+K_1} \ln \frac{1}{A} + \frac{T'_{d0}}{1+K_2} \ln 100A$$

$$K_f = \frac{R_1}{R_f}$$

$$K_2 = \frac{R_1 + R_2}{R_f} \tag{14.20}$$

where

K_1 – initial de-excitation resistor multiple when first-step de-excitation resistor R_1 is put in;

K_2 – de-excitation resistor multiple when second-step de-excitation resistor R_2 is put in.

The rotor current attenuation coefficient A is determined under the constraint that the initial de-excitation overvoltage generated when the first-step de-excitation resistor R_1 is put in is equal to the de-excitation overvoltage generated at the instant when the second-step de-excitation resistor R_2 is put in.

Thus, the excitation overvoltage equation can be listed as follows:

$$K_1 I_{f0} R_f = K_2 A I_{f0} R_f \tag{14.21}$$

The following equation can be obtained from Eq. (14.21):

$$K_2 = \frac{K_1}{A} \tag{14.22}$$

When $K_1 = 5$ and $K_2 = \frac{5}{A}$ are substituted into Eq. (14.20), the following equation can be obtained:

$$t_m = T'_{d0} \left(\frac{1}{6} \ln \frac{1}{A} + \frac{A}{5+A} \ln 100A \right) \tag{14.23}$$

Equation (14.23) shows that de-excitation time $t_m = f(A)$.

So, the rotor current attenuation coefficient A in Eq. (14.23) should be determined first.

To obtain the minimum de-excitation time t_m, the derivative of Eq. (14.23) can be obtained, and $\frac{dt_m}{dA} = 0$. Thus, the rotor current attenuation coefficient A that minimizes t_m can be obtained.

The derivative of Eq. (14.20) is taken and made equal to zero:

$$\begin{aligned}\frac{dt_m}{dA} &= -\frac{1}{6A} + \frac{1}{5+A} + \frac{5}{(5+A)^2} \ln 100A \\ &= \frac{5A^2 + 20A - 25 + 30A \ln 100A}{6A(5+A)^2} = 0\end{aligned} \quad (14.24)$$

Using a numerical method, we find that $A = 0.2197$. When this is substituted into Eq. (14.23), the minimum t_m is obtained:

$$t_m = 0.3825 T_{d0}$$

Compared to the first-step linear de-excitation resistor scheme, the de-excitation time drop is as follows:

$$\Delta t_m = \frac{0.3825}{0.767} = 49.8\%$$

The second-step de-excitation resistor multiple is determined below. The following equation can be obtained from Eq. (14.22):

$$K_2 = \frac{K_1}{A_1}$$

Since $K_1 = 5$,

$$K_2 = \frac{5}{0.2197} = 22.758.$$

$$K_2 = \frac{R_1 + R_2}{R_f} = 22.758$$

$$R_2 = 22.758 R_f - R_1 = 22.758 R_f - 5 R_f = 17.758 R_f$$

The current capacity of the first-step de-excitation resistor is as follows:

$$I_{R1} = (0.1 \sim 0.2) I_{fN}$$

The current capacity of the second-step de-excitation resistor is as follows:

$$I_{R2} = (0.1 \sim 0.2) A I_{fN} = (0.022 \sim 0.044) I_{fN}$$

The current capacity of the field circuit breaker's auxiliary contact S1 and contact S2 is

$$I_{S1,S2} = (0.1 \sim 0.2) I_{fN}$$

But when the second-step linear de-excitation resistor is put in, the breaking current capacity of the contactor S2 is

$$I_{S2} > 0.23 I_{fN}$$

Theoretically, a multi-step linear resistor de-excitation system can also be used to further accelerate the de-excitation. But it is not recommended, since it makes the circuit complex.

14.3.2 Nonlinear Resistor De-excitation System

For a so-called nonlinear resistor, the voltage applied to both ends of the resistor is nonlinearly related to the current flowing through it. Its resistance decreases as the current increases.

The characteristic of a nonlinear resistor is generally represented by the nonlinear resistance coefficient β. The corresponding expression is as follows:

$$U = C I^\beta \quad (14.25)$$

where

U – voltage at both ends of nonlinear resistor;

I – current flowing through nonlinear resistor;

β – nonlinear resistance coefficient, related to resistance valve material;

C – configuration coefficient of nonlinear resistance, related to valve material, geometric dimensioning, and resistance series/parallel combination mode.

Equation (14.25) can also be written expressed as follows:

$$I = HU^\alpha \tag{14.26}$$

where

α – nonlinear resistance coefficient, related only to resistance valve material.

Figure 14.8 shows a comparison of the nonlinear resistance characteristics of silicon carbide SiC and zinc oxide (ZnO). It should be noted that since a ZnO nonlinear resistor has a steeper nonlinearity, the area where the current flowing through the valve is less than 10 mA is defined as the cutoff area and the area where the current is 10 mA or greater is defined as the turn-on area. Such definitions are not required for a SiC nonlinear resistor.

As can be seen from Figure 14.8, the ZnO nonlinear resistor valve has a small leakage current and a steeper nonlinearity.

For a SiC nonlinear resistor, the nonlinear resistance coefficient α is 2–4; that is, β is 0.25–0.5. For a ZnO nonlinear resistor, α is 20–40; that is, β is 0.025–0.05.

14.3.2.1 Expression for Nonlinear De-excitation Time

The connection of a nonlinear resistor de-excitation system is shown in Figure 14.9, where R is a nonlinear resistor.

In the de-excitation state, the following voltage equation is obtained from the generator excitation winding circuit:

$$L_f \frac{di_f}{dt} + i_f R_f + U_R = 0 \tag{14.27}$$

$$U_R = C i_f^\beta$$

Figure 14.8 Comparison of nonlinear resistance characteristics between SiC and ZnO: a – SiC valve; b – ZnO valve.

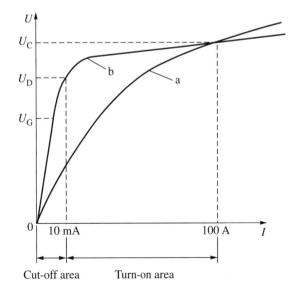

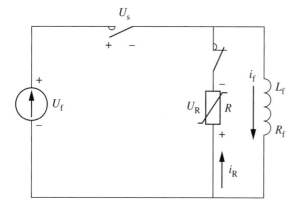

Figure 14.9 Nonlinear resistor de-excitation system.

where

U_R – total voltage drop of nonlinear de-excitation resistor elements.

The excitation current expression in the de-excitation process is obtained from Eq. (14.27):

$$i_f = \left[\left(I_{f0}^{1+\beta} + \frac{C}{R_f}\right) e^{-(1-\beta)\frac{t}{T'_{do}}} - \frac{C}{R_f}\right]^{\frac{1}{1-\beta}} \qquad (14.28)$$

where

I_{f0} – field flashing current before de-excitation.

It is assumed that the reverse voltage established by the field flashing current I_{f0} flowing through the nonlinear resistor at the time of the de-excitation is U_{R0}. Its value is K times the field flashing voltage $I_{f0}R_f$. That is,

$$K = \frac{U_{R_0}}{I_{f0}R_f} = \frac{CI_{f0}^{\beta}}{I_{f0}R_f} = \frac{C}{R_f}I_{f0}^{\beta-1}$$

The equation can also be written as follows:

$$I_{f0}^{\beta-1} = \frac{KR_f}{C}$$

$$I_{f0}^{1-\beta} = \frac{C}{KR_f} \qquad (14.29)$$

When Eq. (14.29) is substituted into Eq. (14.28), the following can be obtained:

$$\left[\left(\frac{C}{KR_f} + \frac{C}{R_f}\right) e^{-(1-\beta)\frac{t}{T'_{do}}} - \frac{C}{R_f}\right]^{\frac{1}{1-\beta}}$$

At the end of the de-excitation, $i_f = 0$ and $t = t_m$. The following equation can be obtained:

$$\frac{C}{KR_f}(1+K)e^{-(1-\beta)\frac{t}{T'_{do}}} - \frac{C}{R_f} = 0 \qquad (14.30)$$

The natural logarithm is taken. Thus, the de-excitation time can be obtained:

$$t_m = \frac{T'_{do}}{1+\beta} L_n \left(1 + \frac{1}{K}\right) \qquad (14.31)$$

14.3.2.2 Nonlinear Resistor De-excitation Commutation Conditions

The connection diagram of a nonlinear resistor de-excitation system is shown in Figure 14.9. When the field circuit breaker is broken, the relation between the circuit voltages must satisfy the following inequality in order

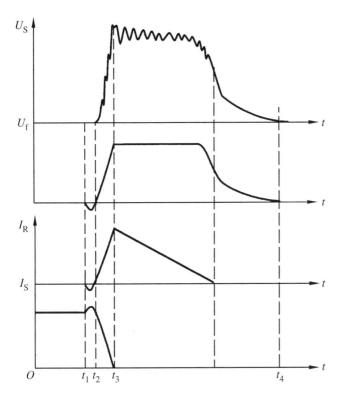

Figure 14.10 Normal commutation waveforms of nonlinear de-excitation resistor: t_1 – main contact field circuit breaker breaking time; t_2 – commutation start time; t_3 – commutation end time; t_4 – de-excitation end time.

to meet the normal commutation conditions to make the initial current I_{f0} in the generator excitation winding smoothly commutated to the circuit connected in parallel by the de-excitation resistor and the generator excitation winding:

$$U_S \geq U_R + U_f \tag{14.32}$$

For the nonlinear resistor de-excitation system, the normal commutation process is shown in Figure 14.10.

When the field circuit breaker is opened, at point t_1, the establishment of the arc voltage U_s is started. The excitation voltage shows a reverse polarity due to the action of the induced emf. At point t_2, Eq. (14.32) is satisfied. The current I_s flowing through the field circuit breaker begins to decrease. The current flowing into the nonlinear resistor R begins to rise. At point t_3, the complete commutation succeeds. If the arc voltage of the field circuit breaker is too low and fails to satisfy Eq. (14.32), the commutation will fail at the time of the de-excitation. All of the excitation current flows through the main contact of the field circuit breaker and is attenuated on the basis of the generator engine time constant T'_{d0}, which may cause a burnout of the main contact of the circuit breaker in severe cases. The failed de-excitation commutation waveforms are shown in Figure 14.11. In the de-excitation process, all of the rotor current flows through the main contact of the field circuit breaker, starting from t_1 and ending at t_4.

14.3.3 Crowbar De-excitation System

14.3.3.1 Overview

In a conventional DC field circuit breaker de-excitation system, the DC field circuit breaker has a main contact and an arc contact to meet the needs of the normal magnetic field circuit breaking/closing and de-excitation operation.

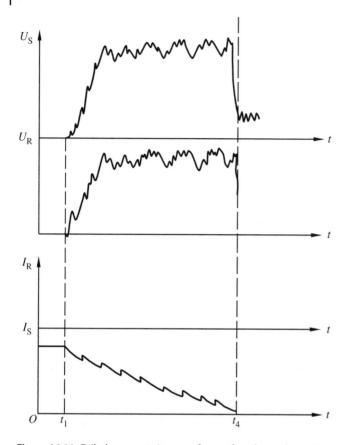

Figure 14.11 Failed commutation waveforms of nonlinear de-excitation resistor.

When such a specially designed and manufactured DC field circuit breaker is used for a nonlinear resistor de-excitation system, due to the strict requirements for the arc voltage generated when the main contact of the field circuit breaker is broken, the DC field circuit breaker structure tends to be complicated. In some cases, the DC field circuit breaker is even required to have two or more series main contacts to establish a higher break voltage, to ensure that the nonlinear resistor is turned on at the time of the de-excitation and the excitation current is commutated to the de-excitation circuit. Such special DC field circuit breakers are subject to high prices and small market size, which has an adverse effect on their production and development. So, in countries other than China, the number of manufacturers of special de-excitation DC field circuit breakers has been gradually decreasing. In de-excitation system applications, general-purpose DC circuit breakers without an auxiliary arc contact are used in place of DC field circuit breakers.

Since a general-purpose DC field circuit breaker has only one normally open main contact and has no normally closed auxiliary arc contact connected to the de-excitation resistor, the de-excitation resistor is connected across the crowbar circuit at both ends of the generator excitation winding in parallel via a power electronic element to meet the conventional de-excitation requirements. The crowbar circuit can also serve as the overvoltage protector for the generator excitation winding circuit.

14.3.3.2 Functions of Crowbar

Figure 14.12 shows the connection of a typical imported crowbar that is widely used in China.

De-excitation In Figure 14.12, the thyristors in Section F02 (V2 and V3) serve as contactless electrical contacts with the de-excitation resistors. When the generator excitation winding circuit reaches the corresponding

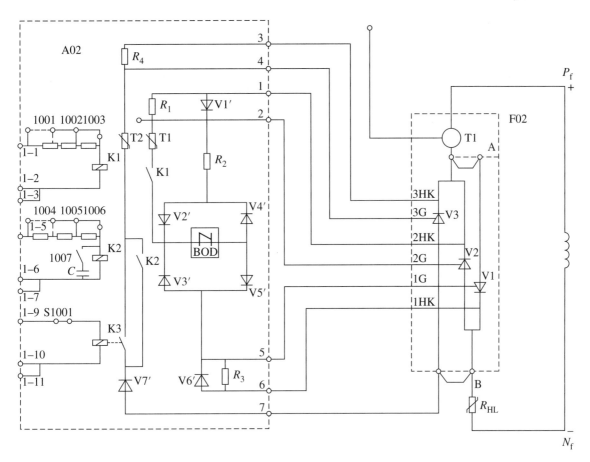

Figure 14.12 Typical crowbar circuit.

forward and inverse overvoltage set values, V2 and V3 are turned on and connected to the nonlinear resistors to limit the overvoltage of the excitation winding circuit. Section A02 dominated by the breakover diode (BOD) is the de-excitation and overvoltage protection value setting area. The attached relays K1, K2, and K3 are relays started by the generator general outlet protection signal. The de-excitation process achieved by the crowbar is introduced first.

In the event of a fault of the external system of a generator-transformer set, the general outlet relay protection signal issues a trip order while acting on the main circuit breaker and field circuit breaker trip circuit. As can be seen from Figure 14.12, the protection trip signals TRIP1 and TRIP2 act on the relays K1 and K2, respectively. When the trip signal is input to K1, the K1 contact is closed. During that period, with the breaking of the main contact of the field circuit breaker, the generator excitation winding generates a reverse excitation voltage. The polarity of the cathode N_f end of the excitation winding is positive. This voltage's path is from the N_f end to R_{NL}–B–1HK–6–V′6–V′3–K1–T1–R1. At this point, Thyristor V2 is triggered by the voltage drop of R_1 ($U_{R1} = U_{21} = U_{2G-2HK}$) and turned on to connect the de-excitation resistors in parallel to both ends of the excitation winding to achieve de-excitation. The current limiting resistor T_1 is used to protect Contact $K1$.

K2 serves as a redundant relay to ensure de-excitation reliability. Since there are capacitors in parallel at both ends of the winding picked up at Contact K2, a delay is caused. After a certain delay after K1 acts, the normally open Contact K2 is closed. The process is similar to the closure process of Contact K1. The rotor reverse voltage's path is N_f–R_{NL}–B–7–V′7–K2–T2–R_4. $U_{R4} = U_{43} = U_{3G-3HK}$. Thyristor V3 is triggered and turned on to connect the de-excitation resistors in parallel to both ends of the excitation winding to achieve the reserve de-excitation. If $R_1 = R_4$ and $R_{T1} = R_{T2}$, the operating values of the connection of the de-excitation resistors to K1 and K2 are the same. As is clear from the above discussion, the function of V2 and V3 is to provide contacts for the connection of the de-excitation resistors.

Overvoltage Protection For a synchronous generator, when the generator is operating in the event of a loss of excitation or a loss of synchronization or a pole slip, depending on the active power, a slip is generated between the generator's stator and rotor, and an overvoltage is generated on the rotor excitation-winding side. In the most serious mode of operation, when the polarity of the overvoltage is the same as that of the rectified voltage outputted by the rectifiers, if the slip induced voltage on the rotor excitation-winding side is high enough, the voltage outputted by the rectifiers will be latched, and the excitation current outputted by it will be blocked to zero, until the current outputted by the rectifiers resumes a positive value. A severe overvoltage will be caused in the period of time when the excitation current is zero.

Figure 14.13 shows the waveform of the rotor overvoltage caused by the blocking of the excitation current for the generator field circuit under the action of the forward overvoltage when the steam turbine loses synchronization at a rotor power angle δ of more than 180°.

The crowbar's overvoltage protection function is described below. In Figure 14.12, A02 is the crowbar overvoltage protection limit value setting area. The U-I characteristic curve of the BOD is shown in Figure 14.14.

Under the action of the forward voltage, when the voltage limiting diode is operating in the turn-on section, the forward leakage current flowing through it is very small. When the applied voltage continues to rise and reaches the voltage limiting diode breakover voltage U_{B0}, the voltage limiting diode is turned on. At this point,

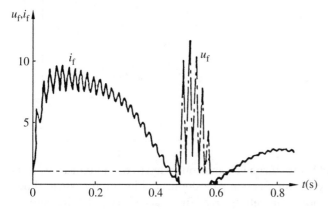

Figure 14.13 Rotor overvoltage waveform when generator excitation current is blocked by forward reverse voltage: u_f – excitation voltage; i_f – excitation current.

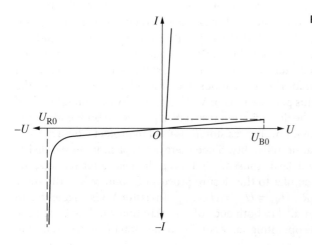

Figure 14.14 U-I characteristic of breakover diode (BOD).

the forward on-state voltage at both ends of the diode is only a few volts and allows the flow of the specified rated current.

When the applied forward voltage disappears, the voltage limiting diode restores the blocking characteristic. When a reverse voltage is applied to the voltage limiting diode and it is greater than the reverse blocking voltage U_{R0}, the voltage limiting diode is reversely broken down, and its operating characteristics cannot be restored.

The following is a description of the crowbar's overvoltage protection action process at any instant when the polarity of the overvoltage caused by the N_f end of the excitation winding is positive in the event of a loss of synchronization and a slip voltage. The polarity of the N_f end of the excitation winding is positive. The overvoltage's path is R_{NL}–B–1HK–6–V'6–V'3–V (BOD)–V'4–R_2–R_1. $U_{R1} = U_{2G-2HK}$. Thyristor V2 is triggered and turned on to connect the voltage limiting resistor R_{NL} that serves concurrently as an overvoltage protector and the excitation winding in parallel to form a closed circuit to limit the overvoltage.

Besides, if the overvoltage positive polarity occurs at the P_f end, the path is P_f–A–2HK–1–V'1–R_2–V'2–V (BOD)–V'5–R_3. At this point, $U_{R3} = U_{56} = U_{1G-1HK}$. Thyristor V1 is triggered and turned on to connect the de-excitation resistor (R_{NL}) and the excitation winding to form a parallel open circuit to limit the overvoltage on the rotor excitation winding side.

As is clear from the above discussion, in the crowbar system, V2 and V3 undertake the de-excitation, and V1 and V2 undertake the overvoltage limiting. The circuit is well designed, in that V2 undertakes both the de-excitation and the overvoltage limiting. This is achieved by selecting different set resistance values in the corresponding de-excitation and overvoltage limiting circuits.

It should be particularly emphasized that in the rotor overvoltage limiting process, once the forward and reverse V1 or V2 are turned on, the nonlinear resistor R_{NL} at both ends of the generator excitation winding will always be in the follow current conducting state, unless the overvoltage energy in the excitation winding circuit is completely absorbed by the de-excitation resistor or the thyristors are blocked by the reverse field negative half wave voltage.

For the de-excitation resistor that serves concurrently as an overvoltage protector in order to prevent the burnout caused by the follow current energy accumulation, a current detection device T1 needs to be connected to the nonlinear resistor circuit. For a hydro-turbine of Three Gorges Hydro Power Station, when the excitation winding circuit overvoltage reaches 1700 V, the overvoltage protection circuit will act. When the current flowing through the crowbar is 300 A, a trip order will be issued.

14.3.4 AC Voltage De-excitation System [36]

An AC voltage de-excitation system is a system that achieves de-excitation by breaking the AC circuit breaker connected to the AC side of the power rectifiers and closing the thyristor pulse.

For a static self-excitation system, priority should be given to the AC voltage de-excitation system.

The basic working principle of the AC voltage de-excitation system is that the trigger pulse of the three-phase fully controlled rectifier bridges should be cut off before the AC circuit breaker connected to the AC side of the power rectifiers is broken, thus enabling the negative peak wave voltage introduced to the secondary side of the field transformer to be utilized for the acceleration of de-excitation.

The typical circuit of an AC voltage de-excitation system with an AC circuit breaker is shown in Figure 14.15. Since the rectifier load is the generator excitation winding, which has a large inductance, the output current can be considered to be a constant current. If the commutation angle is neglected, each phase current on the AC side of the rectifiers is a square wave with a duty ratio of 2/3, and the phase angle difference of each phase current is a 120° electrical angle. So, at any time, in the three-phase currents, one must be zero and the other two numerically equal and opposite in direction. The phase current waveforms of the AC circuit breaker are shown in Figure 14.16. It is assumed that the AC circuit breaker S is broken at t_1. Then, +A and −C are turned on, while the other elements are cut off. If the trigger pulse of the thyristor elements is not cut off before the AC circuit breaker is broken, it will act on the corresponding bridge arm thyristors in order (+A, −C, +B, −A, +C, and −B). At the instant when the AC circuit breaker S is cut off, since +A and −C are turned on and the

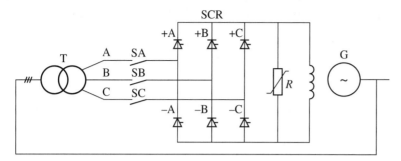

Figure 14.15 Connection of AC voltage de-excitation system.

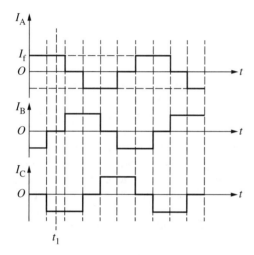

Figure 14.16 Rectifier AC-side current waveforms.

follow current effect of the inductive load plays a part, the current of +A and −C is greater than the holding current and cannot be cut off. The breaks SA and SC maintain the arc burning. The Phase B break SB is easily broken since no current flows through it. Then, +B is triggered. But at this point the Phase B supply voltage has been broken. So, +B cannot be turned on. After a 120° electrical angle or 3.3 ms, −A is triggered. At this point, the anode voltage of −A has become forward. So, −A is immediately turned on, while −C is cut off. The load current flows through +A and −A to form a short-circuit follow current. SA and SC of the AC circuit breaker break the contact for arc quenching. Then, a state of de-excitation with the natural follow current through the thyristor elements is formed. This process is time consuming and cannot achieve fast de-excitation.

An AC circuit breaker achieves breaking primarily by utilizing the zero crossing of the current. However, because the load inductance is large and the excitation current attenuation is slow, it is impossible achieve cut-out by utilizing the zero crossing of the current within 3.3 ms. The most effective measure to address breaking of the AC voltage de-excitation system is to cut out the trigger pulse of the fully controlled rectifier bridges at the instant of the de-excitation. The working principle is illustrated with the equivalent circuit of the AC voltage de-excitation system in Figure 14.17.

In Figure 14.16, it is assumed that the AC circuit breaker is broken and the trigger pulse of the thyristor rectifier bridges is cut off simultaneously at t_1, and the two elements (+A and −C) are always in the follow current conducting state. If the forward voltage drop of the conducting elements is ignored, the voltage applied to both ends of the nonlinear resistor (R) is constrained to be equal to the sum of the power supply line voltage of the field transformer and the break voltage of the AC circuit breaker. It is assumed that $U_S = U_{SA} + U_{SC}$ – i.e., U_S is the sum of the two break voltages. The expression meeting the de-excitation current commutation conditions is as follows:

$$U_{Sm} \geqslant U_{Rm} \pm U_{Tm} \tag{14.33}$$

Figure 14.17 Equivalent circuit of AC voltage de-excitation system.

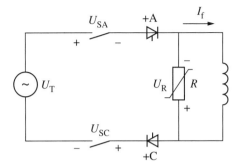

Figure 14.18 Rectifier output voltage and current waveforms.

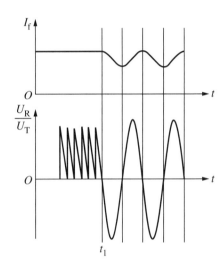

where

U_{Sm} — maximum break voltage of AC circuit breaker;
U_{Rm} — maximum turn-on voltage of nonlinear resistor;
U_{Tm} — maximum line voltage of excitation transformer.

At t_1, the instant of the breaking, the excitation transformer line voltage may be in the positive or negative half wave. For the positive half wave, it can be seen from Eq. (14.33) that the break voltage U_{Sm} of the AC circuit breaker should be greater than or equal to the sum of U_{Rm} and U_{Tm} and should be high in order to meet the commutation conditions. In the negative half cycle, the break voltage required to achieve the commutation conditions is $(U_{Rm}-U_{Tm})$, which can reduce the circuit breaker arc voltage requirements.

In the de-excitation process, the rectifier output current and voltage waveforms are shown in Figure 14.18.

At t_1, an AC circuit breaker breaking order is issued, and the pulse is turned off simultaneously. But there is a certain delay of about tens of milliseconds in the circuit breaker breaking, while the turn-off pulse is completed instantaneously. After the pulse is turned off, the excitation voltage changes from a rectified voltage with an AC component to a sinusoidal AC voltage. After a certain delay, the circuit breaker is broken. The breaking instant is largely random. If U_f happens to be positive at the instant of breaking, the commutation conditions in Eq. (14.33) cannot be met. The break arc will continue to burn, waiting for the arrival of the negative half cycle. For a 50-cycle power supply, each cycle is 20 ms, and the interval is only a dozen milliseconds. If the commutation conditions in Eq. (14.33) are met in the negative half cycle, the commutation will succeed, and the circuit breaker will quench the arc. All of the excitation current will be transferred to the de-excitation resistor for the de-excitation.

14 De-excitation and Rotor Overvoltage Protection of Synchronous Generator

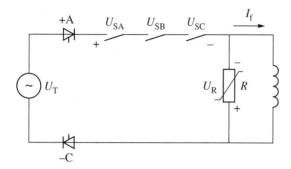

Figure 14.19 Equivalent circuit of DC-side de-excitation.

If the AC circuit breaker is connected to the DC side, when the three circuit breaker contacts are connected in series, de-excitation can also be achieved. De-excitation also requires breaking of the trigger pulse of the thyristor rectifiers. The equivalent de-excitation circuit is shown in Figure 14.19.

There is no essential difference between Figures 14.18 and 14.19. So, Eq. (14.33) also applies to the case where the AC circuit breaker is connected to the DC side.

The following can be observed from a comparison of the two demagnetization methods (connecting the AC circuit breaker to the AC side and connecting the AC circuit breaker to the DC side):

(1) When the circuit breaker is connected to the AC side, the AC zero crossing is utilized for arc quenching. This is AC de-excitation. It should be noted that only two breaks have current flow and establish an arc voltage at the time of de-excitation.
   The commutation condition that should be met at the time of de-excitation is

$$U_{Rm} \leq 2U_{Sm} \pm \sqrt{2}U_{Tm} \tag{14.34}$$

   If the three breaks of the circuit breaker are connected in series to the DC side, they all establish an arc voltage at the time of the de-excitation. Their value will be 1.5 times that when they are connected to the AC side. The commutation conditions can be met more easily.

$$U_{Rm} \leq 3U_{Sm} \pm \sqrt{2}U_{Tm} \tag{14.35}$$

   Obviously, when the AC circuit breaker is connected to the DC side, the AC current zero crossing cannot be utilized for de-excitation. So, strictly speaking, this method cannot be called AC de-excitation.

(2) For the current, if the allowable operating current is 1.0 per unit value when the same circuit breaker is connected to the DC side, the allowable operating current when the circuit breaker is connected to the AC side will be 0.816 per unit value.
   AC circuit breakers have a number of advantages such as low price, mature manufacturing technology, and high operational reliability. It should be noted, however, that their application is subject to certain constraints since AC de-excitation depends on the AC supply voltage. For example, in the event of a short circuit of the generator terminal and a short circuit of the secondary side of the excitation transformer, the AC voltage is zero. The commutation conditions cannot be met. AC de-excitation cannot be achieved. If the short-circuit fault occurs in a rectifier bridge arm or the DC output end, the AC circuit breaker cannot play a role in the breaking. But it is still highly advisable to regard AC negative voltage de-excitation as a mechanical DC de-excitation method alternative scheme.

So, in the de-excitation system of a large hydro-turbine, it should first be ensured that the field circuit breaker is connected to the DC excitation circuit side. For an important unit, another AC air circuit breaker can be provided on the AC side simultaneously. Its role is to switch the power supply of the secondary winding of the excitation transformer to the station service power side to provide an AC-side break for the excitation circuit in the case of electrical braking and achieve de-excitation in the case of a failure of the DC-side field circuit breaker.

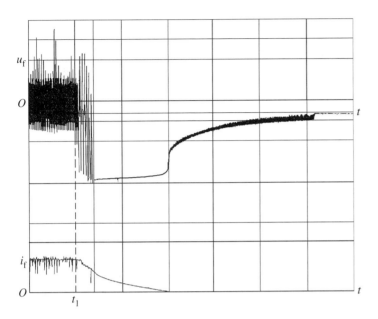

Figure 14.20 Oscillogram of AC negative voltage de-excitation of 300 MW hydro-turbine: u_f – excitation voltage; i_f – excitation current.

Figure 14.20 shows the oscillogram of the AC negative voltage de-excitation of a 300 MW hydro-turbine under load conditions. Before the DC-side field circuit breaker trips, the pulse of the power rectifiers is cut off. At t_1, the second anode sine wave AC voltage of the excitation transformer is introduced into the DC excitation circuit. The de-excitation resistor is turned on through utilization of the superposition of the AC voltage negative half cycle voltage and the field circuit breaker breaking arc voltage. Thus, the de-excitation function or the conversion of the rotor current to the de-excitation resistor is achieved.

14.4 Influence of Saturation on De-excitation [37]

In the magnetic circuit of the generator, only the inductance L of the straight line segment of the magnetization curve is constant. When the magnetic conductor is in the saturated state, the inductance will decrease with the intensification of the saturation. Figure 14.21 shows the variation of the inductance L with the saturation.

It is assumed that point N on the magnetization curve is the normal operating point of the generator. Point b is equivalent to the operating point in the state three times the forced excitation of the generator.

It is assumed that the dynamic inductance of the linear part of the magnetization curve $L = \frac{d\psi}{di_f} = \tan \alpha$. In the forced excitation state, due to the influence of the saturation of the magnetic circuit, the dynamic inductance will significantly decrease to $\frac{\tan \beta}{\tan \alpha} \approx 0.04$. When the de-excitation effect is considered, the saturation should not be ignored. This problem is discussed with an example of linear resistor de-excitation.

For a linear resistor de-excitation system, the de-excitation process can be expressed as follows:

$$\frac{d\psi}{di} + R_i + R_f i = 0 \tag{14.36}$$

The curve of the relation between the flux linkage ψ and the current i_f is represented by the two Line Segments oa and ab, respectively, with the graphical method. The inductance value of each line segment is constant. Thus, on Line Segment oa,

$$L_\alpha = \frac{d\psi}{di_f} = \tan \alpha$$

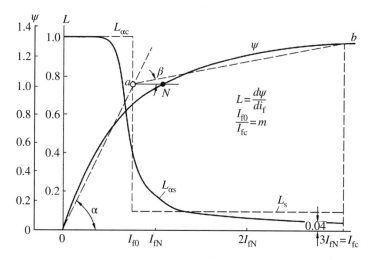

Figure 14.21 Generator magnetization curve and variation of dynamic inductance (L). It is assumed that point N on the magnetization curve is the normal operating point of the generator. Point b is equivalent to the operating point in the state three times the forced excitation of the generator.

On Line Segment ab,

$$L_\beta = \frac{d\psi}{di_f} = \tan\beta$$

Thus, Eq. (14.36) can be replaced by two equations with a constant inductance value.
On Line Segment oa, $0 < i_f < I_{f0}$.

$$L_\alpha = \frac{di_f}{dt} + (R + R_f)i_f = 0 \tag{14.37}$$

On Line Segment ab, $I_{f0} < i_f < I_{fc}$.

$$L_\beta = \frac{di_f}{dt} + (R + R_f)i_f = 0 \tag{14.38}$$

where

I_{f0} — field flashing current before de-excitation;

I_{fc} — forced excitation current.

When Eq. (14.37) is integrated between 0 and I_{f0}, the following is obtained:

$$i_f = I_{f0} e^{-\frac{t}{T_\alpha}}$$
$$T_\alpha = \frac{L_\alpha}{R + R_f} \tag{14.39}$$

where

T_α — equivalent de-excitation time constant.

When $I_{f0}/I_{fc} = m$ is substituted into Eq. (14.39), the current expression for Segment oa of the magnetization curve can be obtained:

$$i_f = mI_{fc} e^{-\frac{t}{T_\alpha}} \tag{14.40}$$

For Segment ab of the magnetization curve, the current varies within the range I_{f0}–I_{fc}. The corresponding current variation expression can be obtained from the integral Eq. (14.38):

$$i_f = I_{fc} e^{-\frac{t}{T_\beta}} \tag{14.41}$$

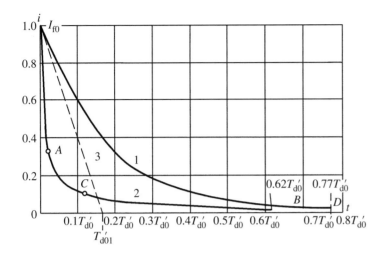

Figure 14.22 Current attenuation characteristics of linear resistor de-excitation system when influence of inductance saturation is taken into account: 1 – inductance is constant $K = 5$; 2 – inductance is variable $K = 5$; 3 – current attenuates linearly $K = 5$.

Under the assumed linearization conditions, the attenuation of the current in the de-excitation process can be divided into two sections: the section from I_{fc} to I_{f0} and the section from I_{f0} to I_{fc}/N at the end of the de-excitation. Generally, $N = 100$, and the total de-excitation time is the sum of the times of the two sections. The time t_1 required for the current's attenuation from I_{fc} to I_{f0} is determined first. From Eq. (14.41), when $i_f = I_{f0}$. When the natural logarithm of both sides of the equation is taken, the corresponding de-excitation time can be obtained:

$$t_1 = T_\beta \ln \frac{1}{m} \tag{14.42}$$

Then, the time t_2 required for the current's attenuation from I_{f0} to I_{fc}/N is determined. The following can be obtained from Eq. (14.40):

$$t_2 = T_\alpha \ln Nm \tag{14.43}$$

The total de-excitation time is as follows:

$$t_1 + t_2 = T_\beta \ln \frac{1}{m} + T_\alpha \ln Nm \tag{14.44}$$

When $N = 100$, $m = 0.38$, $K = 5$ and $\frac{T_\alpha}{T_\beta} = 25$ and $T_\alpha = \frac{T'_{d0}}{K+1}$ is taken into account, the following can be obtained:

$$t_m = \frac{T'_{d0}}{25 \times 6} \times \ln \frac{1}{0.38} + \frac{T'_{d0}}{6} \times \ln(10.0 \times 0.38) = 0.618 T'_{d0}$$

where $T'_{d0} = \frac{L_0}{R_f}$ and L_0 is the inductance corresponding to the linear part of the magnetization curve. The corresponding de-excitation curves are shown in Figure 14.22.

14.5 Influence of Damping Winding Circuit on De-excitation [37]

For an analysis of the de-excitation characteristics of a large synchronous generator, the presence and influence of the damping winding circuit should be taken into account. This is especially true for a steam turbine. In the de-excitation process, part of the energy in the generator excitation winding is transferred to the damping winding circuit. This on the one hand relieves the de-excitation energy burden on the field circuit breaker

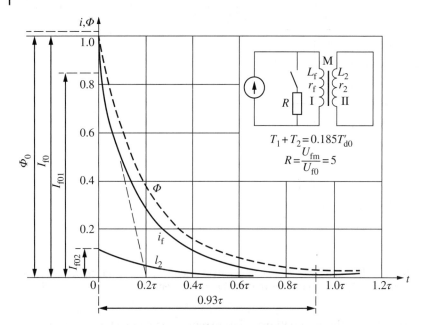

Figure 14.23 Circuit diagram of linear resistor de-excitation system when influence of damping winding is taken into account and current variation curves in excitation winding and damping winding in de-excitation process.

and on the other hand increases the de-excitation time due to the presence of the damping winding. The influence of the damping circuit on the de-excitation characteristics is discussed below on the basis of the specific de-excitation connection.

The de-excitation connection is shown in Figure 14.23. The linear de-excitation resistor R is connected at the time of the de-excitation.

In the normal case, the exciter supplies power to the excitation winding circuit I (L_1 and r_1). The mutual inductance between the excitation winding circuit and the damping winding circuit II (L_2 and r_2) is M. When the coupling coefficient is equal to 1, $M = \sqrt{L_1 L_2}$. Thus, the equations of the de-excitation process are as follows:

$$L_1 \frac{di_1}{dt} + \sqrt{L_1 L_2} \frac{di_2}{dt} + r_1 i_1 = 0 \tag{14.45}$$

$$L_2 \frac{di_2}{dt} + \sqrt{L_1 L_2} \frac{di_1}{dt} + r_2 i_2 = 0 \tag{14.46}$$

At the time of the de-excitation, the de-excitation resistor R is connected. So, r_1 in Eq. (14.45) is the total resistance in Circuit I, including the excitation winding resistance and the de-excitation resistance. That is, $r_1 = r_f + R$. When Eqs. (14.45) and (14.46) are integrated, the variations in the currents (i_1 and i_2) in the excitation and damping windings in the de-excitation process can be obtained:

$$i_1 = I_{f0} \times \frac{T_1}{T_1 + T_2} \times e^{-\frac{t}{T_1 + T_2}} \tag{14.47}$$

$$i_2 = I_{f0} \times \frac{T_1}{T_1 + T_2} \times \sigma e^{-\frac{t}{T_1 + T_2}}$$

$$\sigma = \frac{r_1}{r_2} \sqrt{\frac{L_2}{L_1}}, I_{f0} = \frac{U_{f0}}{r},$$

$$T_1 = \frac{L_1}{r_1}, T_2 = \frac{L_2}{r_2}, T'_{d0} = \frac{L_1}{r} \tag{14.48}$$

where

I_{f0} — field flashing current before de-excitation;
T_1 and T_2 — time constants of Circuit I and II, respectively;
T'_{d0} — time constant of excitation winding.

In addition, the total flux of the generator excitation winding circuit is

$$\Phi = \Phi_0 e^{-\frac{t}{T_1+T_2}} \qquad (14.49)$$

where

Φ_0 — initial total flux.

The parameter selected for the excitation circuit:

$$K = \frac{U_{fm}}{U_{f0}} = \frac{R}{R_f} = 5$$

Thus, the time constant of the excitation winding circuit I is as follows:

$$T_1 = \frac{L_1}{R+R_f} = \frac{L_1}{R_f(K+1)} = \frac{T'_{d0}}{6} = 0.167 T'_{d0}$$

$$T_2 = \frac{L_2}{r_2} = 0.032 T'_{d0}$$

$$T_1 + T_2 = 0.199 T'_{d0}$$

At the instant when the de-excitation resistor is connected and the de-excitation begins, the step change of the current in the excitation winding circuit (I) from the initial value (I_{f0}) is as follows:

$$I_{f01} = I_{f0} \times \frac{T_1}{T_1 + T_2} \qquad (14.50)$$

So, $I_{f01} = I_{f0} \times \frac{0.167 T'_{d0}}{0.199 T'_{d0}} = 0.84 I_{f0}$. Then, the current i_1 attenuates in an exponential curve with a time constant of $(T_1 + T_2)$.

For the damping winding circuit (II), at the instant when the de-excitation begins, the step change of the current from zero is as follows:

$$I_{f02} = I_{f0} \times \frac{T_1}{T_1 \times T_2} \times \sigma \qquad (14.51)$$

Then, it is attenuated exponentially with a time constant of $(T_1 + T_2)$. The flux Φ is also attenuated exponentially with a time constant of $(T_1 + T_2)$ from Φ_0 at the instant of the de-excitation.

As the previous definition of the de-excitation time, it is assumed that the de-excitation process ends when the flux is attenuated from Φ_0 to $\Phi = \frac{1}{100} \times \Phi 0$. Thus, the de-excitation time can be obtained:

$$t_m = (T_1 + T_2) \ln \frac{\Phi_0}{\Phi} = (T_1 + T_2) \ln 100 = (0.166 + 0.032) T'_{d0} \times \ln 100 = 0.93 T'_{d0}$$

It can be obtained from Eq. (14.52) that the de-excitation time t_m of the linear resistor system without a damping winding is $0.76 T'_{d0}$. The de-excitation time with a damping winding is extended by about 21%. The current variations in the excitation and damping windings in the de-excitation process are shown in Figure 14.23.

14.6 Field Circuit Breaker

There are many kinds of field circuit breakers for excitation systems of large synchronous generators, but a trend has virtually set in regarding selection of field circuit breakers through years of accumulated experience in their application and operation.

It should be noted that almost all Chinese and international excitation equipment manufacturers use standardized general-purpose fast DC circuit breakers as field circuit breakers for excitation systems of large hydro, thermal, and nuclear power units. Strictly speaking, such fast DC circuit breakers are not specifically designed for field applications, but when the DC circuit breaker is chosen to be used as the field circuit breaker, the designer should fully integrate the characteristics of both to meet the functional requirements of the field circuit breaker.

14.6.1 DC Field Circuit Breaker

Generally, for large hydropower and thermal units, the following general-purpose fast DC circuit breakers (contactors) are used as field circuit breakers.

14.6.1.1 CEX Series of Modular DC Contactors from France's Lenoir Elec

Structurally, for the CEX series of modular DC contactors, all contacts of the contactor, including the main contact, arc contact, and normally closed contact, are installed on the same shaft, so as to achieve the function of coordinated action. Besides, the DC contactors' multi-break combinations can meet different arc voltage requirements at the time of de-excitation.

The earliest application of the field circuit breaker composed of the CEX series of DC contactors is a 700 MW hydro-turbine of Three Gorges Hydro Power Station. The model is CEX98 5000 4.2. Its rated voltage is 2000 V and rated current is 5000 A. The maximum arc voltage generated at the rated voltage is up to 4000 V. Its configuration structure and main circuit are shown in Figure 14.24 and the elementary wiring diagram in Figure 14.25.

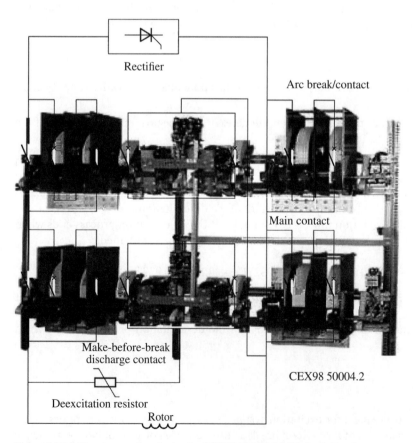

Figure 14.24 Structure and main circuit of CEX98 5000 4.2 field circuit breaker.

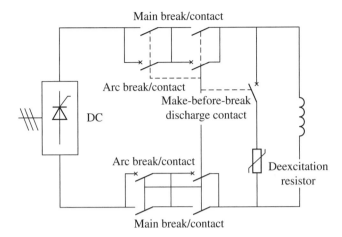

Figure 14.25 Principal wiring diagram of CEX98 5000 4.2 field circuit breaker.

To meet the requirements of the rated excitation voltage and current and reduce the installation space, Three Gorges Hydro Power Station adopts two sets of CEX98-2.1 DC contactors. Each pole of the excitation winding is equipped with a set of CEX98-2.1 DC contactors, which are installed upright to reduce the cabinet width.

On such a field circuit breaker composed of the CEX98 series of DC contactors, the main contact has no arc chute. The arc contact and the discharge contact as normally closed de-excitation contacts have a metal short arc chute.

A shortcoming of a field circuit breaker of the CEX98 series is that the arc chute of the arc contact is small. Thus, if the de-excitation commutation is not initially successful, the arc burning of the break of the main contact will continue or the arc contact will be burned. Another shortcoming of this field circuit breaker is its large size, which necessitates a large installation space.

14.6.1.2 HPB Series of Fast DC Circuit Breakers from Switzerland's Secheron

HPB 81 S or HPB 82 S DC circuit breakers are generally used as field circuit breakers for large hydropower and thermal power units. The corresponding rated voltages are 1000 and 2000 V, rated currents 4500 and 6000 A, and short-circuit current cutoff capacities 100 and 75 kA, respectively. On the basis of practical application experience, their maximum arc voltage is 1.5–2.0 times the rated operating voltage.

The HPB series of DC circuit breakers is widely used in large hydropower and thermal power units. The structure of a HPB series DC circuit breaker is shown in Figure 14.26.

It should be noted that a HPB series field circuit breaker has a large arc extinguish chamber. When the de-excitation commutation conditions are not met initially, the arc striking at the break of the main contact will continue. When the generator rotor excitation current drops to meet the de-excitation commutation conditions, the commutation is successful, and the break of the main contact is cut off. So, when a fast DC circuit breaker is selected as a field circuit breaker, a large arc extinguish chamber is one of the priorities. It should be emphasized that although the HPB series field circuit breaker adopts a metal arc chute (the space between two adjacent chute sheets is 4–5 mm) and works under the short arc working principle, the upper part of the metal arc chute is equipped with an insulating spacer. When a small current flows, the arc passes through the metal sheets. Thus, short arc de-excitation characteristics will be shown. Under high-current conditions, the arc is sucked between the upper insulating chute sheets and elongated. Thus, long arc de-excitation characteristics will be shown.

14.6.1.3 UR Series of DC Circuit Breakers from Switzerland's Secheron

The HPB and UR series from Secheron are fully interchangeable with the arc chute type from the structural standpoint.

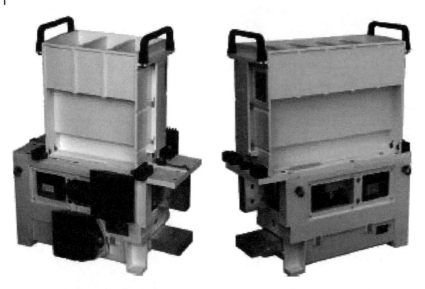

Figure 14.26 Structure of HPB series of DC circuit breakers.

Figure 14.27 Structure of UR series DC circuit breaker.

For the UR series of DC circuit breakers, its rated DC currents include grades such as 2600, 3600, 4000, 6000, and 8000 A. Its rated operating DC voltages include 900 V (for 81S), 1800 V (for 82S arc chute), and 3600 V (for 64 type arc chute).

The structure of a UR series DC circuit breaker is shown in Figure 14.27.

It should be noted that, in DC field circuit breakers for excitation systems of large hydropower and thermal power units, the HPB series is the most widely applied. However, in recent years, some excitation systems of large hydropower and thermal power units have seen many fractures of the 5400 plastic part in the closing

Figure 14.28 Example of fracture of plastic part of closing mechanism of main contact of HPB series fast DC circuit breaker.

mechanism of the HPB series. In a HPB series fast DC circuit breaker, the fracture of the 5400 plastic part will cause inadequate closing of the main contact of the circuit breaker, seriously affecting the operational safety of the excitation system. An example of a typical fracture of a HPB series fast DC circuit breaker is shown in Figure 14.28.

In view of this, some operators have inspected or maintained or replaced their existing HPB series DC circuit breakers.

14.6.1.4 Gerapid Series of Fast DC Circuit Breakers

(1) *Basic parameters and performance characteristics.* The Gerapid series of fast DC circuit breakers achieves high breaking capacity based on new technologies and new materials featuring high insulation under the modular design principle. Its basic technical data are shown in Table 14.2.

The Gerapid series features a compact size. With the same body width size (about 700 mm), it can form DC circuit breakers of different models and specifications. Its rated current is up to 10 000 A, its rated voltage is up to 4000 V, and its performance meets multiple standards such as IEC 947-2, EN 50123-2,

Table 14.2 Technical data of Gerapid series of DC circuit breakers.

Item	Standard operating voltage	Arc chute type	2607	4207	6007	8007
Rated current meeting applicable standard (A)	IEC947-2 EN50123-2 ANSIC37.14	—	2 600 2 600 2 600	4 200 4 200 4 150	6 000 6 000 5 000	8 000 8 000 6 000
Maximum test breaking capacity under specified operating voltage and selected arc chute conditions I_{ccmax} (kA)	DC 1 000 V DC 2 000 V DC 2 000 V DC 3 000 V DC 3 600 V DC 3 600 V	1×2 1×4 2×2 2×3 2×4 EF4–12	244 50 100 50 52 176	244 50 100 50 52 176	200 50 100 50 a) — —	200 — 100 a) — —
Mechanical life (minimum number of maintenance operations)	—	—	50 000–100 000	50 000–100 000	50 000–100 000	30 000–100 000

a) Indicates that this data should be determined in consultation with the manufacturer.

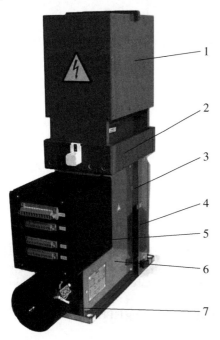

Figure 14.29 External structure of Gerapid series fast DC circuit breaker: 1 – arc chute; 2 – arc chute base; 3 – insulating side panel (optional); 4 – instantaneous trip [adjustable mechanical action (optional)]; 5 – electronic control device; 6 – circuit breaker body with drive mechanism and contact system; 7 – electromagnetic drive mechanism (quick action).

and ANSI C37.14. After the circuit breaker is closed, it is maintained by the mechanical lock, without the need for additional power supply support. Thus, it delivers high reliability.

A Gerapid series fast DC circuit breaker has an electromagnetic trip device applicable to all current directions, whose set value is fixed or adjustable. In terms of accessories, it boasts a perfect built-in secondary control module unit, without the need for additional secondary control units such as closing/opening. In addition, it features a two-stage contact system, where the main contact and the arc contact are separated, as well as a modular arc chute design, which facilitates maintenance and improves arc breaking capacity, enabling a maximum arc breaking voltage of 8000 V.

(2) *Structural characteristics.* The external structure of a Gerapid series fast DC circuit breaker is shown in Figure 14.29, and its internal structure is shown in Figure 14.30.

The structural characteristics of the Gerapid series are as follows:

1) *Two-stage contact system where the main contact and arc contact are separated.* A Gerapid series fast DC circuit breaker features a two-stage contact system where the main contact and arc contact are separated. Since the operation sequence of "closing of arc contact before breaking of main contact" is applied at the time of breaking, the possibility of burnout or welding of the main contact is extremely small. In the contact maintenance, the arc chute can be easily and quickly removed, without the need for movement of the circuit breaker body. If necessary, only the arc contact or the arc strike conductor arc angle or the protective baffle needs to be replaced. The structure of the main contact and the arc contact is shown in Figure 14.31.

2) *Arc chute.* A Gerapid series fast DC circuit breaker features a compact modular arc chute. Structurally, it achieves a whole arc quenching process without the need for an additional blower coil or permanent magnet.

In the case of the same DC circuit breaker body size, depending on the arc chute configuration specifications, the arc voltage can be maintained from 1000 V to a maximum of 8000 V. The corresponding arc chute structure is shown in Figure 14.32.

It should be noted that the arc chute of the Gerapid series fast DC circuit breaker is arranged in two layers (the upper layer and the lower layer). The distance between two adjacent chute sheets is 4–5 mm. The lower part accommodates the metal chute sheets, and the upper part accommodates the insulating spacer. In the case of a small current, the arc passes through the metal arc chute to form a short arc system. In the

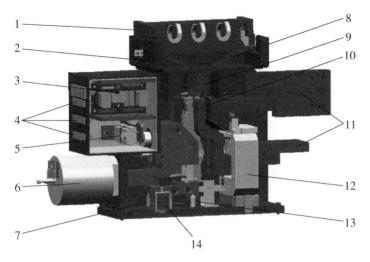

Figure 14.30 Internal structure of Gerapid series fast DC circuit breaker: 1 – arc chute; 2 – base; 3 – electronic control device; 4 – control circuit terminal strip; 5 – auxiliary contact; 6 – electromagnetic drive mechanism; 7 – mechanical forced trip; 8 – arc strike conductor; 9 – arc contact; 10 – main contact; 11 – main circuit terminal; 12 – instantaneous quick trip; 13 – quick trip; 14 – shunt trip (for normal operation).

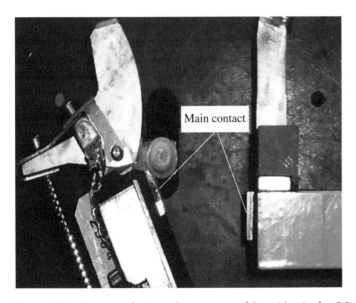

Figure 14.31 Structure of main and arc contacts of Gerapid series fast DC circuit breaker.

case of a high current, the arc goes from the metal arc chute into the insulating spacer to form a long arc system, conductive to arc elongation and extinguishing. The arrangement of the arc chute sheets is shown in Figure 14.33. In the figure, the lower part accommodates the metal arc chute sheets, and the upper part accommodates the insulating spacer.

3) *Electromagnetic drive mechanism and integrated control unit*. The electromagnetic drive mechanism closing time is 150 ms. The minimum action time can be 100 ms. The power is cut off automatically about 400 ms after the closing, without the need for an auxiliary power supply. Meanwhile, the characteristic of trip prevention is delivered. The electromagnetic drive mechanism is applicable to all permitted standard voltage power supply systems.

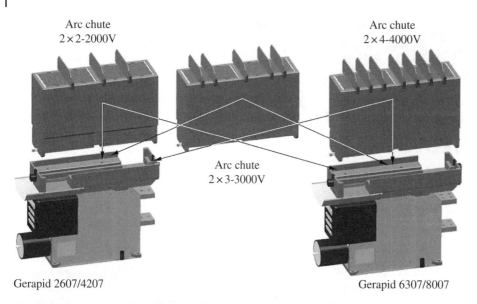

Figure 14.32 Structure of arc chute of Gerapid series fast DC circuit breaker.

Figure 14.33 Arrangement of arc chute sheets of Gerapid series fast DC circuit breaker.

4) *Trip*.
   a. Normal tripping, including shunt tripping, power-off tripping, and power-off tripping with voltage control, whose action time is 20–40 ms;
   b. Protection (overcurrent) instantaneous tripping, whose action time depends on the protection setting;
   c. Emergency instantaneous tripping, whose action time is 3–5 ms in the case of failure of the first protection signal;
   d. Mechanical forced tripping, which is a local manual tripping in the case of loss of all power supplies.
5) *DC current measurement system SEL*. A Gerapid series fast DC circuit breaker features a built-in current measurement system that can replace the conventional shunt. The output signal passing interface is

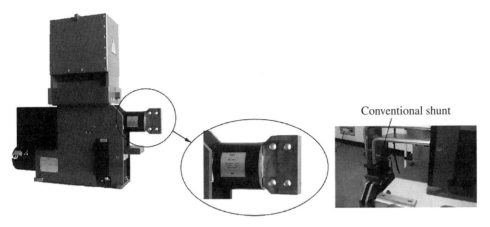

Figure 14.34 Structure of SEL current measurement system of Gerapid series.

applicable to 4–20 mA, ±20 mA, and ±10 V meter systems. The accessory needs no additional assembly accessory or installation space or any additional measurement noise reduction or filtering system element. The current measurement range is up to 6–12 kA and the voltage applied up to 4000 V. The structure of the DC measurement system is shown in Figure 14.34.

It should be noted that the Gerapid series of fast DC circuit breakers has been used as field circuit breakers for the excitation systems of the 1407 MVA units of Sanmen Nuclear Power Station. Since the generator excitation current is greater than 8000 A, two Gerapid series fast DC circuit breakers are used in parallel.

14.6.2 AC Field Circuit Breaker

14.6.2.1 AC Air Circuit Breaker

For AC field circuit breakers between the AC power supply output side of the excitation system and the AC input side of the power rectifiers, the ABB Emax series of air circuit breakers is generally used. The corresponding electrical parameters are shown in Table 14.3. The Emax series of air circuit breakers can be used in AC systems with an AC voltage of up to 1150 V. The types of structure include fixed type and drawout type as well as three-pole and four-pole types. The test voltage of an Emax series circuit breaker is 1250 V.

14.6.2.2 AC Disconnector

On the basis of the Emax AC standard circuit breaker, disconnections applicable to AC voltages of up to 1150 V have been developed. The series product model is Emax/E MS. The structure types also include the fixed type and drawout type as well as the 3-pole and 4-pole types. This accessory is interchangeable with the Emax series of circuit breakers. The standard fixed part is also applicable to 1150 V drawout circuit breakers. The electrical parameters of the Emax/E MS series of AC disconnectors are shown in Table 14.4. The test voltage of the series is 1250 V.

14.6.2.3 DC Disconnector

In some cases, a generator excitation circuit adopts a single-pole DC field circuit breaker, for example, one connected to the anode side of the excitation winding. A DC disconnector can be connected to the cathode, so that the circuit can be simplified. An Emax/E MS AC disconnector can also be used as a DC disconnector. If the disconnector's four-pole contacts are connected in series, it can be used for 1000 V DC/6300 A DC systems. It should be noted that this product can be equipped with the same accessories as the standard circuit breaker, but it should not be equipped with electronic trips, current sensors, or other current measurement or AC application accessories. The electrical parameters of the Emax/E MS series of DC disconnectors are shown in Table 14.5.

Table 14.3 Electrical parameters of Emax series of AC air circuit breakers.

Model		E2B/E		E2N/E				E3H/E				E4H/E		E6H/E	
Rated uninterrupted current (40 °C) I_u (A)		1600	2000	1250	1600	2000	1250	1600	2000	2500	3200	3200	4000	5000	6300
Rated operating voltage U_e (V)		1150	1150	1150	1150	1150	1150	1150	1150	1150	1150	1150	1150	1150	1150
Rated insulation voltage (V)		1250	1250	1250	1250	1250	1250	1250	1250	1250	1250	1250	1250	1250	1250
Rated ultimate short-circuit current breaking capacity I_{cu}	1000 V (kA)	20	20	30	30	30	50	50	50	50	50	65	65	65	65
	1150 V (kA)	20	20	30	30	30	30	30	30	30	30	65	65	65	65
Rated operating short-circuit current breaking capacity I_{cs}	1000 V (kA)	20	20	30	30	30	50	50	50	50	50	65	65	65	65
	1150 V (kA)	20	20	30	30	30	30	30	30	30	30	65	65	65	65
Rated short time withstand current I_{cw} (1 s)	1000 V (kA)	20	20	30	30	30	50	50	50	50	50	65	65	65	65
	1150 V (kA)	20	20	30	30	30	30	30	30	30	30	65	65	65	65
Rated short-circuit making capacity (peak) I_{cm}	1000 V (kA)	40	40	63	63	63	105	105	105	105	105	143	143	143	143
	1150 V (kA)	40	40	63	63	63	63	63	63	63	63	143	143	143	143

Table 14.4 Electrical parameters of Emax/E MS series of AC disconnectors.

Model	E2B/E MS	E2N/E MS	E3H/E MS	E4H/E MS	E6H/E MS
Rated uninterrupted current (40 °C) I_u (A)	1600	1250	1250	3200	5000
	2000	1600	1600	4000	6300
	—	2000	2000	—	—
	—	—	2500	—	—
	—	—	3200	—	—
Number of poles	3/4	3/4	3/4	3/4	3/4
Rated operating voltage U_e (V)	1150	1150	1150	1150	1150
Rated insulation voltage U_i (V)	1250	1250	1250	1250	1250
Rated impulse withstand voltage U_{imp} (kV)	12	12	12	12	12
Rated short time withstand current I_{cw} (1 s) (kA)	20	30	30	65	65
Rated short-circuit making capacity 1000 V AC (peak I_{cm}) (kA)	40	60	105	143	143

Table 14.5 Electrical parameters of Emax/E MS series of DC disconnectors.

Model		E1B/E MS	E2N/E MS	E3H/E MS	E4H/E MS	E6H/E MS
Rated uninterrupted current (40 °C) I_{cu} (A)		800	1250	1250	3200	5000
		1250	1600	1600	4000	6300
		—	2000	2000	—	—
		—	—	2500	—	—
		—	—	3200	—	—
Number of poles		3/4	3/4	3/4	3/4	3/4
Rated operating voltage U_e (V)		750/1000	750/1000	750/1000	750/1000	750/1000
Rated insulation voltage U_i (U)		1000/1000	1000/1000	1000/1000	1000/1000	1000/1000
Rated impulse withstand voltage U_{imp} (kV)		12/12	12/12	12/12	12/12	12/12
Rated short time withstand current I_{cw} (1 s) (kA)		25/20	40/25	50/40	65/65	65/65
Rated short-circuit making capacity I_{cm}	750 V DC (kA)	42/42	52.5/52.5	105/105	143/143	143/143
	1000 V DC (kA)	–/42	–/52.5	–/105	–/143	–/143

The use of an AC vacuum circuit breaker as an AC-side field circuit breaker should be avoided as much as possible, because an AC vacuum circuit breaker has a small clearance between the open main contacts at the time of a break, which will cause extremely high current gradient changes and an extremely high transient overvoltage in the generator excitation winding, endangering the insulation safety.

14.7 Performance Characteristics of Nonlinear De-excitation Resistor [38]

Through years of exploration and practice, large hydropower, thermal, and nuclear power units in China and other countries all adopt SiC nonlinear resistors as de-excitation resistors.

In China, small- and medium-sized hydro and thermal power units, especially 300 MW or smaller ones, generally adopt ZnO nonlinear resistors as de-excitation resistors, with good results.

This section describes the performance characteristics of the widely used Metrosil series of SiC nonlinear resistors from the United Kingdom's M&I.

14.7.1 U-I Characteristic Expression for Single Valve

A nonlinear resistor's U-I characteristic expressions have been shown in Eqs. (14.25) and (14.26).

For an SiC nonlinear resistor, it is fairly inconvenient to represent its characteristics with a U-I coordinate system but more convenient with a logarithmic coordinate system in engineering applications.

The logarithm is taken on both sides of Eq. (14.25). The expression is as follows:

$$\lg U = \lg C + \beta \lg I \tag{14.52}$$

It can be seen from Eq. (14.52) that the U-I characteristic of the nonlinear resistor is approximately a straight line in the double logarithmic U-I coordinate system, as shown in Figure 14.35.

For Eq. (14.52), when $I = 1$ A, the following equation can be obtained:

$$\lg U = \lg C \tag{14.53}$$

or

$$U = C$$

Equation (14.52) indicates that in the U-I double logarithmic coordinate system, Coefficient C is equal to the voltage at which the U-I characteristic curve intersects Axis U when $I = 1$ A.

Similarly, the following equation can be obtained from Eq. (14.26):

$$\lg I = \lg H + \alpha \lg U \tag{14.54}$$

When $U = 1$, in the U-I characteristic curve, the corresponding current $I = H$.

The relation between the coefficients C, H, α, and β in Eqs. (14.25) and (14.26) is discussed below. The following equation can be obtained from Eq. (14.26):

$$U = \sqrt[\alpha]{\frac{I}{H}} = H^{-\frac{1}{\alpha}} I^{\frac{1}{\alpha}} \tag{14.55}$$

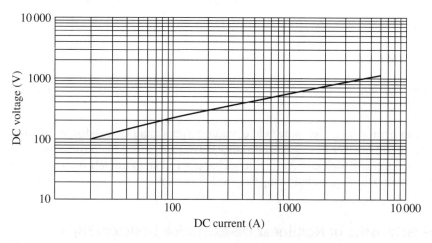

Figure 14.35 Typical U-I characteristic curve of SiC nonlinear resistor.

Table 14.6 Typical C and β of Metrosil series of SiC nonlinear resistors.

Valve thickness (mm)	C (for single valve)	β (for single valve)
20	40–250	0.5–0.3
15	106	0.4
11.25	75	0.4
7.5	53	0.4

When Eq. (14.25) is equal to Eq. (14.55), the following equation can be obtained:

$$C = H^{-\frac{1}{\alpha}}$$

$$\beta = \frac{1}{\alpha}$$

Then, the following can be obtained:

$$\alpha = \frac{1}{\beta}$$

$$C = H^{-\beta} \tag{14.56}$$

Typical C and β values of the Metrosil series of SiC de-excitation resistors are shown in Table 14.6.

Different valve thicknesses, connection modes, and C and β values can be selected to meet the specific requirements for de-excitation resistor parameters under various de-excitation conditions.

14.7.2 U-I Characteristic Expression for Elements

In order to meet the requirements for the total capacity of de-excitation, resistor chips should be connected in series and parallel in practical applications. The U-I characteristic expression for multiple resistors is different from that for a single one.

(1) *In series.* It is assumed that the U-I characteristic expression for a single nonlinear resistor chip is as follows:

$$U = CI^{\beta}$$

When N_S resistor chips are connected in series, their U-I characteristic expression is as follows:

$$U = N_S CI^{\beta} \tag{14.57}$$

(2) *In parallel.* It is assumed that the U-I characteristic expression for a single nonlinear resistor chip is as follows:

$$U = CI^{\beta}$$

When N_P nonlinear resistor chips are connected in parallel, their U-I characteristic expression is as follows:

$$U = C\left(\frac{1}{N_P}\right)^{\beta} \tag{14.58}$$

(3) *In series and parallel.* When N_P nonlinear resistor chips are connected in parallel and N_S ones in series, their U-I characteristic expression is as follows:

$$U = CN_S\left(\frac{1}{N_P}\right)^{\beta} \tag{14.59}$$

14.7.3 Temperature Coefficient of SiC Nonlinear Resistor

For an SiC nonlinear resistor, its material decides its resistance temperature characteristic – its resistance will decrease when its temperature increases. Its voltage at both ends decreases with as its temperature increases at a constant current load or its current increases as its temperature increases at a constant voltage load. In sum, it shows a negative resistance temperature coefficient characteristic. The negative temperature coefficient is that its current increases 0.6% for every 1 °C increase in its temperature for a constant current load, and its voltage decreases 0.12% for every 1 °C increase in its temperature for a constant voltage load.

In fact, there are two applications for the Metrosil series of SiC nonlinear resistors:

(1) A Metrosil series SiC nonlinear resistor as a surge-voltage-absorbing element: The applied voltage is constant. The SiC resistor continuously absorbs the energy. Due to the presence of the leakage current, the temperature will increase. In the design of the SiC resistor cooling scheme, it should be considered that the heat energy generated by the leakage current will be balanced by the natural cooling effect of its element. Thus, the influence of the negative resistance temperature coefficient is compensated for.

(2) A Metrosil series SiC nonlinear resistor as a de-excitation resistor: It is connected to the generator excitation circuit only when the de-excitation circuit acts. So, before the de-excitation acts, the influence of the negative resistance temperature coefficient does not matter. In the few seconds of the de-excitation process, the influence of the negative resistance temperature coefficient is very small and negligible. First, in the de-excitation current attenuation process, the voltage waveform at both ends of the SiC de-excitation resistor does not show a constant voltage characteristic as in the case of a ZnO de-excitation resistor but is a triangular wave that decreases as the de-excitation current decreases. Thus, the influence of the negative resistance temperature coefficient is reduced. Meanwhile, at the time of de-excitation, the energy of the excitation current source is limited, and there is no follow-up energy input. So, the so-called current collapse caused by the influence of the negative resistance temperature coefficient is impossible.

The capacity of an SiC resistor for a de-excitation circuit should be correctly designed to ensure that it absorbs the de-excitation energy while preventing the final temperatures of the elements from exceeding the allowable values under the conditions for natural heat dissipation of the elements.

14.7.4 Calculation of Temperature Rise

For the Metrosil series of SiC de-excitation resistors, the specific heat coefficient is about $0.84\,\text{J}\,(\text{g}\,°\text{C})^{-1}$, and the element density about $2.35\,\text{g}\,\text{cm}^{-3}$. Based on these essential parameters, the temperature rise of the SiC de-excitation resistor in the de-excitation process can be estimated.

For example, for the Metrosil series of SiC 600-A/US16/P/Spec 6298 elements, each element has 16 pieces connected in parallel. For a set of de-excitation resistors, six elements are connected in parallel, and then they are connected in series with a six-element parallel set. The total number of pieces of the set is $16 \times 6 \times 2 = 192$. The element's diameter is 152 mm, thickness is 15 mm, and the center hole size is $\phi 20$. Thus, the total volume of the set of elements can be obtained as follows:

$$V = 192 \times \pi(7.6^2 - 1.0^2) \times 1.5 = 51.33\ (\text{cm}^3)$$

The total mass of the set of elements can be obtained as follows:

$$W = 51.33\,\text{cm}^3 \times 2.35\,\text{g/cm}^3 = 120.6\ \text{kg}$$

The specific heat $\rho = 0.84\,\text{J}\,(\text{g}\,°\text{C})^{-1}$ indicates that the energy required for every 1 °C increase in the temperature per gram of SiC de-excitation resistor is 0.84 J. For the above SiC de-excitation resistor whose total element weight is 118.4 kg, the energy required for every 1 °C increase in its temperature is calculated as follows:

$$W_E = 0.84 \times 122.8 = 103.15\ (\text{kJ}/°\text{C})$$

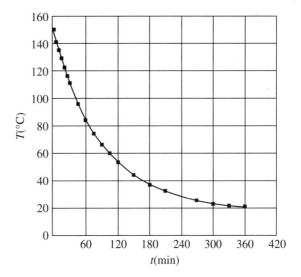

Figure 14.36 Cooling characteristic curve of Metrosil series of SiC nonlinear resistor elements [600A/US14/P (20 mm thick), 600A/US16/P (15 mm thick) and similar sizes].

If the calculation is based on a de-excitation system and the de-excitation capacity in the event of a sudden three-phase short circuit of the generator is 6554 kJ, the final temperature rise of the de-excitation resistor is calculated as follows:

$$\Delta T = 6554 \div 103.15 = 63.54 \; (°C)$$

The room temperature is 25 °C. At this point, the temperature of the de-excitation resistor is 63.54 °C.

The typical cooling curve of the Metrosil series of SiC nonlinear resistor elements is shown in Figure 14.36.

As can be seen from Figure 14.36, it takes approximately 6 h for an element to decrease from the maximum allowable temperature to the room temperature.

14.7.5 De-excitation Time

The de-excitation time expression is shown in Eq. (14.31):

$$t_m = \frac{T'_{d0}}{1-\beta} \ln\left(1 + \frac{1}{K}\right)$$

where K – overvoltage multiple, defined as

$$K \frac{U_{fm}}{U_{f0}} = \frac{I_{f0} R_0}{I_{f0} R_f}$$

In the equation, U_{fm} is the voltage drop generated at the instant of the de-excitation commutation when the initial current I_{f0} flows through the nonlinear resistor.

14.7.6 Parameter Selection for SiC Nonlinear Resistor

A user can select the parameters of a SiC nonlinear resistor on the basis of the following points:

(1) *Maximum de-excitation current*. The rotor non-periodic component amplitude current caused by a sudden three-phase short circuit of the generator can be taken as the maximum de-excitation current. Generally, this current is approximately equal to three times the generator rated excitation current.
(2) *Maximum de-excitation voltage*. It is equal to the residual voltage generated at both ends of the de-excitation resistor when the maximum de-excitation current flows through the de-excitation resistor.
(3) *Maximum de-excitation capacity*. It is usually equal to the magnetic field energy generated in the generator rotor excitation winding in the event of a sudden three-phase short circuit of the generator.

In addition to the above-mentioned method to determine the maximum de-excitation capacity in the event of a sudden three-phase short circuit of the generator, there are de-excitation methods such as generator unloaded, rated, and forced de-excitation states. In the event of tripping of the main circuit breaker of the generator stator circuit, under the principle of conservation of stator and rotor flux linkage, the rotor current will decrease to a level close to the no-load air gap excitation current. The de-excitation capacity is approximately equal to the no-load de-excitation capacity. For the no-load faulty forced de-excitation method, the generator terminal overvoltage protection acts after a 0.3 s delay after the generator voltage rises to 1.3 times the rated voltage and the field circuit breaker trips. At this point, the de-excitation current is higher than the rated excitation current, usually less than the forced excitation current. It should be emphasized that the no-load faulty forced excitation voltage protection set value has been changed from action after 0.3 s delay at a voltage 1.3 times rated voltage to action after 0.2 s delay at a voltage 1.2 times rated voltage for hydro-turbines. This change has greatly improved the operating conditions of field circuit breakers. There have been similar changes in the no-load faulty forced excitation overvoltage set value for thermal power units.

14.7.7 Timeliness of SiC Nonlinear Resistor

The experience in the operation of the Metrosil series of SiC nonlinear resistors produced by M&I shows that there is only a minor change in the U-I characteristic, and the change does not affect the current balance or matching of the branches operating in parallel, or does not affect the overall performance of the whole set of Metrosil series SiC nonlinear resistors. Besides, there is no problem of a decrease in the nominal capacity of the SiC nonlinear resistors after long-term operation. That is, the performance of the Metrosil series of SiC nonlinear resistors is stable and does not change over time. Moreover, the electrical performance of the Metrosil series of SiC nonlinear resistors have stable electrical performance and long time timeliness.

14.7.8 Damage and Failure Forms for SiC Nonlinear Resistor

There are two damage and failure modes for a Metrosil series SiC nonlinear resistor element:

(1) The current flowing through the de-excitation resistor exceeds the current for the specified time limit.
(2) The de-excitation energy absorbed by the de-excitation resistor exceeds the de-excitation capacity of the specified temperature rise.

For the Metrosil series of SiC nonlinear resistors, at the time of de-excitation, the temperature rise depends on the energy absorbed by the elements. For example, the temperature rise of a Spec 6298 600A/US16/P

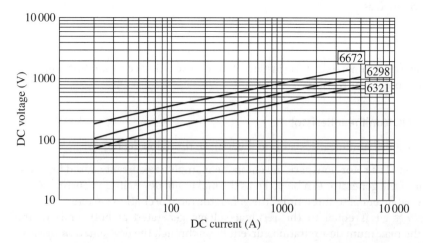

Figure 14.37 U-I characteristic curves of SiC nonlinear resistor elements (600-A/US14/P/Spec 6672, 600-A/US16/P/Spec 6298, and 600-A/US14/P/Spec 6321).

Table 14.7 Specification for Metrosil 600 A series of typical SiC nonlinear resistor elements.

Model	Number of elements in parallel	Element thickness (mm)	Rated current (A)	Maximum voltage at rated current (V)	Maximum voltage at current that is 25% of rated current (V)	Rated energy (kJ)		U-I characteristic curve $U = CI^\beta$	
						For reuse, sufficient cooling time interval should be ensured after each de-excitation (temperature rise is 80 °C at time of de-excitation)	For reuse, sufficient cooling time interval should be ensured after each de-excitation (temperature rise is 105 °C at time of de-excitation)	Occasional (temperature rise at time of de-excitation)	
								C	β
600-A/US14/P/ Spec 6672	14	20	3500	1400	835	780	1000	1250 (130 °C)	60	0.37
600-A/US16/P/ Spec 6298	16	15	4000	1100	635	680	880	1050 (125 °C)	35	0.40
600-A/US14/P/ Spec 6321	16	11	4000	800	460	510	650	820 (128 °C)	25	0.40

The test ground insulation voltage is 5 kV min^{-1}.
The elements' maximum continuous operating temperature is 115 °C, and maximum limit temperature is 160 °C.

nonlinear resistor element will not exceed 80 °C when 680 kJ of de-excitation energy is input, will not exceed 105 °C when 880 kJ of de-excitation energy is input, and will not exceed 130 °C when 1250 kJ of de-excitation energy is input. If the room temperature is 25 °C, the de-excitation resistor operation is safe in any de-excitation mode when the final temperature of the element does not exceed 160 °C.

For the allowable de-excitation energy of a Metrosil series SiC element, its allowable load temperature is taken as a limiting condition. Thus, users can have a clear basis and criterion for selection of the capacity of de-excitation resistors.

If the de-excitation current or de-excitation capacity of a SiC de-excitation resistor element in operation exceeds its limit for some reason, a resistor chip failure may occur. In the form of a "current breakdown," it may break down a resistor chip in the branch and generate an arc at the breakdown point to cause a fault similar to a short circuit. Then, if the current flowing through the fault point is large, the faulty resistor chip will be broken due to overheating. Ultimately, the faulty branch will be put in the open state. This fault process does not affect the normal operation of the resistor chips of the other branches in parallel that are still in the normal state.

However, M&I suggests that in the event of such a fault, the whole set of resistor elements of the faulty parallel branch should be replaced by the original set of spare parts.

It should be noted that for a Metrosil series SiC nonlinear resistor used as a de-excitation resistor, the de-excitation energy injection speed has little effect on the temperature rise under the current de-excitation time (5 s or less) conditions.

14.7.9 Specifications

For a Metrosil series SiC nonlinear resistor, the basic specification is shown in its model name. For example, in the model name 600-A/US14/P/Spec 6672 (1400 V, 3500 A, 1000 kJ), 600 indicates that the diameter of each piece of element of the de-excitation resistor is $\varphi 152$ (6 in.). A indicates that the shape of the element is annular. US14 indicates that each individual element consists of 14 pieces of resistor chips. US indicates that there is no interval between resistor chips. P indicates that 14 resistor chips are connected in parallel in a single element.

Figure 14.37 shows the U-I characteristic curves of de-excitation resistors of three different models.

Table 14.7 shows the specification for the Metrosil 600 A series of typical SiC nonlinear resistor elements.

Figure 14.38 shows the structure of M&I's Metrosil series of SiC nonlinear resistors with a new structure.

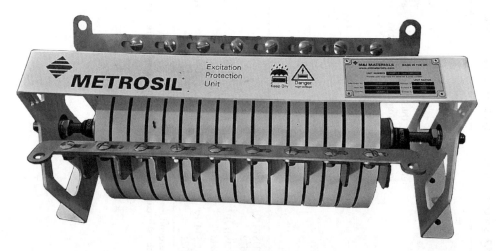

Figure 14.38 Structure of M&I's Metrosil series of SiC nonlinear resistor elements.

15

Excitation System Performance Characteristics of Hydropower Generator Set

15.1 Overview

In recent years, especially since the beginning of the twenty-first century, China has made great progress in hydropower development. In 2002, Three Gorges Hydro Power Station, represented as an iconic hydropower project, put its first 700 MW hydropower unit into operation. As of the end of 2008, a total of 26 units, including 14 on the left bank and the other 12 on the right bank, had been put into operation. At the end of 2010, six sets of 700 MW hydropower units which were located in the under-ground powerhouse were put into operation. By then, Three Gorges Hydro Power Station with a total installed capacity of 22 400 MW had become the world's biggest hydropower station. At the end of July 2014, Xiluodu Hydro Power Station with a total installed capacity of 18×700 MW and Xiangjiaba Hydro Power Station with a total installed capacity of 8×800 MW, which are located in Jinsha River Basin, upstream of Yangtze River, were put into operation as a new batch of hydropower base centers. It should be emphasized that Xiluodu Hydro Power Station has become China's second largest hydropower station second only to Three Gorges Hydro Power Station. Xiangjiaba Hydro Power Station has become China's third largest hydropower station, with the largest per-unit installed capacity (800 MW) among the hydropower units ever built worldwide. In addition, according to the long-term development plan of China Three Gorges Company, Baihetan and Wudongde hydropower stations will also be built upstream of the Yangtze River (downstream of Jinsha River), after Xiluodu and Xiangjiaba hydropower stations. The total installed capacity of Baihetan Hydro Power Station is 16×1000 MW, with the largest per-unit installed capacity (1000 MW) in hydropower units ever built worldwide. Wudongde Hydro Power Station, with a total installed capacity of 12×850 MW, will be put into operation around 2020. At that time, Xiluodu, Xiangjiaba, Baihetan, and Wudongde hydropower stations' total installed capacity will reach 45 200 MW, double Three Gorges Hydro Power Station's total installed capacity (22 400 MW). In order to fully draw lessons from Three Gorges, Xiluodu, and Xiangjiaba hydropower stations' excitation system design and operational experience to facilitate the future selection of excitation system design schemes and parameter optimization for Baihetan and Wudongde hydropower stations, this section describes the design and performance characteristics of excitation systems using Xiangjiaba Hydro Power Station as an example.

15.2 Static Self-Excitation System of Xiangjiaba Hydro Power Station

15.2.1 Overview

Xiangjiaba Hydro Power Station which has eight hydropower units, with a per-unit installed capacity of 800 MW, adopts the SPPA-E3000-SES530 excitation system designed by Siemens. The excitation transformer connected to the generator terminals consists of three single-phase epoxy cast dry-type transformers. The de-excitation and rotor overvoltage protection cabinet are provided by Beijing Sifang under the requirements of tender documents. The power station adopts the CEX series made by France's Lenoir as its DC field circuit breaker and the Metrosil series made by the United Kingdom's M&I as its SiC nonlinear de-excitation resistor. Its complete excitation equipment is provided by Nanjing SPPA.

Design and Application of Modern Synchronous Generator Excitation Systems, First Edition. Jicheng Li.
© 2019 China Electric Power Press. Published 2019 by John Wiley & Sons Singapore Pte. Ltd.

15 Excitation System Performance Characteristics of Hydropower Generator Set

Table 15.1 Parameters of 800 MW hydropower units, Xiangjiaba Hydro Power Station. Authorized by Nanjing SIEMENS power plant automation company.

Parameter	Parameters of units on right bank (Alstom)	Parameters of units on left bank (HEC)	Unit	Remark
Generator model	SF800-84/19990	SF800-80/20400		
Rated capacity	888.9		MVA	
Maximum capacity	840	840	MVA	
PF at maximum capacity	0.9	0.9		
Rated PF	0.9	0.9		
Rated voltage	23	20	kV	
Rated current	22 313	25 660	A	
Rated speed	71.4	75	R/min	
Number of pole pairs	42	40		
Runaway speed	134	150		
Rated frequency	50	50	Hz	
Number of phase	3	3		
Stator winding connection type	7Y	8Y		
Cooling type	All air cooling	All air cooling		
Flywheel moment	⩾490 000	⩾490 000	t·m^2	
Thrust bearing load	4 420	4 420	t	
Stator insulation class	Class F	Class F		
Rotor insulation class	Class F	Class F		
Stator winding temperature	70	69	K	
Stator core temperature rise	60	59	K	
Rotor winding temperature rise	75	74.5	K	
Collecting ring temperature rise	80	78	K	
Machine parts temperature rise (connected with or bordered upon stator winding)	60	60	K	
Thrust bearing surface temperature rise	80	80	K	
Guide bearing surface temperature rise	70	70	K	
Direct-axis synchronous reactance X_d	104/93.2	107.2/94.34	%	Unsaturated/saturated, Rated capacity
Direct-axis transient reactance X'_d	33/28	33/31.2	%	Unsaturated/saturated, Rated capacity
Direct-axis subtransient reactance X''_d	28.4/24	26.3/23	%	Unsaturated/saturated, Rated capacity
Direct-axis transient open-circuit time constant T'_{do}	10.2	9.827	s	95 °C
Direct-axis transient short-circuit time constant T'_d	3.35	2.986	s	95 °C
Stator winding short-circuit time constant	0.151	0.459	s	95 °C
Excitation winding capacitance to ground	0.55	0.55	μF	
Excitation winding self-inductance	1.43	0.983	H	
Stator winding capacitance to ground	3.43	3.846	μF/phase	
Rotor winding resistor	0.143/0.151	0.106 5/0.1129 5	Ω	Under 95 °C/115 °C

Each excitation device is composed of the following main elements: (1) excitation transformer and its accessories (e.g., current transformer (CT), temperature measuring element, terminal cabinet, and so on), (2) thyristor rectifier device, (3) de-excitation and overvoltage protection, (4) excitation regulator, (5) field flashing device, and (6) excitation system controlling, detecting, protecting, measuring, and displaying devices.

Xiangjiaba Hydro Power Station is equipped with four 800 MW hydropower units on each bank. Since the units on the left bank and those on the right bank are different in parameters and excitation equipment configuration, those on the right bank do not adopt electric braking, whereas those on the left bank adopt a 700 kVA electric braking transformer for their excitation equipment.

15.2.2 Generator Parameters and Excitation System Configuration

The parameters of the 800 MW hydropower units of Xiangjiaba Hydro Power Station and its excitation system are shown in Tables 15.1–15.15. The outline drawing of the Siemens E3000-SES530-THYRIPOL excitation system is shown in Figure 15.1. The schematic diagram of the static self-excitation system of the hydropower units on the left bank of Xiangjiaba Hydro Power Station is shown in Figures 15.2 and 15.3.

Table 15.2 Excitation system parameters (SES 530-THYRIPOL). Authorized by Nanjing SIEMENS power plant automation company.

Item		Units on left bank (HEC)	Units on right bank (Alstom)
Excitation device manufacturer		Siemens company	
Model number		Static self-excitation system SES530	
No-load excitation voltage		199.4 V	273 V
No-load excitation current		2310.5 A	1979 A
Excitation voltage at 888.9 MVA and PF = 0.9		512.1 V	540 V
Excitation current at 888.9 MVA and PF = 0.9		4170.4 A	3692 A
Excitation ceiling voltage		1031.4 V	1080 V
Excitation ceiling current		8399.4 A	7384 A
Ceiling voltage time	Permit ceiling voltage time under max. excitation current condition	20 s	
	Permit ceiling voltage time under max. excitation current without rectifier bridge condition	$\geqslant$20 s	
Excitation system voltage response time		$\leqslant$0.03 s	
Rotor overvoltage value under any operation condition of generator (including grid fault disturbance, generator-transformer unit breaker or field breaker trip, etc.)		Excitation winding instantaneous overvoltage should not exceed 70% of the amplitude voltage value (U_s) during factory withstand test (winding to ground). That is $\leqslant$0.7U_s.	
Excitation system unavailability ratio per year		$\leqslant$0.03%	
First fault appear time after excitation put into operation		$\geqslant$50000 h	
Excitation system service life		$\geqslant$50	
Excitation system mean time between failures (MTBF)		$\geqslant$223713 h	

Table 15.3 Auto voltage regulator (AVR) parameters. Authorized by Nanjing SIEMENS power plant automation company.

Item		Parameters
Manufacturer		Siemens
Mode		SES530
AVR regulating range		70%–110%U_N (U_N Stator rated voltage)
ECR regulating range		No-load: 10%–65%I_{fN} (I_{fN} Rotor rated current)
		Load: 30%–110%I_{fN} (I_{fN} Rotor rated current)
AVR regulating precision		±0.2%
AVR regulating time		<1.5 s
Overshoot		<10%
AVR regulating rule		PID + PSS2A (2B)
PSS Inhibition oscillation frequency range		0.1–2.5 Hz
AVR voltage regulation slope range		±15%
AVR hardware configurations	CPU word length	32 bits
	Main frequency	32 MHz
	RAM	256 MB
Communication interface		Modbus or Profibus

Table 15.4 Power rectifier cabinet parameters. Authorized by Nanjing SIEMENS power plant automation company.

Item	Parameters
Thyristor device/manufacturer	SITOR/Siemens
Thyristor manufacturer/model	EUPEC/T1452N52
Thyristor rectifier bridge parallel branches number	5
Series elements number for each branch	1
Parallel branch current sharing coefficient	⩾0.95
Thyristor rectifier cabinet number	5
Thyristor element peak repetitive reverse voltage	5200 V
Thyristor element rated average rectifier forward current	2260 A
Single rectifier bridge load capacity	2200 A
Load capacity without a rectifier bridge	8400 A
Thyristor control angle of ceiling voltage	10°
Thyristor control angle of no-load	78°–81°
Thyristor min. inverse angle	150°
Thyristor cooling type	Air cooling
Cooling fan number	2
Cooling fan rated capacity	1750 W (400 V/three-phase)
Cooling fan noise	<65 db
Thyristor shell temperature of rated conditions	85 °C
Rectifier total losses at rated load	34 kW (left bank)/32 kW (right bank)
Pulse transformer service life	50 yr
Fast acting fuse manufacturer/Type	Siemens/3NE7637-1C
Fast acting fuse rated current	2 × 710 A

Table 15.5 Excitation transformer parameters (Jinpan). Authorized by Nanjing SIEMENS power plant automation company.

Item		Units on left bank (HEC)	Units on right bank (TAH)
Excitation transformer manufacturer		Hainan Jinpan	
Type		Epoxy cast/single-phase	
Rated capacity		3×2500 kVA	3×2300 kVA
Primary-side voltage		$20/\sqrt{3} \pm 5\%$ KV	$23\sqrt{3} \pm 5\%$ KV
Secondary-side voltage		1110 V	1150 V
Insulation class		Class F	Class F
Standard withstand voltage	Primary-side withstand voltage at operation frequency, hang on for 1 min	50 kV	60 kV
	Primary-side impulse withstand voltage (Peak value)	125 kV	150 kV
	Secondary-side withstand voltage at operation frequency, hang on for 1 min	5 kV	5 kV
	Secondary-side impulse withstand voltage (peak value)	10 kV	10 kV
Temperature rise	Winding temperature	80 K	80 K
	Core temperature	65 K	65 K
Short-circuit impedance		8%	8%
Generator loss with max. load	Copper loss (120 °C) (single phase)	19 345 W	19 200 W
	Core loss (single phase)	4 815 W	4 370 W
Connection type		YD11	
Cooling type		AN	
Protection class		IP20	
Winding structure	High-voltage side	Wire-wound	
	Low-voltage side	Wire-wound	
Stalloy type		C130-30	
Stalloy designed field density		1.5 T	
External dimension (Length × Width × Height) (Single phase)		$2\,200 \times 3\,300 \times 3\,100$ mm	
Weight		6 700 kg	6 300 kg

15.2.3 Function and Component of Excitation System

Redundant design concept is adopted in the design of the excitation control system, i.e. two sets of identical control systems (main set and slave set) are adopted in the excitation control system, each set of systems includes: 1 set of SIMATIC S7-300 components, 1 set of SIMOREG CM control module and 1 set of control card T400 integrated in SIMOREG CM. Siemens standard PROFIBUS communication protocol is adopted for external communication of excitation system. Users can also choose other communication protocol if necessary. The redundant structure diagram of excitation control system is shown in Figure 15.3.

15.2.4 Composition of Excitation System

15.2.4.1 Automatic Voltage Regulator

The main functions of Xiangjiaba Hydro Power Station's automatic voltage regulator (AVR) are as follows: it forms the core of the excitation system; serves as the controller, the protector, and the decider of the whole excitation system; and leads the core effects ; besides, according to the operational requirement, it provides

Table 15.6 Power system device parameters (Siemens, Sifang, Jinpan). Authorized by Nanjing SIEMENS power plant automation company.

Item	Voltage (V)	Current (A)	Power (kVA)	Functions	Remark
AC1	380	22/29	Continuous: 1.5 Short-time: 2	AVR auxiliary power supply	Three-phase
AC2	220	23	5	Heater and socket power supply	Single-phase
AC3	380	145/580	Continuous: 10 Short-time: 40	Field flashing power supply	Three-phase
AC4	380	145	10	Cooling fan and auxiliary power supply	Three-phase
DC1	220	7/9	Continuous: 1.5 Short-time: 2	Auxiliary cabinet power supply 1 and AVR power supply 2	
DC2	220	7/9	Continuous: 1.5 Short-time: 2	Auxiliary cabinet power supply 2 and AVR cabinet relays power supply	
DC3	220	7/9	Continuous: 1.5 Short-time: 2	S101/S102/S104/S107 trip power supply 1	
DC4	220	7/9	Continuous: 1.5 Short-time: 2	S101/S102/S104 trip power supply 2	
DC5	220	45	10	S101/S102/S104/S107 closing power supply	

Table 15.7 Field circuit breaker parameters (Sifang). Authorized by Nanjing SIEMENS power plant automation company.

Item	Specification and type
De-excitation cabinet manufacturer	Sifang Jisi
Field circuit breaker manufacturer	France's Lenoir Elec
Field circuit breaker series number	CEX 06 5000 4.2 Ts 2000VD
Field circuit breaker rated voltage	2 000 V
Rated short-circuit voltage	400 V
Rated max. breaking voltage of	3 000 V/4 000 V/5 000 V
Rated continuous current	5 000 A
Rated breaking current at rated short-circuit voltage	2 000 V/30 kA (bipolar)
Max. breaking current at rated max. voltage	4 000 V/21 kA (bipolar)
Instantaneous let-through current at 0.5 s short circuit	50 kA
NC contact breaking current at rated voltage	10 kA
NC contact let-through current for 0.5 s	12 kA
NC contact rated closing current (if it exists)	10 kA
Circuit breaker breaking test standard	ANSI/IEEE C37.18
Opening time	70 ms
Breaking time	90 ms
Closing time	300 ms
Guarantee value of de-excitation breaking arc voltage	4000 V
Max. de-excitation breaking voltage	3 000 V/4 000 V/5 000 V
Max. breaking current at max. de-excitation breaking voltage	2 000 V/30 kA (bipolar)
Breaking current at other de-excitation voltages	4 000 V/21 kA (bipolar)
Insulation voltage	7 500 V
Breaker number	4 NO main breakers, 2 NC auxiliary breakers
Control circuit voltage	220 V dc
Mechanical life	100 000 times
Electrical life	50 000 times
Dimension	1 756 mm × 522 mm × 655 mm

1 closing coil, 2 trip coils, 6 NO, and 6 NC auxiliary contacts.

Table 15.8 AC inlet wire vacuum circuit breaker parameters (Siemens). Authorized by Nanjing SIEMENS power plant automation company.

Item	Parameters	
Manufacturer/model	SIEMENS/3AH3078-8	
Rated current	4000 A	
Rated voltage	7200 V	
Rated short-circuit breaking current	Periodic component	63 kA
	Aperiodic component	36%
Rated short-circuit withstand current	160 kA	
Rated short-circuit stand time	3 s	
Rated short-circuit closing current	160 kA	
Short-circuit breaking current (operation voltage is lower than rated voltage)	75 kA/1250 V	
Opening time	<60 ms	
Breaking time	<80 ms	
Closing time	<80 ms	

Table 15.9 Nonlinear resistor and DC overvoltage protection parameters (Sifang). Authorized by Nanjing SIEMENS power plant automation company.

Item	Parameters
Manufacturer/model	M&I Materials Ltd./FME/600A/US238/119/2S
Material	SiC
Resistor number of parallel branch/series number of each parallel branch	119P/2S
Max. de-excitation allowed current for total group	12 600 A
Max. allowed voltage for total group	2 100 V
Load rate of total group	60%
Nonlinear coefficient of total group	0.39–0.42
Total group capacity	16 MJ
Total group leakage current	2–3 μA
Resistor allowed temperature rise	160 K
Resistor service life	170 000 h
Crowbar action voltage	3 000 V
Max. de-excitation energy calculation value	8.213 MJ (left bank)/8.048 MJ (right bank)
Max. allowed voltage for total group (during de-excitation)	1 700 V
Resistor max. temperature rise (during de-excitation)	108 K

Table 15.10 Anodic overvoltage protection device parameters (Sifang). Authorized by Nanjing SIEMENS power plant automation company.

Item	Parameter
AC side	
Type and mode of connection	RC protection
Excitation transformer secondary-side rated voltage	1 110 V (left bank)/1 150 V (right bank)
Overvoltage limiting value	1 500 V
Surge current	21 000 A
Capacitor parameter (made in Germany)	4 μf/4 000 V AC
Resistor parameters (high-power non-inductive resistor)	40 Ω/500 W
Fuse parameters	FWJ-35A/1 000 V
High-voltage cable (high-voltage low-current)	AGG-4.0/5 000 V, high temperature resistor, 180 °C
AC-side overvoltage multiple	3
Connection mode	Triangle

Table 15.11 Field flashing parameters. Authorized by Nanjing SIEMENS power plant automation company.

Item	Parameter (HEC/TAH)
Field flashing voltage (DC)	25.2 V (left bank)/27.3 V (right bank)
Field flashing current	231.3 A (left bank)/197.9 A (right bank)
Field flashing time	10 s
Field flashing transformer capacity	10 kVA (field flashing transformer rated capacity 15 kVA)
Field flashing transformer primary-side/secondary-side voltage	380 V/25 V (field flashing transformer ratio 400 V/43 V)
Field flashing rectifier manufacturer	SEMIKRON
Field flashing rectifier repetitive reverse peak voltage	2400 V
Field flashing rectifier rated average forward current	430 A

Table 15.12 Electric braking parameters (applicable to only HEC units on left bank). Authorized by Nanjing SIEMENS power plant automation company.

Item	Parameter (HEC)
Field voltage during electric braking	200 V
Field current during electric braking	2313 A
Electric braking time	<600 s

Table 15.13 Braking transformer parameters (applicable to only HEC units on left bank).

Item	Parameter (HEC)
Manufacturer	Hainan JINPAN
Series number	Epoxy-resin filled dry-type transformer (SC9-700/0.38/0.23)
Phase number	Three phase
Rated capacity	700 kVA
Primary-side voltage	380 V
Secondary-side voltage	230 V
Insulation class	Class F
Primary-side power frequency withstand voltage for 1 min	3 kV
Temperature rise	80 K
Short-circuit impedance	6%
Connection type	Yd11
Cooling type	AN
Protection class	IP20
Boundary dimension (length × width × height)	2000 × 1250 × 2200 mm
Weight	2575 kg
Noise	55 dB

Table 15.14 Electric braking vacuum circuit breaker parameters (applicable only to HEC units on left bank). Authorized by Nanjing SIEMENS power plant automation company.

Item	Parameter
Manufacturer	Siemens
Series number	3AH3114-6
Rated current	2 500 A
Rated voltage	12 kV
Rated short-time breaking current (periodic component effective value)	25 kA
Aperiodic component (rated short circuit on-off dc component ratio)	36%
Rated short-circuit withstand current	25 kA
Rated short-circuit stand time	4 s
Rated short-circuit closing current	63 kA
Short-circuit breaking current (operation voltage is lower than rated voltage)	>25 kA
Closing time	40–60 ms
Break time	<75 ms
Opening time	<75 ms
Rated power frequency withstand voltage (to the ground/break)	28 kV
Rated impulse withstand voltage (to the ground/break)	75 kV
Max. braking times	10 000 times

Table 15.15 Excitation cabinet parameters. Authorized by Nanjing SIEMENS power plant automation company.

Item	Parameter		
Numbers	Totally 12 cabinets of PE01-PE12		
Single cabinet boundary dimension (mm) (length × width × height)	Regulator (PE01)		600 × 1200 × 2200
	Auxiliary control cabinet (PE02)		900 × 1200 × 2200
	Rectifier cabinet (PE03-PE07)		600 × 1200 × 2800
	De-excitation switch cabinet (PE08-PE09)		1200 × 1200 × 2800
	De-excitation resistor cabinet (PE10)		1200 × 1200 × 2200
	Excitation transformer cabinet (PE11)		6000 × 3300 × 3100
	AC inlet wire cabinet (EN)	Left bank units (S102, S104)	1200 × 2600 × 3100
		Right bank units (S102)	2000 × 1000 × 3100
	AC overexcitation protection cabinet (EJ)		1000 × 1000 × 3100
Total dimension	Left bank electric braking transformer cabinet		1250 × 2000 × 2200
	Excitation control cabinet 8100 × 1200 × 2800 mm (width × length × height)		
Single cabinet weight (kg)	Regulator		<750
	Auxiliary control cabinet		<750
	Rectifier cabinet		<750
	De-excitation switch cabinet		<750
	De-excitation resistor cabinet		<1000
Total weight (kg)	Right bank excitation transformer, AC inlet wire cabinet, AC overexcitation protection cabinet		6500
	Left bank excitation transformer, AC inlet wire cabinet, AC overexcitation protection cabinet		7000
	Left bank braking transformer		2575

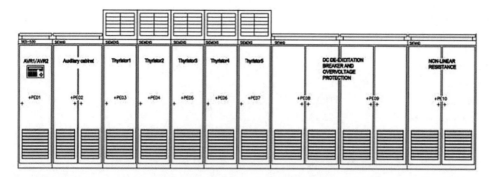

Figure 15.1 External view of E3000 SES530 THYRIPOL® excitation equipment of Xiangjiaba Hydro Power Station. Authorized by Nanjing SIEMENS power plant automation company.

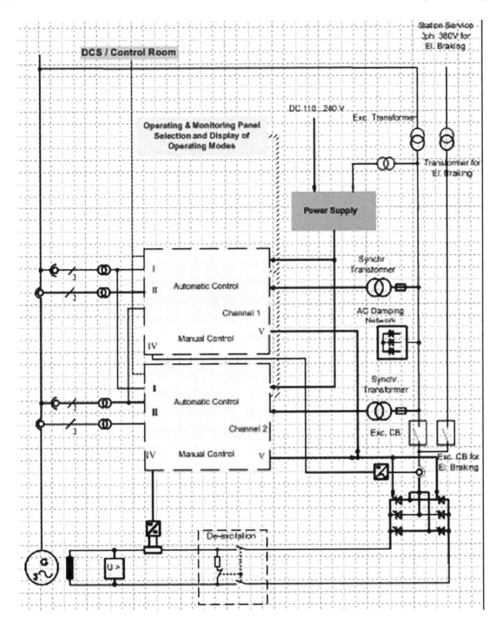

Figure 15.2 Schematic diagram of excitation system employed by 800 MW units of Xiangjiaba Hydro Power Station (left bank). Authorized by Nanjing SIEMENS power plant automation company.

a trigger pulse for the rectifier unit, and supplies corresponding excitation power for the excitation circuit to maintain steady generator operation. Table 15.16 describes the basic parameters of the AVR.

The AVR that Siemens offers is a digital static excitation regulator. Its hardware configuration has the following features:

(1) Two sets of exactly independent SIMOREG.CM+SIMATIC S7 control modules, 100% redundancy design, with every control set having manual mode.
(2) Profibus bus communication system is used between SIMATIC and SIMOREG, which has fast communication rate and high real-time performance.
(3) The Profibus bus communication system is used between SIMATIC and DCS, and also between two sets' controllers in order to switch over with no distribution.

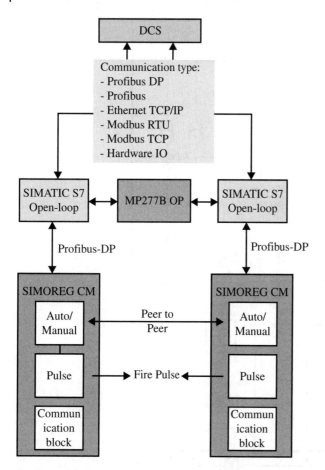

Figure 15.3 Redundancy controller. Authorized by Nanjing SIEMENS power plant automation company.

Table 15.16 Basic parameters of automatic voltage regulator (AVR). Authorized by Nanjing SIEMENS power plant automation company.

Item	Parameter	Remark
AVR regulation range	5%–130%U_n	U_n – stator rated voltage
FCR regulation range	5%–110%I_{fn}	I_{fn} – rotor rated current
Regulator precision	0.2%	
Regulator time	<1.5 s	
Overshoot	10%	
Regulator mode	PID + PSS2A (2B)	
Force control angle	10°	
AVR voltage droop range	±15%	
CPU word length	32 bits	
CPU main frequency	32 MHz	
RAM	256 MB	
Communication interface type	Modbus/Profibus	

(4) There is a control panel on the control cabinet for ease of on-site operation.
(5) Regulator mode of the AVR is proportional–integral–differential (PID) + power system stabilizer (PSS).
(6) Two channels operate in hot standby mode. Both receive input signals, but only the active channel outputs and triggers the rectifier unit. When the active channel is faulty, the backup channel becomes active immediately, and also blocks the faulty channel in order to avoid switchover to the faulty channel.
(7) Under extremely fault conditions, two channels in AVR mode (voltage close loop) are both faulty, and the excitation system will switch over to ECR mode (manual mode of excitation current closed loop control).

15.2.4.2 Thyristor Rectifier

Overview In the SES530 static excitation system, Siemens has adopted the SIMOREG 6RA70 series digital controlled thyristor rectifier device, which is compact and performs control, regulation, monitoring, and additional functions. Connecting SIMOREG rectifier units in parallel can extend rectifier power.

Besides, from the different parameters of the generator, one solution is to combine the controller module and rectifier together as one SIMOREG DC-Master, and the other solution is to use converter equipment made up of the controller module SIMOREG CM and an independent rectifier unit (SITOR or STACK design).

The channel connection between the excitation control system and the main functional units is shown in Figure 15.4.

In the SITOR power unit, the rectifier module is a drawer structure which contains a pulse amplifier, pulse transformer, resistor capacitance absorber, and fast fuse. The branch measure CT is located in the rectifier module, which is easily maintained.

The other rectifier module structure is the Stack, in which the high-power thyristor component and the radiator are fixed into the frame. The Stack structure will help produce more output power.

Siemens has developed a high-power unit standard data sheet that is shown in Table 15.17. The control module SIMOREG CM is installed in the control cabinet independently.

For the 800 MW generators of Xiangjiaba Hydro Power Station, the power unit type of the static excitation system is ERR1150/2200 (see Table 15.4).

Thyristor Rectifier Parameter Selection
(1) *Thyristor rated voltage selection.* From the bidding files, in rated load operation temperature condition, the reverse peak voltage that the thyristor can bear should be larger than 2.75 times the secondary-side maximum peak voltage of the excitation transformer. The repetitive peak reverse voltage of the thyristor should not be lower than 5200 V. The thyristor reverse peak voltage detailed calculation formula is as follows:

$$U_{RRM} \geqslant K\sqrt{2}U_2$$

where

K – Voltage margin factor: 2.75
U_2 – The secondary line voltage of excitation transformer, V

1) For the Alstom generator units,

$$U_{RRM} \geqslant K_U\sqrt{2}U_2 \geqslant 2.75 \times \sqrt{2} \times 1150 \geqslant 4471.8 \text{ V}$$

$U_{RRM} = 5200$ V is taken.

2) For the HEC generator units,

$$U_{RRM} \geqslant K_U\sqrt{2}U_2 \geqslant 2.75 \times \sqrt{2} \times 1110 \geqslant 4316.2 \text{ V}$$

$U_{RRM} = 5200$ V is taken.

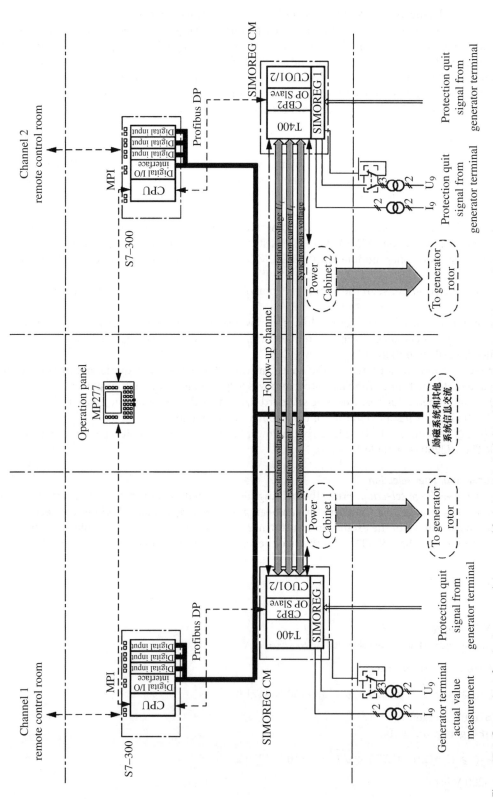

Figure 15.4 Connection of excitation control system and function unit. Authorized by Nanjing SIEMENS power plant automation company.

Table 15.17 SITOR or STACK design standard data sheet. Authorized by Nanjing SIEMENS power plant automation company.

Model	ERR 750/2400 SITOR Module	ERR 550/2900	ERR 900/4500 Stack design	ERR 900/6000
Rated excitation current I_{fn} (A)	2182	2636	4090	5455
Max. continue allowing operation current $I_{Fmax}(=1.1 \times I_{frated})$ (A)	2400	2900	4500	6000
Ceiling current I_p (10 s) (A)	3600	4350	6000	7640
Ceiling current multiple I_p/I_{frated}	1.65	1.65	1.40	1.40
AC max. allowing input voltage of Rectifier bridge (V)	750	550	900	900
Ceiling voltage multiple	2.36	2.31	2.83	2.83
Single channel	1 × 100%	1 × 100%	1 × 100%	2 × 50%
Loss power of rated working condition (kW)	10	11	26	35.5
Cooling air (single bridge) (m³/h)	3200	3200	5760	5760
Noise classes 50 Hz	70 dB (A)	70 dB (A)	72.5 dB (A)	74 dB (A)
60 Hz	73 dB (A)	73 dB (A)	75.5 dB (A)	77 dB (A)
Weight	1300 kg	1300 kg	1660 kg	2750 kg

(2) *Thyristor rated current selection.* From the bidding files, the rectifier cabinets adopt three-phase full control bridge design, meeting excitation system requirements in the generator for every operation condition (include force excitation), there have to be no fewer than four parallel connected rectifier bridges, and each feeder series has one thyristor. The number of parallel rectifier bridges meets the $N-1$ principle, so that when a fault occurs in one bridge, the rectifier bridges perform all functions including force excitation, and when a fault occurs in two bridges, the rectifier bridges perform all functions except force excitation. The current distribution factor should not be less than 0.95.

For Alstom generator units:

The operation with three bridges in parallel meets the requirement of 1.1 times the excitation current, when the generator is working at maximum capacity, and the single bridge output current is as follows:

$$\frac{1.1 \times 3692/3}{0.95} = 1425 \text{ A} < 2{,}200 \text{ A}$$

The operation with four bridges in parallel meets the requirement of twice the normal excitation current during force excitation, and the single bridge output current is as follows:

$$\frac{2 \times 3692/4}{0.95} = 1943.1 \text{ A} < 2{,}200 \text{ A}$$

The average current flowing through a single bridge arm is as follows:

$$I_{T(AV)} = 0.577 \times 2200/1.57 = 808.5 \text{ A}$$

When the current margin factor $K_i = 2.0$, the thyristor forward average current is as follows:

$$I_{T(AV)} = K_i \times I_{T(av)} = 2.0 \times 808.5 = 1{,}617 \text{ A}$$

For HEC generator units:
The operation with three bridges in parallel meets the requirement of 1.1 times the excitation current, when the generator is working at maximum capacity, and the single bridge output current is as follows:

$$\frac{1.1 \times 4170.4/3}{0.95} = 1609.6 \text{ A} < 2{,}200 \text{ A}$$

The operation with four bridges in parallel meets the requirement of twice the normal excitation current during force excitation, and the single bridge output current is as follows:

$$\frac{2 \times 4170.4/4}{0.95} = 2195 \text{ A} < 2{,}200 \text{ A}$$

The average current flowing in the single bridge arm is as follows:

$$I_{T(AV)} = 0.577 \times 2200/1.57 = 808.5 \text{ A}$$

When the current margin factor $K_i = 2.0$, the thyristor forward average current value is as follows:

$$I_{T(AV)} = K_i \times I_{T(av)} = 2.0 \times 808.5 = 1{,}617 \text{ A}$$

(3) *Thyristor commutation overvoltage calculation*
For the Alstom generator units:
$$\frac{0.7}{3}\sqrt{2}U_2 + 2\sqrt{2}U_2 = \frac{0.7}{3}\sqrt{2} \times 1150 + 2\sqrt{2} \times 1150 = 3635.8 \text{ V}$$
For the HEC generator units:
$$\frac{0.7}{3}\sqrt{2}U_2 + 2\sqrt{2}U_2 = \frac{0.7}{3}\sqrt{2} \times 1110 + 2\sqrt{2} \times 1110 = 3505.3 \text{ V}$$

From the above calculation, for the two kinds of generators, we select the same thyristor, T1451N52; the average on-state current is 2260 A (60 °C), and the repetitive peak reverse voltage is 5200 V.

Fast Fuse Parameter Calculation From the thyristor type calculation, including the force excitation operation working condition, a single cabinet may support a maximum current less than 2200 A, and so the fast fuse for an output current of 2200 A is calculated.
The single bridge arm current flow's effective value is as follows:

$$i = 0.577 \times 2200 = 1269.4 \text{ A}$$

The fast fuse rated current is 1420 A.
The Siemens 3NE7637-1C fast fuse is selected. Its rated voltage is 2000 V, and its rated current 2×710 A.

Thyristor Anode Overvoltage Protection Due to the leakage inductance of the excitation transformer, the thyristors' phase switches in the rectifier produce a high peak voltage. In order to suppress the peak overvoltage between any two phases, an RC filter is usually installed in the AC side of the rectifier, in parallel with the thyristor, or in the DC side of the rectifier.

For the 800 MW units of Xiangjiaba Hydro Power Station, Siemens applied a triangle filter circuit in the AC side of the rectifier, producing a superior effect compared the other methods. Figure 15.5 shows the connection, and the related component parameters:
Resistor: 40 Ω/500 W, high-power non-inductive resistor
Capacitor: 4 μF/AC 4000 V, electrolytic capacitor
High-voltage cable: AGG-4.0/5000 V, resistor of temperature 180 °C
Fuse: FWJ-35A/1000 V

The overvoltage protection function can suppress the peak voltage between two phases, and can also suppress the atmospheric overvoltage from the generator or transformer, transient overvoltage during operation, overvoltage during opening or closing of the circuit breaker at the excitation transformer side, operation overvoltage from the excitation transformer coupled capacitor, and so on.

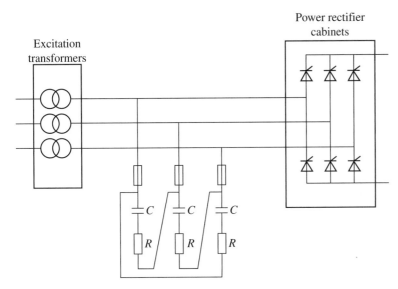

Figure 15.5 RC damper connected to thyristor AC input end. Authorized by Nanjing SIEMENS power plant automation company.

Cooling System of Thyristor Rectifier For Xiangjiaba Hydro Power Station, the ERR 1150/2000 type power unit is used in the excitation system. This type of power unit has forced air cooling. One unit has two cooling fans with two independent power supplies. The station fan power supply is from a 400 V AC power source, and the self-power supply of the fan is from the excitation low-voltage side. In the normal operation mode (AVR mode), two fans of the power unit will switch on alternately during excitation running. In the ECR mode, the fan with the station power supply will be on.

Moreover, during system running, fan-on or fan-off can be controlled by an airflow monitoring sensor in the power unit; the adjusted airflow rate in the sensor can be used to determine whether the fan is working properly. So, when the airflow rate is lower than the adjusted value, the system considers this fan to be faulty and switches on the other fan. If all fans are faulty, system will disable this power unit. Meanwhile, when the number of disabled power units reaches 3, the whole system will be shut down.

15.2.4.3 De-excitation and Overvoltage Protection Device

Xiangjiaba Hydro Power Station, which processes 800 MW hydroelectric units, adopts a de-excitation device equipped with a DC field breaker and also an AC breaker. The constituent parts of the de-excitation device are described as follows.

DC Field Breaker The de-excitation device adopts the CEX 06-5000 4.2 combined breaker from France's Lenoir as the field breaker. The CEX series DC breaker features a mechanical interlock. All the moving contacts are coupled rigidly on the same connecting rod, which ensures synchronization during the opening and closing operations, and the breaker is reliable in all conditions. The main contact (thermal fracture) and the arc contact (arc-blow-out fracture) are separated. Both contacts have an individual de-excitation rupturing pole. By using the technology of the short arc effect, the arc voltage is stabilized. Its modular design ensures ease of maintenance. The CEX series DC breaker has good performance, high technology level, comprehensive functions, and fine operational performance.

The CES 06 5000 4.2 series employs the double rupturing poled DC circuit breaker. The breaker consists of two main fractures; each main fracture contains two thermal fractures and two arc-blow-out fractures in parallel. The rated current is 5500 A, and the rated voltage is 2000 V. The maximum arc voltage is 3200 V. With an included double opening coil, the device achieves full redundancy.

The electric connection of the DC field breaker (main fracture and arc fracture) is designed as in Figure 15.6.

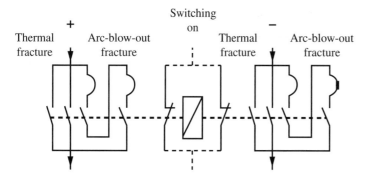

Figure 15.6 DC field breaker fractures' electric connection. Authorized by Nanjing SIEMENS power plant automation company.

Figure 15.7 External view of SIEMENS AH3078-8 vacuum AC circuit breaker. Authorized by Nanjing SIEMENS power plant automation company.

AC Breaker The AC vacuum circuit breaker employs SIEMENS 3AH3078-8 equipment and is installed between the excitation transformer low-voltage side and the converter. This AC breaker is designed as the AC field breaker, which assists the DC field breaker and provides a standby mode for de-excitation operation. When the excitation transformer secondary voltage is maximum, the AC circuit breaker can disconnect the current, if low-voltage side of the transformer is short-circuited, or if the two poles of the rotor are short-circuited directly. Besides, when the electric brake is adopted, this AC circuit breaker can also be used as a disconnector at the time of electric braking.

The external view of a vacuum AC circuit breaker is shown in Figure 15.7. When the metal electrical contact opens, metallic vapor is produced by the interrupted current. The ionized air gives the electricity a path past the breaker until the current reaches zero; at the same time, the arc is extinguished. Then the metallic vapor loses its conductivity within a few microseconds, as a result of which the dielectric strength of the gap between the metal contacts recovers rapidly. At the end, the circuit breaking function is established.

De-excitation and Rotor Overvoltage Protection There are many circumstances that cause overvoltage in excitation systems. On the basis of these causes, we can classify overvoltage as DC-side overvoltage and AC-side overvoltage.

The DC-side overvoltage includes (1) overvoltage caused by field breaker opening, (2) overvoltage caused by open phase closing between the generator and the power grid, (3) overvoltage caused by

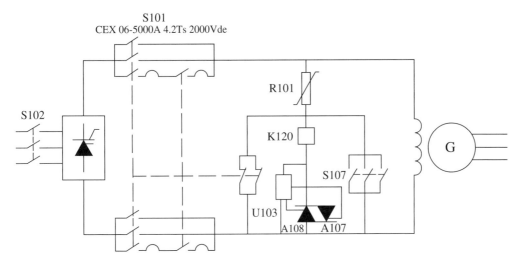

Figure 15.8 Principle of operation: de-excitation and overvoltage protection. Authorized by Nanjing SIEMENS power plant automation company.

three-phase/two-phase short circuit at the transformer high-voltage side, (4) overvoltage caused by asynchronous running, and (5) overvoltage caused by thyristor rectifier commutating. The de-excitation and overvoltage protection for the Xiangjiaba 800 MW hydroelectric generator sets is designed as in Figure 15.8.

As described the figure, the de-excitation device includes the DC magnetic field breaker S101, AC breaker S102 installed at the thyristor's anode, SiC nonlinear resistor connected in parallel with the rotor and breaker S107, which provides a standby mode for de-excitation during an emergency shutdown.

The overvoltage protection device includes a current relay, K120, which is an overvoltage protection crowbar consisting of two thyristor elements A107 and A108 reversely connected in parallel and a flip-flop U103. Combined with a linear resistor, R101, they constitute the rotor forward/reverse overvoltage protection devices. In the event of a forward overvoltage caused by a rotor lightning impulse or asynchronous operation, U103 triggers the forward thyristor A107 to be turned on. R101 is put in to absorb the overvoltage energy and limits the rotor overvoltage. K120 sends an overvoltage pulse signal to the regulator. When the rotor circuit instantaneous overvoltage disappears, the regulator control logic will ensure that A107 automatically resumes blocking to avoid the rectifier bridge's power supply to the de-excitation resistor. If the overvoltage energy is too high or a continuous overvoltage occurs, the regulator will issue a trip order. Similarly, when a reverse overvoltage occurs, U103 triggers the reverse thyristor A108 to be turned on to execute the above process.

During the generator unit normal shutdown, inverter mode de-excitation is adopted. The energy stored in the rotor will be consumed through the inverter bridge. When the field current is zero, the excitation regulator (AVR) will switch off the rectifier bridge trigger pulse and switch on the auxiliary field suppression breaker S107. Then the excitation regulator (AVR) trips off the AC breaker S102, and the de-excitation process ends. At the time, S107 and S102 have no current. The time sequence of the normal shutdown de-excitation process is depicted in Figure 15.9.

In the accident condition, the generator adopts the mechanical de-excitation mode by using the field circuit breaker. When the AVR receives the tripping instruction, the field breaker S101's rupturing pole opens first, and the NC fracture closes. At the same time, the SiC nonlinear resistor is inserted. Then the thermal fracture of the main poles opens. At last, the magnetic arc-blow-out contactor poles open to build up arc voltage. In the meantime, the AVR switches off the rectifier bridge trigger pulse and switches on the auxiliary field suppression breaker S107 to ensure that the SiC nonlinear resistor is put into operation. The closing of S107 will trip off the AC breaker S102. After shutting the trigger pulse, the thyristor bridges freewheel naturally. Within 20 ms, the thyristors assist the field breaker to build up arc voltage by using the negative half-wave of the anode voltage. Then the energy in the rotor is consumed quickly by break-over of the SiC resistor,

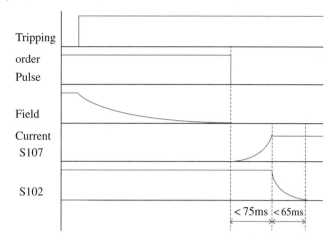

Figure 15.9 Inverter mode de-excitation time sequence. Authorized by Nanjing SIEMENS power plant automation company.

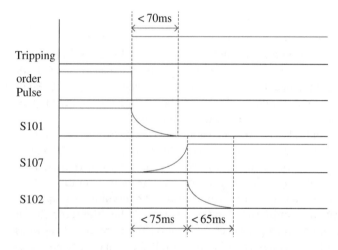

Figure 15.10 Tripping de-excitation action time sequence. Authorized by Nanjing SIEMENS power plant automation company.

and AC field suppression is achieved. The AC breaker S102 and the auxiliary field suppression breaker S107 are standbys for the DC field breaker S101. After receiving the tripping order, under normal circumstances, the DC field breaker S101 switches off within 70–100 ms. If it fails to operate, the auxiliary field suppression breaker S107 will close in 75 ms, and the nonlinear resistor R101 will be put into operation. Then the AC breaker S102 opens within 65 ms. Also, as arc voltage is built up by using the negative half-wave of the anode voltage, speedy energy transfer is realized. This standby design guarantees the reliability of the de-excitation device. The time sequence of the accidental shutdown de-excitation process is shown in Figure 15.10.

SiC Nonlinear Resistor Xiangjiaba Hydro Power Station uses 800 MW hydroelectric generator sets and the Metrosil Varistor Series SiC nonlinear resistor produced by the United Kingdom's M&I Materials Ltd. The type of the whole set of nonlinear resistor is FME/600A/US238/119/2S. The product includes 17 groups, each group consisting of 14 pieces, 2 in series and 7 in parallel. The number of pieces is $17 \times 2 \times 7$ (2 in series and 119 in parallel), that is, a total of 238 pieces. The total de-excitation energy capacity is 16 MJ.

Calculation of De-excitation Resistor Capacity For hydropower generator, the de-excitation resistor capacity could be decided by certain modes of operation, and the maximum value should be selected. The modes of operation include

(1) De-excitation when generator is unloaded;
(2) Generator rated de-excitation;
(3) De-excitation in the event of forced excitation of the generator;
(4) De-excitation in the event of faulty forced excitation caused by loss of control when the generator is unloaded or loaded.

Sudden Generator Three-Phase Short-Circuit De-excitation Practical operation experience shows that the de-excitation capacity is maximum when the generator is three-phase short-circuited suddenly. So the discharge resistor capacity calculated in accordance with this mode could satisfy the needs of all of the above operation modes. According to ANSI/IEEE C37.18, in a three-phase short circuit, the aperiodic component of the field current is suggested as $3\,I_{fN}$. But after determining the resistor capacity, this current should not be referred to as the basis of the rotor's stored energy. When cutting off the external fault, the short-circuit duration is normally within 0.1 s, and the aperiodic component of the attenuated rotor current is close to 60%–70% of the initial value; that is, the maximum value of the aperiodic component could be selected as $0.7 \times 3\,I_{fN} = 2.1\,I_{fN}$.

In addition, when stator short circuits occur, at $t = 0$, due to the constant-flux-linkage theorem, the air-gap voltage is maintained. After the tripping of the main circuit breaker on the generator high-voltage side, the current in the stator, rotor, and amortisseur winding circuits change suddenly to maintain the gap magnetic flux.

The estimation of the resistor capacity depends on the following hypothesis:

(1) When stator short circuits occur, most of the ampere-turns in the rotor consist of triangle flux ampere-turns. The magnetic lines do not pass through the air gaps between rotor and stator magnetic circuits, and so the saturation level is low. For hydropower stations, the saturation coefficient is normally considered to be 0.6.
(2) The energy stored in the rotor is not only consumed by the de-excitation resistor, but also by the rotor resistor, field breaker, damping winding, and the whole iron core. For the consuming coefficient, from experience, values are set as 0.73 for the hydroelectric generator and 0.6 for the steam-turbine generator.

The approximate energy capacity of the resistor is calculated with the following:

$$I_{fm} = 2.1 I_{fN} \tag{15.1}$$

$$W_{fmax} = \frac{1}{2} L'_f I^2_{fm} = \frac{1}{2} T'_d R_{f(115°C)} I^2_{fm} \tag{15.2}$$

$$W_N = K_1 K_2 K_3 W_{fmax} \tag{15.3}$$

I_{fN} — rated field current, A;
W_{fmax} — maximum stored energy of rotor winding, J;
L'_f — rotor winding inductance, H;
T'_d — short-circuit transient time constant of d-axis, s;
$R_{f(115°C)}$ — DC resistor of rotor winding (115 °C);
W_N — de-excitation resistor energy capacity, J;
K_1 — capacity reserve coefficient. The requirement of de-excitation could still be met when 20% of the nonlinear resistors are out of service. $K_1 = 1/0.8 = 1.25$;
K_2 — saturation coefficient, selected as 0.6;
K_3 — energy-consuming distribution coefficient, selected as 0.73;

Table 15.18 Results of de-excitation resistor capacity. Authorized by Nanjing SIEMENS power plant automation company.

Item	Alstom	HEC
Maximum rotor stored energy (MJ)	15	14.7
Energy capacity of de-excitation resistor (MJ)	8.213	8.048
De-excitation residual voltage (V)	1700	1700
Maximum rotor current (A)	7753.2	8819.4

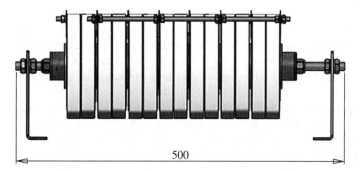

Figure 15.11 External view of each group (600A/US14/7P/2S). Authorized by Nanjing SIEMENS power plant automation company.

The energy capacity of the resistor of the Alstom generator on the right bank is calculated as follows, by using the Formulas (15.1)–(15.3):

$$I_{fm} = 2.1 I_{fN} = 2.1 \times 3692 = 7753.2 \text{ A}$$
$$W_{fmax} = \tfrac{1}{2} L'_f I_{fm}^2 = \tfrac{1}{2} T'_d \cdot R_{f(115°C)} I_{fm}^2 = \tfrac{1}{2} \times 3.35 \times 0.151 \times 7753.2^2 = 15 (\text{MJ})$$
$$W_N = K_1 K_2 K_3 W_{fmax} = 1.25 \times 0.6 \times 0.73 \times 15 = 8.213 (\text{MJ})$$

On the basis of the generator parameters of Alstom and HEC, the calculation results of the energy capacity of resistor are described in Table 15.18.

The nonlinear resistor selection still takes the maximum rotor stored energy into account, so the de-excitation energy capacity is selected as 16 MJ for both types of generator.

The SiC nonlinear resistor adopts United Kingdom's M&I Materials Ltd. product, including 238 pieces in 17 groups (600A/US14/7P/2S). In each group, there are 14 pieces, 2 in series and 7 in parallel. All groups are in parallel. The external view of each group is shown as Figure 15.11. Besides, the characteristic curves of the whole set of nonlinear resistors is shown in Figure 15.12.

The V-A characteristic curves of the whole set of nonlinear resistors are shown in Figure 15.12.

Taking an extreme condition of erroneous excitation forcing no-load into account, the arc voltage built by the field breaker could make the nonlinear resistor break-over credibly, and could also effectively restrict the overvoltage of the rotor. The residual voltage of the SiC is finally set to 1700 V. Normally, the nonlinear resistor is blocked by thyristors A107 and A108 to eliminate the leakage current . When the rotor experiences overvoltage or de-excitation in an emergency shutdown, the nonlinear resistor might be put into operation through A107 or A108, which is triggered by the overvoltage crowbar, or through the rupturing poles of the DC field breaker S101.

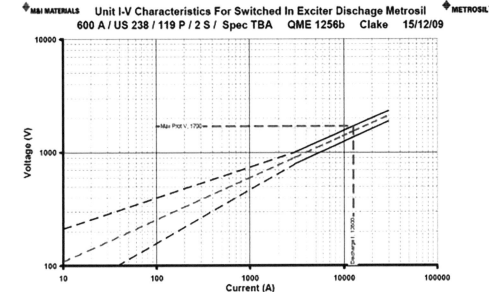

Figure 15.12 Logarithmic U-I characteristic curves of the whole set of the nonlinear resistor. Authorized by Nanjing SIEMENS power plant automation company.

15.2.4.4 Excitation Transformer

Secondary-Side Rated Line Voltage Calculation From the bidding files, the excitation system must guarantee to provide the excitation system the ceiling voltage when the terminal voltage drops to 80% of the rated value. The ceiling voltage is twice the excitation voltage with the generator at maximum capacity. See the following detailed calculation formula:

$$U_{fT2} = \frac{K_u U_{fn}}{0.8 \times 1.35 \times \cos \alpha_{min}}$$

where

K_u – A multiple of the force excitation ($\alpha = 10°$). Here, it is 2 (at 80% U_{GN})

U_{fn} – Excitation voltage when the generator operates at rated condition

(1) For the Alstom generator units,

$$U_{fT2} = \frac{2.0 \times 540}{0.8 \times 1.35 \times \cos 10°} = 1016 \text{ V}$$

Considering the impedance of the transformer and the line voltage loss, the final value is $U_{fN} = 1050$ V.

(2) For the HEC generator units,

$$U_{fT2} = \frac{2.0 \times 515.7}{0.8 \times 1.35 \times \cos 10°} = 970 \text{ V}$$

Considering the impedance of the transformer and the line voltage loss, the final value is $U_{fN} = 1000$ V.

Secondary-Side Rated Line Current Calculation On the basis of the bidding files, the excitation system should be in continuous operation over the long term when the generator has a rated capacity of 888.9 MVA, rated voltage and 1.1 times the excitation current at a power factor of 0.9. The detailed calculation formula is as follows:

$$I_{fT2} = \sqrt{\frac{2}{3}} \times K_a I_{fn}$$

where

K_α – Margin factor 1.1

I_{fn} – Excitation current with generator at rated capacity and rated voltage, and a power factor of 0.9.

(1) For the Alstom generator units,

$$I_{fT2} = \sqrt{\frac{2}{3}} \times 1.1 \times 3692 = 3314 \text{ A}$$

(2) For the HEC generator units,

$$I_{fT2} = \sqrt{\frac{2}{3}} \times 1.1 \times 4199.7 = 3770 \text{ A}$$

Rated Capacity Calculation

(1) For the Alstom generator units,

$$S_{fT2} = \sqrt{3} U_{fT2} I_{fT2} \times 10^{-3} = \sqrt{3} \times 1050 \times 3314 \times 10^{-3} = 6027 \text{ kVA}$$

The single-phase transformer is 2300 kVA.

(2) For the HEC generator units,

$$S_{fT2} = \sqrt{3} U_{fT2} I_{fT2} \times 10^{-3} = \sqrt{3} \times 1000 \times 3770 \times 10^{-3} = 6530 \text{ kVA}$$

The single-phase transformer is 2500 kVA.

The excitation transformer design parameter list is shown in Table 15.19.

15.2.4.5 Electric Braking

Electric Braking Basic Configuration With the expansion of automation, there is also a great requirement for the stop speed of the hydro-turbine. Therefore, nowadays many hydropower stations have the electric braking function. Due to constraints of plant space, the units on the right bank of Xiangjiaba Hydro Power Station do not have electric braking function; only those on the left bank have this function.

The units on the left bank adopt the flexible electric braking mode, where excitation and electric braking share the excitation system thyristor rectifier. At the time of electric braking, the station power AC of 380 V provides the units with the excitation source required for the electric braking through the special braking transformer and the main thyristor rectifier to meet the automatic control and local operation monitoring requirements.

Table 15.19 Excitation transformer design parameters. Authorized by Nanjing SIEMENS power plant automation company.

Item	Alstom units	HEC units
Model number	DC9-2300/23/$\sqrt{3}$	DC9-2500/20/$\sqrt{3}$
Rated capacity	3 × 2300 kVA	3 × 2500 kVA
The primary rated voltage	23/$\sqrt{3}$ kV	20/$\sqrt{3}$ kV
The secondary rated voltage	1150 V	1110 V
The number of phase	Single phase	
Connection type	YD11	
Insulation level	F	
Temperature raise	80 K	
Type	Epoxy cast dry rectifier excitation transformer	

Principle of Electric Braking After the generator is disengaged from the grid, when the generator speed decreases to 50%–60% of the rated speed, the electric braking operation will be executed. The stator windings are short-circuited in three phases. A constant excitation is applied to the rotor windings, so that an electric braking torque is generated and electric braking is achieved.

For large hydropower station, there is high rotational inertia, and stopping process is long. Mechanical braking produces physical deterioration, air pollution, insulation safety problems, and heat dissipation. Thus, if electric braking is included after mechanical braking, the life of the generator and turbine is extended. Furthermore, electric baking has additional benefits, such as high rotation torque, short stop process, no pollution, and so on.

The electric braking stop technique is based on the armature reaction of the synchronous motor and energy conservation theory. When the unit stops, the hydro-turbine's guide vanes shut down, and after de-excitation, the generator rotor only has remanence, which determines the terminal residual voltage. At the same time, the system will monitor whether the electric braker can be enabled. If so, the three-phase short-circuit breaker will close to make the generator stator's three phases short-circuit and then turn the excitation on again. According to the armature reaction theory of the synchronous motor, at this time, an armature reaction will appear. The direct-axis component of this reaction reflects only the decrease or increase in the magnetic torque, not the active power torque. But the quadrature component of this reaction reflects the active power torque that produces an active moment, in the direction opposite to the generator's normal rotation direction, thus increasing the combining braking moment and finally stopping the unit faster.

The left bank of Xiangjiaba Hydro Power Station applies a flexible electric braking configuration, and there is no additional braking rectifier cabinet. AC breakers S102 and S106 are installed inside the excitation breaker cabinet, and the two breakers interlock automatically with the position contact of each other. The braking power supply is from the station power AC of 380 V, through the special braking transformer T100, which is then rectified into the braking current. The electric braking principle is shown in Figure 15.13.

Electrical Braking Operation Process The electrical braking operation process is described below. After the generator goes offline, and after de-excitation, when the unit speed drops to 50%–60% of the rated speed or the voltage drops to 5% of the rated voltage, the excitation system issues an electric braking enable command, checks whether the electric braking short-circuit breaker S101 is closed, and whether the auxiliary de-excitation breaker S107 is open. The excitation system issues a command to open the excitation transformer secondary-side AC breaker S102 and close the braking transformer secondary-side AC breaker S106, and then according to the preset setting point of the excitation current, offers a suitable braking moment. At the same time, the losses produced by the generator stator short-circuit form a braking moment to slow down the generator rotation faster. To strengthen the effect, a preset value is used to make the short-circuit stator current equal to the generator stator rated current.

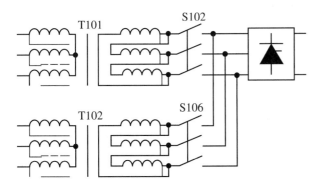

Figure 15.13 Electric braking principle. Authorized by Nanjing SIEMENS power plant automation company.

15.2.5 System Control Logic

The logic of the SES530 excitation system is shown in Figure 15.14.

The figure shows that all complicated logic and programs are processed in the T400 microprocessor. All regulations contain three loops: (1) generator terminal voltage regulation is processed in T400; (2) the value is output to excitation voltage regulation, which is processed in the CUD board; and finally (3) the output goes to the pulse trigger unit, and then, in SIMOREG CM, a pulse is generated to the rectifier. The other excitation current loop is also processed on the CUD board. The outputs of the excitation current loop go directly into the excitation voltage loop, the angle is calculated through a linear function, finally the output goes to the pulse trigger unit, and then, in SIMOREG CM, pulses are generates to the rectifier. In order to satisfy the redundant design requirement, two sets of control systems have the same structure and configuration, and they have independent power supplies.

15.2.5.1 Control Logic Mode

In the open loop and closed loop programs, there are different functional modules, such as fan control, setting value, system parameters, and so on. If the signal value is 1 in the program, it means the real state is the same as the signal name. Generally speaking, programs in the active channel and inactive channel are identical. A few differences will be highlighted in the parameter list. The parameter list final version will be created after commission on-site, the factory parameter list serves as a reference. The signal "ready to start" indicates that the system is ready to start.

Auto mode (generator voltage regulation) and manual mode (excitation current regulation) can be selected in OP panel MP227 on the cabinet front door or by the remote control room.

(1) *Local/manual mode – excitation current regulation.* In this mode, the excitation current setting value is the target regulation value. The setting value is input from the OP panel. The setting value range is 0%–110%, depending on the generator cooling condition and running parameters. Generally speaking, for the external power supply mode, the low limit is 0, and for the shunt mode, the low limit is 35% I_{fn}. In the manual mode, the setting value can also be adjusted by increasing and decreasing the voltage.

Normally, the manual mode is used in test situations or during an emergency. The stator current limiter, overexcitation limiter, underexcitation limiter, and V/Hz limiter are not active in the manual mode. Meanwhile, switching from the manual mode to the auto mode is blocked; such a switch is only allowed manually.

(2) *Local/auto mode generator – voltage regulation.* For this mode, the setting value of the generator voltage is the target regulation value. The setting value can be input on the OP panel. The setting value range is based on the generator running mode, and can be modified in the program. When the generator is online, the regulation range is 95%–100%; when the generator is in the no-load mode, the regulation range is 90%–100%.

(3) *Remote/manual mode – excitation current regulation.* For this mode, the setting value of the excitation current is the target regulation value. The setting value can be input from the control room. If the signal is transferred through communication, an analog value is directly input or an increase and decrease command is issued in the control room; if the signal is transferred through hard wire, only an increase and decrease command is sent. The setting value range is 0%–110%, depending on the generator cooling condition and the running parameters. If the generator is in the no-load mode, the excitation current up limit is 1.15 times the no-load rated excitation current.

Normally, this mode is only used for commissioning or during an emergency. The stator current limiter, overexcitation limiter, underexcitation limiter, and V/Hz limiter are not active in the manual mode.

(4) *Remote/auto mode – generator voltage regulation.* This mode is a totally automatic mode. A feedback signal is recorded in the control room. The setting value of the generator voltage is the target regulation value, and only the generator voltage is adjusted. When the generator is online, regulation of the generator voltage will affect the reactive power and power factor.

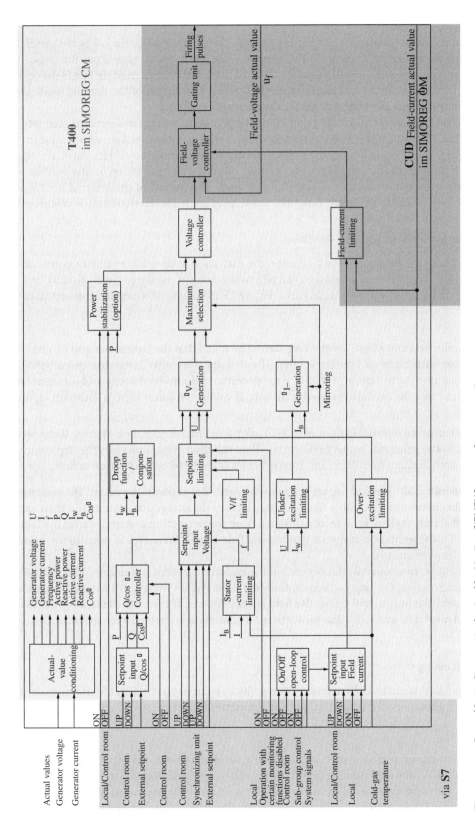

Figure 15.14 Control logic diagram. Authorized by Nanjing SIEMENS power plant automation company.

(5) *Manual/auto mode feature*

In the manual mode (excitation current regulation), the excitation system keeps the generator stable by adjusting the excitation current. In this situation, the generator voltage is not the target regulation value. The excitation current setting value is sent to the T400 control module and is processed by the filter, and then becomes the real excitation setting value. The output of the manual mode loop is sent to the linear function. The trigger angle for thyristors is calculated.

In the auto mode (generator voltage regulation), the generator voltage difference is read into the CUD (SIMOREG CM) through the T400 control module, as the excitation voltage setting value for the internal regulation loop. The up limit of the excitation current regulation loop limits the final output. In this situation, if the excitation current exceeds the limit value, the overexcitation limiter will become active and adjust the excitation current to the allowed range. The output of the internal excitation voltage regulation loop is sent to the linear function. The trigger angle for the thyristors is calculated.

15.2.5.2 Excitation System Start and Stop Condition

Start Condition In both the local and remote modes, the excitation system has to satisfy every condition for start. Meanwhile, the "excitation on" command can be executed only if the signal "ready to start" is 1. The start condition is different because of project differences. Figure 15.15 shows the Xiangjiaba project start condition logic.

Stop Condition In the stop condition, special care should be taken that the stop command can be issued only when the generator output circuit breaker is open. If not, even when the "generator speed ≤90%" signal is not valid, the excitation system cannot stop. But if the generator is in no-load mode and the "generator speed >90%" signal is not valid, the excitation system can stop. The stop condition logic is shown in Figure 15.16.

Protection Trip Whether an internal fault or external fault occurs in the excitation system, the protection trip command is sent by the generator protection system. The excitation system receives the trip command from the protection system through hard wire, and then executes the related protection-off action.

(1) *Excitation internal fault*. The fault signal is sent to the generator protection system. The excitation system opens the field circuit breaker with a feedback signal from the protection system and then executes the protection-off action. If there is no feedback signal from the protection system 3 s after the fault signal is sent by the unit, the excitation system will open the field circuit breaker and execute the protection-off action by itself automatically.
(2) *External fault*. When an external fault occurs but the excitation system is running correctly, the generator protection system sends the trip command, the excitation system inverts until the excitation current is zero, then blocks the pulses, and opens the field circuit breaker. If the invert fails, the system opens the field circuit breaker automatically. The protection-off command is shown in Figure 15.17.

15.2.5.3 Special Running Modes

Unlocked Operation In this mode, the excitation system blocks some protection and limit functions; this mode is only operation by professional engineers. Inputting the administrator password on the OP allows entry into this test mode.

Field Flashing When the excitation system starts under a residual voltage, the input voltage for the rectifier requirement is only 35 V. The field flashing power supply can be selected from AC or DC factory power. When field flashing succeeds and then switches to the self-shunt closed loop excitation mode, the rectifier works normally, and field flashing terminates. All processes are controlled and monitored by the program.

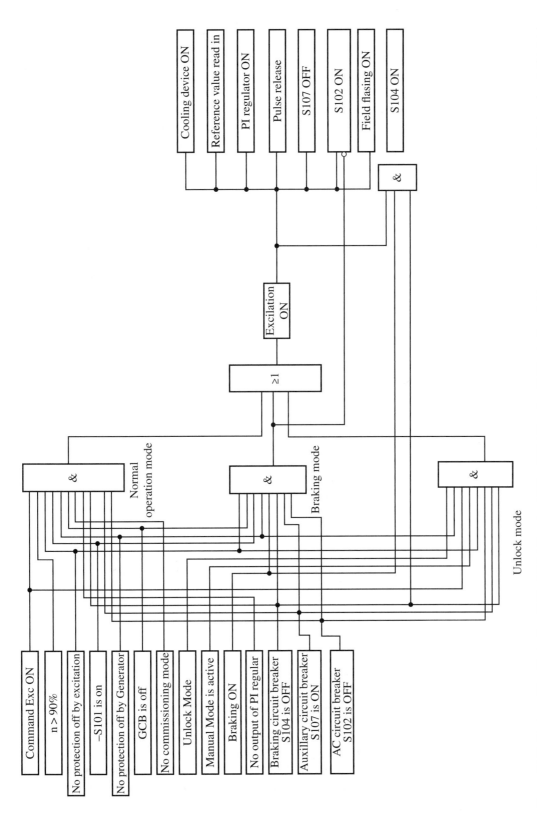

Figure 15.15 Start condition illustration. S101 – DC field breaker; S102 – AC field breaker; S107 – auxiliary field suppression breaker; S104 – electric breaker. Authorized by Nanjing SIEMENS power plant automation company.

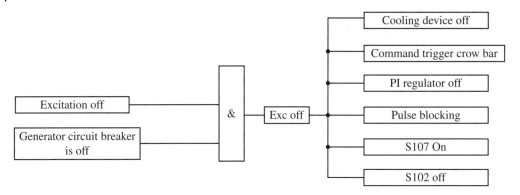

Figure 15.16 Stop condition logic in normal case. Authorized by Nanjing SIEMENS power plant automation company.

De-excitation Mode De-excitation has two operational methods. In the normal stop condition, invert de-excitation is executed as the main method, during which the operational trigger angle is set to 150°, and the DC current returns to the line through the inverter mode. If there is a fault during the normal running condition, the mechanical method is executed: the DC field circuit breaker opens and connects the de-excitation resistor to the circuit to transfer the field energy to heat for de-excitation. If a generator internal fault occurs, protection will send a trip command in order to protect the generator; de-excitation of internal energy should be fast and safe. So the field circuit breaker, de-excitation breaker, and jumper should work together to implement this function. The field circuit breaker cuts off the excitation current and connects the de-excitation resistor to the field circuit to reduce the de-excitation time. When the field circuit breaker opens, with the jumper, the related thyristor is triggered, and the excitation current passes through the resistor. The sudden generator terminal three-phase short-circuit fault will produce the maximum de-excitation energy.

15.2.5.4 Limiters

Limiters keep generator operation safely within the range of the reactive capability. All limiters are integrated in closed loop control. So all limiters are active only in the auto mode; in the manual mode, these limiters are not active.

Stator Current Limiter Stator current limiter function: When the generator stator current persists at a high level, the stator current limiter function prevents the generator from over-temperature insulation damage. Due to the generator's different operating points, if the generator voltage setting value decreases, the limiter active point should be set in the overexcitation area; otherwise, if the generator voltage setting value increases, the limiter active point should be set in the underexcitation area. One point to remember is that the generator current limiter has a dead band for ±10% of the reactive current. The stator current limiter module is shown in Figure 15.18.

V/Hz Limiter The V/Hz limiter limits the ratio of the generator voltage to the frequency in order to prevent overheating from overexcitation of the generator and main transformer. The overexcitation limiter has an inverse time character. The V/Hz limiter calculates the ratio of the generator voltage to the frequency. If this ratio extends the setting point, depending on differences of the exceeding value, the limiter will send decrease voltage signal and alarm trip signals. The V/Hz limiter is enabled during the no-load mode and online. When the frequency changes, the V/Hz limiter will change the generator voltage setting point accordingly, and this relationship is shown in Figure 15.19. Also, parameters can be adjusted in the program, which is also shown in the figure.

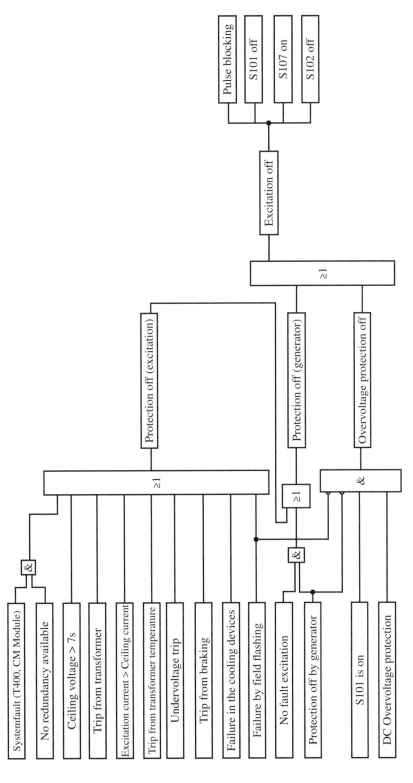

Figure 15.17 Protection trip command. Authorized by Nanjing SIEMENS power plant automation company.

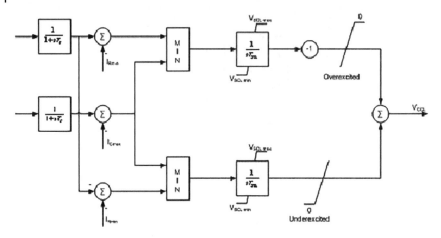

Figure 15.18 Stator current limiter module. Authorized by Nanjing SIEMENS power plant automation company.

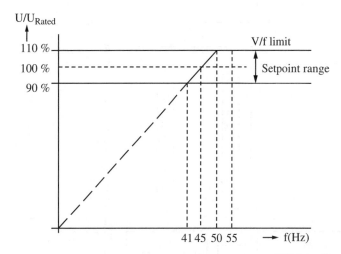

Figure 15.19 V/Hz limiter. Authorized by Nanjing SIEMENS power plant automation company.

Underexcitation Limiter The underexcitation limiter function is to prevent the excitation current from dropping too low, which would make the generator lose static stability. The limiter setting point is a straight linear function which normally contains a multistage line that is traced by 3–4 points in the program. Due to the different allowed levels of the leading phase under the condition of the same generator active power and different voltages, the underexcitation limiter should have the function of modifying the limiter setting point according to the actual voltage variation.

There is a high value selection between the input of the voltage control closed loop and the output of the underexcitation limiter. When the underexcitation limiter is active, the input of the voltage control closed loop is disabled. The generator characteristic curve C, as shown in Figure 15.20, is saved in the program module. According to the input of the active current, the program outputs the related reactive power on the basis of the setting value line. This reactive current can be compared with the actual reactive current to judge whether generator operation point is out of the underexcitation limiter curve. If so, the difference between the actual value and the setting value will increase the excitation current, and meanwhile, the underexcitation limiter active signal will be sent to the control room and OP panel. The underexcitation module is shown in Figure 15.21.

15.2 Static Self-Excitation System of Xiangjiaba Hydro Power Station | 517

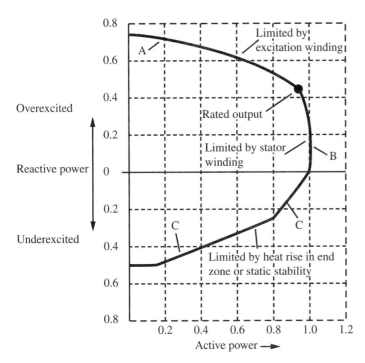

Figure 15.20 Generator operation curve: A – overexcitation limiter; B – stator current limit; C – stator iron core over temperature or underexcitation limiter. Authorized by Nanjing SIEMENS power plant automation company.

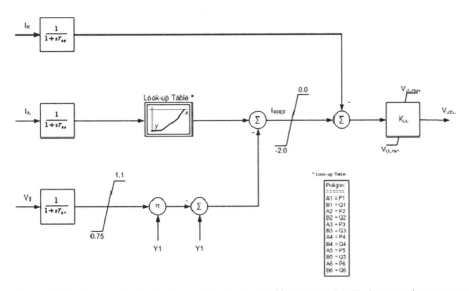

Figure 15.21 Underexcitation limiter module. Authorized by Nanjing SIEMENS power plant automation company.

Overexcitation Limiter The overexcitation limiter function is to prevent the generator rotor winding heat from exceeding the limit heat capacity value. The overexcitation characteristic and generator short-time overload characteristic match and have an inverse time characteristic. The rotor overcurrent time and limiter active returned value can be adjusted on the basis of the rotor overheat performance. Figure 15.22 shows the overexcitation inverse time characteristic curve.

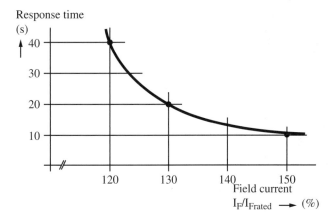

Figure 15.22 Overexcitation limiter inverse time limit characteristic. Authorized by Nanjing SIEMENS power plant automation company.

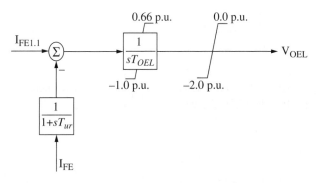

Figure 15.23 Overexcitation limiter module. Authorized by Nanjing SIEMENS power plant automation company. Authorized by Nanjing SIEMENS power plant automation company.

When the excitation current exceeds 1.1 times the maximum allowed rated current, the program begins to calculate the heat accumulated, and if the current exceeds the overheat limit value, the limiter will become active, and then the returned current value should be lower than the maximum allowed current, in order to allow the generator rotor winding and the excitation transformer temperatures to decrease to the normal level. The overexcitation limiter module is shown in Figure 15.23:

Reactive Power/Power Factor Controller The reactive power/power factor controller (Q-, pf-controller) can be switched on or off by operation on OP or in the remote control room. This operating mode can only be activated and operated under the condition that the "Generator on the grid" (generator circuit breaker or online circuit breaker is close) and excitation system operate in the AVR mode. If the generator with activated Q- or pf-control is disconnected from the power network, this upstream control automatically switches back to voltage control. After synchronizing again, the Q or pf-control must be activated manually. The Q- or pf-control can only be activated while the excitation is in the AVR mode. If the operating mode is switched to the manual mode while Q- or pf-control is activated, then "Q-/pf-control OFF" occurs automatically.

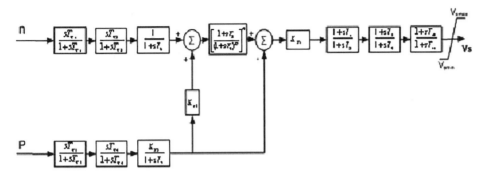

Figure 15.24 PSS 2B model. Authorized by Nanjing SIEMENS power plant automation company.

15.2.5.5 PSS

PSS is an additional control that controls the synchronous motor excitation by means of an AVR to damp the power system power oscillations. The input variables are speed and active power. The Siemens exciter uses the PSS 2B model. The control of the PSS can be achieved by a remote control room or a local operator panel. When the generator active power is less than 20% of the apparent power, the PSS will be automatically turned on. The PSS can be turned on only when the generator is in the automatic synchronization mode. The OP panel displays the action information, and meanwhile the PSS action signal is transmitted to the remote control room via the communication line. Siemens uses the PSS 2B model, as shown in Figure 15.24.

16

Functional Characteristics of Excitation Control and Starting System of Reversible Pumped Storage Unit

16.1 Overview

A reversible pumped storage unit is a motor-driven hydro-turbine with a generator that serves concurrently as a motor and a water turbine that serves concurrently as a pump. During the load valley of the power system (generally at night), it pumps the water stored in the lower reservoir into the upper reservoir with the excess power in the power system, playing the role of load valley filler. During the load peak of the power system (generally in the daytime), when there is a shortage of power, it generates power with the water diverted from the upper reservoir into the lower reservoir, playing the role of peak shaver. Thus, it can serve as a hydro-generator and as a pump motor, acting as not only a power provider but also a power load. In functionality, this is a significant difference between a reversible pumped storage unit and a conventional hydro-turbine.

The past half century has witnessed several stages of development of pumped storage unit structures such as the four-machine pumped storage unit, which has a generating set consisting of a conventional water turbine and a hydro-turbine and a pumping set consisting of a pump and a motor; the three-machine pumped storage unit, which has a generator serving concurrently as a motor but separates the water turbine from the pump; the reversible two-machine pumped storage unit; as well as the variable pole two-speed motor generator set and the motor generator with AC excitation and speed regulation.

This chapter focuses on the reversible two-machine pumped storage unit, which is the most widely used in China.

16.2 Operation Mode and Excitation Control of Pumped Storage Unit

16.2.1 Operation Modes of Pumped Storage Unit

The operational characteristics of a pumped storage unit can be represented in a plane with P-Q coordinates, as shown in Figure 16.1. During the peak load of the power system, the pump storage unit serves as a generator that transmits active power to the system. It can also output reactive power or absorb reactive power. In this case, it operates in the first and second quadrants of the P-Q plane. During the valley load of the power system, the unit serves as a pump motor that utilizes the excess power in the grid, with active power absorbed by the system. It can also send reactive power or absorb reactive power. In this case, it operates in the third and fourth quadrants of the P-Q plane. When the system outputs or absorbs reactive power while absorbing a small amount of active power, it operates in the area of the first and second quadrants near Axis Q in the P-Q plane.

As can be seen from Figure 16.1, for the pumped storage unit connected to the power system, there are four typical stable operation modes: generator lagging phase, generator leading phase, motor lagging phase, and motor leading phase.

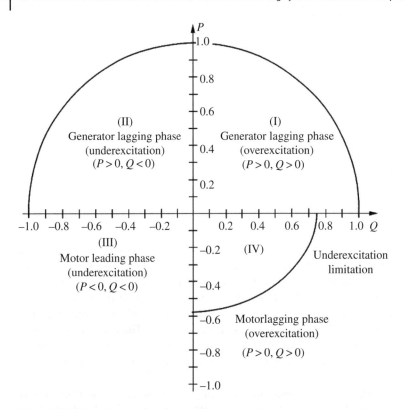

Figure 16.1 Operation modes of pumped storage unit.

16.2.2 Excitation Control Characteristics of Pumped Storage Unit

From the power angle expression of a hydro-turbine,

$$P = \frac{E_q U}{X_d} \sin \delta + \frac{U^2}{2} \times \frac{X_d - X_q}{X_d X_q} \sin 2\delta = P_1 + P_2 \tag{16.1}$$

$$Q = \frac{E_q U}{X_d} \cos \delta - \frac{U^2}{2} \times \frac{X_d + X_q}{X_d X_q} + \frac{U^2}{2} \times \frac{X_d - X_q}{X_d X_q} \cos 2\delta \tag{16.2}$$

It can be seen from Eq. (16.1) that the active power of the synchronous hydro-turbine can be divided into two parts. P_1 is called the basic electromagnetic power, whose value is related to the excitation. P_2 is called the salient pole effect magnetoresistive power, whose value is independent of the excitation and depends on the grid voltage and the motor's vertical and horizontal axis synchronous reactance values.

For Eq. (16.2), when the excitation is zero and $E_q = 0$,

$$Q = -\frac{U^2}{2} \times \frac{X_d + X_q}{X_d X_q} + \frac{U^2}{2} \times \frac{X_d - X_q}{X_d X_q} \cos 2\delta \tag{16.3}$$

When the rotor power angle $\delta = 90°$, that is, $2\delta = 180°$, for the synchronous motor, the reactive power absorbed by the system is the maximum. Its value is as follows:

$$Q_{max} = -\frac{U^2}{X_q} \tag{16.4}$$

Equations (16.1) and (16.2) show that the internal potential E_q of the synchronous motor will change with the excitation current. Thus, the generator can be in the normal excitation, overexcitation, or underexcitation state, and the synchronous motor operates in the generator lagging phase (first quadrant), generator leading phase (second quadrant), motor lagging phase (third quadrant), or motor leading phase (fourth quadrant). It can be seen that a change in the rotor excitation current can change the synchronous motor's reactive power Q, power angle δ, stator current I, and power factor $\cos\varphi$. So, the pumped storage excitation system regulation must meet the requirements of the working conditions in the various operation modes of the unit for excitation regulation.

When the pumped storage unit operates as a generator, the role of the excitation system is generally to keep the voltage constant and improve the stability of the power system operation as required by the system. It regulates the generator terminal voltage and reactive power with the constant voltage regulation method. In the event of a grid failure that causes a voltage drop, the excitation system should have short-term forced excitation capability to ensure the stability of the power system. Besides, to improve the dynamic stability of the power system, the excitation system should be equipped with a power system stabilizer (PSS).

When the pumped storage unit operates as a motor, its terminal voltage and frequency depend on the grid. As a motor of the pump, it primarily inputs electric energy from the system and outputs mechanical energy to drive different loads. In this case, the constant power factor excitation regulation method is generally adopted. The excitation current can be regulated on the basis of the system voltage or load changes. The constant excitation current or constant reactive power regulation method can also be adopted as needed. When the constant excitation current regulation method is adopted, the reactive power decreases with an increase in the active power or increases with a decrease in the active power. So, when the active load trawled by the synchronous motor is stable or does not change much, the constant excitation current regulation method is appropriate. When the load trawled changes drastically, the constant reactive power regulation method can be adopted to make the synchronous motor excitation current increase with an increase in the active load. Thus, the reactive power transmitted by the synchronous motor to the grid remains unchanged, and the grid voltage can avoid large fluctuations. When the pumped storage unit serves as a pump, the system is in the load valley, subject to large voltage changes. Moreover, the load trawled, that is, the pumping head, is slowly changing. In this case, the constant power factor regulation method can increase or decrease the excitation current on the basis of the system voltage or load changes. In the process of reactive power changes, the power factor of the motor can be kept at a constant value, so that the optimum power output can be achieved under economical and energy efficient conditions.

When the unit operates as a synchronous condenser, the constant voltage regulation method can be adopted under both power generation and pumping conditions. Thus, it can automatically increase or decrease the output reactive power as the grid voltage changes in order to maintain the grid voltage. During the deep load valley, it can automatically enter the leading phase operation to absorb the excess reactive power of the grid to prevent the grid voltage from becoming too high.

In the above-mentioned several operation modes, in the event of a grid failure that causes a voltage drop, the excitation system of the pumped storage unit should have the forced excitation function.

For a pumped storage unit operating in the generator or motor state, in the stop process after it is disengaged from the grid, electric braking is required. At the time of electric braking, in order to ensure that the short-circuited stator current is constant, the constant stator current or constant excitation current regulation method can be adopted, so that the requirements of the working conditions in the case of electric braking can be met.

When the static frequency converter (SFC) start or synchronous back-to-back start is adopted for the pumped storage unit, the constant excitation current regulation method is generally adopted. When the SFC starts, the excitation current setpoint remains constant.

16.2.3 Selection of Main Circuit of Excitation System of Pumped Storage Unit

Since a pumped storage unit starts and stops frequently in operation, it does not adopt the main excitation circuit with a rotary exciter but applies the simple self-shunt static excitation connection in order to simplify the structure of the excitation system. The excitation system primarily consists of excitation transformers, three-phase fully controlled rectifier bridges, a field circuit breaker, and an excitation regulator [40]. Some of these devices can serve concurrently as an electrical brake.

When the synchronous start mode is adopted for the pumped storage generator/motor set, the excitation system is required to be able to provide the excitation current when the unit is in the static state, to establish the rotor magnetic potential and make the rotor magnetic potential interact with the stator magnetic potential in order to generate the starting torque. This is fundamentally different from the start of a conventional unit. If the conventional self-shunt excitation system main circuit connection (see Figure 16.2) is adopted, it is impossible to achieve synchronous start. Thus, an external excitation source is required to provide a starting excitation current in the case of static start of the unit. To meet this requirement, there are usually two available excitation main circuit connections, as shown in Figure 16.3.

In Figure 16.3a, the excitation current is outputted by the service power via the starting transformer (braking transformer) through the silicon-controlled rectifier (SCR), so that field flashing, synchronous start, and electric braking of the generator/motor set can be achieved. After the start is completed, the AC power supply is switched to the excitation transformer side of the generator terminal to supply power. This connection is also a self-shunt excitation system when the unit is in normal synchronization. This scheme requires the addition of a starting transformer and switching between two AC circuit breakers. Relatively speaking, it represents a more complex connection and control mode.

In Figure 16.3b, the excitation power is taken from the outside of the outlet circuit breaker of the generator/motor set, that is, the system side. Thus, the excitation transformers are connected with the high-voltage

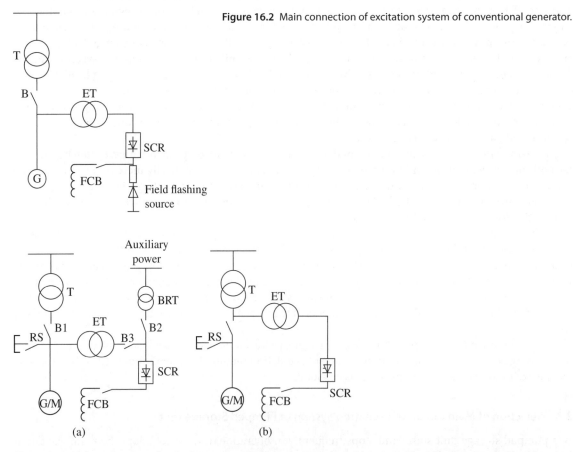

Figure 16.2 Main connection of excitation system of conventional generator.

Figure 16.3 Main connection of excitation system of pumped storage unit: (a) powered by AC power supply of service power, (b) powered by low-voltage power supply of main transformer.

system side. The excitation transformers are always live, without the need for an additional field flashing circuit or starting excitation circuit device. The excitation system adopts the constant excitation current regulation method to directly control the output of the thyristor rectifier bridges in conformity with the start/stop logic control sequence order to achieve field flashing of the generator/motor set, synchronous start under the pump operating conditions, and electric braking at the time of stop. This main circuit features a simple connection and fewer devices and does not require excitation power source switching in the process of start/stop and working condition switching.

The connection in Figure 16.3b can meet the requirements of multi-function excitation systems for integration of start, braking, and operation regulation; improve the operational reliability of excitation systems of pumped storage units subject to frequent start/stop, complex working conditions witching; simplify the device configuration; and facilitate operation control and maintenance. So, it is a preferred connection scheme.

16.3 Application Example of Excitation System of Pumped Storage Unit

A pumped storage unit plays the role of peak shaver, valley filler, and emergency standby in the grid. Its operation modes include generator operating mode (GO), generator condenser operating mode (GCO), pump motor operating mode (PMO), and pump condenser operating mode (PCO). The unit's start modes include SFC start and synchronous back-to-back start. Besides, the unit starts frequently, and the characteristics required for electric braking are required at each stop. In order to meet the above-mentioned requirements of the reversible pumped storage unit, the corresponding requirements for the functionality of the excitation system are described. For ease of description of the various operation modes, the special requirements for the excitation system of the pumped storage unit are briefly described with the excitation system of the units of Tianhuangping Pumped Storage Power Station (see Figure 16.4) as an example. Tianhuangping Pumped Storage Power Station has an installed capacity of 6×300 MW on units manufactured by Alstom.

16.3.1 Excitation System of Units of Tianhuangping Pumped Storage Power Station

At Tianhuangping Pumped Storage Power Station, when a pumped storage unit starts under the pump operating conditions, the excitation system is required to be able to provide a constant excitation current for the rotor circuit when the unit is static [40]. This is a significant difference between a pumped storage unit excitation system and a conventional excitation system.

Under the specific requirements of the project, the main circuit of the excitation system features simple connection and engages fewer devices. The start/stop process does not involve excitation source switching.

In view of the extremely low probability of an interphase short-circuit fault enabled with the enclosed phase-isolated busbar (including the excitation transformers) adopted for the generator terminal, the scheme shown in Figure 16.3b is adopted for the main connection of the excitation system. It should be emphasized that the effect of any grid failure on the excitation system should be noted when the main circuit connection scheme shown in Figure 16.3b is adopted. For example, a grid failure will cause tripping of the circuit breaker of the unit, which will lead to a loss of power supply for the excitation transformers and loss of control of the thyristor rectifiers. Thus, the thyristor elements of the two arms in the same phase turned on last will be in an out-of-control on state. That may cause burnout of the thyristor elements due to the overload. So, in the design of such an excitation system with excitation transformers connected to the outside of the generator terminal circuit breaker, the thyristor elements' ability to withstand the impact of the overload caused by the loss of control in the above-mentioned fault state should be verified. The excitation parameters of the pumped storage unit of Tianhuangping Pumped Storage Power Station include a no-load excitation voltage of 126 V, no-load excitation current of 953 A, rated excitation voltage of 233 V, rated excitation current of 1764 A, forced excitation ceiling voltage of 583 V, forced excitation ceiling current of 3528 A, excitation transformer rated capacity of 3×470 kVA, primary rated voltage of 18 kV, and secondary rated voltage of 510 V.

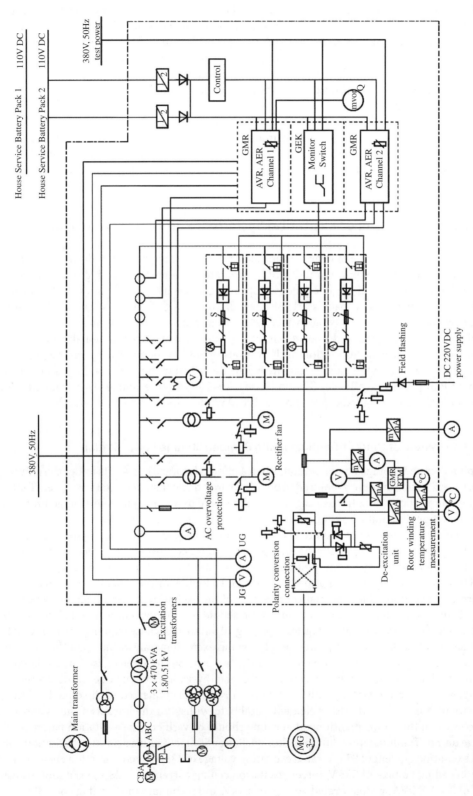

Figure 16.4 Connection of excitation system of 300 MW pumped storage units of Tianhuangping Pumped Storage Power Station.

16.3.2 Composition of Excitation System

The excitation system of the pumped storage units of Tianhuangping Pumped Storage Power Station is provided by Austria's VA TECH ELEN.

16.3.2.1 Excitation Transformers
The excitation transformers consist of three single-phase epoxy cast dry-type transformers. Each single-phase transformer is installed in a separate enclosed metal cabinet. The transformers' rated capacity is 3×470 kVA. Their primary and secondary rated voltages are 18 kV and 0.51 kV, respectively. They feature H-class insulation and AN mode. Their allowable temperature rise is 80 K. In order to ensure the operational safety of the equipment, each single-phase transformer is equipped with a thermosensitive element for transformer temperature measurement. Depending on the temperature setting, two temperature sensors can be used for the winding temperature alarm and overheating trip.

16.3.2.2 Power Rectifiers
Four thyristor rectifiers are connected in parallel. The three-phase fully controlled bridge connection scheme is adopted. Each rectifier bridge and its accessories such as fast fuse, pulse amplifier, bushing-type current sharing core, and R-C absorption circuit are installed in one cubicle. Each rectifier bridge consists of six thyristor elements with an inverse peak voltage of 2600 V. Each rectifier can continuously output 1000 A at an ambient temperature of 45 °C. Each rectifier bridge's DC side and AC side are equipped with a disconnector for ease of its throw-in and cut-out. Even if one rectifier bridge quits the operation, the other three can meet the full load operation of the excitation system.

The top of each rectifier is equipped with a sound-muffling ventilation casing with two cooling fans that serve as standbys for each other. Each fan is equipped with a pressure switch that monitors the pressure of the air duct. In the event of a failure of the fan in service, the standby fan starts automatically. The power supply of the fans is taken from the low-voltage side of the excitation transformers. In the event of a black start of the unit, the power supply of the fans can be taken from the service power.

16.3.2.3 Initial Field Flashing Circuit and De-excitation System
In the normal case, the high-voltage side of the excitation transformers has a power supply. So, the initial field flashing current can be provided directly by the excitation power rectifiers. When the unit needs a black start, the field flashing current is taken from the 220 V DC system of the power station via the field flashing contactor. The field flashing current is 140 V DC, and the field flashing time is 4 s. When the field flashing current makes the generator terminal voltage rise to 5% of the generator rated voltage, the thyristor rectifiers begin to work to make the unit voltage rise to the rated value. When the excitation current reaches 20% of the no-load excitation current, the field flashing contactor is automatically disconnected to quit the field flashing circuit.

The de-excitation system consists of a DC contactor (CEX2000) with a main contact and an arc contact and a SiC nonlinear de-excitation resistor. The DC contactor's maximum breaking current is 18 000 A, and its maximum breaking voltage is 1500 V. The maximum current allowed to flow through the nonlinear resistor is 5000 A. In that case, the maximum voltage at both ends of the de-excitation resistor is 1100 V, and the absorbable energy is 4200 kJ. When a trip signal is input into the de-excitation system, the DC contactor as a field circuit breaker will immediately disconnect the excitation source, and its arc contact will connect the nonlinear resistor to the rotor circuit of the unit for fast de-excitation. In order to improve the reliability of the de-excitation in the event of an accident, the DC contactor is equipped with a double trip coil. Besides, in view of the frequent starts and stops of the pump storage unit, when the unit is normally stopping, the thyristor rectifiers are switches to the inversion mode for de-excitation first. Then, the DC contactor is disconnected in the absence of a load. Thus, the service life of the DC contactor can be improved.

16.3.2.4 Overvoltage Protection Devices
The AC and DC sides of each rectifier are equipped with overvoltage protection devices. A selenium overvoltage limiter is adopted for the AC side to limit the spike voltage. The DC-side overvoltage protection circuit

is a crowbar consisting of two thyristor elements in parallel which have opposing polarity. The nonlinear de-excitation resistor is connected when the crowbar is turned on. If the DC-side overvoltage exceeds the set value (1500 V) of the thyristor trigger module, the corresponding thyristor element will be turned on to connect the de-excitation resistor to the rotor winding circuit. It can suppress the overvoltage on the DC side in the case of a forward or reverse overvoltage. Meanwhile, the monitoring relay on the corresponding overvoltage side will act and send a signal. If the overvoltage does not disappear within the set time, the overvoltage protection device will act on the unit for trip and stop.

16.3.2.5 Excitation Regulator

The excitation regulator features a dual-channel structure. The two channels are independent of each other. Each channel consists of a GMR3 digital excitation regulator device. It is capable of automatic voltage regulation (AVR) and artificial excitation current regulation (AER). The hardware configuration of the GMR3 excitation regulator (see Figure 16.5) is as follows:

(1) The digital circuit power supply (NGT) supplies (5 ± 15) V power for the regulator's internal digital circuit.
(2) The main processor unit (MRB) is primarily used to complete AVR, control of various limiters and auxiliary regulators, and sequential logic and processing of input/output signals.
(3) The subprocessor unit PIM has three subprocessors A, B, and C. The function of Subprocessor A is to form the trigger pulse of the six thyristor elements of the rectifier bridge on the basis of the thyristor trigger control angle signal provided by Subprocessor B and the synchronization signal of the rectifier power supply. The function of Subprocessor B is to convert the regulation signal after the main processor into a trigger angle signal and send it to Subprocessor A. Another function of Subprocessor B is automatic excitation current regulation to meet the requirements of manual operation, electric braking, and pump start. The function of Subprocessor C is to receive the measured signals such as the unit voltage and

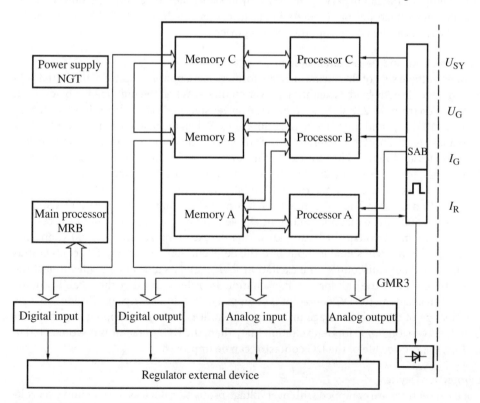

Figure 16.5 Structure of GMR3 digital excitation regulator.

current from the signal processing board (SAB), calculate the various parameters required for the voltage regulation, and supply them to the main processor (MRB).

(4) The signal processing board SAB filters and matches the signals required for the excitation regulator, including the electric variables outputted by the electric variables isolation plate CE130 such as the unit voltage/current and the excitation current and the thyristor voltage outputted by the power panel GEA30, sends them to the subprocessor board PIM, and pre-amplifies the thyristor trigger pulse signal outputted by the subprocessor board.

(5) The digital input card DE32 is used to input the control signals and the status signals of the external devices.

(6) The analog output card AA8 (for regulation of Channel I only) is used to output analog signals such as the rotor temperature and the unit reactive power.

(7) The electric variables isolation plate GE130 is used for electrical isolation and pre-processing of the signals required for the excitation regulator such as the unit voltage/current and the excitation current.

(8) The regulator power panel GEA30 supplies 24 V power to the digital circuit power supply NGT and provides the thyristor a synchronous voltage signal for the signal processing board SAB. In normal operation, GEA30 is powered by the AC bus of the excitation system. In test, it can be powered by the AC 380 V power supply of the service power through switch selection.

The trigger pulse signals outputted by the two regulation channels are outputted to the four power rectifiers via the shared pulse switching unit UMPI, the pulse monitoring unit IMU1, and the pulse distribution unit GEV26, respectively.

16.3.3 Excitation Regulator Software

The excitation regulator software mainly includes the operating system, the regulator program, and the subprograms for the subprocessors. Among these, the regulator program logic programmable software operating in the main processor uses the functional module language, that is, the software module optimally based on the execution time included in the operating system, for programming. The subprograms operating in the subprocessors and the operating systems operating in the main processor are hardware. The operating system is responsible for input and output conversion, regulator program coordination, and regulator serial port communication, and it can provide some self-diagnosis functions. The subprograms are used in the subprocessors to process relevant information required to be rapidly processed such as the thyristor trigger angle calculation, the trigger pulse formation, and the measured value calculation.

The regulator program can perform such functions as AVR, automatic excitation current regulation, constant excitation current regulation, maximum excitation current limitation, limitation of overexcitation current with time limit, minimum excitation current limitation, difference adjustment based on generator current, stator current limitation, load limitation, voltage/frequency limitation, PSS, power factor regulation, system voltage tracking, rotor temperature measurement, and sequential logic control.

The regulator program is divided into several functional modules which have different execution cycles. For example, the voltage regulator program module has an execution cycle of 2 ms, the maximum excitation current limitation program module 4 ms, the logic control program module 50 ms, and the parameter exchange program module up to 300 ms (representing the longest execution cycle). All the modules are stored in an electrically erasable programmable read-only memory (EEPROM) for easy modification.

The GMR3 digital regulator is equipped with eltermGMR, a special handheld terminal which can be easily used to read the measured analog variables inputted into the regulator and the set values inside the regulator and modify all of the regulator's set values and operation-related parameters. In addition, the handheld terminal has a fault diagnosis function, identifying the faulty module and displaying the cause. In order to avoid faulty modification of the regulator, passwords at a number of levels are provided for the handheld terminal, so that its operation range can be identified on the basis of the operation and maintenance personnel levels. Moreover, this function can also be achieved via a compatible PC equipped with special software. In that case, the application program can be modified.

16.3.4 Excitation Regulator Software Flow Diagram [41]

16.3.4.1 Overview

The structure of the GMR3 digital AVR is shown in Figure 16.6. As can be seen from the figure, the excitation regulator includes two control elements. The main element is the AVR automatic voltage control element (VCON), including the PID control element and the PI control element (CCON), that is, the AER excitation current control subordinate element.

Since the dual-loop structure features good dynamic response in all operation modes and applies to a wide frequency range, it has been widely used in pumped storage units.

In the PID voltage regulator, the control law is as follows:

$$F_{UREG}(s) = V_{PU}\left(1 + \frac{1}{sT_{NU}} + \frac{sK_{DU}}{1 + sT_{DU}}\right) \quad (16.5)$$

where

$F_{UREG}(s)$ – transfer function of voltage regulator;
V_{PU} – proportional magnification factor;
K_{DU} – differential magnification factor;
T_{DU} – differential damping coefficient
T_{NU} – stabilization time.

For the PI current regulator,

$$F_{IREG}(s) = V_{PI}\left(1 + \frac{1}{sT_{NI}}\right) \quad (16.6)$$

where

$F_{IREG}(s)$ – transfer function of current regulator;
V_{PI} – proportional magnification factor;
T_{NI} – stabilization time.

The position of Switch S can be changed to select any of the two operation modes: AVR (automatic operation mode) and AER (manual operation mode).

In the automatic mode, both the AVR channel and the AER channel are active. The generator voltage U_{GK} is controlled through the voltage setpoint variable U_{GSW}. In this mode, reactive power difference adjustment and other limiting functions are all operating effectively.

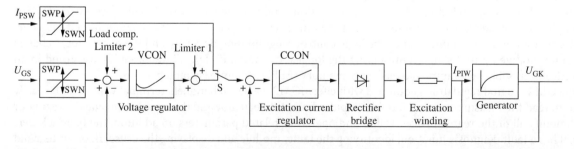

Figure 16.6 Structure of GMR3 digital excitation regulator: U_{GK} – actual stator voltage; U_{GS} – set stator voltage; I_{PIW} – actual excitation current; I_{PSW} – set excitation current; Limiter 1 – signal outputted by excitation current limiter without delay; Limiter 2 – total output signal of several limiters; Load comp. – active and reactive power difference adjustment output signal; S – operation mode selector switch.

In the manual mode, the current loop CONT controls the excitation current on the basis of the manual operation mode setpoint I_{PSW}. In the automatic mode, the AVR VONT is active. The generator voltage U_{GK} is controlled on the basis of the setpoint U_{GSW} in the automatic operation mode. In this mode, the output of the AVR acts as a setpoint of the AER.

16.3.4.2 Switching Between Automatic Mode and Manual Mode

The automatically tracking regulators allow switching from one mode to another without the need for a balance sheet. Switching from the AVR (automatic mode) to the AER (manual mode) can be done manually or automatically. In the event of a failure of the AVR, switching from the AVR to the AER will be done automatically while switching from the AER to the AVR needs to be done manually.

If the operation mode has been switched to the manual mode, this manual mode will be active in the stop and restart process, unless the operation mode is manually switched to the automatic mode. If the setpoint exceeds the setpoint range of the AVR in the manual mode (AER), switching to the automatic mode will be latched.

16.3.4.3 Reactive Power/Power Factor Regulator

Regulation of the reactive power regulation factor can be done with a reactive power regulator. The reactive power is regulated to the setpoint by changing the generator voltage setpoint. In this mode, the AVR is always active. Thus, the transient voltage fluctuations caused by the load can be regulated. The structure of the reactive power regulator is shown in Figure 16.7.

The reactive power regulator feedback is as follows:

$$F_{QRF}(s) = \frac{dK_{PQRF}}{1 + sT_{IQRF}} \tag{16.7}$$

where

$F_{QRF}(s)$ – transfer function of reactive power feedback regulator;

K_{PQRF} – differential magnification factor;

T_{IQRF} – differential damping coefficient.

The setpoint of the reactive power regulator depends on the selected reactive power Q or the power factor $\cos\phi$. The symbol of the setpoint can be positive or negative. A positive sign represents inductive reactive power, while a negative sign represents capacitive reactive power. The setpoint can be limited with the limiter of the selected reactive power regulator.

The reactive power regulator contains a three state controller: outputting the BHQ (order: raise, reactive power regulation), outputting the BTQ (order: lower, reactive power regulation), and outputting no order. These orders act on the voltage setpoint, changing the reactive power of the synchronous motor.

In order to stabilize the regulation, the regulator feeds the differentiated setpoint back to the actual values.

Figure 16.7 Structure of reactive power regulator: Q_{ist} – actual reactive power (reactive power or power factor); Q_{SW} – reactive power setpoint; U_{GSW} – generator voltage setpoint; BHQ – order: raise, reactive power regulation; BTQ – order: lower, reactive power regulation.

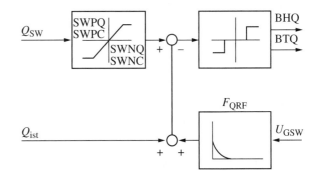

The actual values of the reactive power/power factor regulator are as follows:

$$Q = U_G I_Q$$
$$\tan \varphi = \frac{I_Q}{I_P}$$

where

I_P and I_Q – active and reactive current of generator.

Switching between the voltage regulator and the reactive power regulator is possible only after the synchronous generator is synchronized.

16.3.4.4 Reactive Power Capacity Characteristic Curves and Limiters of Generator

The reactive power capacity characteristic curves of the generator are shown in Figure 16.8, which also shows the limitation functions of the limiters. The function of each limiter is as follows:

(1) The minimum excitation current limiter without delay increases the excitation current to prevent operation at a minimum excitation current value less than the setting value.
(2) The maximum excitation current limiter without delay decreases the excitation current to limit the maximum allowable ceiling excitation current.
(3) The maximum excitation current limiter with delay decreases the excitation current to prevent the unit from overheating caused by an excessive excitation current. It has an inverse time lag characteristic.
(4) The stator current limiter with delay decreases or increases the excitation current based on the operation mode (overexcitation or underexcitation) to prevent the unit from overheating caused by an excessive stator current. It has an inverse time lag characteristic.
(5) The power angle limiter without delay increases the excitation current to prevent the synchronous motor's loss of synchronization.
(6) The voltage/frequency limiter without delay decreases the excitation current to prevent the magnetic density of the generator and the main transformer from exceeding the allowable range.
(7) The generator voltage limiter with delay decreases/increases the excitation current to prevent the generator voltage from exceeding the allowable operating value.

16.3.4.5 Limiters

(1) *Excitation current limiter without delay.* The excitation current limiter without delay includes two (maximum and minimum) PI regulators. On the basis of the reactive power capacity characteristic curves of the generator, the minimum limiter limits the excitation current to the minimum allowable value, and

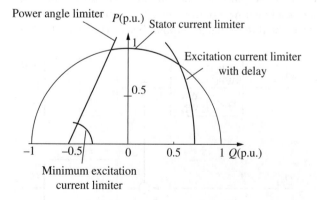

Figure 16.8 Reactive power capacity characteristic curves of generator.

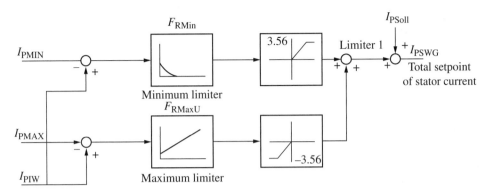

Figure 16.9 Block diagram of excitation current limiter without delay: I_{PIW} – actual excitation current; I_{PMIN} – minimum excitation current limit; I_{PMAX} – maximum excitation current limit; Limiter 1 – limiter output signal (see the block diagram of the regulator); I_{PSoll} – output signal of voltage regulator; I_{PSWG} – total setpoint of excitation current in automatic mode.

the maximum limiter controls the maximum excitation current of the generator below the forced excitation ceiling current. The two limiters can be individually thrown in or cut out. The block diagram of the excitation current limiter without delay is shown in Figure 16.9.

The transfer function of the PI regulator of the minimum excitation current limiter is as follows:

$$F_{RMin}(s) = K_{PMin} + \frac{1}{sT_{IMin}} \tag{16.8}$$

where

$F_{RMin}(s)$ – transfer function of minimum excitation current limiter;

K_{PMin} – proportional magnification factor;

T_{IMin} – integration time.

The transfer function of the PI regulator of the maximum excitation current limiter is as follows:

$$F_{RMaxU}(s) = K_{PMaxU} + \frac{1}{sT_{IMaxU}} \tag{16.9}$$

where

$F_{RMaxU}(s)$ – transfer function of maximum excitation current limiter;

K_{PMaxU} – proportional magnification factor;

T_{IMaxU} – integration time.

The output signals of the limiter are superposed in the signals outputted by the voltage regulator. So, the limiter can directly affect excitation current regulation. When the actual excitation current is below the minimum limit I_{PMIN}, the limiter will start and the minimum excitation current limiter will increase the excitation current. When the actual excitation current is above the maximum limit I_{PMAX}, the limiter will start and the maximum excitation current limiter will decrease the excitation current.

(2) *The maximum excitation current limiter with delay.* The current outputted by the maximum excitation current limiter has an inverse time lag characteristic to ensure that the excitation current operates at the allowable temperature of the rotor excitation winding. The block diagram of the maximum excitation current limiter with delay is shown in Figure 16.10.

The actual excitation current IPIW is sent to a comparator via a delay unit for comparison.

If the actual excitation current after the delay is greater than the maximum allowable value I_{PMAXV}, the comparator will start. Meanwhile, the limit integrator begins integration in the negative direction. The output of the integrator will reduce the setpoint of the stator voltage.

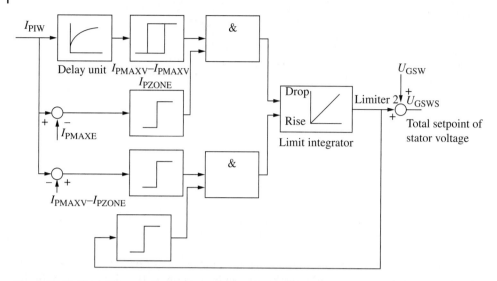

Figure 16.10 Block diagram of maximum excitation current limiter with delay: I_{PIW} – actual excitation current; I_{PMAXV} – maximum excitation current; I_{PZONE} – lagging zone; Limiter 2 – output signal of Limiter 2; U_{GSW} – setpoint of generator voltage; U_{GSWS} – total setpoint of generator voltage.

When the excitation current drops to below the maximum allowable value I_{PMAXV}, the limit integrator will be removed, and the output signal will remain unchanged.

When the excitation current drops to $I_{PMAXV} - I_{PZONE}$, the limit integrator's control output signal will be reset to zero, and the effect of limitation will be reduced.

The transfer function of the delay unit is as follows:

$$F(s) = \frac{1}{1 + sT_{VIPB}} \tag{16.10}$$

The response time t_{an} of the comparator can be obtained from Eq. (16.11):

$$t_{an} = -T_{VIPB} \ln \frac{I_{P2} - I_{PMAXV}}{I_{P2} - I_{P1}} \tag{16.11}$$

where

T_{VIPB} – delay time;
I_{P1} – operating current before overcurrent;
I_{P2} – overcurrent;
I_{PMAXV} – comparator threshold limit.

Figure 16.11 shows the excitation current response characteristics of the maximum excitation limiter with delay.

For Figure 16.11a, $I_{P1} = 0.2$ p.u., $I_{PMAXV} = 1.1$ p.u., and $I_{P2} = 1.4$ p.u.

When the data are substituted into Eq. (16.11), the following equation is obtained:

$$\frac{t_{an}}{T_{VIPB}} = \ln \frac{1.4 - 1.1}{1.4 - 0.2} = 1.386$$

For Figure 16.11b, $I_{P1} = 0.8$ p.u., $I_{PMAXV} = 1.1$ p.u., and $I_{P2} = 1.4$ p.u.

$$\frac{t_{an}}{T_{VIPB}} = \ln \frac{1.4 - 1.1}{1.4 - 0.8} = 0.693$$

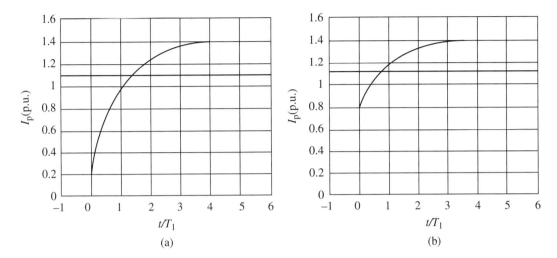

Figure 16.11 Step response of excitation current: (a) $t_{an}/T_1 = 1.3863$, (b) $t_{an}/T_1 = 0.6931$.

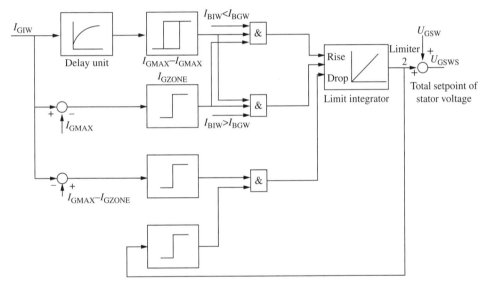

Figure 16.12 Block diagram of stator current limiter: I_{GIW} – actual stator current; I_{GMAX} – maximum stator current; I_{GZONE} – lagging zone; Limiter 2 – output signal of Limiter 2; I_{BIW} – stator current output; I_{BGW} – reactive current start limit; U_{GSW} – setpoint of stator voltage; U_{GSWS} – total setpoint of stator voltage.

16.3.4.6 Stator Current Limiter

The stator current limiter has a current inverse time lag delay characteristic to ensure that the stator current is operating at the allowable temperature. It limits the stator current on the basis of the working state of the unit (overexcitation or underexcitation) and increases or decreases the excitation current accordingly. The block diagram of the stator current limiter is shown in Figure 16.12.

The actual stator current I_{GIW} is sent via a delay unit to a comparator for comparison. When the actual stator current after the delay is greater than the maximum allowable value I_{GMAXV}, the comparator will start, and meanwhile the limit integrator will receive a start signal. The integration direction depends on the operating point (overexcitation or underexcitation). The two parameters – I_{BGWP} (positive reactive current limit) and I_{BGWN} (positive reactive current limit) – decide that the unit is in the lagging zone, where the unit operates at a power factor of 1. In the lagging zone, the stator current limiter will remain latched. When the stator current drops to below the maximum allowable value I_{GMAXV}, the limit integrator will be removed, and the output signal will remain unchanged. When the stator current value drops to $I_{GMAX}-I_{GZONE}$, the control output signal of the limit integrator will be reset to zero, and the effect of limitation will be reduced.

The transfer function of the delay unit is as follows:

$$F(s) = \frac{1}{1 + sT_{VIGB}} \tag{16.12}$$

The response time of the comparator can be calculated from Eq. (16.13):

$$t_{an} = -T_{VIGB} \ln \frac{I_{G2} - I_{GMAX}}{I_{G2} - I_{G1}} \tag{16.13}$$

where

T_{VIGB} — delay time;
I_{G1} — operating current before overcurrent;
I_{G2} — overcurrent;
I_{GMAX} — comparator threshold limit

16.3.4.7 Rotor Power Angle Limiter

The power angle limiter is used to limit underexcitation. When the unit's power angle exceeds the maximum value and endangers its operation stability, the power limiter will act without delay. The block diagram of the rotor power angle limiter is shown in Figure 16.13.

The transfer function of this limit controller is as follows:
For the PI unit:

$$F_{RPI}(s) = K_{PUEB} + \frac{1}{sT_{IUEB}} \tag{16.14}$$

For the DTI unit:

$$F_{RDT1}(s) = \frac{sK_{DUEB}}{1 + sT_{DUEB}} \tag{16.15}$$

where

K_{PUEB} — proportional magnification factor;
T_{IUEB} — integration time;
K_{DUEB} — differential magnification factor;
T_{DUEB} — differential damping coefficient.

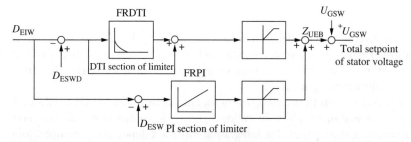

Figure 16.13 Block diagram of rotor power angle limiter: D_{EIW} – actual power angle; D_{ESWD} – power angle differential limit; D_{ESW} – static power angle limit; Z_{UEB} – additional signal of underexcitation limiter; U_{GSW} – setpoint of stator voltage; U_{GSWS} – total setpoint of stator voltage.

The control variable equation is as follows:

$$Z_{UEB}(s) = \left[\frac{D_{ESWD}}{s} - D_{EIW}(s)\right][F_{RDT1}(s) + 1]$$
$$+ \left[\frac{D_{ESW}}{s} - D_{EIW}(s)\right]F_{RPI}(s) \tag{16.16}$$

The underexcitation limiter is tuned with the PIDTI. The DTI component of the regulator is used to quickly intervene if abrupt changes occur in the power angle while working in the case of a small magnification factor taken for the PI unit. Generally, when the power angle of a synchronous motor is involved, what should be first talked about should be the internal power angle or the external power angle, as shown in Figure 16.14.

The internal power angle δ_i is between the rotor air gap emf e_p direction (i.e., the rotor horizontal axis) and the stator voltage U. The external power angle δ_a is between e_p and the system voltage U_n. The external power angle δ_a plays an important role in the operation stability of the unit.

$$\delta_a = \arctan\left|\frac{P}{Q + \frac{U^2}{X_q}}\right| + \arctan\left|\frac{P}{Q - \frac{U^2}{X_n}}\right|$$

To calculate δ_a, $\dot{e}_p$ is replaced by a phasor in the same phase as $\dot{e}$.
The internal emf is calculated as follows:

$$\dot{e}(t) = \dot{u}(t) + j\dot{i}(t)X_q \tag{16.17}$$

The system voltage is calculated as follows:

$$\dot{u}(t) = \dot{u}(t) - j\dot{i}(t)X_n \tag{16.18}$$

16.3.4.8 Voltage/Frequency Limiter Generating Additional Frequency Signals
The voltage/frequency limiter determines the maximum and minimum values of the stator voltage on the basis of the frequency of the synchronous motor or keeps the ratio of the stator voltage to the frequency of the

Figure 16.14 Power angle calculation phasor diagram for synchronous motor: e – internal voltage; U – stator voltage; U_n – system voltage; i – stator current; δ_i – internal power angle; δ_a – external power angle; X_d – motor vertical axis synchronous reactance; X_q – motor horizontal axis synchronous reactance; X_n – system impedance.

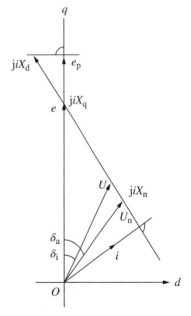

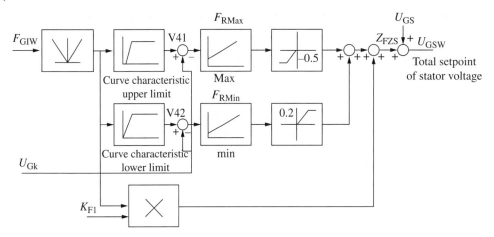

Figure 16.15 Block diagram of voltage/frequency limiter: F_{GIW} – stator voltage frequency; U_{GK} – actual stator voltage; Z_{FZS} – additional signals of voltage/frequency limiter; U_{GSW} – setpoint of stator voltage; K_{FI} – frequency magnification factor when generator is operating on separate grid.

synchronous motor constant. In addition, it generates frequency-dependent additional signals during voltage regulation. These two functions can be individually thrown in or cut out.

The block diagram of the voltage/frequency limiter is shown in Figure 16.15.

The transfer function of the PI regulator for regulation of the maximum and minimum values of the stator voltage is as follows:

$$F_{RMin}(s) = F_{RMax}(s) = K_{PF} + \frac{1}{sT_{IF}} \tag{16.19}$$

where

$F_{RMax}(s)$ – transfer function of maximum stator voltage limiter;

K_{PF} – proportional amplifier;

$F_{RMin}(s)$ – transfer function of minimum stator voltage limiter;

T_{IF} – integration time.

The actual frequency F_{GIW} of the generator is introduced into the upper and lower limit characteristic curve elements for comparison to form the upper and lower limits of the generator stator voltage. That is, the generator voltage can only operate between the two limits. The output, through the subtractor, determines the difference between the stator voltage U_{GK} and the allowable limits and outputs the differences to the two PI regulators respectively to achieve the maximum and minimum limits of the voltage/frequency ratio.

Figure 16.16 shows the operating range of the voltage/frequency limiter.

The slope and the maximum limit of the generator voltage are determined by F_{GMNFG} and F_{GMNUG}.

For the frequency-related additional signals shown in Figures 16.15, their frequency is linearly proportional to the generator stator voltage. This function is applicable to the case where the generator has a resistive load and operates on a separate grid. A linear proportional relation between the generator voltage and the frequency will be more conducive to the governor's speed regulation.

The voltage/frequency upper and lower limits and the three frequency-related functional signals, that is, the U/f limiter's additional signals, are all outputted to the generator voltage setting value.

16.3.4.9 PSS

The PSS can provide additional signals to damp the rotor power oscillations. It generates auxiliary signals with the calculated active value, finds its differential, and breaks it down into components proportional to the

Figure 16.16 Operating range of voltage/frequency limiter: F_{GMXUG} and F_{GMNUG} – upper limit and lower limit of generator voltage, respectively; F_{GMXFG} and F_{GMNFG} – upper limit and lower limit of generator frequency, respectively; F_{GMXFO} – base point determining upper limit curve.

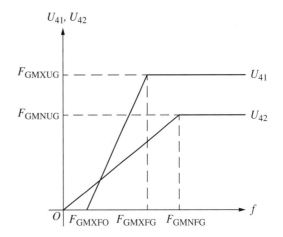

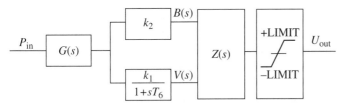

Figure 16.17 Block diagram of transfer function of PSS.

angular velocity and the angular acceleration of the generator rotor. The two components can form additional signals with regulable phase and amplitude, acting on the set point of the voltage regulator. These signals can limit the positive direction and negative direction values (BEGR+ and BEGR−) to ensure that the voltage of the control generator fluctuates within the allowable range. The transfer function of the PSS is shown in Figure 16.17.

The transfer function of each of the PSS signals is now given.

The input filter is as follows:

$$G(s) = \frac{1}{1+sT_1} \times \frac{1}{1+sT_2} \times \frac{sT_3}{1+sT_3} \times \frac{sT_4}{1+sT_4} \times \frac{sT_5}{1+sT_5}$$

The acceleration signal is as follows:

$$B(s) = G(s)k_2$$

The speed signal is as follows:

$$V(s) = G(s) \times \frac{k_1}{1+sT_6}$$

$$T = T_2 = 0.02$$

$$T_3 = T_4 = T_5 = 4.22$$

$$T_6 = 2.15$$

where

- T_1–T_6 — filter time constant;
- k_1 — speed signal magnification factor;
- k_2 — acceleration signal magnification factor.

The integrated PSS signal $Z(s)$ consists of speed and acceleration signals:

$$Z(s) = +a[B(s)(1-p) + V(s)(1-|p-1|)] \text{ (when } p = 0\text{--}2)$$

$$Z(s) = +a[B(s)(3-p) + V(s)(1-|p-3|)] \text{ (when } p = 2\text{--}4)$$

where

a – magnification factor (AMPL), ranging from 0 to 15.9995;

p – phase (PHASE), ranging from 0 to 4 = 0–360°.

In the PSS, the two regulable parameters a and p allow the phase and the magnification factor to be individually set and added to the generator voltage setting value.

16.3.5 Additional Functions of Excitation Regulator

16.3.5.1 Limitation Protection

In order to ensure the operational safety of the excitation device and the unit, the following protection functions are provided in the regulation control software of the excitation regulator as described above:

the maximum excitation current limiter, used to limit the output of the maximum allowable excitation current;
the excitation overcurrent limiter, used to limit the output of the maximum allowable continuous current;
the minimum excitation current limiter, used to limit the output of the minimum allowable excitation current;
the stator current limiter, used to limit the maximum allowable stator current of the generator motor;
the load power angle limiter, used to prevent the load power angle between the generator voltage and the current from being too large;
the voltage/frequency limiter, used to prevent the voltage/frequency ratio from being too large when the generator is operating;
the PSS, used for low-frequency power oscillations of the damping system;
and the forced excitation limiter, used to limit the excitation voltage and current multiples at the time of forced excitation in the event of a significant drop of the generator terminal voltage caused by a circuit remote fault.

16.3.5.2 Operation and Switching of Excitation Regulator

The excitation regulator has remote and local control modes and auto, manual, and test operation modes.

In the local control mode, non-disturbance switching between Channel 1 and 2 and between the auto operation mode and the manual operation mode as well as start/stop of the excitation device, increase or decrease of the load, and switching between the fan in service and the standby can be achieved.

16.3.5.3 Debugging and Self-Test of Excitation Regulator

Each channel main processing board (MRB) is equipped with an RS232 communication port. A PC or handheld terminal can be used to not only modify some of the parameters or control functions of the software in operation but also to monitor some important operating data and control variables during the test operation of the unit. Meanwhile, each channel has a certain self-test function that monitors the working status of the main excitation elements such as the excitation input and output card, control program, excitation source, and thyristors in real time and issues alarms or trip signals in a timely manner.

16.3.6 Excitation Regulation in Different Operation Modes

16.3.6.1 Working Conditions Including GO, GCO, PMO, and PCO Provided for Pumped Storage Units of Tianhuangping Pumped Storage Power Station

For the PMO mode, SFC start is adopted, with back-to-back start as a standby. For stopping the unit, an electric brake is provided. So, the excitation system should not only have the functions of conventional excitation systems but also meet the synchronous start, operational, and electric brake requirements of the PMO mode.

16.3.6.2 Simultaneous Start under PMO Conditions

During the SFC or back-to-back start, the excitation regulator operates in the AER mode. When the pump start conditions are in place (the SFC or back-to-back start mode has been selected; the excitation system has been ready for the start of the unit; and the unit's speed is lower than 1%), the "excitation throw-in" order can be given via the power station monitoring system or the excitation local control interface to close the field circuit breaker. During the first 10 s of the start process, the excitation regulator provides the excitation current at the setpoint. In the following 20 s, the excitation current will increase to the determined excitation current. This excitation current will be maintained until the unit's speed rises to 90% of the rated speed. Then, the regulator will be automatically switched from the AER mode to the AVR mode.

16.3.6.3 Pump Operation Mode

A pumped storage unit operates in the form of a pump load in the power system. The pump unit will become a high-value load of the power system. In order to reduce the line reactive power loss, the pump unit should operate in a state of a power factor $\cos\varphi \approx 1$ as far as possible. The excitation system of the units of Tianhuangping Pumped Storage Power Station is equipped with a power factor regulator. When the unit is set to in the constant power factor $\cos\varphi \approx 1$, the excitation system will control the excitation current and make the unit operate in a state of a set constant power factor $\cos\varphi \approx 1$. Thus, the loss caused by the power system's transmission of reactive power can be reduced.

16.3.6.4 Electric Braking

Since the excitation source is taken from the low-voltage side of the main transformer, the excitation system is still able to provide the excitation current required for the electric braking during the unit braking although the unit circuit breaker is in the open position. At the time of the electric braking, the braking current is controlled by the rotor current. At this point, the excitation system works in the AER mode. The amplitude of the braking current is determined by the excitation current. When the excitation throw-in conditions are in place for the electric braking (e.g., the excitation system is accident-free and there is no stop order; the unit electric braking signal has been given; the unit speed is greater than 1% and less than 95%; and the generator motor circuit breaker is in the open position), the "excitation throw-in" order will be given via the power station monitoring system or the excitation cubicle local control panel, so that the unit short-circuit switch and the field circuit breaker are first closed, after which the unit enters the electric braking state. When the unit speed is 1%, the electric braking process will end. When the unit stops rotating, the excitation system can automatically cut off the excitation current through the internal generator current monitoring.

In summary, the excitation system of the pumped storage units of Tianhuangping Pumped Storage Power Station is a multifunctional excitation system which integrates such functions as field flashing, SFC start, back-to-back start, generator/motor operation, and electrical braking and meets the operational requirements of pumped storage units. In terms of functionality, it is significantly different from conventional excitation systems.

16.3.6.5 SFC or Back-To-Back Operation

When the unit control system selects SFC start or back-to-back start in the PMO or PCO mode, the field circuit breaker is closed after the field flashing. The excitation current regulator controls the output excitation current. As the unit speed increases, the excitation current remains constant. When the speed is greater than 90% of the rated speed, the excitation system is automatically switched to the AVR mode.

16.3.6.6 Test Operation Mode

The excitation test operation mode is used to check the characteristics of the power elements of the excitation system with the unit short-circuit test and the generator no-load test. When the excitation system is switched to the test mode, the excitation operates in the manual mode. The field flashing current set manually is zero. The start/stop of the excitation system and the increase/decrease of the excitation current can be controlled through the excitation regulator panel.

16.4 Working Principle of SFC

16.4.1 Connection of SFC of Pumped Storage Unit

Generally, a reversible pumped storage unit has five steady-state operation modes, including GO, GCO, PMO, PCO, and Stop.

The PMO mode must be started by an external power supply. Generally, SFC start is the main start mode, with synchronous pair trawling (back-to-back start, one trawling unit + one trawled unit) as a standby.

In essence, the SFC control system is a device based on the speed regulator. The SFC applies torque to the pump motor and obtains the desired speed in turn.

This section focuses on the working principle and connection of SFCs.

(1) On the basis of power supply for main circuit, SFC categories include high-low-high SFC with power supply by step-down transformer and high-high SFC without power supply by step-down transformer.
(2) On the basis of coupling between rectifier and inverter circuits, SFC categories include AC-DC-AC SFC and AC-AC SFC.
(3) On the basis of intermediate DC coupling combination, SFC categories include reactor coupling current-type SFC and capacitor coupling voltage-type SFC.

Since pumped storage units pose no special requirements for dynamic response performance of SFC, China's pumped storage power stations generally adopt AC-DC-AC current source SFCs composed of a thyristor and a smoothing reactor.

For a high-low-high SFC, a step-down transformer is adopted on its input side and a step-up transformer on its output side. The purpose is to match the SFC voltage with the grid- and generator-side voltage. Since generator motors of pumped storage power stations have large capacity, investment in high-low-high SFCs is large if the transformer and reactor configuration and continuous conductor costs of the system are considered. However, in view of the current level of manufacturing of power electronic components and voltage withstand capacity, the high-low-high SFC connection is still the mainstream choice.

In order to suppress the influence of the harmonic components of a high-high SFC, generally an isolation transformer is adopted on its input side or a harmonic filter is added and connected to it. The output side is directly connected with the generator-motor set stator side, bypassing the step-up transformer. Compared to a high-low-high transformer, this scheme boasts of lower equipment investment and operation and maintenance costs and higher reliability.

It should be noted that a general transformer is generally a 6-pulse wave rectifier, generating large harmonic components. The difference between transformer secondary winding connection modes such as star connection and triangular connection creates a 30° electrical angle between the two three-phase AC power supplies of the secondary winding in the same phase. Thus, the number of pulse waves of the rectifier circuit can be increased from 6 to 12, so that the low-order harmonic current components can be reduced.

So, there are two schemes for the above two SFC main circuits: the 6-pulse wave and 12-pulse wave schemes. Figure 16.18 shows several typical SFC connection modes classified by function.

16.4.2 Composition of SFC

The functionality of an SFC is described with the 6-pulse wave high-low-high SFC of the 300 MW pumped storage units of Tianhuangping Pumped Storage Power Station as an example.

Tianhuangping Pumped Storage Power Station has 6300 MW pumped storage units, representing the largest total capacity in power stations in China.

The SFC is supplied by France's CEGELEC. The six units share two SFC sets. The main advantages of this SFC are stepless speed change, smooth start, and good dynamic response performance.

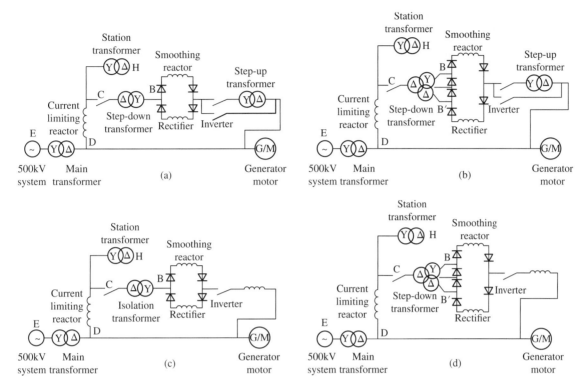

Figure 16.18 Typical connection of SFC: (a) high-low-high SFC 6-pulse wave scheme, (b) high-low-high SFC 12-pulse wave scheme, (c) high-high SFC 6-pulse wave scheme, (d) high-high SFC 12-pulse wave scheme.

For an AC-DC-AC coupling current-type SFC, its rated voltage is 18 kV, maximum input capacity 33 MVA, and rated output power 22 MW. Its grid-side input power's rated voltage is $18 \times (1 \pm 10\%)$ kV, rated frequency 50 Hz, and rated current 120 A. Its output voltage is 0–19.8 kV, output frequency 0–52.5 Hz, and output rated current 120 A. Its design starting acceleration time is 210 s. It allows eight continuous starts (including two standby starts) of the unit. Thus, it can be determined that the short-time duty is continuous operation of 30 min and a stop of 30 min at the rated current.

The functionality of the SFC is described below on the basis of its component units.

16.4.2.1 Power Unit

The power unit includes a current limiting reactor that stabilizes and limits the SFC input current. Circuit breakers connected to the SFCs input and output circuits cut off the circuit current in the event of a fault in the unit start process or after the synchronization or in the start circuit. During the start of the SFC, the input step-down transformer connected between the grid and the network bridge will play the role of a harmonic filter by suppressing the large number of high-order harmonics generated in the start process and will improve the operating power factor and the voltage and current waveforms of the SFC.

In the SFC circuit, the thyristor bridge connected to the grid side is called a rectifier bridge or a simple recurrent network (SRN). Its role is to rectify the low voltage of the secondary winding of the step-down transformer into a DC voltage.

The role of the thyristor bridge connected to the pumped storage unit side is to invert the DC current into an AC current. The thyristor bridge is called an inverter bridge or simple recurrent machine (SRM). The connection is shown in Figure 16.19.

544 | *16 Functional Characteristics of Excitation Control and Starting System of Reversible Pumped Storage Unit*

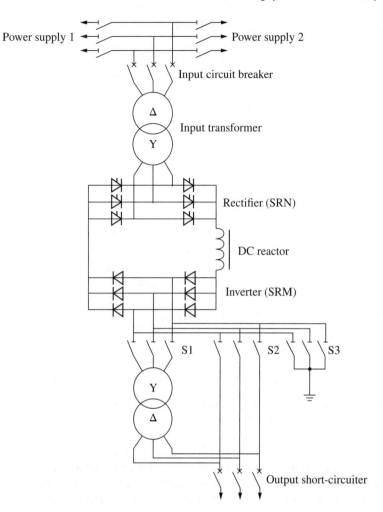

Figure 16.19 Composition of SFC.

16.4.2.2 Control and Protection Units

The control units include the measurement unit, the pulse unit, the programmable numerical controller (PNC), and the programmable logic controller (PLC). The functionality of each unit is described as follows:

(1) The measurement unit is used to measure the various variables required for the SFC regulation control, such as the SRM-/SRN-side voltage, the voltage provided by the current transformer, the current signal, and the rotor position sensor.
(2) The pulse unit provides such functions as thyristor trigger pulse and photoelectric control conversion.
(3) The PNC is used for closed-loop regulation and control of the SFC. Through regulation of the thyristor control angle, it can change the unit's current and power factor to achieve control of the unit's torque and speed.
    The PNC controls and monitors the SRN and the SRM through an external processing unit (PU). The PNC sends six trigger signals through the fiber. Each arm receives one trigger signal, after which the trigger card distributes the pulse signal to each arm. Each arm has 18 thyristor elements.
(4) The monitoring card sends the status information of each thyristor, including fault signals, to the PLC through the serial connection port.
(5) The PLC is used to monitor the input and output variables of the SFC system as well as the fault status handling.

The PLC also processes breaking/closing of the input and output circuit breakers (OCBs) as well as handling of the auxiliaries (cooling water pump, oil pump, and fan). It also processes the SFC start sequence and the frequency reference values.

When the selector switch is "Follow-up on," the grid frequency will be followed up. When the selector switch is "Follow-up off," the frequency will be regulated in accordance with the "Speed up" or "Speed down" order issued by the synchronizer.

16.4.2.3 SFC Control Mode

The control flow of the SFC circuit is described briefly below, with Tianhuangping Pumped Storage Power Station as an example. As mentioned earlier, the fully controlled bridge in the SFC circuit has two operation states: rectification and inversion. The DC output end of the fully controlled bridge can be equivalent to a series circuit consisting of an emf or cemf and a diode. The diode specifies the direction of the DC current. The amplitude and the direction of the DC voltage U_d are determined by the closed-loop control device. When U_d is positive, it is a rectifier bridge, which rectifies the AC power inputted by the grid or the motor and outputs it to the DC load circuit in the form of DC. When U_d is negative, it is an inverter, which inverts the DC power inputted by the DC circuit and feeds it back to the grid or motor side in the form of AC power at the corresponding frequency.

As is well known, there must be an appropriate AC voltage for automatic commutation of the fully controlled bridge. When the AC voltage is too low, automatic commutation of the fully controlled bridge is impossible. In the SFC start process, the excitation current applied to the motor rotor is constant, and the terminal voltage of the unit being trawled is proportional to its speed (U/F is constant). At the beginning of the start, the unit is in the static state, with a terminal voltage of zero. Automatic commutation of the thyristor bridge is impossible. So, the SFC start process is divided into two stages: the low-speed operation stage (pulse coupling operation mode) and the high-speed operation stage (synchronous operation mode). Thus, the low-voltage commutation and trigger problems can be solved in different ways:

(1) *Low-speed start stage.* In order to achieve commutation of the thyristor bridge under the low-voltage low-speed start conditions, the position of the rotor and the direction of the starting torque must be measured first. The rotor position is measured with a magneto-sensitive mechanical position sensor and the torque direction with a PNC closed-loop control device. The working principle of PNC closed-loop control is shown in Figure 16.20.

The current and voltage signals measured from the SRM- and SRN-side current and voltage transformers are converted by the *I/U* and *U/U* converters into voltage signals and sent into the SCN203 card of the PNC. The SCN203 card converts these voltage signals into digital variables and sends them into the central PUs of the SCN824 card (for SRN control) and the SCN825 card (for SRM control) of the PNC, respectively. Meanwhile, at the beginning of the trawling, the rotor position sensor sends the rotor position signal into the SCN825 card. These signals are calculated via the internal software in the SCN824 card and the SCN825 card, and then the thyristor trigger pulse signals are outputted by the SCB302 card. The trigger signals outputted are current signals, which are converted by the SCN954 card and the SCN653 card of the photoelectric converter into optical signals and then sent via the fiber into the 20AM/20AE trigger cards. The trigger signals are sent to the pulse unit. The 20AM/20AE trigger cards are powered by 220 V AC (5 V/15 V DC). The 18 thyristor elements of each arm use one 20AM card and two 20AE cards. The pulse signal passes through the 20AM/20AE cards to form 19 pulse signals. One of these pulse signals is sent into the 20RM/20RE cards as the synchronization signal of the pulse monitoring unit, while the other 18 signals are sent into the OEXA31 cards of the 18 thyristors. Each OEXA31 card converts the optical trigger signals inputted into square wave signals that control the thyristors. When the thyristor forward bias voltage is greater than 45 V, the OEXA31 card will generate a 15 μs pulse and send it to the thyristor trigger pole, thus controlling the thyristor turn-on. An OEXA31 card will send a pulse feedback signal to the 20RM/20RE card to identify whether the thyristor or pulse trigger circuit is faulty. If the thyristor or pulse trigger circuit is faulty, the 20RM/20RE card will convey information regarding the fault of the 20RS card. Then, the information will be transmitted by the 20RS card to the PLC and displayed by the PLC through the display unit.

16 Functional Characteristics of Excitation Control and Starting System of Reversible Pumped Storage Unit

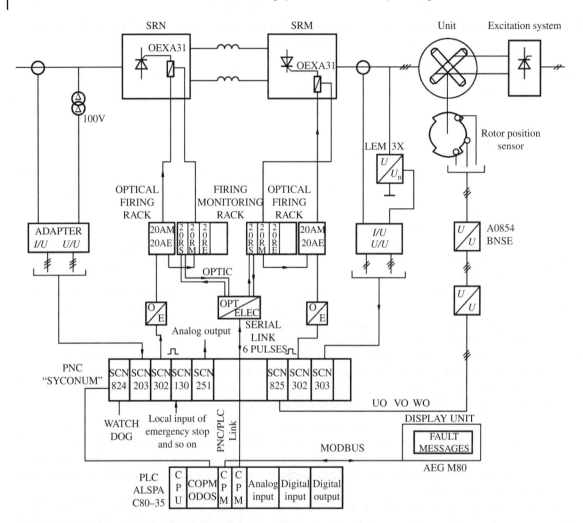

Figure 16.20 Working principle of PNC closed-loop control.

The following is a brief description of the start process of triggering the thyristor elements with the pulse coupling mode when the unit goes from the static state to the low-speed stage.

When the unit is in the static state, first the actual position of the rotor is measured with the shaft angle sensor mounted on the shaft. Then, the maximum starting torque that can be obtained when currents are added to two of the three phases of the winding of the stator is determined. For the operation of the shaft angle sensor, a tooth plate of the simulated rotor pole is installed which rotates synchronously with the main shaft on the shaft of the generator. Three sensors that simulate the positions of the three phases of the stator windings are installed on the racks. The three electromagnetic sensors are distributed at an electrical angle of 120°. The center line of the magnetic poles of the three phases R, S, and T of the stator windings coincides with the center line of the square wave voltage waveforms induced by the sensors. Each time the rotor turns around a pole position sensor, it receives a square wave voltage with a constant amplitude and a pulse width of 180°. These square waves are consistent with the rotor magnetic field in phase. After being processed, the square waves form six trigger pulses with a width of 120° at intervals of 60°, which respectively control the six arm thyristor elements of the SRM. The element currents that are turned on will be sent to the corresponding two of the three phases of the stator windings respectively to generate a starting torque.

The SFC is unlocked immediately after the unit obtains a constant excitation current and calculates the initial position of the rotor. First, it provides currents for two of the three phases of the stator windings that can generate the maximum forward acceleration torque in accordance with its decision logic. At this point, the commutation SRN is working in the constant current rectification control state. In terms of the selection of the SRM control angle, in order to make the unit establish a terminal voltage as soon as possible at the early stage of the start of the pump, the SRM adopts the constant control angle mode during the pulse operation control to ensure that no failure occurs in the commutation process. After the pump begins to rotate from the static state under the action of the initial electromagnetic torque, on the unit stator windings, three alternating voltages corresponding to its speed will be induced. Although the magnitude of the voltages at this point is low and insufficient to provide the commutation current of the SRM arm, it can be taken as a criterion for the phase for the SRM's commutation control. In order to make the SRM arm complete the commutation in the absence of commutation voltage support, the forced current interruption method is adopted. That is, at the moment when the need for the arm set's commutation is identified, first the trigger angle of the SRN is quickly moved to the inversion state (greater than 90°). The DC of the DC connection circuit is rapidly eliminated by virtue of the reverse action of the DC output voltage of the SRN. When the SFC circuit current is forced to decrease to zero, all the thyristors of the SRM are in the off state. After it is verified that the stator current has been interrupted, the trigger pulses are transmitted to the SRM. They get to a set of thyristor elements to be triggered in turn. Correspondingly, the SRM's full inversion function settings are canceled, and a commutation current is provided for the SRM. Then, the SRN is restored to the rectified state of the constant current control, while the SRM is unlocked immediate after the unit establishes an excitation current and thus calculates the initial position of the rotor. Thus, the continuous DC circuit generates a DC current. Meanwhile, the unlocked SFC first provides currents for two of the three phases of the stator that can generate the maximum forward acceleration torque. Since the output current of the SFC is intermittent in the commutation process described above, the time interval is generally kept at about 5 ms to ensure that the SRM can complete the commutation reliably. So, with the acceleration of the pump and the unit started, the SRM's commutation interval is shorter and shorter. When the speed reaches 10% of the rated speed and the unit stator windings' induced potential has the ability to provide an SRM current, the SRM will be automatically switched from pulse operation control to unit voltage natural commutation control. With the increase in the rotor speed, sinusoidally changing voltage waveforms in different phases are induced in the three phases R, S, and T of the stator, respectively. At this point, the three-phase induced potential not only acts as a basis for the SRM to identify the automatic commutation phase sequence but also provides commutation currents for commutation of the SRM arms. The commutation mode changes from artificial forced pulse commutation to natural commutation, that is, transition to a normal excitation control mode with the terminal voltage of the generator as the control object.

It should be emphasized that at the low-speed stage, each commutation period is equal to the sum of the time for the cancellation of the original circuit current and the non-commutation time and the circuit current recovery time. The time required for the cancellation of the circuit current is at least 4 ms. With the increase in the frequency of the unit trawled, the time taken to cancel and recover the circuit current will be shorter and shorter. Ultimately, the forced commutation cannot be sustained. So, an upper limit must be set for the operating frequency of the pulse coupling mode. Generally, this frequency is set to 5 Hz or so.

(2) *High-speed trawling stage.* When the unit speed is at the high-speed start phase when $f > 5\,\text{Hz}$, with the rise in the speed of the pump unit trawled, the terminal voltage directly proportional to the frequency increases. With the rise in the terminal voltage, the SRM thyristor output AC voltage increases. At this point, the elements enter a working state for natural commutation. That is, any thyristor element current that quits commutation will be automatically cut off. So, at this stage, no rotor position sensor signal is needed for the trigger. The PNC will control the trigger current of the SRM and the SRN by measuring the SRM- and SRN-side voltages and currents on the basis of the torque setpoint and the frequency benchmark value, regulate the starting current outputted by the SFC, and trawl the unit to 49.5 Hz (the frequency

benchmark value). Through the PLC and monitoring information exchange, the unit synchronizer gives the "Speed up" or "Speed down" order as required and ultimately synchronizes the unit.

As mentioned above, the upper limit of the operating frequency of the pulse coupling mode should be higher than the lower limit of the operating frequency of the synchronous operation mode in order to enable the entire SFC to operate normally within the entire frequency range. For effective switching and connection between the two operation modes, the switching frequency is just the switching frequency between the two stages of the SFC. Generally, this switching frequency should be 2.5–8 Hz. For Tianhuangping Pumped Storage Power Station, this switching frequency is 3.75 Hz.

16.4.3 SFC Start Procedure

When the SFC is in the remote automatic control mode, its start procedure can be divided into five processes.

16.4.3.1 SFC Auxiliaries Start Process

The PSCS sends the "SFC auxiliaries starting order" to the SFC. The SFC automatically starts the auxiliaries such as the input transformer oil pump, the cooling unit deionized water pumps and the SRN and SRM fans, and issues an order to close the harmonic filter switch. After the auxiliaries are started, the SFC detects normal input transformer oil flow, deionized water flow and conductivity, and SRN and SRM air temperatures and then sends the "SFC auxiliaries started" signal to the PSCS.

16.4.3.2 SFC Preparation Process

After sending the "SFC auxiliaries started" signal, the SFC issues two orders including "Operation water flow for input transformer on" and "Operation water flow for input bridges on" to the PSCS. The PSCS opens the SFC external cooling water drain valve to ensure that the SFCI has an external cooling water flow. Then, the SFC automatically closes the input circuit breaker. After the SFC input breaker is closed, the SFC sends three signals including "Input CB closed," "SFC ready," and "SFC ready for on command" to the PSCS. The SFC enters the "standby" state. If the PSCS does not issue any order to the SFC at this point, the SFC will remain in the "standby" state. If the PSCS issues the "SFC on command" order to the SFC, the SFC will automatically close the OCB. After the OCB is closed, the SFC sends the "Output CB closed" signal to the PSCS. Then, the PSCS sends a "Unit (1–6) selected" signal to the SFC. The SFC selects the rotor position sensor circuit of the corresponding unit for measurement on the basis of this signal.

16.4.3.3 SFC Start Process

After selecting and connecting the rotor position sensor circuit of the corresponding unit, the SFC issues the "Excitation start" order to the PSCS. The PSCS makes the excitation system apply an excitation to the rotor windings of the corresponding unit. Then, the PSCS sends the "Excitation ready" feedback signal to the SFC. At this point, the SFC releases the pulse latch and establishes a current circuit. The generator/motor starts, with its speed rising. The excitation system provides a constant excitation current for the generator/motor. The ratio of the stator voltage to the unit speed rises (i.e., U/F regulation). As the unit speed rises to 495 r min^{-1} ($f \geqslant 49.5$Hz), the SFC sends the "SFC ready for synchronizing" signal to the PSCS. At this point, there are two possibilities:

(1) The frequency follow-up selector is switched to "OFF."
(2) The frequency follow-up selector is switched to "ON."

If the frequency follow-up selector is switched to "OFF," the SFC will regulate the unit speed in accordance with the "Raise" or "Lower" order issued by the synchronizer. If the frequency follow-up selector is switched to "ON," the SFC will take the grid frequency it measures as a frequency benchmark value and regulate the unit speed based on it. Once the unit meets the synchronization conditions, it will immediately close the generator circuit breaker (GCB) and send the "GCB closing order" signal to the SFC. After the SFC receives the "GCB closing order" signal, it immediately latches the regulator to prevent the formation of a loop circuit between the grid, the generator/motor, and the SFC. When the starting current of the SFC is zero, the SFC will automatically open the OCB immediately. After the output breaker of the SFC is opened, the SFC sends two signals including "Output CB Opened" and "SFC regulation stopped" to the PSCS. At this point, the SFC automatically enters the next process.

16.4.3.4 SFC Standby Process

After the above process is completed, the SFC automatically enters the standby state. The SFC's input circuit breaker is in the close position. The filter switch is in the close or open position. The input transformer oil pump, the cooling unit deionized water pumps, and the SRN and SRM fans are in the operation state. At this point, there are two possibilities:

(1) Another unit is started.
(2) The SFC enters the stop state.

16.4.3.5 SFC Auxiliaries Stop Process

When the SFC is in the standby state, the PSCS issues the "SFC auxiliaries stopping order" to the SFC. After the SFC receives the order, it immediately opens the input circuit breaker. After the input circuit breaker is opened, the SFC gives a signal to the PSCS. Five minutes after the SFC receives the "SFC auxiliaries stopping order," the SFC issues two orders including "Operation water flow for bridge off" and "Operation water flow for input transformer off" to stop the operation of the input transformer oil pump, the cooling unit deionized water pumps, and the SRN and SRM fans. When the SFC auxiliaries all stop, the SFC will send the "SFC auxiliaries stopped" signal to the PSCS, and it will be able to act as a standby again. When the SFC is in the local control state, its start program can also be divided into six processes:

(1) STEP 1. Auxiliaries starting and harmonic filter, circuit breaker, input transformer oil pump, cooling unit deionized water pumps, and SRN and SRM fans starting.
(2) STEP 2. Input CB closing order.
(3) STEP 3. ON command.
(4) STEP 4. Unit selection.
(5) STEP 5. Not used.
(6) STEP 6. Auxiliaries stopping order.

When the SFC is in the "MAINTENANCE" state, it cannot start the unit and can only start/stop the input transformer oil pump, the cooling unit deionized water pumps (1 and 2), and the SRN and SRM fans in the GA03 enclosure in order to check whether the cooling unit deionized water flow and the pump and fan working conditions are normal.

16.4.4 Establishment of SFC Electrical Axis

The SFC system is related to the unit only when the PMO or PCO mode starts. The two start modes are virtually the same at the beginning but different after the unit synchronization.

The electrical axis is an electrical connection between the SFC system and the unit established through the electrical devices. The control of different devices may be done by different systems, but the operation flow is basically constant.

The electrical axis establishment process is described below, with Unit No. 4 of Tianhuangping Pumped Storage Power Station as an example. The establishment process of the electrical axis of the unit: The unit's commutation disconnector (PRD04) is closed → The starting bus connection disconnector SBI712 is closed → The SFC1 output disconnector OPI81 is closed → The disconnector trawled SBI41 is closed → The SFC1 output circuit breaker OCB81 is closed → The excitation system field circuit breaker FCB04 is closed. The connection of the electrical primary circuit is shown in Figure 16.21.

When the above electrical axis is established, SFC1 will unlock the regulation and establish a current, and the trawling unit starts rotating. It should be noted that the fling-in and cut-out of the starting bus connection disconnector (SBI712) depends on the SFC selection of the unit start. If the start of Unit No. 4 selects SFC2, SBI712 will remain open.

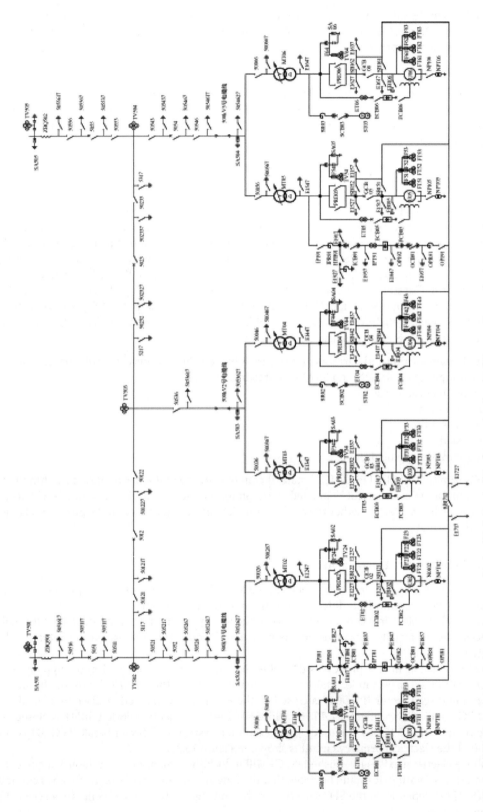

Figure 16.21 Connection of electrical primary circuit of Unit No. 4 of Tianhuangping Pumped Storage Power Station.

The six 300 MW pumped storage units of Tianhuangping Pumped Storage Power Station share two SFC sets. When the unit synchronization is finished, the unit circuit breaker GCB04 is closed → The SFC01 output circuit breaker OCB81 is immediately opened → The SFC1 output disconnector OPI81 is cut out → The trawled disconnector SBI41 of the unit is cut out → The starting bus connection disconnector SBI712 is opened.

In fact, when the GCB completes the synchronization, the SFC will immediately latch the regulation function while issuing an order to open the OCB. At this point, the electrical axis has been released, but the other devices need to automatically restore their state, and the preparations for the start of the other units need to be made.

When the unit is operating in the start process in the PCO mode, the monitoring system is the common interface of the SFC system, the excitation system, and the synchronizer. The ultimate goal of the coordination between the three is to achieve the unit synchronization.

When the unit selects the SFC start in the synchronization phase modulation mode, it will start its auxiliaries and send a signal to start the SFC's auxiliaries in accordance with the monitor's system sequential control logic. Meanwhile, the SFC operating mode is selected for the excitation system. Then, under the coordination between the SFC and the excitation system, the electrical axis of the SFC start is established. After the SFC system identifies the unit rotor position, it issues an order to start the excitation system. After the order (excitation current established) is given, the SFC starts establishing a current to start the unit and divide the process into two stages (low-speed stage and high-speed stage). The unit is started directly at a speed of 49.5 Hz. At this point, the excitation system regulates the unit terminal voltage to 95% U_N. The monitor will send a signal to start the synchronizer. The synchronizer will follow up the frequency and voltage of the system, continuously issuing voltage and frequency regulation orders to the excitation system and the SFC. When the synchronizer detects that the three elements of synchronization meet the specified conditions, it will issue an order to close the unit outlet circuit breaker GCB. After the synchronization is completed, the established electrical axes will be released in the sequence described above. The SFC system will stop the regulation and automatically enter the standby state. Meanwhile, the excitation system will work normally in the AVR mode. The specific sequence is shown in Figure 16.22.

It should be noted that, in the unit start process, the SFC is actually equivalent to a governor which regulates the unit speed to make it meet the synchronization conditions. When the PCO starts in the unit synchronization phase modulation mode, the governor system will receive an order from the monitor and enter the phase modulation mode. Thus, the governor's opening limit will be set to the close position, and the unit's load reference value is set to zero. During the unit's transition from PCO to PMO, when the governor detects that the ball valve opening is greater than 40%, the governor system will enter the PMO mode and open the guide vanes.

It should be noted that when the unit adopts synchronized pair trawling, that is, back-to-back start, the establishment process of the electrical axis of the electrical primary main connection is the same.

16.4.5 Starting System of SFC of Units of Tiantang Pumped Storage Power Station [42]

Tiantang Pumped Storage Power Station has a total installed capacity of 2 × 35 MW on pumped storage units. For the units, SFC start is adopted, with back-to-back start as a standby. The SFC is provided by Hungary's GANX ANSALDO. Its rated output power is 4 MW, rated input voltage 1300 V, output voltage 0–1450 V, and output current 0–1380 A. Its design allowable start time is 200 s and actual set start time is 128 s. It allows four consecutive starts.

16.4.5.1 Composition of SFC System

The SFC can provide a variable frequency power source from zero to rated frequency, synchronously trawling the unit to synchronization. There are two SFC categories: separate control SFC and self-control SFC. A separate control SFC powers the pump motor unit with a separate SFC circuit, while a self-control SFC controls the SFC circuit trigger pulse with the rotor position detector. Tiantang Pumped Storage Power Station adopts the former category. The SFC system consists primarily of two 12-phase pulse wave rectifier-inverter (LSC-MSC) units (SFC221) connected in parallel, two sets of reactors, step-down and step-up transformers,

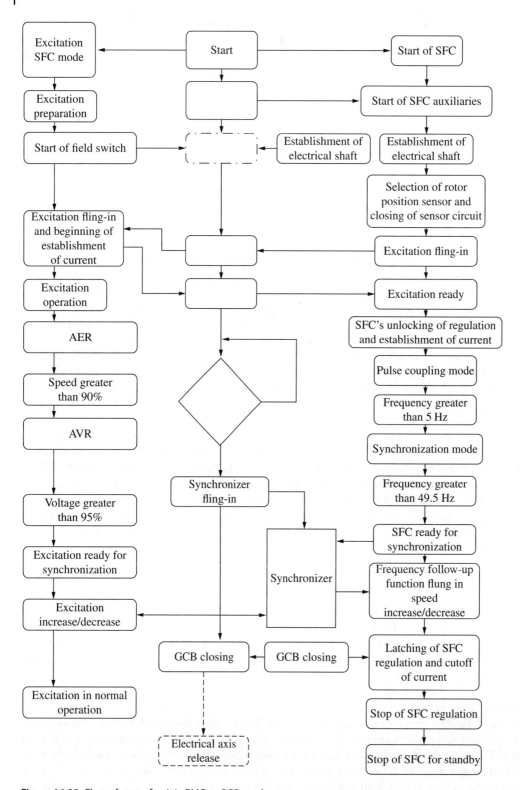

Figure 16.22 Flow of start of unit in PMO or PCO mode.

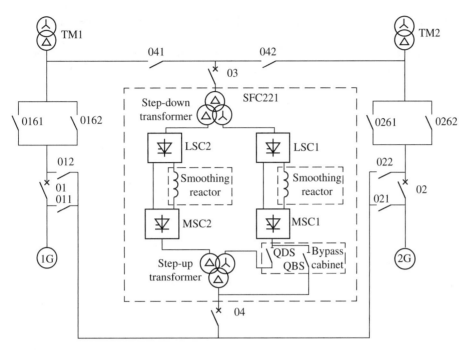

Figure 16.23 Connection of main circuit of SFC system.

bypass cabinets, and relevant 10 kV disconnectors. The connection of the main circuit is shown in Figure 16.23. This circuit is a high-low-high SFC.

SFC221 is a 12-phase pulse wave SFC with two 6-pulse wave rectifier-inverter units (LSC1-MSC1 and LSC2-MSC2) connected in parallel. Each arm has only one thyristor. LSC1 and LSC2 are three-phase grid-side SRNs, respectively connected to the secondary winding of the triangularly connected three-winding step-down transformer on the main power supply. MSC1 and MSC2 are three-phase unit-side SRMs. The MSC1 output can be connected to the starting bus or the star secondary winding of the step-up transformer via the bypass cabinet QBS or QDS, respectively, while the MSC2 output is connected directly to the triangularly connected secondary winding of the step-up transformer.

The secondary windings of the step-up and step-down transformers are connected in triangular and star shapes, respectively. Their phasor phase difference is 30°. The role of the transformers is to match the grid voltage with the generator-motor set voltage, reducing the input and output voltages of the SFC and the number of thyristor elements in series. Besides, the primary and secondary windings adopt a triangular connection, thus reducing the influence of the harmonics generated by the rectifiers on the grid and the generator and playing the role of fault current limiter and DC isolator.

The smoothing reactor connected to the SRN output end suppresses the pulse wave of the DC circuit, thus improving the operating conditions of the inverters and playing the role of limiter of the initial growth rate of fault currents.

The bypass disconnector QBS guarantees the supply of a great starting electromagnetic torque to the unit. When the unit starts, the QBS is closed. LSC1-MSC1 is put into operation, with its output directly connected to the generator motor windings. Thus, the unit gets a large starting current. When the unit speed is greater than 5 Hz, the QBS is cut off and the QDS is flung in. The two LSC-MSC sets were put into operation simultaneously.

16.4.5.2 Back-To-Back Start

In the back-to-back start mode, one unit (conventional unit or pumped storage unit) of the power station or an adjacent power station supplies the starting power. Before the start, the starter unit and the unit being started are connected to the starter bus through the electrical axis and properly excited. At the time of the start, the guide vanes of the starter unit are opened. The low-frequency voltage induced by its stator winding side is applied to the unit being started and a starting torque is generated, so that the unit being started rotates synchronously with the starter unit. With the gradual opening of the guide vanes, the rise of the speed and the increase in the starter unit terminal voltage, their respective excitation regulators are flung in when the speed rises to about 80% of the rated speed. After it rises to the rated value synchronously, synchronization is achieved.

The excitation regulators of the two units (the starter unit as a generator and the pump motor unit as the unit being started) are chosen to operate in the manual constant excitation current state. This value is generally chosen as the no-load excitation current.

There are different back-to-back start schemes classified by starting bus position. The schemes where the starting bus is located on the low- and high-voltage side of the main transformer are called low- and high-voltage back-to-back start, respectively. Back-to-back start schemes are also classified by starting bus connection direction. The scheme where a specific unit acts as a starter unit to start the several other units is called 1−n back-to-back start. The scheme where any unit can start the other units and be started by any other unit is called interconnected back-to-back start. The characteristics of back-to-back start are as follows:

(1) The start-up process does not receive power from the system and has no effect on the system operation.
(2) For the runner chamber pressurization start, the capacity of the starter unit is only 15%−20% of the capacity of the unit being started. As long as the capacity of the starter unit is sufficient, it can start two units of the same type simultaneously.
(3) A small- or medium-capacity unit of the adjacent power station can also act as a starter unit as long as it meets the start conditions (e.g., capacity, starting circuit impedance). Thus, the scope of application of back-to-back start is expanded.
(4) Since a unit needs to be first excited before being started, the excitation transformer is required to be connected to the outside of the GCB or powered by the service power.
(5) A starting bus and switchgear are needed, and the electrical connection or layout is complex.
(6) The interconnected back-to-back start scheme can improve the flexibility but also complicate the operation and control circuits.
(7) The corresponding protection device is required to be able to reliably act within a range of 0–50 Hz.
(8) Since back-to-back start needs a unit that acts as a generator to start the other units, it needs to be used in coordination with other start schemes such as SFC start when all the units are required to enter the PMO mode.

Back-to-back start is applicable to units with various capacities. It can be used for both the runner chamber pressurization start and start in water without runner chamber pressurization. The additional cost of back-to-back start is generally lower than that of coaxial small-capacity motor start. When the number of units is large, back-to-back start boasts higher economy.

Back-to-back start has been widely used in China's pumped storage power stations. The key to its success is the opening speed and degree of the guide vanes of the starter unit as well as the excitation current of the starter unit (generator) and the unit being started (motor).

For back-to-back start, the phase of the current flowing into the corresponding two-phase stator windings of the unit being started should be selected. The basis for the selection is that the direction of the rotating magnetic field generated by the two-phase stator current and the direction of the magnetic field generated by the constant excitation current of the rotor windings should be orthogonal at 90° in order to obtain the maximum starting torque.

16.4.6 Electromagnetic Induction Method for Identification of Initial Position of Rotor [43]

The electromagnetic induction method is a method introduced by Alstom in recent years, which identifies and measures the initial position of the rotor with electromagnetic phase in a pumped storage unit. The working principle: First, the SFC sets and gives an excitation current value that changes in a step function. Although the pump unit is in the static state, the excitation current of the unit changes during its increase in accordance with the given order, and thus the changing excitation current will induce voltages in the three phases of the stator windings. Obviously, the phase and magnitude of these voltages are related to the initial position of the rotor. The amplitude of the line voltage of the two phases with the strongest electromagnetic induction must be the maximum.

Thus, if a current is applied to these two phases first, the rotor will generate the maximum forward acceleration torque. Under the above physical concept, a calculation program for determination of the initial position of the rotor is set during the pulse control of the unit at the low-speed stage. Based on the three-phase stator voltage detected during the establishment of the excitation current of the unit, the initial position of the rotor before the start of the pump unit can be calculated, and the two phases of the stator windings where the current flowing in at the moment of the start of the pump can generate the maximum forward acceleration torque can be identified. This method for identification of the initial position of the rotor has been applied to such pumped storage power stations as Langyashan and Zhanghewan.

The electromagnetic induction method differs from the above-mentioned method of measuring the initial position of the rotor with a mechanical position sensor in that it does not require a mechanical sensor.

The basic working principle of the electromagnetic induction method is as follows.

At the beginning of the start of the unit, the rotor is in the static state, and the stator and rotor relative motion mechanism cannot be used to identify the rotor position. However, at the instant when the excitation current is applied to the rotor excitation windings, emfs in different phases will be induced in the three-phase windings of the stator. The position of the rotor can be calculated on the basis of these emfs.

When the excitation current is applied, the fluxes generated due to the mutual inductance in the stator three-phase windings can be represented by Eq. (16.20). The stator and rotor winding fluxes are shown in Figure 16.24.

$$\begin{aligned} \Phi_U &= Mi_f \cos\gamma \\ \Phi_V &= Mi_f \cos(\gamma + 120°) \\ \Phi_W &= Mi_f \cos(\gamma - 120°) \end{aligned} \quad (16.20)$$

where

$\Phi_U, \Phi_V,$ and Φ_W — fluxes generated by rotor current in three phases of stator windings;

M — mutual inductance between stator and rotor windings;

i_f — rotor current;

γ — angle of Phase U axis of rotor axis.

Figure 16.24 Initial position of rotor of synchronous motor and six fan sections.

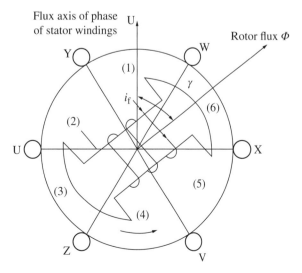

The rotor current can be expressed by Eq. (16.21)

$$i_f = \frac{u_f}{r_f}\left(1 - e^{-\frac{r_f}{L_f}t}\right) \tag{16.21}$$

where

u_f — voltage applied to rotor windings;

r_f and L_f — resistance and inductance of rotor windings.

The emfs induced in the stator three-phase windings can be expressed by Eq. (16.22):

$$\left.\begin{aligned}e_U &= -\frac{d\Phi_U}{dt} = -\frac{d}{dt}(Mi_f \cos\gamma) = -M\frac{u_f}{r_f}\cos\gamma\frac{d}{dt}\left(1 - e^{-\frac{r_f}{L_f}t}\right) = -M\frac{u_f}{L_f}\cos\gamma\, e^{-\frac{r_f}{L_f}t} \\ e_V &= -\frac{d\Phi_V}{dt} = -M\frac{u_f}{r_f}\cos(\gamma + 120°)e^{-\frac{r_f}{L_f}t} \\ e_W &= -\frac{d\Phi_W}{dt} = -M\frac{u_f}{r_f}\cos(\gamma - 120°)e^{-\frac{r_f}{L_f}t}\end{aligned}\right\} \tag{16.22}$$

The maximum emfs induced in the three phases of the stator windings appear at the moment when the voltage is applied to the rotor windings or when t is 0, as shown in Eq. (16.23):

$$\left.\begin{aligned}e_{U0} &= -M\frac{u_f}{L_f}\cos\gamma = -k\cos\gamma \\ e_{V0} &= -M\frac{u_f}{L_f}\cos(\gamma + 120°) = -k\cos(\gamma + 120°) \\ e_{W0} &= -M\frac{u_f}{L_f}\cos(\gamma - 120°) = -k\cos(\gamma - 120°)\end{aligned}\right\} \tag{16.23}$$

where

$e_{U0}, e_{V0},$ and e_{W0} — maximum emfs induced in three phases of stator windings;

k — coefficient it equals $k = \frac{Mu_f}{L_f}$.

From the trigonometric functions, the following can be obtained from Eq. (16.23):

$$\left.\begin{aligned}\cos\gamma &= -\frac{1}{k}e_{U0} \\ \sin\gamma &= \frac{e_{V0} - e_{W0}}{\sqrt{3}k} \\ \tan\gamma &= \frac{e_{W0} - e_{V0}}{e_{U0}} \\ \tan\gamma &= \frac{u_{W0} - u_{V0}}{u_{U0}} \\ \gamma &= \arctan\frac{u_{W0} - u_{V0}}{u_{U0}}\end{aligned}\right\} \tag{16.24}$$

In the case where the stator windings are unloaded, $e_{U0}, e_{V0},$ and e_{W0} and $u_U, u_V,$ and u_W are equal. Since the latter can be measured, it is easy to obtain γ and thus identify the initial position of the rotor. Calculating γ on the basis of $\tan\gamma$ can prevent errors from influencing the motor parameters.

There may be an infinite number of possible initial positions of the rotor, but the SRM has only six possible turn-on arm combinations. So, the infinite number of possible initial positions of the rotor must be merged into six to meet the SRM control requirements.

The space in the motor stator is divided into six 60° fan sections. The origin axis of each fan section is the axis of the magnetic field of the stator windings. The rotor axis must be in one of the six fan sections.

At the instant of the application of the current to the rotor windings, the range of γ when the rotor is in each position (see Line A in Figure 16.25) is shown in Line B in Figure 16.25.

16.4.7 Capacity Calculation of SFC

The relationship between the trawling torque T_M and the resistance torque T_R during the start of the SFC is as follows:

$$T_M - T_R = J \frac{d\Omega}{dt} \tag{16.25}$$

where

T_M – trawling torque of SFC in N m;
T_R – resistance torque of unit, which is a function speed, in N m;
J – moment of inertia of rotating element of unit in t m²;
Ω – angular velocity of unit in rad s⁻¹.

The relationship between T_M and T_R is shown in Figure 16.26.

The relationship between J and the moment of flywheel GD^2 is as follows:

$$J = \frac{GD^2}{4g} \tag{16.26}$$

where

GD^2 – measured in N m²;
g – gravity acceleration in m s⁻².

The relationship between Ω and the number of revolutions per minute n is as follows:

$$\Omega = \frac{2\pi n}{60} \tag{16.27}$$

Equations (16.28) and (16.29) can be obtained from the above differential equation:

$$dt = \frac{1}{T_M - T_R} J \, d\Omega \tag{16.28}$$

$$t = \int_0^N \frac{1}{T_M - T_R} J \, d\Omega = \int_0^N \frac{1}{\frac{P_{SFC} - P_R}{\Omega}} J \, d\Omega \tag{16.29}$$

$$P_{SFC} = T_M \Omega$$
$$P_R = T_R \Omega$$

where

P_{SFC} – capacity of SFC W;
P_R – power loss equivalent to resistance torque during start of unit.

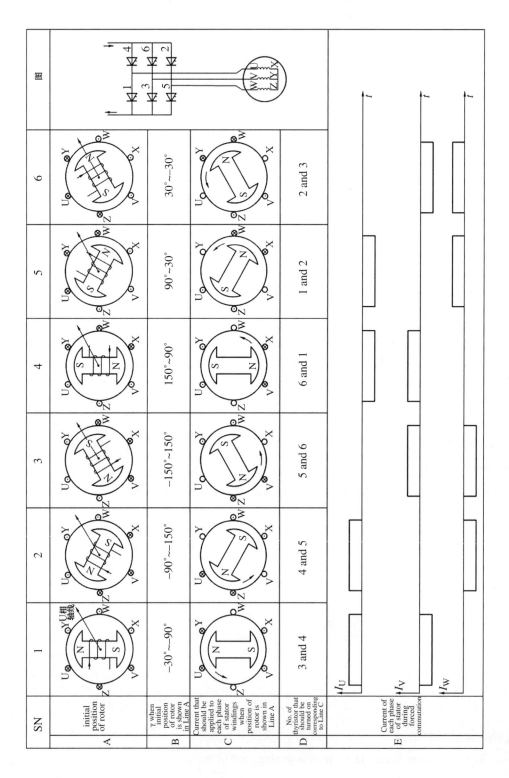

Figure 16.25 Identification of initial position of rotor and phase current during forced commutation.

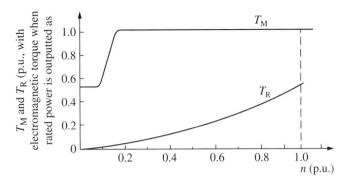

Figure 16.26 Variation curves of trawling torque T_M, resistance torque T_R and speed n.

Table 16.1 Expressions relating power loss and speed.

Power loss item	Expression relating power loss and speed
Resistance loss of runner in air	$P_1 = an^3$
Resistance loss of motor in air	$P_2 = bn^3$
Thrust bearing friction loss	$P_3 = cn^{1.5}$
Guide bearing friction loss	$P_4 = dn^2$
Iron loss of motor when unloaded	$P_5 = en^2$
Total loss	$P_R = P_1 + P_2 + P_3 + P_4 + P_5$

As an example, the relational expressions between the power loss and the speed are shown in Table 16.1.

It can be seen that the capacity of the SFC depends on the required unit acceleration time, the moment of inertia, and the resistance torque. In order to reduce the capacity of the SFC, the water in the runner chamber should be drained into the draft tube with the compressed air before the start to disconnect the runner from the water.

P_{SFC} can be obtained from this equation. P_R is a function of speed. The equation needs to be solved with a special program. P_{SFC} is generally 6%–10% of the unit capacity (measured in megawatts). A relationship similar to an inverse proportion exists between P_{SFC} and the acceleration time. Increasing P_{SFC} can shorten the acceleration time. However, when P_{SFC} increases to a certain extent, the effect of shortening the acceleration time will no longer be significant. So, when the power system's requirement for the pump start time is not too high, too large an SFC capacity should be avoided.

16.4.8 SFC Power Supply Connection

Since the SFC generates harmonic components that affect the power supply quality of the grid during the start of the unit, the impact on the grid should be taken into account when the SFC connection point is considered. However, since the SFC is for short-time duty and the SFC connection point is not a characteristic of the public grid, there is usually no need to require a demanding harmonic current level for pumped storage power stations as specified by the international standards. The conventional SFC connection schemes are shown in Figure 16.27.

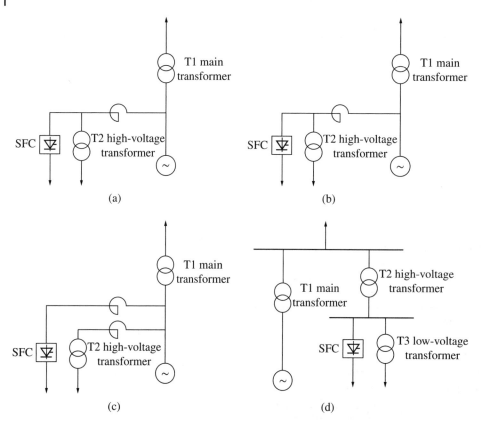

Figure 16.27 Several SFC-station transformer connection schemes: (a) connection via common reactor, (b) connection without reactor, (c) connection via reactor, (d) connection via high-voltage station transformer.

16.5 SFC Current and Speed Dual Closed-Loop Control System [44]

The schematic diagram of the current and speed dual closed-loop control system of the SFC is shown in Figure 16.28. In terms of control function, the SFC should be able to complete the control of the entire process of the unit operating as a pump from the start in the static state to the synchronization. Figure 16.28 shows the block diagram of the control of the SFC. It primarily includes the speed control, the generator terminal voltage control, and the control of the synchronization of the pump's grid connection during the start of the unit.

16.5.1 Speed Control of Pump Motor Unit during SFC Start

During the start of the pump unit, the SRM in the inversion mode adopts the fixed angle open loop control mode. In order to prevent the inverse commutation failure, the control angle setpoint is generally a 140°–150° electrical angle. Thus, the corresponding DC outlet voltage U_{d2} will be directly proportional to the AC voltage U_{srm} of the unit. The SRN, which is in the rectification mode, is directly controlled by the current and speed dual closed-loop system. At the time of the SFC start of the pump, the SFC sets and optimizes a speed rise curve in advance. Then, the speed setting unit outputs a speed base value which is the equivalent of the grid frequency and corresponds to 50 Hz. Under the action of the dn/dt element (limiting the speed base value rise rate), a speed set value n_ω is generated and compared to the actual speed of the unit n_s. The deviation acts as an input control signal of the speed regulator (outer closed loop) and the current regulator (inner closed loop). Finally, the control angle of the SRN is regulated to control the operating current of the SFC circuit, so that the unit can quickly follow up the changes in the speed orders. When n_ω is greater than (or smaller than) n_s,

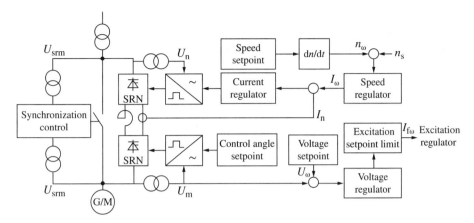

Figure 16.28 Schematic diagram of current and speed double closed-loop control system of SFC.

the SFC will automatically increase (or decrease) the electromagnetic torque supplied to the unit. When n_ω is equal to n_s, the SFC maintains the original operation state. That is, when the electromagnetic torque provided by the SFC is equal to the mechanical load resistance torque of the unit, the unit will maintain a certain speed. This procedure is just the working principle of the speed control of the SFC during the start of the pump unit.

The main factors that affect the speed characteristics of the pump during the SFC start are the parameters of the speed regulator and the current regulator. So, during the debugging of the pump to be started, the parameters of these two regulators should undergo an optimization test, so that the unit can obtain fast and smooth optimal start control characteristics. Normally, in order to prevent the pump's speed fluctuations during the acceleration, the response of the current regulator in the inner closed-loop control should be faster than that of the speed regulator in the outer closed-loop control. Besides, the right pump start speed (dn/dt) can also effectively reduce the unit vibration and improve the start success rate.

16.5.2 Unit Voltage Control during SFC Start of Pump Motor Unit

During the start of the pump, the unit voltage increases proportionally with the speed. That is, the constant excitation current control mode is adopted. The measured unit voltage U_m is compared to the voltage setpoint U_ω (a function of speed) set in the SFC. The deviation enters the voltage regulator and then generates the excitation current setpoint $I_{f\omega}$ and orders the unit's excitation control system to perform the corresponding control. Thus, voltage regulation in the SFC and excitation regulation of the unit are combined to constitute a complete unit terminal voltage control system. The characteristics of the unit terminal voltage control system not only affect the start process of the pump but also directly concern the synchronization process of the unit. So, it is important to select appropriate regulation parameters for them on the basis of the coordination between the unit terminal voltage regulator and the excitation current regulator.

16.5.3 Synchronization Control of Pump Motor Unit

Synchronization is an important step in the SFC start process of the pump. It achieves the pump's transition from variable frequency acceleration to synchronization with the grid, thus ending the whole SFC start process of the pump. The SFC synchronization control is put in when the unit started is close to the grid frequency. First, the synchronization measurement system orders the synchronization regulation system to perform the corresponding regulation on the basis of the frequency and amplitude deviations between the unit voltage U_{srm} and the grid voltage U_{srn}, makes the frequency deviation signal replace the speed deviation signal adopted in the pump start acceleration control process, and sends it directly into the speed regulator. Meanwhile, it makes the grid voltage U_{srn} replace the voltage setpoint adopted in the pump start acceleration control process and takes it as a setpoint of the unit voltage regulator of the pump synchronization control. When the SFC speed, the unit voltage, and the frequency deviation all meet the synchronization conditions, the synchronization logic decision circuit in the synchronization measurement system is put in and issues a synchronization order under the principle of phase angle equality. The SFC is immediately latched to close the synchronization circuit breaker. After the SFC is latched, the DC current is attenuated. The SFC OCB is opened to cut off the SFC output circuit. The pump's SFC start and synchronization process is completed.

16.6 Influence of SFC Start Current Harmonic Components on Power Station and Power System [45]

With the increased share of large-capacity thermal and nuclear power units in the power system, large-sized pumped storage power stations have become one of the most effective measures of addressing peak shaving, frequency modulation, and operational economy of the power system. Since pumped storage units undergo frequent starts and stops, it is necessary to equip them with high-performance and high-reliability starting devices. SFCs characterized by high operational reliability, low energy consumption, and operational flexibility have become a starting device widely used by large- and medium-sized pumped storage power stations. Currently, the three-phase fully controlled bridge circuit composed of thyristor elements is widely used in pumped storage units in China. It performs commutation with the grid voltage and obtains the required frequency and amplitude-regulable sinusoidal voltage output waveform with the phase control method. An SFC circuit with three-phase AC input and single-phase DC output can be seen as a circuit composed of two antiparallel three-phase rectifier circuits. One of the two circuits provides a forward output current, but its trigger control angle is modulated, and thus its output is not a constant DC. The other provides a reverse output current, but its trigger control angle is modulated, and thus its output is not a constant DC but a voltage close to sine. Three circuits are connected in a certain way to supply power to the three-phase load, thus constituting the SFC with a three-phase output.

Since phase control is adopted, the input end of the SFC requires the system to provide a lagging reactive power, which will reduce the input power factor of the system. Besides, due to the input current modulation by the output waveform, the asymmetry between the three-phase circuit and the magnetic circuit, and the thyristor control angle error, the input current contains the characteristic harmonics of a general rectifier circuit (P-pulsating rectifier, whose characteristic harmonics are $K \cdot P \pm 1$, $K = 1, 2, 3, \ldots$) and may also contain harmonics of other frequencies, that is, non-characteristic harmonics. Generally, the non-characteristic harmonics are numerically small and can be neglected. Only when the non-characteristic harmonics resonate will their values increase and cause harm. Resonance is a hard-to-predict phenomenon which cannot be found without debugging. On the other hand, rational connections help eliminate resonance.

16.6.1 Harmfulness and Characteristics of Harmonics

Harmonics are harmful to the power system. Harmonic currents may cause additional loss of devices and aging of insulation performance, and voltage harmonics may affect the reliability of actions of electrical devices. So, applicable standards have been developed to specify and limit harmonics. For example, IEEE Std. 519 recommends converter harmonic current limits, as shown in Table 16.2.

The power access point of the SFC is usually on the low-voltage side (13.8–18 kV) of the main transformer. Obviously, the harmonic content of this access point is the highest and the most greatly distorted. However, this access point is not a public grid. The only electrical equipment that may be connected to this point is a high-voltage station transformer. Harmonics may cause the transformer to generate additional heat. However, it takes time for the heat to accumulate. The SFC's operation time is only 5 min or so. The additional harmonic loss generated by the SFC in the transformer is much smaller than the heating loss caused by the continuous harmonics. Some data indicate that the available capacity of the power transformer is 67% of the rated value when the ratio of the additional loss caused by the harmonics to the basic loss is 0.5. So, according to the most

Table 16.2 Converter harmonic current limits recommended by IEEE Std. 519.

Characteristic frequency	5	7	11	13	17	19	23	25	Harmonic factor HP
6-pulse wave I_h/I_1 (%)	17.5	11.0	4.5	2.9	1.5	1.0	0.9	0.8	1.39
12-pulse wave I_h/I_1 (%)	2.6	1.6	4.5	2.9	0.2	0.1	0.9	0.8	0.71

Note: h stands for the order of harmonics $h = 5, 7, 11, 13, \ldots$ for three-phase bridge converter connection; I_h stands for the harmonic current corresponding to h harmonic; I_1 stands for the rated current; and HP stands for harmonic factor.

conservative estimate, in general, the available capacity of the station transformer will not be less than 70% of the rated value. Moreover, in the capacity selection for a station transformer, unforeseen circumstances are allowed for. According to the test results of Phase I of Guangzhou Pumped Storage Power Station, the station transformer's rated current is 135 A, but the measured current is only 16.5–37.5 A – just 12%–28% of the rated value. Furthermore, it is a fact that a dry-type transformer can operate for a one hour event at 120% load. So, it can be considered that the adverse effects of the harmonics on the station transformer are negligible.

Thus, only two access points including the point of common coupling (PCC) and the 400 V service power bus of the power station need to take the harmonic level into account.

It should be noted that it is not appropriate to use the SFC connection point as a PCC in the engineering design, because that will raise the harmonic limit standard to an impracticable level. In fact, only high-voltage (220–500 kV) buses fall within the scope of the public grid. Meanwhile, the test results of Phase I of Guangzhou Pumped Storage Power Station show that the harmonic level of the high-voltage bus is always much lower than the national standard.

As for the harmonic level of 400 V buses, it is stipulated internationally that the total harmonic distortion (*THD*) of the voltage shall be no greater than 5%. International standards generally stipulate that the index should be 4%–6%. It should be noted that this is a limitation on continuous harmonics. There has been no special provision for the case where pumped storage power stations act as short-time harmonic load, though the national standard developer has explained that the international provision targets continuous harmonics; short-time harmonics are not subject to it. IEEE Std. 519 stipulates that the short-time harmonic limit (during start or in special cases) can be extended to 50%. Some engineering calculations show that the voltage *THD* of a 400 V bus is far less than 5% in the most unfavorable circumstances. So, even if the limit is not extended, it is lower than the internationally stipulated level.

16.6.2 Simplified Calculation of Harmonic Components

In engineering design, it is necessary to understand and estimate the influence of the harmonic components generated by the SFC on the pumped storage power station and the relevant systems.

Figure 16.29 shows the equivalent system used in the calculation of the harmonic components of the SFC.

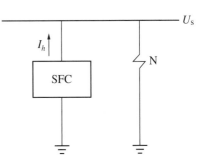

Figure 16.29 Equivalent system for calculation of harmonic components of SFC: U_S – equivalent power system network; N – equivalent load system; I_h – harmonic components generated by SFC harmonic current source.

The harmonic factor HF is defined as follows:

$$HF = \frac{\sqrt{\sum(h^2 I_h^2)}}{I_1} \quad (16.30)$$

where

h — order of harmonic, depending on SFC SRM connection $h = 5, 7, 11, 13, \ldots$ for three-phase bridge 6-pulse wave circuit;

I_h — harmonic current corresponding to h harmonic;

I_1 — power frequency rated power supply of SFC.

The THD at the point where the SFC is connected to the bus is

$$THD = \frac{\sqrt{\sum U_h^2}}{\dfrac{U_1}{\sqrt{3}}} \quad (16.31)$$

where

U_h — h harmonic voltage;

U_1 — power frequency rated line voltage of point where SFC is connected to bus.

$$U_h = I_h h X_S \quad (16.32)$$

where

X_S — power frequency reactance reduced to side of bus system to which SFC is connected.

When Eq. (16.32) is substituted into Eq. (16.31), the following equation is obtained:

$$THD = \frac{\sqrt{\sum(h^2 I_h^2) X_S}}{\dfrac{U_1}{\sqrt{3}}} \quad (16.33)$$

When the definition of harmonic factor in Eq. (16.30) is substituted into Eq. (16.33), the following is obtained:

$$THD = \frac{HF X_S}{\dfrac{U_1}{\sqrt{3}}} \quad (16.34)$$

or

$$THD = HF U_1 I_1 X_S \times \frac{\sqrt{3}}{U_1^2} = \frac{(\sqrt{3} U_1 \, I_1)\, HF}{\dfrac{U_1^2}{X_S}}$$

$$SSFC = \sqrt{3} U_1 I_1$$

$$SCC = \frac{U_1^2}{X_S} \quad (16.35)$$

where

SSFC – rated capacity of SFC MVA;

SCC – short-circuit capacity of bus to which SFC is connected MVA.

The THD generated by the SFC can be estimated on the basis of Eq. (16.35). GB/T 14549-1993 *Quality of Electric Energy Supply Harmonics in Public Supply Network* specifies that the allowable THD is 3% for 35 and 66 kV public supply networks, 4% for 10 kV ones, and 5% for 400 V ones.

It should be emphasized that if the *THD* at the SFC connection point estimated with the foregoing method does not exceed the national standard, the corresponding *THD* at the PCC will be certainly smaller than the *THD* generated at the harmonic source.

16.6.3 Improvement of Harmonic Operation State of SFC

Through the SFC operation state analysis and the harmonic component estimation in the above sections, the following conclusions can be drawn:

(1) The SFC of a pumped storage power station is an SFC start device for short-time duty, and the harmonic loss it generates is short-time. It is expected that the *THD* generated by the harmonics can meet the national standard. If it fails to meet the national standard, the limit can be extended appropriately. For example, the limit can be increased by reference to international standards.

(2) In general, the SFC access point is not the PCC and should not be taken as a criterion for compliance with the harmonic level standard. For the preliminary estimation, simplified calculation of the harmonic at the point can be conducted (the simplified calculation is only applicable to the SFC access point). If the harmonic at the point meets the standard, the harmonics at the other connection points will meet the standard. If the harmonic at this point does not meet the standard, a detailed analysis and calculation will be needed. For example, a calculation with a computer program will be needed to identify whether the harmonics at other relevant points meet the standard.

(3) A high-voltage bus is an actual PCC. The *THD* here does not exceed the limit provided in the national standard in most cases.

(4) As long as the main connection is appropriately chosen, the harmonics generated by the SFC will not have a noteworthy harmful effect on the equipment of the power station. For example, the electrical distance between the SFC access point and the service power should be increased. The power supply of the SFC should be directly connected from the low-voltage side of the main transformer, and its connection from the secondary side of the station transformer should be avoided as far as possible. If it must be connected to the secondary side of the station transformer for some reason, a detailed analysis will be needed. If necessary, a 12-pulse converter or a filter system can be adopted to improve the harmonic operation state. If conditions are available, the SFC and the high-voltage station transformer should be separately connected to the low-voltage side of the main transformer of the different sets. As far as possible, the service power should be connected to the station transformer that is not in parallel with the SFC. Thus, the harmonic level of the service power can be greatly reduced.

(5) An SFC with a 12-pulse rectifier can reduce the 5th or 7th harmonic to a minimum (it can completely eliminate them theoretically). According to the data provided by Alstom, the price and volume of the 12-pulse scheme is only 1.05 times that of the 6-pulse scheme. If the harmonic level goes beyond the standard due to the connection limitation, the 12-pulse wave scheme is one of the effective solutions to this problem.

(6) In most cases, it is unnecessary to add a filter system to an SFC. Complete filter systems require additional investment and a larger plant area. Besides, a filter is a capacitive load for the power frequency, which is detrimental to the operation of the power station, and sometimes an associated reactor becomes necessary by way of compensation. The long-term put-in of a filter system will increase the power loss of the power station. Moreover, the frequent put-in/cut-off of the filter system may cause circuit breaker restrike to reduce the operational reliability of the whole station.

(7) In order to block the possible zero-order harmonics and increase the electrical distance between the harmonic source and the other points, an input transformer should be considered for the SFC.
(8) Resonance and non-characteristic harmonics are hard to predict to some extent. Under the requirement for elimination of characteristic harmonics, the filter of a device often cannot eliminate the non-characteristic harmonic values. Sometimes, increasing the electrical distance between the harmonic source and the load may be effective.
(9) Relay protection is prone to faulty action caused by harmonic interference, especially harmonic resonance. Requirements for harmonic interference protection should be put forward for relay protection equipment for pumped storage units.

16.7 Local Control Unit (LCU) Control Procedure for Pumped Storage Unit

The main roles of pumped storage units are system load peak shaver and valley filler as well as phase modulator and backup load provider.

For pumped storage units, generally distributed automatic control system is adopted for power station monitoring.

The entire monitoring system is divided into four control levels: scheduling control level, remote control room control level, local control room control level, and local unit control level, with control authority from high to low. The scheduling control level is directly controlled by the provincial grid dispatching control center and connected to it via a fiber channel; the AGC mode is adopted. The remote control room control level is located in the city/county and connected to the local control network via a fiber channel. The local unit control level is connected to each LCU via an optical fiber ring network. For ease of elaboration, the 2×35 MW unit monitoring system of Tiantang Pumped Storage Power Station is taken as an example to illustrate its control process.

16.7.1 LCU Configuration

The LCU of the units of Tiantang Pumped Storage Power Station is a complete computer monitoring system equipped with dual CPUs [46]. The LCU is responsible for the station's communication with the upper computer and other stations as well as processing of its key information such as logical control and numerical calculations. Online monitoring can be achieved between the two CPUs. If one of the CPU fails, the other can lead the unit to a safe state.

16.7.2 LCU Control Program

The operating conditions of Tiantang Pumped Storage Power Station are complex. The LCU control program divides the entire sequential control process into several transition transient states and operation steady states, as shown in Figure 16.30. Transition transient states include standstill, no-load generation, trawler in back-to-back operation mode, and PCO. Operation steady states include stop, GO, GCO, PMO, and black start.

On the basis of the above-mentioned transition transient states and operation steady states, the corresponding transitions are defined. For example, the conversion from stop to standstill is defined as transition. Each transition is divided into several steps. Each step controls the corresponding specific device as required. In the execution of each step, a single step operation time is set on the basis of the actual actions of the on-site equipment under the requirements for equipment operation safety. Within the operation time, when the corresponding equipment does not meet the operation requirements, the control program will give a timeout alarm and automatically execute the corresponding stop sequence flow. After the control program is gradually broken down step by step as described above, the structure of the control program is very clear. GO and PMO are the main operation modes of a pumped storage unit. The following is a description of the control process and characteristics of the unit in the two operation modes.

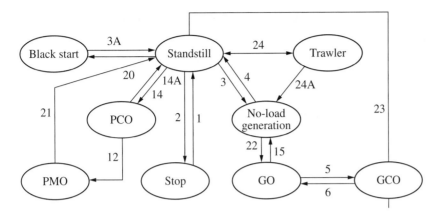

Figure 16.30 Unit control structure flow.

16.7.3 GO Control Process

When the unit LCU control program receives the "GO order," the control program first determines whether the unit start conditions are met. The start conditions include the 220 kV system-related circuit breaker or disconnector, the state of the water inlet gate, 10 kV system-related disconnector, and the state of the unit automation elements. When the start conditions are met, the control program will execute the following operations in the sequence of Stop → Transition 1 → Transition 3 → Transition 22. The flow is shown in Figure 16.29.

STEP1-1 (90 s). The generator cooling water is opened. The mechanical braking is put in. The air shroud is put out; the guide vanes are locked and put out; the generator fans are put in; and the bearing oil vapor extraction fans are put in. The time in parentheses indicates the waiting time when this step is executed.

STEP3-1 (10 s). The governor and the excitation are in GO mode. A commutation disconnector 0162 or 0262 is closed; and the turbine guide bearing low oil level alarm is latched.

STEP3-2 (180 s). The mechanical braking is put out. The butterfly valve is opened at the 100% opening; and the high-pressure filling pump is put in.

STEP3-3 (45 s). The governor opening limiter is opened in the starting position. The guide vanes are opened; and the unit speed is >95%.

STEP3-4 (30 s). The excitation is put in; and the high-pressure filling pump is put out. The turbine guide bearing low oil level alarm signal is put in.

STEP22-1 (210 s). The unit speed is >98%; the unit terminal voltage is >90%; the synchronization is started; and the GCB is closed.

STEP22-2 (200 s). The governor opening limiter is opened in the maximum position. The guide vanes reach the set opening; the unit reaches GO; and the required active and reactive loads are set.

16.7.4 PMO Control Flow

In the PMO mode, there are two start modes: SFC start and back-to-back trawling. The former is the normal start mode, while the former is a backup start mode. When the unit start conditions are met, if the unit LCU control program receives the "SFC" order to start the pumping, the control program will execute the following operations in the sequence Stop → Transition 1 → Transition 14A → Transition 12.

STEP1-1 (90 s). The generator cooling water is opened; and the mechanical braking is put in. The air shroud is put out. The guide vanes are locked and put out, and the generator fans are put in. The bearing oil vapor extraction fans are put in; and the turbine cooling water is opened.

STEP14A- (10 s). The governor is in the PMO mode. The SFC start device conducts a self-check.

STEP14A-2 (10 s). A commutation disconnector 0161 or 0261 is closed; and a trawling disconnecting link 011 or 021 is closed.

STEP14A-3 (60 s). The air inflation and water pressurization valves and the air inflation, water pressurization, and re-inflation valves are opened. The water bypass valve is opened, and the air bypass valve is opened.

STEP14A-4 (20 s). The air inflation and water pressurization valves close the water level; the air inflation and water pressurization valves and the air inflation, water pressurization, and re-inflation valves are closed; the mechanical braking is put out; and the high-pressure filling pump is put in. The excitation is in the SFC mode.

STEP14A-5 (200 s). The SFC is started; and the trawling unit speed is >95%.

STEP14A-6 (30 s). The high-pressure filling pump is put out.

STEP14A-7 (110 s). The unit terminal voltage is >90% after the unit speed is >98%; the synchronization system is started; and the GCB is closed.

STEP14A-8 (10 s). 011 or 021 is opened. The SFC mode is put out. The SFC stands still.

STEP12-1 (10 s). The air bypass valve is closed.

STEP12-2 (220 s). The water bypass valve is closed; the guide vane opening limiter is opened in the maximum position; the butterfly valve is opened at the 100% opening; and the air inflation, water pressurization, and re-inflation valves are closed.

STEP12–3 (60 s). The top cover exhaust valve is opened.

STEP12-4 (6 s). The governor is started, and the guide vanes are opened after the pressured is established. The pump head optimization method is put in. The top cover exhaust valve is closed after a 90 s delay.

STEP12-5 (105 s). The water turbine cooling water is closed when the top cover exhaust valve is closed; and the unit reaches PMO.

16.8 Pumped Storage Unit Operating as Synchronous Condenser

As mentioned above, pumped storage units play the role of grid load peak shaver, frequency modulator, synchronous condenser, and emergency standby. So, the synchronous condenser mode is more complex than the GO mode. This section focuses on the particularity of the synchronous condenser mode.

16.8.1 GCO Mode

GCO start includes the start and operation process from stop steady state to stop transient state and then from stop transient state to GCO. It is similar to the start process of a conventional hydropower unit. Firstly, the unit cooling equipment is started to cool the unit. The, the ball valve and guide vanes are opened at a certain opening to make the unit rotate in the direction of power generation. At the rated speed, the excitation is put in. After the synchronization conditions are met and the unit is synchronized, the ball valve and guide vanes are closed and the water pressurization valve is opened, so that the pump turbine runner rotates in the gas. In the water pressurization process, the pump turbine's upper and lower impeller rings, water supply cooling, water ring drain valve, and volute air bypass valve are opened, so that the runner can rotate in the gas of a certain depth sealed by the water ring.

16.8.2 PCO Mode

The PCO mode is also a necessary process of the start of the PMO mode. For a pump unit working in the synchronous condenser start mode, SFC start or back-to-back start can be adopted. In the PCO start process, first a common electrical axis should be established between the starter device (the trawler or the SFC) and the device being started, and the device being started should be trawled for rotation under the action of the electromagnetic torque. In order to prevent too large a starting current and reduction in the capacity of the starting device, the device being started can be trawled directly in the water at the initial stage of the start when the speed is more than 15% of the rated value. When the unit speed is greater than 15% of the rated value, the water level in the runner chamber is lowered with compressed air to make the trawling power loss equivalent to 6%–8% of the main unit's rated power. At this point, the unit runner is virtually rotating in the air. After the trawled unit is synchronized, the unit enters the PCO mode.

16.8.3 PCO Start Mode and Coordination

The PCO mode not only requires the water pressurization system to form a water ring of a certain depth in the runner chamber but also has the following requirements for the other related devices:

(1) The requirements of PCO for the SFC starting power: China's large pumped storage units generally coordinate air inflation and water pressurization in the runner chamber with SFC start. As mentioned above, in the case of the PMO and PCO start, if the runner is started in the water, the water resistance torque is large. For a mixed flow reversible unit with a medium or low specific speed, when the guide vanes close the runner's starting power in the water, 40%–50% of the unit rated output may be achieved. If the runner is started through rotation in the air, the starting power generally does not exceed 6%–8% of the rated power. For a pump unit with a rated power of 336 MW, the starting power will be 336 MW × 6% = 20.16 MW on the basis of a starting power of 6% of the rated power in the synchronous condenser start mode.

(2) The requirements of PCO start and operation for the functionality of the excitation system: In the PCO start mode, in the case of SFC start or back-to-back start, the excitation regulator operates in the manual constant ECR mode before the unit is synchronized. In the synchronization process, the AVR automatic channel replaces the manual channel automatically to regulate the excitation in the AVR mode to change the synchronous condenser power.

16.9 De-excitation System of Pumped Storage Unit

There are two de-excitation systems used in large pumped storage units: the DC field circuit breaker de-excitation system composed of a DC field circuit breaker and an AC de-excitation system composed of an AC air circuit breaker connected to the secondary side of the excitation transformer [47].

In China, the former has been widely used, and the latter has also been applied to pumped storage units in recent years. For example, Phase II of Guangzhou Pumped Storage Power Station uses the Siemens THYRIPOL automatic excitation system and AC de-excitation system composed of an AC air circuit breaker, linear de-excitation resistors, and a crowbar for its 4 × 300 MW pumped storage units.

China started early and has gone deep in terms of AC de-excitation system study, with many achievements. For example, an AC de-excitation system has been applied to the 5 × 300 MW units of Baishan Hydro Power Station in northeast China and the 125 and 120 MW hydro-turbines of Gezhouba Hydro Power Station.

In this section, the performance of AC de-excitation system is presented, with the AC de-excitation system used in the pumped storage units of Guangzhou Pumped Storage Power Station as an example.

16.9.1 Composition of AC De-excitation System

The schematic diagram of the AC de-excitation system is shown in Figure 16.31.

The key parameters of the excitation system of Phase II of Guangzhou Pumped Storage Power Station include a rated excitation voltage of 255 V, rated excitation current of 2334 A, ceiling excitation voltage of 638 V, ceiling excitation current of 4668 A; a no-load excitation voltage of 101 V, no-load excitation current of 1320 A, an excitation transformer voltage of 18 000 V/510 V, and excitation transformer capacity of 3 × 62 kVA.

The thyristor crowbar in the de-excitation system consists of a forward overvoltage protection element A107, a reverse overvoltage protection element A106, a BOD overvoltage monitor, a thyristor trigger unit U121, a de-excitation contact K611, and an overcurrent alarm relay K112. The de-excitation program of the

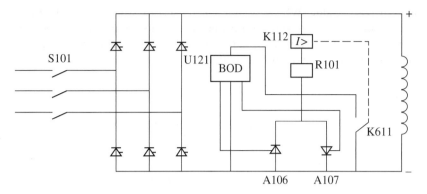

Figure 16.31 Schematic diagram of AC de-excitation system.

AC de-excitation system: First, the thyristor trigger pulses are blocked. The AC field circuit breaker S101 is broken, and meanwhile the de-excitation contact K611 is switched on. The generator conducts de-excitation through the linear de-excitation resistors. If the overvoltage generated when the switch is broken reaches the trigger voltage of the voltage limiting BOD, the thyristor crowbar can be turned on and connected to the de-excitation resistors to protect the rotor winding insulation from breakdown by the overvoltage.

16.9.2 Functional Characteristics of AC De-excitation System

16.9.2.1 AC Field Circuit Breaker Application Characteristics

Compared with an ordinary AC switch, an AC field circuit breaker must have a certain DC breaking capacity. Although it is installed on the secondary AC side of the excitation transformers, the thyristor trigger pulses are cut off while it is broken. So, the two-phase thyristor is in the follow current state. Thus, the breaking process of the AC switch is similar to the breaking process on the DC side. Although there is a zero point on the AC voltage, the current is still a constant DC current under the action of a large inductance.

Besides, in a de-excitation system with nonlinear resistors, it is required that the AC field circuit breaker's breaking arc voltage should be appropriately improved to meet the de-excitation commutation requirements. So, an ordinary AC switch needs renovation such as a change in the arc chute structure design to appropriately increase its breaking arc voltage to make it applicable to AC de-excitation systems.

16.9.2.2 Coordination of Thyristor Rectifier Bridge Trigger Pulse AC Field Circuit Breaker

As mentioned above, all the trigger pulses of the thyristor rectifier bridges must be blocked while the AC field circuit breaker is broken. If the AC field circuit breaker is broken but the thyristor trigger pulses are not blocked, the thyristor elements of the two bridge arms in the same phase will go out of control according to the conclusion drawn from the analysis of the thyristor rectifier bridge trigger process. At this point, the excitation current will pass through the out-of-control conductive thyristor elements of the two bridge arms and the rotor excitation winding circuit to form a natural follow current attenuation state, thereby greatly extending the de-excitation time.

16.9.2.3 De-excitation System Composed of AC Switch and Linear Resistor Crowbar

The crowbar composed of the forward and reverse thyristors in parallel are connected to the linear de-excitation resistors in series and then connected to both ends of the excitation set. Such a circuit integrating de-excitation and overvoltage protection s often referred to as a crowbar circuit internationally and is called a crowbar in China. It should be noted that, after the pulses are blocked, the conductive excitation transformer secondary winding two-phase AC sinusoidal voltage will be introduced into the DC excitation circuit. Meanwhile, when the AC sinusoidal voltage is negative, its polarity and the break voltage of the AC field circuit breaker are added. This relieves the requirements for the break voltage of the AC field circuit breaker to a certain extent, especially in the case of de-excitation with nonlinear de-excitation resistors.

The composition of the de-excitation system is shown in Figure 16.31. The de-excitation system includes the following:

(1) The AC field circuit breaker S101, which is a 3WN1671 extended arc chute sheet-type three-phase AC air circuit breaker manufactured by Siemens. Its key parameters are as follows: rated operating AC voltage: 690 V, rated current: 2500 A, rated short-circuit breaking capacity: 80 kA (rms), peak short-circuit closing capacity: 176 kA (rms), closing DC voltage: 220 V, opening DC voltage: 220 V/48 V, mechanical life: opening/closing 100 000 times, electrical life: rated current breaking 1000, and minimum opening/closing time interval: 120 ms.
(2) The de-excitation resistors R101, which are linear resistors, serving concurrently as rotor winding overvoltage protection. Their material is cast iron, and the model used is 3PR3201-1B. A single resistor is 0.12 Ω. Two resistors are connected in series to ensure that their energy consumption capacity in the most serious de-excitation case does not exceed 80% of their operating energy consumption capacity.
(3) The overvoltage protection. Rotor field overvoltages include transient overvoltage of AC power supply caused by switch operations, open-circuit overvoltage of rotor field circuit, and overvoltage caused by loss of synchronization of generator or interphase short circuit.

The overvoltage protection consists of the forward and reverse thyristors A106 and A107 in parallel and the trigger control unit 121. U121 consists of a BOD and a resistor. The overvoltage setpoint is 1883 V. When the overvoltage at both ends of the rotor reaches the setpoint, a thyristor (A106 or A107) will be triggered by the BOD trigger element. A larger current generated by the overvoltage flows through the relay K112 in series with A106 and A107. The relay is started, and the auxiliary contactor K611 is closed to limit the rotor overvoltage to a low level.

16.9.2.4 Characteristics of AC De-excitation System

The AC air circuit breaker S101 serves as a de-excitation circuit breaker and a thyristor rectifier bridge overvoltage protection device concurrently. Compared with a DC switch, model selection is easier for the AC switch:

(1) The de-excitation mode where the linear resistor R101 and the auxiliary contact K611 are coordinated is slower but has the advantage of simple and reliable operations.
(2) If the line resistor and the forward and reverse thyristors A106 and A107 in parallel are coordinated as forward and reverse overvoltage protection, the rotor winding and thyristor bridge overvoltage can be limited below the insulation withstand level.
(3) In the AC voltage de-excitation mode in the emergency state where the AC de-excitation circuit breaker is tripped while the normal de-excitation broken by the thyristor inverter and the AC de-excitation circuit breaker in coordination and the latched thyristor bridge pulses are introduced into the AC negative half cycle voltage, AC de-excitation functionality and speed are improved.

16.9.3 Operation Time Sequence of AC De-excitation System

16.9.3.1 De-excitation for Stop in Normal Case

Inversion de-excitation is adopted for the normal stopping of the unit. At this point, both ends of the rotor windings whose thyristor control trigger angle $\alpha = 150°$ are trawled and a reverse de-excitation voltage is added that changes with the unit terminal voltage, so that the excitation current is attenuated with time. In this case, the de-excitation time depends primarily on the reverse de-excitation voltage amplitude.

It should be emphasized that the de-excitation speed in a self-excitation system is slower than that in a separate excitation system because the inverse voltage at the time of the de-excitation is attenuated with the unit terminal voltage in the self-excitation system.

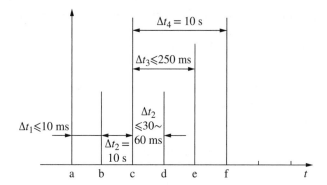

Figure 16.32 Time sequence of normal stop de-excitation process: a – giving of AC field circuit breaker S101 breaking order; b – start of inversion de-excitation (time set to 10 s); c – end of inverter de-excitation, breaking of AC field circuit breaker S101 and closing of K611; d – breaking of AC field circuit breaker S101 and signal feedback; e – closing of K611 and signal feedback; f – release of pulse latch time.

Ten seconds after the start of the de-excitation, the pulses are latched, and meanwhile the AC field circuit breaker S101 breaks the auxiliary contact to close K611. The magnetic field energy is consumed through the de-excitation resistor R101. The specific time sequence of the de-excitation process is shown in Figure 16.32.

16.9.3.2 De-excitation for Stop in Case of Fault

When the excitation system receives an external fault trip order, the AC field circuit breaker S101 is immediately broken, and its auxiliary contact closes K611. Meanwhile, all the thyristor element trigger pulses are latched, and the magnetic field energy is absorbed by the linear de-excitation resistor R101. At the same time, after the pulses are latched, a sinusoidal single-phase AC voltage is introduced. This voltage can conduct de-excitation in coordination with the break voltage of the AC field circuit breaker in the negative half cycle. The magnetic field current is reduced in an exponential function. If K611 cannot be normally closed due to some accident, the voltage protection BOD element will be turned on to limit the rotor overvoltage and conduct the de-excitation.

The de-excitation time t for the stop in the case of the fault is determined by the following equation:

$$t = \frac{TR_f}{R_c + R_f}$$

where

T – generator D-axis open or short-circuit time constant in s;

R_c – de-excitation resistor R101;

R_f – rotor winding resistance.

16.10 Electric Braking of Pumped Storage Unit [48]

Since pumped storage units play the role of grid load peak shaver and valley filler, they undergo frequent starts and stops. For example, the 300 MW units of Guangzhou Pumped Storage Power Station register more than 1000 starts and stops per year on average. Conventional mechanical braking for stopping units may accelerate damage and consumption of brake parts.

Electric braking is an ideal braking mode for the hydro-turbine. Its biggest advantage is large braking torque, which can significantly reduce stop time, avoid generator terminal contamination, facilitate maintenance, eliminate the need for regular observation of mechanical braking system shoe damage, effectively improve operating conditions of hydro-turbines, and meet the requirement for rapid mode transition. Electric braking can be achieved in different ways. For example, unit braking torque can be provided by copper loss generated by the short-circuit current of short-circuited stator windings. Braking can also be achieved when stator windings are connected to a low-voltage circuit in reverse phase sequence. Besides, the DC current can be input into stator windings to achieve braking. Studies show that the braking torque provided by the copper loss generated by the short-circuiting stator windings is the most effective braking method. In the operation process, the generator is disengaged from the grid for stop de-excitation. The water turbine guide mechanism is closed, and the unit begins to slow down naturally. When the unit speed decreases to 50%, the generator outlet short-circuit switch is turned on. The generator stator three-phase windings are short-circuited. Meanwhile, the service power serves as an AC power supply for electric braking. The rectified DC is supplied to the generator excitation winding. Generally, this DC excitation current is equal to the current that makes the stator short-circuit current the rated value. The electric braking torque generated by the unit, the water resistance torque caused by the rotation of the water turbine runner in the water, the air resistance torque caused by the ventilation loss of the generator, and the unit's total resistance torque composed of the resistance torque caused by the bearing friction loss and the copper loss are balanced, so that the unit can stop in the shortest possible time.

When the unit speed decreases to below 10% of the rated value, the mechanical braking will be put in until the unit stop. However, for a pumped storage unit, the speed is sometimes 1% when the mechanical braking system is put in.

In general, for a unit in a rotating state, the energy balance equation can be written as follows:

$$\frac{1}{2}I\omega^2 = \frac{1}{2}I\omega_0^2 - \int \sum P \, dt \tag{16.36}$$

where

I — inertia torque of unit;
ω — angular velocity of unit;
ω_0 — initial angular velocity of unit;
$\sum P$ — sum of various losses including mechanical loss, copper loss, and iron loss.

When both sides of Eq. (16.36) are differentiated, the following equation can be obtained:

$$I\omega\frac{d\omega}{dt} = -\sum P \tag{16.37}$$

When $I = \frac{GD^2}{4} \times 10^3$ (kg·m²) is substituted into Eq. (16.37), the following equation can be obtained:

$$\frac{GD^2}{4} \times 2\pi\frac{np}{60} \times \frac{2\pi p}{60} \times \frac{dn}{dt} = -\sum P \tag{16.38}$$

or it can be written as follows:

$$GD^2 \times \left(\frac{2\pi}{60}\right)^2 \times p^2 n \times \frac{dn}{dt} = -\sum P \tag{16.39}$$

where

GD^2 — rotational inertia of unit kg·m² × 10³
n — number of revolutions of unit;
$\sum P$ — stator winding copper loss, loss of bearings, ventilation loss, and loss of water turbine runner;
p — number of pole pairs.

When Eq. (16.39) is integrated, the braking time can be obtained:

$$t = GD^2 \times \left(\frac{2\pi}{60}\right)^2 p^2 \int_0^{n_1} n\,dn \tag{16.40}$$

Table 16.3 Relation between braking torque and speed.

Loss category	Braking power	Braking torque
Water turbine runner friction	$\propto n^3$	$\propto n^2$
Generator wind friction loss	$\propto n^3$	$\propto n^2$
Bearing friction loss	$\propto n^{1.5}$	$\propto n^{0.5}$
Stator winding copper loss	Constant	$\propto n^{-1}$
Excitation winding copper loss	Constant	$\propto n^{-1}$

The braking time of the hydro-turbine is determined from Eq. (16.40) and the simulation calculation program. From the given speed, each resistance torque can be obtained. The curve of the relationship between the speed and the time in the stop process can be obtained point by point.

According to studies, the braking torque is a function of speed. The approximate relationship is shown in Table 16.3.

As can be seen from the above relationship, when the unit is operating at a high speed, the water resistance torque and the air resistance torque play a major role. When the unit is operating at a low speed, the stator winding copper loss plays a major role. The other braking torques decline sharply with the decrease in the speed. The electromagnetic power is unrelated to the speed. The electric braking torque increases with the increase in the speed. So, electric braking is a very good braking mode.

16.11 Shaft Current Protection of Pumped Storage Unit [49]

Protection of the shaft current of the pumped storage unit is descried below, with Tianhuangping Pumped Storage Power Station as an example. As a pure pumped storage power station capable of daily regulation, Tianhuangping Pumped Storage Power Station has six 300 MW reversible units, adopting a vertical shaft suspension synchronous motor designed, manufactured, installed, and debugged by Canada's GE.

Since the first unit was put into operation in September 1998, a number of unit start failures caused by shaft current protection action have been seen.

The units of Tianhuangping Pumped Storage Power Station feature a long axis, many sections, high speed ($N_r = 500$ r min^{-1}), and high stator ($H = 3.05$ m). Compared with conventional hydro-turbines, such a unit sees a greater potential difference induced on the large shafts on both sides of the rotor during normal operation (varying slightly with the operation mode and the load). If there is an insulation resistance decrease or short circuit of the insulation layer when the unit is operating normally, a large shaft current will be formed (Circuit 2 in Figure 16.33). In general, if the density of the shaft current passing through the bushing surface exceeds 0.2 A cm^2, bushing corrosion, oil film damage, bushing heating, and even bushing surface burnout may be caused, which will endanger the operational safety of the unit. So, in unit protection, it is particularly important to appropriately configure and install reliable shaft current protection.

16.11.1 Configuration of Shaft Current Protection

A thrust bearing and upper guide bearing are arranged on the upper part of the generator-motor set of Tianhuangping Pumped Storage Power Station. The thrust bearing oil basin upper cover plate is sealed with Babbitt metal pad. So, insulation pads are arranged in three positions including the thrust pad, the thrust oil basin cover, and the upper guide bearing to prevent the shaft current from constituting a circuit. The insulation pad of the thrust pad consists of two parts, one arranged between the thrust head and the runner plate (two layers, each 1 mm thick) and the other arranged between the thrust base plate and the upper rack (four layers, each

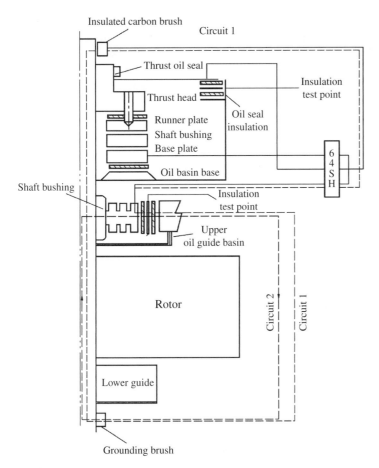

Figure 16.33 Schematic diagram of generator shaft current protection. *Note*: Circuit 1 and 2 are upper guide shaft current action circuits.

1.2 mm thick). The insulation pad of the thrust oil basin cover is arranged on the joint face between itself with the thrust oil basin (two layers, each 1.2 mm thick). The insulation pad of the upper guide bearing is arranged between the upper guide shaft bushing rack and the upper rack support (two layers, each 1.2 mm thick). One insulation test point is provided for each of the two layers of the insulation pad of the thrust oil cover and the upper guide bearing for daily inspection. In order to prevent the suspension potential, the emf induced by the large shaft when the unit is operating normally, from being too high and discharging to the bushing surface, a large shaft grounding carbon brush is provided under the lower guide. To minimize the impact of the shaft current on the busing surface, a large shaft insulation carbon brush is provided above the thrust cover. In the normal case, if one of the three insulation pads is damaged, the shaft current generated by the large shaft-induced potential will form a shaft current circuit through the large shaft, the insulation carbon brush, the protection relay, the bushing rack, the damaged insulation pads, the support, the earth, and the large shaft grounding carbon brush since the bushing surface oil film impedance is much larger than the large shaft insulation carbon brush impedance (oil film impedance is up to $50\,000\,\Omega$ at the rated speed). This can not only protect the bushing surface but also issue an alarm and even achieve stop.

On the units of Tianhuangping Pumped Storage Power Station, the separate current relay 64SH is used as an integrated-circuit-type three-phase two-stage overcurrent relay for shaft current protection. One of its sides is connected to the insulation parts of the thrust bearing, the thrust cover, and the upper bearing. The other side is short-circuited in three phases and then connected as a public terminal to the insulation carbon brush of the large shaft. Compared with the conventional shaft current protection scheme with a large shaft current transformer TA where the external electromagnetic or load current changes interfere with the secondary-side output of the current transformer TA, its biggest advantage is that it can avoid such interference. When the shaft current protection relay 64SH detects that the current is greater than the protection setpoint, it will act on the alarm and trip the unit after a certain delay.

Setpoints of Section 16.1 of shaft current protection are $I = 0.06$ A and $t = 2$ s. Setpoints of Section 16.2 of shaft current protection are $I = 0.1$ A and $t = 1$ s.

16.11.2 Problems and Handling

16.11.2.1 Shaft Current Protection Action Caused by Metal Lap Joint Between Upper Bushing Back and Oil Cooler Upper Cover Plate (See Figure 16.34)

When a unit of Tianhuangping Pumped Storage Power Station was debugged after an overhaul, it tripped many times because of the action of the upper guide shaft current protection. The main reason was that the manufacturer provides a filler strip between the bushings in the design of the upper guide bearing and fastens it with a screw in order to improve the lubrication effect. If any improper tool or method is used during the reassembly, a small amount of scrap iron can easily fall when the screw is fastened. Since the gap between the upper oil cooler upper cover plate and the upper guide bushing back cannot be absolutely even, the scrap iron falling into the upper oil guide basin can easily cause a lap joint between the shaft bushing and the oil cooler upper cover plate. That will lead to the grounding of the upper guide bushing, resulting in the action of the shaft current protection. For such an accident, generally the bidirectional rotation characteristic of the reversible unit can be utilized. When the unit rotates in the opposite direction, the scrap iron which is short-circuited between the upper guide bushing back and the oil cooler upper cover plate can fall off. It is required that the scrap iron in the oil guide basin should be sucked out with a magnet during an overhaul. For a unit, if the gap between the bushing back and the upper oil guide cooler upper cover plate is small, the inner circumference of the upper oil guide cooler upper cover plate can be ground off (by about 2 mm on a single side) to increase the gap between the shaft bushing and the oil basin.

16.11.2.2 Decrease in Upper Insulation of Thrust Bearing Caused by Impurities in Lubricating Oil

An oil-lubricated Babbitt metal bushing surface mechanical seal and an insulated carbon brush are provided on the upper part of the thrust bearing in turn. The mechanically sealed bushing surface and the insulated

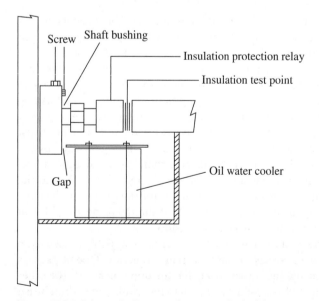

Figure 16.34 Schematic diagram of upper guide shaft bearing current renovation.

carbon brush are in direct contact with the large shaft. When the unit is in normal operation, the carbon dust generated by the carbon brush will enter the thrust oil basin along the gap between the large shaft and the thrust upper cover and enter, together with the metal powder generated by the Babbitt metal bushing surface, the oil circulation in the thrust oil basin. Some oil streams containing conductive impurities will get into the insulation pad between the thrust head and the runner plate through the four positioning pin holes between the thrust head and the runner plate. For the unit of Tianhuangping Pumped Storage Power Station, through a long period of operation, there was carbon dust or Babbitt metal powder around the pin holes, which decreased the upper insulation of the thrust bearing.

To prevent a decrease in the upper insulation pad of the thrust bearing and burnout of the thrust bushing caused by the impurities in the lubricating oil, the following measures were taken:

(1) The Babbitt metal powder generated by the Babbitt metal mechanical seal is closely related to the large shaft pre-tightening force. Test results showed that the pre-tightening force could be appropriately lowered to 3/4 of the original design value on the premise that the normal operation of the mechanical seal of the Babbitt metal bushing surface is not affected.
(2) Since the original design does not provide any online filter device, the thrust bearing lubricating oil contamination intensifies with time. So, an online filter was installed on the thrust oil basin.
(3) Since the large shaft insulated carbon brush is too close to (about 120 mm away from) the thrust oil basin upper cover plate, the large amount of carbon powder generated can easily fall into the thrust oil basin. When the unit was overhauled, the insulated carbon brush was moved upward to deviate from the thrust oil upper cover plate as far as possible (approximately 750 mm). Besides, a dustband was added to the upper part of the thrust upper cover plate.

16.11.2.3 Equipment Grounding in Insulation Measurement Area

During an overhaul, if the insulation of the wire at the insulation test point is damaged, the resistance to ground measured with a multimeter may be normal, but the value measured with a 500 V insulation resistance meter is zero. So, it is generally required that the insulation resistance of an insulation pad should be measured with a 500 V insulation resistance meter.

16.12 Application Characteristics of PSS of Pumped Storage Unit

The application characteristics of the PSS in a pumped storage unit are described below, with Guangzhou Pumped Storage Power Station as an example [50].

With a total installed capacity of 8×300 MW, Guangzhou Pumped Storage Power Station is the world's largest-capacity pumped storage power station. The functions of the pumped storage units in a power system are as follows:

(1) Peaking shaver – A pumped storage power station can utilize the surplus power in the system during the load valley to make its pumped storage units operate as pump motors to pump the water in the lower reservoir to the upper reservoir for power generation at the load peak of the system.
(2) Emergency standby – With a flexible and rapid start – the transition from start to full load within only 1 to 2 min and mode transition from PMO to GO within only 3 to 4 min – a pumped storage power station can be used as an emergency power supply of the power system.
(3) Frequency modulator – With good adaptability to rapid changes in load and frequency modulation performance, a pumped storage power station can be used as a flexible and reliable frequency modulation power supply.
(4) Synchronous condenser – Since a pumped storage power station is generally close to the load center and can be easily controlled, it can be used as a synchronous condenser to undertake the synchronous condenser task in the power system. Thus, the parameter design study and dynamic stability analysis for PSS of pumped storage units are of great significance for improvement of the stability level of the grid.

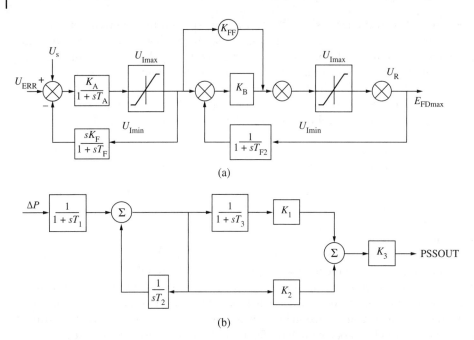

Figure 16.35 Block diagrams of excitation system and PSS of 300 MW units of Facility B of Guangzhou Pumped Storage Power Station: (a) block diagram of transfer function of excitation system, (b) block diagram of transfer function.

The block diagrams of the excitation system and the PSS of the 300 MW units of Facility B of Guangzhou Pumped Storage Power Station are shown in Figure 16.35.

In Figure 16.35a, $K_A = 400$, $T_A = 3$ ms, $K_F = 4000$, $T_F = 3$ ms, $T_{F2} = 100$ s, $U_{AMAX} = 5.853$, $U_{AMIN} = 0$, $U_{Rmax} = -4.679$, and $E_{FDmax} = 5.853$.

$$W(s) = \frac{1}{1+sT_1} \times \frac{sT_2}{1+sT_2} \times \left(\frac{K_1}{1+sT_3} + K_2 \right) \times K_3$$

where

$T_1 = 20$ ms, $T_2 = 1500$ ms, $T_3 = 20$ ms, $K_1 = 0$, $K_2 = -0.750$, and $K_3 = 1$.

A PSS is generally designed for the GO mode. In the design operation mode, it can effectively suppress low-frequency oscillations. Since the dynamic stability of a pumped storage motor is symmetrical in regard to the active power, a power direction PSS designed on the basis of the GO mode that can identify the power direction can effectively suppress low-frequency oscillations in the PMO mode as well to ensure that the pumped storage motor has dynamic stability in the PMO mode consistent with that in the GO mode.

There should be a difference of 180° in the PSS compensation phase between the GO mode and the PMO mode, so as to ensure that the PSS can provide positive damping for the electromechanical mode in both operation modes. A power direction PSS is just one of such solutions. A power direction PSS can be made by adding a power direction element to a conventional PSS. This is the main characteristic of the application of the PSS to pumped storage units.

17

Performance Characteristics of Excitation System of 1000 MW Turbine Generator Unit

17.1 Introduction of Excitation System of Turbine Generator of Malaysian Manjung 4 Thermal Power Station

This chapter takes the potential source static excitation system as an example to describe the performance characteristics of the excitation system used by the 1000 MW turbine generator. The potential source static excitation system is supplied by France's Alstom for the 1270 MVA turbine generator of the Malaysian Manjung 4 Thermal Power Station.

The system diagram for the potential source static excitation system of the turbine generator used by the Malaysian Manjung 4 Thermal Power Station is depicted in Figure 17.1.

For the excitation system, the latest generation of CONTROGEN-type excitation regulator is adopted by Alstom. There are two sets of completely independent and fully redundant double-automatic channels, designed for the standard configuration of the excitation regulation system, and the system consists of one channel comprising an automatic voltage regulator (AVR) with an integrated FCR (field current regulator).

In addition, each channel is equipped with an independent power supply and the firing of the rectifier module. A failure of the active channel causes the system to switch over to the standby channel. Two-way follow-up provides smooth transfer between the regulators. This redundant structure guarantees the best availability of power generation, with automatic regulation and integrated FCR. Each power converter is designed to flow continuously 110% of the rated current and the transient ceiling current every 15 min.

Each channel can allow parallel operation of eight power thyristor bridges with independent pulse amplification and an isolation circuit. In addition, each group of rectifier bridges has an independent signal detection function. Once a fault signal is detected, the power thyristor bridge in the circuit will be automatically locked.

AVR and the power cabinet communicate via double-Ethernet. The pulse generator installed in power cabinet can reduce hardwire and interference among cabinets to ensure reliability of equipment effectively. In addition, the system can conduct hardware fault detection, communication fault detection, pulse monitoring, and conduction monitoring.

For redundant modular design, the control card can be exchanged during operation of the unit, and the software download and restoration can be re-operated for intelligent modular and controller.

In the communication function, the CONTROGEN-type excitation regulator offers a wide choice of communication, such as communication with DCS, and the on-site control unit can be implemented through the TCP communication protocol of Modbus. The communication between two sets of regulators can be implemented through the bus interface with EPL and the CPU card, which is equipped with two serial ports that can support serial read and writing and the Modbus communication protocol, and so on.

In the excitation system, the REDEX-300 series rectifier cabinet with more than 30 years of operation experience is adopted by Alstom for the power rectifier cabinet, there are totally six cabinets operating in parallel,

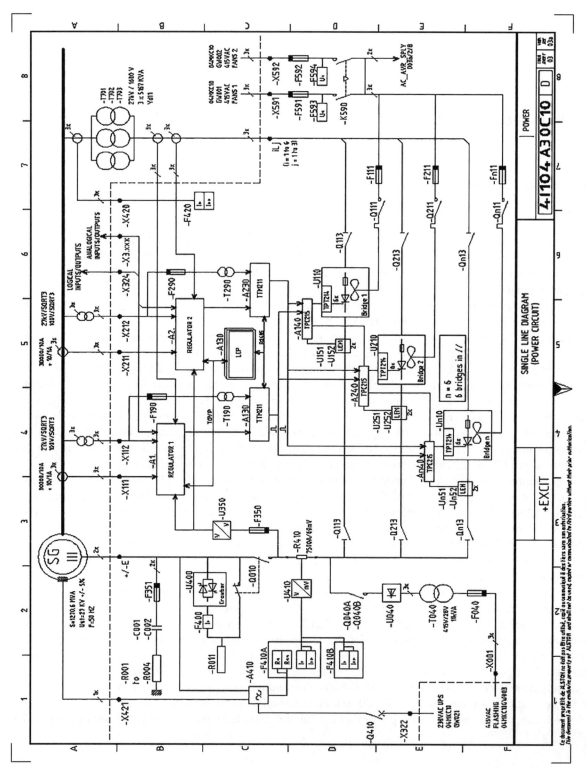

Figure 17.1 System diagram for static excitation system of 1270 MVA turbine generator of Manjung 4 Thermal Power Station: LCP – local control panel; TTM – thyristor trigger module; TPC – thyristor pulse controller; LEM – current transducer; TPT – thyristor pulse transformer. Authorized by Beijing G.E. Automation Engineering Co., Ltd.

and the maximum continuous current output of the excitation system can reach 6721 A. In the design of the fan, the double-fan and double-power supply air cooling system can be adopted for the REDEX-300 series rectifier cabinet.

17.2 Key Parameters of Turbine Generator Unit and Excitation System

17.2.1 Key Parameters of Generator Unit

The key parameters of the generator unit [51] are shown in Tables 17.1–17.7.

17.2.2 Key Parameters of Excitation System

17.2.2.1 Data of Generator Unit
Generator rated power: 1080 MW;
  Generator rated voltage: 27 kV;
  Generator rated voltage range: ±5%;
  Rated power factor: 0.85
  Rated current: 27 169.7 A;
  Rated frequency: 50 Hz;
  Rated frequency variation range: +3%/−5%;
  Rated speed: 3000 r min^{-1};

Table 17.1 Main parameters of generator unit. Authorized by Beijing G.E. Automation Engineering Co., Ltd.

Parameter	Sign	Value	Unit
IEC60034			
Temperature rise class		B	
Rated apparent power	S_N	1270.6	MVA
Rated active power	P_N	1080.0	MW
Rated terminal voltage (+5.0%/−5.0%)	U_N	27 000	V
Rated phase current	I_N	27 170	A
Rated power factor	$\cos\phi_N$	0.85	p.u.
Rated frequency (+3.0%/−5.0%)	f_N	50	Hz
Rated speed	n_N	3000	r min^{-1}
Generator field current at no load, rated terminal voltage	I_{fo}	1843	A
Generator field voltage at no load, rated terminal voltage	U_{fo}	211	V
Generator field current at rated output	I_{fN}	6110	A
Generator field voltage at rated output	U_{fN}	763	V
Ceiling factor		2	p.u.
Ceiling voltage		1526	V
Ceiling time		10	s
Short-circuit ratio	SCR	0.426	p.u.
Generator output with one cooler out of service		80.0	%

Table 17.2 Generator reactance and resistance. Authorized by Beijing G.E. Automation Engineering Co., Ltd.

Parameter	Sign	Value	Unit
Rated impedance	Z_N	0.547	Ω
Direct-axis synchronous reactance (unsaturated)	x_d	2.49	p.u.
Direct-axis transient reactance (unsaturated)	x'_d	0.321	p.u.
Direct-axis sub-transient reactance (unsaturated)	x''_d	0.246	p.u.
Direct-axis transient reactance (saturated)	x'_{dv}	0.305	p.u.
Direct-axis sub-transient reactance (saturated)	x''_{dv}	0.214	p.u.
Quadrature-axis synchronous reactance (unsaturated)	x_q	2.439	p.u.
Quadrature-axis transient reactance (unsaturated)	x'_q	0.512	p.u.
Quadrature-axis sub-transient reactance (unsaturated)	x''_q	0.256	p.u.
Negative-sequence reactance (unsaturated)	x_2	0.247	p.u.
Zero-sequence reactance (unsaturated)	x_0	0.103	p.u.
Negative-sequence reactance (saturated)	x_{2v}	0.212	p.u.
Zero-sequence reactance (saturated)	x_{ov}	0.089	p.u.
Potier reactance	x_P	0.396	p.u.
Leakage reactance (stator)	x_Q	0.216	p.u.
Positive-sequence resistance	r_1	0.0031	p.u.
Negative-sequence resistance at (95 °C)	r_2	0.0188	p.u.
Zero-sequence resistance at (95 °C)	r_o	0.00203	p.u.
Stator resistance per phase at (95 °C)	R_a	1.167	mΩ
Rotor resistance at (95 °C)	R_f	0.1188	Ω

Table 17.3 Time constant (unsaturated). Authorized by Beijing G.E. Automation Engineering Co., Ltd.

Parameter	Sign	Value	Unit
Direct-axis transient open-circuit time constant at (95 °C)	T'_{do}	6.01	s
Direct-axis transient short-circuit time constant at (95 °C)	T'_d	0.77	s
Direct-axis sub-transient open-circuit time constant at (95 °C)	T'_{do}	0.022	s
Direct-axis sub-transient short-circuit time constant at (95 °C)	T''_d	0.017	s
Quadrature-axis transient open-circuit time constant at (95 °C)	T'_{qo}	0.81	s
Quadrature-axis transient short-circuit time constant at (95 °C)	T'_q	0.17	s
Quadrature-axis sub-transient open-circuit time constant at (95 °C)	T''_q	0.39	s
Quadrature-axis sub-transient short-circuit time constant at (95 °C)	T''_q	0.034	s
Short-circuit time constant of armature winding at (95 °C)	T_a	0.017	s

Table 17.4 Miscellaneous (electrical). Authorized by Beijing G.E. Automation Engineering Co., Ltd.

Parameter	Sign	Value	Unit
Three-phase stator winding capacitance ($_{3xCphase}$ to ground)	C	0.982	µF
Continuous unbalance load, maximum	$I_{2\infty}$	0.052	p.u.
Short-time capability for unbal. faults, max.	$I_2^2 t$	5.0	p.u. s
Saturation factor $If_0/If_{air\text{-}gap}$ (according to IEEE100)		1.06	p.u.
Voltage increase at sudden load rejection and rated $\cos\phi = 0.85$ (without AVR action)		46.3	%
Voltage increase at sudden load rejection and $\cos\phi = 1$ (without AVR action)		36.7	%
Three-phase short-circuit peak current value	I_{p3}	323	kA
Three-phase steady short-circuit current (rms value)	I_{k3}	38 399	A

Table 17.5 Torque and inertia. Authorized by Beijing G.E. Automation Engineering Co., Ltd.

Parameter	Sign	Value	Unit
Nominal torque	M_N	3 438	kN m
Pull-out torque	M_{kipp}	5 935	kN m
Maximum two-phase short-circuit torque	M_{k2}	24 587	kN m
Moment of Inertia (generator + exciter only)	J	17 138	kg m^2
Constant of Inertia (generator + exciter only)	H	0.666	s

Table 17.6 Critical speed of generator only. Authorized by Beijing G.E. Automation Engineering Co., Ltd.

Parameter	Value	Unit
1 Critical speed (for information only)	575	r min^{-1}
2 Critical speed (for information only)	1500	r min^{-1}
3 Critical speed (for information only)	3930	r min^{-1}
Revolutions per minute at Overspeed test of (2 min)	3600	r min^{-1}

Table 17.7 Generator losses at rated load. Authorized by Beijing G.E. Automation Engineering Co., Ltd.

Parameter	Sign	Value	Unit
Core losses	P_{Fe}	945	kW
Copper losses of stator (95 °C)	P_{cu1}	2 585	kW
Stray load losses	P_{sup}	1 387	kW
Rotor copper losses at 95 °C	P_{cu2}	4 463	kW
Excitation losses	P_{exc}	138	kW
Windage losses	P_{ven}	1 570	kW
Bearing losses	P_{lag}	960	kW
Total losses	P_{tot}	12 048	kW

Generator reactance:

$$x_d = 2.492 \text{ p.u.},$$
$$x'_d = 0.323 \text{ p.u.},$$
$$x''_d = 0.246 \text{ p.u.};$$

Generator time constant:

$$T'_{do} = 6.011 \text{ s},$$
$$T'_d = 0.775 \text{ s},$$
$$T_a = 0.39 \text{ s},$$

17.2.2.2 Data of Excitation System
Field winding resistance: $0.1188 \, \Omega$ (95 °C);
No-load excitation current for rated voltage at air-gap line: $I_{fg} = 1700$ A;
No-load rated excitation voltage of generator: $U_{fo} = 211$ V;
No-load rated excitation current: $I_{fo} = 1843$ A;
Field voltage of generator at rated load: $U_{fN} = 762.8$ V (75 °C);
Filed current at rated load: $I_{fN} = 6109.6$ A;
In 110% rated load and 110% rated voltage of generator:
Maximum field voltage: $U_{fd} = 839.08$ V,
Maximum field current: $I_{fd} = 6720.56$ A;
Permissible forced excitation time (two times the rated ceiling forced excitation voltage): 10 s;
Permissible forced excitation time (two times the rated ceiling forced excitation current): 10 s;
Maximum ambient temperature indoor excitation cubicle: 40 °C;
Maximum ambient temperature indoor excitation transformer: 45 °C;
Maximum humidity: 80%–100%

According to IEC 60034-16-1-2011, continuous field current must be at least equal to the maximum steady-state field current of the generator at 1080 MVA and 27 kV + 5%.

$$I_{EN} = I_{fd} = 6720.56 \text{ A}$$

According to IEC 60034-16-1, the continuous excitation voltage must be at least equal to the direct voltage at the excitation system output terminals which the excitation system can provide when delivering excitation system current.

$$U_{EN} = U_{fd} = 839.08 \text{ A}$$

When the generator terminal voltage is 80% of the rated voltage, the excitation system can output two times the rated forced excitation voltage. The permissible time is 10 s. The ceiling forced excitation voltage of the excitation device is calculated as follows:

$$U_c = 2U_{fN} = 2 \times 762.8 = 1525.6 \text{ V}$$

The ceiling current is $2I_{fN}$ for 10 s with a maximum of four accesses per hour from and with return to the permanent output current (in the case of more than four accesses per hour, some elements will trip owing to over-temperature). The ceiling current of the excitation device is calculated as follows:

$$I_c = 2I_{fN} = 2 \times 6109.6 = 12\,219.2 \text{ A}$$

17.3 Parameter Calculation of Main Components of Excitation System

17.3.1 Excitation Transformer

Parameter calculation is performed in accordance with IEC60146-1-1, IEC60146-1-2, and IEC60146-1-3 [52].

17.3.1.1 Excitation Transformer Output Current

Permanent output current:

$$I_{tN} = \sqrt{\frac{2}{3}} \times I_{EN} = 5483.97 \text{ A}$$

Transient output current:

$$I_{tNs} = \sqrt{\frac{2}{3}} \times I_p = 9970.8 \text{ A}$$

where

I_p — forced excitation current (two times rated excitation current).

Short-time overloading output current $I_{tNs} = 9970.8$ A, during 10 s (four times per hour).

17.3.1.2 Transformer Output Voltage

When the generator terminal voltage is at 80%, the minimum control angle of the thyristor is 10°, and the effective value for the line voltage at the secondary side of the excitation transformer is

$$U_{rms} = \frac{\pi}{3\sqrt{2}} \times \frac{U_p/0.8}{\cos 10°} = 1433 \text{ V}$$

where

U_p — forced excitation voltage (two times rated excitation voltage).

Ideal output power:

$$S_i = \sqrt{3} \times U_{rms} \times I_{tN} = 13610 \text{ kVA}$$

Inductive voltage drop:

$$U_x = \frac{3}{\pi} \times x_r \times \frac{U_{rms}^2}{S_i} \times I_p = 141 \text{ V}$$

where

x_r — short-circuit impedance of excitation transformer 8%.

Thyristor and excitation circuit cable drop voltage:

$$U_f = 2U_{t0} + 0.05 U_{rms} = 73 \text{ V}$$

where

U_{t0} — forward voltage drop of thyristors 0.94 V.

The factor 0.05 corresponds to the 5% voltage drop in the cables. On the basis of this value, the total output voltage of the excitation transformer can be obtained.

$$U_{tN} = U_{rms} + \frac{\pi}{3\sqrt{2}}(U_f + U_x) = 1591 \text{ V}$$

When the voltage at the generator terminal drops to 80% of the rated value, it is also possible to provide two times the ceiling voltage value at forcing, and the corresponding secondary voltage value U_{2N} of excitation transformer will be

$$U_{2N} \geq U_{tN} = 1591 \text{ V}$$

The value $U_{2N} = 1600$ V is taken.

17.3.1.3 Excitation Transformer Capacity
Continuous capacity:

$$S_{tN} = \sqrt{3}U_{2N}I_{tN} = \sqrt{3} \times 1600 \times 5483.97 = 15197 \text{ kVA}$$

The value $S_{tN} = 15\,500$ kVA is taken. When the generator terminal voltage drops to 80% of the rated value, the short-time capacity is as follows:

$$S_{ts} = \sqrt{3}U_{2N}I_{tNs} = \sqrt{3} \times 1600 \times 9970.8 = 27631 \text{ kVA}$$

The value $S_{ts} = 28\,000$ kVA is taken.

17.3.1.4 Design Values of Excitation Transformer
The ambient temperature shall be subject to the internal temperature in the cabinet of excitation transformer.
Design capacity: $S_{tN} = 15\,500$ kVA.
Secondary rated output voltage: $U_{2N} = 1600$ V
Primary rated primary voltage: 27 kV
The permitted overload capacity for 10 s is 28 000 kVA (four times per hour permitted for forced excitation)
Rated frequency: 50 Hz (+3/−5)%
Installation altitude: 0–1000 m

17.3.2 DC Field Circuit Breaker

17.3.2.1 Rated Operating Voltage of Field Circuit Breaker
The maximum excitation voltage of the generator under the normal operating conductions is as follows:

$$U_{EN} = 839.08 \text{ V}$$

1000 V is taken as the field circuit breaker breaking voltage U_{BN}:

$$U_{BN} > U_{EN}$$

In the forced excitation state:

$$U_{EP} = 2U_{fN} = 2 \times 762.8 = 1525.6 \text{ V}$$

The operating voltage of the field circuit breaker at the time of forced excitation is as follows:

$$U_{BP} = 1.5U_{EP} = 1.5 \times 1525.6 = 2288.4 \text{ V}$$

$U_{BP} = 2250$ V is taken.

17.3.2.2 Rated Current of Main Contact of Field Circuit Breaker

$I_{BN} \geq I_{EN}$

$I_{EN} = I_{fd} = 6720.56 \text{ A}$

$I_{BN} \geq 6720.56 \text{ A}$

8500 A is taken as the design value of I_{BN}.

17.3.2.3 Design Values of Field Circuit Breaker

The field current rms value, taking into account four ceiling accesses per hour is one ceiling access for 15 min (900 s).

The effective value of the current flowing through the main contact within 1 h is

$$I_{rms} = \sqrt{\frac{890 I_{EN}^2 + 10 I_{EC}^2}{900}}$$

$$= \sqrt{\frac{890 \times 6720.56^2 + 10 \times 12219.2^2}{900}} = 6806 \text{ A}$$

where

I_{EN} – maximum continuous excitation current value;

I_{EC} – ceiling forced excitation current value $I_{EC} = 2I_{fN} = 12\,219.2 \text{ A}$.

The insulation voltage of the field circuit breaker is 6250 V.

We choose DC breaker CEX 06 8500 3.1, which is produced by France's LENOIR ELEC.

17.3.3 Field Discharge Resistor

The field discharge resistor will limit the field overvoltage at field breaker opening. This overvoltage must be lower than the field breaker maximum interrupting voltage.

Maximum field current at field breaker opening:

Excitation ceiling current:

$I_{EC} = 12\,219.2 \text{ A}$

Field breaker maximum interrupting voltage:

$U_{BP} = 2250 \text{ V}$

According to IEC 60034-1, dielectric rotor is given by

Case $U_{fN} > 500 \text{ V}$,

$U_{text} = 2U_{fN} + 4000 = 2 \times 762.8 + 4000 = 5600 \text{ V}$

When the residual voltage U_{max} produced on the discharge resistance during field voltage is limited to 1500 V, the resistance value for the flowing discharge resistance is

$$R_d \leq \frac{U_{max}}{I_{EC}} = \frac{1500}{12219.2} = 0.123 \text{ }\Omega$$

Discharge resistance design:

$R_d = 0.123 \text{ }\Omega$

When a three-phase short circuit suddenly occurs on the generator, the redundancy coefficient for the stored capacity in field winding is 1.2; then

$$W_F = \frac{1}{2} \times 1.2 \times T'_d \times R_f \times I_{EP}^2 = \frac{1}{2} \times 1.2 \times 0.77 \times 0.1188 \times 12219.2^2 = 8194.9 \text{ kJ}$$

The value $W_F = 8400$ kJ is taken.

17.3.4 Crowbar

The trigger voltage should be lower than 75% of the voltage value in the withstand voltage test for the generator excitation winding, and when $U_{fN} > 500$ V, the test voltage is

$$U_t = 2U_{fN} + 4000 = 2 \times 762.8 + 4000 = 5600 \text{ V}$$

Lower than 75% dielectric rotor:
$U_{th} < 0.75 \times 5600 = 4200$ V
Lower than reverse repetitive peak voltage of thyristor bridges: $U_{th} < U_{RRM} = 6500$ V
Greater than maximum spike overvoltage on DC side:

$$U_{th} > 1600 \times 1.3 \times 1.414 = 2910 \text{ V}$$

The BOD trigger voltage is 3200 V.

When the crowbar is turned on, the current flowing in the thyristors should be greater than the forced excitation ceiling current.

Thyristor VRRM to be chosen to be just greater than the BOD ignition value:

17.3.5 Field Flashing

The field flashing circuit is designed to provide current to the rotor field during the generator start-up sequence.

17.3.5.1 Excitation Data

The excitation rated flashing current is equal to 15% of the rated no-load field current of the generator. Excitation rated flashing current:

$$I_{et} = 0.15 I_{f0} = 0.15 \times 1843 = 276 \text{ A}$$

Rated excitation flashing current:

$$I_{etN} = 280 \text{ A}$$

Excitation rated flashing voltage:
Given field resistor:

$$R_f = 0.1188 \text{ }\Omega \text{ } (95°C)$$

Rated excitation voltage:

$$U_{etN} = I_{et}R_f = 280 \times 0.1188 = 33.26 \text{ V}$$

Design of the equipment:

$$U_{etN} = 33 \text{ V}.$$

17.3.5.2 AC Flashing – Design of the Transformer

Transformer permanent output current:

$$I_{ttN} = \sqrt{\frac{2}{3}} \times I_{etN} = 228.5 \text{ A}$$

Transformer ideal output voltage:
$$U_{rms} = \frac{\pi}{3\sqrt{2}} \times U_{etN} = 24.4 \text{ V}$$

Ideal output power:
$$S_{et} = \sqrt{3} \times U_{rms} \times I_{ttN} = 9.66\text{k VA}$$

Inductive voltage drop:
$$U_x = \frac{3}{\pi} \times x_r \times \frac{U_{rms}^2}{S_{et}} \times I_{ttN} = 1.14 \text{ V}$$

where

x_r — transformer short-circuit impedance (8.5%).

External voltage drop (diodes, cables, etc.):
$$U_f = 2U_{t0} + 0.05 U_{rms} = 3.22 \text{ V}$$

where

U_{t0} — threshold voltage of the diodes 1 V.

The factor 0.05 corresponds to the 5% voltage drop in the cables. Output voltage of the transformer:
$$U_{ttn} = U_{rms} + \frac{\pi}{3\sqrt{2}} \times (U_f + U_x) = 27.6 \text{ V}$$

Transformer permanent power:
$$S_{ttN} = \sqrt{3} \times U_{ttN} \times I_{ttN} = \sqrt{3} \times 27.6 \times 228.5 = 11 \text{ kVA}$$

Rated primary voltage: 415 V.
Maximum permanent primary voltage: $1.1 \times 415 = 456.5$ V.
Rated frequency: 50 Hz, (+3/−5)%.
Altitude: <1000 m.

17.3.6 Current Transformer

17.3.6.1 Current Transformer on Primary Side of Excitation Transformer

The current transformer is used for over-current protection measurement

Bi-phase short-circuit current:

Rated excitation current: $I_{tN} = \sqrt{\frac{2}{3}} \times I_{EN} = 5438.97$ A

The primary-side current value for the current transformer of the excitation transformer corresponding to the above current is

$$1.2 \times 5483.97/(27000/1600) = 390 \text{ A}$$

The primary-side current of the selected current transformer is 400 A
The specification is 400 A/1 A, 5P20, 15 VA (P121 power consumption 0.025 VA)

17.3.6.2 Current Transformer on Secondary Side of Excitation Transformer

It should be able to read a ceiling current of 12 220 A ≥ back in AC side: 10 000 A ac.
Current transformer to be chosen: 10 000 A/1 A, Cl 0.5, 30 VA (PMF175, power consumption 15 W)

17.3.7 Parameters and Configuration of Excitation System

The parameters and configuration of the excitation system are shown in Table 17.8.

Table 17.8 Parameters and configuration of excitation system. Authorized by Beijing G.E. Automation Engineering Co., Ltd.

SN	Description	Unit	Design value
1	Excitation transformer		
	Type		Dry type
	Model		Cast resin
	Rated capacity	kVA	3 × 5 167
	Supplier		China
	Rated voltage		
	Primary	kV	27
	Secondary	kV	1.6
	Frequency	Hz	50
	Phases		3
	Connection		Y/d11
	Earthing methods		Direct ground
	Insulation class		F
	Insulation Withstand voltage	kV	35
	Pulse voltage		
	Primary	kV	170
	Secondary	kV	20
	50 Hz withstand voltage of power (1 min) frequency		
	Primary	kV	70
	Secondary	kV	10
	Cooling method		AN/AF
	Losses		
	Copper loss	kW	3 × 15.17
	Iron loss	kW	3 × 8.38
	Addition loss	kW	3 × 5.31
	Total loss	kW	3 × 28.86
	Efficiency	%	99.4
	Protection class		IP23
	Voltage regulation	%	0.375 2 (single phase)
	Impedance voltage	%	10
	Zero-sequence impedance	%	7.2
	Primary coil resistance	Ω	0.0448 (single-phase)
	Exciting current	A	0.7
	Noise level (AN)	dBA	60
	Overload capacity	kVA	AN = 1.1
	Transformer temperature signal		Alarming, and trip
	Allowable maximum temperature rise of transformer	K	100
	Dimension (W × D × H)	mm	7 650 × 4 410 × 3 200
	Shipping dimension (W × D × H)	mm	10 393 × 7154 × 3530
	Weight	t	16.8 (single-phase)

Table 17.8 (Continued)

SN	Description	Unit	Design value
2	Converter cabinet		
	Type		Three-phase full – controlled thyristor bridge
	Rated current for elements of thyristor	A V	2 220
	negative voltage		6 500
	Rated voltage	V	762.8
	Rated current	A	6 109.6
	Cooling mode		AF
	Serial components number/arm		1
	Parallel path number		1
	Noise of converter cabinet	dB	<80
	Average time of fan without failure	h	30 000
	Quantity of converter cabinet		6
	Average-current coefficient		0.9
3	FCB (field current breaker)		
	Rated voltage	V	1 500
	Rated current	A	8 500
	Main contact interrupting current (effective value), $L/R = 2$ ms	kA	80
	DC control voltage	V	110
	Maximum extinction voltage	V	2 250
	Supplier		LENO1R
	Type		CEX 068500 3.1
4	AVR characteristic		
	Range of automatically regulated voltage	%	30–110
	Range of manually regulated voltage	%	20–110
	Static differential ratio	%	<±0.5
	Transient resolution ratio	ms	5
	Record length of each time	s	<300
	Total record length	s	300
	Response time	s	0.1
	Control cycle	ms	5
	Sampling method		AC
	Phase margin	(°)	40
	Gain margin	dB	6
	Control rule		PI + PSS
	Manufacture		ALSTOM
5	Parameters of field flashing circuit		
	Field flashing voltage	V (AC)	415
	Field flashing current	A (AC)	13.5
	field flashing power losses	kVA	11
6	Overvoltage protection of rotor		
	Linear resistance capacity of overvoltage protection	MJ	8.4
	Operant voltage value of overvoltage protection	V	≤3 200
7	Measuring device to rotor insulating to earth		MiCOM P342 + MiCOM P391

17.4 Block Diagram of Automatically Regulated Excitation System

For the potential source static excitation system for the 1270.6 MVA generator of the Malaysian Manjung 4 Thermal Power Plant, the Alstom CONTROGEN latest generation of automatically regulated excitation systems is adopted, and the base value for per unit value in the cabinet of the automatically regulated excitation system is as follows:

(1) The active power and reactive power are referred to the rated apparent power: S_N, $S_N = 1270.6$ MVA
(2) The generator stator voltage is referred to the rated stator voltage: U_N, $U_N = 27\,000$ V
(3) The generator field current is referred to the generator field current necessary to produce the rated terminal voltage at no-load on the air-gap line.
(4) The base generator field voltage is linked to the base generator field current through the generator rotor resistance at 100 °C (0.121 Ω) according to IEEE Std 421.1: 209.57 V.

17.4.1 AVR Reference Voltage

The block diagram of the AVR reference voltage [53] is shown in Figure 17.2. The input quantities of the unit include the droop setting, regulation of reactive or power factor, input of the reference voltage, frequency, limit of magnetic flux, and so on.

Reactive power and the power factor (RPPF) regulator output are connected to summing point instead of to the voltage reference if the RPPF mode is selected (RPPF_sld = 1). The typical AVR reference voltage is shown in Table 17.9.

17.4.2 AVR Reference Voltage Limitation

See Figure 17.3 for the AVR reference voltage limitation, and Table 17.10 for the typical parameters.

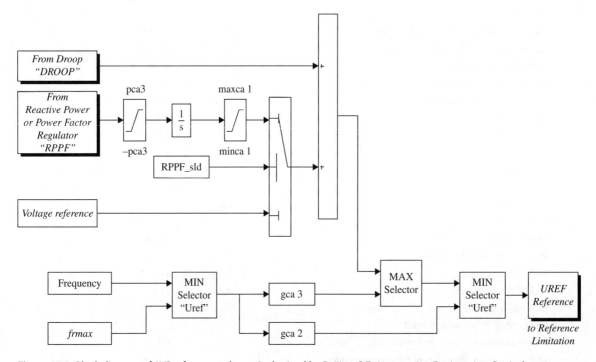

Figure 17.2 Block diagram of AVR reference voltage. Authorized by Beijing G.E. Automation Engineering Co., Ltd.

17.4 Block Diagram of Automatically Regulated Excitation System

Table 17.9 Typical parameter values of AVR reference voltage. Authorized by Beijing G.E. Automation Engineering Co., Ltd.

Parameter	Value	Unit	Description	Range
Pca3	0.0033	p.u. s^{-1}	Slope limiter parameter	[0, 1]
Limitation flux				
f_{rmax}	1.8	p.u.	Maximum frequency limit	[0, 1.8]
gca2	1.06	p.u.	Maximum U/F limit	[0, 1.5]
gca3	0	p.u.	Minimum U/F limit	[0, 1]

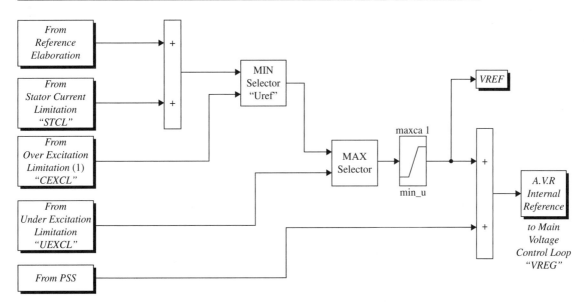

Figure 17.3 Diagram of AVR reference voltage limitation. Authorized by Beijing G.E. Automation Engineering Co., Ltd.

Table 17.10 Typical parameter values for AVR reference voltage limitation. Authorized by Beijing G.E. Automation Engineering Co., Ltd.

Parameter	Value	Unit	Description	Range
maxca1	1.05	p.u.	Maximum voltage reference	[1.05, 1.1]
min_u	0.95	p.u.	Minimum voltage reference	[0.9, 0.95]

17.4.3 Main Voltage Control Loop for Static Excitation System - VREG

See Figure 17.4 for the main voltage control loop and Table 17.11 for the typical parameters.

17.4.4 Under-Excitation Limit – UEL

See Figure 17.5 for the block diagram of the low excitation limit, and Table 17.12 for the typical parameters.

17.4.5 Overexcitation Limitation for Static Excitation System – OEL

See Figure 17.6 for the block diagram of the overexcitation limiter (OEL), and Table 17.13 for the typical parameters.

17 Performance Characteristics of Excitation System of 1000 MW Turbine Generator Unit

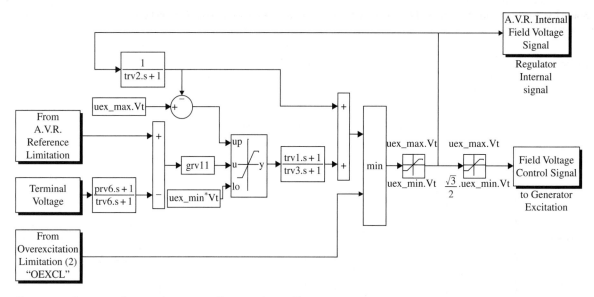

Figure 17.4 Diagram of main voltage control loop. Authorized by Beijing G.E. Automation Engineering Co., Ltd.

Table 17.11 Typical parameters of AVR. Authorized by Beijing G.E. Automation Engineering Co., Ltd.

Parameter	Value	Unit	Description	Range
g_{rv11}	40	p.u.	Main loop proportional gain	[1.05, 1.1]
t_{rv2}	4	s	Integral time constant	[0.9, 0.95]
t_{rv1}	1	s	Phase lead/lag filter numerator time constant	
t_{rv3}	1	s	Phase lead/lag filter denominator time constant	
p_{rv6}	1	s	Voltage feedback numerator time constant	
t_{rv6}	1	s	Voltage feedback denominator time constant	
u_{ex_max}	7.28	p.u.	Positive ceiling voltage with nominal generator voltage	
u_{ex_min}	7.28	p.u.	Negative ceiling voltage with nominal generator voltage	

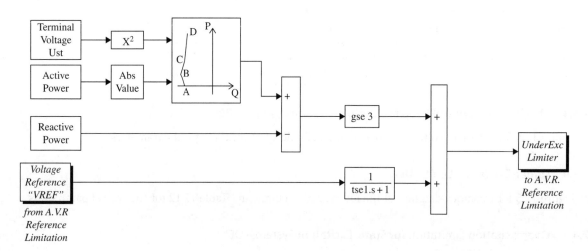

Figure 17.5 Under-excitation limitation. Authorized by Beijing G.E. Automation Engineering Co., Ltd.

17.4 Block Diagram of Automatically Regulated Excitation System

Table 17.12 Typical parameter values of low excitation limit. Authorized by Beijing G.E. Automation Engineering Co., Ltd.

Parameter	Value	Unit	Description	Range
g_{se3}	0.25	p.u.	Gain	[0, 0.5]
t_{se1}	0.5	s	Integral time constant	[0.5, 5]
var_A	−0.28	p.u.	Point A: Reactive power at nominal voltage	
pwr_A	0	p.u.	Point A: Active power	
var_B	−0.28	p.u.	Point B: Reactive power at nominal voltage	
pwr_B	0.25	p.u.	Point B: Active power	
var_C	−0.28	p.u.	Point C: Reactive power at nominal voltage	
pwr_C	0.5	p.u.	Point C: Active power	
var_D	−0.28	p.u.	Point D: Reactive power at nominal voltage	
pwr_D	0.85	p.u.	Point D: Active power	

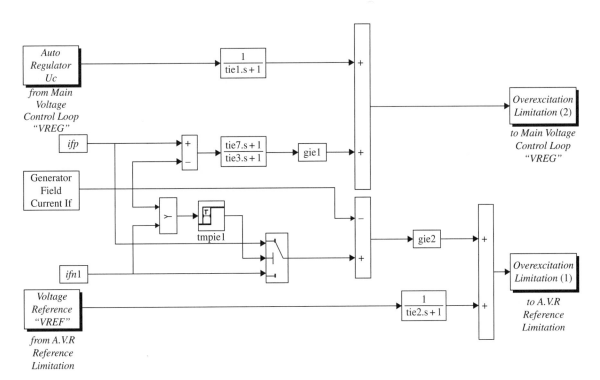

Figure 17.6 Block diagram of overexcitation limit. Authorized by Beijing G.E. Automation Engineering Co., Ltd.

17.4.6 Dual-Input PSS

See Figure 17.7 for the dual-input PSS, and Table 17.14 for the typical parameters.

17.4.7 Stator Current Limitation – STCL

See Figure 17.8 for the diagram of STCL, and Table 17.15 for the typical parameters.

Table 17.13 Typical parameter values for diagram of overexcitation limitation. Authorized by Beijing G.E. Automation Engineering Co., Ltd.

Parameters	Values	Unit	Description	Range
Ceiling limitation				
i_{fp}	7.05	p.u.	Ceiling limitation set point	[1, 10]
g_{ie1}	4	p.u.	Gain (ceiling loop)	[0.1, 2]
t_{ie1}	1	s	Integral time constant of (ceiling loop)	[0.1, 5]
t_{ie7}	1	s	Phase leading filter numerator time constant	[0.05, 3]
t_{ie3}	1	s	Phase leading filter denominator time constant	
Thermal limitation				
i_{fp1}	3.88	p.u.	Thermal limitation set-point	[0.1, 10]
g_{ie2}	0.155	p.u.	Terminal limitation gain	[0.1, 2]
t_{ie2}	1	s	Thermal limitation time constant	[5, 30]
t_{mpie1}	10	s	Ceiling limitation set-point time delay (with reset)	

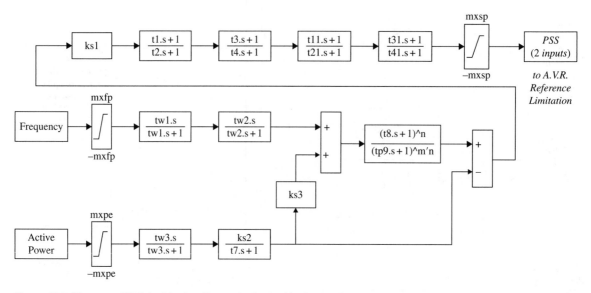

Figure 17.7 Diagram of PSS double-signal input. Authorized by Beijing G.E. Automation Engineering Co., Ltd.

17.4.8 Reactive Power or Power Factor Regulators – RPPF

See Figure 17.9 for the diagram of RPPF regulators and Table 17.16 for the typical parameters.

17.4.9 Droop

See Figure 17.10 for the block diagram of the difference regulation unit, and Table 17.17 for the typical parameters.

17.4.10 Potential – Source Excitation System ST7B

See Figure 17.11 for the block diagram of the static excitation system of voltage source-ST7B, and Table 17.18 for the typical parameters.

Table 17.14 Typical parameter values of dual-input PSS. Authorized by Beijing G.E. Automation Engineering Co., Ltd.

Parameter	Value	Unit	Description	Range
k_{s3}	1	p.u.	Mechanical channel gain	[0 or 1]
t_{w1}	3	s	Speed channel wash-out time constant	[1, 10]
t_{w2}	3	s	Speed channel wash-out time constant	[1, 30]
t_{w3}	3	s	Power channel wash-out time constant	[1, 10]
k_{s2}	0.426	p.u.	Power channel gain	[0, 1]
t_7	3	s	Phase lead/lag filter high frequency gain phase	[1, 30]
t_8	0.6	s		[0, 2]
t_9	0.15	s		[0.1, 0.5]
m	4	integer	Ramp track filter parameter	[1, 5]
n	1	integer	Ramp track filter parameter	[1, 4]
k_{s1}	6[a]	p.u.	PSS gain	[1, 150]
t_1	0.390	s	Phase lead/lag filter numerator time constant (1)	[0, 10]
t_2	0.065	s	Phase lead/lag filter denominator time constant (1)	[0.015, 3]
t_3	0.156	s	Phase lead/lag filter numerator time constant (2)	[0, 10]
t_4	0.026	s	Phase lead/lag filter denominator time constant (2)	[0.015, 3]
t_{11}	1	s	Phase lead/lag filter numerator time constant (3)	[0, 10]
t_{21}	1	s	Phase lead/lag filter denominator time constant (3)	[0.015, 3]
t_{31}	1	s	Phase lead/lag filter numerator time constant (4)	[0, 10]
t_{41}	1	s	Phase lead/lag filter denominator time constant (4)	[0.015, 3]
mxfp	0.1	p.u.	Frequency deviation Input limit	[0, 0.2]
mxpe	1.5	p.u.	Active power Input limitation	[0, 2]
mxsp	0.05	p.u.	Output limitation	[0, 0.1]

a) Must be defined from on-site tests and (or) stability study.

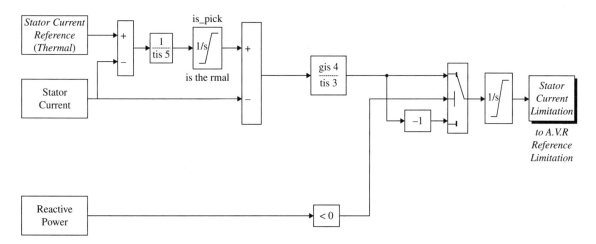

Figure 17.8 Stator current limitation diagram. Authorized by Beijing G.E. Automation Engineering Co., Ltd.

Table 17.15 Stator current limitation. Authorized by Beijing G.E. Automation Engineering Co., Ltd.

Parameter	Value	Unit	Description	Range
g_{is4}	1	p.u.	Gain	[0.1, 10]
t_{is3}	5	s	Integral time constant	[1, 200]
t_{is5}	10	s	Time constant of the inverse time current correction	[5, 50]
is_pick	1.16	p.u.	Maximum permitted stator current threshold (inverse time)	[1, 2]
is_thermal	1.06	p.u.	Stator current reference	[1, 1.5]

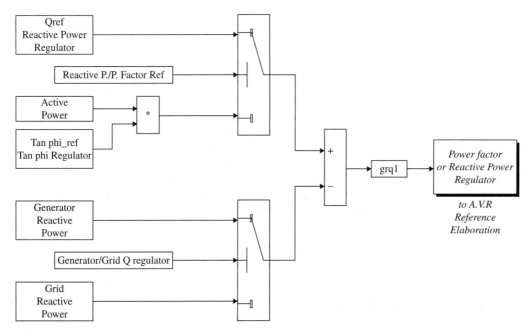

Figure 17.9 Block diagram of regulators of reactive power and power factor. Authorized by Beijing G.E. Automation Engineering Co., Ltd.

Table 17.16 Reactive power or power factor regulator. Authorized by Beijing G.E. Automation Engineering Co., Ltd.

Parameters	Values	Unit	Description	Range
g_{rq1}	0.1	p.u.	Proportion gain for the operation mode	[0.01, 1]

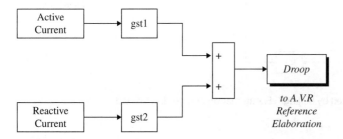

Figure 17.10 Diagram of droop. Authorized by Beijing G.E. Automation Engineering Co., Ltd.

17.4 Block Diagram of Automatically Regulated Excitation System

Table 17.17 Droop. Authorized by Beijing G.E. Automation Engineering Co., Ltd.

Parameter	Value	MoU	Description	Range
g_{st1}	0	p.u.	Active current compensation Gain	[0, 0.2]
g_{st2}	0	p.u.	Reactive current compensation Gain	[−0.2, 0.2]

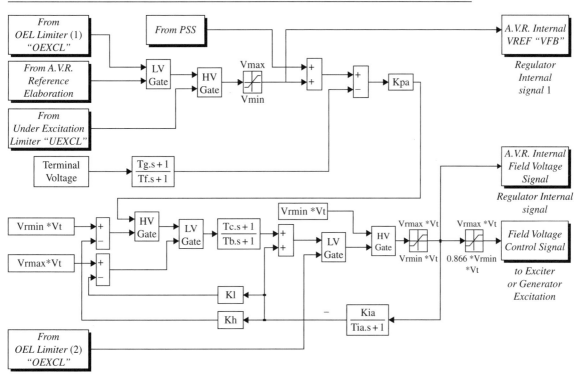

Figure 17.11 Diagram of static excitation system of voltage source-ST7B. Authorized by Beijing G.E. Automation Engineering Co., Ltd.

Table 17.18 Typical parameters for voltage source-ST7B. Authorized by Beijing G.E. Automation Engineering Co., Ltd.

Parameter	Value	MoU	Description
K_{pa}	$=g_{rv11}$	p.u.	Voltage regulator proportional gain
K_{ip}	$=1$	p.u.	Gain of integral term
T_{ia}	$=t_{rv2}$	s	Voltage regulator integral time constant
T_c	$=t_{rv1}$	s	Lead/lag filter numerator time constant
T_b	$=t_{rv3}$	s	Lead/lag filter denominator time constant
T_g	$=p_{rv6}$	s	Voltage feedback numerator time constant
T_1	$=t_{rv6}$	s	Voltage feedback denominator time constant
V_{max}	$=maxca1$	p.u.	Positive limit of internal voltage reference
V_{min}	$=minca1$	p.u.	Negative limit of internal voltage reference
K_I	1	–	Integral term validation for inner positive limitation
K_h	0	–	Integral term validation for inner negative limitation
V_{rmax}	$=u_{ex\_max}$	p.u.	Positive ceiling voltage with nominal generator voltage
V_{rmin}	$=-u_{ex\_max}$	p.u.	Negative ceiling voltage with nominal generator voltage

18

Performance Characteristics of 1000 MW Nuclear Power Steam Turbine Excitation System

This chapter will focus on describing the performance and features of the excitation system in the steam turbine generator unit of Fuqing Nuclear Power Station and Sanmen Nuclear Power Station.

18.1 Performance Characteristics of Steam Turbine Generator Brushless Excitation System of Fuqing Nuclear Power Station [54]

18.1.1 Basic Parameters

The base value of the relevant parameters [54] in the brushless excitation system of the steam turbine generator of Fuqing Nuclear Power Station is selected according to the following definition:

(1) *Generator active and reactive power base value.* The generator rated apparent power of 1278 MVA.
(2) *Generator stator rated voltage base value.* 24 kV.
(3) *Generator excitation current base value.* The value of the excitation current is required when the generator voltage is the no-load rated voltage value, which is 2189 A.
(4) *Generator excitation voltage base value.* When the generator excitation current is the per-unit base value through the generator excitation winding, the temperature is 100 °C, and the winding resistance is 0.08127 Ω, according to IEEE std. 421.1 it is specified as 175 V.
(5) *Exciter excitation current base value.* When the generator generates the no-load rated voltage, the excitation current required for the exciter is 59.94 A.
(6) *Exciter excitation voltage base value.* When the excitation current flowing through the exciter field winding is equal to the per-unit base value and the field winding temperature is 100 °C, the winding resistance is 1.0822 Ω. According to IEEE std. 421.1, the exciter excitation voltage base value is specified as 53.06 V.

18.1.1.1 Generator Parameters
The parameters of the generator set are shown in Table 18.1.

18.1.1.2 Exciter Parameters
The exciter parameters are shown in Table 18.2:

18.1.1.3 Excitation Transformer Parameters
The excitation transformation parameters are shown in Table 18.3.

18.1.2 Description of Excitation System

The 1150 MW turbine generator excitation system of the M310 + reactor type Fuqing Nuclear Power Station uses the brushless excitation system, whose power is supplied by excitation transformers connected to the generator terminal. The corresponding excitation system single-line diagram is shown in Figure 18.1.

Design and Application of Modern Synchronous Generator Excitation Systems, First Edition. Jicheng Li.
© 2019 China Electric Power Press. Published 2019 by John Wiley & Sons Singapore Pte. Ltd.

18 Performance Characteristics of 1000 MW Nuclear Power Steam Turbine Excitation System

Table 18.1 Generator parameters. Authorized by Beijing G.E. Automation Engineering Co., Ltd.

Parameter	Symbol	Rated value	Unit	Remark
Rated apparent power	S_N	1278	MVA	
Rated active power	P_N	1150	MW	
Rated terminal voltage	U_N	24	kV	
Rated stator current	I_N	30 739	A	
Rated frequency	f_N	50	Hz	
Rated power factor	P_F	0.9	N/A	
Generator no-load excitation current	I_{f0}	2189	A	
Generator no-load voltage excitation current	U_{f0}	169 (84 °C)	V	
Rated excitation current	I_{fn}	5795	A	
Rated excitation voltage	U_{fn}	461	V	
Direct-axis transient short-circuit time constant	T_d	1.83	s	
Direct-axis transient open-circuit time constant	T'_{do}	9.33	s	
Direct-axis synchronous reactance	X_d	195.5	%	
Direct-axis transient reactance (saturation value)	X'_d	35.4	%	

Table 18.2 Brushless exciter parameters (Exciter model: TKJ167-45). Authorized by Beijing G.E. Automation Engineering Co., Ltd.

Parameter	Symbol	Rated value	Unit		Remark
Exciter field resistance 21 °C	R_{fex}	0.8270	Ω		
Exciter field resistance 95 °C	R_{fex}	1.066	Ω		
Exciter field resistance 120 °C	R_{fex}	1.147	Ω		
Open-circuit transient time constant	T'_{do}	0.426	s		
Exciter time constant	T_e	0.17	s		
Generator no-load rating	Generator field voltage U_{f0}	169 V	Generator field current I_{f0}	2189 A	
	Exciter field voltage U_{fexo}	53.8 V	Exciter field current I_{fexo}	46.9 A	
Generator rating	Generator field voltage U_{fN}	461 V	Generator field current I_{fN}	5795 A	
	Exciter field voltage U_{fexN}	129.9 V	Exciter field current I_{fexN}	113.2 A	

Note: The excitation voltage of the exciter is measured under the rated operating conditions of the generator under no load. $U_{fexo} = 39.9$ V, excitation current $I_{fexo} = 39.9$ A, generator rated load conditions under the excitation magnetic field excitation voltage $U_{fexN} = 87.7$ V, excitation current $I_{fexN} = 84.7$ A.

18.1.2.1 Automatic Regulation Excitation System

The steam turbine generator uses the Alstom P320 V2 type excitation regulator. The excitation regulator is configured in a fully redundant dual automatic channel. In normal operation, one automatic channel will operate with 100% load and the other channel serves as the standby.

The power plant control system developed for Alstom by P320 includes the DCS function, steam turbine regulator, generator excitation, monitoring, synchronism, and other functions. P320AVR is a three-channel excitation control system. Two automatic/manual regulators control a rectifier bridge; the third channel having only a manual regulator can control two bridges at the same time. In normal operation, an automatic regulator operates with a rectifier bridge, and the other rectifier bridge does not have an output. When both automatic channels are exited, the third channel controls both rectifier bridges at the same time. The rated

Table 18.3 Excitation transformer parameters. Authorized by Beijing G.E. Automation Engineering Co., Ltd.

Parameter	Value	Unit
Rated power	54	kVA
Overload power	88	kVA/10 s/1 time/15 min
Rated secondary voltage	300	V
Rated primary voltage	24	kV
Rated frequency	50	Hz
Impedance voltage	6	%
Altitude	<1000	m
Load type	6-phase full-controlled rectifier bridge	

output current is about 100 A, so the rectifier bridge can serve as a cold standby; in addition, the AC and DC sides of each rectifier bridge are equipped with disconnectors. The adjustment function of the regulator is the same as that of the conventional project. It has a power system stabilizer (PSS), rotor current limit, V/Hz limit, reactive power/active compensation, excitation limit, and so on.

18.1.2.2 External Communication Network

The external communication network still adopts the integrated design of turbine-generator-control (TGC) and automatic voltage regulator (AVR), and its communication network is relatively complicated. F8000 is the Worldfip field bus, and the operating data are exchanged between AVR, TGC, and TGC controllers. S8000 uses Ethernet, exchanging information with the engineering station and external DCS. The office network also uses Ethernet, which is used to connect to the client computer at the engineering station during debugging, without affecting the operation. In the actual work, the integration of communication design means that debugging involves different professional staff, and coordination is an inconvenience, so Alstom has canceled the integrated communication design structure in the other project after the Fuqing Nuclear Power Stations project. DCS communicates with AVR directly, and TGC and AVR do not have horizontal contact.

18.1.2.3 Excitation System Monitoring and Protection

The following excitation system monitoring and protection devices are part of the excitation system of the steam turbine generator of China's domestic CPR1000 reactor nuclear power plant.

(1) Grounding Rotor Earthing Protection Device.
   Slip rings and carbon brushes for detection are equipped at the excitation end of the generator. Three slip rings are respectively connected to the positive, negative pole of the field winding and at the rotating shaft of the generator rotor to perform grounding detection for the generator rotor winding. Because the carbon brushes and slip rings are installed inside the armature drum of brushless exciter, in order to facilitate maintenance and extend service life, short-time duty rather than continuous duty is adopted by this device, which should be inspected once daily by contacting with the carbon brushes automatically for 10 s. The corresponding signal is sent to the rotor earthing protection relay located in the excitation regulation cubicle. In addition, a manual rotor earthing detection button is provided to facilitate inspection by operation personnel at any time.
   The grounding detection device designed by superimposing the AC voltage principle is injected with a 40 V power frequency AC voltage, and a 3.3 kΩ resistor is connected in the measuring circuit. The alarm

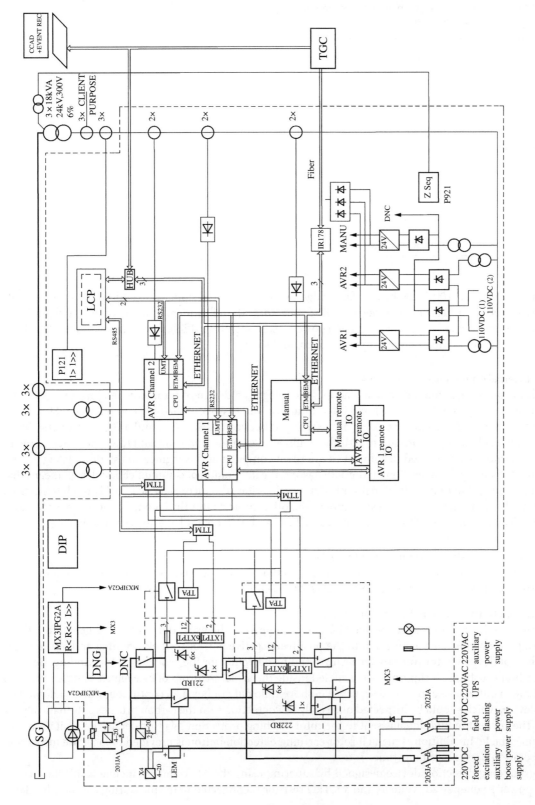

Figure 18.1 Single-line diagram for brushless excitation system of 1150 MW steam-generator unit. Authorized by Beijing G.E. Automation Engineering Co., Ltd.

rings when the measuring resistance is less than 4 kΩ, and the machine shuts down when the value is less than 2 kΩ. As the generator type uses the brushless excitation mode, the rotor lacks the collector ring. In order to install a rotor grounding protection device, a small slip ring is added to excitation end of the generator rotating shaft. The device structure schematic diagram is shown in Figure 18.2.

The structure chart of slip rings and brush gear for the rotor earthing protection device is shown in Figure 18.3.

The monitoring mode is designed for periodic monitoring. The brush is periodically lifted by a separate lifting brush device (electromagnet), and a superimposed alternating voltage is introduced between the brush and the slip ring. The lifting brush device is powered by 125 V DC. The lifting brush device should exit automatically after completing one measurement, and then the lifting brush is used again for automatic measurement and exit in the next cycle. The general measurement cycle is set to monitor every 24 hours. The advantage of this design is reduction in the wear caused by long-term high-speed rotating friction between the slip ring and the brush, which increases the service life. But the interval period cannot be set too long, otherwise unit ground faults may not be detected. The protection device can also manually send measuring signals of the lifting brush. When the generator set is running, if the generator set operating parameters are abnormal, the brush must be promptly raised manually to detect whether rotor winding ground faults occur.

The superimposing AC voltage protection principle is shown in Figure 18.4. C_{mE} is the capacitance of the field winding to ground; R_{mE} is the resistance of the field winding to ground; $U_\sim$ is the AC component of the excitation winding sets. C_i is the isolated DC capacitor, and its role is to prevent the DC current generated by the excitation voltage from flowing through the AC power supply circuit. K is the relay, and its impedance is R_K. UV is the voltage converter, and its main role is to separate the AC power from the rotor circuit and reduce the AC voltage to the value required for the protecting device.

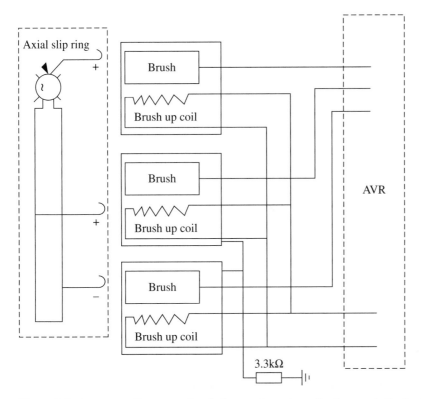

Figure 18.2 Rotor grounding protection device structure schematic diagram. Authorized by Beijing G.E. Automation Engineering Co., Ltd.

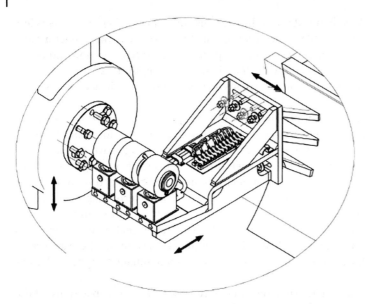

Figure 18.3 Structure chart of slip rings and brush gear for rotor earthing protection devices. Authorized by Beijing G.E. Automation Engineering Co., Ltd.

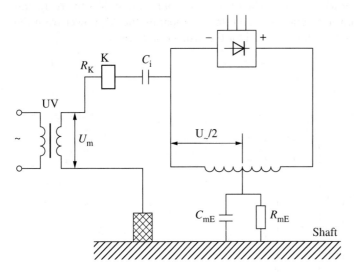

Figure 18.4 Superimposed AC voltage protection schematic diagram. Authorized by Beijing G.E. Automation Engineering Co., Ltd.

In normal operation, regardless of the contact resistance between the brush and the shaft, the current flowing through the relay is

$$I_{-} = \frac{U_m + \frac{1}{2}U_{\sim}}{R_x - jX_{ci} - j\frac{R_{mE}X_{mE}}{R_{mE} - jX_{mE}}}$$

where $j\frac{R_{mE}X_{mE}}{R_{mE} - jX_{mE}}$ is the impedance of the C_{mE} capacitive reactance in parallel with R_{mE} resistance. When the field winding to the ground insulation is reduced, the current flowing through the relay will increase, and the protection works when the current reaches the set value.

(2) Excitation AC Overcurrent Protection.

The relay monitors the excitation current. The first stage fault causes the regulator exchange to enter standby when the excitation current exceeds a reference threshold, then issues a tripping order if the fault persists. The second stage fault protects the excitation transformer when a direct short-circuit fault occurs at the output of the thyristor, and issues a tripping order as soon as the excitation current exceeds a reference threshold.

(3) Excitation Transformer Overcurrent Protection.

For the potential source excitation system, CTs usually are installed on the HV side and LV side of the excitation transformer for differential, overcurrent, and overload protection. However, an appropriate CT is not available due to the rated current of the excitation transformer HV winding being less than 1 A; so a zero sequence overvoltage relay is adopted to detect a fault in the excitation transformer primary winding.

(4) Rotating Diode Non-Conduction Detection System (DNC).

The rotating diode non-conduction detection system is composed of three Hall sensors installed on the inductor support toward the conductor of the rotating armature to monitor the conducting condition. The signals from the three sensors are connected to the DNC for processing; then the testing results are send to the relevant AVR circuit where corresponding alarm or tripping signals are sent out. The rotating diode non-conducting monitoring protection circuit is shown Figure 18.5.

(5) Shaft Current Detection Device.

The shaft current detection device detects possible shaft current and protects the turbo-generator shaft. This protection is obtained by de-energizing a switch which shunts the fault current to the grounding without passing through the shaft current detection equipment.

(6) Boosting System.

The boosting system is implemented by connecting another boosting thyristor bridge in parallel with the thyristor bridge. At the specified terminal voltage of the generator, the boosting forcing is supplied by

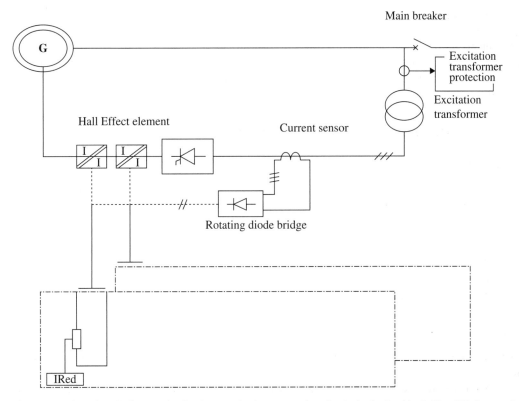

Figure 18.5 Rotating diode non-conducting monitoring protection circuit. Authorized by Beijing G.E. Automation Engineering Co., Ltd.

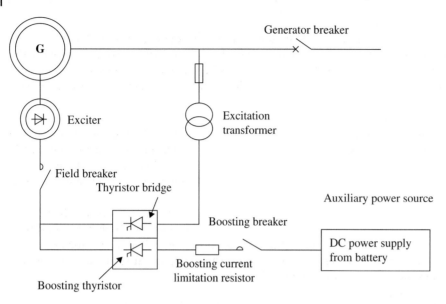

Figure 18.6 Schematic diagram of boosting system. Authorized by Beijing G.E. Automation Engineering Co., Ltd.

another boosting thyristor bridge. Boosting is initiated by generating pulses (close pulses) to the excitation thyristor when the generator voltage falls below the pre-set threshold, like $U_{min} = 0.7$ p.u. Meanwhile, the thyristor bridge will be blocked. The schematic diagram of the boosting system is shown in Figure 18.6. The boosting circuit breaker will be closed when the generator is connected to the grid. Boosting is initiated by generating pulses (close pulses) to the excitation thyristor when the generator voltage falls below the pre-set threshold U_{min}. Then, as the generator terminal voltage rises, the boosting circuit breaker opens; the boosting circuit breaker closes again after the generator terminal voltage recovers to the normal value, to get ready for the next boosting. After stopping, boosting is inhibited during a pre-set time period that depends on different cooling times considering the rotor thermal capability. Fuqing Nuclear Power Station allows boosting four times in an hour; each cooling time interval is 15 min, and the boosting time is 10 s.

18.2 Structural Characteristics of Brushless Excitation System

18.2.1 Basic Parameters

The TKJ167-45 type brushless exciter is divided into two main parts: static components and rotating parts [55]. The main parameters are as follows.

Rated capacity	4000 kVA	Rated magnetic field voltage	93 V
Rated voltage	454 V	Rated field current	158 A
Rated current	5863 A	Armature winding number	117
Pole pair	11	Armature winding phase	39

18.2.2 Structural Characteristics

TKJ167-45 is a suspended type brushless exciter, and the simplified wiring diagram of the excitation system is shown in Figure 18.7.

In terms of mechanical structure, though the main exciter with a brushless structure is used in the unit excitation system, from the electric standpoint, because the exciting power of the main exciter is taken from

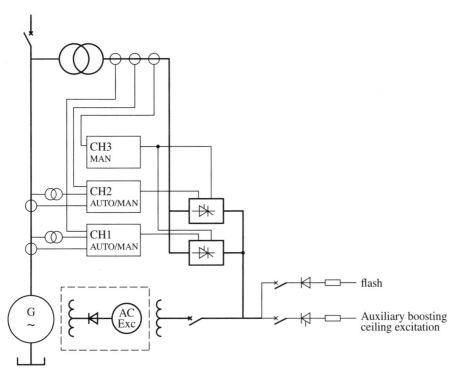

Figure 18.7 Simplified wiring diagram of brushless excitation system. Authorized by Beijing G.E. Automation Engineering Co., Ltd.

the exciting transformer connected to the generator terminal, this excitation system is a self-excited excitation system.

The TKJ167-45 suspended type exciter is the proprietary technology of Alstom; at present, the suspended type exciter unit can be produced for the units including Ningde, Fang Jiashan, Hongyanhe, and Fuqing nuclear power units in the domestic Dongfang Electrical Machinery Factory through technology transfer. The structure chart of the TKJ167-45 suspended type exciter is shown in Figure 18.8 [56].

The structural features of TKJ167-45 type brushless exciter are as follows:

(1) A single-terminal suspended rotating armature is used. The rotating armature and rotating rectifying ring are fixed on the rotation rotor of the exciter. The bottom of the cylinder is hung and fixed on the exciter end tail of the generator spindle, and a backup bearing does not need to be added. In addition, the armature winding and diode of the exciter are fixed on the inner side of the rotating armature cylinder. The advantage of this structure is that it can directly connect the armature winding in the cylinder with the rectifying module of the rectifying ring, simplifying the mechanical structure of the brushless exciter and improving the reliability of unit operation.
(2) The exciting winding of the exciter is fixed on the static pedestal, and the excitation current needed is supplied by the automatic excitation regulator brushless exciter.
(3) The exciter uses the 39-phase polygon multi-phase armature winding, and the number of pole pairs is 11, which makes the ripple wave of the rotating voltage output by the multi-phase exciter armature smaller. The exciter is far superior to the common three-phase type or six-phase star-shaped rectifying circuit from the standpoint of rectifying power conversion and ripple wave factor reduction. The corresponding wiring mode is shown in Figure 18.9. The brushless exciter armature winding consists of 39-phase windings connected in series to form a polygonal closed loop. The phase outgoing line connects to two diodes through the fast fuse, and forms a 39-phase full-wave bridge rectifier circuit.

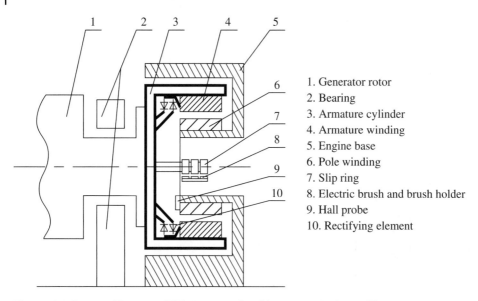

Figure 18.8 Structural features of TKJ167-45 type brushless exciter. Authorized by Beijing G.E. Automation Engineering Co., Ltd.

1. Generator rotor
2. Bearing
3. Armature cylinder
4. Armature winding
5. Engine base
6. Pole winding
7. Slip ring
8. Electric brush and brush holder
9. Hall probe
10. Rectifying element

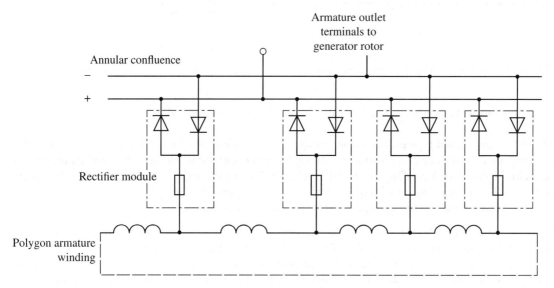

Figure 18.9 Wiring diagram of rotating armature winding and diode of brushless exciter. Authorized by Beijing G.E. Automation Engineering Co., Ltd.

18.2.3 Rotary Diode Rectifier

The rotary diode rectifier and fast fuse combination unit structure is shown in Figure 18.10. The relevant structure is as follows:

(1) The radiator base is made of aluminum material, and its transverse surface heat dissipation groove and two cooling holes can cool the radiator base sufficiently. The radiator base is mounted in the exciter armature through the mounting seat. The upper surface of the radiator base is machined to a different gradient to ensure contact and fixation with the fuse and with the diode.

(2) Each phase winding outgoing line is equipped with a fuse and two diodes. The stud is screwed with the glue into the bottom of the fuse, and the fuse is secured to the bottom of the heat sink. The polarity of the two parallel diodes connected to the rotating rectifier ring is shown in Figure 18.10.

The rotary diode rectifier and AC armature winding connection are shown in Figure 18.11.

18.2 Structural Characteristics of Brushless Excitation System

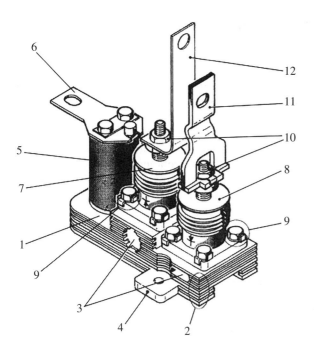

1. Radiator base
2. Heat dissipation slot
3. Cooling holes
4. Mounting seat;
5. Fuse
6. Electric connector
7. Anode of diode
8. Cathode of diode
9. Diode fixing bolt
10. Diode connecting piece fixing bolt
11. Copper connector
12. Copper connector

Figure 18.10 Structure diagram of rotating diode rectifier and quick fuse combination unit. Authorized by Beijing G.E. Automation Engineering Co., Ltd.

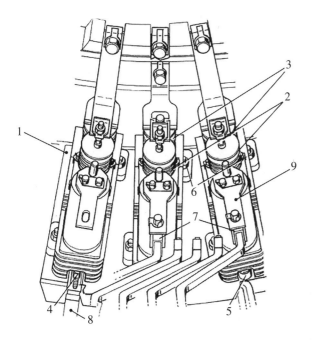

1. Insulated arc plate
2. Anode diodes
3. Cathode diodes
4. Fixing bolts
5. Insulated cover
6. Self-locking nut
7. AC winding connection terminals.
8. Locating ribs
9. Fuse connection

Figure 18.11 Connection of rotating diode rectifier to AC armature winding. Authorized by Beijing G.E. Automation Engineering Co., Ltd.

18.3 Analysis of Working State of Multi-Phase Brushless Exciter

18.3.1 Overview

For rectifier circuits of rotary diodes for large-scale steam-generator brushless excitation systems, two kinds of wiring methods are used widely at present, which are the three-phase bridge and the multi-phase bridge rectifier circuit [57–59].

For the three-phase bridge rectifier circuit, the AC voltage supplied by the brushless exciter rotary armature has a sinusoidal voltage waveform, the wiring is simple, and the energy conversion ratio is higher.

For the multi-phase bridge rectifier circuit, the AC voltage supplied by the rotating armature has a trapezoidal voltage waveform, which can significantly reduce the ripple coefficient of the rectified voltage and further improve the utilization of the brushless exciter capacity. Besides, in the rectifier line powered by a multi-phase armature winding, the branch armature winding can be directly connected with the rotary diode rectifier of a separate branch, and there is no uneven current distribution problem in the multi-branch diode parallel circuit.

18.3.2 Working State Analysis of Odd-Numbered Diodes' Rectifier Circuit

A 39-phase bridge rectifier circuit applied in the brushless exciter in Fuqing Nuclear Power Station is shown in Figure 18.12. The armature windings are connected in series by a 39-phase winding to form a polygonal closed loop.

For the 39-phase brushless exciter, normally, each positive and negative conduction angle is $\varphi = \frac{2\pi}{78} = \frac{\pi}{39}$. Since the armature winding is the inductive load, as the winding current increases, the commutation time and its commutation angle between the two commutator windings will also increase. Depending on the value of the commutation angle γ, the condition of commutation can be divided into the following three working conditions:

(1) When $0 \leqslant \gamma \leqslant \frac{\pi}{39}$, since the commutation angle $\gamma \leqslant \frac{\pi}{39}$ (equal to the $\gamma \leqslant \frac{\pi}{3}$ in the three-phase-bridge rectifier circuit), we have the first phase commutation state or to the two- to three-element commutation state, in which two rectifier components alternate conduction normally and three rectifier components conduct during commutation.

(2) When $\gamma \leqslant \frac{\pi}{39}$, we have the second commutation state – that is, three rectifier components are in the on state at the same time – in which the commutation angle γ is equal to the conduction angle ϕ. With the increase in current, the commutation moment changes in each phase winding. In the second commutation state, the angle will be a fixed value when it increase to $\frac{\pi}{39}$. But in this commutation process, the forced hysteresis control angle α will begin to appear and grow from 0 to $\frac{\pi}{2\times 39}$. So, for the 39-phase rectifier circuit

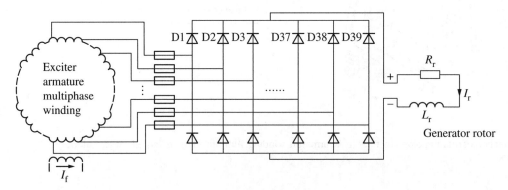

Figure 18.12 39-phase brushless exciter bridge diode rectifier circuit. Authorized by Beijing G.E. Automation Engineering Co., Ltd.

operating in the second commutation state, the online detection of the diode operating state increases due to the presence of a forced hysteresis control angle α' during the commutation process.

(3) When $\frac{\pi}{39} \leqslant \gamma \leqslant \frac{2\pi}{39}$, we have the third type of working state, in which components are under three to four working states; that is, three components or four components alternately turn on the way to work. It should be noted that the third type of operation is abnormal, and it will not transition to this working state until the rectifier seriously overloads, the AC side of the power supply voltage drops, and a short circuit occurs in the DC side.

A distinct operating characteristic of the third commutation work state is that there is an inherent forced hysteresis control angle $\alpha = \frac{\pi}{39}$ in the rectified line. The control angle set does not work when it is lower than $\alpha_p \leqslant \frac{\pi}{39}$, and the operating state of the rectifier line is determined by the inherent hysteresis control angle α of the line. Only if the control angle $\alpha_p > \frac{\pi}{39}$ will the operating state of the rectifier line be determined by the control angle α_p. The relationship between the commutation angle γ and forced lag angle α for different conditions is shown in Figure 18.13.

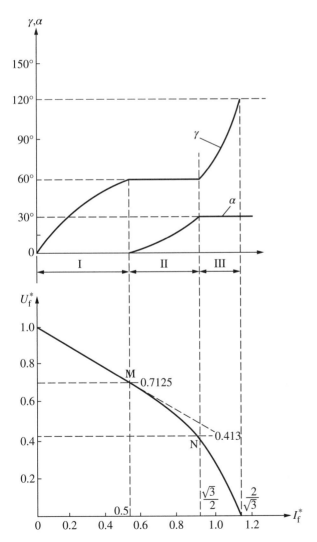

Figure 18.13 Commutation angle γ and forced lag angle α curve of rectifier under different working conditions. Authorized by Beijing G.E. Automation Engineering Co., Ltd.

18.3.3 Double-Layer Fractional Slot Variable Pitch Wave Winding Connection

Fixed excitation pole winding of the Fuqing Nuclear Power Station 39-phase brushless excitation generator (inside) and the rotating armature winding AC (lateral) diagram are shown in Figure 18.14. The number of magnetic poles of the exciter is $P = 11$.

The armature winding is a 117-slot double-layer fractional slot variable pitch wave winding structure, along the counterclockwise direction; the line rod is from the lower layer to the upper layer for a long pitch, the slot pitch between 6 slot pitches is $y_1 = 6$, the short pitch is from the upper layer to the lower layer of the winding bar, and the slot pitch between 5 slot pitches is $y_2 = 5$. The space slot angle is

$$\alpha = \frac{360°}{Z} = \frac{360°}{117} \approx 3.077°$$

In the formula, Z is the number of slots.

Slot electric angle:

$$\alpha_1 = P\alpha = \frac{P360°}{117} = \frac{11}{117} \times 360° \approx 33.846°$$

where P is the number of pole pairs.

Polar distance:

$$\tau = \frac{Z}{2P} = \frac{117}{22} \approx 5.318$$

Number of slots per phase per pole:

$$q = \frac{Z}{2Pm} = \frac{117}{22 \times 39} = \frac{3}{22}.$$

where m is the phase number.

Output frequency of armature winding:

$$f = \frac{nP}{60} = \frac{1500 \times 11}{60} = 275 \text{ (Hz)}$$

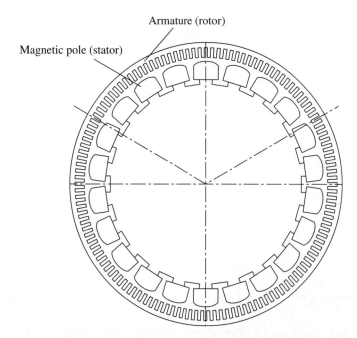

Figure 18.14 Brushless exciter, excitation, and armature winding. Authorized by Beijing G.E. Automation Engineering Co., Ltd.

18.3.4 Phase Angle of 39-Phase Outgoing Lines Emf

Because the armature windings use a triangular waveform winding structure, the entire armature forms a large closed-loop wiring. The phase of the triangular wave windings is determined by the position of the taps of the bar connection points. The 39-phase exciter is taped from every 3 of the 177 bar connection points in the armature slot to form a total of 39 taps and 39 outgoing lines, so the phase between two adjacent taps' outgoing lines is $3\alpha_1 = \frac{3P360°}{117} = \frac{3\times 11}{117} \times 360° = 101.538°$. Any tap can be defined as the first phase outgoing line, and the phase angle is zero; then the next tap phase angle is $-\frac{3\times 11}{117} \times 360°$, and then the next is $-\frac{6\times 11}{117} \times 360°$......

The tap phase angle followed is

$$\varphi_i = \frac{11(i-1)}{39} \times 360°$$

where i is the phase number, ranging from 1 to 39.

From the above analysis, we can obtain the following conclusions:

(1) In the absence of a parallel branch, 39 taps' outgoing lines are 39-phase windings' outgoing lines.
(2) The armature winding 39 outgoing lines and 78 rotary diodes constitute a 39-phase full-wave bridge rectifier circuit, and 39-phase outgoing lines produce an equal amplitude, symmetrical phase emf. The emf phases respectively are $0, -\frac{1}{39} \times 360°, -\frac{2}{39} \times 360°,...,-\frac{38}{39} \times 360°$.
(3) It should be noted that the two phases are structurally adjacent but are not electrically adjacent. The emf star chart composed of 39-phase armature windings is shown in Figure 18.15.

18.3.5 Phase Angle of Emf in Single Winding

As mentioned above, based on the symmetry of the structure, any phase extraction point can be taken as a reference point. The first phase winding starts from the 117-slot bar; then through the 111-slot upper wire bar connects to the 106-slot lower bar, then connects to the 100-slot upper bar, 95-slot lower bar to the 89-slot upper wire bar to form the first phase winding. Figure 18.16 show the single-phase wave winding expansion, each phase consisting the three windings in series.

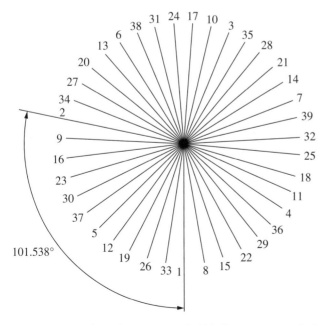

Figure 18.15 Emf star chart composed of 39-phase armature windings. Authorized by Beijing G.E. Automation Engineering Co., Ltd.

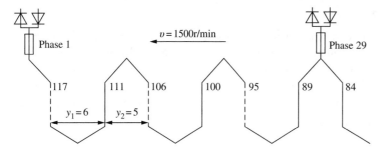

Figure 18.16 Wave winding single-phase expansion diagram. Authorized by Beijing G.E. Automation Engineering Co., Ltd.

18.3.6 Phase Analysis of Single Winding Emf

According to the synthesis electromotive force star phase diagram of the exciter armature winding in Figure 18.15, it can be seen that the electromotive force amplitude of each slot bar is the same, the induction emf amplitude is the same, and the induction emf phase of the space adjacent bar line is a slot electrical angle $\frac{11}{117} \times 360° = 33.846°$.

Assuming that the 117th slot outgoing line is the zero phase reference, there are 6 slots between the 111th slot and the 117th slot, and therefore the phase of the 111th slot is $\frac{11 \times 6}{117} \times 360°$. It is assumed that e_{117} is the synthesis emf of the single-turn winding of e_{111} and e_{117}. We see that the electric phasor e_{111} lags behind electric phasor e_{117} $b = \frac{11 \times 6}{117} \times 360° = 203.076°$. Analysis shows that $e_{k1} = e_{117} - e_{111}$; the phase relationship between the single-turn synthetic emf e_{k1} and e_{117} is shown in Figure 18.17. The emf e_{k1} lags behind the electrical angle a of e_{117}, which is given by $a = \frac{180° - (360° - b)}{2} = \frac{3.75}{117} \times 360° = 11.5°$.

18.3.7 Phase of Phase Emf

From Figure 18.16, we can see that each phase of the armature winding consists of three windings: e_{k2}, e_{k3} were the second and third winding emfs of phase 1, and e_{y1} is the synthetic emf of phase 1. There are 11 slots between the second winding (106–100 slot) and the first winding (117–111 slot), and the angle of e_{k2} lagging behind e_{k1} is

$$c = \frac{11 \times 11}{117} \times 360° = \frac{121}{117} \times 360° = \frac{4}{117} \times 360° = 12.3°$$

Similarly, the angle of e_{k3} lagging behind e_{k2} is c. e_{y1} is superimposed by e_{k1}, e_{k2}, and e_{k3}, and the phasor diagram is shown in Figure 18.18.

e_{k1}, e_{k2}, and e_{k3} amplitudes are equal; from Figure 18.18, we can see that the angle between e_{y1} and e_{k1} is also c, so the electrical angle of e_{y1} lagging behind e_{k1} is $c = 12.3°$. From the above, we see that the angle of the armature winding phase 1 emf phase, e_{y1}, is lagging behind the initial slot, that is, the 117th slot, and the phase of the armature winding emf e_{117} is

$$\theta = a + c\frac{121}{117} \times 360° = 11.5° + 12.3° = 23.8°$$

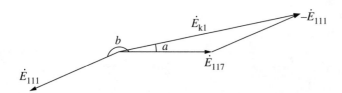

Figure 18.17 Wave winding single-turn emf phasor diagram. Authorized by Beijing G.E. Automation Engineering Co., Ltd.

Figure 18.18 Armature winding phase emf phasor diagram. Authorized by Beijing G.E. Automation Engineering Co., Ltd.

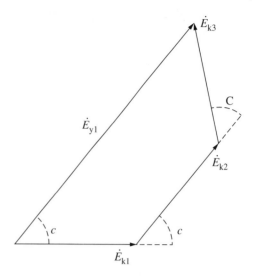

18.3.8 Armature Winding Phase Emf Phase Diagram

Each phase of the armature winding is composed of three coils, each coil slot pitch is 11, and so the slot pitch between the two-phase winding is $11 \times 3 = 33$ slots; the phase of the two-phase winding is

$$\delta = \frac{11 \times 3 \times 11}{117} \times 360° = 36.9°$$

18.3.9 Phasor Diagram of Total Emf Composed of 39-Phase Windings

The phasor diagram of the total emf composed of 39-phase polygonal windings is shown in Figure 18.19.

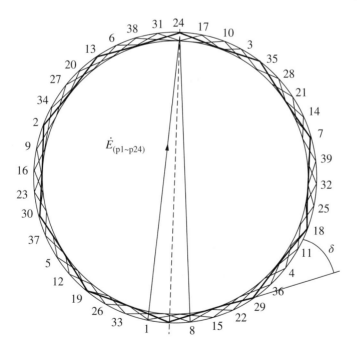

Figure 18.19 Phasor diagram of 39-phase polygonal winding. Authorized by Beijing G.E. Automation Engineering Co., Ltd.

18.4 Calculation of Excitation System Parameters of Fuqing Nuclear Power Station

18.4.1 Ceiling Parameters

According to the technical specification (Table 18.2), the ceiling current will be

$$I_{fexc} = 2.076 \times I_{fexN} = 2.076 \times 113.2 = 235 \text{ A}$$

Exciter limitation: $I_{fexc} = 207.4$ A.
The converter design current will be

$$I_{fexd} = 1.1 \times I_{fexN} = 1.1 \times 113.2 = 124.52 \text{ A}$$

The exciter transformer design current will be

$$I_{fexd} = 1.1 \times I_{fexN} = 124.52 \text{ A}$$

The positive voltage ceiling will be

$$U_{fexc} = 2.0785 \times U_{fexN} = 2.0786 \times 129.9 = 270 \text{ V}$$

Exciter limitation: $U_{fexl} = 237.9$ V.

18.4.2 Excitation Transformer

(1) *Transformer secondary voltage design.* In the ceiling condition, when the network voltage is at 0.8 p.u., the rectifier voltage will be

$$U_{fc} = 1.35 U_{20} \cos \alpha - \Delta U$$

where

- U_{20} The rectifier voltage under no load;
- α The lowest firing angle of the thyristor bridge, $\alpha = 10°$;
- U_{fc} The positive exciter voltage limitation under ceiling conditions, $U_{fc} = U_{fexl} = 237.9$ V;
- ΔU The largest voltage drop at the ceiling current when the output current of the rectifier bridge is 207.4 A, $\Delta U = 35$ V;

$$U_{20} \geq \frac{237.9 + 35}{1.35 \times \cos 10° \times 0.8} \approx 257.5 \text{ V}$$

So, $U_{20} = 300$ V.

(2) *Transformer power design.* The transformer rated power is calculated according to the design current IFD and the second voltage,

$$I_{fexd} = 1.1 \times I_{fexN} = 116.6 \text{ A}$$

The transformer transient power id calculated according to the defined ceiling current, and so

$$S_{ntr} = \sqrt{2} \times 116.6 \times 300 \approx 49.5 \text{ kVA}$$

Choice: $S_{ntr} = 3 \times 17.7 = 53.1$ kVA.
The transformer transient power is calculated according to the defined ceiling current, and so $I_{fexc} = 207.4$ A,

$$S_{ctr} = \sqrt{2} \times 207.4 \times 300 \approx 88 \text{ kVA}$$

Table 18.4 Single-phase excitation transformer technical specifications. Authorized by Beijing G.E. Automation Engineering Co., Ltd.

Parameter	Value	Unit
Rated power	54	kVA
Overload power	88	kVA/10 s/1 times/15 min
Rated no-load secondary voltage	300	V
Rated primary voltage	24 + 10%	kV
Rated frequency	50 ± 5%	Hz
Impedance voltage	0.06	
Altitude	<1000	m
Full-controlled bridge rectifying circuit	6-pulse rectifier	

Two overload conditions shall be taken into account for the actual power calculation of the transformer:

- Ceiling access, 10 s every 15 min. The duty cycle, once every 10 s, is our basis for calculating the temperature rise of the power component.

Table 18.4 shows the technical specifications of the single-phase excitation transformer.

18.4.3 Thyristor Bridge

(1) *Rated current*. According to the contract, the design current for the rectifier bridges is 1.1 times the rated current, $I_{\text{fexn}} = 113.2$ A. So,

$$1.1 I_{\text{fexn}} = 124.52 \text{ A}$$

(2) *Repetitive reverse voltage*. According to the contract, the design ratio of the thyristor voltage will be 2.7:

$$U_{\text{RRM}} \geq 2.75 \times U_{20} \times \sqrt{2}$$
$$U_{\text{RRM}} \geq 2.75 \times 300 \times \sqrt{2} \approx 1166.6 \text{ V}$$

so, $U_{\text{RRM}} = 1200$ V.

The SKT100 Semikron type thyristor components are selected, and their parameters are shown in Table 18.5.

The thyristor thermal calculations are shown in Table 18.6.

In the rectification bridge thermal calculation, the following basic assumptions are used:
1) The ambient temperature inside the cubicle is 45 °C.
2) The temperature if the thyristor ceiling air is 45 °C.
3) The calculation relationships are according to IEC60747-6 2000.

Thyristor rectifier bridge technical specifications:
1) Six pulse thyristor bridges.
2) Repetitive reverse voltage: 1200 V.
3) Cooling: Natural air.
4) Overvoltage protection: By RC snubbers.

(3) *Transient impedance*. The transient thermal impedance is given by the following curve.

Losses and permanent or ceiling temperature results are shown in Table 18.7 (see Figure 18.20).

(4) Calculation of Voltage Drop of Thyristor Rectifier Bridge.

Rectifier bridge voltage drop factors are:

Table 18.5 SKT100 Semikron parameters. Authorized by Beijing G.E. Automation Engineering Co., Ltd.

Parameter	Value
Selected power component	SKT100 Semikron
V_{RRM} (V)	1200
I_{TAVM}, permanent current (A)	100
I_{TSM} (kA)	2
(without reverse voltage) (KA² s)	20 000
Rt (mΩ)	2.4
Vto (V)	1
Rth (j–h) (K/W)	0.31 + 0.08 K/W, @120° DC
Zth (j–h) /5 s (K/W)	0.31 + 0.08 K/W, @120° DC
Junction (°C)	130
Heat sink	P1/200 Semikron
Rth (h–a) (K/W)	0.6
Zth (h–a) /5 s (K/W)	0.05
Cooling	Air natural

Table 18.6 Formulas for thyristor thermal calculation. Authorized by Beijing G.E. Automation Engineering Co., Ltd.

Thyristor losses	$P_{OL} = V_{to} \times \frac{I_F}{3} + rt \times \frac{I_{F2}}{3}$
Junction temperature rise, thermal transient impedance	$\Delta T_1 = \sum_{q=1}^{Q}(P_q - P_{q-1})Z_{th}(t - t_Q - t_{q-1}) Z'_{th}(t = t_Q - t_{q-1})$
Field voltage	$V_{fex} = RF(@100°C) \times I_{fex}$
Thyristor junction temperature	$T_J = T_A + [Rth(j-h) + Rth(h-a)] \times P_{OL}$
Heat sink temperature	$T_S = T_A + Rth(h-a) \times P_{OL}$

Table 18.7 SKT100 thyristor temperature rise under different working conditions. Authorized by Beijing G.E. Automation Engineering Co., Ltd.

Item	Rated	Current for long time	Ceiling
Field current (A)	113.2	116.6	207.4
Number of bridge	1	1	1
Times (s)	∞	∞	10
Field resistance (Ω)	1.147	1.147	1.147
Field voltage (V)	129.9	133.7	237.9
Ambient temperature (°C)	40	40	40
POL (W)/thyristor	47.9	49.75	103.5
Junction temperature (°C)	—	—	—
Heat sink temperature (°C)	—	—	—

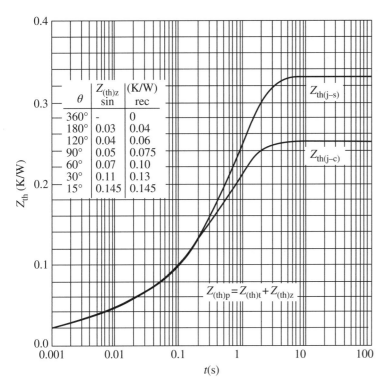

Figure 18.20 SKT100 thyristor transient thermal resistance curve. Authorized by Beijing G.E. Automation Engineering Co., Ltd.

1) Commutation voltage drop due to leakage of excitation transformer.
2) Impedance drop of excitation transformer.
3) The forward voltage drop of the thyristor element.
   Under different load conditions, the thyristor rectifier bridge voltage drop calculation results are shown in Table 18.8.

(5) Fuse Preliminary Design
   According to the design current, the rms current through the thyristor is

$$I_{fus,\ rms,\ N} = \frac{I_{fexN} \times 1.1}{n \times k} \sqrt{\frac{2}{3}}$$

Table 18.8 Thyristor rectifier bridge voltage drop under different working states. Authorized by Beijing G.E. Automation Engineering Co., Ltd.

Item	Rated	Ceiling	Current for long time
No-load transformer secondary voltage (V)	300	300	300 × 0.8
Transformer rated power	54	54	54
Impedance voltage (p. u.)	0.06	0.06	0.06
Thyristor forward voltage drop U_{to} (V)	1.50	1.50	1.50
Thyristor slope resistance value (μΩ)	200	200	200
Exciter excitation current (A)	113.2	116.6	207.4
Reactor voltage drop (V)	7.5	8	22.5
Resistance voltage drop (V)	3.5	4	10.5
Thyristor voltage drop (V)	1.2	1.2	1.2
Total pressure drop (V)	12.2	13.2	35

where $k = 0.8$, and the coefficient n is the number of parallel bridges; that is, $n = 1$.

$$I_{\text{fus, rms, N}} = \frac{113.2 \times 1.1}{1 \times 0.8}\sqrt{\frac{2}{3}} = 127.0877 \text{ A}$$

So, the rated current of fuse = 200 A;

According to the ceiling current, the fuse will be able to accept the ceiling current for 10 s:

$$I_{\text{fus, rms, c}} = \frac{I_{\text{fexc}} \times 1.1}{1 \times 0.8} = \frac{207.4 \times 1.1}{1 \times 0.8} = 285.175 \text{ A}$$

The technical specifications of the fuse are shown in Table 18.9.

18.4.4 Other Thyristor Protections

(1) *Temperature monitoring*. The temperature of the bridge is controlled with three thermal switches.
(2) The rotor insulation monitoring is carried out by the insulator resistor shown in Table 18.10.
(3) Under the ceiling current, the rotor insulation monitoring is carried out by the MX3IPG2A device and insulator resistor, and the value set is 126 A/10 s.
(4) The overcurrent protection is carried out by the P121 device shown in Table 18.11.

18.4.5 Flashing Circuit

No-load field current $I_{\text{fex0}} = 46.9$ A.
No-load field voltage $U_{\text{fex0}} = 53.8$ V.

Table 18.9 Fuse technical specifications. Authorized by Beijing G.E. Automation Engineering Co., Ltd.

Selected fuse	69 URD 30 TTF 200/FERRAZ
Permanent rated current (A)	200
Max voltage (V)	690
Ceiling current 10 s	546
I^2t (kA$^2 \cdot$ s)	3

I^2t design: The design criterion: $\alpha \cdot (I^2t)_T \leq (I^2t)_F$ ($I^2t)_F$ – Thyristor heat capacity; $(I^2t)_T$ – Fuse heat capacity; α – Redundancy factor, $1.5 < \alpha < 2$. Thyristor $(I^2t)_F = 20\text{kA}^2 \cdot$ s, 130°C; $(I^2t)_T = 3\text{kA}^2 \cdot$ s, 290 V.

Table 18.10 AC exciter field winding to ground insulation resistance monitoring. Authorized by Beijing G.E. Automation Engineering Co., Ltd.

Preliminary – Field winding to ground resistance monitoring		
First stage field to ground resistance (alarm)	Ω	4000
Delay time	s	5
Second stage field to ground resistance (trip)	Ω	2000
Delay time	s	0

Preliminary – Field winding to ground resistance monitoring, First stage field to ground resistance (alarm) = 4000 Ω; Delay time = 5 s; Second stage field to ground resistance (trip) = 2000 Ω; Delay time = 0 s.

Table 18.11 P121 overcurrent protection device. Authorized by Beijing G.E. Automation Engineering Co., Ltd.

Preliminary – Overcurrent monitoring		
First stage overcurrent*	A	126.838
Delay time	s	0.5
Second stage overcurrent**	A	128.676
Delay time = ceiling time	s	11
Second stage trip	A	238.97
Delay time	s	1

* First stage = alarm for display only, ** second stage = trip.

The flashing power supply is 220 V dc power supply for the auxiliary supply. In addition, if the flashing power supply voltage drops to 80% rated, it can still meet the requirement that the flashing current be greater than 20% of I_{fex0}.

The required flashing current will be in the range 20%–30%.

The needed extra resistance is $0.8 \times 220/(53.8/46.9 + R) = 46.9 \times 0.2, R = 17.616\,\Omega$.

18.4.6 De-excitation Circuit

(1) *De-excitation resistor residual voltage.* An SiC nonlinear de-excitation resistor is adopted to conduct de-excitation on the excitation winding side of the AC exciter. Under the most serious de-excitation conditions, the de-excitation resistor residual voltage should meet the following requirements.

At the time of de-excitation, the residual voltage on the nonlinear resistor should be smaller than 0.75 of the rotor excitation winding factory test value. When $U_{fexN} = 129.9$ V, the rotor test voltage is 1500 V. So, the residual voltage generated at the time of de-excitation is as follows:

$$U_{NR} \leq 0.7 \times 1500 = 1050 \text{ V}$$

$$U_{NR} = 1000 \text{ V is taken.}$$

(2) *De-excitation resistor capacity.* The capacity of the de-excitation resistor is determined on the basis of the capacity of the forced excitation state of the generator. The field energy stored in the excitation winding is expressed as follows:

$$W = \frac{1}{2} \times 1.8 \times T'_{d0} \times R_{fex,120°} \times I^2_{fexc}$$
$$= \frac{1}{2} \times 1.8 \times 0.426 \times 1.147 \times 207.4^2$$
$$= 18.9 \text{ kJ}$$

The technical specification for the de-excitation resistor selected is shown in Table 18.12.

18.4.7 Field Circuit Breaker

The French LENOIR EL EC OEX 57B field circuit breaker is selected:

$$U_{sN} = 500 \text{ V} > U_{fexN} = 129.9 \text{ V}$$
$$U_{fexcl} = 270 \text{ V} < U_{sN} = 500 \text{ V}$$

(1) $I_{SN} = 250$ A, with redundancy considered.

$$I_{sN} = \frac{I_{sN}}{1.08} = \frac{250}{1.08} = 231.4 > I_{fexd} = 211 \text{ A}$$

Table 18.12 Technical specification for de-excitation resistor. Authorized by Beijing G.E. Automation Engineering Co., Ltd.

Item	Value
Reference	C21/330 V LANGLADE&POCARD
Voltage limit	1500 V
Initial current	90 A
Energy	90 kJ
Insulation	1.1–3 kV
Ambient temperature	50 °C

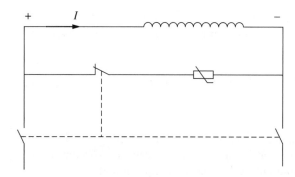

Figure 18.21 Field circuit. Authorized by Beijing G.E. Automation Engineering Co., Ltd.

(2) Main contact interrupted currents, S/C operation

The S/C to be considered is a two-phase short-circuit, located downstream of the DC circuit breaker. In that case, the operating voltage is 0.

$$I_{sc} = \frac{1.1 \times 100e^3}{0.06 \times 340 \times \sqrt{3}} \times \sqrt{\frac{3}{2}} = 2.696 \text{ A}$$

The DC breaker breaking current, at 1000 V, is: 5 kA > 2.696 kA
The short-time current capability is 10 kA (1 s) > 2.696 kA

(3) De-excitation contacts, rated short-time current.
Rated making current of the discharge contacts:
– DC breaker rated making current of the discharge contacts = 6 kA > 207.4 A.

The field circuit is shown in Figure 18.21.

18.5 Static Excitation System of Sanmen Nuclear Power Station [60]

18.5.1 Introduction

Sanmen Nuclear Power Station 1407 MVA half-speed steam-generator set is the first implementation of the AP1000 pressurized water reactor (PWR). The generator excitation system used is UNITROL®6800

18.5.2 Basic Parameters

18.5.2.1 Available Synchronous Machine Data

The basic parameters of the 1407 MVA steam-generator unit of Sanmen Nuclear Power Station are shown in Table 18.13.

18.5 Static Excitation System of Sanmen Nuclear Power Station

Table 18.13 Basic parameters of Sanmen Nuclear Power Station. Authorized by ABB.

Description	Sign	Value	Unit
Rated apparent power	S_N	1 407 000.0	kVA
Rated stator voltage	U_N	24 000.0	V
Rated frequency	f_N	50.0	Hz
Rated power factor	p.f.	0.900	
Rated field current	I_{fN}	9 265.0	A
Rated field voltage	U_{fN}	510.0	V
Rotor test voltage	U_{test}	5 100.0	V
Field resistance	R_{fN}	0.054 8	Ω
Direct-axis synchronous reactance	X_d	1.830 0	p.u.
Direct-axis transient reactance (unsaturated)	$X_{d'}$	0.434 0	p.u.
Direct axis sub-transient reactance (unsaturated)	$X'_d\,d$	0.348 0	p.u.
Quadrature axis synchronous reactance (unsaturated)	X_q	1.790 0	p.u.
Direct axis short-circuit transient time constant	T'_d	1.810 0	s
Direct-axis short-circuit sub-transient time constant	T''_d	0.015 0	s
Direct-axis open-circuit transient time constant	T'_{d0}	8.710 0	s
Armature time constant	T_a	0.180 0	s

Table 18.14 Rated output value of excitation system of Sanmen Nuclear Power Station. Authorized by ABB.

Description	Sign	Value	Unit
Rated system continuous DC output current	I_{eN}	10 192.5	A
Rated system continuous DC output voltage	U_{eN}	560.8	V
System ceiling current	I_p	18 530.0	A
System ceiling application time	t_p	20.00	S
On-load system ceiling voltage	U_{PL}	1 021.0	V
Undervoltage factor for ceiling condition	kU_{min}	0.70	
Ambient temperature	T_{amb}	45	°C
Altitude above sea level	H	1 000	m

Table 18.15 Excitation transformer parameters of Sanmen Nuclear Power Station. Authorized by ABB.

Description	Value	Unit
Rated apparent power	3 × 6500	kVA
Rated primary voltage	24 000	VAC
No-load secondary voltage	1328	VAC
Rated secondary current	8321	AAC
Rated frequency	50	Hz
Short-circuit reactance	0.08	p.u.
Short-circuit resistance	0.01	p.u.

18.5.2.2 Excitation System Rated Output Value
The excitation system rated output value is shown in Table 18.14.

18.5.2.3 Excitation Transformer Parameters
The excitation transformer supplied from the machine terminals (shunt supply) is shown in Table 18.15.

18.5.2.4 Thyristor Rectifier Parameters
The thyristor rectifier technical specifications of Sanmen Nuclear Power Station excitation system are shown in Table 18.16.

Table 18.16 Thyristor rectifier technical parameters of Sanmen Nuclear Power Station. Authorized by ABB.

Description	Value	Unit
Maximum ambient temperature	45	°C
Maximum altitude above sea level	1 000	m
Converter type	UNL 143xy0Vab2cd	
4 in. Si IP31 1 + 1 fans in N-1 converter configuration and 1 fan unit(s) for normal operation		
5-pole isolator for online maintenance	NO	
Minimum number of parallel converters	6	
Total number of parallel converters	7	
Current sharing factor	0.95	
Cooling air temperature at cubicle outlet	65	°C
Thyristor type	5STP 34Q5200	
Thyristor blocking voltage	5 200	V
Minimum thyristor blocking voltage	5 149	V
Thyristor blocking voltage security factor (min./actual)	2.75/2.78	
Fuse type	170M7255	
Positive on load ceiling voltage at ceiling current	1 549.2	V
Negative ceiling voltage at rated field current	−1 389.4	V
Limited ceiling current	18 530.0	A
Ceiling application time	20.00	s
Junction temperature after application of ceiling current	109.4	°C
Unlimited ceiling current	26 630.3	A
Rated continuous current output (pre-load)	10 191.5	A
Current per thyristor at pre-load condition during conduction	1 788.0	A
di/dt per thyristor at pre-load condition	5.1	A/μs
Junction temperature at pre-load condition	95.6	°C
Maximum admissible junction temperature	125	°C
Pre-load converter losses (6 converters + 7 OVP)	70.358	kW
When the self-excitation system output rated field current of the generator I_{fN} = 10 191.5 A, thyristor control angle α = 67.7°. Under these electrical angle conditions, the thyristor output is directly short-circuited.		
Peak current	53.9	kA
Time to current peak value	2.5	ms
Total actual I^2t	9.5	MA^2 s
Total I^2t fuse limit	180.7	MA^2 s
Total short-circuit duration	5.85	ms

Table 18.17 Calculation parameters in case of DC-side short circuit. Authorized by ABB.

Description	Value	Unit
Short circuit at slip rings		
Peak current	155.0	kA
Time to current peak value	5.5	ms
Total actual I^2t	139.5	MA^2 s
Total I^2t fuse limit	20.5	MA^2 s
Total short-circuit duration	11.48	ms
Short circuit at converter output:		
Peak current	160.6	kA
Time to current peak value	5.5	ms
Total actual I^2t	149.5	MA^2 s
Total I^2t fuse limit	20.5	MA^2 s
Total short-circuit duration	11.46	ms
Short circuit FROM PRE-LOAD condition, at rated system current IEN = 10 191.5 A and at firing angle = 67.7° electr.		
Peak current	53.9	kA
Time to current peak value	2.5	ms
Total actual I^2t	9.5	MA^2 s
Total I^2t fuse limit	180.7	MA^2 s
Total short-circuit duration	5.85	ms

18.5.2.5 DC Short Circuit

The calculation parameters of a variety of DC-side short-circuit cases are shown in Table 18.17.

18.5.2.6 Nonlinear Field Suppression

The parameters of nonlinear field suppression are shown in Table 18.18.

18.5.3 Functional Description of Static Excitation System

18.5.3.1 Overview

A static excitation system regulates the terminal voltage and the reactive power flow of the synchronous machine by direct control of the field current using thyristor converters. Figure 18.22 below shows an overview of the excitation system and its typical environment.

The type code used in Sanmen Nuclear Station is T6S-O/U571-S12800, which is described as follows:

T	–	Triple – Dual AUTO channel + Backup controller for both AUTO channels
S	–	Standard (N-1 redundancy)
6	–	UNITROL 6000 digital system
–O/	–	Without additional functions
U5	–	ABB type UNL14300 thyristor converter depending on application requirements
7	–	Number of forward running bridges
1	–	1 thyristor per bridge arm, 3 phases, 6 pulse
S	–	DC breaker, single pole rupturing
12 800	–	Rated breaker current in amperes

The UNITROL 6800 excitation system single-line wiring diagram is shown in Figure 18.23.

Table 18.18 Parameters of nonlinear field suppression. Authorized by ABB.

Description	Value	Unit
Field breaker		
Field breaker type (placed at DC side of converter)	2 × GERAPID 8007 2 × 2	
Arcing voltage	2 800.0	V
Rated current	12 800.0	A
Maximum breaking current capacity	1 50 000.0	A
Resistor		
Number of parallel resistor assemblies	5	
Resistor type 1	HIER 464797	
Number of series assemblies of resistor type 1	4	
Total number of assemblies	20	
Total energy limit	20 000.0	kW s
Field suppression from no-load condition		
Calculated energy on resistors during field suppression	19 102	kW s
Field suppression after three short circuits at stator terminals		
Total field discharge time	0.74	s
Calculated energy on resistors during field suppression	12 868.7	kW s
Thyristor $i^2 t$ (0.2 s) actual	39.4	MA^2 s
Crowbar		
Thyristor type	HUEL 412328	
Thyristor minimum blocking voltage	3 945	V
Nominal voltage of BOD element	3 000	V

18.5.3.2 Control Electronics

(1) *Hardware configuration.* For the UNITROL 6800 excitation system, the hardware configuration is shown in Figure 18.24.

The AC800PEC high-speed processor is used in the UNITROL 6800 system and has the following characteristics:

1) It can quickly perform analog and digital I/O interface conversion; the typical cycle is 400 μs.
2) The same controller can quickly achieve closed-loop and conventional process logic control.
3) Fast analog to digital conversion. AC800P C can be fully integrated in the ABB Control^{IT} software environment and seamlessly connected in different ABB products.

(2) *Regulator basic functions.*

1) The excitation system consists of three channels: two sets of equivalent AVR automatic channels provide 100% automatic redundancy, and another backup manual channel FCR (field current regulator) provides 100% manual redundancy. In normal operation, only one set of three channels is running.
2) AVR automatic operation can provide all the control, limiting the monitoring and protection functions; the manual control system FCR is a backup of the AVR operation, and is used mainly for system debugging and maintenance.
3) Once a fault in the normal operation of the AVR1 automatic channel occurs, if the fault continues, the following procedure will be used for channel switching: AVR1 → AVR2 → FCR2 → FCR1 → BFCR → Trip, and then the corresponding logic is removed.

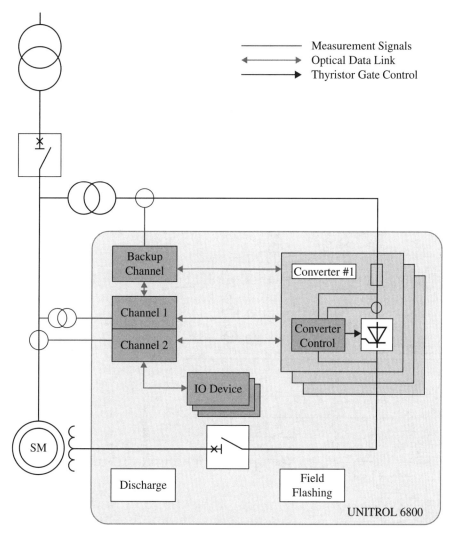

Figure 18.22 Excitation system overview. Authorized by ABB.

The diagram of the dual-channel AVR, FCR automatic excitation regulator is shown in Figure 18.25. When a fault occurs in the two automatic channels, the backup manual magnetic field current adjustment channel will be provided by the BFCR. This channel also includes inverse time overcurrent protection, and the signal is taken from the primary transformer or secondary-side current transformer.

4) *Closed-loop control function*. This feature includes an AVR with a PID filter, a manual automatic FCR with a PI filter, and a maximum and minimum field current limiter, a P/Q limiter and a V/Hz limiter. According to the user requirements, it can provide standard PSS2A/2B and PSS 4B and other PSS functions.

Figure 18.26 shows the generator real-time power running characteristic curve. The curve can determine the area of the corresponding limited characteristics.

5) *The communication between the control channel and the rectifier*. The optical fiber data link between the control channels of the excitation regulator and between the interface with the power rectifier and the rectifier communication link are shown in Figure 18.27. A full range of communication links can be used to meet the needs of logic programs under various operating conditions.

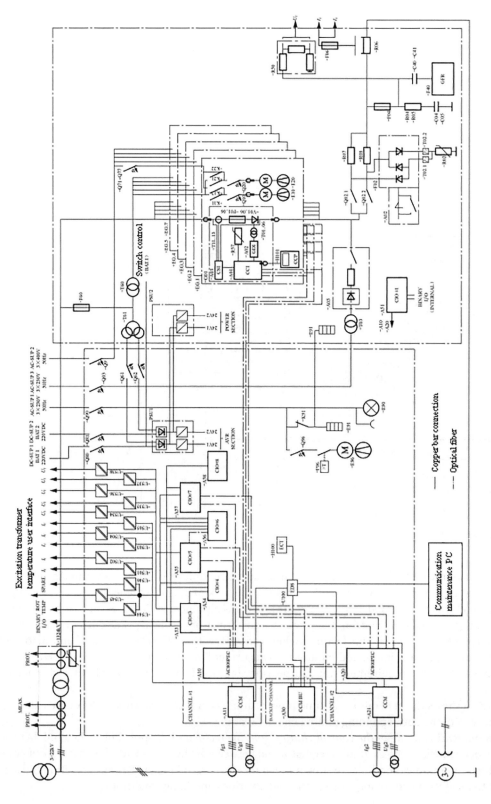

Figure 18.23 UNITROL 6800 excitation system single-line wiring diagram: CCI – thyristor control interface; CIO – combination of input and output; EDS – Ethernet device switch; CCM – communication control measuring device; CSI – thyristor signal interface; GDI – trigger device interface; CCMBU – CCM backup channel; ECT – excitation control terminal; CCP – thyristor control panel; SU – power supply unit. Authorized by ABB.

18.5 *Static Excitation System of Sanmen Nuclear Power Station* | 631

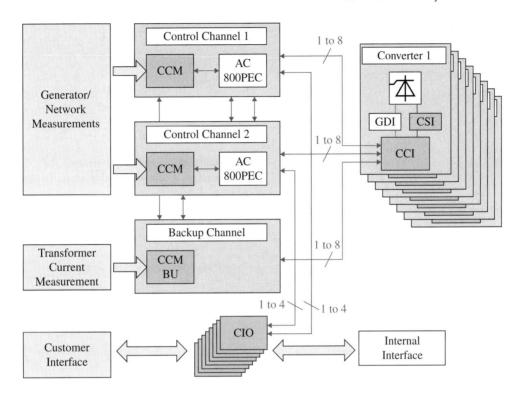

— Analog measurements signals
— Optical data transmission
— Power semiconductor control signals
⟹ General data path

Figure 18.24 Hardware layout of UNITROL 6800 excitation system. Authorized by ABB.

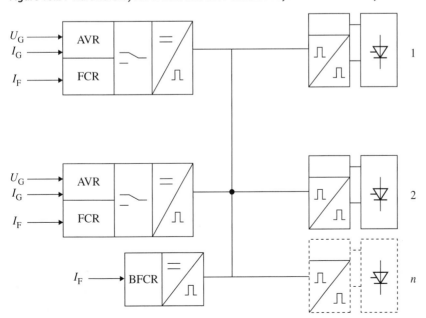

Figure 18.25 Dual-channel AVR, FCR automatic excitation regulator diagram: AVR – automatic voltage regulator; FCR – field current regulator; BFCR – backup field current regulator. Authorized by ABB.

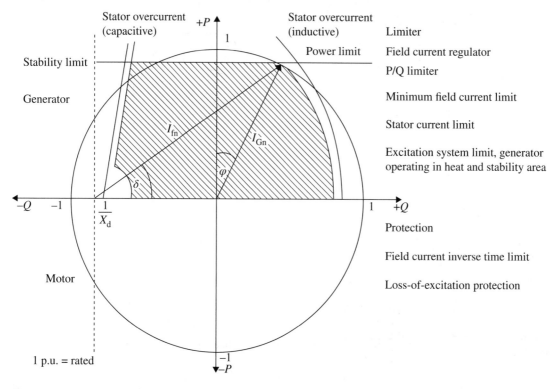

Figure 18.26 Generator real-time power running characteristic curve. Authorized by ABB.

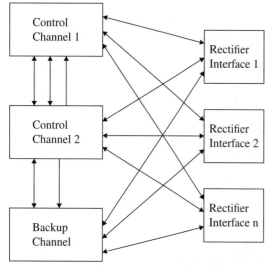

Figure 18.27 Communication links between control channels and rectifiers. Authorized by ABB.

(3) *Converters*. The thyristor in UNL14300 for the UNITROL 6800 Static Excitation is high reverse voltage and high current controllable, which diameter is 4 inches. Each power rectifier cabinet has 1 bridge, and each arm has 1 thyristor.

The thyristor rectifier can be operated in four quadrants and can be continuously operated by the rectified power supply and in fully open mode. The thyristor rated parameter is 4200 V/4275 A, and the single cabinet long-term maximum output current can reach 3994 A. The thyristor du/dt is $2\,\text{kV}\,\mu\text{s}^{-1}$, which

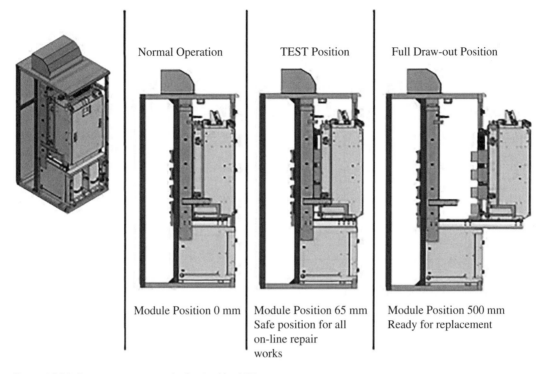

Figure 18.28 Draw-out converter. Authorized by ABB.

can effectively prevent the breakdown caused by mis-conducting accidents and conduction instantaneous overheating.

The basic functions of the UNL14300 rectifier cabinet are as follows:
1) The power rectifier cabinet is $1+1$ redundant (dual bridge), or is NP-redundant ($N-m \leqslant 2$, $m=1$).
2) The thyristor uses a 120° pulse column trigger signal, the typical pulse frequency is 62.5 kHz, the front steepness is 1.25 A μs^{-1}, and the first pulse trigger current peak is more than 3 A to satisfy the strong trigger requirements. While conducting, the thyristor carriers can be quickly spread over the entire silicon to limit the local overheating caused by the moment of conduction, and the maximum voltage of the pulse transformer is 7.5 kV.
3) The fast fuse in series with the thyristor components has a high breaking capacity, which can effectively cut off the impact short-circuit current generated by faults. Fuse and thyristor A/S characteristics have a good combination, which can protect the thyristor components to the maximum extent.
4) The rectifier cabinet uses the blocking snubber connected in the DC output side.
5) The thyristor radiator is installed in the cooling air duct. It is possible to replace the damaged parts directly in front of the cabinet after removing the front duct.
6) The power cabinet cooling system uses a redundant configuration, with two sets of the fan 1 group for operation, and another group for standby. At runtime, if the fan fails, the standby fan will start automatically. In addition, each fan is supplied by the plant power and excitation transformer power.
7) The rectifier cabinet can use IP20-IP54 protection.
8) The rectifier cabinet uses an intelligent dynamic current sharing device, and the current sharing coefficient can go up to 0.95. Through the converter control interface (CCI) on the control panel, the ventilation and temperature of the rectifier cabinet and the on-line conduction state of each arm element can be monitored and judged. The corresponding alarm and fault signals are transmitted to the excitation regulator through high-speed connected optical fiber for adjustment.

18 Performance Characteristics of 1000 MW Nuclear Power Steam Turbine Excitation System

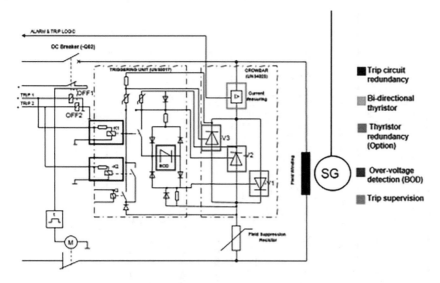

Figure 18.29 De-excitation and overvoltage protection circuits. Authorized by ABB.

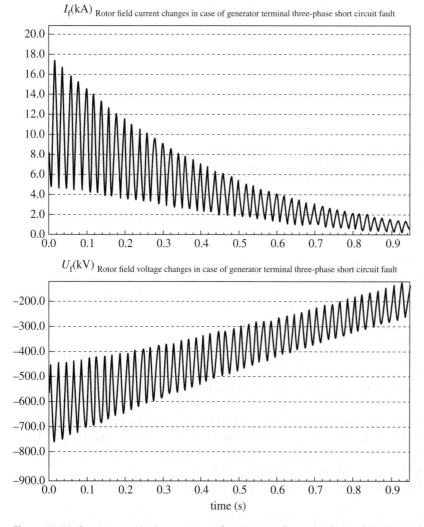

Figure 18.30 Generator excitation current and excitation voltage simulation diagram. Authorized by ABB.

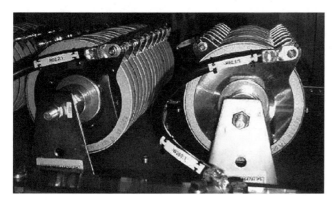

Figure 18.31 De-excitation resistance, external view. Authorized by ABB.

The UNL 14300 power rectifier cabinet diagram is shown in Figure 18.28. Typical data of UNL14300 power rectifier cabinet are as follows:
(1) Input voltage 720–1500 V_{AC}
(2) Output current 2600–900 A_{DC}
(3) Double redundant cooling fan
(4) Rectifier cabinet working state rapid diagnosis
(5) Modular design, ease of parallel connection, system current up to 10 000 A

In addition, due to the UNL14300 rectifier cabinet with a drawer structure, with the help of the joystick, the rectifier cabinet unit can be pulled 65 mm to a safe position for online maintenance, removed to a 500 mm position for maintenance, and online repair and function checks without pole isolation.

(4) *Excitation transformer*. Excitation transformer basic performance characteristics are as follows:

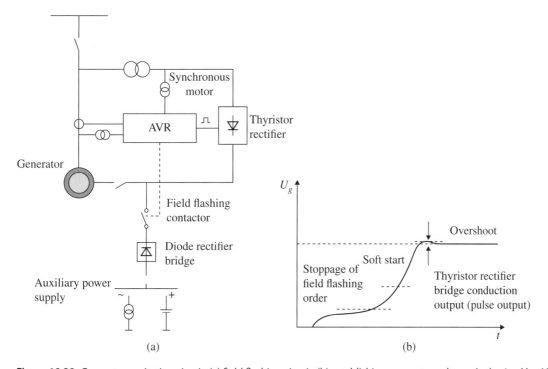

Figure 18.32 Generator excitation circuit: (a) field flashing circuit, (b) establishing generator voltage. Authorized by ABB.

1) Epoxy cast single-phase dry-type transformers.
2) Two-level PT100-type winding temperature measurement device.
3) The electrostatic ground shield between HV and LV high- and low-voltage winding.
4) Per phase of HV high-voltage winding has three current transformers 500/1 A, 50 VA.
5) Per phase of LV low-voltage winding has two current transformers 10 000 A/1 A, 50 VA.
6) Connection technical specifications of HV side of the high-voltage winding and the upper part of the flat copper busbar.
7) Rated apparent power is 3×6500 kVA.
8) Rated voltage is 24 kV.
9) No-load secondary rated voltage is 1324 V.
10) Rated frequency is 50 Hz.
11) Short-circuit impedance is 8%.
12) Impact withstand voltage test standard is in accordance with IEC60 726.
13) AN cooling standard is in accordance with IEC60 726.
14) Winding temperature meter (LV/HV): Class B.
15) Insulation class (LV/HV): Class F.
16) Winding connection (Yd11) (HV/LV).

(5) *Crowbar/Overvoltage Protection.* The Sanmen Nuclear Power Station de-excitation and overvoltage protection system has many special features, because the generator excitation current Ip can reach 18 530 ADC; hence this equipment was selected.

1) *Field breaker.* Sanmen nuclear turbines generator rated excitation current $I_{fN} = 9265$ ADC. At present, the only fast DC circuit breaker worldwide that can be used as a magnetic circuit breaker application is GERAPID produced by GE, but this series of products has a maximum rated current of 8000 A. ABB uses two DC circuit breakers parallel in the Sanmen Nuclear Power Station de-excitation system: $2 \times$ GERAPID8007 2×2, DC circuit breaker rated current 12 800 A, arc voltage 2000 V, maximum breaking capacity 1 50 000 A.

2) *De-excitation and overvoltage protection.* The de-excitation and overvoltage protection circuits are shown in Figure 18.29, where V1 and V2 thyristors are positive and negative overvoltage protection, V2 and V3 thyristors are the de-excitation branch, and the turning diode BOD can limit the generator forward rotor overvoltage protection action value according to the set voltage. The de-excitation resistance uses the British M&I Metrosil series SiC nonlinear de-excitation resistance.

The whole component: Five groups in parallel for each group with four components in series constitute the total de-excitation resistance, and the total resistance power is 20 MJ. When a three-phase short circuit occurs on the generator side, the corresponding generator excitation current and excitation voltage simulation diagram are shown in Figure 18.30. Figure 18.31 shows the external view of the de-excitation resistance.

(6) *Build-up field circuit.* For generators with static self-excited excitation systems, it is usually necessary to supply a smaller initial field current to establish an initial voltage.

The procedure for establishing the generator voltage from the excitation circuit is as follows:
1) Close generator magnetic field breaker.
2) Release pulses to thyristors.
3) If the generator voltage does not rise, close the excitation circuit contactor.
4) The generator voltage starts to rise.
5) Trigger the thyristor in the generator power rectifier to switch.
6) Switch off excitation contactor, and exit the excitation circuit.
7) Soft-start control software makes the generator-side voltage reach the rated value.

The corresponding process is shown in Figure 18.32.

(7) *Shaft current protection.* The large-scale turbo-generator sets usually use the static self-excited system. Because of the high-frequency harmonic components in the excitation voltage, a high suspension emf will appear in the shaft due to the influence of the ground winding on the ground capacitance. It is possible

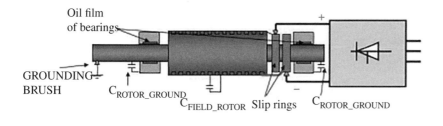
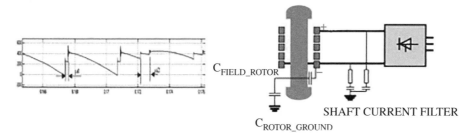

Figure 18.33 Generator shaft current protection principle diagram. Authorized by ABB.

for a shaft current to flow through the oil film of the bearing and cause discharge of the bearing surface, resulting in corrosion damage. To prevent shaft current damage, we usually use the grounding carbon brush and other bypass measures. The corresponding generator shaft current protection schematic diagram is shown in Figure 18.33. Once a high-frequency axis voltage is generated, it can be bypassed by a grounding carbon brush circuit.

References

1 Lu, Q., Mei, S., and Sun, Y. (2008). *Power System Nonlinear Control. Version 2*. Beijing: Tsinghua University Press.
2 Koishikawa, S., Michgam, T., and Onizuk, N. (1990). Advanced control of reactive power supply enhancing voltage stability of a bulk power transimission system and new scheme of monitor on voltage security. *Group 39 of CIGRE* 38/39-01.
3 Large synchronous generator set NR-PSS principle and engineering implementation. China Hydropower Association of Power System Automation Specialized Committee, Annual meeting and academic exchange. Guangzhou, 2008.
4 Michigami, T., Onizuka, N., and Kitamura, S. (1990). Development of a new generator excitation system for improving voltage stability (PSVR). *Proceedings of the Japanese Electrical Society, Part B* 110 (11).
5 Zhou, D. (2004). *Synchronous Generator Operation Technology and Practice. Version 2*. Beijing: China Power Press.
6 Li, J. (1987). *Modern Synchronous Generator Rectifier Excitation System*. Beijing: Hydraulic Power Press.
7 Zhdanov, P.C. (1948). *Power System Stabilization*. Moscow. Soviet Union: Energy Press.
8 Winnikov, B.A. (1985). *Electromechanical Energy Transient Processes in Power Systems*. Soviet Union: Higher Education Press.
9 Liu, Z. (1998). Synchronous generator excitation control tasks and design ideas. China Electrical Engineering Society of excitation work group, Annual meeting. Beijing.
10 Specification and characteristics of Synchronous electrical machine. Japanese Electrical Society Technical Report No. 536, part II, 1995.
11 China Zhejiang University Power Generation Teaching and Research Group (1985). *DC Transmission*. Beijing: Hydraulic Power Press.
12 Grepov, E.A. (1960). *Synchronous Generator Excitation System with Controllable Converter*. Literature Press. Soviet Academy of Sciences.
13 Grepov, E.A. (1982). *Hydro Generator*. Moscow, Soviet Union: Energy Press.
14 Huang, Y. and Li, X. (1998). *Modern Excitation System and Control of Synchronous Electric Machine*. Chengdu: Chengdu University of Science and Technology Press.
15 Morozova, Y.A. (1976). *Parameters and Characteristics of the Thyristor Excitation Systems of Powerful Synchronous Generator*. Moscow. Soviet Union: Energy Press.
16 Bluehine, G.N., Egnatof, P.E., and Kowalinkof, G.A. (1973). Study on forced excitation process of synchronous generator. *Soviet Union. Jou. Electricity*. (6).
17 Gao, J. and Zhang, L. (1986). *Basic Theory and Analysis Method of Electric Machine Transition Process*. Beijing: Science Press.
18 Li, J. (1999). *Training Materials for 900MW Unit Brushless Excitation System of Daya Bay Nuclear Power Plant (Application)*. Shenzhen: Daya Bay Nuclear Power Plant Training Center.
19 Investigation and study on brushless exciter of synchronous machine. Japan Society of Electricity Technical Report No. 652, 1997.

Design and Application of Modern Synchronous Generator Excitation Systems, First Edition. Jicheng Li.
© 2019 China Electric Power Press. Published 2019 by John Wiley & Sons Singapore Pte. Ltd.

20 Chen, X. (1997). Discussion on protection mode of excitation transformer at machine end. *Electrical Design Technology* (1).
21 Meyer, A. (1988). Shaft voltage in Turbo generators. Recent development of a new grounding design to improve the reliability of bearings. CIGRE 11–10.
22 Candeior, R. (1988). Shaft voltage in large Turbo generators with static excitation. CIGRE 11–04.
23 Dr. Hans-Joachhim Herrmann and Dijun Gao. Generator loss-excitation protection based on admittance measurements — very close to the generator operational limit diagram. The first international conference on Hydropower technology: volume 1. Beijing. China power press, 2006.
24 Chen, Y., Liu, G., and Zhong, M. (2008). Design and application of electrical braking for hydropower units. China Hydropower Association of Power System Automation Specialized Committee, Annual meeting. and academic exchange meeting of the discipline group of generator excitation system. Guangzhou.
25 Cui, J. (2000). *Digital Excitation Device Handout*. China Tianjin: Hebei University of Industrial.
26 Li, J. (1988). Specification features of excitation transformers for modern static excitation systems. *Automation of Hydropower Plants* 20 (2): 53–59.
27 Zhang, H. (2001). HTC resin insulation excitation transformer design and analysis. China Electric Engineering Society excitation sub - Committee. Excitation academic Seminar. Beijing.
28 Geafol (1985). *Castresin Transformers 50 to 2500 kVA Planning Guides*. Geafol Factory user guidelines: Germany.
29 Xie, T. and Lu, J. (2002). Three gorges project excitation transformer design. *Electro Mechanical Engineering Technology* (2).
30 Wang, B., Zhang, J., and Zhou, Y. Research on high power rectifier based on loop heat pipe radiator. Proceedings of the international conference on electrical generation technology: volume 2. Beijing. China Power Press, 2006.
31 Wang, W., Shi, L., and Ma, Q. Factors affecting shunt operation of thyristor excitation rectifier bridge. Proceedings of the international conference on electrical generation technology: volume 1. Beijing. China Power Press, 2006.
32 Zang, J., Chen, X., and Hu, S. (2007). Excitation overvoltage protection device and its application in three gorges power plant. *Mechanical & Electrical Technique of Hydropower Station* (6).
33 Kaili Electric company. Bai mountain hydro-power plant AC peak voltage absorption device test report. China Hefei, 2006.
34 He, C. (2005). The fault analysis and judgment of the thyristor in static excitation system. *Hydraulic Power Generation* (9).
35 Zhang, D. (1988). *Linear Resistance Grading de-Excitation System*. Beijing: Beijing Hydropower Planning and Design Institute.
36 Li, Z. (2006). AC de-excitation discussion. *Large Electric Machine Technology* (6).
37 Bloon, O.B. (1961). Automatic de-excitation device. Soviet union.
38 Li, J. (2008). Performance characteristics of Metroosil nonlinear resistance. *Automation of Hydropower Plants* (1).
39 Hu, Y. (1998). Main circuit scheme selection and equipment introduction of Tianhuang ping pumped storage unit excitation system. *Automation of Hydropower Plant* (2).
40 Jing, G. (2002). Tianhuang ping pumped storage unit excitation system operation. *Hydropower Station Electro-Mechanical Technology* (2).
41 VA TCH HYDRO. THYNE6 Static excitation system. Beijing. 2006.
42 Yang, H. (2004). Analysis of frequency conversion system of the Tiantan pumped storage power plant. *Automation of Hydropower plant* (2).
43 Convertteam. Langya mountain static frequency converter training course. Paris, 2006.
44 Lu, H. (2006). *Large Pumped Storage Power Plant Automatic Control System Research Report*. National Grid Nanjing Institute of Automation: Nanjing.

45 Jiang, S. and Jiang, F. (2002). Langya mountain pumped storage power plant harmonic analysis. Hydropower plant automation information network, the eighth session of the network conference to exchange information. Beijing.
46 Yang, H. and Liu, S. (2004). The Tiantan pumped storage power plant unit LCU control procedures. *Automation of Hydropower Plant* (2).
47 Zhang, M., Zeng, G., and Liu, M. (2002). The application of ac voltage de-excitation in large pumped storage unit. Proceedings of the symposium on power system stability and synchronous generator excitation system. Beijing.
48 Yamamoto (1975). Recent control unit for hydropower plant. *Mitsubishi Electric Journal of Technology* 149 (9).
49 Zhang, Y. and Zeng, H. (2002). Improvement of shaft current protection for Tianhuang ping hydropower plant. *Hydropower Station Electromechanical Technology* (2).
50 Zhang, M. and Zeng, G. (2007). PSS parameter design and field test of pumped storage units. *Hydropower Station Electromechanical Technology* (3).
51 Mj4 Generator Technical Data. Alstom-Power, 2001.
52 Mj4 Sommaire Table of Contents. Alstom-Power, 2011.
53 Controgen transter diagram static excitation Mj4. Alstom-Power, 2011.
54 Generator Technical Data of Fu-Qing Nuclear Power Plant. Alstom-Power, 2011.
55 Exciter TKJ 167–45 O & M Manual. Alstom-Power, 2011.
56 Zhang, L. and Yi, J. (2012). *Application of P320 AVR and TKJ Brushless Exciter Units in Domestic Nuclear Power Units*. China Dongfang Electric Machinery factory.
57 Jinyuan, Liu, G., and Chen, X. (2013). Hall element for brushless exciter rotary diode fault on-line monitoring. *China Electric Machinery Technology* 5 (1): 22–27.
58 Jia, X. and Chen, S. (1999). An odd number of phase brushless DC generator commutation process. *China Journal of Xi'an Jiaotong University* 12: 18–22.
59 Michael Liwschitz-Garik, Dr-ing, Assisted by Clyde C. Whipple. E.E. Electric Machinery, Volume A-C, Machines, Second Printing, 1948.
60 Sanmen NPP, Unitrol. 6000. ABB. 2016.

Index

a

AC disconnectors   495
AC field circuit breaker   495
AC harmonic value   153
AC-1 model   238
AC-2 model   243
AC sampling   337
AC separate exciter type excitation system   113
AC voltage de-excitation   459
Admittance measurement principle and derivation   316
AF radiator   407
Air gap line   52
Analysis of working state of Multi-phase Brushless Exciter   612
AN radiator   407
Application characteristics of PSS of pumped storage unit   577
Armature reaction and power factor   267
Atmospheric overvoltage   377
Automatic reactive power regulator   118
Automation excitation regulator   329
Average rectifier voltage   144
AVR gain   128

b

Back to back start   553
Basic expression for electric braking   329
Basic functions of PSVR   31
Basic control law of excitation system   97
Boosting system   607
Bridge inverter circuit   98
Brushless excitation system   215

c

Calculation of rectifier loss   406
Calculation of reverse repetitive peak voltage   400
Calculation of Thyristor Element Loss   401
Capacity of power rectifier operating in parallel   433
Characteristics of digital excitation system   383
Characteristics of digital phase shift   347
Characteristics of SCR excitation system for steam turbine generator   258
Characteristics of separately excited SCR excitation system   255
Characteristics of static self-excitation system   288
Characteristics of synchronous generator   35
Characteristics of typical digital excitation system   353
Classical Control Theory   1
Commutation process   414
Comparison between digital and conventional current sharing   415
Comparison of impedance and admittance measurement   319
Composition of SFC   542
Composition of brushless excitation system   221
Control characteristics of brushless excitation system   227
Control of electric braking system   326
Cooling of large capacity power rectifier   407
Coordination between excitation restriction and loss of excitation protection   311
Criterion of stability level   68
Current sharing factor   271
Current sharing of power rectifier   413

d

Damage and failure forms for SiC nonlinear resistor   482
DC field circuit breaker   467
De-excitation and overvoltage protection device   501
De-excitation and rotor overvoltage protection   439
De-excitation system classification   447
De-excitation system of pumped storage unit   569
De-excitation time   481

Design and Application of Modern Synchronous Generator Excitation Systems, First Edition. Jicheng Li.
© 2019 China Electric Power Press. Published 2019 by John Wiley & Sons Singapore Pte. Ltd.

Design values of excitation transformer   568
Differential control law   101
Digital current sharing   415
Digital discrete technology   330
Digital of high-power heat pipe rectifier   411
Digital phase shift trigger   346
Digital sampling and signal conversion   337
Diode brushless excitation system   215
Discretization of continuous system   332
Dynamic stability   67

e
Effective element current value   147
Electric braking   321
Electromagnetic induction method for identification of initial position of rotor   554
Electrostatic shield   387
Energy conversion by reverse current recovery effect   378
Energy conversion caused by commutation voltage   377
Establishment of SFC Electrical   49
Evaluation of performance of de-excitation system   443
Evolution and development of excitation control   1
Excitation regulator software   529
Excitation system monitor and protection   603
Excitation system start and stop condition   512
Excitation transformer   365
Excitation voltage response for AC exciter with rectifier load   201
Expression for nonlinear de-excitation time   453
External characteristic curve for rectifier   168
External Communication Network   601

f
Field circuit breaker   467
Fourier Algorithm for AC Sampling   338
Fundamental Wave and harmonic value for alternating current   152
Fuse preliminary design   621

g
General external characteristics of exciter   189
Generator operating limit diagram   316
Gerapid series of fast DC circuit breaker   471
Go Control Process   567
Grounding rotor earthing protection device   603

h
Harmfulness and characteristics of harmonics   562
Harmonic calculation   392
Harmonic current analysis   389
Heat dissipation and cooling   368
HIR excitation system   88

i
Impedance voltage   371
Improvement of harmonic of operation state of SFC   565
Improving dynamic stability   74
Improving transient stability   74
Influence of damping winding circuit on de-excitation   465
Influence of harmonic current   391
Influence of harmonic current load on electromagnetic characteristics of auxiliary generator   260
Influence of saturation on de-excitation   463
Integral Control Law   99
Integral time constant   129
Inversion de-excitation of self thyristor excitation system   300

k
Key points of design of electric braking circuit   325

l
Linear multivariable control   7
Linear multivariable total controller   20
Linear optimal control system   11
Linear phase shift element   348
Linear resistor de-excitation commutation condition   447
Local control unit control procedure for pumped storage unit   566
Loop heat pipe radiator   409

m
Mathematical model for brushless excitation system   232
Mathematical model of the excitation system   108
Maximum excitation current limiter   358
Mechanical braking   321
Modern control theory   7

n
Nonlinear multivariable excitation control   20
Numerical description the PID control law   105
Noise   382

Noise of epoxy dry-type excitation transformer   382
Noise reduction measures   382

o

Operating capacity characteristic curve of synchronous generator   41
Operating characteristic curve of generator   51
Operating principle of three-phase bridge inverter   170
Operating principle of three-phase bridge rectifier   139
Operation characteristic of thyristor   395
Operation mode and excitation control of pumped storage unit   521
Operation modes of digital excitation system   354
Over current protection   376
Over Excitation Limiter   121
Over voltage protection   376

p

Parallel correction-dynamic feedback   108
Parallel correction-proportional feedback   107
Parallel feedback correction of excitation control system   107
Parameterization of high-low voltage bridge rectifier   276
Parameterization of power rectifier   400
Parameter selection for SIC nonlinear resistor   481
Performance characteristics of excitation system of 1000MW turbine generator unit   679
Performance characteristics of nonlinear de-excitation resistor   477
Performance of Fuqing Nuclear Power Station   601
Per-unit value setting   345
Phase compensation   131
PI type brushless excitation system   114
Power angle characteristics of synchronous generator   38
Power factor of rectifying devices excitation system   268
Power rectifier   395
Power rectifier anode overvoltage suppressor   417
Power system stability   67
Power system voltage regulator   23
Principle for parameter selection for commutation snubber   421
Proportional control law   99
PSS parameter setting   131
PSVR control mode   29
PSVR simulation test   33

Pumped storage unit operating as synchronous condenser   568

r

R-C protection of power rectifier thyristor elements   421
Rectifier current ratio   184
Rectifier voltage ratio   184
Resistance to sudden short-circuit current   375
Rotary diode rectifier   610
Rotating diode non-conduction detection system   607

s

Salient pole generator   35
SCR brushless excitation system   218
Selection of braking system   321
Selection of braking transformer   321
Selection of fan model   403
Selection of main circuit of excitation system of pumped storage   521
Selection of quick fuse parameters   404
Selection of radiator   401
Separately excited SCR excitation system   255
SFC control mode   545
SFC current and speed dual closed-loop control system   560
SFC power supply connection   559
SFC start process   548
Shaft current protection of pumped storage unit   574
Shaft grounding system for large steam turbine generator   309
Short circuit characteristic curve   51
Short-circuit current calculations for AC exciter   205
Short-circuit current of self-excited generator   291
Single Input, Single Output (SISO)   3
Source and protection of shaft voltage   307
Specification and essential parameters of thyristor rectifier elements   395
Speed control of pump motor unit during SFC start   561
Stability analysis of excitation system   99
Static characteristic of excitation system   77
Static excitation system of Sanmen Nuclear Power Station   624
Static self-excitation system   109
Static voltage droop ratio of generator   80

Structural characteristics of resin cast dry-type excitation transformer 367
Synchronization control of pump motor unit 561

t

Technical specification for brushless excitation system 219
Test voltage 381
Theoretical basis of digital control 330
Three-phase bridge rectifier circuit 137
Three-Phase one-point algorithm 339
Thyristor damage and failure 442
Thyristor damage and failure analysis and identification 430
Time constant compensation in case of small deviation signal 229
Timeliness of SIC nonlinear resistor 447
Transfer function of control system 98
Transfer function of holder 333
Transient characteristic of synchronous generator 54
Transient equation of rotor excitation circuit 59
Transient process of separately excited 276
Transient response of large disturbance signal 87
Transient response of small deviation signal 90
Transient state characteristic of excitation system 87
Transient state process of AC exciter with rectifier load 191
Type I commutation state 139
Type II commutation state 167
Type III commutation state 161

u

Uncertainty of parallel operation of double-bridge power rectifiers 437
Unit step response 9
Unit voltage control during SFC start of pump unit 561

v

V/Hz over-excitation limiter 357
Voltage response characteristics of AC exciter 224
Voltage response ratio of excitation system 88
V-shaped curve of hydro-generator 52

w

Working principle of heat pipe 408